Urban Storm Water Management

Second Edition

Urban Storm Water Management

Second Edition

Hormoz Pazwash

CRC Press
Taylor & Francis Group
Boca Raton London New York

CRC Press is an imprint of the
Taylor & Francis Group, an **informa** business

CRC Press
Taylor & Francis Group
6000 Broken Sound Parkway NW, Suite 300
Boca Raton, FL 33487-2742

© 2016 by Taylor & Francis Group, LLC
CRC Press is an imprint of Taylor & Francis Group, an Informa business

Printed on acid-free paper
Version Date: 20150601

International Standard Book Number-13: 978-1-4822-9895-6 (Hardback)

Library of Congress Cataloging-in-Publication Data

Pazwash, Hormoz.
  Urban storm water management / Hormoz Pazwash. -- Second edition.
    pages cm
  Includes bibliographical references and index.
    ISBN 978-1-4822-9895-6 (hardcover : alk. paper) 1. Storm sewers. 2. Urban runoff--Management.
    I. Title.

TD665.P46 2016
628'.21--dc23   2015020431

Visit the Taylor & Francis Web site at
http://www.taylorandfrancis.com

and the CRC Press Web site at
http://www.crcpress.com

To humanity and my family (Zaman, Haleh, and Hooman)

Contents

Preface..xvii
Author xix
Abbreviations..xxi
Notations xxv

Chapter 1	Urbanization Impact on Runoff 1	

1.1   Impacts on Storm Water Quantity.. 1
1.2   Impacts on Water Quality... 3
1.3   NPS Pollutants and Their Impacts 4
          1.3.1   Floatables 6
          1.3.2   Sediment... 7
          1.3.3   Nutrients and Pesticides 9
          1.3.4   Heavy Metals.. 11
          1.3.5   Pathogens, Fecal Coliform 11
          1.3.6   Road Salt 12
          1.3.7   Petroleum Hydrocarbons... 14
          1.3.8   Atmospheric Dust... 15
1.4   Management of Storm Water Runoff 16
Problems... 18
Appendix 1A: NRC Report Summary, October 15, 2008 19
References 23

Chapter 2	Pipe and Open Channel Flow: A Review... 25	

2.1   Flow Classifications.. 25
2.2   Energy Equation 25
2.3   Specific Energy; Critical Flow 26
          2.3.1   Critical Depth 27
          2.3.2   Critical Flow in Rectangular Channels....................................... 28
          2.3.3   Critical Flow in Trapezoidal Channels 29
          2.3.4   Critical Flow in Partly Full Circular Pipes 31
2.4   Normal Depth.. 34
          2.4.1   Chezy Equation 35
          2.4.2   Manning Formula... 35
2.5   Calculation of Flow Depth 42
          2.5.1   Circular Sections 42
          2.5.2   Trapezoidal Sections 44
2.6   Energy Losses in Pipes and Culverts 50
          2.6.1   Friction Losses 50
          2.6.2   Local Losses... 51
                  2.6.2.1   Entrance and Exit Losses 52
                  2.6.2.2   Sudden Expansions or Contractions 53
                  2.6.2.3   Bend Losses 54
                  2.6.2.4   Head Loss at Transitions... 54
                  2.6.2.5   Junction Losses 55
                  2.6.2.6   Losses at Access Holes and Inlets 55

Problems 59
Appendix 2A: Hydraulic Properties of Round and Elliptical Pipes 62
References 66

Chapter 3 Hydrologic Calculations 67
3.1   Rainfall Process 67
       3.1.1   Intensity–Duration–Frequency Curves 67
       3.1.2   Rainfall Data 73
       3.1.3   Rainfall Hyetograph 75
3.2   Initial Abstractions 76
       3.2.1   Interception 78
       3.2.2   Depression Storage 79
3.3   Infiltration 82
       3.3.1   Green-Ampt Model 83
       3.3.2   Horton Equation 87
       3.3.3   Philip Infiltration Model 90
       3.3.4   Infiltration Indexes 91
3.4   Measurement of Infiltration and Permeability 92
       3.4.1   Infiltrometers 92
       3.4.2   Permeameters 93
       3.4.3   Soil Gradation Analysis 94
3.5   Hydrographs 96
       3.5.1   Time of Concentration Equations and Nomographs 97
                3.5.1.1   Kirpich Equation 97
                3.5.1.2   Izzard Equation 97
                3.5.1.3   Kirby Equation 98
                3.5.1.4   Garden State Parkway Nomograph 99
       3.5.2   Other Methods of Time of Concentration Calculation 101
                3.5.2.1   SCS Method 101
                3.5.2.2   FHWA Method 103
       3.5.3   Sheet Flow Length Analysis 108
3.6   Runoff Calculation Methods 109
       3.6.1   Rational Method 110
       3.6.2   Limitations of Rational Method 112
       3.6.3   Modified Rational Method 113
       3.6.4   SCS TR-55 Method 115
       3.6.5   SCS Peak Discharge Calculations 119
                3.6.5.1   Graphical Method 119
                3.6.5.2   Tabular Method 120
       3.6.6   SCS Unit Hydrograph Method 121
       3.6.7   Limitations/Drawbacks of TR-55 Method 123
       3.6.8   WinTR-55 Method 126
3.7   Universal Runoff Method 131
       3.7.1   Lag Time between Rainfall and Runoff 131
       3.7.2   Runoff Volume and Discharge from Pervious Surfaces 133
       3.7.3   Lag Time and Runoff Equations for Impervious Surfaces 134
       3.7.4   Equations for Composite Surfaces 135
                3.7.4.1   Universal Runoff Equations for Composite Surfaces 135
       3.7.5   Universal Method Application to Nonuniform Rainfall 139

3.8   Storm Water Management Models 140
Problems 141
Appendix 3A: The Sheet Flow Time of Concentration in TR-55 Method: A
            Commentary 145
Appendix 3B: Snyder Unit Hydrographs 148
Appendix 3C: USGS Nationwide Urban Hydrograph 150
Appendix 3D: USGS Regression Equations for Urban Peak Discharges 151
Appendix 3E: USGS StreamStats Program: A Case Study 152
References 153

Chapter 4   Design of Storm Drainage Systems 155

4.1   Introduction to Roadway Drainage Analysis 155
        4.1.1   Gutter Flow 155
4.2   Types of Inlets 160
4.3   Inlet Design 165
        4.3.1   Grate Inlets at Grade 165
        4.3.2   Curb Opening Inlets 171
        4.3.3   Slotted Inlets 172
        4.3.4   Combination Inlets 173
        4.3.5   New Jersey Inlets 173
        4.3.6   Grates on Sag 175
4.4   Inlets Spacing 176
4.5   Inlets on Roadways at 0% Grade 181
4.6   Design of Storm Drains 184
4.7   Hydraulic Design of Culverts 191
4.8   Erosion Control at Outfalls 206
        4.8.1   Riprap Aprons 207
        4.8.2   Preformed Scour Holes 209
4.9   Drainage Channels 213
        4.9.1   Permissible Velocity Concept 213
        4.9.2   Tractive Force Method 215
        4.9.3   Bare Soil and Stone Lining 217
        4.9.4   Side Slope Stability 219
        4.9.5   Grass Lining 221
        4.9.6   Manning's Roughness Coefficient Variation with Lining 224
        4.9.7   Channel Bends 234
        4.9.8   Composite Lining 237
4.10   Other Linings 238
        4.10.1   Gabion Baskets and Mattresses 239
        4.10.2   Turf Reinforcement Mats (TRMs) 241
        4.10.3   Erosion Control Blankets (ECBs) 242
        4.10.4   Properties of ECBs and TRMs 243
        4.10.5   Design of RECP Lined Channels 244
Problems 246
Appendix 4A: Derivation of Gutter Flow Equation 249
Appendix 4B: Derivation of Flow Equations for Inlets on Roadways at 0% Grade 251
Appendix 4C: Hydraulic Design Charts for Inlets 253
Appendix 4D: Critical Flow Charts for Round and Elliptical Pipes 254
Appendix 4E: Permissible Shear Stress of Cohesive Material in HEC-15 255

Appendix 4F: Propex Turf Reinforcement Mats and ArmorMax Properties 256
Appendix 4G: Landlok Erosion Control Blankets and SUPERGRO Properties
      by Propex, Inc..260
References 264

Chapter 5   Storm Water Management Regulations...265

    5.1   Introduction, Federal Regulations 265
        5.1.1   NPDES, Phase I Program...265
        5.1.2   NPDES, Phase II Program 266
    5.2   An Overview of Current Storm Water Management Regulations 268
        5.2.1   EISA Section 438 271
    5.3   NJDEP Storm Water Management Regulations...271
        5.3.1   Runoff Quantity Requirement..272
        5.3.2   Storm Water Quality Standards 273
        5.3.3   Groundwater Recharge Standards...275
        5.3.4   Runoff Calculation Methods 278
        5.3.5   Standards for Storm Water Management Structures.........................278
        5.3.6   Nonstructural Storm Water Strategies 278
        5.3.7   Municipal Storm Water Management Review 279
        5.3.8   Suggestions for Improving the NJDEP Regulations 279
    5.4   State of Maryland Storm Water Management Regulations...........................280
        5.4.1   Water Quality Volume, WQ_v 281
        5.4.2   Recharge Volume Criteria, Re_v 282
        5.4.3   Channel Protection Storage Volume Criteria, Cp_v...........................284
        5.4.4   Overbank Protection Volume Criteria, Q_p 286
        5.4.5   Extreme Flood Volume Criteria, Q_f 286
        5.4.6   BMP Design 286
        5.4.7   Environmental Site Design (ESD) 290
              5.4.7.1   ESD Storm Water Management Requirements 292
        5.4.8   Addressing ESD 294
              5.4.8.1   Alternative Surfaces 294
              5.4.8.2   Nonstructural Practices 295
              5.4.8.3   Microscale Practices...296
        5.4.9   Redevelopment 304
              5.4.9.1   Introduction 304
              5.4.9.2   Redevelopment Policy...304
        5.4.10   Special Criteria...305
              5.4.10.1   Sensitive Waters...305
              5.4.10.2   Wetlands, Waterways, and Critical Areas 306
    5.5   State of New York Storm Water Regulations 306
        5.5.1   Introduction 306
        5.5.2   Water Quality Volume...307
        5.5.3 WQ_v Treatment Practices...308
        5.5.4   Stream Channel Protection Volume Requirement (Cp_v).................309
        5.5.5   Overbank Flow Control Criteria (Q_p)... 311
        5.5.6   Extreme Flood Control Criteria (Q_f).. 312
        5.5.7   Downstream Analysis 313
        5.5.8   Conveyance System Design Criteria 314
        5.5.9   Storm Water Hotspots 315

5.5.10 Redevelopment Projects 316
5.5.11 Enhanced Phosphorus Removal Standards 316
Problems 316
Appendix 5A: NJDEP-Approved TSS Removal Rate for Vegetated Filter Strips ... 318
5A.1 Required Filter Strip Length 318
Appendix 5B: Maryland's Reduced Curve Numbers for ESD Sizing Requirement 321
References 323

Chapter 6 Manufactured Water Treatment Devices 325
6.1 An Overview 325
6.2 Certification of Water Quality Devices 327
6.2.1 NJCAT Certification 328
6.3 Types of Manufactured Devices 332
6.3.1 Catch Basin Inserts 332
6.3.2 Hydrodynamic Separation Water Quality Devices 337
6.3.3 Media Filtration Water Quality Devices 343
6.4 Bioretention Cells 350
Problems 362
Appendix 6A: Verification of Terre Kleen by EPA and NSF 363
6A.1 Technology Description 363
6A.2 Verification Testing Description 364
6A.2.1 Methods and Procedures 364
6A.3 Verification of Performance 365
6A.3.1 Test Results 365
6A.3.2 System Operation 366
6A.3.3 Quality Assurance/Quality Control 366
6A.3.4 Note for This Revision 366
6A.3.5 Availability of Supporting Documents 367
Appendix 6B: NJCAT 2013 Procedure: Appendix A—MTD Verification Process 368
Appendix 6C: NJDEP Certification Letter for Filterra Bioretention Systems,
Formerly a Division of Americast, Inc. Now a Part of Contech 370
Appendix 6D: Dimension and Capacity of CDS Models 372
References 373

Chapter 7 Structural Storm Water Management Systems 375
7.1 Detention Basins/Wet Ponds 375
7.1.1 Flow Routing through Detention Basins 377
7.1.2 Outlet Structure Design 379
7.1.2.1 Orifice 379
7.1.2.2 Rectangular Weir 379
7.1.2.3 Triangular Weir 380
7.1.2.4 Cipolleti Weir 381
7.1.2.5 Broad-Crested Weir 382
7.1.2.6 Overflow Grates 382
7.1.2.7 Stand Pipes 383
7.1.2.8 Hydro-Brake, Fluidic-Cone 383
7.1.2.9 Thirsty Duck 383

7.2   Preliminary Sizing of Detention Basins...386
       7.2.1   Rational and Modified Rational Methods Estimation.....................386
       7.2.2   SCS TR-55 Method Estimation 387
       7.2.3   Universal Method of Storage Volume Estimation...........................388
       7.2.4   Adjusting Detention Storage Volume Estimation 388
7.3   Extended Detention Basins...392
7.4   Underground Detention Basins 411
       7.4.1   Solid and Perforated Pipes 412
       7.4.2   Chambers...415
       7.4.3   Plastic and Concrete Vaults 420
7.5   Water Treatment Structures...440
       7.5.1   Vegetative Swales 440
       7.5.2   Sand Filters..442
7.6   Infiltration Basins 448
7.7   Retention–Infiltration Basins...459
       7.7.1   Dry Wells 459
Problems..474
References 477

Chapter 8   Newer Trends in Storm Water Management (Green Infrastructure)........................479

8.1   Introduction/Source Reduction/Control 479
       8.1.1   Introduction 479
       8.1.2   Source Reduction 479
       8.1.3   Source Control...480
       8.1.4   Source Reduction Benefits 481
8.2   Low-Impact Development 482
8.3   Smart Growth 483
8.4   Green Infrastructure...484
8.5   LEED and Green Buildings 485
8.6   Porous Pavements 488
       8.6.1   Open Cell Paving Grids 488
       8.6.2   Porous Asphalt 489
       8.6.3   Pervious Concrete 493
       8.6.4   Glass Pave..495
       8.6.5   Concrete Pavers 495
       8.6.6   Open Cell Pavers 500
       8.6.7   Nonconcrete Pavers 502
8.7   Green Roofs..505
       8.7.1   Green Roof Construction 505
       8.7.2   Storm Water Management Analysis of Green Roofs 508
8.8   Blue Roofs 511
8.9   Storm Water Wetlands... 511
8.10   Subsurface Gravel Wetlands..514
8.11   Filter Strips 516
       8.11.1   Application 516
       8.11.2   Design Criteria 519
8.12   Bioretention Basins, Swales, and Cells 520
       8.12.1   Bioretention Basins 521
       8.12.2   Bioswales...522
       8.12.3   Bioretention Cells..523

8.13   Rain Gardens...525
8.14   Cost Effectiveness of BMPs 530
8.15   Other Nonstructural Measures 531
       8.15.1   Common (General) Measures 531
       8.15.2   Clustered Developments..532
8.16   Minimal Impact Developments 532
8.17   Storm Water Fees 537
Problems...538
Appendix 8A: Drivable Grass® Technical Specification Guide 540
References 543

Chapter 9   Installation, Inspection, and Maintenance of Storm Water
Management Systems..545
9.1   Soil Erosion and Sediment Control Measures.................................545
9.2   Installation of Pipes..545
       9.2.1   Round Reinforced Concrete Pipes 547
       9.2.2   Elliptical Concrete Pipes...551
       9.2.3   Prestressed Concrete Pipes..551
       9.2.4   Concrete Box Culverts 554
       9.2.5   HDPE Pipes...554
       9.2.6   Dewatering 559
9.3   Watertight Joints...561
       9.3.1   Pipe Joints 561
       9.3.2   Pipe Connection to Manhole/Inlet 563
       9.3.3   Infiltration/Exfiltration Testing 565
9.4   Construction of Detention Basins/Ponds...565
       9.4.1   Detention Basins..565
       9.4.2   Infiltration Basins 566
       9.4.3   Wet Ponds 567
       9.4.4   Grass Swales..567
       9.4.5   Dry Wells and Infiltration Chambers 568
       9.4.6   Outlet Structures 569
9.5   Slope Stabilization...570
9.6   Inspection and Maintenance...571
       9.6.1   Objectives of Inspection and Maintenance 571
       9.6.2   Maintenance of Vegetative and Paved Areas 573
              9.6.2.1   Lawns/Landscapes..573
              9.6.2.2   Pavements 574
       9.6.3   Maintenance of Storm Water Drainage Systems 574
              9.6.3.1   Restoration of Grass- and Riprap-Lined Swales 574
              9.6.3.2   Snow and Ice Removal 575
              9.6.3.3   Removal of Sediment and Floatables from Drainage
                      Systems 575
              9.6.3.4   Control of Potential Mosquito Breeding Habitats 575
       9.6.4   Maintenance of Ponds/Detention Basins 575
              9.6.4.1   Algae and Weed Control..................................575
              9.6.4.2   Underground Detention Basins........................576
              9.6.4.3   Wet Ponds 576
              9.6.4.4   Outlet Structures...576

9.6.5   Maintenance of Water Treatment Devices 577
            9.6.5.1   Catch Basin Inserts 577
            9.6.5.2   Manufactured Water Treatment Devices........................577
9.6.6   Repair of Storm Water Management Facilities................................577
9.6.7   Neglect in Maintenance...579
9.7   Inspection, Operation, and Maintenance Manual 579
Problems..586
Appendix 9A: Installation Of HDPE Pipe by Advanced Drainage System (ADS)...587
9A.1   Backfill Envelope Construction..587
9A.2   Backfill Placement..588
9A.3   Mechanical Compaction Equipment 588
9A.4   Joints...589
9A.5   Construction and Paving Equipment..589
9A.6   Joining Different Pipe Types or Sizes 589
9A.7   Curvilinear Installations 590
9A.8   Vertical Installations 590
9A.9   Steep Slope Installations 590
9A.10   Cambered Installations..591
9A.11   Sliplining 591
Appendix 9B: Installation Guidelines For Armormax 592
Appendix 9C: Construction Inspection Checklist: An Overview...........................596
9C.1   Soil Erosion and Sediment Control Measures 596
9C.2   Excavation 596
9C.3   Pipe Installation..596
            9C.3.1   Trenching 596
            9C.3.2   Pipe Laying..596
            9C.3.3   Manholes/Inlets 596
            9C.3.4   Backfilling 596
            9C.3.5   Repairs 597
9C.4   Site Restoration 597
Appendix 9D: General Specifications for Maintenance of FloGard+Plus
                Catch Basin Insert Filters 598
References 599

Chapter 10 Water Conservation and Reuse 601
10.1   Trends in Supply and Demand 601
10.2   Water Conservation 602
10.3   Indoor Conservation 603
            10.3.1   Residential Buildings 603
            10.3.2   Urinals in Nonresidential Buildings...................................605
            10.3.3   Other Indoor Saving Tips...606
            10.3.4   Economy of Water-Saver Fixtures 606
10.4   Outdoor Conservation 607
            10.4.1   An Overview 607
            10.4.2   Conservation of Outdoor Water Use: A Summary 612
            10.4.3   Other Water Conservation Measures 613

10.5 Water Reuse 613
  10.5.1 Wastewater Reuse 616
  10.5.2 Recycled Wastewater Market 617
  10.5.3 Reuse of Grey Water 619
  10.5.4 Treatment of Wastewater and Grey Water 620
10.6 Reuse of Rainwater and Storm Water Runoff 621
  10.6.1 Quantity of Urban Runoff 623
10.7 Rainwater Harvesting 625
  10.7.1 Harvesting Roof Rain 626
  10.7.2 Problems with Rain Barrels 630
10.8 Suggested Actions for Widespread Conservation and Reuse 631
  10.8.1 Public Education 631
  10.8.2 Task Force 631
  10.8.3 Reaching Out 631
  10.8.4 Reward 631
  10.8.5 Block Programs 631
  10.8.6 Enforcement 631
  10.8.7 Pilot Projects 631
  10.8.8 Organizations/Alliances for Water Reuse 632
  10.8.9 Benefits of Water Conservation and Reuse 632
Problems 633
Appendix 10A: List of Programs and Nonprofit Organizations for Water
  Conservation and Reuse 634
  10A.1 Water Conservation 634
  10A.2 Water Conservation and Reuse—Nonprofit Organizations 634
Appendix 10B: EPA 2012 Guidelines for Water Reuse 636
References 639

Glossary 641

Appendix A: System International (SI) 651

**Appendix B: Unified Soil Classification System and Nominal Sizes of Coarse
and Fine Aggregates** 655

Index 663

Preface

Urbanization has had a drastic impact on the natural process of storm water runoff. It has increased both the peak and the volume of runoff, has reduced infiltration, and has caused water pollution. Traditionally, runoff used to be conveyed by storm drains directly into streams and lakes. To avoid increased flooding, towns and states adopted regulations that initially mandated maintaining the peak rates of runoff from urban developments. However, since this did not fully address the flooding problem, some regulatory agencies later required certain reductions in the peak rates of runoff.

To address the concerns on substantial pollution from nonpoint sources, the US Environmental Protection Agency (EPA) promulgated the National Pollutant Discharge Elimination System (NPDES), phase II, permitting program under the 1987 Water Quality Act. This act applies to all municipal separate storm and sewer systems (MS4s) and construction sites disturbing over 1 acre of land. Many states have since adopted regulations to extend the NPDES to smaller sites. New Jersey, among others, has also adopted regulations that include groundwater recharge. To address the evolving regulations, storm water management practices have been continually changing during the past 30 years, and they continue to change.

Aside from teaching and practicing in various domains of hydraulic engineering since receiving my PhD in February 1970 under the supervision of the late Dr. Ven Te Chow, I have been extensively involved in the field of storm water management since 1985. This experience has included design of drainage and storm water management systems for hundreds of projects and teaching pertinent courses, including "Urban Storm Water Management," "Drainage Design," "Watershed Modeling," and "Advanced Hydraulics" at the Stevens Institute of Technology.

In addition, I have reviewed drainage design by others. In this respect, I have reviewed thousands of plans and storm water management calculation reports submitted by consulting engineers to the municipalities that my employer has served as a municipal engineer. In my experience, a significant number of civil engineers need more knowledge to accurately perform runoff calculations and properly design storm water management systems, including detention/retention basins. Thus, there is a great need for a practical, concise, yet thorough book that will guide practicing engineers and municipal planners in the design of storm water management elements. This book is intended to meet this objective and serves as a "cookbook" on urban storm water management, which is one of the most challenging and dynamic fields of engineering.

The book covers all the subject matter needed to guide practitioners to design drainage and storm water management systems efficiently. It includes numerous examples of hydrologic and hydraulic calculations involved in this field. The book also contains ample case studies that exemplify the methods and procedures for the design of drainage networks and structural and nonstructural storm water management systems, such as extended detention basins, infiltration basins, underground retention/infiltration basins, rain gardens, pervious pavements, and vegetative buffers.

Since a vast majority of practicing engineers in the United States has yet to become familiar with the System International (SI) system of units, English units are used throughout the book. However, all equations and a large number of examples and case studies are presented in SI units as well, so the book can be used internationally.

The book is divided into 10 chapters. Chapter 1 provides an introduction to the impacts of development on the quality and quantity of storm water runoff. Chapter 2 presents an overview of pipe and open channel flow equations, supplemented with charts and tables to simplify hydraulic design of runoff conveyance systems.

Chapter 3 covers the elements of the rainfall-runoff process and includes several examples that guide practitioners to perform runoff calculations accurately. Included in this chapter is a universal rainfall-runoff model that I have developed. Chapter 4 covers the design of inlets, storm drains, culverts, vegetative swales, and erosion control systems. The examples in this chapter show that roadway inlets have far less capacity in intercepting runoff than many engineers perceive.

Chapter 5 contains an overview of the EPA and sample states' storm water management regulations, including those of New Jersey and Maryland, which have among the most stringent rules in the nation. This chapter also covers the shortcomings of the regulations and suggestions for improvements. Chapter 6 provides a description of various types of manufactured storm water treatment devices that are increasingly employed to address the applicable water quality requirements.

Chapter 7 covers design of various types of structural storm water management systems such as detention/retention and infiltration basins. It provides examples and case studies to guide the practitioner to design structural storm water management systems effectively. Chapter 8 provides a description of various types of nonstructural, source-reduction measures such as porous pavements, rain gardens, green roofs, blue roofs, and filter strips, among others, some of which are generally more effective and far less expensive than structural systems. These measures, also referred to as green infrastructure, have become increasingly popular in recent years and are forming the future trend in storm water management. This chapter also provides a comparative cost analysis of various non-structural practices.

Chapter 9 includes an overview of the installation methods of drainage and storm water management facilities. It also presents suggested maintenance measures for storm water management elements. While maintenance is crucial for the proper functioning of a system, it has been, and is, generally being neglected. Also, the cost of maintaining systems as required by applicable regulations will most likely exceed anyone's expectations.

Chapter 10 shows how storm water runoff is a resource that can be conserved and used in an efficient and cost-effective manner. This view of storm water runoff as a resource is contrary to a general view that the runoff is a waste to be disposed of in a regulated manner. The chapter further discusses the use of rainwater from roofs and suggests the sizing of rain tanks to collect it. This concept, which I initially introduced in 1994, is now vastly being marketed as rain barrels, which are smaller than the rain tanks I had suggested.

This book is targeted at a large readership. The subject matter is of interest to all professionals involved in the design of drainage and storm water management systems for urban developments and roadway projects, as well as those engaged in urban planning, including municipal engineers and officials and water resources planners. In addition, it can be used as a textbook in an upper undergraduate/graduate-level course on storm water management. Because of numerous examples and case studies and the inclusion of approximately 220 problems/questions together with a solutions manual, it would be an ideal selection for classroom teaching. Due to an ever-increasing demand for storm water management practitioners, more and more universities around the country are offering courses in storm water management.

Due to my busy schedule, it took many evenings, weekends, and holidays to prepare the first edition of this book. It also took months to update the book, cover additional material, and add many more worked examples as well as over three times as many problems/questions in this edition. I am indebted to Dr. Stephen T. Boswell, president of Boswell Engineering, for his moral support and appreciation of my publication activities alongside engineering consultation. I would also like to thank Kathy Chwiej, who undertook the word processing of both the original manuscript and this edition, volunteering her free time.

Author

 **Hormoz Pazwash, PhD,** earned his BS, CE with highest honor among the entire graduating class of 1963 from Tehran University. He continued his graduate studies under the supervision of the late Dr. Ven Te Chow at the University of Illinois in Urbana-Champaign, earning an MS and a PhD in civil engineering. In 1970 he joined the Faculty of Engineering at Tehran University and, in the next 7 years, held the positions of assistant professor, associate professor, and chairman of the Department of Civil Engineering. His other academic appointments have included visiting professorships at the University of California, Berkeley; the University of Akron in Ohio; and an associate professorship at Northeastern University in Boston. He has also served as an adjunct professor at Stevens Institute of Technology in Hoboken, New Jersey. Dr. Pazwash has received various academic awards, including a fellowship at Tehran University and a Fulbright Scholarship at the University of California, Berkeley.

He is listed in the 1992–1993 premier edition of Marcus Who's Who in Science and Engineering and the International Who's Who of Professionals. Since 1985, Dr. Pazwash has been practicing as a consulting engineer. For nearly 30 years now, Dr. Pazwash has held the position of project manager and director of hydraulic/hydrologic engineering at Boswell Engineering in South Hackensack, New Jersey. He has also taught a number of senior- and graduate-level courses in the field of water resources and storm water management at Stevens Institute of Technology.

Dr. Pazwash is the author of nearly 60 papers, five books and a chapter of the Encyclopedia of Environmental Management, published September 2014. His professional experience covers a broad range of disciplines in water resources and hydrologic/hydraulic engineering. He has been involved in projects that include evaluation of regional water resources; the design of pipelines, channels, and culverts; hydrologic and hydraulic analysis of rivers and streams; flood control projects; reservoir and dam safety studies; and the design of drainage and storm water management systems. Dr. Pazwash is well respected in the field of engineering due to his professional accomplishments.

Dr. Pazwash holds professional engineering licenses in New Jersey and New York. He is a life member and Fellow of the American Society of Civil Engineers (ASCE) and a Diplomat of the American Academy of Water Resources Engineers (AAWRE), D.WRE. He is also a member of the American Water Resources Association (AWRA).

Abbreviations

AASHTO	American Association of State Highway Officials
AAWRE	American Academy of Water Resources Engineers
ac	acres = 43,560 square feet
ACIS	Applied climate information system
ANSI	American National Standards Institute
ASCE	American Society of Civil Engineers
ASTM	American Society for Testing and Materials
AWRA	American Water Resources Association
AWWA	American Water Works Association
BDF	Basin development factor
BMP	Best management practice
°C	Celsius, temperature in metric units
CAFRA	Coastal Area Facility Review Act
CCF	A unit of tap water use, equal to 100 cubic feet
Cd	Candela: unit of luminous
cf	cubic feet
cfs	cubic feet per second
CGS	Construction general permit
CLSM	Controlled low-strength material
cm	Centimeter
COD	Chemical oxygen demand
csm	Cubic feet per second per square mile (in TR-55 method)
CSO	Combined sewer overflow
CU	Customary units (English units)
CWA	Clean Water Act
DPW	Department of Public Works
D.WRE	Diplomat Water Resources Engineer
ECB	Erosion control blanket
ECOS	Environmental Council of States
EGL	Energy grade line
EISA	Energy Independence and Security Act
EPA	Environmental Protection Agency
ESD	Environmental site design
°F	Fahrenheit; temperature in English units
FEMA	Federal Emergency Management Agency
FHWA	Federal Highway Administration
ft	foot, feet
gpcd	gallons per capita per day
gpd	gallons per day
gpf	gallons per flush
gpm	gallons per minute
GSR-32	Geological survey report-32 (specific to New Jersey)
h	Hour
ha	Hectares
Hg	Mercury
HGL	Hydraulic grade line
in.	Inch, inches

kg	Kilogram
km	Kilometer = 1000 m
kN	KiloNewton
L	Liter
LEED	Leadership in energy and environmental design
LOD	Limit of development, limit of disturbance
Lpcd	Liters per capita per day
m	Meter
mgd	Million gallons per day
mi	Mile = 5280 ft
μm	Micron, micrometer (0.001 mm)
mm	Millimeter
mol	Mole
MOU	Memorandum of understanding
MS4	Municipal separate storm sewer system
MSGP	Multisection general permit
MTD	Manufactured treatment device
N	Newton (unit of force in metric units)
NAHB	National Association of Home Builders
NAS	National Academy of Sciences
NGVD	National Geodetic Vertical Datum
NJAC	New Jersey Administrative Code
NJCAT	New Jersey Corporation for Advanced Technology
NJDEP	New Jersey Department of Environmental Protection
NJPDES	New Jersey Pollutant Discharge Elimination System
NOAA	National Oceanic and Atmospheric Administration
NPDES	National Pollutant Discharge Elimination System
NPS	Nonpoint source
NRC	National Research Council
NRCS	Natural Resources Conservation Service
NSF	National Science Foundation
NSPS	Nonstructural strategies point system
NURP	National Urban Runoff Program
OSHA	Occupational Safety and Health Administration
Pa	Pascal (N/m^2)
RCP	Reinforced concrete pipe
RECP	Rolled erosion control products
ROW	Right of way
s	second
SCS	Soil Conservation Service (now NRCS)
SI	System International = metric units
SRI	Solar reflectance index
SWMM	Storm Water Management Model
SWPPP	Storm water pollution prevention plan
TARP	Technology acceptance and reciprocity partnership
TCLP	Toxicity characteristic leach procedure
TN	Total nitrogen
TP	Total phosphorous
TRM	Turf reinforcement mat
TSS	Total suspended solids
USCS	Unified Soil Classification System

USDA	US Department of Agriculture
USDOT	US Department of Transportation
USGBC	US Green Building Council
USGS	US Geological Survey
WEF	Water Environment Federation
WQE	Water quality event
WS	Water surface
ZLD	Zero liquid discharge

Notations

α	alpha: angle
γ	gamma: specific gravity
θ	theta: angle, moisture content
μ	mu: dynamic viscosity
ν	nu: kinematic viscosity
π	pi: 3.1415
ρ	rho: fluid density
Σ	sigma: sum
ψ	psi: soil suction
A	Area; cross section area of flow in pipe or open channel: Amperes
C	A constant; runoff coefficient
CN	Soil curve number in the SCS/NRCS runoff calculations methods
d	Depth of flow in channels
D	Pipe diameter
g	Acceleration of gravity: 9.81 m/s^2, 32.2 ft/s^2
I	Rainfall intensity (in./h, mm/h)
I_a	Initial abstractions
K	Kelvin; a pipe/channel flow factor
n	Manning's roughness coefficient
P	Wetted perimeter
q	Discharge per unit width
Q	Discharge, peak discharge (cfs, m^3/s)
R	Runoff depth (in., mm)
R, r	Hydraulic radius
S_c	Critical slope
S_g	Specific gravity
t	Time
V	Velocity
V_c	Critical velocity
y	Depth of flow
y_c	Critical dept

1 Urbanization Impact on Runoff

Urban development alters the natural hydrologic process. It increases and accelerates runoff, reduces infiltration, and deteriorates water quality. Neglecting these impacts in the past has aggravated flooding, caused stream pollution, and lowered water tables. These impacts are discussed in this chapter. In addition, urbanization causes thermal pollution from dark impervious surfaces, such as roofs and streets. This latter impact, which is linked to global warming, is, however, beyond the scope of this book.

1.1   IMPACTS ON STORM WATER QUANTITY

Rain falling on a virgin land such as forests, grasslands, and wetlands is partly intercepted by vegetation, partly retained by surface depressions and puddles, and partly infiltrates into the ground. Only a small fraction flows overland. Urbanization disturbs land and replaces natural vegetation with impervious surfaces such as roads, driveways, parking areas, and building roofs and compact soils. Thus, it eliminates interception, reduces surface retention, and diminishes infiltration.

Urban development begins with clearing woods and grading the ground. The trees that intercept rainfall are removed and natural depressions and puddles that temporarily impound rainfall are flattened to uniform grade. The humus layer of forest and the organic matter that act like a sponge and absorb rainfall are scraped off. The soil which allowed infiltration is also severely compacted. The land can no longer prevent rainfall from being rapidly turned to runoff. These significant impacts of development during construction are often overlooked in practice.

The impact of developments on runoff worsens after construction as roads, parking lots, driveways, rooftops, and other impervious surfaces stop the infiltration of rainfall into the ground. The result is increased flooding and a significant reduction in groundwater recharge. Because of soil compaction by heavy construction equipment, the impacts are more profound in areas where the soil has high permeability or has high organic matter. Healthy soil, containing 5% to 6% organic matter, can hold water up to 40% by volume (Hudson, 1994).

In addition, construction of drainage systems comprising inlets, pipes, and drainage channels to collect and convey the runoff from developments turns overland flow to concentrated runoff. The result is a shorter time of concentration (i.e., the time it takes for the runoff from the entire watershed area to reach its outlet point). These changes, namely reduction of rainfall abstraction and shortening of the time of concentration, create a significant increase in both the peak and volume of runoff but a reduction in ground water recharge. Unnaturally high storm water discharges overwhelm the capacity of streams and rivers and cause flooding. According to the EPA, approximately 24% of rivers and streams, nearly 458,000 km (271,000 miles), suffer from decreased vegetation cover, which leads to more erosion and water pollution (EPA, 2013a,b).

The impact of an urban development on the runoff and infiltration is depicted by Figure 1.1. The resulting change in the runoff hydrograph (a graph of runoff discharge with time) is shown schematically in Figure 1.2. These figures indicate that the urbanization increases the peak and volume of runoff, shortens the time of concentration, and also reduces the infiltration. By lowering the water table and reducing base flow, the hydrograph of a developed watershed has a sharper peak and lower base flow during dry weather. This impact is more pronounced in areas with soils of high permeability. In deep, sandy soils, rainfalls, however intense, infiltrate the ground and there is little runoff.

Before the eighteenth century, over 95% of the world's population lived in rural areas and farmed in the absence of any machinery. In the United States, the population was approximately four million people (1.3% of today's population) and 95% lived in rural areas in 1800. The world's

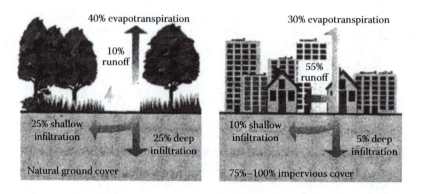

FIGURE 1.1   Impact of urbanization on infiltration and runoff. (From US Environmental Protection Agency, Protecting water quality from urban runoff, EPA 841-F-03-003, National Research Council, October 2008, *Urban Stormwater Management in the United States*, National Academy Press, Washington, DC.)

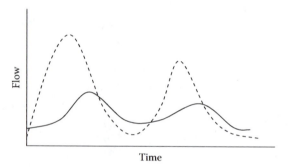

FIGURE 1.2   Effect of urbanization on runoff hydrograph.

population was tenfold less than that of today.* There was even far less production and consumption of foods and goods on a per-capita basis. As such, there were insignificant adverse impacts to the environment in general and storm water in particular. The Industrial Revolution, which developed in England between 1750 and 1850 and rapidly spread to other European countries, resulted in a change in living style and stimulated the growth of cities. Throughout Europe, the percentage of people who lived in cities rose from 17% in 1801 to 54% in 1891.

In the United States urbanization was a long and gradual process. Massachusetts and Rhode Island were among the northeastern states that already had an urban majority in 1850, but the majority of populations in southern states were still rural until after World War I, which ended in 1918.

Traditionally, storm water management practices considered runoff as a waste to be removed and disposed of quickly from developments. Thus, the runoff from urban and suburban developments and municipalities used to be collected and conveyed by drainage systems and directly discharged into lakes and streams. The result, apart from importing impurities to receiving water bodies, was increased occurrence of flash floods, elevated flood levels, and expansion of flood-prone areas, which adversely impacted properties along and adjacent to streams and lakes.

To offset these adverse impacts on runoff quantity, municipalities and states have adopted storm water management regulations. Earlier regulations (and practices as well) were aimed at maintaining

---

* According to the United Nations 1999 estimate, the world population grew from 980 million in 1800 to 2.52 billion in 1950. This implies a 0.6% annual growth in population within that period, which seems too low. Estimating the annual population growth during the identified period at 1%, the world population in 1800 would be calculated at 566 million. Note that the population has grown at a rate of approximately 1.7% from 1950 to 2010, when the world population reached 7 billion.

the peak rate of runoff for a selected storm frequency, commonly a 10- or 25-year storm event. However, a single storm criterion was found to be rather ineffective since it would not attenuate runoff sufficiently to maintain predevelopment peaks for other storm frequencies. The regulations were subsequently amended to cover multiple storm events, covering a large range in frequency, such as 1- or 2-, 10- and 100-year storms. To address these regulations, detention basins and ponds were employed in practice. These practices, too, were found not to fully address the impact of a development. Because of having a larger volume than predevelopment conditions, the outflow hydrograph from a detention basin/pond was more prolonged than that of predevelopment. As a result, the composite discharges from detention basins/ponds in a watershed exceeded the predevelopment runoff in that watershed. To compensate for these effects, as will be discussed in a later chapter, some states, New Jersey included, adopted regulations that require certain reductions in the peak rates of runoff.

1.2 IMPACTS ON WATER QUALITY

Storm water has long been regarded as a major source of urban flooding; however, its role in degrading streams, lakes, and rivers has been given attention in the past three decades. Still, many people do not believe that rain falling on pavement and carried by a storm drain is polluted enough to need treatment.

The porous natural terrain of varied landscapes like forests, grasslands, and wetlands absorbs rainfall and allows it to filter into the ground. Urbanization replaces native soil cover with pavements, such as roofs, roads, and driveways, and eliminates both absorption and filtration processes. Most of the pure rain and snow melt mix with man-made or natural pollutants on the ground, flow into drainage systems, and are quickly transported into streams and lakes. The forceful power of this flow erodes the bed and banks of the receiving water body and creates more sediment pollution. It also results in destruction of interstitial spaces within substrate where invertebrates live. Sensitive species, such as stonefish and caddis flies, have begun to disappear.

It is to be noted that mining, septic systems, agriculture, and airborne sediment also contribute to storm water pollution. Among these, agriculture is the leading source of contamination of streams and lakes in the United States. However, unlike golf courses and landscaping, agriculture is exempt from the Clean Water Act in the United States and will remain a source of water pollution until it is regulated.

According to the US Environmental Protection Agency (EPA), storm water runoff is the most common source of water pollution (EPA, 2000). The pollutants that are present in urban areas come from various diffuse or nonpoint sources. A nonpoint source pollutant is contrasted with a point source pollutant where discharge to a water body occurs at a single location, such as outfalls from a chemical factory or sewage treatment plant. Because of originating from many different sources and spreading out overland, nonpoint source (NPS) pollution is more difficult to control or regulate than point source pollution, such as municipal waste. Also because of the nonlinear relationship between land uses and pollutant loading, a conclusive cause–effect relationship cannot be established.

What adds to this difficulty is the varied nature of storm water runoff. The composition and magnitude of urban runoff is highly time dependent. Unlike municipal sewage, which is continuous and which does not vary more than a few fold daily, the flow of storm water is intermittent. Both the flow rate and the pollutant loads can vary by several orders of magnitude during a storm period. Consequently, the traditional centralized water quality treatment practices, also referred to as end-of-pipe practices, are far less effective than sewage treatment plants. Most common pollutants present in urban storm water originate from

- Soil erosion from bare land—an acute problem during construction
- Lawn chemicals such as fertilizers and pesticides
- Road salt and other de-icing substances
- Household products (paints, thinners, solvents, cleaning agents, etc.)
- Heavy metals from roof shingles, motor vehicles

TABLE 1.1

Median Event Mean Concentration for Urban Land Uses

Pollutant	Units	Residential		Mixed		Commercial		Open/Nonurban	
		Median	COV	Median	COV	Median	COV	Median	COV
BOD	mg/L	10	0.41	7.8	0.52	9.3	0.31	–	–
COD	mg/L	73	0.55	65	0.58	57	0.39	40	0.78
TSS	mg/L	101	0.96	67	1.14	69	0.85	70	2.92
Total lead	µg/L	144	0.75	114	1.35	104	0.68	30	1.52
Total copper	µg/L	33	0.99	27	1.32	29	0.81	–	–
Total zinc	µg/L	135	0.84	154	0.78	226	1.07	195	0.66
Total Kjeldahl nitrogen	µg/L	1900	0.73	1288	0.50	1179	0.43	965	1.00
Nitrate + nitrite	µg/L	736	0.83	558	0.67	572	0.48	543	0.91
Total phosphorus	µg/L	383	0.69	263	0.75	201	0.67	121	1.66
Soluble phosphorus	µg/L	143	0.46	56	0.75	80	0.71	26	2.11

Source: US Environmental Protection Agency, *Results of the Nationwide Urban Runoff Program: Vol. 1—Final Report*, Water Planning Division, Washington, DC, December 1983, National Technical Information Service (NTIS) publication no. 83-185552.

Note: COV: coefficient of variation.

- Oil leaks and illicit disposal
- Dust—atmospheric and automobiles (tire and pavement wear)
- Failing septic systems and illicit sewer connections

In addition, runoff from developed areas can raise temperature in streams. Table 1.1 presents a comparison of pollutants in urban runoff of various land uses.

Causes of urban pollution are readily detectable and include

- Water decolorization
- Excessive plant growth in streams and lakes
- Scum and algae floats near lake shores
- Unpleasant odors
- Fewer fish and wildlife
- Fish kill
- Sediment accumulation in storm drains and ditches

1.3 NPS POLLUTANTS AND THEIR IMPACTS

As indicated, a large number of pollutants are present in urban runoff. The sources and impacts of pollutants have been researched by many organizations including the US EPA. The EPA in cooperation with the US Geological Survey (USGS) conducted a comprehensive research study of urban storm water pollution across the United States between 1979 and 1983. The result of this research project, called National Urban Runoff Program (NURP), was published in *Vol. 1 Final Report* (EPA, 1983a) and *Executive Summary* (EPA, 1983b). Among the conclusions of the study are the following:

- "Heavy metals [especially copper, lead, and zinc] are by far the most prevalent priority pollutant constituents found in urban runoff.... Copper is suggested to be the most significant threat of the three."

TABLE 1.2
Sources of Pollutants in Urban Runoff

Pollutant	Source
Floatables	Shopping centers, streets, parking lots, parks and recreational areas
Sediment	Construction sites, roads, lawns and gardens
Nitrogen and phosphorus	Lawn fertilizers, detergents, pet wastes, automobile deposition
Organic materials	Lawns and gardens, parks, golf courses, leaves and animal wastes
Pesticides and herbicides	Lawns and gardens, roadside channels, parks, golf courses
Metals	Atmospheric deposition, automobiles, industrial sites, steel bridges corrosion
Oil and grease	Parking lots, truck stops, roads, driveways, car washes and gas stations, car and truck service stations, and illicit dumping
Bacteria, coliform	Lawns, roads, septic systems, leaky sanitary sewers, pet wastes, Canadian geese droppings

- "Coliform bacteria are present at high levels in urban runoff."
- "Nutrients are generally present in urban runoff, but … [generally] concentrations do not appear to be high in comparison with other possible discharges."
- "Oxygen demanding substances are present in urban runoff at concentrations approximating those in secondary treatment plant discharges."
- "The physical aspects of urban runoff, e.g., erosion and scour, can be a significant cause of habitat disruption and can affect the type of fishery present."
- "Detention basins … [and] recharge devices are capable of providing very effective removal of pollutants in urban runoff."
- "Wet basins (basins which maintain a permanent water pool) have the greatest performance capabilities."
- "Wetlands are considered to be a promising technique for control of urban runoff quality."
- "Organic priority pollutants in urban runoff do not appear to pose a general threat to freshwater aquatic life."

In 1987, the EPA amended the Clean Water Act of 1972* requiring states, local governments, and industry to address pollution sources indicated in the 1983 report. This amendment mandated any construction activity that disturbs 5 acres (±2.0 ha) or more and all municipal separate sewer storm water systems (MS4s) to obtain a National Pollutant Discharge Elimination System (NPDES) permit. As will be discussed in Chapter 5, the NPDES permit has been amended to cover smaller sites.

Major pollutants generated by urban storm water runoff are sediment, lawn fertilizers and nutrients, heavy metals, hydrocarbons, and coliform. Table 1.2 lists sources of contaminants in urban storm water runoff and Table 1.3 shows typical concentration of storm water pollutants.

In a 2000 report to Congress, the EPA cited diffused (nonpoint) sources of pollution as the top reason for those national waterways that were too polluted for swimming and fishing. The EPA's "National Water Quality Inventory; 2002 Report to the Congress" identified urban runoff as one of the leading sources of water quality impairment both in surface water and groundwater (http://www.epa.gov/305b/2002report-catched). In March 2013, the EPA released the "National Rivers and Streams Assessment Report." This report was based on a survey conducted in 2008 and 2009 covering a total of 1924 randomly selected sites across the contiguous 48 states. Of these, half were large streams and rivers. The study indicates that 55% of 1.19 million miles (1.92 million kilometers) of the nation's river and stream lengths are in poor biological condition, 23% in fair condition, and only

---

* The Clean Water Act of 1972 was established to regulate discharges of point source pollutants into waters of the United States. This Act did not affect individual homeowners; however, it required municipal, industrial, or other facilities to obtain an NPDES permit from the EPA if their discharges went directly to surface waters (Haugton, 1987).

TABLE 1.3

Typical Pollutant Concentration in Urban Storm Water

Pollutant	Typical Concentration	Unit
Total suspended sediment	80	mg/L
Total phosphorus[a]	0.30	mg/L
Total nitrogen[b]	2.0	mg/L
Total organic carbon	12.7	mg/L
Fecal coliform bacteria	3800	MPN/100 mL
Escherichia coli bacteria	1450	MPN/100 mL
Oil and grease	3	mg/L
Petroleum hydrocarbons	3.5	mg/L
Cadmium	2	μg/L
Copper	10	μg/L
Lead	30	μg/L
Zinc	140	μg/L
Chlorides (winter only)	200	mg/L
Insecticides	0.1 to 2.0	μg/L
Herbicides	1 to 5.0	μg/L

Source: New Jersey Department of Environmental Protection (NJDEP), *Stormwater Best Management Practices Manual*, February 2004, Table 1-1, Trenton, NJ; *The State of New York Stormwater Management Design Manual*, August 2010, Table 2-1; Article 63, Chapter 1 of the Maryland Department of Environmental Protection Manual, Table 1.

Note: MPN: most probable number.

[a] Average total phosphorus concentration in residential and commercial sites are reported as 0.38 and 0.20 mg/L, respectively, by US EPA (1983); see Table 1.1.

[b] Total nitrogen in residential and commercial sites is on average 2.6 and 1.75 mg/L, respectively, US EPA (1983).

21% in good condition. Biological condition, which is the most comprehensive indicator of water body health, is related to total nitrogen, total phosphorous, and acidification, of which phosphorous and nitrogen are by far the most widespread. Of the nation's rivers and streams, 40% have high levels of phosphorous and 28% have high levels of nitrogen. Acidification, although a problem in less than 1% of the surveyed rivers and stream lengths, has a significant impact on biological conditions; poor biological condition is 50% more likely in waters affected by acidification.

To lessen adverse impacts of the urban runoff pollutants on the environment, proper measures should be taken to control them through source reduction and removal. The primary urban runoff pollutants and means of their control are briefly discussed in the following sections. More detailed information on the runoff quality can be found in Schueler (1987, 1997), US EPA (1983a, 2008), Walker (1987), Terrene Institute (1994), and Caltrans (2010).

1.3.1 FLOATABLES

Floatables include cans, bottles, jars, nylon bags, paper, cardboard, leaves, and branches. These materials are encountered in surface water and are no concern for groundwater contamination. Plastic materials are generally nondegradable and may last for centuries, so they build up behind culverts and clog storm drains; see Figure 1.3. The most effective means of controlling these materials is public education. If everyone recycled and no one littered, there would be little floatable waste.

FIGURE 1.3   Bottles and floatables trapped behind culverts. (Photo by the author.)

1.3.2 SEDIMENT

Sediment, which is one of the most prevalent pollutants in urban runoff, is generated due to developments. The largest amount of sediment load is created during the construction phase of a development project. Thus, it is imperative to install and maintain adequate sediment and erosion control measures during construction to avoid discharge of large quantities of sediment in the form of muddy water into downstream drainage systems, waterways, and lakes.

The sediment load discharged from a development increases significantly with the rainfall intensity (Pazwash, 1982b). A large storm event may deposit more sediment in a siltation basin than the cumulative load during the balance of the construction period. The same phenomenon is evidenced for lakes and reservoirs. A case in point was Sefidrud Dam in northern Iran, which was constructed during the 1954 to 1962 period. Sediment measurements during the 1954 to 1976 period indicated a range in the annual sediment inflow of 14 million metric tons to 218.3 metric tons in the water years 1955 and 1969, respectively.* The river discharge during the same period was measured at 3 billion m³ (106 million ft³) in 1955 and 14 billion m³ (494 billion ft³) in 1969. A major flood on March 10, 1969, carried 15.55 million tons of sediment in the reservoir, more than the sediment inflow during the entire water year of 1955. The data also revealed that the ratio of the sediment load to the stream discharge in the 1969 water year (a wet year) was over 3.3-fold greater than that in 1955, which was a dry year (Pazwash, 1982a).

A feasible means of removing sediment from storm water runoff is to route the runoff through a detention basin or pond (wet basin). A detention basin functions like a sediment basin in a water supply treatment plant. It retards the flow and allows the sediment to settle to the bottom. The sediment removal effectiveness (also called trap efficiency) of a pond or detention basin depends on the length and depth of the pond and the residence time, namely the time it takes for the runoff to be discharged from the basin. More importantly, this efficiency depends on the size, shape, and type of sediment material. The falling velocity of a particle varies exponentially with its size and can be calculated from the following equation:

$$C_D \left(\frac{\pi d^2}{4} \right) \left(\frac{\rho V_f^2}{2} \right) = \left(\frac{\pi d^3}{6} \right) (\gamma_s - \gamma_w) \tag{1.1}$$

---

* A water year begins October 1 and ends September 30 of the following year.

The term on the left side of the equation represents drag force on the particle as it falls in water and the term on the right side is the net weight of the particle in water. The parameters in this equation are

C_D = drag coefficient, function of the Reynolds number ($R_e = Vd/\upsilon$)
d = particle size, diameter for spherical particles
ρ = density of fluid
V_f = falling velocity
γ_s = unit weight of particle
γ_w = unit weight of water

Rewriting the preceding equation in terms of the falling velocity yields

$$V_f = 2 \left[\frac{g(S-1)}{3} \right]^{1/2} \left[\frac{d}{C_D} \right]^{1/2} \tag{1.2}$$

where S is the specific weight of the particle. Approximating $S = 2.65$, the preceding equation further simplifies as

$$V_f = 0.147 \left(\frac{d}{C_D} \right)^{1/2} \quad \text{SI (metric unit)} \tag{1.3}$$

$$V_f = 2.43 \left(\frac{d}{C_D} \right)^{1/2} \quad \text{CU (customary unit)} \tag{1.4}$$

In the preceding equations, d is expressed in millimeters and inches, respectively. For very small particles where $R_e \leq 1$, the drag coefficient is given by Stokes's law:

$$C_D = \frac{24}{R_e}; \quad R_e = \frac{Vd}{\upsilon} \tag{1.5}$$

where υ = the kinematic viscosity of water.
Combining Equations 1.2 and 1.5 yields

$$V_f = \frac{g(S-1)d^2}{18\upsilon} \tag{1.6}$$

where V_f = settling velocity m/s (ft/s) and d = particle size, m (ft).
For larger Reynolds numbers, a number of equations have been proposed. The following simple equation is proposed by Pazwash (2007, Chapter 7):

$$C_D = \left(\frac{24}{R_e} \right) \left(1 + R_e^{2/3}/6 \right) \tag{1.7}$$

This equation is in good agreement with experimental data for $1 < R_e < 1000$. For large granular particles such as coarse sand and gravel, where $R_e > 1000$, the drag coefficient ranges from 0.4 to 0.45 up to $R_e = 2 \times 10^5$, at which the flow becomes turbulent, and the drag coefficient drops to 0.2. Table 1.4, prepared by the author, lists settling velocity of spherical particles in water.

Table 1.4 implies that sand particles settle in a matter of minutes and silt in hours in a typical 1.5 m (5 ft) deep pond; however, clay particles ($d < 0.004$ mm) take days to settle. Therefore, the suspended sediment removal rate of a detention basin/pond is controlled by the type of sediment carried by runoff. This also implies that, contrary to some publications (NJDEP, 2004), the total

TABLE 1.4
Settling Velocity of Spherical Particles in Water[a]

Particle Diameter		Fall Velocity		Particle Size		Fall Velocity[b]	
mm	in	m/h	ft/h	mm	in	m/s	ft/s
0.001	4×10^{-5}	3.2×10^{-3}	1.05×10^{-2}	0.20	0.008	0.02	0.07
0.002	8×10^{-5}	1.3×10^{-2}	4.3×10^{-2}	0.30	0.012	0.04	0.13
0.005	2×10^{-4}	0.08	0.26	0.4	0.016	0.06	0.20
0.01	4×10^{-4}	0.32	1.05	0.5	0.020	0.08	0.26
0.02	8×10^{-4}	1.3	4.27	0.6	0.024	0.10	0.33
0.05	0.002	8.1	26.6	0.8	0.03	0.13	0.43
0.10	0.004	32.4	106.3	1.0	0.04	0.15	0.49

[a] At 20°C (68°F).
[b] Rounded to second decimal place.

suspended solids (TSSs) removal of a detention basin is not merely a function of detention time, but more importantly depends on the size and type of sediment material (fine sand, silt, or clay) as well as its density and therefore is site specific.

1.3.3 Nutrients and Pesticides

A comprehensive national study of nutrients in streams and groundwater was conducted by the US Geological survey from 1992 through 2004. The study results were published in the USGS (2010a) circular 1350, titled "The Quality of Our Nation's Water—Nutrients in the Nation's Streams and Groundwater, 1992–2004," and highlighted in the USGS fact sheet (2010-3078; USGS 2010b). The results indicate that excessive nutrient concentration is a widespread cause of ecological degradation and that despite major federal, state, and local regulations to control point and nonpoint sources and transport of nutrients, concentrations of nutrients have remained the same or increased in many streams and aquifers nationwide since the early 1990s.

The USGS study finds that nitrate levels in the Illinois River decreased by 21% between 2000 and 2010; however, the level continued to increase in the Mississippi and Missouri Rivers. The increase between 2000 and 2010 was 29% in the upper Mississippi River and 43% in the Missouri River. At the Mississippi River outlet to the Gulf of Mexico, the increase was 12% in the same period. Excessive nitrate and other nutrients from the Mississippi River strongly affect the extent and severity of the hypoxic zone, known as the dead zone, that forms in the northern Gulf of Mexico every summer. The dead zone is characterized by extremely low oxygen levels in the bottom or near-bottom waters, degraded water quality, and impaired marine life. The 2013 Gulf dead zone covered 15,120 km[2] (5840 square miles), an area the size of the state of Connecticut.

Phosphorus and nitrogen are the essential nutrients for plants and are also the prime nutrients in urban storm water runoff. These substances are mostly inorganic, comprising orthophosphates, nitrates, and ammonia. Nitrogen is usually added to a watershed as organic-N or ammonia (NH_3) and stays attached to the soil until oxidation converts it to nitrate. So nitrogen is most often transported by water as nitrate (NO_3). In rural and residential areas, significant amounts of nutrients originate from fertilizers, manure, or dairy farming. Pet wastes, Canadian geese droppings, detergents, and raw sanitary wastes also add to nutrient loading. It has been reported that urban watersheds typically generate 5 to 20 times as much phosphorus per unit volume of runoff per year as compared to underdeveloped watersheds in a given region (Walker, 1987).

The sources, dispersion, transport, and fate of phosphorus in the environment are highly complex. In urban and suburban storm water runoff, phosphorus sources include detergents, fertilizers, lubricants, household cleaners, paints, and, of course, natural soil. Lawns are significant contributors of nutrients in storm water runoff—four times higher than other land uses such as roads, driveways, and streets. So fertilizers from lawn and landscaping are major sources of nutrients. Soil tests in New Jersey indicate that most soils have plenty of phosphorus for plant growth. Therefore, the use of fertilizer containing phosphorus has been banned in many municipalities in New Jersey (http://www.nj.gov/dep/watershedmgt).

Phosphates often attach to fine soil particles and remain in the soil until it is either utilized by plants or carried away with the soil as suspended sediment. Nitrates, however, are much more soluble and during late winter and after heavy rainfalls may penetrate into the water table and contaminate groundwater. High levels of nitrogen and phosphorous increase algae growth, which harms water quality, food resources, and habitats, and decrease oxygen, which adversely affects the fish and other aquatic life. At high concentration, nitrates can also cause a public health hazard to drinking water.

According to the EPA's 2013 "National Rivers and Streams Assessment Report," over 27% of the streams and rivers have excessive levels of nitrogen and 40% have excessive phosphorous. The concentration of phosphorus and nitrogen in a river, lake, or estuary increases biological productivity, resulting in nuisance algae growth and eutrophic conditions. Eutrophication mostly occurs in small, semistagnant agriculture ponds and urban lakes where the water is retained for over 2 weeks. In the growing season, these water bodies experience chronic eutrophication whose symptoms are water discoloration, unpleasant odors, algae scum, low oxygen levels, release of toxins, and ill effects on fish life. Eutrophication is also a major environmental problem in all coastal areas in the United States.

Barnegat Bay in New Jersey and Chesapeake Bay in Maryland are the two worst basins for eutrophication in the nation. The watershed boundaries of Barnegat Bay almost fully coincide with the jurisdictional boundaries of Ocean County, which has been one of the fastest growing development areas in New Jersey. The vast development in this county since 2000 has been a prime reason for degradation of Barnegat Bay. According to a 2010 summit meeting at Rutgers University, Barnegat Bay absorbs an estimated 1.4 million lb (60,000 kg) of nitrogen load annually. Over two-thirds of this load is believed to originate from storm water runoff. The occurrence of high permeability soils in this area aggravates the impact of development on runoff, as not only the impervious area, but also soil compaction reduces the infiltration within disturbed land. While the natural soil was found to have a density specific gravity of 1.5, the specific gravity of turf in residential developments that were constructed in the 1970s was measured to range from 1.75 to 1.9. These measurements indicate a loss of soil porosity from less than 30% to over 40%. To restore soil permeability, the compacted soil should be loosened and amendments such as lime, organic matter, gypsum, and proper fertilizer should be added to the soil layers.

Total nitrogen (TN) and total phosphorous (TP) are expensive to remove from storm water. The average cost of removing TN and TP using a number of storm water management systems are reported at approximately $530/lb ($1175/kg) and $2750/lb ($6100/kg), respectively (England and Listopad, 2012).

The total phosphorus and total nitrogen concentrations in urban runoff, listed in Table 1.1, are significantly less than treated wastewater concentrations. However, it is to be noted that storm water volumes are far greater than sanitary flows during wet weather conditions.

Pesticides, which include insecticides, herbicides, rodenticides, and fungicides, are routinely used in urban areas and in agricultural lands. These substances can contaminate soil, water, and air and have toxic effects on the ecosystem and human life. Pesticides decrease aquatic populations either directly by damaging the food chain or indirectly by reducing the population of phytoplankton, which lowers oxygen levels in the water. Highly carcinogenic pesticides (such as DDT, dieldrin, and chlordane) have been removed from the market; the EPA banned diazinon and chlorpyritos for urban use in 2004 and 2005, respectively. These pesticides have been replaced by others, especially the pyrethroid pesticides in urban areas. A number of these pesticides, however, are more toxic to fish and zooplankton than the phased-out pesticides. Morcover, many of the pyrethroid pesticides tend to bond strongly to soil particles and accumulate in sediments (Lee and Jones-Lee, 2005). Urban waterways in California are found to experience widespread toxicity caused by application of insecticides currently registered by the EPA. Other commonly used pesticides, such as malathion, are suspected carcinogens through direct contact.

Upon application, pesticides leave the site by becoming dissolved in storm water runoff or by binding to suspended sediment carried in runoff. Pesticides can also contaminate groundwater through infiltration. The transport and fate of pesticides depend on their physical and chemical interaction with soil and water. The effect of fertilizer and pesticides on the environment depends on our gardening and lawn-care habits. Our gardening habits have significantly improved since decades ago when homeowners burned autumn leaves and applied harmful pesticides and too much synthetic fertilizer to their lawns and garden plants. But we still overfertilize and have a long way to go to minimize adverse effects on the environment. It is time to realize that lawns, apart from requiring repeated fertilizer applications, use tremendous amounts of water. Thus, we should consider downsizing lawn area, and instead use low-maintenance native plants and landscaping to form greenery.

1.3.4 HEAVY METALS

The main sources of metals in urban storm water runoff are automobiles and industry. According to the aforementioned NURP study, copper, lead, and zinc are by far the most prevalent heavy metals found in urban runoff. Atmospheric deposition, both wet and dry, can also make a significant contribution in some parts of the country. Metals can also occur naturally in soil and the quantities of metals leaching into water from natural sources elevates as water pH drops.

Zinc is one of the most prevalent heavy metals found in storm water runoff here in the United States and abroad. In New Zealand, for example, zinc is among the primary contaminants in urban runoff and is generated from galvanized roofing (Tveten and Williamson, 2006). Copper, zinc, and mercury can cause health problems, but lead is the prime concern in toxicity and public health. Lead tends to precipitate in aquatic systems and accumulate in soils and sediment (Lee and Jones-Lee, 2006). It has cumulative neurological adverse effects and is particularly harmful to children. Tetraethyl lead in gasoline had been one of the principal sources of lead in storm water runoff. However, with the use of unleaded gasoline, the pollution from this source has diminished.

The effective control of heavy metals is not simple. Although better gas mileage will reduce the production of heavy metals, a much more significant reduction can only be brought up by a change in our lifestyle, discussed later in this chapter.

1.3.5 PATHOGENS, FECAL COLIFORM

Pathogens are dangerous microscopic viral or bacterial organisms that can cause some form of illness. Fecal wastes from pets, birds, and Canadian geese are the prime sources of pathogens in urban storm water runoff. However, overflow from combined sewers is by far the largest source of pathogens where storm and sewer systems are combined. Many older, more intensely developed cities and municipalities in the United States do not have separate storm and sanitary sewers. Examples include Fort Lee, parts of North Bergen, and Ridgefield in the state of New Jersey and the boroughs of Brooklyn and Queens in New York City. Improperly sized or located septic systems contaminate groundwater, especially in areas of very permeable soil and/or high water tables and where fractured rocks and well casings create entry routes.

Storm water runoff poses little threat to human health if it does not come in contact with domestic wastewater or fecal waste. However, when inadequately treated combined sewer flows are discharged to beaches and lakes, there is a public health risk due to pathogen contamination. Likewise, the risk exists where runoff comes into contact with shellfish beds and in swimming ponds where large flocks of Canadian geese may roam.

Runoff contaminated with fecal sources can transmit a number of human diseases. Some well-known bacteria agents include the *Salmonella* group, which causes typhoid fever and intestinal fever, and the Shigella genus group which causes bacterial dysentery. Other bacterial agents include *E. coli* and *Vibrio cholerae*, the latter of which can cause cholera. Gastroenteritis, which is caused by *E. coli*, is the leading waterborne human infectious disease in the United States.

Fecal coliform concentrations were studied by the EPA (1983a,b) at 17 sites for 156 storm events. The study showed that coliform bacteria are present at high levels in urban runoff and can be expected to exceed EPA water quality criteria during and immediately after storm events. Coliform counts during warmer periods were found to be approximately 20 times greater than those found during colder months. This could be partly due to washout of pets' and birds' wastes with runoff, which is far greater from rain in summer than snow in winter.

Concentrations of coliform bacteria and *E. coli* bacteria serve as indicators for the presence of pathogenic bacteria. The bacteria TMDL (total mean daily load) adopted by Ventura County in California in 2006 includes both a summer dry weather single day maximum of 235 MPN (most probable number) per 100 milliliters and a 30-day geomean at 126 MPN. Monitoring bacterial TMDL at locations within the city of Thousand Oaks in California exceeded summer dry weather standard 30% to 40% of the time (Carson and Sercu, 2013).

In a study of impaired waters and total maximum daily loads, the EPA found that pathogens are by far the leading cause of impairment in 303(d)* listed waters nationwide with the most common being fecal coliform or *E. coli* (Kaspersen, 2009). The aforementioned 2013 EPA study finds that about 9% of streams and rivers have high levels of bacteria. A number of attempts have been made to chase away Canadian geese from the pond perimeters. None have been successful. The author suggests taking away the birds' fresh eggs to reduce Canadian geese populations.

1.3.6 ROAD SALT

Road salt, namely common salt (sodium chloride), which is readily available, inexpensive, and effectively depresses the freezing point of ice, is the most widespread de-icing substance used in many states. New Hampshire was the first state in the United States to use salt on an experimental basis to melt snow on roadways in 1938 (Richardson, 2012). Three years later, nearly 5,000 tons of salt were spread on US highways. According to the EPA more than 13 million tons (11.8×10^6 metric tons)[†] of salt are applied to roads annually throughout the country (*Civil Engineering*, 2005). Now over 18 million tons (16.3 million metric tons) are used for de-icing in the United States each year.

The use of road salt for de-icing has doubled during the past 20 years. Massachusetts, New Hampshire, and New York have higher than average use; Massachusetts is highest at 19.94 tons per lane-mile (11.2 metric tons/lane km) per year. New York City's road salt use averages 500,000 tons/year or 16.6 tons/lane-mile/year (9.4 metric tons/lane-km/year). The New York State Department of Transportation requires an application rate of 225 lb/lane-mile (63 kg/lane-km) for light snow and 270 lb/lane-mile (76 kg/lane-km) for each application during heavy snow (Wegner and Yaggi, 2001). Since the roads are salted no more than 10 to 20 times each winter, these figures indicate that a large portion of the used salt is unaccounted for.

Studies by the US Geological Survey and private agencies, such as the Cary Institute (a nonprofit environmental research and education organization in Millbrook, NY), have shown that salt does not disperse as quickly as was previously perceived and that concentrations of 100 to 200 mg/L, which are common, have a considerable adverse effect on aquatic life. The high use of salt has resulted in widespread contamination in the northeast United Sates. Chloride concentrations have

---

* Under Section 303(d) of the 1972 Clean Water Act, states, territories, and authorized tribes were required to develop lists of impaired waters (those that do not meet water quality standards).
† Ton = 2000 lb = 907 kg; metric ton = 1000 kg.

at times exceeded 5 g/L (0.5% by weight) in urban areas. This is nearly one-fourth of the salt concentration in sea water.

Salt enters drainage systems from precipitation falling on salt stockpiles and from salt application to roadways, driveways, and parking areas. Salt in runoff infiltrates into the ground and contaminates groundwater. Due to long residence time, salt concentrations tend to build up in groundwater, reaching their highest level in summer. Excessive concentrations of salt in fresh waters can retard springtime mixing due to density gradient, and also decrease oxygen levels causing high mortality among fish and bottom-living organisms. It can also alter natural saline concentrations in estuaries and bays, disrupting shellfish reproduction. Increased salt concentrations produce a displeasing water taste, which may require expensive treatment for domestic use.

Salt can affect the water chemistry miles downstream of where it is applied. According to research by Stephen Norton, PhD, a professor of geological sciences at the University of Maine, salt causes certain minerals to leach out of soils in the streams. At high concentration, salt can increase the acidity of water, having effects similar to those of acid rain (*Civil Engineering*, 2005). Other studies of road salt have found ecological consequences of salt including adverse impacts on macroinvertebrates. High levels of salt in drinking water can result in high blood pressure and kidney malfunctioning as well as damage to eggs and embryos of wood frogs and salamanders.

Sodium chloride is one of the emerging contaminants of great concern in urban runoff. Road salt is often overapplied, so there is a potential for reduction. Its use can be reduced as follows:

- Plow before applying salt. Snow needs a lot of salt to melt.
- Use wet salt before freezing occurs. This reduces the salt loss from 25% for dry salt to only 4%.
- Mix sand and salt and apply it in parallel strips.
- Move trucks at moderate speed. At 50 km/h (30 miles/h) speed, 30% of salt is spread out beyond the roadway.
- Monitor salt application with temperature. One kilogram (pound) of salt melts 40 kg (40 lb) of snow at $-1°C$ (30°F) but approximately 5 kg (5 lb) at $-12°C$ (10°F).
- Never salt before snow falls. Weather forecasts are not always reliable. Snow may not fall or may turn to rain.

A number of de-icing materials have been tested as an alternative to road salt. Among these, calcium chloride (CaCl), calcium magnesium acetate (CMA), and potassium acetate (KA) are the most notable. Calcium chloride requires special handling and is more expensive than salt. However, it is very effective and fast acting at temperatures below $-18°C$ (0°F). CMA is found to be slower acting than salt when applied during or after snow. Also, its effectiveness diminishes below $-5°C$ (23°F) (WIDOT, 1987). In addition, CMA costs $660 to $770 per metric ton ($600 to $700 per ton) compared with $25 to $45 per metric ton ($35 average) for road salt. Potassium acetate (KA) is often used as a base for chloride-free liquid de-icing material. Advantages of KA include low corrosion, relatively high performance, and, above all, low environmental impact. However, KA costs many times more than salt and its cost is comparable with CMA. Studies show that among all de-icers, sodium chloride has the most deteriorating effect on concrete surfaces and is significantly more harmful than calcium chloride (Mishra, 2001).

Calcium chloride also has been employed in a pavement called Verglimit as an anti-icing agent (Clines, 2003). Verglimit is a bituminous concrete pavement containing calcium chloride pellets encapsulated in linseed oil and caustic soda. It is most suited for bridge decks, steep slopes, and shaded areas, which are more prone to icing. Verglimit has been used in Europe since 1974, in North America since 1976, and in Japan since 1978. Tests in New Jersey have found the material to be effective at $-4°C$ (24°F).

In airports where managing storm water is critical to their operations, glycol is used as a de-icing agent. The impact of glycol fluid on the environment is related to the high oxygen demand it exerts when discharged into streams and rivers. With the EPA ruling limiting the discharge of glycol and other pollutants from airports, the traditional de-icing process is changing. In some airports, including Buffalo Niagara International Airport, the changes have already been made. In this airport an underground wetland system has been designed to treat glycol. The storm water management system in each airport will need to be designed based on the existing drainage infrastructure and de-icing operations.

1.3.7 PETROLEUM HYDROCARBONS

Petroleum hydrocarbons include oil and grease, the "BTEX" compounds (benzene, toluene, ethyl benzene, and xylene), and a variety of polynuclear aromatic hydrocarbons (PAHs). These pollutants enter storm water runoff from garages, parking lots, roadways, leaky storage tanks, auto emissions, and improper or illicit disposal of waste oil. It may be hard to believe that there are still people who dump their used car oil (and antifreeze) into municipal storm drains. Petroleum hydrocarbons have acute toxicity to humans at low density (Schueler, 1987). They also render water unsuitable for designated uses.

A study involving measurement of petroleum hydrocarbon concentrations in urban runoff from a variety of impervious areas in the District of Columbia and suburban Maryland has shown that the amount of car traffic affects the hydrocarbon concentration in runoff with the medium concentration ranging from 0.6 to 6.6 mL/L (Shepp, 1996). These concentrations exceed the maximum concentrations recommended for the protection of drinking water supplies and fisheries production—in the range of 0.01 to 0.1 mL/L.

Hydrocarbons are harmful to water supply in a liquid state; however, they are absorbed and adsorbed onto sediment and solid particles so rapidly that they are found mainly as particulate in water. Only considerable masses of oil, such as oil spills, will remain in a liquid state. Petroleum hydrocarbons are biodegradable in an aerobic environment, although at a very slow rate. The rate can be significantly hastened by forced aeration.

Building more efficient cars, avoiding unnecessary trips, and, more importantly, car pooling will reduce the hydrocarbon concentration in urban runoff. Carpooling lanes were employed in New Jersey in the mid-1990s. However, the program was abandoned after nearly 3 years as the high-occupancy car lanes were not effectively utilized. The lack of success was partly due to low gas prices, which did not outweigh the convenience of having one's own car. More importantly, the program failed because it required a minimum of three passengers in a car, which limited the number of people with common commutes. In Southern California and other heavy traffic areas where an express lane is provisioned for cars with two or more passengers, the carpool program has been fairly successful.

Removing hydrocarbons and fuels at airports and seaports has been given special attention in recent years. Albany International Airport has installed a treatment system to address the water quality of runoff from fuel facility expansion discharged to Shaker Creek, which leads to a downstream drinking water intake in Albany County. The system includes trench-drains in refueling areas, conveying the runoff to a storm water lift station, where the runoff is pumped to a vault with a bank of Smart Sponge absorption filters. The filters supplied by Clear Water Solutions have been found to produce superior effluent quality at a much lower capital cost than traditional activated carbon filter media. These filters do not leach contaminants; rather, they transform them into manageable solid waste. The filters are easy to maintain and replace and can be disposed of as solid waste at a reportedly relatively low cost (Shane, 2007b).

A number of other airports in the Northeast use similar filters to treat runoff. These include Newark International Airport in New Jersey and Westchester County Airport in New York State. In 2002, Westchester County Airport installed Ultra Urban Filters (manufactured by AbTech Industries) in 54 selected critical inlets at the airport (Shane, 2007a). The infiltration media, called Smart Sponge, comprises a blend of polymers and has been shown to effectively absorb contaminants from water. Polymers are composed of molecules that are nonleaching and bond gasoline, oil, and grease, transforming them to a gel-like material and saturating at two- to fivefold their dry weight. The filters come in two standard shapes or modular units, one for curb inlets and a single unit for drop inlets. Based on the experience on this project, the filters were good for 2 years, in the absence of any spill, before they needed to be replaced. The removed filters were tested to ensure that they had no hazardous substances before they were hauled away for recycling.

1.3.8 Atmospheric Dust

Atmospheric dust is defined as minute particles suspended by slight currents and slowly settling in calm air. Generally, less dust exists at high elevations over the ocean and more at low levels over cities. Dust is created by activities such as overgrazing, deforestation, and improper agricultural and construction practices. However, most of the atmospheric dust is produced by natural causes, namely climatic conditions and in particular wind. Arid and semiarid regions generate significantly more dust than areas of high precipitation where the ground surface is protected with vegetation. Wind erosion is the largest source of dust pollution; it erodes bare ground and detaches and entrains solid particles in the air. Strong winds can create a thick cloud of dust in the sky. Figure 1.4 depicts the dust clouding the air during a strong wind.

Dust not only is an air pollutant, but airborne dust also collects on urban surfaces such as roads, driveways, roofs, and pools and washes away with rain to form water pollution. Large dust particles quickly settle on the ground, but large quantities of fine dust particles remain suspended in the air for a long time. The effects of the August 1883 volcanic eruption of Mt. Krakatau in Indonesia were observed years after its occurrence. Generally, dust particles larger than 1 μm (0.001 mm) settle as dry deposition because of their larger settling velocity; smaller particles are primarily removed by wet deposition. Specifically, the small particles adhere to rain and fall with raindrops. This is best observed by dirt spots on cars and pavements due to a rain that falls after days of dry weather. The 2013 EPA study indicates that a little over 1% of streams and rivers have excessively high levels of mercury. Atmospheric deposition of coal ash is the prime source of mercury in streams.

Recent studies estimate global dust-emission rates within a range of 1000 to 3000 teragrams (10^{12} g = 10^6 metric tons) per year. About 80% of the dust is from the Northern Hemisphere, in a dust belt extending from North Africa to the Middle East to Central and South Asia and to China. The largest source of dust is the Sahara Desert with an estimated range of 160 to 760 Tg (teragrams) per year (*Encyclopedia of Earth*, 2007). It is worthy to note that winds, apart from carrying dust, also transport plant seeds, bacteria, viruses, fungi, and various pollutants, not all of which are harmful; some are food for earth.

FIGURE 1.4 Dust caused by wind erosion clouding the sky and forming a dark shade. (Photo by the author.)

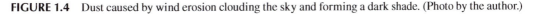

1.4 MANAGEMENT OF STORM WATER RUNOFF

Current national and state storm water management regulations govern the quantity and the quality of runoff from the site of developments beyond certain threshold limits. To address these regulations, a variety of measures have been developed. Because of the diversity of measures, a generic term of best management practices (BMPs) was introduced to the field of storm water management (see EPA, 1999). BMPs refer to measures or structures that under given conditions control the quantity and improve the quality of storm water runoff in a most cost-effective manner. BMPs may be divided into either preemptive measures aimed at lowering the production of runoff and pollutants from a site or corrective measures to reduce the amount of runoff and its pollutants once generated. These approaches are also referred to as nonstructural BMPs and structural (or end-of-pipe) BMPs, respectively.

Structural measures include detention/retention basins, ponds, infiltration basins, and the like. Nonstructural measures, on the other hand, are those measures that tend to reduce the runoff volume and pollutant generation. Reducing imperviousness, disconnecting impervious areas, and avoiding excessive use of lawn chemicals are examples of nonstructural measures. In recent years a large number of manufactured devices have been introduced to the market for the treatment of storm water quality. Structural, nonstructural, and manufactured water-quality devices will be discussed in separate chapters in this book.

Addressing adverse impacts of developments on the environment requires challenging storm water management actions in the future. A concern with storm water runoff is that it becomes more polluted as it traverses through street gutters and drainage systems to get to a facility to receive treatment. The traditional approach to storm water management, namely the end-of-pipe methods, will be neither feasible nor cost effective.

A single BMP may not, and often cannot, address all storm water problems. Each BMP has its own limitations depending on its intended objectives and specific site conditions. In general, the best solution to control pollution is to reduce its production in the first place, noting that prevention is a more effective solution than cure.

The National Research Council (NRC) under a contract with the EPA released a report titled "Urban Storm Water Management in the United States" on October 15, 2008. This report, which is the product of a 26-month study by a 15-member committee, includes a description of the history of storm water management in the United States and an overview of storm water regulations and the federal regulatory program. A 500 plus page report, which was prepared for the EPA, concludes that radical changes to the EPA storm water program are necessary to reverse degradation of freshwater resources. Among the key findings and general recommendations of the study are

- A change from the EPA's current piecemeal regulatory system to a new watershed-based permitting system that focuses on all discharges to streams and water bodies.
- Focusing more on the increased volume of water and less on chemical pollutants.
- The area of disturbance due to urban land use grows faster than the population. The impact of this trend must be considered in the EPA's storm water regulatory program.

A summary of the NRC's report, in four sheets, is included as Appendix 1A. Following the NRC's recommendation, the EPA has selected a number of projects for pilot study. The EPA has also incorporated certain findings and recommendations of the study in the current construction general permit adopted in 2012.

As indicated, a most cost-effective and environmentally friendly solution to manage water quality is to reduce pollution in the first place. The following is a list of some measures to achieve this goal:

- Pollution prevention/reduction/good housekeeping
- Reducing imperviousness

- Retaining a large portion of rainfall
- Directing runoff from impervious surfaces (roofs and driveways) to lawns/landscaping
- Public education
- Illicit discharge detection and elimination
- Improved soil erosion control during construction
- Postconstruction runoff and erosion control
- Avoid littering and dispose of trash and recyclables properly
- Reducing application of lawn chemicals (fertilizers and pesticides)
- Proper disposal of paint and household chemicals

Some of these measures will be explored later in this book. It is to be noted, however, that most important of all is to redefine the criterion of the impact of a development on the environment. The state-of-the-art practices, such as low-impact development, green infrastructure, and sustainable development, aim at reducing impacts on the per-lot or -site basis. Rating the impact on the basis of disturbance per unit area of land is a misleading criterion because the land is disturbed to provide people with housing and other amenities. Therefore, the criterion for impact on the environment should be based not on how much disturbance a project creates, but rather on the amount of disturbance on a per-capita basis (Pazwash, 2011, 2012). On this basis, as will be shown later in this book, compact developments such as condominiums, multifamily residential buildings, and, above all, city living have a far less adverse impact on the environment than single-family homes.

Therefore, to reduce impacts of urbanization on storm water runoff, our urban planning should be based on compact mixed-use developments, building around existing transport stations, and, above all, building cities having streets with walkways and bikeways. This will minimize the impervious coverage per capita, resulting in a proportionate reduction in the quantitative and qualitative impacts on storm water. Compact developments and, in particular, city living significantly lessen, if not nearly eliminate, the use of private cars, which in our society form the general means of commuting to work and moving around by the public, and which generate traffic jams and also create air and water pollution. This will significantly reduce air pollution and greenhouse gas emission, road rage, and countless hours of nonproductive daily life. In addition, they will reduce energy consumption and overall adverse impacts to the environment. City living, of course, entails a drastic change from our suburban lifestyle and may not be acceptable by many people, but is the best solution to reduce our adverse impacts on storm water runoff and the environment as a whole.

As indicated earlier, land disturbance due to urbanization is growing faster than population growth. Thus, our adverse impacts to the environment in general and storm water in particular will not diminish; rather, they will grow with time. It is a misperception to believe that sustainable developments will avoid compromising the ability of future generations to meet their needs. True sustainability is an unrealistic goal that cannot be achieved. The best we can hope for is to minimize our ill impacts on the environment.

Man's adverse impacts on the environment are not limited to storm water runoff and water pollution. We pollute the air, contaminate the earth, and change the natural processes. Although technology has reduced the amount of per-car emission per kilometer (mile), there has been a 250% increase in vehicular travel distance since 1970. Also, the number of cars is expected to double from one billion to two billion in the next 20 years.

Ill impacts of our activities on the natural world are enormous and beyond imagination. There should be a greater concern regarding the destruction of numerous species of insects and microbial organisms that may be vital to the maintenance of the life cycle on earth. Dr. Edward Wilson, professor emeritus of Harvard University and two-time Pulitzer Prize winner, thinks that it is quite possible that we could destroy the rest of the natural world and with it as many as one-half of plants and animals on earth by the end of the century (Wilson, 2006; Schulte, 2006).

PROBLEMS

1.1   Why does urbanization impact storm water?
1.2   What are the main impacts of urbanization on storm water runoff?
1.3   What are the impacts of urbanization on runoff quantity?
1.4   Why does urbanization increase storm water volume?
1.5   Does urbanization increase the peak rate of runoff? If your answer is yes, what are the reasons?
1.6   Does urbanization impact storm water quality? If so, what are the changes?
1.7   Does urbanization increase or decrease the sediment load? Explain your answer.
1.8   Why does sediment contribute to storm water contamination?
1.9   Is the sediment a problem during construction? If so, is it more of a problem than that of postconstruction? Explain your answer.
1.10 What does the settling velocity of a particle in water depend on? How is it related to particle size?
1.11 How is sediment commonly removed from storm water?
1.12 Does urbanization increase nutrients in runoff?
1.13 Name the prime nutrients present in storm water.
1.14 What are the sources of nutrients in urban storm water?
1.15 Do nutrients have adverse impacts on water bodies? If so, list the impacts.
1.16 Are there any heavy metals in urban runoff? If so, what are they?
1.17 Where does the heavy metal in urban runoff originate?
1.18 Are pathogens bacterial or viral organisms? What are their sources in storm water runoff?
1.19 Can pathogens cause human diseases? Name an infective disease associated with pathogen bacteria.
1.20 What is the main source of sodium chloride (salt) in urban runoff?
1.21 Can salt in drinking water affect our health?
1.22 Outline the most effective measures to reduce the use of salt on roads.
1.23 Are petroleum hydrocarbons present in urban runoff? If so, where do they come from?
1.24 Do single-family homes or compact developments have a smaller impact on runoff? Explain your reason.
1.25 To properly compare impacts of developments on runoff, what should the impacts be based on?

APPENDIX 1A: NRC REPORT SUMMARY, OCTOBER 15, 2008

THE NATIONAL ACADEMIES

REPORT IN BRIEF

Urban Stormwater Management in the United States

The rapid conversion of land to urban and suburban areas has profoundly altered how water flows during and following storm events, putting higher volumes of water and more pollutants into the nation's rivers, lakes, and estuaries. These changes have degraded water quality and habitat in virtually every urban stream system. The Clean Water Act regulatory framework for addressing sewage and industrial wastes is not well suited to the more difficult problem of stormwater discharges. This report calls for an entirely new permitting structure that would put authority and accountability for stormwater discharges at the municipal level. A number of additional actions, such as conserving natural areas, reducing hard surface cover (e.g., roads and parking lots), and retrofitting urban areas with features that hold and treat stormwater, are recommended.

Stormwater has long been regarded as a major culprit in urban flooding, but only in the past 30 years have policymakers appreciated its significant role in degrading the streams, rivers, lakes, and other waterbodies in urban and suburban areas. Large volumes of rapidly moving stormwater can harm species habitat and pollute sensitive drinking water sources, among other impacts. Urban stormwater is estimated to be the primary source of impairment for 13 percent of assessed rivers, 18 percent of lakes, and 32 percent of estuaries—significant numbers given that urban areas cover only 3 percent of the land mass of the United States.

Photo by Roger Bannerman

Urbanization—the conversion of forests and agricultural land to suburban and urban areas—is proceeding at an unprecedented pace in the United States. Stormwater discharges have emerged as a problem because the flow of water is dramatically altered as land is urbanized. Typically, vegetation and topsoil are removed to make way for buildings, roads, and other infrastructure, and drainage networks are installed. The loss of the water-retaining functions of soil and vegetation causes stormwater to reach streams in short concentrated bursts. In addition, roads, parking lots, and other "impervious surfaces" channel and speed the flow of water to streams. When combined with pollutants from lawns, motor vehicles, domesticated animals, industries, and other urban sources that are picked up by the stormwater, these changes have led to water quality degradation in virtually all urban streams.

In 1987 Congress wrote a new section into the Clean Water Act's National Pollutant Discharge Elimination System to help address the role of stormwater in impairing water quality. This system, which is enforced by the U.S. Environmental Protection Agency (EPA), has focused on reducing pollutants from industrial process wastewater and municipal sewage discharges—"point sources" of pollution that are relatively straightforward to regulate. Under the new "stormwater program," the number of permittees in the National Pollutant Discharge Elimination System has ballooned from about 100,000 to more than 500,000, to include stormwater permittees from municipal areas, industry, and construction sites one acre or larger. Not only do stormwater permittees vastly out number wastewater permittees, it is much more difficult to collect and treat stormwater than wastewater.

In light of these challenges, EPA asked the National Research Council to review its stormwater program, considering all entities regulated under the program (i.e., municipal, industrial, and construction). The report finds that the stormwater program will require significant changes if it is to improve the quality of the nation's waters. Fortunately, there are a number of actions that can be taken. The report concludes that the course of action most likely to halt and reverse degradation of the nation's waterways would be to base all stormwater and other wastewater discharge permits on watershed boundaries instead of political boundaries, which is a radical shift from the current structure.

The Challenges of Regulating Stormwater

One of the problems in managing stormwater discharge is that it is being addressed so late in the development of urban areas. Historically, stormwater management has meant flood control—by moving water away from structures and cities as fast as possible. Ideally, stormwater discharges would be regulated through direct controls on land use, strict limits on both the quantity and quality of stormwater runoff into surface waters, and rigorous monitoring of adjacent waterbodies to ensure that they are not degraded by stormwater discharges. Future land use development would be controlled to minimize stormwater discharges. Products or sources that contribute pollutants through stormwater—like de-icing materials, fertilizers, and vehicular exhaust—would be regulated by EPA at a national level to ensure that the most environmentally benign materials are used.

The current regulatory scheme lacks many of these attributes. EPA's program has monitoring requirements that are so benign as to be of little use for the purposes of program compliance. Most dischargers have no measurable, enforceable requirements. Instead, the stormwater permits leave a great deal of discretion to the regulated community to set their own standards, develop their own pollution control schemes, and to self-monitor. Current statistics on the states' implementation of the stormwater program, compliance with stormwater requirements, and the ability of states

High volumes of stormwater discharge have badly damaged this stream near Philadelphia, which is suffering from Urban Stream Syndrome. Photo by Chris Crockett, City of Philadelphia Water Department.

and EPA to incorporate stormwater permits with pollution limits are uniformly discouraging.

Significant changes to the current regulatory program are necessary to provide meaningful regulation of stormwater dischargers in the future. One idea is to focus the stormwater program less on chemical pollutants in stormwater and more on problems associated with increased volumes of water. Some states have used flow volumes as a metric for controlling and reducing stormwater discharge; other regulators have used the extent of hard surfaces (impervious cover) as a proxy for stormwater pollutants. These substitutes for the traditional focus on the "discharge" of "pollutants" have great potential as stormwater management tools because they provide specific and measurable targets. At the same time, they focus regulators on the problems of increased water volume, which include a condition known as Urban Stream Syndrome (see image above).

In addition, the federal government should provide more financial support to state and local efforts to regulate stormwater. Today, the stormwater program still receives much less funding than the wastewater program despite having many more permittees.

The Case For Watershed Permitting

The report concludes that the most likely way to halt and reverse damage to waterbodies is through a substantial departure from the status quo—namely a watershed permitting structure that bases all stormwater and other wastewater discharge permits on watershed boundaries instead of political boundaries. Watershed-based permitting is not a new concept, but it has been attempted in only a few communities.

The proposed watershed permitting structure would put both the authority and accountability for stormwater discharges at the municipal level. A municipal lead permittee, such as a city, would work in partnership with other municipalities in the watershed as co-permittees. Permitting authorities (designated states or, otherwise, EPA) would adopt a minimum goal in every watershed to avoid any further loss or degradation of designated beneficial uses in the watershed's component waterbodies and additional goals in some cases aimed at recovering lost beneficial uses. Permittees, with support by the states or EPA, would then conduct comprehensive impact source analyses as a foundation for targeting solutions.

The approach gives municipal co-permittees more responsibility, with commensurately greater authority and funding, to manage all of the sources discharging to the waterbodies comprising the watershed. The report also outlines a new monitoring program structured to assess progress toward meeting objectives, diagnosing reasons for any lack of progress, and determining compliance by dischargers. The proposal further includes market-based trading of credits among dischargers to achieve overall compliance in the most efficient manner, and adaptive management to determine additional actions if monitoring demonstrates failure to achieve objectives.

As a first step to taking the proposed program nationwide, a pilot program is recommended that will allow EPA to work through some of the more predictable impediments to watershed-based permitting, such as the inevitable limits of an urban municipality's authority within a larger watershed.

Short of adopting watershed-based permitting, other smaller-scale changes to the EPA stormwater program are possible. The report recommends that EPA integrate the three different permitting types so that construction and industrial sites come under the jurisdiction of their associated municipalities.

Stormwater Management Approaches

Even in the absence of regulatory changes, there are many stormwater management approaches that can be used to prevent, reduce, and treat stormwater flows. Central to the EPA Stormwater Program is the requirement for permittees to develop stormwater pollution prevention plans that include stormwater control measures. When designed, constructed, and maintained correctly, stormwater control measures have been demonstrated to reduce runoff volume and peak flows and to remove pollutants. A classic example is the removal of lead from gasoline, which has reduced lead concentrations in stormwater by at least a factor of four.

Stormwater control measures are grouped in two categories: nonstructural and structural. Nonstructural stormwater control measures include a wide range of actions that can reduce the volume of runoff and pollutants from a new development. Examples include the use of products that contain less pollutants; improved urban design, for example, of new developments that have fewer hard surfaces; the disconnection of downspouts from hard surfaces to instead connect with porous surfaces; the conservation of natural areas; and improved watershed and land use planning.

Structural stormwater control measures are designed to reduce the volume and pollutants of small storms by the capture and reuse of stormwater, the infiltration of stormwater into porous surfaces, and the evaporation of stormwater. Examples include rainwater harvesting systems that capture runoff

There are many innovative approaches to stormwater management that can be applied in urban and suburban areas. Chicago's City Hall (left) was retrofitted with a "green roof" to capture stormwater. Photo courtesy CDF Inc. The downspoutings on the house (right) drain onto a porous surface instead of onto a driveway. Photo by William Wenk.

from roofs in rain barrels, tanks, or cisterns; the use of permeable pavement; the creation of "infiltration trenches," into which stormwater can

Data on Stormwater Discharges

Thanks to a 10-year effort to collect and analyze monitoring data from municipal separate storm sewer systems nationwide, a lot is known about the quality of stormwater from urbanized areas. Residential land use has been shown to be a relatively smaller source of many pollutants, but it is the largest fraction of land use in most communities, typically making it the largest stormwater source on a mass pollutant discharge basis. Freeway, industrial, and commercial areas can be very significant sources of heavy metals, and their discharge significance is usually much greater than their land area indicates. Construction sites are usually the overwhelming source of sediment in urban areas, even though they make up very small areas of most communities. These results come from many thousands of storm events, systematically compiled. These data make it possible to accurately estimate the concentration of many pollutants for any given storm.

seep or is piped; the planting of rain gardens on both public and private lands, and the planting of "swales" along the roadside that capture and treat stormwater.

The report recommends that nonstructural stormwater control measures be considered first before structural practices, because their use reduces the reliance on and need for structural measures. The report discusses the characteristics, applicability, goals, effectiveness, and cost of nearly 20 different broad categories of stormwater control measures, organized as they might be applied from the roof top to the stream.

There is an opportunity to retrofit urban areas with stormwater control measures. Promoting growth in these areas is a good thing because it can take pressure off the suburban fringes, thereby preventing sprawl, and because it minimizes the creation of new impervious surfaces. However, it can be more expensive because there is existing infrastructure and limited availability and affordability of land. Both innovative zoning and development incentives, along with careful selection of stormwater control measures, are needed to achieve fair and effective stormwater management in these areas.

This traffic island has a "bioinfiltration" system to capture water. Photo courtesy Villanova Urban Stormwater Partnership.

Committee on Reducing Stormwater Discharge Contributions to Water Pollution: Claire Welty, (*Chair***),** University of Maryland, Baltimore County; **Lawrence E. Band**, University of North Carolina, Chapel Hill; **Roger T. Bannerman**, Wisconsin Department of Natural Resources; **Derek B. Booth**, Stillwater Sciences, Inc.; **Richard R. Horner**, University of Washington, Seattle; **Charles R. O'Melia**, Johns Hopkins University; **Robert E. Pitt**, University of Alabama; **Edward T. Rankin**, Midwest Biodiversity Institute; **Thomas R. Schueler,** Chesapeake Stormwater Network; **Kurt Stephenson**, Virginia Polytechnic Institute and State University; **Xavier Swamikannu**, California EPA, Los Angeles Regional Water Board; **Robert G. Traver**, Villanova University; **Wendy E. Wagner**, University of Texas School of Law; **William E. Wenk**, Wenk Associates, Inc.; **Laura Ehlers** (Study Director), National Research Council.

This report brief was prepared by the National Research Council based on the committee's report. For more information or copies, contact the Water Science and Technology Board at (202) 334-3422 or visit http://nationalacademies.org/wstb. Copies of *Urban Stormwater Management in the United States* are available from the National Academies Press, 500 Fifth Street, NW, Washington, D.C. 20001; (800) 624-6242; www.nap.edu.

REFERENCES

Caltrans, 2010, Storm water quality handbooks project planning and design guide (PPDG), State of California, Dept. of Transportation, CTSW-RT-10-254.03, July.

Carson, R.A. and Sercu, B., 2013, Efforts to achieve compliance with coliform plan objectives, *Stormwater*, July/August, 10–19.

Civil Engineering, 2005, News brief: Salting road during winter dangers neighboring ecosystems, May, 36–37.

Clines K., 2003, Bid list materials for deicing and anti-icing, *Better Roads Magazine*, April.

Encyclopedia of Earth, 2007, Global dust budget.

England, G. and Listopad, C., 2012, Use of TMDL credits for BMP comparisons, *Stormwater*, May, 38–43.

EPA (US Environmental Protection Agency), 1983a, *Results of the nationwide urban runoff program: Vol. 1— Final report.* Water Planning Division, Washington, DC, National Technical Information Service (NTIS) publication no. 83-185552.

———— 1983b, *Results of the nationwide urban runoff program, executive summary*, Water Planning Division, Washington, DC, National Technical Information Service (NTIS), accession no. PB84-185545.

———— 1999, Preliminary data summary of urban stormwater best management practices, EP-821-R-99-012, August.

———— 2000, National water quality inventory, 1998 report to Congress, USEPA 841-R-00-001, Washington, DC.

———— 2008, Protecting water quality from urban runoff, EPA 841-F-03-003, National Research Council, October 2008, *Urban stormwater management in the United States*, National Academy Press, Washington, DC.

———— 2013a, National rivers and streams assessment 2008–2009, a collaborative survey, draft, February 28, EPA/841/D-13/001.

———— 2013b, March 26, Water headlines, http://water.epa.gov/about/owners/waterheadlines/2013.

Haugton, M., 1987, The Clean Water Act of 1987, US Bureau of National Affairs, Arlington, VA.

Hudson, B.E., 1994, Soil organic matter and available water capacity, *Journal of Soil and Water Conservation*, 49 (2): 189–194.

Kaspersen, J., 2009, The great bug hunt, editor's comments, *Stormwater*, March/April, 6.

Lee, G.F. and Jones-Lee, A., 2005, Urban storm water runoff quality issues, *Water encyclopedia: Surface and agricultural water*, Wiley, Hoboken, NJ, pp. 432–437.

———— 2006, Lead as a storm water runoff pollutant, *Stormwater*, September, 88–91.

Mishra, S.K., 2001, A mixture for snow and ice, roads and bridges, December, pp. 18–21. http://www.roads bridges.com.

NJDEP (New Jersey Department of Environmental Protection), 2004, *Storm water best management practices manual*, Table 1-1, February, Trenton, NJ.

Pazwash, H., 1982a, Sedimentation in reservoirs, case of Sefidrud Dam, in *Proceedings of 3rd Congress of the Asian and Pacific Regional Division of the I.A.H.R.* Aug. 24–26, 1982, Bandung, Indonesia, Vol. C, pp. 215–223.

———— 1982b, Erosion and sedimentations, Effect of reservoirs, in *Proceedings of 1982 International Symposium on Surface Mining Hydrology, Sedimentology, and Reclamation,* University of Kentucky, Lexington, Dec. 5–10, pp. 457–461.

———— 2007, *Fluid mechanics and hydraulic engineer*, Tehran University Press, Iran.

———— 2011, *Urban storm water management*, 1st ed., CRC Press, Boca Raton, FL.

———— 2012, Development sustainability, proper basis, presented at OIDA International Conference on Sustainable Development, Montclair State University, Montclair, NJ, August 1.

Richardson, D.C., 2012, Ice school, melding the science and craft of winter road maintenance, *Stormwater*, January/February, 14–21.

Schueler, T., 1987, *Controlling urban runoff: A practical manual for planning and designing urban BMPs.* Metropolitan Washington Council of Governments, Washington, DC.

———— 1997, Comparative removal capability of urban BMPs: A reanalysis. *Watershed Protection Techniques*, 1 (2): 515–520.

Schulte, B., 2006, Q&A: Edward Wilson, *U.S. News & World Report*, September 4, http://www.usnews.com.

Shane, J.I., 2007a, Westchester County Airport meets tough international standards, *Stormwater*, October, 72–82.

———— 2007b, Airport support Albany International protects local waterways from fuel facility expansion runoff using absorption filter media, *Stormwater Solutions*, November/December, 30–33.

Shepp, D.L., 1996, Petroleum hydrocarbon concentrations observed in runoff form discrete, urbanized automotive-intensive land uses. Metropolitan Washington Council of Governments, Washington, DC.

State of New York Stormwater Management Design Manual, 2010, Prepared by Center for Watershed Protection, Maryland, for New York State Department of Environmental Conservation, August.

Terrene Institute, 1994, Urbanization and water quality, a guide to protecting the urban environment, in cooperation with EPA, Terrance Institute, Washington, DC, March.

Tveten, R. and Williamson, B., 2006, Zinc found as one of the primary contaminants in New Zealand's urban storm water, *ASCE EWRI* (Environmental and Water Resources Institute), 8 (1), 2.

USGS (US Geological Survey), 2010a, Circular 1350: Nutrients in the nation's streams and groundwater.

———— 2010b, Nutrients in the nation's streams and groundwater, national findings and implications, fact sheet, 2010-3078, http://pubs.usgs.gov/fs2010/3078/.

———— 2013, News release, October 30, http://www.usgs.gov/newsroom/article.asp?ID=3715.

Walker, W.W., 1987, Phosphorus removal by urban runoff detention basins, *Lake and Reservoir Management*, III: 314–326.

Wegner, W. and Yaggi, M., 2001, Environmental impacts of road salt and alternatives in the New York City watershed, *Stormwater*, May/June.

WIDOT (Wisconsin Department of Transportation), 1987, Field deicing tests of high quality calcium magnesium acetate (CMA).

Wilson, E.O., 2006, *The creation: An appeal to save life on earth*, W. W. Norton, New York.

2 Pipe and Open Channel Flow
A Review

Hydraulic principles including energy equations, specific energy, and critical flow are briefly discussed in this chapter. Also presented in this chapter is flow in pipes and open channels. This chapter also discusses losses in manholes and junctions.

2.1 FLOW CLASSIFICATIONS

Since the design equations for pipes and channels have been developed for specific flow conditions, it is first necessary to classify the types of flow. Depending on temporal and spatial variations, the flow may be classified as steady or unsteady, uniform or nonuniform, gradually varied or rapidly varied. In the design of pipes and channels in urban storm water management systems, the flow is generally considered steady and uniform, in that the flow velocity, discharge, and depth are assumed to remain unchanged with time or distance along a reach of a conduit. The depth of uniform flow is called the "normal" depth.

Uniform flow can occur as pressure flow, such as full flow in pipes, or nonpressure (also called free surface flow), such as flow in channels and pipes when partially full. Nonuniform flow is a flow with changing depth and velocity along the channel or conduit. This type of flow can be gradually varied, if the changes in depth and velocity are gradual and occur over a considerable length, or rapidly varied flow where the changes in flow are abrupt and occur over a very short distance. Overland flow on paved surfaces, gutter flow along roadways, and flow in natural streams are examples of gradually varied flow. Flow over emergency spillways, hydraulic jumps, and flow under sluice gates are examples of rapidly varied flow.

The uniform flow equations can be (and commonly are) applied to short distance intervals in gradually varied flows; however, these equations are not applicable to rapidly varied flows.

2.2 ENERGY EQUATION

A flowing fluid has potential, pressure, and kinetic energies at any given point. The sum of these three, interchangeable energy components is called total energy. The principle of conservation of energy states that for an ideal fluid with no external energy sources or sinks, the total energy remains constant along the flow, though the contribution from the three components changes. In a real fluid, energy losses including frictional and form losses (also called local or minor losses) should be accounted for in the energy equation. In a system with a pump, the head added by a pump also enters into the equation. In a channel flow, shown by Figure 2.1, the total energy head, defined as the total energy per unit weight at a point, is given by

$$H = z + d + \frac{p}{\gamma} + \frac{V^2}{2g} \tag{2.1}$$

where
- H is the total energy head
- z is the elevation of channel bed above an arbitrary, or fixed, datum (such as the National Geodetic Vertical Datum [NGVD, 29])

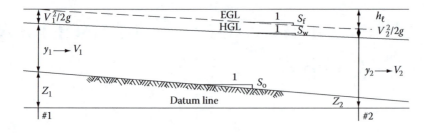

FIGURE 2.1   Energy components in an open channel flow.

    d is the distance of the point above the bed
    p is the fluid pressure at the point
    V is the flow velocity
    γ is the fluid unit weight

It is evident that H, which represents energy per unit weight, has units of length and thus is termed the energy head or total head. The term p/γ is the pressure per unit weight and called the pressure head, and $V^2/2g$ is the kinetic energy per unit weight and is known as the velocity head.

The sum of $p/\gamma + d$, which represents the pressure head at the streambed, can be approximated by the flow depth, y, in channels of mild slope. In this case, the energy equation applied between sections 1 and 2 is expressed in the following simple form:

$$H_1 = H_2 + h_\ell \tag{2.2}$$

where

$$H_1 = z_1 + y_1 + \frac{V_1^2}{2g}; \quad H_2 = z_2 + y_2 + \frac{V_2^2}{2g}$$

and

$$h_1 = \text{head loss between sections 1 and 2}$$

The line representing the sum of terms $z + y$ is known as the hydraulic grade line (HGL). In open channel flow, the hydraulic grade line coincides with the water surface profile, and the line lying at a distance equal to the velocity head above the HGL is called the energy grade line or simply energy line (EL). The slope of the EL in a uniform channel represents the energy loss (due to friction) per unit length of channel. In closed conduit flows such as pipes flowing full, the hydraulic grade line generally does not coincide with the crown of the pipe. In this case, the HGL, which is the sum of $p/\gamma + d$ (d being the pipe diameter), may lie above or below the crown of the pipe depending on whether the pipe is flowing under a pressure larger or smaller than the atmospheric pressure, respectively. In a storm drain, running full and under pressure, the hydraulic grade line lies above the crown of the pipe.

2.3   SPECIFIC ENERGY; CRITICAL FLOW

If the datum of elevation is taken at the channel bed or invert of a partially full pipe, Equation 2.2 simplifies as

$$E = y + \frac{V^2}{2g} \tag{2.3}$$

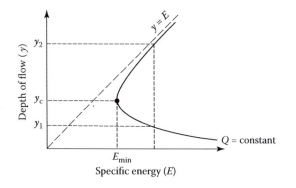

FIGURE 2.2 Variation of specific energy with depth.

where E is known as the specific energy. As the depth of fluid increases, the velocity head $V^2/2g$ decreases. At a certain depth, E, which is the sum of flow depth and velocity head, becomes minimal. This is illustrated by Figure 2.2, which is a plot of E as a function of y. The depth at which the energy becomes minimal is known as the critical depth and the velocity associated with this depth is called the critical velocity, V_c. This type of flow is discussed in more detail in the following section.

2.3.1 CRITICAL DEPTH

In a channel of any geometry, the specific energy equation is expressed as

$$E = y + \frac{V^2}{2g} = y + \frac{Q^2}{2gA^2} \tag{2.4}$$

where
  A = flow area m² (ft²)
  Q = discharge m³/s (cubic feet per second [cfs])

The critical depth associated with minimum specific energy satisfies the following equation:

$$\frac{dE}{dy} = 1 - \frac{Q^2}{gA^3} \times \frac{dA}{dy} = 0 \tag{2.5}$$

Since $dA = Tdy$, where T is the top width, Equation 2.5 simplifies as

$$\frac{Q^2T}{gA^3} = 1 \tag{2.6}$$

or

$$Q = \left(\frac{gA^3}{T} \right)^{1/2} \tag{2.7}$$

Substituting $Q = AV$ and $T = A/D$ in Equation 2.6 results in

$$\frac{V_c^2}{gD} = 1 \tag{2.8}$$

or

$$\frac{V_c^2}{2g} = \frac{D}{2}$$

(2.9)

and

$$F_r = \frac{V_c}{\sqrt{gD}} = 1$$

(2.10)

where $D = A/T$ is called the mean flow depth, and F_r is the Froude number.

Thus, for any channel including partly full pipes, the critical flow occurs when the Froude number, F_r, equals one or velocity head equals one-half of mean flow depth and

$$E = y_c + \frac{D}{2}$$

(2.11)

The critical depth and critical discharge for any section can be calculated from Equation 2.6 through an iterative procedure.

2.3.2 CRITICAL FLOW IN RECTANGULAR CHANNELS

In rectangular channels (and partly full rectangular culverts) the area is related to the depth of flow by

$$A = by$$

(2.12)

where b is the width of the channel and y is the depth of flow. For a rectangular channel, the general Equation 2.6 simplifies as

$$\frac{q^2}{gy_c^3} = 1$$

(2.13)

or

$$y_c = \left(\frac{q^2}{g}\right)^{1/3}$$

(2.14)

where $q = Q/b$ represents the discharge per unit width of the channel.

Substituting $q = Vy_c$ in Equation 2.14 and rearranging gives

$$y_c = \frac{V^2}{g}$$

(2.15)

and

$$V_c = \sqrt{gy_c}$$

(2.16)

Combining Equations 2.3 and 2.15 yields

$$E = 1.5y_c$$

(2.17)

or

$$y_c = \left(\frac{2}{3}\right)E$$

(2.18)

Since the flow in box culverts (rectangular flumes) with free water surface resembles an open channel flow, Equations 2.12 through 2.18 are equally applicable to box culverts.

Rearranging Equation 2.14, the critical depth can be calculated directly from the following equations:

$$y_c = 0.467 \left(\frac{Q}{b} \right)^{2/3} \quad \text{SI} \tag{2.19}$$

$$y_c = 0.314 \left(\frac{Q}{b} \right)^{2/3} \quad \text{CU} \tag{2.20}$$

where
   Q = discharge, m³/s (cfs)
   y_c = critical depth, m (ft)
   b = bottom width, m (ft)

and the coefficients 0.467 and 0.314 are founded to the third decimal place.

2.3.3 CRITICAL FLOW IN TRAPEZOIDAL CHANNELS

In a trapezoidal channel, the area, A, and the top width, T, are presented by the following equations:

$$A = by + my^2 \tag{2.21}$$

$$T = b + 2my \tag{2.22}$$

where b, y, and m are the bottom width, depth, and side slope (horizontal to vertical ratio) of the channel, respectively. Combining Equations 2.7, 2.21, and 2.22 and simplifying results in

$$Q_c = \sqrt{g} \, K_c b y^{3/2} \tag{2.23}$$

where

$$K_c = \left\{ \frac{\left[1 + m \left(\dfrac{y}{b} \right) \right]^3}{\left[1 + 2m \left(\dfrac{y}{b} \right) \right]} \right\}^{1/2} \tag{2.24}$$

is a dimensionless number.

    Table 2.1, prepared by the author, lists values of K_c for a wide range in the y/b ratio and side slope, m. Since K_c is dimensionless, this table is equally applicable to metric and English systems of units.

Example 2.1

Calculate the critical depth for a discharge of 120 cfs in a 4-foot wide box culvert.

Solution

Using Equation 2.14,

$$q = \frac{120}{4} = 30$$

$$y_c = (30^2/32.2)^{1/3}$$

$$y_c = 3.035 \approx 3.04 \text{ ft}$$

TABLE 2.1
K_c Values for Trapezoidal Channels[a]

y/b	1	1.5	2.0	3.0
			m	
0.1	1.053	1.082	1.111	1.172
0.2	1.111	1.172	1.235	1.364
0.3	1.172	1.267	1.364	1.565
0.4	1.235	1.364	1.498	1.770
0.5	1.299	1.464	1.633	1.976
0.6	1.364	1.565	1.770	2.185
0.7	1.431	1.667	1.907	2.394
0.8	1.498	1.770	2.046	2.603
0.9	1.565	1.873	2.185	2.813
1.0	1.633	1.976	2.324	3.024
1.2	1.770	2.185	2.603	3.445
1.4	1.907	2.394	2.883	3.868

[a] $K_c = \{[1 + m(y/b)]^3/[1 + 2m(y/b)]\}^{1/2}$.

or Equation 2.20,

$$y_c = 0.314(120/4)^{2/3} = 3.032 \text{ ft}$$

Example 2.2

Calculate the critical discharge in a trapezoidal channel of 1.5 m bottom width and 2:1 side slope for a depth of 1 m.

The solution is prepared using both Equations 2.23 and 2.24 and Table 2.1.

Using Equation 2.24,

$$\frac{y}{b} = \frac{1}{1.5} = 0.667$$

$$K_c = [(1 + 2 \times 0.667)^3/(1 + 2 \times 2 \times 0.667)]^{1/2}$$

$$K_c = 1.861$$

$$Q = 1.861 \times \sqrt{9.81} \times 1.5 \times 1^{3/2} = 8.74 \text{ m}^3/\text{s}$$

Using Table 2.1, interpolate K_c between y/b = 0.6 and y/b = 0.7 under the m = 2 column

$$\text{For } \frac{y}{b} = 0.6, K_c = 1.770$$

$$\text{For } \frac{y}{b} = 0.7, K_c = 1.907$$

Interpolate:

$$K = 1.77 + \frac{(1.907 - 1.770)}{(0.7 - 0.6)} \times (0.67 - 0.6) = 1.852$$

$$Q = \sqrt{9.81} \times 1.862 \times 1.5 \times 1^{1.5} = 8.755 \text{ m}^3/\text{s}$$

Example 2.3

Water flows at a rate of 2 m³/s in a trapezoidal channel lined with Reno mattress. The channel has 1 m bottom width and 1:1 side slopes. Calculate the critical depth in this channel.

Solution

$$A = y + y^2$$

$$T = 1 + 2y$$

$$f(y) = \frac{Q^2 T}{gA^3} = \frac{4(1+2y)}{9.81(y+y^2)^3} = 0.40775 \frac{(1+2y)}{(y+y^2)^3}$$

The critical depth can be calculated by solving the equation $f(y) = 1$, by trial and error. The calculations are tabulated below:

y	1.00	0.50	0.600	0.610	0.602
$f(y)$	0.15	1.93	1.014	0.959	1.002

Thus, $y_c = 0.602$ m.

2.3.4   CRITICAL FLOW IN PARTLY FULL CIRCULAR PIPES

For circular sections flowing partly full (see sketch in Table 2.2), A and T are expressed by

$$A = \left(\frac{D^2}{8}\right)(2\alpha - \sin 2\alpha) \tag{2.25}$$

and

$$T = D \sin \alpha \tag{2.26}$$

Substituting these equations in the general Equation 2.6 gives

$$\frac{(64Q^2 \sin \alpha)}{\left[gD^5\left(\alpha - \frac{1}{2}\sin 2\alpha\right)^3\right]} = 1 \tag{2.27}$$

This equation simplifies as

$$Q = K_c g^{1/2} D^{5/2} \tag{2.28}$$

where

$$K_c = \frac{(\alpha - 0.5 \sin 2\alpha)^{3/2}}{8(\sin \alpha)^{1/2}} \tag{2.29}$$

In the preceding equation, the angle α is expressed in radians and is related to the flow depth by

$$\frac{y_c}{D} = \frac{(1 - \cos \alpha)}{2} \tag{2.30}$$

TABLE 2.2

Critical Flow in Partly Full Circular Pipes

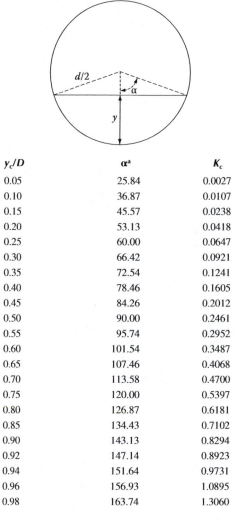

y_c/D	α^a	K_c
0.05	25.84	0.0027
0.10	36.87	0.0107
0.15	45.57	0.0238
0.20	53.13	0.0418
0.25	60.00	0.0647
0.30	66.42	0.0921
0.35	72.54	0.1241
0.40	78.46	0.1605
0.45	84.26	0.2012
0.50	90.00	0.2461
0.55	95.74	0.2952
0.60	101.54	0.3487
0.65	107.46	0.4068
0.70	113.58	0.4700
0.75	120.00	0.5397
0.80	126.87	0.6181
0.85	134.43	0.7102
0.90	143.13	0.8294
0.92	147.14	0.8923
0.94	151.64	0.9731
0.96	156.93	1.0895
0.98	163.74	1.3060

[a] $\alpha = \cos^{-1}(1 - 2\,y/D)$.

or

$$\alpha = \cos^{-1}\left(\frac{1 - 2y_c}{D}\right) \tag{2.31}$$

Calculating the angle α from Equation 2.31 and substituting in Equation 2.29, K_c can be calculated for any given y/D. Table 2.2, prepared by the author, lists K_c for a wide range in y_c/D. It is to be noted that K_c is dimensionless; therefore, Table 2.2 can be used in metric units and customary units as well.

Figure 2.3a presents the critical discharge as a function of flow depth for circular pipes ranging in size from 300 to 4500 mm in diameter. A similar relation for pipes of 1.0 ft (12 in.) to 15 ft (180 in.) in diameter is plotted in Figure 2.3b. Figures 2.3a and b, though less accurate

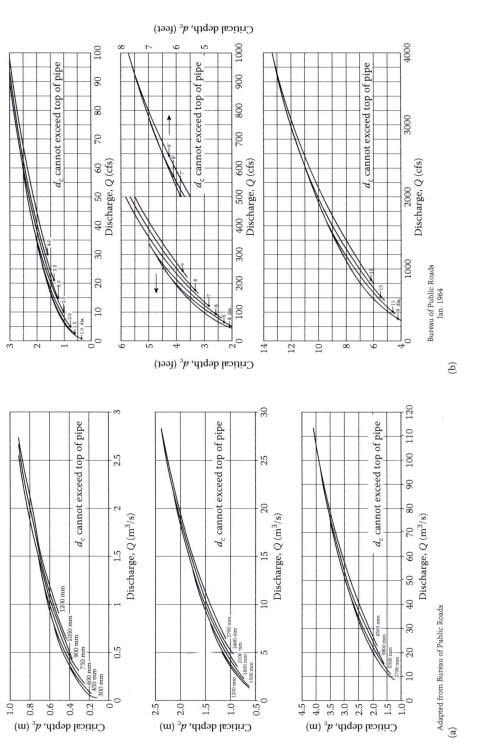

Adapted from Bureau of Public Roads

(a)

FIGURE 2.3 (a) Critical depth-discharge relation for circular pipes in metric units. (b) Critical depth-discharge relation for circular pipes in customary units. Note: The critical depth, y_c, is designated as d_c in this figure.

than Table 2.2, allow for a direct estimation of critical depth versus discharge or vice versa for various pipe sizes.

Example 2.4

Calculate the critical discharge for a 48-inch RCP pipe, if $y_c = 2.5$ ft.

Solution

$$\frac{y_c}{D} = \frac{2.5}{4} = 0.625$$

Calculate the angle α using Equation 2.31.

$$\cos^{-1}(1 - 1.25) = \cos^{-1}(-0.25)$$

$$\alpha = 104.48° = 104.48 \times \frac{\pi}{180} = 1.824 \text{ radians}$$

Then calculate K_c using Equation 2.29:

$$\sin \alpha = 0.968$$

$$\sin 2\alpha = -0.484$$

$$K_c = 0.377$$

$$Q = K_c \times \sqrt{g}D^{5/2} = 0.377 \times (32.2)^{1/2} (4)^{5/2} = 68.5 \text{ cfs}$$

Alternatively, interpolate α and K_c between $d_c/D = 0.6$ and 0.65 in Table 2.2:

$$\alpha = (101.54 + 107.46)/2 = 104.50°$$

$$K_c = (0.3487 + 0.4068)/2 = 0.3778$$

$$Q = 68.6 \text{ cfs}$$

Exercise: Calculate the critical discharge in a 1200 mm pipe for $y_c = 750$ mm.

2.4   NORMAL DEPTH

Normal depth is defined as the depth of flow for a constant discharge and under the force of gravity. Specifically, normal depth is associated with a free surface flow and is differentiated from pressure flow where the pressure at the top of conduit deviates from atmospheric pressure. In a uniform flow, the slope of the channel, the slope of HGL, and the slope of EL are equal and the HGL and EL are parallel to the channel bed.

A number of equations have been introduced for analyzing the normal flow in channels. Two of the more popular equations are discussed next.

2.4.1 CHEZY EQUATION

In 1775 Chezy presented the following equation for the average flow velocity in a channel:

$$V = C \sqrt{RS} \tag{2.32}$$

where
  R is the hydraulic radius, defined as the area divided by the wetted perimeter ($R = A/P$)
  S is the energy gradient, which is the same as channel slope in uniform flow
  C is the Chezy coefficient, which has the dimension of $L^{1/2}/T$

For C a number of formulae have been proposed, some of which are as follows:

a. Pavlosky formula: Pavlosky presented the following formula for C:

$$C = \left(\frac{1}{n}\right) R^{\alpha} \tag{2.33}$$

  where $\alpha = 1.5\sqrt{n}$; $R \leq 1$ m and $\alpha = 1.3\sqrt{n}$; $R > 1$ m.
  In this formula, n is the Manning's roughness coefficient, to be defined in the next section.

b. Prague Hydraulics Institute formula: This institute has proposed C to be a function of the hydraulic radius, R, and the mean stone diameter, d_{50}, on the channel bed as follows:

$$C = 18 \log\left(\frac{R}{d_{50}}\right) + 3 \tag{2.34}$$

c. Bazin formula: Bazin developed the following relation between Chezy's C and hydraulic radius, R:

$$C = \frac{87}{\left(1 + \dfrac{m}{\sqrt{R}}\right)} \tag{2.35}$$

In this formula, m is the Bazin coefficient, which varies from 0.07 for very smooth concrete channels to 1.90 for rock lined channels. These coefficients correspond to Manning's n values of 0.011 and 0.04, respectively. Table 2.3 provides the relation between the Manning's n and Bazin's m for lined channels and natural streams (Pazwash, 2007).

2.4.2 MANNING FORMULA

In 1889, Manning presented the following formula for the normal flow velocity in a channel in English units:

$$V = \left(\frac{1.486}{n}\right) R^{2/3} S^{1/2} \tag{2.36}$$

TABLE 2.3

Typical Values of the Manning's *n* and Bazin's *m* (Chezy's Equation) for Lined Channels and Natural Streams

Type of Cover	Manning's, *n*	Bazin's, *m*
Concrete, very smooth	0.010	0.07
Concrete, normal	0.015	0.25
Brick/mortar stone lined	0.016	0.35
Earth channel, smooth	0.023	0.85
Earth channel, semismooth	0.027	1.30
Rocky channels	0.040	1.90
Streams/rivers, normal	0.030	1.65

In metric units, the Manning formula, which is also known as Gauckler-Manning-Strickler formula in Europe, reads as

$$V = \left(\frac{1}{n}\right) R^{2/3} S^{1/2} \tag{2.37}$$

where
V = flow velocity, m/s (ft/s)
R = hydraulic radius, m (ft)
S = energy slope, m/m (ft/ft)
n = Manning's roughness coefficient

The coefficient n varies with the surface roughness of channel. From studies of streams with gravel beds in Switzerland in 1923, Strickler proposed the following equation relating n with the gravel size:

$$n = 0.0417 d_{50}^{1/6} \tag{2.38}$$

where d is the median size of the bed material in meters. In customary units, with d expressed in feet, this equation becomes (Henderson, 1966)

$$n = 0.034 d_{50}^{1/6} \tag{2.39}$$

Equating Equations 2.32 and 2.39 will relate the Chezy's coefficient, C, with the Manning's coefficient, n, as follows:

$$C = \frac{1.49 R^{1/6}}{n} \tag{2.40}$$

The constant 1.486 in the Manning formula in English units represents the ratio of meter to foot to power one-third. More accurately, this constant would be

$$(3.2808)^{1/3} = 1.4859$$

In practice, however, this constant is commonly rounded to 1.49.

Thus, it appears that this formula must have been originally developed in metric units. It was suggested that the Manning formula was first proposed in metric units by Hagen in 1876 (Chow, 1959).

The literature generally ascribes a dimension $L^{1/6}$ to n and $L^{1/2}/T$ to 1.49. In practice, however, the same n value is used in the CU and SI versions of the Manning formula. For this reason, it is more appropriate to treat n as a dimensionless number and ascribe dimension $L^{1/3}/T$ to the coefficients 1.49 and 1.0 so that the Manning formula is dimensionally homogeneous. This statement is verified by the relation $(1 \text{ m/s})^{1/3} = (3.28 \text{ ft/s})^{1/3} = 1.486 \text{ (ft/s)}^{1/3}$.

In terms of discharge, the Manning formula is given as

$$Q = \left(\frac{1}{n} \right) AR^{2/3} S^{1/2} \tag{2.41}$$

$$Q = \left(\frac{1.49}{n} \right) AR^{2/3} S^{1/2} \tag{2.42}$$

in metric and English units, respectively.

In uniform channel flow where the energy slope, S, and the slope of channel bed (or pipe flowing partially full), S_o, are equal, the Manning formula may be expressed as

$$Q = \left(\frac{1}{n} \right) AR^{2/3} S_o^{1/2} \quad \text{SI} \tag{2.43}$$

$$Q = \left(\frac{1.49}{n} \right) AR^{2/3} S_o^{1/2} \quad \text{CU} \tag{2.44}$$

where
  Q = discharge, m^3/s (cfs)
  A = flow area, m^2 (ft^2)
  S_o = channel slope, m/m (ft/ft)

It is to be noted that the Manning formula is also employed for pipe flow calculations and design of storm drains.

Table 2.4 provides equations for areas and hydraulic radius of pipe and channel geometries commonly used in practice. The Manning formula is used not only for analyzing uniform flow, but also for gradually varied flow in pipes and open channels. Table 2.5 lists typical values of "n" for different makes of pipes and types of linings of man-made or natural channels. A more complete listing of n values can be found in Barnes (1987), Brater et al. (1996), Chow (1959), Henderson (1966), and FHWA (2012, 2013).

Figures 2.4a and b present nomographs for the solution of the Manning formula for full flow in pipes in metric and customary units, respectively.

Tables 2A.1 to 2A.3 in Appendix 2A provide hydraulic properties for round and elliptical concrete pipes. These tables facilitate performing design calculations for storm drains.

TABLE 2.4
Geometric Properties of Channel Sections

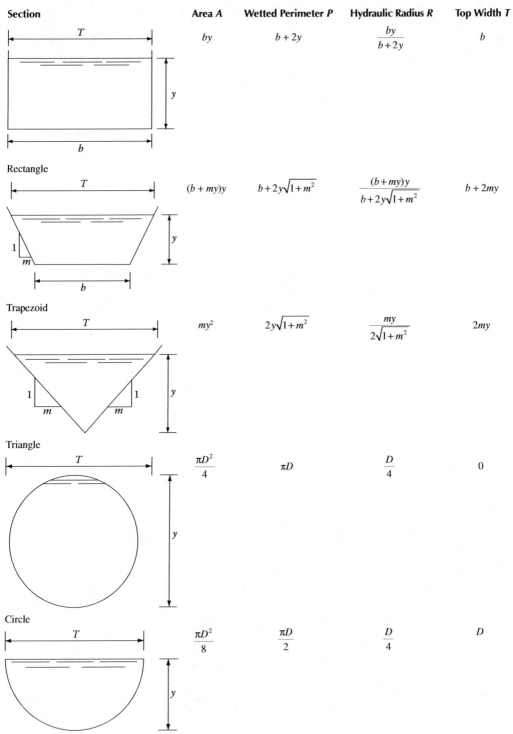

Section	Area A	Wetted Perimeter P	Hydraulic Radius R	Top Width T
Rectangle	by	$b + 2y$	$\dfrac{by}{b+2y}$	b
Trapezoid	$(b + my)y$	$b + 2y\sqrt{1+m^2}$	$\dfrac{(b+my)y}{b+2y\sqrt{1+m^2}}$	$b + 2my$
Triangle	my^2	$2y\sqrt{1+m^2}$	$\dfrac{my}{2\sqrt{1+m^2}}$	$2my$
Circle	$\dfrac{\pi D^2}{4}$	πD	$\dfrac{D}{4}$	0
Semicircle	$\dfrac{\pi D^2}{8}$	$\dfrac{\pi D}{2}$	$\dfrac{D}{4}$	D

(Continued)

TABLE 2.4 (CONTINUED)
Geometric Properties of Channel Sections

Section	Area A	Wetted Perimeter P	Hydraulic Radius R	Top Width T
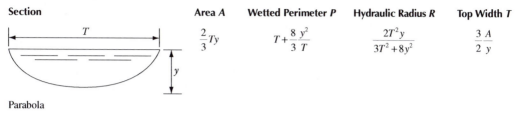	$\dfrac{2}{3}Ty$	$T+\dfrac{8}{3}\dfrac{y^2}{T}$	$\dfrac{2T^2y}{3T^2+8y^2}$	$\dfrac{3}{2}\dfrac{A}{y}$

Parabola

Note: When $T > 4y$.

TABLE 2.5
Manning's Roughness Coefficient, *n*, for Pipes and Channels

Conduit Material	Manning's *n*
Concrete pipes	0.012–0.013
Ductile iron pipes	0.011–0.015
PVC pipes	0.010–0.011
HDPE pipes	0.011–0.012[a]
Vitrified clay pipes	0.011–0.015
Corrugated metal pipes (1/2–3 in. [12.5–7.6 mm] corrugation)	0.022–0.028
Old brick lined conduits	0.013–0.017
Open Channels (Lined)	
Brick	0.012–0.018
Concrete	0.011–0.016
Riprap	0.025–0.040
Grass	0.030–0.300[b]
Dredged Channels	
Earth, straight and uniform	0.020–0.03
Earth, winding	0.025–0.04
Rock, 2–12 in. (5–30 cm)	0.025–0.06
Natural Channels and Streams	
Fairly uniform	0.030–0.05
Irregular	0.040–0.08
Overbanks	0.080–0.12

[a] ADS, the largest manufacturer of HDPE pipes in the United States, recommends using $n = 0.012$.

[b] Varies with grass height, depth, and velocity of flow (see Chapter 4).

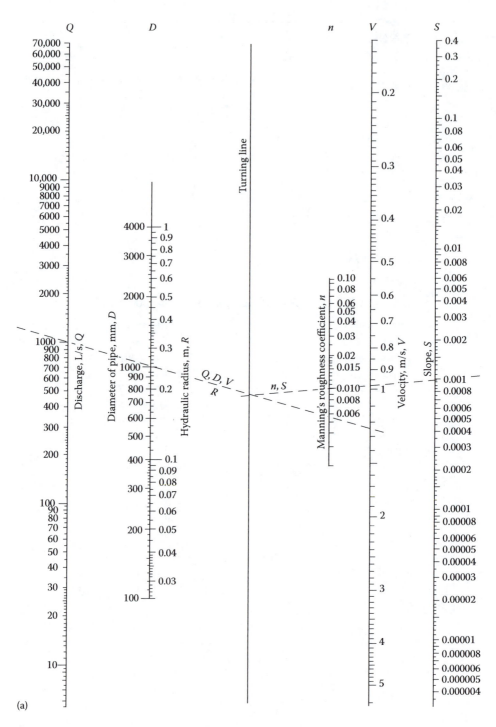

FIGURE 2.4 (a) Manning formula nomograph (SI).

(*Continued*)

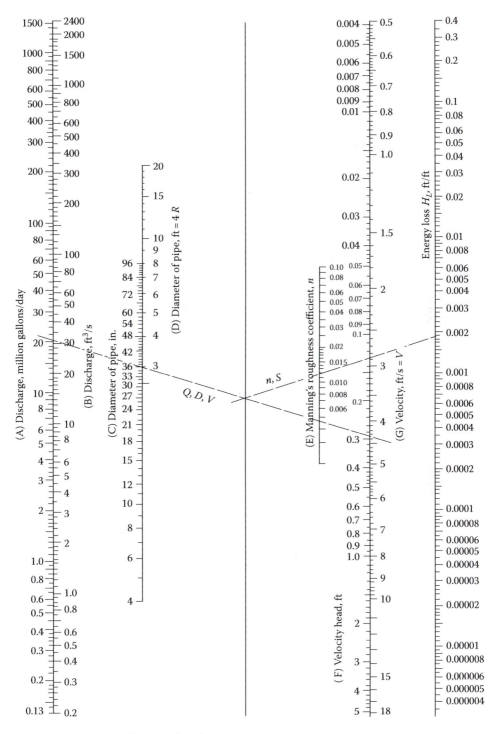

Alignment chart for energy loss in pipes, for Manning's formula
Note: use chart for flow computations, $H_L = S$

(b)

FIGURE 2.4 (CONTINUED)   (b) Manning formula nomograph (CU).

2.5   CALCULATION OF FLOW DEPTH

Discharge in a conduit depends on both the area and hydraulic radius, each of which not only is a function of depth but also depends on the conduit geometry. Therefore, a simple relationship cannot be established between discharge and flow depth. In practice, the depth of flow is commonly calculated using an iterative process. Simplified procedures for calculating the flow depths in circular pipes (flowing partly full) and trapezoidal channels are derived in the following sections.

2.5.1   CIRCULAR SECTIONS

For circular sections, such as drainage pipes flowing partly full, the Manning formula may be written as

$$Q = \left(\frac{1}{n}\right) KD^{8/3} S^{1/2} \quad \text{SI} \tag{2.45}$$

$$Q = \left(\frac{1.49}{n}\right) KD^{8/3} S^{1/2} \quad \text{CU} \tag{2.46}$$

where D is pipe diameter in m (ft) and K is a dimensionless number, given by the following equation:

$$K = \frac{AR^{2/3}}{D^{8/3}} \tag{2.47}$$

Hydraulic properties A and R may be expressed in terms of pipe geometrical parameters in the following dimensionless forms:

$$\frac{A}{D^2} = \frac{(2\alpha - \sin 2\alpha)}{8} \tag{2.48}$$

$$\frac{P}{D} = \alpha \tag{2.49}$$

$$\frac{R}{D} = \frac{1}{4}\left[1 - \frac{(\sin 2\alpha)}{2\alpha}\right] \tag{2.50}$$

where

$$\alpha = \cos^{-1}(1 - 2y/D) \tag{2.51}$$

Therefore,

$$K = \frac{(2\alpha - \sin 2\alpha)^{5/3}}{32\alpha^{2/3}} \tag{2.52}$$

In the preceding equations α is in radians (degrees · π/180).

These parameters together with the ratio of average depth, D_a, to pipe diameter, D, are listed in Table 2.6. This table shows that the maximum discharge in pipes flowing partly full occurs at approximately $y/D = 0.94$, representing the highest K value. More accurate calculations show that the maximum discharges occur at $y/D = 0.938$. Since K is dimensionless, it can be applied directly to Equations 2.45 and 2.46. A graphical representation of dimensionless area, velocities, and discharges is presented in Figure 2.5. Using either Figure 2.5 or Table 2.6 greatly simplifies pipe flow

TABLE 2.6

Hydraulic Properties of Partly Full Circular Sections

y/D	T/D	R/D	D_a/D	A/D^2	K
0.05	0.4359	0.0326	0.0337	0.0147	0.0015
0.10	0.6000	0.0635	0.0681	0.0409	0.0065
0.15	0.7141	0.0929	0.1034	0.0739	0.0152
0.20	0.8000	0.1206	0.1398	0.1118	0.0273
0.25	0.8660	0.1466	0.1773	0.1535	0.0427
0.30	0.9165	0.1709	0.2162	0.1982	0.0610
0.35	0.9539	0.1935	0.2568	0.2450	0.0820
0.40	0.9798	0.2142	0.2994	0.2934	0.1050
0.45	0.9950	0.2331	0.3445	0.3428	0.1298
0.50	1.0000	0.2500	0.3927	0.3927	0.1558
0.55	0.9950	0.2649	0.4448	0.4426	0.1826
0.60	0.9798	0.2776	0.5022	0.4920	0.2094
0.65	0.9539	0.2881	0.5665	0.5404	0.2358
0.70	0.9165	0.2962	0.6407	0.5872	0.2610
0.75	0.8660	0.3017	0.7296	0.6319	0.2842
0.80	0.8000	0.3042	0.8420	0.6736	0.3047
0.85	0.7141	0.3033	0.9963	0.7115	0.3212
0.90	0.6000	0.2980	1.2409	0.7445	0.3322
0.92	0.5426	0.2944	1.3933	0.7560	0.3345
0.94	0.4750	0.2895	1.6131	0.7662	0.3353
0.96	0.3919	0.2829	1.9771	0.7749	0.3339
0.98	0.2800	0.2735	2.7916	0.7816	0.3294
1.00	0.0000	0.2500	∞	0.7854	0.3117

$\alpha = \cos^{-1}(1 - 2y/D)$

$T = D \sin \alpha = 2(y(D - y))^{0.5}$, top width

$A/D^2 = \dfrac{(2\alpha - \sin 2\alpha)}{8}$

$P = \alpha D$

$R = A/P = \dfrac{D(2\alpha - \sin 2\alpha)}{8\alpha}$

$K = (AR^{2/3})/D^{8/3} = \dfrac{[2\alpha - \sin 2\alpha]^{5/3}}{32[\alpha]^{5/3}}$

$D_a = A/T = \dfrac{D(2\alpha - \sin 2\alpha)}{8(\sin \alpha)}$, average depth

$D_a/D = \dfrac{(2\alpha - \sin 2\alpha)}{8(\sin \alpha)}$

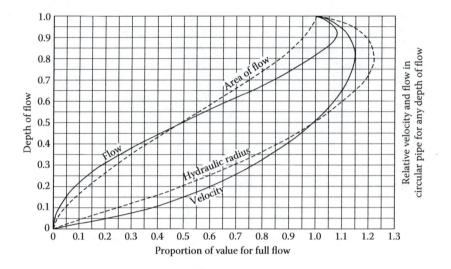

FIGURE 2.5 Variation of relative velocity and flow in partly full circular pipe.

calculations (Pazwash, 2007). Figure 2A.1 in the appendix at the end of this chapter shows dimensionless partly full flow parameters for elliptical pipes.

2.5.2 TRAPEZOIDAL SECTIONS

In a trapezoidal section, the Manning formula may be written as

$$Q = \left(\frac{K}{n}\right) b^{8/3} S^{1/2} \tag{2.53}$$

$$Q = 1.49 \left(\frac{K}{n}\right) b^{8/3} S^{1/2} \quad \text{CU} \tag{2.54}$$

where

$$K = \frac{AR^{2/3}}{b^{8/3}} = \frac{\left[\left(\dfrac{y}{b}\right) + m\left(\dfrac{y}{b}\right)^2\right]^{5/3}}{\left[1 + 2\left(\dfrac{y}{b}\right)\sqrt{\left(1 + m^2\right)}\right]^{2/3}} \tag{2.55}$$

b = bottom width
y = flow depth
m = side slope (H/V)

TABLE 2.7

Hydraulic Parameter K for Rectangular and Trapezoidal Channels

$$K = \frac{AR^{2/3}}{b^{8/3}} = \frac{[(y/b)+m(y/b)^2]^{5/3}}{\left[1+2(y/b)\sqrt{(1+m^2)}\right]^{2/3}}$$

y/b	$m = 0$	$m = 1/4$	$m = 1/2$	$m = 3/4$	$m = 1.0$	$m = 3/2$	$m = 2.0$	$m = 3.0$
0.00	0.0000	0.0000	0.0000	0.0000	0.0000	0.0000	0.0000	0.0000
0.01	0.0005	0.0005	0.0005	0.0005	0.0005	0.0005	0.0005	0.0005
0.02	0.0014	0.0014	0.0015	0.0015	0.0015	0.0015	0.0015	0.0015
0.03	0.0028	0.0028	0.0028	0.0029	0.0029	0.0029	0.0029	0.0029
0.04	0.0044	0.0045	0.0046	0.0046	0.0047	0.0047	0.0048	0.0049
0.05	0.0064	0.0065	0.0066	0.0067	0.0067	0.0069	0.0070	0.0071
0.06	0.0085	0.0087	0.0089	0.0090	0.0091	0.0093	0.0095	0.0098
0.07	0.0109	0.0112	0.0114	0.0116	0.0118	0.0121	0.0123	0.0128
0.08	0.0135	0.0139	0.0142	0.0145	0.0147	0.0152	0.0155	0.0162
0.09	0.0162	0.0167	0.0172	0.0176	0.0179	0.0185	0.0190	0.0199
0.10	0.0191	0.0198	0.0204	0.0209	0.0214	0.0221	0.0228	0.0241
0.15	0.0356	0.0376	0.0394	0.0409	0.0422	0.0445	0.0466	0.0504
0.20	0.0547	0.0589	0.0627	0.0659	0.0687	0.0737	0.0783	0.0868
0.25	0.0757	0.0832	0.0898	0.0956	0.1007	0.1099	0.1182	0.1340
0.30	0.0983	0.1100	0.1205	0.1298	0.1382	0.1532	0.1669	0.1928
0.35	0.1220	0.1392	0.1547	0.1686	0.1812	0.2038	0.2246	0.2641
0.40	0.1468	0.0167	0.1922	0.2118	0.2297	0.2621	0.2919	0.3486
0.45	0.1723	0.0198	0.2330	0.2596	0.2840	0.3283	0.3693	0.4472
0.50	0.1984	0.2390	0.2770	0.3119	0.3440	0.4027	0.4571	0.5606
0.55	0.2251	0.2761	0.3243	0.3688	0.4100	0.4856	0.5558	0.6896
0.60	0.2523	0.3150	0.3748	0.4304	0.4822	0.5773	0.6660	0.8350
0.65	0.2799	0.3557	0.4286	0.4968	0.5605	0.6781	0.7880	0.9974
0.70	0.3079	0.3980	0.4856	0.5681	0.6453	0.7884	0.9222	1.1777
0.75	0.3361	0.4421	0.5460	0.6442	0.7367	0.9083	1.0692	1.3766
0.80	0.3646	0.4879	0.6096	0.7254	0.8347	1.0383	1.2293	1.5946
0.85	0.3934	0.5354	0.6767	0.8118	0.9297	1.1785	1.4030	1.8325
0.90	0.4223	0.5846	0.7471	0.9033	1.0516	1.3292	1.5907	2.0911
0.95	0.4514	0.6354	0.8210	1.0002	1.1708	1.4908	1.7927	2.3708
1.00	0.4807	0.6879	0.8984	1.1024	1.2973	1.6636	2.0095	2.6725

where

$$K = \frac{Q \cdot n}{b^{8/3} S^{1/2}} \quad \text{SI}$$

$$K = \frac{Q \cdot n}{1.49\, b^{8/3} S^{1/2}} \quad \text{CU}$$

Table 2.7, prepared by the author, lists dimensionless K values for a wide range in the (y/b) ratio and side slope, m. The column $m = 0$ in this table represents a rectangular section such as a box culvert. Since K is dimensionless, this table equally applies to SI and CU units. Figure 2.6 depicts a graphical representation of K as a function of y/b and m. The use of either Table 2.7 or Figure 2.6 eliminates the need for an iterative solution and greatly simplifies open channel flow calculations.

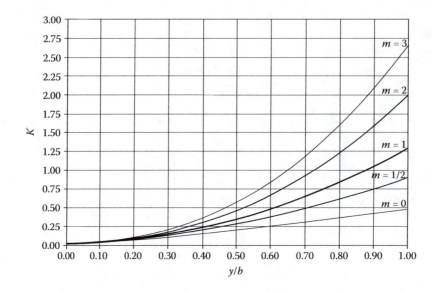

FIGURE 2.6   Variation of K with y/b for rectangular and trapezoidal channels.

Example 2.5

A roadside channel is to carry a discharge of 1 m³/s. The channel is lined with 5 cm stone, has a 1 m bottom width, 2:1 (2H:1V) side slopes, and 0.75% longitudinal slope. Calculate the normal depth of flow in this channel.

Solution

The calculations are performed using both the iterative procedure and Table 2.7.

a. Iterative solution:
   Express the Manning formula as

$$AR^{2/3} = \frac{nQ}{S^{1/2}}$$

$$A = by + my^2 = y + 2y^2$$

$$P = b + 2y\sqrt{1+m^2} = 1 + 2\sqrt{5}y$$

$$R = \frac{A}{P} = \frac{(y+2y^2)}{(1+2\sqrt{5}y)}$$

$$AR^{2/3} = \frac{\left(y+2y^2\right)^{5/3}}{\left(1+2\sqrt{5}y\right)^{2/3}}$$

Using Table 2.5, estimate n value at 0.025

$$\frac{nQ}{\sqrt{S}} = \frac{0.025 \times 1}{(0.0075)^{1/2}} = 0.2887$$

Then

$$AR^{2/3} = \frac{[(y(1+2y)]^{5/3}}{\left[1+2\sqrt{5}y\right]^{2/3}} = 0.2887$$

Raise both sides of the preceding equation to 3/2 power:

$$\frac{[y(1+2y)]^{5/2}}{(1+4.472y)} = (0.2887)^{1.5} = 0.1551$$

Now solve the equation $f(y) = 0.1551$; where

$$f(y) = \frac{[y(1+2y)]^{5/2}}{(1+4.472y)}$$

First try $y = 0.5$ m

$$f(y) = 0.2975 > 0.1551$$

Second try $y = 0.4$ m

$$f(y) = 0.1577 > 0.1551$$

Third try $y = 0.39$ m

$$f(y) = 0.1413 < 0.1551$$

Fourth try $y = 0.395$ m

$$f(y) = 0.1467 < 0.1551$$

Interpolate y between 0.395 and 0.4 to get 0.398 m.
b. Use of Table 2.7:
Under the $m = 2$ column, search for $K = 0.2887$
y/b lies between 0.35 and 0.4 at

$$\frac{y}{b} = 0.35 \quad K = 0.2248;$$

$$\frac{y}{b} = 0.4 \quad K = 0.2919$$

Calculate depth, y, by interpolating y/b between these ratios, as follows:

$$\frac{y}{b} = 0.35 + \frac{0.2887 - 0.2248}{0.2919 - 0.2248} \times (0.4 - 0.35) = 0.35 + 0.952 \times 0.05$$

$$\frac{y}{b} = 0.3976, \text{ say, } \frac{y}{b} = 0.398$$

$$y = 0.398 \times 1 = 0.398 \text{ m}$$

Example 2.6

To carry storm water runoff at a rate of 15 cfs, an 18 in. reinforced concrete pipe (RCP) is to be used. Calculate the minimum slope required in order that the pipe carries this flow under normal flow conditions. If the pipe is laid at 3% slope, what would be the depth of flow?

Solution

Use $n = 0.012$ for smooth pipe

$$A = \frac{\pi D^2}{4} = 1.767 \text{ sf}$$

$$R = \frac{D}{4} = 0.375 \text{ ft; see Table 2.4}$$

Full flow:

$$Q = \frac{149}{0.012} \times 1.767 \times (0.375)^{2/3} \times S^{1/2}$$

$$Q = 114 \times S^{1/2}$$

Note: Alternatively, we can use Table 2A.1, to read $K = 114.0$.

$$S = \left(\frac{15}{114}\right)^2 = 0.017; \ 1.7\%$$

Partly full flow:
Instead of performing iterative calculations, we use Table 2.6 to facilitate the solution.
Calculate $K = AR^{2/3}/D^{8/3}$.
Using Equation 2.46,

$$K = \frac{Qn}{1.49(D)^{8/3} S^{1/2}}$$

$$K = \frac{15 \times 0.012}{1.49(1.5)^{8/3} \times (0.03)^{1/2}} = 0.2366$$

According to Table 2.6, y/D lies between 0.65 and 0.7, though much closer to the former. At

$$\frac{y}{D} = 0.65 \quad K = 0.2358$$

$$\frac{y}{D} = 0.7 \quad K = 0.2610$$

By interpolation:

$$\frac{y}{D} = 0.65 + \frac{0.2366 - 0.2358}{0.2610 - 0.2358} \times (0.7 - 0.65)$$

$$\frac{y}{D} = 0.6516$$

$y = 0.6516 \times 15 = 9.78$ in., say, 9.8 in.

Alternatively: solve this problem using Figure 2.5.
First, calculate the full flow capacity at 3% slope:

$$Q_F = K\sqrt{S} = 114 \times (0.03)^{1/2} = 19.75 \text{ cfs}$$

Then calculate the Q/Q_F ratio:

$$\frac{Q}{Q_F} = \frac{15}{19.75} = 0.76$$

For this flow ratio, Figure 2.5 gives

$$\frac{y}{D} = 0.65$$

$y = 9.75$ in., say, 9.8 in.

Example 2.7

A 38 in. × 60 in. horizontal elliptical reinforced concrete pipe is to convey a design discharge of 130 cfs at 70% of full depth. What should the pipe slope be?

Solution

For $y/d = 0.7$, Figure 2A.1 in the appendix to this chapter shows $Q/Q_F = 0.86$.
   Then $Q_F = 130/0.86 = 151.2$ cfs
   For $n = 0.012$, Table 2A.3 gives $K = 1565$

$$Q_F = 1565 \, (S)^{1/2}$$

$$S = \frac{(151.2)^2}{1565} = 0.0093$$

$$S = 0.93\%$$

Example 2.8

During a large storm, the depth of water in the pipe of the previous example is measured at 31.5 in. Calculate the discharge.

Solution

$$\frac{y}{D} = \frac{31.5}{38} = 0.83$$

Entering in Figure 2A.1 with relative depth = 0.83 gives

$$\frac{Q}{Q_F} = 1.05$$

$$Q = 1.05 \times 145.3 = 152.6 \text{ cfs}$$

Note: Commonly, drainage pipes are designed to carry a discharge less than or equal to their normal capacity. However, flow in culverts may exceed the normal capacity due to inlet control condition.

The inlet control capacity of circular pipes is shown in Figure 2A.2a and b in Appendix 2A at the end of this chapter. More information on this matter is provided in Chapter 4.

2.6   ENERGY LOSSES IN PIPES AND CULVERTS

Energy losses, also called "head losses," occur due to friction along flow and also at locations where there is a rapid change either in the cross section or the flow direction. In design of drainage systems, these losses must be calculated in order to define the hydraulic grade line and/or energy line along the flow. While frictional losses are gradual and fairly uniform, the losses at junctions are abrupt and localized. These losses are commonly termed "minor losses" in practice. However, it is to be noted that, in a reach of pipe or culvert with many junctions and bends and/or partial obstructions, the latter losses may surpass frictional losses. Thus, these losses are termed local losses rather than minor losses in this book. Example 2.9 later in this section exemplifies the matter. The following sections present equations for estimating head losses in drainage systems.

2.6.1   FRICTION LOSSES

The lead loss due to friction is calculated as follows:

$$h_f = S_f L \tag{2.56}$$

where
  h_f = friction loss, m (ft)
  S_f = friction slope, m/m (ft/ft)
  L = length of pipe, m (ft)

The frictional slope, S_f, in this equation is the same as the hydraulic grade line. In uniform flow, as previously indicated, this slope is also equal to the slope of the pipe or channel. The friction slope in uniform flow can be calculated using the Manning formula, which may be rearranged as follows:

$$S_f = c \frac{n^2 V^2}{R^{1.33}} \tag{2.57}$$

where
  R = hydraulic radius
  c = 1.0 metric (0.45 CU)

Expressing the frictional slope in terms of the velocity head:

$$S_f = \frac{19.62 \, n^2}{R^{1.33}} \frac{V^2}{2g} \quad \text{SI} \tag{2.58}$$

$$S_f = \frac{29 \, n^2}{R^{1.33}} \frac{V^2}{2g} \quad \text{CU} \tag{2.59}$$

Combining Equations 2.56 and 2.58/2.59 gives

$$h_f = \frac{19.62 \, n^2 L}{R^{1.33}} \frac{V^2}{2g} \tag{2.60}$$

$$h_f = \frac{29 \, n^2 L}{R^{1.33}} \frac{V^2}{2g} \tag{2.61}*$$

2.6.2 LOCAL LOSSES

In practice, local losses are commonly expressed in terms of the velocity head, as follows:

$$h_\ell = k \frac{V^2}{2g} \tag{2.62}$$

where
  h_ℓ = local head loss
  k = dimensionless loss coefficient
  $V^2/2g$ = velocity head

In water pipes, local losses are also expressed in terms of equivalent pipe length producing the equal amount of loss. This length is calculated as follows:

$$L_\ell = \frac{h_L}{S_f} \tag{2.63}$$

Although not commonly practiced, the same procedure may be used in culverts flowing full. This matter is illustrated by Example 2.9 later in this chapter.

When there is a sudden change in cross section, such as an abrupt contraction or expansion, Equation 2.62 may be written in terms of the difference in the velocity head across the junction, as follows:

$$h_L = k\Delta \frac{V^2}{2g} \tag{2.64}$$

where $\Delta V^2/2g$ represent the change in the velocity head.

---

* Note: Using 1.486 rather than 1.49 in the Manning's formula, the coefficient 29 will read 29.2.

The loss coefficient, k, depends on the type of flow, namely open channel or pressure flow, and also varies from subcritical to supercritical. More information on this matter can be found in Chow (1959), Henderson (1966), French (1985), Brater et al. (1996), and FHWA (2013). Head loss equations for some of the common drainage structures are presented in the following sections.

2.6.2.1 Entrance and Exit Losses

At the inlet face of a pipe or culvert, the flow velocity increases; as a result, the pressure drops. Neglecting the velocity head at the section just before the entrance, Equation 2.64 simplifies as

$$h_\ell = k_c \frac{V^2}{2g} \qquad (2.65)$$

where V is the flow velocity in the pipe. The entrance loss coefficient, k_e, for pipes varies from 0.1 for rounded entrance to 0.7 for mitered to slope connections. Table 2.8 lists entrance loss coefficients for pipes and culverts under various wing wall conditions.

At the exit section from a pipe or culvert to a body of water, the flow undergoes a sudden expansion. The exit loss coefficient at this section depends on the flow condition of the receiving water body. For a stagnant body of water, the entire velocity head is lost at the outlet, and therefore

$$h_o = \frac{V_o^2}{2g} \qquad (2.66)$$

TABLE 2.8
Entrance Loss Coefficient, k_e, for Pipes and Culverts

Type of Structure	k_e
Concrete Pipe	
Mitered to conform to slope	0.7
End section	0.5
Projecting, square cut	0.5
Projecting, groove end	0.2
Square-edge at headwall	0.5
Rounded at headwall	0.2
Groove end	0.2
Beveled edges	0.2
Box Culvert	
Wing walls parallel (extending at sides)	0.7
Wing walls at 30°–75° to culvert	0.5
Headwall along embankment (no wing walls)	0.5
Rounded entrance at headwall	0.2
Tapered to slope	0.2
Plastic Pipes	
End section	0.2
Square cut	0.5

However, if the body of water is in motion, such as a channel or stream, the exit loss is

$$h_{\mathrm{o}} = \frac{V_{\mathrm{o}}^2}{2g} - \frac{V_{\mathrm{d}}^2}{2g} \qquad (2.67)$$

where
   V_{o} = outlet velocity
   V_{d} = channel velocity in the direction of pipe flow

When the outlet is perpendicular to the receiving channel, $V_{\mathrm{d}} = 0$ and the preceding equations simplify as

$$h_{\mathrm{o}} = V_{\mathrm{o}}^2 / 2g \qquad (2.68)$$

reflecting an exit loss coefficient, $k_{\mathrm{o}} = 1$.

2.6.2.2 Sudden Expansions or Contractions

The loss in a sudden expansion results from separation of the flow from the wall within a distance from the expansion section. This is illustrated in Figure 2.7. Applying the momentum and energy equations between sections 1 and 2 results in the following equation for head loss:

$$h_{\mathrm{e}} = \frac{(V_1 - V_2)^2}{2g} \qquad (2.69)$$

This equation may be expressed as

$$h_{\mathrm{e}} = \frac{kV_1^?}{2g} \qquad (2.70)$$

where

$$k = \left[1 - \left(\frac{D_1}{D_2} \right)^2 \right]^2 \qquad (2.71)$$

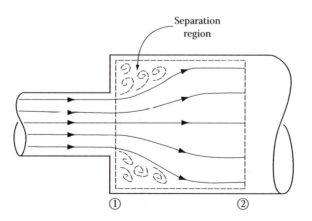

FIGURE 2.7   Sudden expansion.

TABLE 2.9
Sudden Contraction Values, k_c, for Pipes

D_2/D_1	0.90	0.80	0.60	0.40	0.20	0.10
k_c	0.05	0.15	0.28	0.36	0.42	0.45

A special case is a pipe ending at a reservoir. In this case, velocity V_2 is negligible compared to V_1 and the preceding equation simplifies to $h_1 = V_1^2/2g$, which is the same as the exit loss Equation 2.68.

In a sudden contraction, the equation for head loss is expressed as

$$h_1 = \frac{k_c V_2^2}{2g} \tag{2.72}$$

where the coefficient k depends on the ratio of the areas before and after the contraction section. Typical values of k_c are listed in Table 2.9.

2.6.2.3 Bend Losses

At a pipe bend, the head loss is expressed by the following equation:

$$h_b = k_b \frac{V^2}{2g} \tag{2.73}$$

where

$$k_b = 0.003\,(\Delta\theta) \tag{2.74}$$

and $\Delta\theta$ = angle of curvature in degrees.

For a 90° bend, the head loss coefficient would be approximately 0.3. The head loss in a culvert flowing full may be estimated using Equation 2.73 with a bend factor given as

$$k_b = 0.5\left(\frac{\Delta\theta}{90}\right)^{1/2} \tag{2.75}$$

For a 90° bend, this equation gives $k_b = 0.5$. The preceding equations are for sharp bends. When using a circular bend having a radius four times or greater than the culvert diameter, the head loss can be neglected.

2.6.2.4 Head Loss at Transitions

In gradual expansions and contractions, the head loss depends on the angle of expansion/contraction. It also depends on the ratio of pipe diameters, which reflects the length of the transition. In general, the head loss is given by

$$h_1 = k_t\left(\frac{V_1^2}{2g} - \frac{V_2^2}{2g}\right) \tag{2.76}$$

where k_t depends on the angle of expansion/contraction as depicted on Figure 2.8 and the pipe diameters.

Table 2.10 provides the expansion loss coefficient for angles ranging from 5° to 60°. For more information, the reader is referred to FHWA (2013) and ASCE and WEF (1992).

In a gradual contraction, the lead loss is smaller than that of an expansion for a given angle. The head loss coefficient in this case may be approximated as one-half of the expansion coefficients listed in Table 2.10.

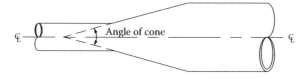

FIGURE 2.8 Sketch of an expansion transition.

TABLE 2.10
Gradual Expansion Loss Coefficients for Angles Ranging from 5° to 60°

	Angle of Cone						
D_2/D_1	10°	20°	45°	60°	90°	120°	180°
1.5	0.17	0.40	1.06	1.21	1.14	1.07	1.00
3	0.17	0.40	0.86	1.02	1.06	1.04	1.00

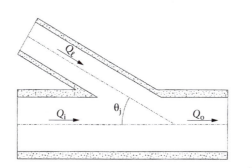

FIGURE 2.9 Pipe junction definition.

2.6.2.5 Junction Losses

At a pipe junction, such as that shown in Figure 2.9, the head loss can be calculated applying the momentum equation as follows:

$$h_j = \frac{\left[(Q_o V_o - Q_i V_i) - (Q_\ell V_\ell)\cos\theta_j\right]}{\left[0.5g(A_o + A_i)\right]} + h_i - h_o \tag{2.77}$$

where

Q_o, Q_i, and Q_ℓ = outlet, inlet, and lateral flows, respectively, m³/s (ft³/s)
V_o, V_i, and V_ℓ = outlet, inlet, and lateral velocities, respectively, m/s (ft/s)
h_o, h_i = outlet and inlet velocity heads, m (ft)
A_o, A_i = outlet and inlet cross-sectional areas, m² (ft²)
θ_j = angle between the trunk pipe and lateral pipe

2.6.2.6 Losses at Access Holes and Inlets

At an access hole or junction hole (commonly referred to as manholes), there is a loss of head due to velocity change, especially if the incoming and outgoing pipes differ in size. In addition, a loss occurs when the flow direction changes and this loss can be greater than the former. To reduce the loss at a bending access hole, a curved vane or deflector may be installed in the direction of flow.

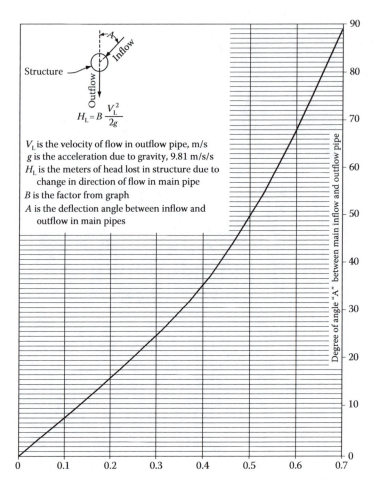

FIGURE 2.10 Bend loss factor for access holes.

Losses due to a bend at an access hole may be estimated using Figure 2.10. The head loss on this figure is calculated based on the outflow velocity head. The velocity head differential of inflow and outflow pipes may be added to account for the velocity change.

The head loss at an inlet depends largely on the change in the flow direction between the inflow and outflow pipes. The head loss at inlets can be estimated using the following equation:

$$h_i = k_i \left(\frac{V_i^2}{2g} \right) \tag{2.78}$$

where
  V_1 = inflow velocity
  k_i = head loss coefficient

The head loss coefficients for a number of common inlet arrangements are listed here:

- Straight run–square edge, $k_i = 0.5$
- Angled through 90°, $k_i = 1.5$

For intermediate angle of deflection in flow direction, k_i may be interpolated between 0.5 and 1.5.

Example 2.9

A drainage system consisting of a series of box culverts is shown in Figure 2.11. This system is aged and in part is located under a building in a town in northern New Jersey. Express local losses in each section in terms of its equivalent length.

Solution

First relate frictional losses in each section to the velocity head using Equation 2.59:

$$S_f = \left(\frac{29.0 \times n^2}{R^{1.33}} \right) \frac{V^2}{2g}$$

Estimate $n = 0.015$ for the old culvert:

$$S_f = \left(\frac{6.53 \times 10^{-3}}{R^{1.33}} \right) \frac{V^2}{2g}$$

Then equate local losses to frictional losses:

$$h_\ell = \frac{kV^2}{2g} = S_f L_{eq}$$

where L_{eq} is the equivalent length of culvert having the same frictional loss as the local loss. Combining these equations yields

$$L_{eq} = \frac{k}{\left(\dfrac{6.53 \times 10^{-3}}{R^{1.33}} \right)}$$

or

$$L_{eq} = 153.1 \times R^{1.33} \times k$$

Now estimate total local head loss coefficients for each section and calculate the equivalent length:

a. 5 ft × 3.5 ft box culvert; $L = 223$ ft; local losses in this section include exit loss at chamber and a bend loss. Estimate exit loss coefficient at chamber at 0.5.
    The contraction loss from 9.5 ft × 4.5 ft culvert to 5 ft × 3.5 ft

$$A_1 = 5 \times 3.5 = 17.5 \text{ ft}^2$$

$$A_2 = 9.5 \text{ ft} \times 4.5 \text{ ft} = 42.75 \text{ ft}^2$$

$$\frac{A_2}{A_1} = 0.41$$

Table 2.9 gives $k = 0.26$

$$\Sigma k = 0.5 + 0.26 = 0.76$$

$$R = \frac{(5 \times 3.5)}{[2(5 + 3.5)]} = 1.03$$

$$L_{eq} = 153.1 \times (1.03)^{1.33} \times 0.76 = 121.0 \text{ ft}$$

Total equivalent length = 223 + 121.0 = 344.0 ft.

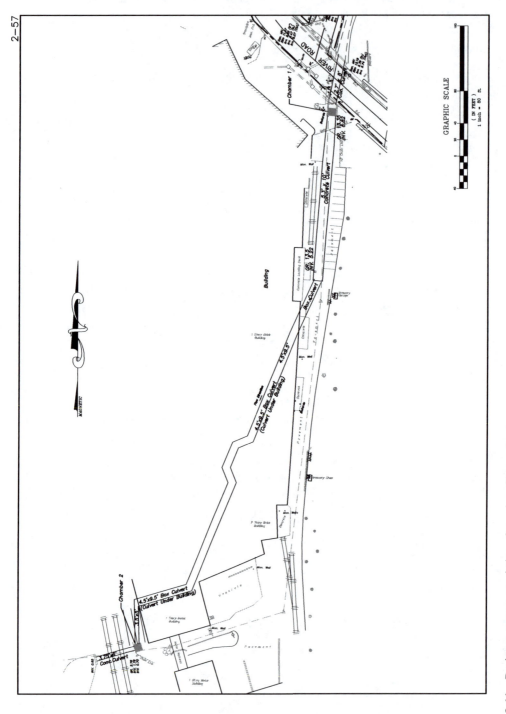

FIGURE 2.11   Drainage system consisting of a series of box culverts.

b. 9.5 ft × 4.5 ft culvert; this section includes five 45° bend and a ±60° bend. Using Equation 2.75:

$$\sum k_b = 5\left[0.5\times\left(\frac{45}{90}\right)^{1/2}+0.5\left(\frac{60}{90}\right)^{1/2}\right]=2.18$$

Also, there is a contraction from the upstream 10 ft × 5.3 ft culvert to this culvert:

$$\frac{A_2}{A_1}=\frac{(9.5\times4.5)}{(10\times5.3)}=\frac{42.75}{53}=0.8$$

$$D_2/D_1 = (0.8)^{1/2} = 0.9$$

$$k_c = 0.05; \text{ see Table 2.9}$$

$$\Sigma k = 2.18 + 0.05 = 2.23$$

$$R=\frac{(42.75)}{[2(9.5+4.5)]}=1.527 \text{ ft}$$

$$L_{eq} = 153.1 \times (1.527)^{1.33} \times 2.18 = 586 \text{ ft}$$

Total equivalent length = (70 + 140 + 20 + 30 + 20 + 210) + 586 = 1077 ft.

c. 10 ft × 5.3 ft culvert; local losses in this section are due to a 20° bend at the connection to the downstream section and an entrance loss:

$$k_b = 0.5\left(\frac{20}{90}\right)^{0.5}=0.24$$

$$k_e = 0.5 \text{ entrance loss coefficient}$$

$$\Sigma k = 0.74$$

$$R=\frac{(10\times5.3)}{[2(10+5.3)]}=1.73 \text{ ft}$$

$$L_{eq} = 153.1 \times 1.73^{1.33} \times 0.74 = 234.9 \text{ ft}$$

$$\text{Total} = 223 + 234.9 = 457.9 \text{ ft; say, } 458 \text{ ft}$$

PROBLEMS

2.1 For a wide channel, assume a velocity distribution of

$$\frac{v}{u*} = 5.75\log\frac{d}{k}+8.5$$

where v is the flow velocity at a distance of d from the bed, k is the surface roughness, and $u* = \sqrt{\tau_o/\rho}$ is the shear velocity. Show that the velocity at 0.6 depth from the water surface approximates the average velocity in a section.

Also show that the mean of the velocity readings at the relative depths of 0.2 and 0.8 along a vertical line from the water surface approximates the average velocity in the section.

2.2   A riprapped trapezoidal channel with a longitudinal slope of 0.3%, bottom width of 2 m, side slopes of 2:1 (2H,1V) is to carry 10 m³/s. Calculate the normal depth of water in this channel. Use: $n = 0.03$.

2.3   Solve Problem 2.2 for a channel width of 6 ft and 350 cfs discharge.

2.4   Calculate the normal capacity and full flow velocity for reinforced concrete pipes (RCPs) of 15, 18, 24, 30, and 36 in. diameter. The pipes are at 2% grade. Use Manning's $n = 0.012$.

2.5   Solve the preceding problem in metric units for RCP pipes of 300, 375, 450, 600, 750, and 900 mm diameter.

2.6   Calculate the normal flow depth in a 24 in. pipe laid at 2% slope for a discharge equal to 65% of the pipe capacity.

2.7   Calculate the normal depth of flow in a 600 mm concrete pipe for a discharge equal to 70% of the pipe capacity. The pipe is at 1.5% slope. Base your calculations on $n = 0.013$.

2.8   The design flow for a 24 in. RCP is calculated at 30 cfs. The pipe is at 2% slope and emanates from a manhole 3.5 ft deep. The invert of the pipe is set at the bottom of the manhole. Is this pipe adequate for the design flow?

2.9   A 500 mm concrete pipe is employed for conveying storm water at a rate of 400 L/s. The pipe is at 2% slope and emanates from an inlet 1.0 m deep. The invert of the pipe is set at the bottom of the inlet. Perform calculations to determine whether or not this pipe is adequate for the design flow. Base your calculations on $n = 0.013$.

2.10   A semicircular concrete channel 1.5 m in diameter is laid on a 0.5% slope. Calculate the maximum discharge, in m³/s, that this channel can carry when the brim full. Take $n = 0.013$.

2.11   Redo Problem 2.10 for a semicircular channel of 60 in. diameter.

2.12   Calculate the normal depth of flow in a 120 cm concrete culvert ($n = 0.012$) on a slope of 0.5% for a discharge of 1 m³/s.

2.13   Redo Problem 2.12 for a 48 in. culvert and 35 cfs flow.

2.14   Select the size of pipe needed to carry a discharge of 10 m³/s at no more than 65% full. The pipe is at 1% grade. Use $n = 0.013$. Note that large pipe comes in 150 mm diameter intervals.

2.15   A circular concrete pipe culvert is to carry a discharge of 450 cfs on a slope of 0.008 and is to run no more than 60% (by depth) full. Concrete pipes are available in internal diameters that are multiple of 6 in. Choose a suitable culvert diameter. Use $n = 0.012$.

2.16   The critical depth in a storm drain, 60 in. diameter, is 2.2 ft. What is the critical discharge?

2.17   Redo Problem 2.16 for a critical depth of 66 cm in a 150 cm pipe.

2.18   Calculate the critical depth in Example 2.4 in this chapter using Table 2.2.

2.19   Water flows at a rate of 3.0 m³/s in a 3 m wide rectangular concrete channel at a depth of 0.5 m.
       a. Calculate the slope for $n = 0.012$.
       b. What slope would be required to produce the critical flow for the given discharge? Also, what is the critical depth?

2.20   Redo Problem 2.19 for a discharge of 100 ft³/s in a 10 ft wide rectangular channel, and at a depth of 1.5 ft.

2.21   Water discharges at a rate of 8 m³/s in a trapezoidal channel of bottom width of 2 m and side slopes of 2:1. Calculate the critical depth and the critical velocity for this channel. What will be the critical slope, if the channel is riprap lined ($n = 0.035$)?

2.22   Solve Problem 2.21 for a discharge of 280 cfs in a trapezoidal channel of 6 ft bottom width.

2.23 Calculate the normal depth of flow in a roadside trapezoidal shaped channel for a 1.5 m³/s design discharge. The channel is at 1% slope, lined with 100 mm stone riprap, and has a bottom width of side slopes of 1.2 m and 2:1, respectively. Estimate $n = 0.04$.

2.24 Solve Problem 2.23 for the following flow parameters:
   a. $Q = 50$ cfs
   b. $b = 4$ ft
   c. Riprap size = 4 in.

2.25 Solve Example 2.6 in this chapter for a discharge of 0.5 m³/s and 450 mm pipe.

2.26 Flow parameters in a junction similar to that shown in Figure 2.9 are
   a. $Q_o = 0.8$ m³/s
   b. $Q_\ell = 0.2$ m³/s
   c. $D_i = 0.6$ m; $D_o = 0.675$ m; $D_\ell = 0.3$ m
   d. $\theta_j = 30°$
   Calculate the head loss in this junction.

2.27 Solve Problem 2.26 for the following case:
   a. $Q_o = 30$ cfs
   b. $Q_\ell = 15$ cfs
   c. $D_i = 24$ in., $D_o = 30$ in., $D_\ell = 18$ in.

2.28 Calculate the head loss in the drainage system of Example 2.9 for a discharge of 250 cfs.

2.29 Solve Problem 2.28 for a discharge of 7.5 m³/s.

APPENDIX 2A: HYDRAULIC PROPERTIES OF ROUND AND ELLIPTICAL PIPES

TABLE 2A.1
Hydraulic Properties of Circular Pipes Flowing Full, SI Units

Pipe Diameter (mm)	Area (m²)	Hydraulic Radius (m)	$K = A \times R^{2/3}/n$[a]			
			n = 0.010	n = 0.011	n = 0.012	n = 0.013
150	0.018	0.0375	0.20	0.18	0.16	0.15
200	0.031	0.0500	0.43	0.39	0.36	0.33
250	0.049	0.0625	0.77	0.70	0.64	0.59
300	0.071	0.0750	1.26	1.14	1.05	0.97
375	0.110	0.0938	2.28	2.07	1.90	1.75
450	0.159	0.1125	3.71	3.37	3.90	2.85
525	0.216	0.1313	5.59	5.08	4.66	4.30
600	0.283	0.1500	7.98	7.26	6.65	6.14
675	0.358	0.1688	10.93	9.93	9.11	8.14
750	0.442	0.1875	14.47	13.16	12.06	11.13
825	0.535	0.2063	18.66	16.96	15.55	14.35
900	0.636	0.2250	23.53	21.39	19.61	18.10
1050	0.866	0.2625	35.50	32.27	29.58	27.31
1125	0.994	0.2813	42.67	38.79	35.56	32.82
1200	1.131	0.3000	50.68	46.07	42.23	38.99
1350	1.431	0.3375	69.38	63.08	57.82	53.37
1500	1.767	0.3750	91.89	83.54	76.58	70.69
1800	2.545	0.4500	149.43	135.85	124.53	114.95

[a] K = conveyance factor; $Q = KS^{1/2}$.

TABLE 2A.2
Hydraulic Properties of Circular Pipes Flowing Full, CU

D Pipe Diameter (in.)	A Area (sq. ft)	R Hydraulic Radius (ft)	$K = \dfrac{149}{n} \times A \times R^{2/3}$[a]			
			n = 0.010	n = 0.011	n = 0.012	n = 0.013
8	0.349	0.167	15.8	14.3	13.1	12.1
10	0.545	0.208	28.4	25.8	23.6	21.8
12	0.785	0.250	46.4	42.1	38.6	35.7
15	1.227	0.312	84.1	76.5	70.1	64.7
18	1.767	0.375	137.0	124.0	114.0	105.0
21	2.405	0.437	206.0	187.0	172.0	158.0
24	3.142	0.500	294.0	267.0	245.0	226.0
27	3.976	0.562	402.0	366.0	335.0	310.0
30	4.909	0.625	533.0	485.0	444.0	410.0
33	5.940	0.688	686.0	624.0	574.0	530.0
36	7.069	0.750	867.0	788.0	722.0	666.0
42	9.621	0.875	1308.0	1189.0	1090.0	1006.0
48	12.566	1.000	1867.0	1698.0	1556.0	1436.0
54	15.904	1.125	2557.0	2325.0	2131.0	1967.0
60	19.635	1.250	3385.0	3077.0	2821.0	2604.0
66	23.758	1.375	4364.0	3967.0	3636.0	3357.0
72	28.274	1.500	5504.0	5004.0	4587.0	4234.0

Source: American Concrete Pipe Association, *Concrete Pipe Design Manual*, 17th printing, 2005.

[a] K = conveyance factor; $Q = KS^{1/2}$.

TABLE 2A.3
Hydraulic Parameters Elliptical Concrete Pipe, Flowing Full CU

Pipe Size $R \times S$ (HE) $S \times R$ (VE0) (in.)	Approximate Equivalent Circular Diameter (in.)	A Area (sq. ft)	R Hydraulic Radius (ft)	Value of $C_1 = \dfrac{1.486}{n} \times A \times R^{2/3}$			
				$n = 0.010$	$n = 0.011$	$n = 0.012$	$n = 0.013$
14 × 23	18	1.8	0.367	138	125	116	108
19 × 30	24	3.3	0.490	301	274	252	232
22 × 34	27	4.1	0.546	405	368	339	313
24 × 38	30	5.1	0.613	547	497	456	421
27 × 42	33	6.3	0.686	728	662	607	560
29 × 45	36	7.4	0.736	891	810	746	686
32 × 49	39	8.8	0.812	1140	1036	948	875
34 × 53	42	10.2	0.875	1386	1260	1156	1067
38 × 60	48	12.9	0.969	1878	1707	1565	1445
43 × 68	54	16.6	1.106	2635	2395	2196	2027
48 × 76	60	20.5	1.229	3491	3174	2910	2686
53 × 83	66	24.8	1.352	4503	4094	3753	3464
58 × 91	72	29.5	1.475	5680	5164	4734	4370
63 × 98	78	34.6	1.598	7027	6388	5856	5406
68 × 106	84	40.1	1.721	8560	7790	7140	6590
72 × 113	90	46.1	1.845	10,300	9365	8584	7925
77 × 121	96	52.4	1.967	12,220	11,110	10,190	9403
82 × 128	102	59.2	2.091	14,380	13,070	11,980	11,060
87 × 136	108	66.4	2.215	16,770	15,240	13,970	12,900
92 × 143	114	74.0	2.340	19,380	17,620	16,150	14,910
97 × 151	120	82.0	2.461	22,190	20,180	18,490	17,070
106 × 166	132	99.2	2.707	28,630	26,020	23,860	22,020
116 × 180	144	118.6	2.968	36,400	33,100	30,340	28,000

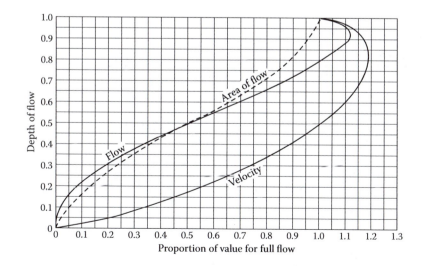

FIGURE 2A.1 Relative flow parameters versus flow depth in horizontal elliptical concrete pipes; CU and SI units.

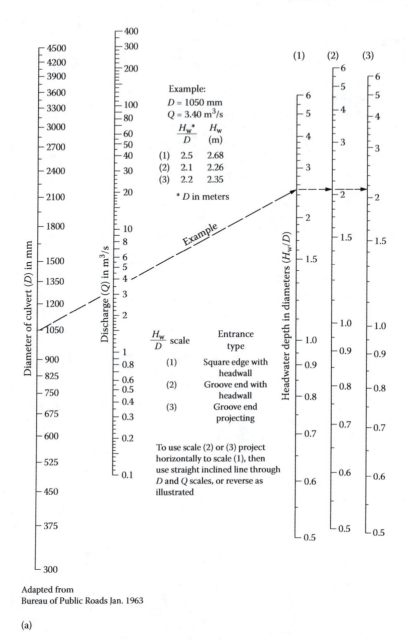

Adapted from
Bureau of Public Roads Jan. 1963

(a)

FIGURE 2A.2   (a) Inlet control headwater depth for round pipes, in metric units.

(Continued)

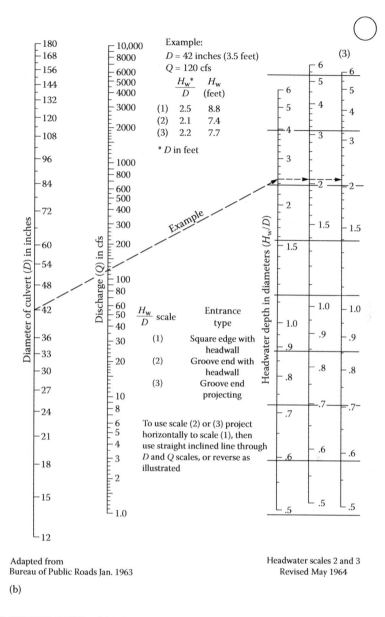

Adapted from
Bureau of Public Roads Jan. 1963

Headwater scales 2 and 3
Revised May 1964

(b)

FIGURE 2A.2 (CONTINUED)   (b) Inlet control headwater depth for round pipes, in English units.

REFERENCES

American Concrete Pipe Association, 2005, *Concrete pipe design manual*, 17th printing.

ASCE and WEF, 1992, *Design and construction of urban storm water management systems*, American Society of Civil Engineers Manuals and Reports of Engineering Practice no. 77 and Water Environment Federation manual of practice FD-20.

Barnes, H.H., 1987, Roughness characteristics of natural channels, water supply paper 1849, US Geological Survey.

Brater, E.F., King, H.W., Lindell, J.E., and Wei, C.Y., 1996, *Handbook of hydraulics*, 7th ed., McGraw-Hill, New York.

Chow, V.T., 1959, *Open channel hydraulics*, Chapter 5, McGraw-Hill, New York.

FHWA (US Department of Transportation), April 2012, Hydraulic design of highway culverts, hydraulic design series no. 5 (HDS-5), 3rd ed.

——— 2013, Urban drainage design manual, hydraulic engineering circular (HEC) no. 22, 3rd ed., September 2009, revised August 2013.

French, R.H., 1985, *Open channel hydraulics*, Chapter 4, McGraw-Hill, New York.

Henderson, F.M., 1966, *Open channel flow*, Chapter 4, Macmillan, New York.

Pazwash, H., 2007, *Fluid mechanics and hydraulic engineering,* Tehran University Press, Iran.

3 Hydrologic Calculations

This chapter covers the principles of hydrologic calculations as they relate to storm water management. Discussed in this chapter are those elements of the hydrologic cycle that affect rainfall–runoff relations. These include rainfall process, vegetal interception, surface retention, and infiltration. Also discussed in this chapter are runoff calculation methods and their limitations. A physically based model developed by the author is also presented in this chapter. Ample examples and case studies are included in the chapter to illustrate proper application of hydrologic methods.

3.1 RAINFALL PROCESS

Precipitation is a dynamic process. It has both spatial and temporal variation. Not only does it vary from one location to another, but at a given location, rainfall also has a varied pattern. It can fall at a faster rate at the beginning, in the middle, or at the end of storm duration. A rain storm can also have more than one maximum or minimum. The falling rate of rain with time is called rainfall intensity.

3.1.1 Intensity–Duration–Frequency Curves

To simplify rainfall–runoff relations, some methods of calculating the peak rate of runoff assume a constant rainfall intensity equal to its mean during the storm period. This intensity, which represents the rainfall depth divided by the rainfall duration, is expressed in millimeters per hour (mm/h) or inches per hour (in./h).

Rainfall intensity is dependent on the rainfall duration. It also varies with the frequency of the rainfall event; the less frequent the storm is, the larger its intensity will be. Curves representing the rainfall intensity–duration–frequency (IDF) relation have been developed using long-term rainfall data from precipitation stations in several states and large cities in the United States and many other countries.

Figure 3.1a and b present rainfall IDF curves in New Jersey. These figures show that, for example, the intensity of a 10-year, 60-minute duration storm is approximately 50 mm/h (2 in./h). Regional IDF curves in the United States are available in the Department of Transportation drainage manuals and/or Department of Environmental Protection/Conservation publications in some states. Based on the analysis of precipitation stations, NOAA (National Oceanic and Atmospheric Administration) has developed point precipitation frequency estimates for the entire United States. These frequency estimates are published in NOAA Atlas 14 and can be downloaded for any observation site free of charge from their website (http://hdsc.nws.noaa.gov). This website allows the user to print the table or the table and color graph of rainfall intensity (or rainfall depth)–duration–frequency estimates for any precipitation station in the United States. The website also gives the option of downloading the data in English or metric units. Table 3.1, for example, provides rainfall intensity–duration–frequency estimates for San Francisco, California, in metric units, while Table 3.2 lists rainfall depth–duration–frequency data in English units for Atlantic City in New Jersey.

A comparison of Tables 3.1 and 3.2 reveals a significant difference in rainfall pattern between the East Coast and the West Coast in the United States. For the 10-year, 60-minute storm, for example, the tables indicate 55 mm (2.16 in.) rainfall in Atlantic City, while only 22 mm (0.866 in.) precipitation in San Francisco.

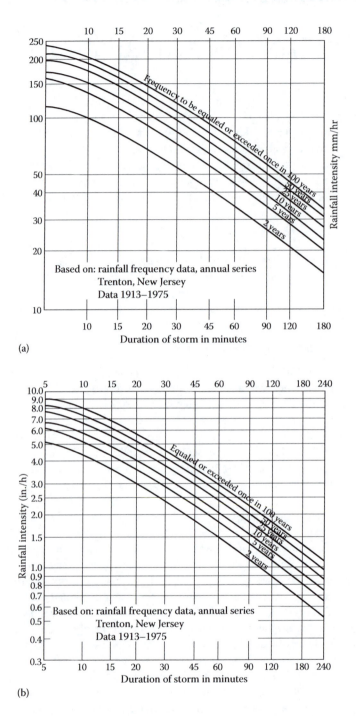

FIGURE 3.1 Rainfall intensity-duration-frequency curves in New Jersey: (a) metric units and (b) customary units.

TABLE 3.1
Rainfall IDF Data for San Francisco

PDS-Based Point Precipitation Frequency Estimates with 90% Confidence Intervals (in millimeters/hour)

Duration	Average Recurrence Interval (Years)									
	1	2	5	10	25	50	100	200	500	1000
5-min	43	54	67	79	95	107	120	134	153	168
	(39–49)	(48–61)	(60–77)	(69–91)	(80–114)	(88–132)	(96–152)	(103–176)	(112–210)	(118–241)
10-min	31	38	48	57	68	77	86	96	110	120
	(28–35)	(34–44)	(43–55)	(50–65)	(57–81)	(63–95)	(69–109)	(74–126)	(80–151)	(85–173)
15-min	25	31	39	46	55	62	69	77	88	97
	(22–28)	(28–35)	(35–44)	(40–52)	(46–66)	(51–76)	(55–88)	(60–101)	(65–122)	(68–139)
30-min	17	21	27	31	38	43	48	53	61	66
	(15–19)	(19–24)	(24–30)	(27–36)	(32–45)	(35–52)	(38–60)	(41–70)	(44–83)	(47–95)
60-min	12	15	19	22	26	30	34	37	43	47
	(11–14)	(13–17)	(17–21)	(19–25)	(22–32)	(25–37)	(27–43)	(29–49)	(31–59)	(33–67)
2-hr	9	10	13	15	18	21	23	26	29	32
	(8–10)	(9–12)	(12–15)	(13–17)	(15–22)	(17–25)	(18–29)	(30–34)	(22–40)	(23–46)
3-hr	7	9	11	13	15	17	19	21	24	27
	(6–8)	(8–10)	(10–12)	(11–14)	(13–18)	(14–21)	(15–24)	(16–28)	(18–34)	(19–38)
6-hr	5	6	7	9	10	12	13	15	17	19
	(4–5)	(5–7)	(7–8)	(8–10)	(9–12)	(10–14)	(11–17)	(11–19)	(12–23)	(13–27)
12-hr	3	4	5	6	7	8	9	10	12	13
	(3–4)	(3–4)	(4–6)	(5–7)	(6–8)	(7–10)	(7–12)	(8–13)	(9–16)	(9–19)
24-hr	2	3	3	4	5	5	6	7	8	9
	(2–2)	(2–3)	(3–4)	(3–4)	(4–6)	(5–6)	(5–8)	(6–9)	(6–11)	(7–12)
2-day	1	2	2	2	3	3	4	4	5	5
	(1–1)	(1–2)	(2–2)	(2–3)	(2–3)	(3–4)	(3–5)	(3–5)	(4–6)	(4–7)
3-day	1	1	1	2	2	2	3	3	4	4
	(1–1)	(1–1)	(1–2)	(2–2)	(2–3)	(2–3)	(2–3)	(2–4)	(3–5)	(3–5)

(Continued)

TABLE 3.1 (CONTINUED)
Rainfall IDF Data for San Francisco

PDS-Based Point Precipitation Frequency Estimates with 90% Confidence Intervals (in millimeters/hour)

Duration	Average Recurrence Interval (Years)									
	1	2	5	10	25	50	100	200	500	1000
4-day	1	1	1	1	2	2	2	3	3	3
	(1–1)	(1–1)	(1–1)	(1–2)	(2–2)	(2–2)	(2–3)	(2–3)	(2–4)	(2–4)
7-day	1	1	1	1	1	1	2	2	2	2
	(1–1)	(1–1)	(1–1)	(1–1)	(1–1)	(1–2)	(1–2)	(1–2)	(2–3)	(2–3)
10-day	0	1	1	1	1	1	1	1	2	2
	(0–1)	(1–1)	(1–1)	(1–1)	(1–1)	(1–1)	(1–2)	(1–2)	(1–2)	(1–2)
20-day	0	0	0	1	1	1	1	1	1	1
	(0–0)	(0–0)	(0–0)	(0–1)	(1–1)	(1–1)	(1–1)	(1–1)	(1–1)	(1–1)
30-day	0	0	0	0	0	1	1	1	1	1
	(0–0)	(0–0)	(0–0)	(0–1)	(0–1)	(0–1))	(1–1)	(1–1)	(1–1)	(1–1)
45-day	0	0	0	0	0	0	1	1	1	1
	(0–0)	(0–0)	(0–0)	(0–0)	(0–1)	(0–1)	(0–1)	(0–1)	(0–1)	(0–1)
60-day	0	0	0	0	0	0	0	1	1	1
	(0–0)	(0–0)	(0–0)	(0–0)	(0–0)	(0–1)	(0–1)	(0–1)	(0–1)	(0–1)

Source: Google Maps; NOAA National Weather Service, Silver Spring, Maryland, NOAA Atlas 14, volume 2, version 3, point precipitation frequency estimates.

Note: Location name: San Francisco, California; latitude: 37.7694°; longitude: −122.4333°; elevation: 55 m; PF tabular.

TABLE 3.2
Rainfall IDF Data for Atlantic City

PDS-Based Point Precipitation Frequency Estimates with 90% Confidence Intervals (in inches)

Duration	Average Recurrence Interval (Years)									
	1	2	5	10	25	50	100	200	500	1000
5-min	0.361 (0.324–0.401)	0.429 (0.385–0.475)	0.501 (0.448–0.554)	0.566 (0.506–0.627)	0.639 (0.568–0.709)	0.696 (0.617–0.774)	0.752 (0.663–0.837)	0.804 (0.703–0.899)	0.867 (0.750–0.977)	0.923 (0.789–1.05)
10-min	0.577 (0.518–0.641)	0.686 (0.615–0.759)	0.802 (0.718–0.888)	0.905 (0.809–1.00)	1.02 (0.906–1.13)	1.11 (0.983–1.23)	1.19 (1.05–1.33)	1.27 (1.11–1.43)	1.37 (1.19–1.55)	1.45 (1.24–1.65)
15-min	0.721 (0.647–0.801)	0.862 (0.773–0.954)	1.02 (0.908–1.12)	1.15 (1.02–1.27)	1.29 (1.15–1.43)	1.40 (1.25–1.56)	1.51 (1.33–1.68)	1.61 (1.41–1.80)	1.73 (1.49–1.95)	1.82 (1.56–2.07)
30-min	0.989 (0.887–1.10)	1.19 (1.07–1.32)	1.44 (1.29–1.60)	1.66 (1.48–1.84)	1.91 (1.70–2.12)	2.11 (1.88–2.35)	2.31 (2.04–2.58)	2.50 (2.19–2.80)	2.75 (2.38–3.10)	2.95 (2.53–3.35)
60-min	1.23 (1.11–1.37)	1.49 (1.34–1.65)	1.85 (1.65–2.05)	2.16 (1.93–2.40)	2.54 (2.26–2.82)	2.87 (2.54–3.19)	3.19 (2.81–3.55)	3.51 (3.07–3.93)	3.94 (3.41–4.44)	4.31 (3.69–4.90)
2-hr	1.56 (1.38–1.78)	1.90 (1.68–2.15)	2.37 (2.08–2.68)	2.78 (2.45–3.16)	3.31 (2.89–3.76)	3.75 (3.27–4.28)	4.21 (3.63–4.82)	4.68 (4.01–5.39)	5.32 (4.49–6.17)	5.88 (4.92–6.88)
3-hr	1.73 (1.52–1.97)	2.10 (1.85–2.39)	2.62 (2.29–2.99)	3.10 (2.70–3.54)	3.70 (3.21–4.25)	4.24 (3.64–4.86)	4.78 (4.08–5.51)	5.36 (4.52–6.19)	6.15 (5.12–7.17)	6.87 (5.63–8.05)
6-hr	2.11 (1.87–2.44)	2.54 (2.25–2.95)	3.17 (2.79–3.66)	3.76 (3.29–4.35)	4.54 (3.95–5.26)	5.25 (4.52–6.08)	5.98 (5.11–6.95)	6.78 (5.72–7.91)	7.91 (6.55–9.28)	8.93 (7.28–10.6)
12-hr	2.49 (2.21–2.86)	3.00 (2.66–3.44)	3.75 (3.32–4.29)	4.50 (3.96–5.14)	5.53 (4.83–6.33)	6.48 (5.60–7.43)	7.50 (6.40–8.64)	8.63 (7.25–10.0)	10.3 (8.44–12.0)	11.8 (9.51–13.9)
24-hr	2.73 (2.48–3.05)	3.32 (3.01–3.71)	4.32 (3.91–4.82)	5.18 (4.67–5.78)	6.49 (5.81–7.21)	7.65 (6.79–8.45)	8.94 (7.87–9.85)	10.4 (9.08–11.4)	12.6 (10.8–13.8)	14.6 (12.4–15.9)
2-day	3.09 (2.79–3.46)	3.77 (3.40–4.22)	4.89 (4.41–5.48)	5.87 (5.27–6.55)	7.33 (6.53–8.16)	8.61 (7.64–9.57)	10.0 (8.84–11.1)	11.7 (10.2–12.9)	14.1 (12.1–15.6)	16.2 (13.8–18.0)
3-day	3.24 (2.93–3.61)	3.93 (3.56–4.38)	5.08 (4.60–5.66)	6.08 (5.47–6.74)	7.55 (6.76–8.36)	8.84 (7.88–9.77)	10.3 (9.08–11.3)	11.9 (10.4–13.1)	14.3 (12.3–15.7)	16.3 (14.0–18.0)
4-day	3.38 (3.07–3.75)	4.10 (3.73–4.54)	5.28 (4.79–5.84)	6.29 (5.68–6.94)	7.78 (6.99–8.55)	9.07 (8.11–9.96)	10.5 (9.32–11.5)	12.1 (10.6–13.2)	14.5 (12.6–15.8)	16.5 (14.2–18.1)

(Continued)

TABLE 3.2 (CONTINUED)
Rainfall IDF Data for Atlantic City

PDS-Based Point Precipitation Frequency Estimates with 90% Confidence Intervals (in inches)

Duration	Average Recurrence Interval (Years)									
	1	2	5	10	25	50	100	200	500	1000
7-day	3.91 (3.58–4.29)	4.72 (4.31–5.18)	5.98 (5.46–6.58)	7.04 (6.42–7.73)	8.62 (7.81–9.44)	9.97 (8.99–10.9)	11.4 (10.3–12.5)	13.1 (11.6–14.3)	15.5 (13.6–16.9)	17.5 (15.2–19.1)
10-day	4.37 (4.02–4.76)	5.24 (4.83–5.71)	6.53 (6.00–7.12)	7.60 (6.98–8.30)	9.15 (8.36–9.96)	10.4 (9.51–11.3)	11.8 (10.7–12.8)	13.3 (12.0–14.5)	15.7 (13.9–17.0)	17.6 (15.6–19.2)
20-day	5.87 (5.48–6.30)	6.99 (6.53–7.50)	8.45 (7.88–9.07)	9.64 (8.98–10.3)	11.3 (10.5–12.1)	12.6 (11.7–13.5)	14.0 (12.9–15.0)	15.5 (14.2–16.5)	17.4 (15.9–18.7)	19.0 (17.2–20.4)
30-day	7.30 (6.83–7.83)	8.66 (8.09–9.28)	10.3 (9.64–11.0)	11.6 (10.9–12.4)	13.4 (12.5–14.4)	14.8 (13.8–15.9)	16.3 (15.0–17.4)	17.8 (16.3–19.0)	19.7 (18.1–21.1)	21.3 (19.4–22.8)
45-day	9.23 (8.70–9.81)	10.9 (10.3–11.6)	12.8 (12.0–13.6)	14.2 (13.4–15.1)	16.1 (15.1–17.1)	17.5 (16.4–18.6)	18.9 (17.7–20.1)	20.3 (18.9–21.5)	22.0 (20.5–23.4)	23.3 (21.6–24.8)
60-day	10.9 (10.3–11.5)	12.8 (12.2–13.5)	14.9 (14.0–15.7)	16.3 (15.4–17.3)	18.3 (17.2–19.3)	19.7 (18.5–20.8)	21.0 (19.7–22.2)	22.3 (20.9–23.5)	23.9 (22.3–25.2)	25.0 (23.2–26.4)

Source: Google Maps; NOAA National Weather Service, Silver Spring, Maryland, NOAA Atlas 14, volume 2, version 3, Atlantic City Marina, Station ID: 28-0325, point precipitation frequency estimates.

Note: Location name: Atlantic City, New Jersey; latitude: 39.3833°; longitude: −74.4333°; elevation (station metadata): 10 ft; PF tabular.

3.1.2   RAINFALL DATA

Daily amounts of precipitation are measured at gauging stations operated by the National Oceanic and Atmospheric Administration throughout the United States. Some universities also collect rainfall data for research purposes. In New Jersey, for example, Rutgers University collects and analyzes rainfall and snowfall data. The NOAA, which was formerly called the US Weather Bureau and is run by the US Department of Commerce, operates over 6700 stations throughout the country. Table 3.3 exemplifies the type of climate data available through the NOAA. This table, which shows daily data for the month of February, 2014, at Newark International Airport in New Jersey, is

TABLE 3.3

Climate Record for Newark International Airport in New Jersey, February 2014

NOWData—NOAA Online Weather Data

NEWARK INTL AP (286026)
Observed Daily Data
Month: February 2014

Day	MaxT	MinT	AvgT	HDD	CDD	Pcpn	Snow	Snwg
1	44	28	36.0	29	0	0.00	0.0	T
2	55	28	41.5	23	0	0.01	0.0	T
3	42	24	33.0	32	0	0.90	7.7	1
4	35	18	26.5	38	0	0.00	0.0	7
5	34	29	31.5	33	0	1.44	4.6	10
6	31	21	26.0	39	0	0.00	0.0	9
7	22	22	27.0	38	0	0.00	0.0	8
8	29	19	24.0	41	0	0.00	0.0	8
9	29	17	23.0	42	0	0.09	1.2	7
10	29	20	24.5	40	0	0.00	0.0	9
11	26	16	21.0	44	0	0.00	0.0	9
12	25	7	16.0	49	0	0.00	0.0	8
13	35	22	28.5	36	0	1.35	9.4	10
14	40	28	34.0	31	0	0.22	2.5	18
15	37	26	31.5	33	0	0.37	2.8	15
16	30	19	24.5	40	0	0.00	0.0	16
17	33	15	24.0	41	0	0.00	0.0	16
18	40	27	33.5	31	0	0.16	1.8	16
19	43	25	.34.0	31	0	0.24	0.0	15
20	49	33	41.0	24	0	0.02	0.0	13
21	45	35	40.0	25	0	0.11	0.0	10
22	52	31	41.5	23	0	0.00	0.0	9
23	54	33	43.5	21	0	T	0.0	4
24	42	27	34.5	30	0	T	0.0	1
25	35	24	29.5	35	0	T	T	T
26	34	17	25.5	39	0	0.03	0.3	T
27	34	13	23.5	41	0	T	T	0
28	25	9	17.0	48	0	0.00	0.0	0
Smry	37.1	22.6	29.9	977	0	4.94	30.3	7.8

Note: Official data and data for additional locations and years are available from the regional climate centers and the National Climatic Data Center.

downloaded from the website: http://nowdata.rcc-acis.org/PHI/PubACIS_results. The applied climate information system (ACIS) website of the NOAA includes over 6700 stations throughout the United States. The rainfall data in the northeastern United States can also be downloaded for the current and the previous month from the Cornell University Northeast Regional Climate Center website (http://www.nrcc.cornell.edu/page_nowdata.html).

Daily precipitation data are also collected by the Community Collaborative Rain, Hale & Snow (CoCoRaHS) Network. The data for any station operated by the CoCoRaHS throughout the United States can be downloaded from their website (http://www.cocorahs.org/). Figure 3.2 exemplifies the information that can be accessed using this website.

Hourly rainfall depths are measured at selected recording precipitation stations in the United States. The hourly rainfall records in the northeastern states are kept at a climate center of Cornell

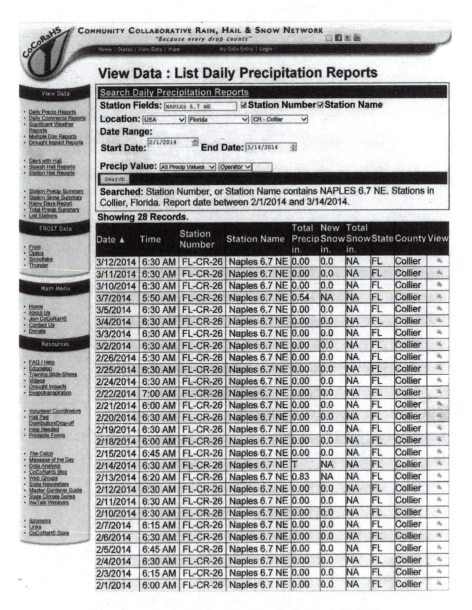

FIGURE 3.2   Daily rainfall records for Naples, Florida, February 1–March 14, 2014.

University. These data may be purchased from the website Northeast Regional Climate Center of the University (http://www.nrcs.cornell.edu/page_nowdata.html).

3.1.3 RAINFALL HYETOGRAPH

In a given storm, the instantaneous rainfall intensity is the slope of the cumulative (mass) rainfall depth at that time. The variation of the instantaneous rainfall intensity with time, called hyetograph, is a continuous curve. To simplify hydrologic analysis, this curve is divided into discrete segments, each representing the average rainfall intensity over a time increment. Figure 3.3 presents a rainfall hyetograph. Hyetographs are more accurate than the average intensity during the storm period and are used in calculating runoff hydrograph (variation of runoff rate with time). Actual storm hyetographs can be constructed from those precipitation stations that either measure hourly rainfall depth or have a recorder continuously measuring rainfall.

Since rainfall hyetographs vary from one storm to another, runoff hydrographs are usually constructed based on synthetic rainfall distribution. The SCS 24-hour rainfall distributions are among the most widely used synthetic hyetographs in the United States. These hyetographs were developed by the US Department of Agriculture, Soil Conservation Service (SCS), which is now the National Resources Conservation Service (NRCS). SCS has four types of 24-hour unit hyetographs designated as: type I, type IA, type II, and type III. Figure 3.4 presents these rainfall types and Figure 3.5 shows the geographic location in the United States where each rainfall type applies.

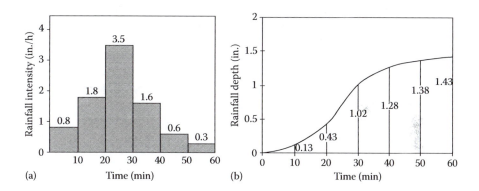

FIGURE 3.3 Rainfall hyetograph. (a) Discrete hyetograph. (b) Cumulative rainfall.

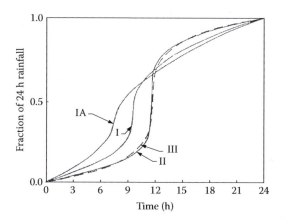

FIGURE 3.4 SCS 24-hour rainfall distributions.

FIGURE 3.5 Approximate geographic locations for SCS rainfall distribution.

Figure 3.4 indicates that type IA and type II are the least and the most intense storms, respectively. An inspection of this table also indicates that nearly 50% of the 24-hour rainfall depth occurs within the middle 2 hours of the 24-hour storm period in type II and type III storms. The SCS 24-hour storm distributions apply to all storm frequencies, each of which has a different total rainfall depth. This assumes that the shape of the 24-hour rainfall hyetograph is independent of the return period. Figure 3.6 shows the 24-hour rainfall depths with a return period of 10 years in various parts of the United States. Similar graphs for 24-hour storm events of 2-, 25-, 50-, and 100-year frequencies may be found in TR-55 (1986). In some states, the 24-hour rainfall depths have been refined on a regional basis. Table 3.4, for example, lists the 24-hour rainfall depths in various counties in New Jersey.

3.2 INITIAL ABSTRACTIONS

Rain falling on a pervious area is partly intercepted by tree canopies and leaves and vegetation, partly fills the surface depressions and puddles, and partly infiltrates into the ground. These retention elements in the beginning of a rainfall are known as initial abstraction. If the rain ends before the initial abstraction is satisfied, there will be no excess water to flow overland. However, when the rainfall exceeds the initial losses, water builds up on the catchment surface; flows overland, completely filling surface depressions on its path; and gradually concentrates in gullies or swales. Through its route, the runoff continues to infiltrate into the ground. A part of the infiltrated water percolates downward through the soil to contribute to base flow and groundwater recharge. Figure 3.7 is a schematic representation of disposition of storm water during a uniformly distributed rainfall.

The initial abstraction of a vegetative cover is significant during light and moderate, short-duration storms. In fact, during such storms, the vegetation may fully retain the rainfall. This behavior is evidenced by observations that no runoff occurs from lawns during storms of short duration. On an annual basis, a large portion of the rainfall is infiltrated into the ground and retained by vegetation that is lost to evapotranspiration, and therefore does not contribute to runoff.

Storm abstraction by vegetation continues until the vegetation becomes saturated. Beyond this point, vegetation has little retention effect; however, other losses, namely surface retention in puddles or depressions and infiltration into the ground, continue to occur. In addition to these, evaporation also adds to losses; this effect is, however, negligible for short-duration storms.

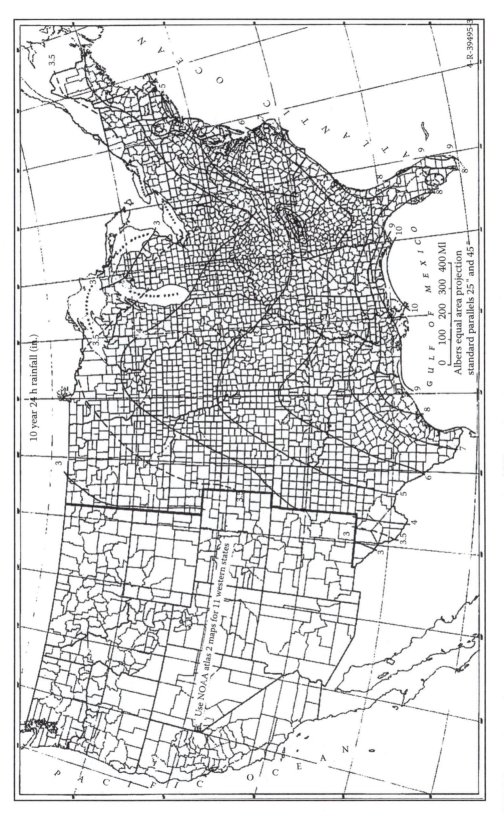

FIGURE 3.6 Twenty-four-hour rainfall depths with a return period of 10 years in various parts of the United States.

TABLE 3.4
New Jersey 24-Hour Rainfall Frequency Data[a]

County	1 Year	2 Years	5 Years	10 Years	25 Years	50 Years	100 Years
Atlantic	2.8	3.3	4.3	5.2	6.5	7.6	8.9
Bergen	2.8	3.3	4.3	5.1	6.3	7.3	8.4
Burlington	2.8	3.4	4.3	5.2	6.4	7.6	8.8
Camden	2.8	3.3	4.3	5.1	6.3	7.3	8.5
Cape May	2.8	3.3	4.2	5.1	6.4	7.5	8.8
Cumberland	2.8	3.3	4.2	5.1	6.4	7.5	8.8
Essex	2.8	3.4	4.4	5.2	6.4	7.5	8.7
Gloucester	2.8	3.3	4.2	5.0	6.2	7.3	8.5
Hudson	2.7	3.3	4.2	5.0	6.2	7.2	8.3
Hunterdon	2.9	3.4	4.3	5.0	6.1	7.0	8.0
Mercer	2.8	3.3	4.2	5.0	6.2	7.2	8.3
Middlesex	2.8	3.3	4.3	5.1	6.4	7.4	8.6
Monmouth	2.9	3.4	4.4	5.2	6.5	7.7	8.9
Morris	3.0	3.5	4.5	5.2	6.3	7.3	8.3
Ocean	3.0	3.4	4.5	5.4	6.7	7.9	9.2
Passaic	3.0	3.5	4.4	5.3	6.5	7.5	8.7
Salem	2.8	3.3	4.2	5.0	6.2	7.3	8.5
Somerset	2.8	3.3	4.3	5.0	6.2	7.2	8.2
Sussex	2.7	3.2	4.0	4.7	5.7	6.6	7.6
Union	2.8	3.4	4.4	5.2	6.4	7.5	8.7
Warren	2.8	3.3	4.2	4.9	5.9	6.8	7.8

Source: USDA Natural Resources Conservation Service New Jersey State Office and rounded to the first decimal place.

[a] Rainfall amounts in inches.

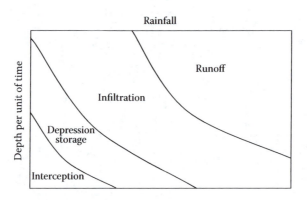

FIGURE 3.7 Schematic diagram of disposition of a uniform rainfall.

3.2.1 Interception

Interception, as indicated, is a portion of abstraction by tree canopies, shrubs, and vegetation before the rain reaches the ground. This occurs as wetting of vegetation surface and retention of rain drops on the tree leaves. Interception is small in urban areas; however, it is quite significant in vegetated and forested land. Development projects, including housing and roadway projects that involve clearing of woods, eliminate interception, thereby increasing runoff.

A number of empirical formulae have been suggested for interception, and many of these are similar to that originally proposed by Horton in 1919, as follows:

$$I = a + b\, P^n \tag{3.1}$$

where a, b, and n are constants and P is the precipitation depth. In spite of its simplicity, the parameters of a, b, and n are vegetation and soil dependent and cannot be readily determined. This severely limits application of this equation, in practice. A more sophisticated interception equation is suggested by Brooks et al. (1991):

$$I = S(1 - e^{-P/S}) + KE_t \tag{3.2}$$

where
  I = interception depth, mm (in.)
  S = storage capacity of vegetation per unit of projected area, mm (in.)
  P = rainfall depth, mm (in.)
  E = evaporation rate, mm/h (in./h)
  t = storm duration, h
  K = leaf index area; the ratio of the upper surface area of intercepting leaves to the projected area of plant (or trees) on the ground

Storage capacity, S, is typically in the range of 1 mm (0.04 in.) for bare woods to 7 mm (0.3 in.) for evergreen trees and spruce. For crops, the interception during a 25 mm (1 in.) rain falling in 1 hour varies from 0.8 mm (0.03 in.) for corn to 4.1 mm (0.15 in.) for small grains and 8.4 mm (0.33 in.) for cotton. Therefore, trees and plants can intercept a large portion of rain during light or moderate short-duration storms.

For P/S ratios of 3 or larger, the second term in the parentheses becomes increasingly small. Thus, interception depths can be approximated to be equal to S for rainfalls exceeding 4 mm to 25 mm (0.15 to 1 in.), depending on the vegetative cover.

Example 3.1

Estimated parameters in Equation 3.2 for a wooded area are $S = 5$ mm, $K = 3$, and $E = 0.25$ mm/h. Calculate the interception depth during a 30-minute storm of 50 mm/h (2 in./h) intensity.

Solution

$$P = 50 \times 30 \text{ min}/60 \text{ min} = 25 \text{ mm}$$

Using Equation 3.2,

$$I = 5\,(1 - e^{-25/5}) + 3 \times 0.25 \times 30 \text{ min}/60 \text{ min} = 5.34 \text{ mm } (0.21 \text{ in.})$$

This example demonstrates that interception by trees and vegetation takes a large portion of the rain; the shorter the storm duration and the lighter the storm, the larger the interception to rainfall ratio will be.

3.2.2 DEPRESSION STORAGE

Depression storage represents the water that is retained in surface depressions during a storm. Surface depressions act like miniature reservoirs that hold rainfall until they are filled. Depressions vary widely in size, namely area and depth. After the interception is satisfied, the rain falling on the ground begins to fill the depressions on the surface and infiltrate into the ground. The small depressions are filled first and overflow to larger depressions; once those are filled, the water begins to flow overland.

The water retained in depressions does not contribute to runoff; rather, it infiltrates into the ground and evaporates after the storm. Depression storage depends on the type of soil cover, natural topography, and ground slope; its magnitude is commonly expressed in terms of average depth of water over the drainage basin in millimeters or inches. Typical values of depression storage are reported to vary from 1 to 3.0 mm (0.04 to 0.12 in.) for paved areas, and up to 7.5 mm (0.3 in.) in forested land (ASCE, 1992). However, the storage may be significantly higher on flat, uneven land and where the land has blind drainage, namely large depressions. For lawns, the depression storage varies from 3.0 to 5.0 mm (0.12 to 0.2 in.).

The volume of water in depression storage V_s can be expressed as (Linsley et al., 1982)

$$V_s = S_d \left(1 - e^{-P_e/S_d}\right) \tag{3.3}$$

where

V_s = depression storage per unit area, mm (in.)
S_d = depression storage capacity, mm (in.)
P_e = depth of precipitation in excess of interception and infiltration

Equation 3.3 neglects any evaporation. In this equation the terms in the parentheses imply that if all depressions are filled, $P_e = 0$ and the depression storage would be equal to its capacity. This, of course, assumes that there will be no overland flow until all depressions are full. However, this condition only occurs when the largest depressions are located downstream. The assumption that depression storage subtracts the volume of rainfall from the initial storm period has been used with satisfactory results under normal conditions.

Experiments indicate that depression storage of an impervious area depends on its slope and varies from 3 mm (0.12 in.) for 1% slope to 1.3 mm (0.05 in.) for 3% slope. In natural basins, S_d varies from 10 to 50 mm (0.4 to 2 in.). For lawn and turf, depression storage as indicated is about 5 mm (0.20 in.).

The rate of depression storage during rainfall, V_s, can be expressed by

$$V_s = e^{-P_e/S_d}(I - f) \tag{3.4}$$

where

I = rainfall intensity, mm/h (in./h)
f = infiltration rate, mm/h (in./h)

Example 3.2

A 10-year, 10-min storm has an average intensity of approximately 4.5 in./h (114 mm/h) in New Jersey. Calculate the portion of rainfall that turns into runoff from a mildly sloped, semismooth paved area.

Solution

Calculations are performed in both CU and SI units, and the depression storage is estimated at 2.5 mm (0.1 in.).
The rainfall depth is

$$\frac{4.5 \times 10}{60} = 0.75 \text{ in.}$$

$$\frac{114 \times 10}{60} = 19 \text{ mm}$$

The net rainfall becoming runoff is

$$0.75 - 0.1 = 0.65 \text{ in.}$$

$$19 - 2.5 = 16.5 \text{ mm}$$

The ratio of runoff to rainfall depth is

$$C = \frac{0.65}{0.75} = 0.87$$

$$C = \frac{16.5}{19} = 0.87$$

This ratio, as will be discussed later, is called the runoff coefficient. Since depression storage is independent of rainfall depth, the runoff coefficient is not constant; rather, it varies with the rainfall intensity and duration. The shorter the storm duration and the lighter the storm, the smaller the runoff coefficient will be. In practice, however, a constant runoff coefficient is employed to perform runoff calculations, which is unrealistic.

Example 3.3

A 0.5 acre (2025 m²) subdivided lot contains 5000 ft² (465 m²*) of impervious surfaces and the remainder is wooded/landscape. Estimate the percentage of rainfall that turns into runoff for a 10-min storm of 5.7 in./h (145 mm/h) intensity. Assume the amounts of interception at 0.15 in. (3.8 mm) and depression storages for pervious and impervious areas at 0.4 in. (10 mm) and 0.1 in. (2.5 mm), respectively. Also, conservatively neglect evaporation and infiltration.

Solution

Calculations are presented both in CU and SI units.
Rainfall depth is

$$\frac{5.7 \times 10}{60} = 0.95 \text{ in.}$$

$$\frac{145 \times 10}{60} = 24.1 \text{ mm}$$

Abstraction and depression storage amount to

$$\frac{[(43,560 \times 0.5 - 5000) \times (0.15 + 0.40) + 5000 \times 0.1]}{12} = 810.8 \text{ ft}^3$$

$$\frac{[(2025 - 465) \times (3.8 + 10) + 465 \times 2.5]}{1000} = 22.69 \text{ m}^3$$

Abstraction depth is

$$\frac{810.8}{(43,560 \times 0.5)} = 0.037 \text{ ft} = 0.45 \text{ in.}$$

$$\frac{22.69}{2025} = 0.0112 \text{ m} = 11.2 \text{ mm}$$

---

* For practicality, the lot area and the impervious surfaces are rounded to full number in metric units.

Net rainfall becoming runoff is

$$0.95 - 0.45 = 0.50 \text{ in.}$$

$$24.1 - 11.2 = 12.9 \text{ mm}$$

The ratio of runoff to rainfall is

$$\frac{0.5}{0.95} = 53\%$$

$$\frac{12.9}{24.1} = 53\%$$

Note: Infiltration during rainfall is ignored in this problem. Accounting for infiltration, the runoff to rainfall ratio would be far smaller than that calculated herein. See Example 3.4.

3.3   INFILTRATION

Infiltration is the process of passage of water through the surface soil. Many factors affect infiltration. These factors may be classified as natural factors and surface factors. Natural factors are related to natural processes, such as precipitation, freezing, season, temperature, moisture, and, above all, soil texture. Surface factors are associated with soil cover. A bare soil forms a crust under the impact of raindrops and this, in turn, reduces infiltration. By preventing the soil from crust formation, a grass cover increases infiltration.

The infiltration process is different from percolation, which represents downward flow of water through soil due to gravity. Although different, the two processes are closely related, since infiltration cannot continue indefinitely unless percolation removes infiltrated water from the surface soil. Percolation occurs through the flow of water in noncapillary channels. The capillary water, namely the water absorbed to the soil particles, does not flow downward by gravity. The capillary suction distinguishes permeable soils, such as sand, from impermeable soils, such as clay, and is much smaller for the former than the latter. Typical capillary suction can be less than 1 cm (0.4 in.) for sand but over 5 m (15 ft) for clay.

The infiltration rate is equal to the percolation rate just below the ground surface, where the soil is saturated. The movement of water through soil is governed by Darcy's law:

$$q = K \frac{dh}{dz} \tag{3.5}$$

where
  K is called permeability or hydraulic conductivity and is a function of soil texture and moisture content
  h is the piezometric head of pore water
  z is the vertical coordinate taken positive downward in the preceding equation

The piezometric head is the total of pore water pressure and depth, z, as follows:

$$h = \frac{p}{d} + z \tag{3.6}$$

A negative pore water pressure indicates tension or suction. This occurs for an unsaturated soil, where due to capillary effect, the soil possesses a negative pore pressure.

Because of capillary (suction) effect, both piezometric head and permeability are at a maximum when a soil is dry. The maximum rate at which the water can enter soil under a given set of conditions is called the infiltration capacity, f_p. The actual infiltration, f, equals f_p only when the effective rainfall intensity, namely the rainfall intensity less the rate of interception and the rate of depression storage, equals or exceeds f_p. As infiltration continues, the soil pores become filled with water, the capillary suction diminishes, and the infiltration reaches its lower limit, which is governed by the gravity flow alone. Under this condition, the infiltration rate becomes equal to the percolation rate, which is also called the hydraulic conductivity, K. If the soil is stratified, the least pervious subsoil layer limits the infiltration.

A number of equations or models are available for estimating infiltration. The Horton equation, Green-Ampt method, and Philip model are the most widely used in engineering practice. The validity of each of these models should be based on its consistency with the actual infiltration process.

3.3.1 GREEN-AMPT MODEL

This physically based model was originally introduced by Green and Ampt in 1911 and was placed on a firm basis by Philip in 1954. The Green-Ampt model, also called delta function model, is one of the most realistic models for infiltration. This model is employed in such widely continuous simulation models as the Environmental Protection Agency's (EPA's) SWMM (Storm Water Management Model). However, as will be shown in this section, this model involves implicit equations and tedious iterative calculations, and it breaks down if improperly applied.

Consider water is impounded to a depth of H_o over the ground surface. When the infiltration begins, the soil below the ground becomes saturated with water, but the soil is unsaturated further down. This produces a sharp moisture gradient near the interface of moist and dry soil, resulting in a high infiltration rate. Figure 3.8 depicts the infiltration process for this case. As the infiltration continues, the interface, called wetting front, moves downward. If the rain is sustained, the wetting front eventually reaches the water table.

Using the straight line approximation for the saturated soil between the soil surface and the wetting front and neglecting ponding depth, H_o, the Darcy's equation (Equation 3.5) becomes

$$q = f_p = \frac{K[0-(L+\psi)]}{(0-L)}$$

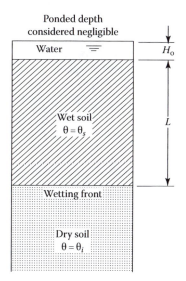

FIGURE 3.8 Schematic diagram of Green-Ampt infiltration model.

or

$$f_p = K\left(1 + \frac{\psi}{L}\right)$$ (3.7)

where
f_p = potential infiltration rate
ψ = soil suction at the wetting front

Since suction is in a downward direction, it would be positive in this equation, if the z direction is also downward. The cumulative infiltration, F, is equal to the product of depth to the wetting front, L, and the initial moisture deficit, that is,

$$F = L(\alpha - \theta_i) = L\Delta\theta$$ (3.8)

where
θ_i = initial moisture content in the dry soil
$\Delta\theta$ = the soil moisture deficit
α = soil porosity

Eliminating L between Equations 3.7 and 3.8 yields

$$f_p = K\,[1 + (\psi\Delta\theta/F)]$$ (3.9)

or, inversely,

$$F = \frac{\psi\Delta\theta}{(f_p/K - 1)}$$ (3.10)

Table 3.5 gives typical values of Green-Ampt parameters for the US Department of Agriculture (USDA) soil texture classes. The preceding equations indicate that the initial infiltration capacity is larger than the hydraulic conductivity. However, as infiltration progresses, ψ diminishes and the infiltration capacity decreases, eventually approaching the hydraulic conductivity. It is to be noted that the listed values of permeability in Table 3.5 are several fold smaller than reliable information (see, e.g., Freeze and Cherry, 1979; Linsley et al., 1982; Todd, 1980).*

The preceding equations for f and F are valid only when water is impounded on the ground and/or the rainfall intensity exceeds the infiltration capacity. Since Equations 3.9 and 3.10 have two variables, they cannot be solved directly. However, noting that

$$f_p = \frac{dF}{dt}$$ (3.11)

and combining Equations 3.8 and 3.10 and separating variables gives

$$\left[\frac{F}{(F + \psi\Delta\theta)}\right]dF = K\,dt$$ (3.12)

---

* The values in Table 3.5 are also far smaller than those in Tables 3.6 and 3.8 in this text.

TABLE 3.5

Green-Ampt Infiltration Parameters for USDA Soil Texture Classes[a]

USDA Soil Classification	Porosity θ	Effective Porosity	Permeability mm/h (in./h)	Wetting Front Suction Head Ψ mm (in.)
Sand	0.44	0.42	117.8 (4.64)	49.5 (1.95)
Loamy sand	0.44	0.40	29.9 (1.18)	61.3 (2.41)
Sandy loam	0.45	0.41	10.9 (0.43)	110.1 (4.33)
Loam	0.46	0.43	3.4 (0.13)	88.9 (3.50)
Silt loam	0.50	0.49	6.5 (0.26)	166.8 (6.57)
Sandy clay loam	0.40	0.33	1.5 (0.06)	218.5 (8.60)
Clay loam	0.46	0.31	1.0 (0.04)	208.8 (8.22)
Silty clay loam	0.47	0.43	1.0 (0.04)	273.0 (10.75)
Sandy clay	0.43	0.32	0.6 (0.02)	239.0 (9.41)
Silty clay	0.48	0.42	0.5 (0.02)	292.2 (11.50)
Clay	0.48	0.39	0.3 (0.01)	316.3 (12.45)

Source: Rawls, W. J. et al., 1983, *Journal of Hydraulic Division, ACSE*, 109 (1): 62–70; condensed and rounded to the second decimal place.

Note: Actual θ and $\Delta\theta$ values vary by nearly 30% from the average values listed in this table; ψ values vary by up to 25-fold from the listed values.

[a] See Figure 3.8.

Integrating from $t = 0$ to $t = t$ results in the following equation:

$$Kt = F - \Delta\theta\psi \ln\left(1 + \frac{F}{\psi\Delta\theta}\right) \qquad (3.13)$$

This equation can be solved through an iterative process to calculate F at any time during the storm period when the rainfall intensity exceeds the infiltration capacity. Having F (the infiltration depth), f_p can be calculated using Equation 3.9.

If the rainfall intensity is initially less than the infiltration capacity, all the rainfall infiltrates until the initial moisture deficit is satisfied. In this case, the ponding does not occur at time $t = 0$, but at a time $t = t_p$, where

$$t_p = F_p/I \qquad (3.14)$$

and

$$f = I \text{ for } t \leq t_p \qquad (3.15)$$

It is to be noted that F_p, which is the infiltrated water depth before ponding occurs, cannot be calculated using Equation 3.10. Substituting $f_p = I$ in that equation yields negative infiltration volume when $I < K$ and breaks down when $K = I$, both of which are unrealistic.

In the preceding case, the variation of infiltrated depth with time can be written as

$$K(t - t_p + t_p') = F - \psi\Delta\theta \ln\left[1 + \frac{F}{\psi\Delta\theta}\right] \qquad (3.16)$$

where, as indicated, t_p is the time at which ponding occurs and t_p' is the equivalent time to infiltrate volume F_p, under initial surface ponding condition. The time t_p' can be calculated by substituting F_p for F in Equation 3.13.

Under this condition, Equations 3.14 and 3.16 are to be employed all together to calculate F in an iterative process. To expedite the calculation process, the Green-Ampt model is generally applied by incrementing F and solving for t in Equation 3.16 and then using Equation 3.9 to calculate f_p. The following example illustrates the tediousness of the calculation process for a relatively simple rainfall distribution.

Example 3.4

A loamy sandy soil has the following properties:

    $K = 40$ mm/h
    $\theta = 0.45$
    $\theta_i = 0.15$
    $\psi = 50$ mm water

Using the Green-Ampt method, calculate the time for soil surface to become saturated for a 60-minute rainfall with the following distribution:

    $I = 30$ mm/h; 0–15 minutes
    $I = 60$ mm/h; 15–60 minutes

Solution

$$\Delta\theta = 0.45 - 0.15 = 0.3$$

$$\psi\Delta\theta = 50 \times 0.3 = 15$$

During the first 15 minutes, the rainfall intensity is less than the minimum infiltration capacity, namely K, therefore all the rainfall infiltrates:

$$F_{15} = \frac{30 \times 15}{60} = 7.5\,\text{mm}$$

In the $t = 15$- to 60-minute period, since the rainfall intensity is larger than hydraulic conductivity, ponding is possible. Use Equation 3.9 to relate f_p and F:

$$f_p = 40 + \frac{600}{F}$$

The time to surface ponding is calculated by equating the infiltration capacity to the rainfall intensity in the preceding equation:

$$60 = 40 + \frac{600}{F_p}$$

$$F_p = 30\,\text{mm}$$

Time to surface ponding is

$$t_p = \frac{F_p}{I} = \frac{30}{60} = 0.5\,\text{h} = 30\,\text{min}$$

Therefore, the ponding starts at

$$t_p = 30 + 15 = 45 \text{ min} = 0.75 \text{ h}$$

from the beginning of rainfall.
Cumulative infiltration at this time is

$$F = 7.5 \text{ mm} + 30 \text{ mm} = 37.5 \text{ mm}$$

Next, calculate t_p', namely the time that it would take $F = 37.5$ mm to infiltrate at potential rate from $t = 0$; using Equation 3.13:

$$40 \times t_p' = 37.5 - 15\ln\left(1 + \frac{37.5}{15}\right)$$

$$t_p' = \frac{18.71}{40} = 0.468 \text{ h} = 28.1 \text{ min.}$$

Express F as a function of time using Equation 3.16:

$$40(1 - t_p + t_p') = F - 15\ln\left(1 + \frac{F}{15}\right)$$

Input the previously calculated t_p and t_p' values in this equation:

$$40(1 - 0.75 + 0.468) = F - 15\ln\left(1 + \frac{F}{15}\right)$$

Simplify:

$$F - 15\ln\left(1 + \frac{F}{15}\right) = 28.72$$

Solve iteratively:

$$F = 51 \text{ mm}$$

Since the amount of rainfall at $t = 1$ h is

$$7.5 + \frac{60 \times 45}{60} = 52.5 \text{ mm,}$$

the ponding depth at the end of the storm ($t = 1$ h) is

$$52.5 \text{ mm} - 51 \text{ mm} = 1.5 \text{ mm}$$

3.3.2 HORTON EQUATION

In 1939/1940 Horton presented the following empirical equation for the infiltration capacity at a given time:

$$f_t = f_c + (f_o - f_c) e^{-\alpha t} \tag{3.17}$$

where

f_t is the infiltration capacity at time t

f_o and f_c are the initial and ultimate (or equilibrium) infiltration rates, respectively

e is the naperine base

α is a decay constant, which depends on soil

t is time from beginning of rainfall

Integrating Equation 3.17 gives the cumulative infiltration depth:

$$F_t = f_c t + \left[\frac{(f_o - f_c)(1 - e^{-\alpha t})}{\alpha} \right] \tag{3.18}$$

Eliminating t between Equations 3.17 and 3.18 yields a direct relationship between f and F:

$$F_t = \frac{f_c \ln(f_o - f_c)}{\alpha} - \frac{f_c \ln(f - f_c)}{K} + \frac{(f_o - f)}{\alpha} \tag{3.19}$$

To solve the preceding equation, parameters f_o, f_c, and α need to be known. These parameters may be determined from a graph of measured infiltration rate over an extended period of time. The tail end of the graph can be extrapolated to calculate f_c. Two sets of f and t data can be taken from the plot and inputted into Equation 3.19 to solve for f_o and α. Though apparently a simple process, the procedure is tedious and somewhat inconclusive.

Similarly to the Green-Ampt method, Equations 3.18 and 3.19 are valid only when the net rainfall intensity exceeds the infiltration capacity. Specifically:

$$f = f_p \qquad I \geq f_p \tag{3.20}$$

If the rainfall intensity is smaller than the infiltration capacity, $(I < f_p)$, the infiltration occurs at the rate of $f = I$.

Experience shows that the infiltration capacity for a given soil varies with the initial moisture content, organic matter of soil, vegetative cover, and season (Linsley et al., 1982). These effects result in a large range of permeability of soil. Consequently, reported values of f_o, f_c, and α in technical literature do not generally agree, but rather vary widely from one researcher to another. In addition, some of the values do not appear to be consistent with field observations. This is exemplified by values of f_o and f_c for clay reported by Butler and Davies (see Chin, 2006) at 75 mm/h and 3 mm/h, respectively; both of which are exaggerative. The same researchers report $f_c = 12$ mm/h for medium textured soil and $f_c = 25$ mm/h for coarse textured soils. While the former results grossly exaggerate the permeability of clay, the latter figures underestimate the permeability of medium and coarse textured soils. Also, the reported values of α by Rawls et al. (referenced in several publications including Chin, 2006) include a value of $\alpha = 0.64$ min^{-1} for a loamy sand. This implies the soil loses 25% of its initial infiltration capacity within just 1 minute after rain and reaches its ultimate limit in less than 10 minutes, both of which appear unrealistic. Because of these inconsistencies, no reliable values for f_o and α are presented here. Only typical values of f_c, which are primarily soil dependent, are provided for the USDA soil textures in Table 3.6.

An uncertainty in the Horton equation, as indicated, is that the reported α values exaggerate the decay of the infiltration with time. To adjust for this deficiency, Viessman et al. (1989) suggest relating f as a function of the cumulative infiltration, rather than time. This form of the Horton equation has been employed in the EPA's SWMM. Figure 3.9 depicts typical variation of infiltration with time.

TABLE 3.6

Typical Values of $f_c = K$ for USDA Soil Textures

Soil Type	f_c mm/h	f_c in./h
Sand	>500	>20.0
Loamy sand	250	10.0
Sandy loam	100	4.0
Loam	40	1.6
Sandy clay loam	30	1.2
Silt loam	10	0.4
Clay loam	7.5	0.3
Sandy clay	5.0	0.2
Silt	2.5	0.1
Silty clay loam	2.0	0.08
Silty clay	1.0	0.04
Clay	<0.5	<0.02

Note: Values of f_c are derived from soil permeability/textural triangle (Figure 3.12). For sandy loam through sandy clay soil covered with turf, the f_c values may be two- to threefold larger than those shown in this table.

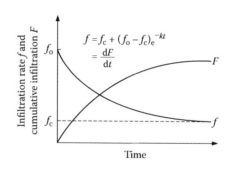

$$f = f_c + (f_o - f_c)e^{-kt}$$
$$= \frac{dF}{dt}$$

FIGURE 3.9 Infiltration curves.

Example 3.5

Assuming $f_o = 55$ mm/h, $f_c = 25$ mm/h, and $\alpha = 1.2$ hour^{-1}, calculate infiltration rates at the end of 1, 2, and 3 hours during a 3-hour storm having intensity larger than infiltration. Also calculate the cumulative infiltration at the end of the 3-hour period.

Solution

$$f_p = f_c + (f_o - f_c)\,e^{-\alpha t}$$

$$f_c = 25 + 30\,e^{-1.2t}$$

 f_p at the end of each hour interval is

$$f_1 = 34.04 \text{ mm}; \; f_2 = 27.72 \text{ mm}; \; f_3 = 25.82 \text{ mm}$$

$$F(t) = f_c t + \left(\frac{f_o - f_c}{\alpha} \right)(1 - e^{-\alpha t})$$

$$= 25 \times 3 + \left(\frac{30}{1.2} \right)(1 - e^{-3.6}) = 99.3 \, \text{mm}$$

Note: Under initially wet ground surface conditions, the infiltration rate would be nearly equal to hydraulic conductivity, f_c, and the cumulative infiltration $= 3 \times 25 = 75$ mm.

Example 3.6

Rain falls at a rate of 1.5 in./h for 1 hour on a lawn in a silty loam soil. Calculate the amount of runoff within the rainfall period. Conservatively estimate $f = K$ and use Table 3.6.

Solution

From Table 3.6, $K = 0.4$ in./h.

Since the soil is covered with lawn, estimate $K = 1.0$ in./h (2.5 times larger than bare soil). Estimate interception and depression storage for lawn at 0.1 and 0.25 in., respectively. Runoff depth = net rainfall = 1.5 – 1.0 – (0.1 + 0.25) = 0.15 in.

3.3.3   Philip Infiltration Model

Philip in 1958 presented the following model:

$$f = \frac{1}{2} s t^{-1/2} + K \tag{3.21}$$

in which

f = instantaneous infiltration rate
s = an empirical parameter related to the progression of the water front
K = hydraulic conductivity at the surface
t = time

In this equation, $f = \infty$ for $t = 0$; however, it gradually decreases with time until $f = K$ at $t = \infty$. Integrating this equation results in

$$F = s t^{1/2} + Kt \tag{3.22}$$

in which F is cumulative depth of infiltration. The parameter "s" in this equation, similar to the exponent α in Horton's equation, depends on soil properties and ground cover and varies widely.

Example 3.7

A 1000 m² residential lot comprises 350 square feet of impervious area, and the remainder is covered with grass. Assume:

- Lawn interception = 2 mm
- Depression storage = 7 mm for grass and 1.5 mm for pavement
- Initial and equilibrium infiltration rates of 75 mm/h and 25 mm/h and a Horton decay constant of $\alpha = 1.0^{-1}$ hour

Calculate the portion of rainfall that turns into runoff for a 30-minute duration rainfall having 100 mm/h intensity, using Horton's equation. Also, calculate the runoff to rainfall ratio.

Solution

The overall initial abstractions and depression storage for the site are calculated as follows:

$$\text{Lawn area} = 1000 - 350 = 650 \text{ m}^2$$

The combined depth of initial abstraction and depression storage is

$$[650 \times (2+7) + 350 \times 1.5]/1000 = 6.4 \text{ mm}$$

Infiltration rate, Horton equation, is

$$f = 25 + (75 - 25) \ e^{-1} = 25 + 50 \ e^{-1}$$

The infiltration amount is

$$F(t) = f_c t + \left(\frac{f_o - f_c}{\alpha} \right)(1 - e^{-\alpha_t})$$

For $t = 30$ minutes $= 0.5$ h,

$$F = 25 \times 0.5 + 50 \ (1 - e^{-0.5}) = 12.5 + 50 \ (1 - 0.607) = 32.2 \text{ mm}$$

Average infiltration over the entire lot is

$$\frac{(650 \times 32.2 + 350 \times 0.0)}{1000} = 20.9 \text{ mm}$$

Total losses = 6.4 + 20.9 = 37.3 mm
Rainfall depth = 100 × 30 min/60 min = 50 mm
Runoff depth = rainfall depth − combined losses

$$= 50 - 37.3 = 12.7 \text{ mm}$$

$$\frac{\text{Runoff}}{\text{Rainfall}} = \frac{12.7}{50} = 0.25$$

As seen, a large percentage of rainfall is dissipated through initial abstraction and infiltration. The losses are significantly larger for soils of higher permeability.

3.3.4 INFILTRATION INDEXES

In practice, infiltration indexes are commonly employed to simplify calculations. Infiltration indexes generally assume that infiltration occurs at a constant rate during the storm. As such, the indexes underestimate the initial rate of infiltration and exaggerate its ultimate value. The best application

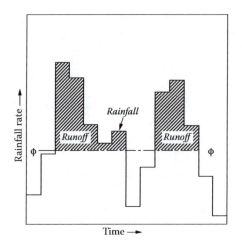

FIGURE 3.10   The φ infiltration index.

of an index is to a large storm on wet soil where infiltration rate may be assumed to be relatively constant.

The most common index is termed as φ index. In this index the total volume of rainfall abstraction is estimated and is distributed uniformly during the storm period (Figure 3.10). The depth of precipitation above the index line then represents the overland runoff. A variation of this index is the W index in which the initial abstractions are deducted from the early storm period.

To calculate the φ index for a given storm, the amount of observed runoff from the storm hydrograph is deducted from the total precipitation and the difference is divided by the storm duration. It is to be noted, however, that a φ index obtained from a single storm may not be applicable to other storms.

3.4   MEASUREMENT OF INFILTRATION AND PERMEABILITY

Infiltration and permeability can be measured directly through in situ percolation tests. Soil samples also can be sent to a laboratory for measurement of permeability. Permeability can also be estimated indirectly through soil gradation. These methods are discussed in this section.

3.4.1   INFILTROMETERS

Infiltration rates into soils can be measured with a ring infiltrometer, consisting of a metal tube 8 to 12 in. (20 to 30 cm) in diameter. This tube is driven 18 to 24 in. (45 to 60 cm) into the soil with 4 in. (10 cm) or so projecting above the ground. Water is poured into the tube and the rate of drop is measured as an indicator of infiltration rate. As water enters the soil, the air escapes around the tube; as such, the measured water drop exaggerates actual infiltration rate. To minimize this effect, a double ring infiltrometer is used, and both rings are filled with water. In this device, the water that enters the inner ring tends to move down with a minimal lateral spread. Because of spatial variation of permeability, several measurements spread over the area of interest should be made to determine the average value of infiltration rate.

Figure 3.11 shows a double ring infiltrometer manufactured by Rickly Hydrological Company in Columbus, Ohio. This instrument consists of two concentric rings, a driving plate with handles for inner and outer rings. The outer and inner rings are 24 and 12 in. diameter, respectively. The rings are driven into the soil and partially filled with water and the volume of water poured in the inner tube to maintain the water level for a specific period of time is measured using a Mariotte tube. The information is converted into the infiltration rate using the data sheet provided by the manufacturer.

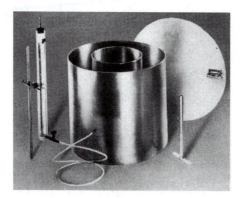

FIGURE 3.11 Rickly double-ring infiltrometer.

3.4.2 PERMEAMETERS

A permeameter is a laboratory device for measuring hydraulic conductivity. In this device, flow is maintained through a small sample of soil material, and flow rate and head are measured. Two types of permeameters are available: constant head and falling head.

The constant-head permeameter can measure hydraulic conductivity of consolidated or unconsolidated soils under low head (Figure 3.12a). Water flows upward through the sample; by measuring its rate, the hydraulic conductivity can be obtained from Darcy's law, as follows:

$$K = \frac{(Q/\pi R^2)}{(H/L)} \tag{3.23}$$

where
Q = rate of flow (volume measured divided by time)
L = height of sample
H = constant head
R = radius of sample in permeameter

The soil should be fully saturated before taking any measurement. In practice, several measurements are made to obtain a reliable result.

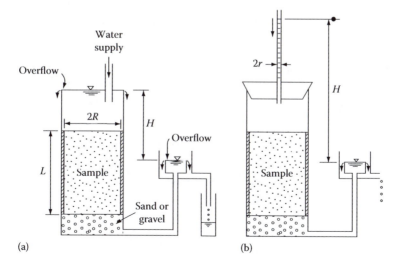

(a)   (b)

FIGURE 3.12 Permeameters for measuring K: (a) constant head and (b) falling head.

In the falling head permeameter (Figure 3.12b), a slender tall tube is filled with water, and the rate of fall of the water level in the tube is measured. The hydraulic conductivity can be obtained by noting that the measured rate of flow is

$$Q = \pi r^2 \frac{dH}{dt} \tag{3.24}$$

and that the flow rate through the sample follows Darcy's law:

$$Q = \pi R^2 K \frac{H}{L} \tag{3.25}$$

By equating the preceding equations and integrating, K can be obtained as follows:

$$K = \left(\frac{r}{R}\right)^2 \left(\frac{L}{t}\right) \ln\left(\frac{H_1}{H_2}\right) \tag{3.26}$$

where
   $t =$ is the time interval for the water level in the tube to fall from H_1 to H_2
   $r =$ radius of tall tube
   $\ln =$ natural logarithm
   R and $L =$ are as previously defined

The laboratory measure of permeability is more accurate than in situ percolation testing; however, it has a drawback of disturbing soil structure and stratification.

3.4.3   SOIL GRADATION ANALYSIS

The percolation rate depends primarily on the soil particle size. Soil porosity has little effect on this rate. In fact, clayey soils, having a higher porosity than both silt and sand, are far less permeable than either of them. Soils are commonly classified by their particle size distribution. Table 3.7 presents USDA soil classification. In nature, the soil may be loam, silt, clay, or a mixture thereof. Thus, percolation and infiltration are controlled by the soil texture, which is defined as the proportions by weight of clay, silt, loam, and sand after granular material larger than 2 mm is removed.

The USDA has proposed a soil texture triangle, which is shown in Figure 3.13. This figure, for example, shows that a soil composed of 30% clay, 50% sand, and 20% silt is classified as sandy clay loam. If more than 15% of soil is larger than 2 mm, a prefix, such as gravely or stony, is added to the soil texture names on this figure.

Gradation analysis of a soil sample in a laboratory provides an indirect method of estimating permeability. In this method, the percentages of sand, silt, and clay are measured. Based on this analysis, the soil texture is determined using the USDA Soil Texture Triangle. Then, the soil permeability is estimated using the Soil Permeability/Textural Triangle in Figure 3.14. As shown in this figure, the soils are classified into six permeability ratings or classes, identified as K_0 through K_5;

TABLE 3.7
USDA Soil Classification

Soil Name	Particle Size (mm)
Clay	<0.002
Silt	0.002–0.05
Sand	0.05–2.0
Gravel	>2.0

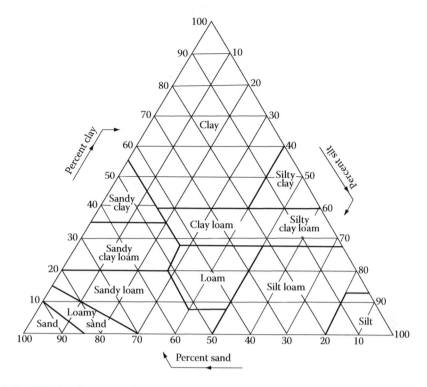

FIGURE 3.13 USDA soil texture triangle.

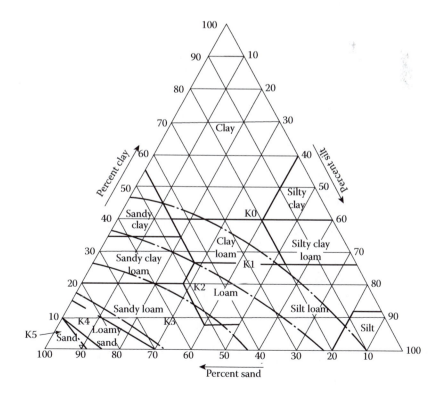

FIGURE 3.14 Soil permeability/textural triangle.

TABLE 3.8
Permeability of USDA Soil
Texture Triangle

| Permeability Class | Permeability | |
	in./h	(mm/h)
K5	>20	(500)
K4	6–20	(150–500)
K3	2–6	(50–150)
K2	0.6–2	(15–50)
K1	0.2–0.6	(5–15)
K0	<0.2	(5)

K_0 has the lowest permeability and K_5 the highest. The permeability rating for these soil classes, which ranges from less than 5 mm/h (0.2 in./h) for K_0 soils to well over 500 mm/h (20 in./h) for K_5 soils, is provided in Table 3.8.

3.5 HYDROGRAPHS

A hydrograph is defined as the graph showing variation of discharge or stage with time. In storm water management, a hydrograph represents temporal variation of runoff rate or discharge. As discussed in a previous section, only a portion of rainfall contributes to runoff; that is the portion that is not intercepted by tree leaves and vegetation, retained in surface depression, or infiltrated into the ground. The relation between rainfall and runoff is an integral part of storm water management calculations.

A hydrograph of a single storm is shown in Figure 3.15. A hydrograph is characterized by a rising limb, peak, and a falling limb, also known as a recession. The rising limb reflects increased runoff as the rainfall is stored overland and the recession represents the withdrawal of stored water after a storm is ended. The time to peak, also referred to as time of concentration, represents the time it takes for the runoff from the entire catchment area to reach the point of discharge. According to Figure 3.15, a hydrograph is characterized by three parameters: peak flow, time to peak (also referred to as time of concentration, T_c), and time base, T_B. These parameters are discussed in the following sections.

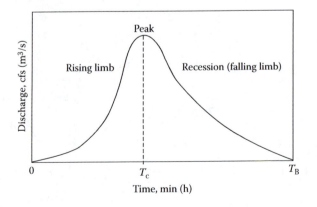

FIGURE 3.15   A typical storm hydrograph.

3.5.1 TIME OF CONCENTRATION EQUATIONS AND NOMOGRAPHS

Time of concentration is defined as the time for the runoff from a catchment area to reach equilibrium under a steady rainfall. It is also defined as the longest travel time it takes the runoff to reach the discharge point of a catchment area. The travel time is a parameter most often used to characterize the response of a catchment area to rainfalls. This parameter is a function of length scale, L, average catchment slope, S, and the catchment surface condition. The time of concentration is the sum of the overland flow time and the travel time in drainage channels along the flow route to the outlet.

A number of empirical equations and nomographs have been proposed for the time of concentration calculations; the most popular of which are the following.

3.5.1.1 Kirpich Equation

Kirpich, in 1940, presented the following equation in customary units (Chow et al., 1988):

$$T_c = \frac{0.0078 L^{0.77}}{S^{0.385}} \tag{3.27}$$

In metric units, this equation becomes

$$T_c = \frac{0.21 L^{0.77}}{S_o^{0.385}} \tag{3.28}$$

where
T_c = time of concentration, in minutes
L = flow length, ft (m)
S = average slope of flow path, ft/ft (m/m)

Kirpich's equation was originally developed from Soil Conservation Service (currently NRCS) data for seven rural basins in Tennessee with well-defined channels and slopes ranging from 3% to 10%. Figure 3.16 shows a graphical solution of Kirpich's equation, which is widely used in practice. For overland flow on lawn, the T_c obtained from this figure should be multiplied by 2. The author finds that Kirpich's nomograph (Figure 3.15) gives unreasonably short times of concentrations and therefore does not recommend its use. See Example 3.8.

3.5.1.2 Izzard Equation

Based on laboratory experiments by the Bureau of Public Works on roads and turf surfaces, Izzard, in 1946, developed the following equation in customary units:

$$T_c = \frac{41.025(0.0007I + C_r)L^{1/3}}{S^{1/3} \cdot I^{2/3}} \tag{3.29}$$

In metric units, this equation reads as

$$T_c = \frac{527(2.8 \times 10^{-5} \times I + C_r)L^{1/3}}{S^{1/3} \times I^{2/3}} \tag{3.30}$$

where
T_c = time of concentration, in minutes
I = rainfall intensity, in./h (mm/h)

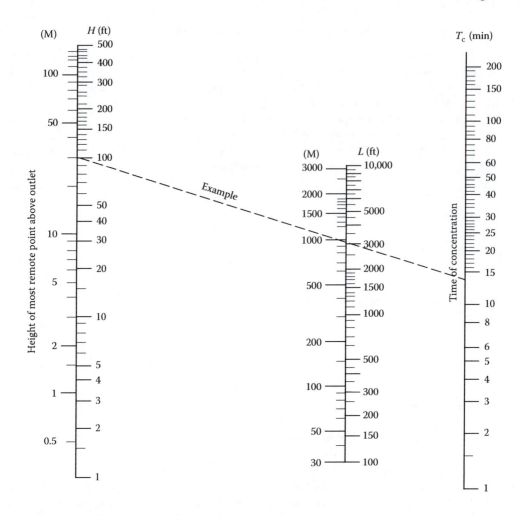

FIGURE 3.16 Kirpich time of concentration nomograph. (Notes: Use nomograph for natural catchment areas with well-defined channels, for overland on bare soil, and for roadside grass swales. For overland flow on lawn, multiply T_c by 2. For overland flow on pavement, multiply T_c by 0.4. For concrete channel, multiply T_c by 0.2.)

L = length of flow path, ft (m)
S = slope of flow, ft/ft (m/m)
C_r = retardance coefficient, ranging from 0.007 for very smooth pavements to 0.06 for dense turf (see Table 3.9)

3.5.1.3 Kirby Equation

Kirby, in 1959, proposed the following equation in customary units:

$$T_c = 0.83 \left(\frac{Lr}{S^{1/2}} \right)^{0.467} \tag{3.31}$$

This equation in metric units becomes

$$T_c = 1.45 \left(\frac{Lr}{S^{1/2}} \right)^{0.467} \tag{3.32}$$

TABLE 3.9
Values of C_r in Izzard Equation

Surface	C_r
Very smooth asphalt	0.007
Tar and sand pavement	0.008
Concrete	0.012
Closely clipped sod	0.016
Tar and travel pavement	0.017
Dense blue grass	0.060

TABLE 3.10
Retardance Roughness Coefficient, r, in Kirby Equation

Surface	r
Smooth pavement	0.02
Asphalt/concrete	0.05–0.15
Smooth, bare, packed soil, free of stones	0.10
Light turf	0.20
Poor grass on moderately rough ground	0.20
Average grass	0.40
Dense turf	0.17–0.80
Dense grass	0.17–0.30
Bermuda grass	0.30–0.48
Deciduous timberland	0.60
Conifer timberland, dense grass	0.60

Source: Westphal, J. A., 2001, in L.W. Mays, ed. *Storm water collection system design handbook*, McGraw-Hill, New York.

where

T_c = time of concentration, in minutes
L = flow length, ft (m)
r = retardance roughness coefficient, varying from 0.02 to 0.06 (see Table 3.10)
S = slope of catchment in ft/ft (m/m)

3.5.1.4 Garden State Parkway Nomograph

In 1957, the New Jersey Highway Authority—Garden State Parkway adopted the time of concentration nomograph shown in Figure 3.17. In this figure T_c is shown as a function of the flow length, type of surface cover, and slope of the land. The surface cover ranges from woodland to pavement and anything in between. Based on the experience of the author, this figure yields a more realistic result than Kirpich's equation for natural land surfaces. This matter is illustrated by the following example.

It is to be noted that neither in Figure 3.16 nor in Kirpich's, Izzard's, and Kirby's equations is the time of concentration proportional with the overland flow length. Therefore, contrary to a common practice, the flow reach should not be divided into reaches, as this would exaggerate the time of concentration.

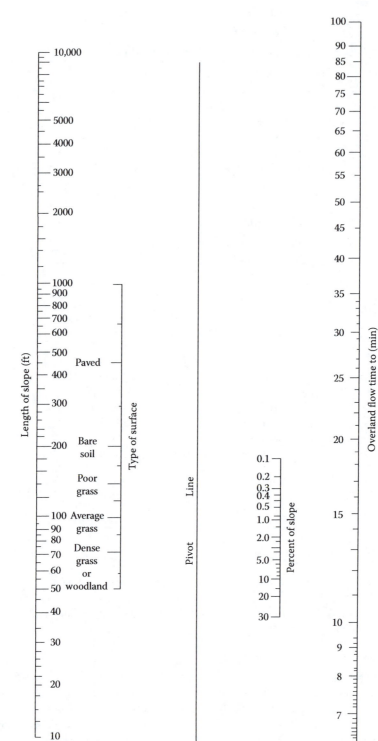

FIGURE 3.17 New Jersey Highway Authority—Garden State Parkway (1957). Time of concentration nomograph.

Example 3.8

A 5 acre (2.0 hectares) watershed is covered by 15% pavement and 85% dense grass. The watershed main flow path is 1100 ft (335 m) long and has an average slope of 3%. Assuming that the flow path is entirely vegetated, calculate the time of concentration using Kirpich, Izzard, and Kirby equations and the New Jersey Garden State Parkway nomograph (Figure 3.15). Use a rainfall intensity of 4 in./h (100 mm/h).

Solution

a. Kirpich equation (Equation 3.27)

$$T_c = \frac{0.0078 \times 1100^{0.77}}{0.03^{0.385}} = 6.6 \text{ min}$$

Multiply the calculated t_c by 2 for grass cover:

$$T_c = 2 \times 6.6 = 13.2 \text{ min}$$

b. Izzard equation
   For dense grass, $C_r = 0.06$ (see Table 3.9)
   Inputting $I = 4.0$ in./h and $C_r = 0.06$ in Equation 3.29.

$$T_c = \frac{41.025(0.007 \times 4 + 0.06) \times 1100^{1/3}}{(0.03^{1/3} \times 4^{2/3})} = 34 \text{ min}$$

c. Kirby equation
   From Table 3.10, $r = 0.3$ (maximum for dense grass).

$$T_c = 0.83 \left(\frac{1100 \times 0.3}{0.03^{1/2}} \right)^{0.467}$$

$$= 28.2 \text{ min, say, } 28 \text{ min}$$

d. Figure 3.17
   Draw a straight line connecting 1100 ft on the length scale to dense grass on the "type of surface" line and extend to the pivot line. Then, from the point of intersection draw another line to $S = 3\%$ and extend to the time of concentration line to read $T_c = 34$ minutes.

Note that the time of concentration obtained from Izzard and Kirby equations is in fair agreement and nearly 2.5 times longer than that of the Kirpich equation (Figure 3.16).

3.5.2   OTHER METHODS OF TIME OF CONCENTRATION CALCULATION

3.5.2.1   SCS Method

The Soil Conservation Service, in 1975, presented a method for calculating runoff hydrographs. This method, which is presented in technical release no. 55 (1986), includes a procedure for the time of concentration calculation as follows. The flow path is divided into three segments: sheet flow, shallow concentrated flow, and channel flow. The sheet flow time of concentration is calculated by

$$T_t = \frac{0.007(nL)^{0.8}}{\left(P^{0.5} \times S^{0.4} \right)} \tag{3.33}$$

In metric units this equation becomes

$$T_t = \frac{0.091(nL)^{0.8}}{\left(P^{0.5} \times S^{0.4}\right)} \tag{3.34}$$

The parameters in the preceding equations are

n = Manning's roughness coefficient, varying from 0.011 for smooth pavements to 0.4 for
  wooded areas (see Table 3.11)
L = Length of sheet flow, ft (m); limited to 150 ft (±50 m) in pervious surfaces and 100 ft
  (30 m) for pavements
P = 2 year-24 hour precipitation, in. (mm)
S = land slope, ft/ft (m/m)
T_t = sheet flow travel time, hours

The shallow concentrated flow lies between sheet flow and channel flow, if any. The average
flow velocity for this segment is derived from Figure 3.18, and the shallow concentrated flow time
of concentration is calculated by

$$T_s = \frac{L}{60V} \tag{3.35}$$

where
  T_s = time of concentration, in minutes
  L = length of this flow segment, in ft, m
  V = shallow concentrated flow velocity in ft/s, m/s

Channel flow begins where the flow concentrates into swales or pipes. The flow velocity for this
segment is calculated using the Manning formula, which, as noted in Chapter 2, is expressed as

$$V = \frac{R^{2/3} S^{1/2}}{n} \quad \text{SI} \tag{3.36}$$

and

$$V = \frac{1.49 R^{2/3} S^{1/2}}{n} \quad \text{CU} \tag{3.37}$$

TABLE 3.11
Values of Manning's *n* for Sheet Flow

Surface Description	*n*
Smooth asphalt	0.011
Fallow (no residue)	0.050
Cultivated soil	0.06–0.17
Grass	
Short grass	0.150
Dense grass	0.240
Bermuda grass	0.400
Range (natural)	0.130
Woods	0.400

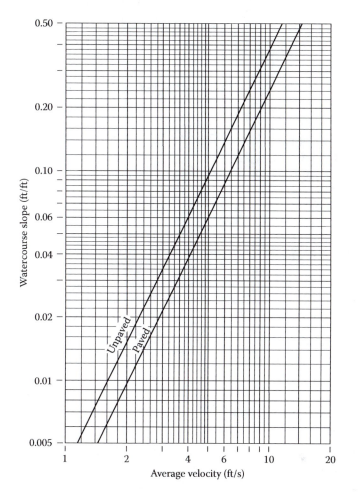

FIGURE 3.18   Shallow concentrated flow velocity (SCS, TR-55 method).

where

   V = flow velocity, ft/s (m/s)

   R = hydraulic radius, ft (m)

   S = energy slope (same as the channel slope in uniform flow), ft/ft (m/m)

   n = Manning roughness coefficient, ranging from 0.025 to 0.04 for gravel and rock channel to over 0.10 for grass swales (see Chapter 4 for more detail)

The channel flow time of concentration, in minutes, can be calculated from Equation 3.35, where L is the length of channel flow segment.

3.5.2.2   FHWA Method

The Federal Highway Administration, US Department of Transportation, has presented a method of calculating time of concentration components in Hydraulic Engineering Circular no. 22 (2013) and Hydraulic Design Series no. 4 (2001). In this method, similar to that of SCS, the flow path is divided into three segments. The sheet flow component of time of concentration relates to the shallow mass of runoff with a uniform depth across the sloping surface. Its length is generally longer than 25 m (80 ft) but rarely over 130 m (400 ft). Applying the kinematic wave model, a derivative of the Manning formula, for estimating the time of equilibrium for fully developed turbulent flow (representing the limit of sheet flow segment) results in the following equation HDS-2 (FHWA, 2002):

$$T_t = \left(\frac{K_n}{I^{0.4}}\right)\left(\frac{nL}{S^{1/2}}\right)^{0.6}$$ (3.38)

where
T_t = sheet flow travel time, minutes
L = flow length, m (ft)
I = rainfall intensity of the design storm, mm/h (in./h)
S = surface slope, m/m (ft/ft)
K_n = empirical coefficient = 6.92 SI, 0.933 CU
n = Manning's roughness coefficient

Equation 3.38 may be applied to calculate the overall time of concentration in small catchment areas.

After a distance of 100 m (300 ft), sheet flow tends to concentrate in rills and then gullies. The flow velocity in this reach, which is referred to as shallow concentrated flow, can be estimated from the following equation:

$$V_s = K_u k S^{0.5}$$ (3.39)

where
K_u = 1 SI; 3.28 CU
k = intercept coefficient (see Table 3.12)
S = slope, percent

The flow velocity in the channel/pipe flow reach is calculated using the Manning formula (Equations 3.36 and 3.37), the same as in the SCS method. The overall time of concentration is then calculated by adding travel time in the previously indicated three segments as follows:

$$T_c = T_t + \left(\frac{L_s}{60V_s}\right) + \left(\frac{L_c}{60V_c}\right)$$ (3.40)

TABLE 3.12

Intercept Coefficient, k, in Shallow-Concentrated Flow Velocity[a]

Land Cover/Flow Regime	k
Forest with heavy ground litter; hay meadow (overland flow)	0.076
Trash fallow or minimum tillage cultivation; contour or strip cropped; woodland (overland flow)	0.152
Short grass pasture (overland flow)	0.213
Cultivated straight row (overland flow)	0.274
Nearly bare and untilled (overland flow); alluvial fans in western mountain regions	0.305
Grassed waterway (shallow concentrated flow)	0.457
Unpaved area (shallow concentrated flow)	0.491
Paved area (shallow concentrated flow; small upland gullies	0.619

Source: FHWA Hydraulic Design Series no. 2, 2nd ed., 2002, publication no. FHWA-NHI02-001, Chapter 6; HEC-22, 3rd ed., 2009 (revised August 2013), publication no. FHWA-NHI-10-009, Chapter 3.

[a] Equation 3.39.

where L_s and L_c are the lengths of the shallow concentrated flow and channel/pipe flow segments, respectively, and other parameters are as defined previously.

It is to be noted that the sheet flow time of concentration equation in this method, though seemingly similar to the SCS method, is different from that method. In the SCS method the 2-year rainfall depth is used to calculate the sheet flow travel time for all storm frequencies. However, in the FHWA method, like the Izzard equation, T_t is calculated using the storm intensity for a design storm, and as such, the sheet flow travel time varies with the storm frequency.

Since the rainfall intensity, I, depends on T_c, which is not initially known, the calculation of T_t requires an iterative process. Specifically, an initial estimate of T_c is assumed to obtain I from the rainfall intensity–duration–frequency (IDF) curve for the locality. The T_c is then calculated by adding the time of concentration components for sheet flow, shallow concentrated flow, and channel flow using Equations 3.38 through 3.40, respectively; it is checked against the initial value of T_c. If they differ, the process is repeated until the two successive T_c estimates become equal. The following example illustrates the calculation process.

Like other methods presented in a previous section, the sheet flow time of concentration in the TR-55 and FHWA methods is not a linear function of the flow length. Therefore, dividing the sheet flow reach into subreaches, which is commonly practiced, will result in an overestimation of the sheet flow travel time. This matter will be exemplified for the TR-55 method later in this chapter.

Example 3.9

Flow path characteristics for a 4 ha catchment area are as follows:

Flow Segment	Length, m	Slope	Segment Cover
Sheet flow	50	0.7%	Short grass
Shallow concentrated	85	1.2%	Short grass
Pipe flow	180	1.0%	450 mm (18 in.) concrete pipe

Calculate the time of concentration, using:

a. TR-55 method
b. FHWA method (HEC-22)

The local 2 year-24 hour storm is 84 mm (3.3 in.) and the rainfall intensity–duration relation for the design storm is listed below:

T_c, min	20	30	45	60	80
I, mm/h	110	82	63	50	42

Solution

a. TR-55 method
- Sheet flow, using Equation 3.34 and $n = 0.24$ (see Table 3.11):

$$T_t = \frac{0.091(0.24 \times 50)^{0.8}}{(84^{0.5} \times 0.007^{0.4})} = 0.527 \, \text{hr} = 31.7 \, \text{min}$$

- Shallow concentrated flow: For $S = 1.2\%$, read the flow velocity for unpaved surface on Figure 3.18:

$$V = 1.8 \text{ ft/s} = 0.55 \text{ m/s}$$

$$T_s = 85/(60 \times 0.559) = 2.6 \text{ min}$$

- Pipe flow:
  Assuming full flow in pipe:

$$V = \frac{R^{2/3} S^{1/2}}{n}$$

$$R = \frac{D}{4} = \frac{450}{4} = 112.5 \text{ mm} = 0.113 \text{ m}$$

$n = 0.012$; see Table 2.5 in Chapter 2

$$V = \frac{0.113^{2/3} \times (0.01)^{1/2}}{0.012} = 1.95 \text{ m/s}$$

$$T_{ch} = \frac{180}{60 \times 1.95} = 1.5 \text{ min}$$

- Time of concentration:

$$T_c = 31.7 + 2.6 + 1.5 = 35.8 \text{ min, say, 36 min}$$

b. FHWA method
   Assume $T_c = 30$ min as a trial; refine if necessary:
- Sheet flow travel time (Equation 3.38):
  Based on the IDF table, $I = 82$ mm/h

$$T_t = \left[\frac{6.92}{(82)^{0.4}}\right]\left[\frac{0.24 \times 50}{(0.007)^{0.5}}\right]^{0.6} = 23.4 \text{ min}$$

- Shallow concentrated flow:

$$V = kS^{0.5}$$

$k = 0.213$, short grass pasture overland flow (see Table 3.12)

$$S = 1 \ (1\%)$$

$$V = 0.213 \times (1)^{0.5} = 0.213 \text{ m/s}$$

$$T_s = 85/(60 \times 0.213) = 6.7 \text{ min}$$

- Pipe flow travel time:
  Same as TR-55 method

$$T_c = 1.5 \text{ min}$$

- Time of concentration:

$$T_c = 23.4 + 6.7 + 1.5 = 31.6 \text{ min, round to 32 minutes}$$

The calculated T_c is not significantly different from the assumed value, so no further trial is necessary.

Note: The results from the two methods are not markedly different; however, the FHWA method gives more reasonable answers for shallow concentrated flow velocity than the SCS TR-55 method.

Example 3.10

Using the TR-55 method, calculate the time of concentration for a 10 acre watershed, having

150 ft sheet flow in woods, at 2% slope
375 ft shallow concentrated flow in grass, at 3% slope
500 ft channel flow: Manning's $n = 0.06$; flow area $= 10$ ft^2; wetted perimeter $= 12$ ft, slope $= 1\%$
The 2 year-24 hour storm is 3.5 in.

Solution

a. Sheet flow, $n = 0.4$ woods

$$T_t = \frac{0.007(n \times L)^{0.8}}{\left(P_2^{0.5} \times S^{0.4}\right)} = \frac{0.007(0.4 \times 150)^{0.8}}{[3.5^{0.5} \times (0.02)^{0.4}]}$$

$$= 0.47 \text{ hour}$$

b. Shallow concentrated flow

For $S = 0.03$, Figure 3.17 reads $V = 2.8$ fps for shallow concentrated flow

$$T_t = \frac{L}{3600V} = \frac{375}{3600 \times 2.8} = 0.04 \text{ hour (rounded to second decimal place)}$$

c. Channel flow

$$R = \frac{A}{P} = \frac{10}{12} = 0.83 \text{ ft}$$

$$V = \left(\frac{1.49}{n}\right) R^{2/3} S^{1/2}$$

$$V = \left(\frac{1.49}{0.06}\right) \times (0.83)^{2/3} \times (0.01)^{1/2} = 2.2 \text{ fps}$$

$$T_t = \frac{500}{(2.2 \times 3600)} = 0.06 \, \text{hour}$$

$T_c = 0.47 + 0.04 + 0.06 = 0.57$ hour, round to 0.6 hour

3.5.3 SHEET FLOW LENGTH ANALYSIS

Rain falling on a surface flows at a shallow depth with the rate and velocity of flow increasing along the flow. At a point located at distance, L, from the onset, the flow rate and velocity are given by the following parametric equations:

$$q \approx I \cdot L \tag{3.41}$$

$$V \approx I \cdot L/d \tag{3.42}$$

where
  q = flow rate per unit width
  L = flow length from the high point
  I = rainfall intensity
  d = depth of flow

the sign "≈" represents the proportionality factor, which depends on the system of units (SI, CU).
    The Reynolds number at the point is

$$R_e = Vd/\nu \tag{3.43}$$

Combining Equations 3.42 and 3.43 results in

$$R_e \approx I \cdot L/\nu \tag{3.44}$$

The original Izzard equation (1946) for the time of concentration placed a limit of laminar sheet flow based on the product of rainfall intensity and length as follows:

$$I \cdot L < 500 \tag{3.45}$$

where I is the rainfall intensity in inches/hour and L is the sheet flow length, in feet. This equation reflects a Reynolds number of approximately 1100. Chow (1964) reported that Izzard's equation is in reasonable agreement for sheet flow across very wide airport aprons. For a rainfall intensity of 1.5 in./h, which approximates a 2-year, 60-minute storm in New Jersey, the Izzard equation gives a maximum sheet flow length of 330 ft. In metric units, expressing I in mm/h and L in meters, Equation 3.45 becomes

$$I \cdot L \leq 3870 \tag{3.46}$$

McCuen and Spiess (1995) report that $nL/\sqrt{s}$ is a better parameter than either L or IL as a limiting criterion for sheet flow. This parameter was later expanded to Equation 3.38 for sheet flow time of concentration (McCuen et al., 2002). Factors affecting sheet flow are identified in this book as follows.
    To relate the sheet flow length with the surface slope and roughness, the flow velocity in Equation 3.42 is equated to velocity in the Manning's formula in which the hydraulic radius is replaced by the flow depth in a broad sheet flow:

$$I \cdot L/d = (1/n)d^{2/3} S^{1/2} \tag{3.47}$$

Solving for L:

$$L \approx (d^{5/3} \sqrt{S})/(n \cdot I) \qquad (3.48)$$

This equation indicates that the sheet flow length is lumped in the parameter $(nL/\sqrt{S}) \cdot (I/d^{5/3})$. This parameter includes not only $nL/\sqrt{S}$, but also the flow depth, d, and rainfall intensity, I. The proportionality factor depends on the system of units. In metric units, Equation 3.48 is expressed as

$$L = 36(d^{5/3} \sqrt{S})/(n \cdot I) \qquad (3.49)$$

Rewriting this equation in terms of flow depth:

$$d = 0.116(nI/\sqrt{S})^{0.6} \times L^{0.6} \qquad (3.50)$$

In these equations:

L = sheet flow length, m
d = flow depth, mm
I = rainfall intensity, mm/h
S = surface slope, m/m

The term in the parentheses on the right-hand side is identical to the second term in the parentheses of the FHWA equation for sheet flow time of concentration (Equation 3.38). This demonstrates that both depth of sheet flow and time of concentration are related to rainfall intensity to power 0.6 and the slope of surface to power 0.3.

Combining Equation 3.46, which is based on the Izzard's experiment, and Equation 3.50 yields the following equation for limiting sheet flow depth:

$$d_{sf} = 16.5(n/\sqrt{S})^{0.6} \qquad (3.51)$$

This equation indicates that the sheet flow depth increases with increase in surface roughness and reduction in slope.

In customary units, Equation 3.51 is expressed as

$$d = 0.65(n/\sqrt{S})^{0.6} \qquad (3.52)$$

where d is in inches and I is in inches/hour.

For a smooth surface ($n = 0.015$) at 2% slope, Equations 3.51 and 3.52 give $d = 4.3$ mm and 0.17 in., which appear reasonable. Substituting this depth in Equation 3.49, the sheet flow length for a rainfall intensity of 75 mm/h (3 in./h) is calculated at 51.5 m (169 ft).

3.6   RUNOFF CALCULATION METHODS

In storm water management practices, runoff calculations are generally performed based on two methods: the rational method and the TR-55 method. A brief review and comparison of the rational and the SCS methods are presented by Pazwash (1989). An in-depth discussion of these two methods is presented in the following sections. Also presented in this book is a universal method developed by the author.

3.6.1 RATIONAL METHOD

In 1898, Emil Kuichling introduced a simple relation for estimating peak rates of runoff for a catchment area. This method, known as the rational formula, is as follows:

$$Q = CIA* \qquad \text{CU units} \tag{3.53}$$

$$Q = CIA/360 \qquad \text{SI units} \tag{3.54}$$

where
    C = runoff coefficient, dimensionless
    I = rainfall intensity, mm/h (in./h)
    A = catchment area, hectares (acres)
    Q = discharge, m³/s (cfs)

The rational method is based on the assumptions of linearity and proportionality, in that

a. Runoff is proportional to rainfall intensity.
b. Abstractions and losses vary linearly with rainfall and are incorporated in the runoff coefficient.
c. The duration of storm must be equal to or longer than the time of concentration of catchment area.

Runoff coefficient varies from less than 0.1 for forested lands of mild slope to 0.95 for paved areas. Table 3.13 lists typical values of runoff coefficient for various land surface covers and developments. The US Department of Transportation recommends using the coefficients in this table for storms of less than 25-year frequency and applying an adjustment factor, C_r, to less frequent storms as follows:

$$T_r < 25 \text{ years,} \qquad C_r = 1.0$$

$$T_r = 25 \text{ years,} \qquad C_r = 1.1$$

$$T_r = 50 \text{ years,} \qquad C_r = 1.20$$

$$T_r = 100 \text{ years,} \qquad C_r = 1.25$$

Where there is more than one land use, the weighted runoff coefficient is calculated and the resulting C value is used in the rational formula. The calculation process is illustrated by Example 3.11.

The rainfall intensity in the rational formula is the average intensity of rainfall having a duration equal to the time of concentration of the catchment area. This storm has a larger intensity than longer storm durations and produces the largest runoff for that intensity. As noted previously, the rainfall intensity depends on the storm frequency. For a given frequency, the peak flow is calculated using the IDF curves for the locality.

The rational formula might yield a larger peak runoff for a portion of a drainage area than the entire area if the subarea is mostly paved and has a much shorter time of concentration than the entire area. Therefore, when using the rational formula it should be checked if the entire area or a downstream subarea, which is highly developed and has a connected drainage system, would give a

---

* This equation inherits an approximation as follows: 43,560 sf/acre × 1/(12 × 3600) = 1.088.

TABLE 3.13

Typical Values of Runoff Coefficient, C[a]

Description of Area	Runoff Coefficient
Business	
Downtown areas	0.70–0.95
Neighborhood areas	0.50–0.70
Residential	
Single-family areas	0.30–0.50
Multiunits, detached	0.40–0.60
Multiunits, attached	0.60–0.75
Residential, suburban	0.25–0.40
Apartment dwelling areas	0.50–0.70
Industrial	
Light areas	0.50–0.80
Heavy areas	0.60–0.90
Parks, cemeteries	0.10–0.25
Railroad yard areas	0.20–0.35
Unimproved areas	0.10–0.30
Pavement	
Asphalt or concrete	0.70–0.95
Brick	0.70–0.85
Roofs	0.75–0.95
Lawns, sandy soil	
Flat, 2%	0.05–0.10
Average, 2%–7%	0.10–0.15
Steep, 7% or more	0.15–0.20
Lawns, heavy soil	
Flat, 2%	0.13–0.17
Average, 2%–7%	0.18–0.22
Steep, 7% or more	0.25–0.35

Source: ASCE and WEF, 1992, Design and construction of urban storm water management systems, *ASCE Manuals and Reports of Engineering Practice*, no. 77, and *WEF Manual of Practice* FD-20.

[a] For storms of up to 10-year frequency. A somewhat higher value for C may be used for larger design storms.

greater discharge. This may occur in pipe design where an upstream pipe reach receives runoff from a mostly paved area while a downstream pipe has a larger but mostly pervious drainage area with a longer time of concentration than the upstream pipe.

For a storm duration equal to the time of concentration, the runoff hydrograph reaches its peak at the time of concentration, T_c, and recedes to zero at a time base, T_b. The rational hydrograph is commonly plotted as a triangular hydrograph with a time base equal to 2.67 longer than the time of concentration. Figure 3.19 shows a typical rational method hydrograph. It is evident that for storm durations shorter than the time of concentration, the runoff from all of the catchment area cannot reach the point of outlet.

It should be noted that for fully paved areas such as parking lots and building roofs, the use of the preceding indicated hydrograph results in a runoff volume greater than rainfall amount, which is unrealistic. Pazwash (1992) suggests selecting a time base according to the impervious coverage or simply on the runoff coefficient. Table 3.14 presents a modification of this relation.

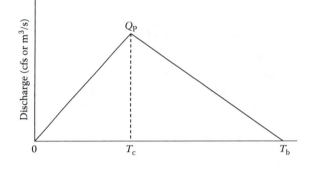

FIGURE 3.19   Rational method hydrograph.

TABLE 3.14
Time Base in Rational Method

C	T_b/T_c
0.75–1.0	2.00
0.5–0.75	2.33
0.30–0.5	2.50
<0.3	2.67

3.6.2   LIMITATIONS OF RATIONAL METHOD

The rational method assumes that the runoff is directly related to and is proportional to the rainfall rate, C being the proportionality factor. As discussed earlier, such an assumption is unrealistic. For this reason, many jurisdictions limit the use of the rational method to small areas. ASCE (1992), for example, recommends that the rational method be used for areas less than 80 ha (200 acres). However, the New Jersey Department of Environmental Protection limits application of the method to 20 acres (8 ha). The author suggests the use of the rational method up to 100 ha (250 acres). The rational method is best suited and most accurate for mostly paved areas where interception is nonexistent, infiltration is negligible, and surface retention small.

In an attempt to account for the effect of rainfall intensity on the losses, some jurisdictional agencies specify adjusting runoff coefficient based on storm frequency. The New Jersey Soil Erosion and Sediment Control Standards (1999), for example, recommend using the same runoff coefficient for 2- and 10-year storms and applying a 1.20 factor and a 1.25 factor to the runoff coefficient for the 2-year storm to get the 50- and 100-year storms, respectively. However, as will be discussed later, not only the rainfall intensity but also the storm duration affect the initial abstractions and hence the runoff–rainfall ratio. Therefore, applying an adjacent factor to the runoff coefficient will not fully correct the inadequacy of the method for calculating peak rates of runoff. This is evidenced by Table 3.15, which lists some of the suggested runoff coefficients in Denver, Colorado (see Maidment, 1993). While these coefficients may not be applicable to all localities, the C values for smaller rainfall depths are in good agreement with visual observations, which indicate that the runoff coefficient varies with rainfall depth.

Example 3.11

Calculate the peak rates of runoff for 2- and 10-year frequency storms for the catchment area of Example 3.10 based on the rational formula. This area is 90% grass covered and 10% paved. Base

---

TABLE 3.15

Suggested Runoff Coefficients for Denver Area, CO

Land Use/Surface Cover	2-h Rainfall Depth, in. (mm)			
	1.2 (30)	1.7 (38)	2.0 (51)	3.1 (79)
Lawns, sandy soil	0	0.05	0.10	0.20
Lawns, clayey soil	0.05	0.15	0.25	0.50
Paved areas	0.87	0.88	0.90	0.93
Gravel pavement	0.15	0.25	0.35	0.65
Roofs	0.80	0.85	0.90	0.93

Note: Values in this table may be applied for an area averaging; the coefficients may not be valid for areas over 200 acres.

---

your calculations on the New Jersey IDF curves and the calculated T_c using the FHWA method in Example 3.9.

Solution

For this site, select $C = 0.15$ for vegetated land and $C = 0.95$ for pavement (see Table 3.13). Calculate the composite C value as follows:

$$C = 0.15 \times (90/100) + 0.95(10/100) = 0.23$$

Round T_c in Example 3.9 to 32 minutes. Interpolate rainfall intensities for a storm duration of 32 minutes in Figure 3.1b.

$$I_2 = 2.2 \text{ in./h}$$

$$I_{10} = 3.0 \text{ in./h}$$

Input these parameters in the rational formula:

$$Q = 0.23 \times 10 \times I = 2.3 \, I$$

$$Q_2 = 5.1 \text{ cfs (rounded to the first decimal place)}$$

$$Q_{10} = 6.9 \text{ cfs}$$

3.6.3 MODIFIED RATIONAL METHOD

The rational method is widely used for estimating the peak flow for small drainage basins, urban developments, and roadway drainage systems. In this method the peak runoff for a given storm frequency is calculated for storm durations longer than the time of concentration. Storms lasting longer than the time of concentration have smaller intensity and produce smaller peak flows than the triangular hydrograph; however, they generate greater volumes of runoff. Therefore, such storms may create a larger discharge when routed through a detention basin. Thus, routing computations for a detention basin should be performed for a range of storm durations to determine the critical storm, namely, the one that produces the largest discharge. This will be illustrated by Example 3.12.

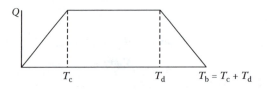

FIGURE 3.20 Modified rational hydrograph.

A modified rational hydrograph is represented by a trapezoid peaking at the time of concentration, T_c, having a uniform discharge extending to the storm duration T_d, and falling to zero at a time equal to $T_d + T_c$. Figure 3.20 presents a modified rational hydrograph. The peak discharge in modified rational hydrographs is calculated using the rational formula with a rainfall intensity associated with the storm duration. The runoff volume in a modified rational hydrograph is calculated as follows:

$$V = \frac{1}{2}T_cQ + (T_d - T_c)Q + \frac{1}{2}T_cQ$$

This equation simplifies as

$$V = T_dQ \qquad\qquad (3.55)$$

This equation shows that the runoff volume is independent of the time of concentration and depends solely on the storm duration. For a storm duration equal to T_c, the runoff hydrograph becomes a triangle, as was shown in Figure 3.19.

Example 3.12

Calculate the runoff volume from a 45,200 sq. ft flat roof for 10-year storms of the following durations:

  a. 10-minute storm duration, representing the time of concentration of roof runoff
  b. 30-minute duration
  c. 60-minute duration

Use the IDF curves in Figure 3.1b.

Solution

1. Rational formula
       Use:

$$C = 0.95 \text{ for roof area (Table 3.13)}$$

$$A = 45,200/43,560 = 1.038 \text{ acres}$$

$$Q = CIA = 0.95 \times 1.038 \times I = 0.986I$$

    a.   10-minute storm

$$I = 5.8 \text{ in./h}$$

$$Q = 0.986 \times 5.8 = 5.72 \text{ cfs}$$

Use an isosceles triangular hydrograph to calculate runoff volume:

$$V = 2 \times T_c \times Q/2 = 10 \times 60 \times 5.72 = 3430 \text{ cf}$$

b. 30-minute storm

$$I = 3.2 \text{ in./h}$$

$$Q = 0.986 \times 3.2 = 3.15 \text{ cfs}$$

$$V = Q \times T_d = 3.15 \times 30 \times 60 = 5678 \text{ cf}$$

c. 60-minute storm

$$I = 2.0 \text{ in./h}$$

$$Q = 0.986 \times 2.0 = 1.97 \text{ cfs}$$

$$V = 1.97 \times 60 \times 60 = 7098 \text{ cf}$$

2. Direct volume calculations

Alternatively, the runoff volumes can be calculated from the product of area times rainfall depth, accounting for losses. For the 30-minute storm, the calculations are exemplified as follows:

$$\text{Rainfall depth} = 3.2 \text{ in./h} \times 30/60 = 1.2 \text{ in.}$$

$$\text{Effective rainfall depth} = 1.2 \times 0.95 = 1.14 \text{ in.}$$

where 0.06 in. (1.2–1.14) represents the initial losses, namely depression storage (surface retention).

$$\text{Runoff volume} = 45200 \times 1.14/12 = 4294 \text{ cf}$$

This is more accurate than the former result; the difference is due to an inherent approximation in the rational formula.

EXERCISE

Solve the preceding example for a 4200 m² roof of a commercial building for the following rainfall events:

a. 10-minute duration storm of 145 mm/h
b. 30-minute duration storm of 80 mm/h
c. 60-minute duration storm of 50 mm/h

3.6.4 SCS TR-55 Method

Based on tests on small agricultural plots, the Soil Conservation Service (presently the NRCS) had developed a method for relating the rainfall–runoff relation. This method was presented in a report titled "Technical Release No. 55 (TR-55)," which was originally issued in 1975 and revised in 1986. The method, which is based on 24-hour storm events and time of concentrations, T_c, ranging from 0.1 hour to 10 hours, is applied to small and midsized catchment areas up to 5 square miles (13 km²). The SCS method separates the rainfall into three components: initial abstraction, I_a; retention storage, F; and runoff depth, R (denoted as Q in the TR-55).

Figure 3.21 depicts these three components. The initial abstraction includes interception, initial infiltration, and depression storage. If the amount of rainfall is less than the initial abstraction, no runoff occurs. The retention, F, represents the continuing losses after the initial abstraction has occurred and is primarily due to the infiltration into the ground surface.

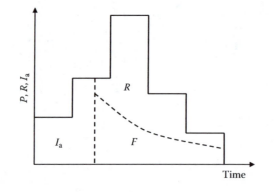

FIGURE 3.21   Components of the SCS model.

The SCS method is based on the hypothesis that the ratio of actual retention, F, to potential (maximum) retention, S, is equal to the ratio of runoff, R, to the potential runoff, namely $P - I_a$. This assumption is presented by the following equation:

$$\frac{F}{S} = \frac{R}{(P - I_a)} \tag{3.56}$$

The potential retention, S, does not include I_a. According to Figure 3.21,

$$F = P - R - I_a \tag{3.57}$$

Combining Equations 3.56 and 3.57 yields

$$R = \frac{(P - I_a)^2}{(P - I_a) + S} \tag{3.58}$$

Using field data, the SCS method further assumes the initial abstraction, I_a, to be equal to 2/10 (0.2) of the maximum retention, S, namely:

$$I_a = 0.2\,S \tag{3.59}$$

Studies indicate that the preceding assumption exaggerates peak rates of runoff, especially for small to medium storms (Pazwash, 1989; Schneider and McCuen, 2005).

Eliminating I_a between Equations 3.58 and 3.59 results in

$$R = \frac{(P - 0.2S)^2}{(P + 0.8S)} \tag{3.60}$$

Since this equation is dimensionally homogeneous, it is equally applicable to SI and customary units.

This equation relates R and P, provided that S is known. To establish a relation for S, the SCS introduces the soil curve number, CN, by the following equation:

$$CN = \frac{1000}{(10 + S)} \tag{3.61}$$

where CN varies with soil group (to be defined later) and soil cover.

Inversely:

$$S = \frac{1000}{CN} - 10,\ \text{in.} \tag{3.62}$$

In SI units, these equations become

$$CN = \frac{1000}{(10 + 0.0394S)} \tag{3.63}$$

and

$$S = 25.4\left(\frac{1000}{CN} - 10\right), \text{ mm} \tag{3.64}$$

The curve number, CN, as calculated by Equation 3.61 or 3.63, applies to normal antecedent moisture conditions, known as AMC II. In drier than normal soil conditions (AMC I) and in wetter than normal soil conditions (AMC III), the curve number would be adjusted as follows (Chow et al., 1988):

$$CN\,(I) = \frac{CN}{(2.38 - 0.014\ CN)} \tag{3.65}$$

$$CN\,(III) = \frac{CN}{(0.43 + 0.0057\ CN)} \tag{3.66}$$

Table 3.16 lists runoff depths for AMC I and AMC III relative to AMC II for a given storm event and CN value.

For the purpose of calculating runoff, the SCS classifies the soils into four hydrologic groups. These are group A through group D, where group A is the most permeable and group D the least. Table 3.17 presents a description of these soils and their permeability.

As indicated, the SCS relates the soil curve number to soil cover and hydrologic soil groups. Table 3.18 lists soil curve numbers for residential developments and Table 3.19 presents soil curve numbers for virgin lands. CN tables for agricultural and forested lands may be found in the TR-55 manual (1986).

TABLE 3.16

Relative Amounts of Runoff for Different AMCs[a]

AMC Type	CN	Runoff in. (mm)	Runoff/Rainfall	Relative to AMC II
II	72	2.2 (55.9)	44%	100%
I	52	0.8 (20.3)	16%	36%
III	86	3.47 (88.1)	69%	158%

[a] For 5 in. (127 mm)—24-hour storm.

TABLE 3.17

Description of SCS Soil Groups

Group	Soil Type	Minimum Infiltration Rate in./h (mm/h)
A	Deep sand; deep loess; aggregated silts	0.30 (7.6)
B	Shallow loess; sandy loam	0.15–0.30 (3.8–7.6)
C	Clay loams, shallow sandy loam; organic soils; soils of high clay concentration	0.05–0.15 (1.3–3.8)
D	Soils that swell significantly when wet; heavy plastic clays, certain saline soils	0–0.05 (0–1.3)

TABLE 3.18
SCS Curve Numbers for Urban Areas

Cover Description		Curve Numbers for Hydrologic Soil Group			
Cover Type and Hydrologic Condition	Average Percent Impervious Area	A	B	C	D
Fully Developed Urban Areas (Vegetation Established)					
Open space (lawns, parks, golf courses, cemeteries, etc.)					
Poor condition (grass cover < 50%)		68	79	86	89
Fair condition (grass cover 50% to 75%)		49	69	79	84
Good condition (grass cover > 75%)		39	61	74	80
Impervious areas					
Paved parking lots, roofs, driveways, etc. (excluding right-of-way)		98	98	98	98
Streets and roads					
Paved; curbs and storm sewers (excluding right-of-way)		98	98	98	98
Paved; open ditches (including right-of-way)		83	89	92	93
Gravel (including right-of-way)		76	85	89	91
Dirt (including right-of-way)		72	82	87	89
Western desert urban areas					
Natural desert landscaping (pervious areas only)		63	77	85	88
Artificial desert landscaping (impervious week barrier, desert shrub with 1- to 2-in. sand or gravel mulch and basin borders)		96	96	96	96
Urban districts					
Commercial and business	85	89	92	94	95
Industrial	72	81	88	91	93
Residential districts by average lot size					
1/8 acre or less (town houses)	65	77	85	90	92
1/4 acre	38	61	75	83	87
1/3 acre	30	57	72	81	86
1/2 acre	25	54	70	80	84
1 acre	20	51	68	79	84
2 acres	12	46	65	77	82
Developing Urban Areas					
Newly graded areas (pervious areas only, no vegetation)	77	86	91	94	
Idle lands (CNs are determined using cover types similar to those in Table 3.19)					

Note: Average runoff condition and Ia = 0.2S. The average percent impervious area shown was used to develop the composite CNs. Other assumptions are as follows: Impervious areas are directly connected to the drainage system, impervious areas have a CN of 98, and pervious areas are considered equivalent to open space in good hydrologic condition. CNs for other combinations of conditions may be computed using Figures 3.22a or b. CNs shown are equivalent to those of pasture. Composite CNs may be computed for other combination of open space cover type. Composite CNs for natural desert landscaping should be computed using Figures 3.22a or b based on the impervious area percentage (CN = 98) and the pervious area CN. The pervious area CNs are assumed equivalent to desert shrub in poor hydrologic conditions. Composite CNs to use for the design of temporary measures during grading and construction should be computed using Figure 3.22a or b based on the degree of development (impervious area percentage) and the CNs for the newly graded pervious areas.

TABLE 3.19

Runoff Curve Numbers for Virgin Lands[a]

	Curve Numbers for Hydrologic Soil Group			
Cover Type	A	B	C	D
Brush—brush-week-grass mixture with brush the major element	30	48	65	73
Woods—grass combination	32	58	72	79
Woods	30	55	70	77

[a] Listed CN values are for soil cover in good condition. Conservatively, this condition should be assumed in performing runoff calculations for the predevelopment conditions.

When impervious areas are not wholly connected, the runoff is partly dissipated through losses in pervious areas. In such a case, the composite CN would be somewhat smaller than that of fully connected impervious surfaces and may be calculated from Figure 3.22. This figure is applicable for cases where total impervious coverage in a catchment area is less than or equal to 30%.

3.6.5 SCS Peak Discharge Calculations

In the TR-55 method, the peak runoff, Q_p, may be calculated using two different methods: the graphical method and tabular hydrograph method.

3.6.5.1 Graphical Method

The peak runoff in the graphical method is calculated using the following equation:

$$Q_p = q_u \times A \times R \times F_p \tag{3.67}$$

where

q_u = unit peak discharge in cfs/mi^2, abbreviated as csm
A = catchment area, in mi^2
R = runoff depth from a 24-hour storm in inches
F_p = pond and swamp adjustment factor, dimensionless
Q_p = peak runoff, in cfs

The unit peak discharge, q_u, depends on the time of concentration, T_c, and travel time, T_t, and is calculated using the following equation:

$$\log (q_u) = C_0 + C_1 \log T_t + C_2 \log T_c - C_3 \tag{3.68}$$

where C_0, C_1, and C_2 are related to I_a/P. The initial abstraction, I_a, in turn is a function of the soil curve number, and the relation is given in Table 3.20. The time of concentration calculation in the TR-55 method was discussed in Section 3.5.2.

To simplify calculations, the TR-55 manual includes a graphical representation of Equation 3.68 for rainfall distribution types I, IA, II and III. These storm hyetographs and a geographic map, where they apply, were presented in a previous section. A graph of q_u versus T_c and I_a/P for type III storms, which is typical of the East Coast of the United States. including New England, New Jersey, Delaware, and eastern portions of the Carolinas, is shown in Figure 3.23. For graphs of q_u for other storm distributions, the reader is referred to the TR-55 manual. The SCS method of calculating the time of concentration was presented in Section 3.5 and this matter will be discussed further in the next section. Table 3.21 lists values of F_p against the percentage of a catchment area covered by ponds and swamps.

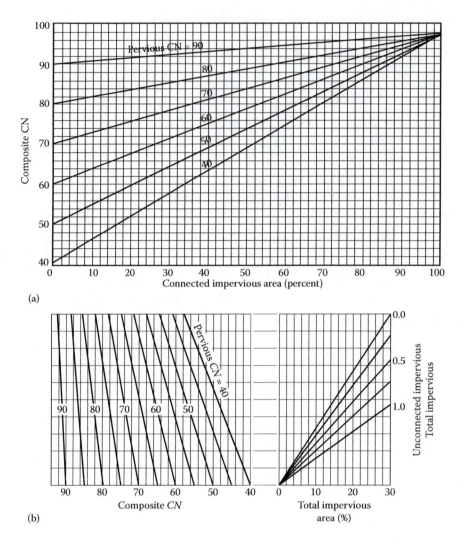

FIGURE 3.22 (a) Composite CN with connected impervious area. (b) Composite CN for partly connected impervious area (total impervious < 30%).

3.6.5.2 Tabular Method

The tabular hydrograph method, as its name implies, gives variation of runoff discharge with time. Included in the TR-55 are tables listing the unit discharges in cubic feet per section (cfs) per square mile per inch of runoff versus the time of concentration and travel time for values of I_a/P ranging from 0.1 to 1.0, T_c ranging from 0.1 to 2.0 h, and travel time of 0 to 3 h. Multiplying the unit hydrograph ordinates in these tables by the watershed area in square miles and the depth of runoff, R, in inches will produce runoff hydrographs. A number of computer programs that facilitate calculations are available commercially. One such earlier program was called Quick TR-55 in MS DOS by Haestad Methods* in Waterbury. A new version of this program in Windows is available at Bentley. StormCad is another software package available for performing the TR-55 calculations.

---

* Haestad Methods was acquired by Bently in 2009.

TABLE 3.20
I_a Values for Soil Curve Numbers

Curve Number	I_a (in.)	Curve Number	I_a (in.)
40	3.000	70	0.857
41	2.878	71	0.817
42	2.762	72	0.778
43	2.651	73	0.740
44	2.545	74	0.703
45	2.444	75	0.667
46	2.348	76	0.632
47	2.255	77	0.597
48	2.167	78	0.564
49	2.082	79	0.532
50	2.000	80	0.500
51	1.922	81	0.469
52	1.846	82	0.439
53	1.774	83	0.410
54	1.704	84	0.381
55	1.636	85	0.353
56	1.571	86	0.326
57	1.509	87	0.299
58	1.448	88	0.273
59	1.390	89	0.247
60	1.333	90	0.222
61	1.279	91	0.198
62	1.226	92	0.174
63	1.175	93	0.151
64	1.125	94	0.128
65	1.077	95	0.105
66	1.030	96	0.083
67	0.985	97	0.062
68	0.941	98	0.041
69	0.899		

3.6.6 SCS Unit Hydrograph Method

The SCS has developed a synthetic hydrograph which is widely used. This hydrograph, similar to the rational hydrograph, has a triangular distribution as shown on Figure 3.24. In this figure,

D = excess rainfall period (not the same as storm duration)
L_a = lag time of catchment area; time from center of rainfall excess to the time to peak
T_p = time to peak = 0.5 D + 0.6 T_c (T_c = time of concentration)
Q_p = peak runoff rate, cfs

The peak runoff rate, Q_p, is given by the following equations:

$$Q_p = 0.208 \frac{A \times R}{T_p} \quad \text{SI*} \tag{3.69}$$

---

* In HECC-22 (2013), pp. 3–33, a coefficient of 2.08 is noted for 1 mm of unit hydrograph. This coefficient should read 0.208.

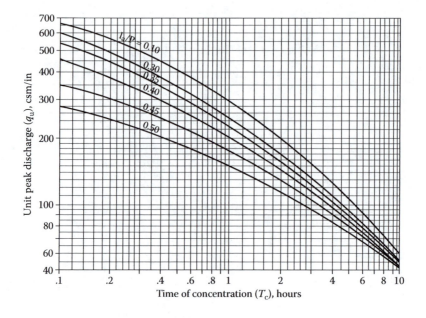

FIGURE 3.23 Unit peak discharge (q_u) for SCS type III rainfall.

TABLE 3.21
Pond and Swamp Adjustment Factor, F_p

Percentage of Pond and Swamp Area	F_p
0.0%	1.00
0.2%	0.97
1.0%	0.87
3.0%	0.75
5.0%[a]	0.72

[a] If the percentage of pond and swamp area exceeds 5%, then it is recommended to route the runoff through ponded areas.

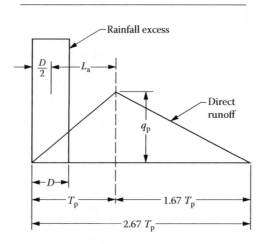

FIGURE 3.24 SCS synthetic triangular hydrograph.

$$Q_p = 484 \frac{A \times R}{T_p} \quad \text{CU} \tag{3.70}$$

where

Q_p = peak flow, m³/s (ft³/s)
A = catchment area in km² (mi²)
R = direct runoff depth (1 for unit hydrograph), mm (in.)
T_p = time to peak, hours

The constant 0.208 (484) is an average value obtained from studies. This method requires calculating T_p. In practice, T_p is approximated at $T_p = (2/3)T_c$.

In flat coastal areas of Maryland, Delaware, Virginia, and New Jersey, where sandy soils are prevalent, the preceding equations exaggerate the peak runoff. In these areas, the SCS unit hydrograph is modified as Delmarva Unit Hydrograph in which the coefficient 284 is substituted for 484 in Equation 3.70. In metric units (Equation 3.69), the coefficient 0.208 is replaced by 0.122.

3.6.7 LIMITATIONS/DRAWBACKS OF TR-55 METHOD

The TR-55 method of runoff calculations is based on a number of assumptions, the most notable of which are

a. The proportionality of retention storage and runoff depth expressed by Equation 3.55. This assumption is contradictory to the physical process in that the larger the retention storage is, the smaller is the runoff.
b. An initial abstraction equal to 2/10 (0.2) of the potential retention is a highly conservation assumption in that it implies that the soil has already lost 80% of its retention capacity from a prior storm before a storm event occurs.
c. The assumption that infiltration rates are far smaller than actual values. Specifically, the actual hydraulic conductivities, which are the minimum value of infiltration rates, are many times greater than the infiltration rates in Table 3.17. This is especially the case for soils of high and moderate permeability.
d. The SCS classifies natural soils into four groups with a small variation in infiltration from one group to another. In reality, the soil permeability varies very widely for different soils (refer to Figure 3.14).
e. Many developed areas are classified as urban land. This soil classification has no defined hydrologic group. As such, a soil group has to be assumed in performing runoff calculations in these areas.
f. There is no justification for sheet flow length, which is limited to 100 ft in the latest time of concentration worksheet, dated June 2004.
g. A comparison of the kinematic wave equations (Equation 3.38) with Equation 3.33 of the TR-5 method indicates the following discrepancies:
   - The exponent $(nL/\sqrt{s})$ in the TR-55 method deviates from the kinematic wave equation.
   - The time of concentration is a function of the rainfall intensity, I, in the kinematic wave equation but of the 2-year rainfall depth, P_2, in the TR-55 method. The exponents for I and P in these equations also differ.
   - The 2-year storm is hypothetical. Its relation with sheet flow length cannot be determined.
h. In the TR-55 method, the same time of concentration is applied for any storm frequency. This is contradictory to the kinematic wave equation and observations that indicate that, as rainfall intensity increases, the flow rate (and sheet flow velocity) increases and the time of concentration shortens.

   i. Worksheet 3 in the TR-55 method allows dividing the sheet flow reach to two segments. Since in Equations 3.33 and 3.34 the sheet flow time of concentration is a nonlinear function of the sheet flow length, such segmenting results in overestimation of the sheet flow time. This is exemplified by the first and second worksheets in Appendix 3A. The former table shows the time of concentration calculations for 100 ft reach separated to 60 ft at 2% slope and 40 ft at 1% slope and the latter relates a 100 ft reach at a uniform slope of 2%. As shown, the calculated sheet flow time of concentration on the latter table, which represents a steeper surface, is 0.16 h shorter than the segmented reach of the smaller slope, which is unrealistic. Using a composite slope equal to 0.016% yields a time of concentration of 0.65, which is more realistic (see the third worksheet in Appendix 3A).

   j. In the TR-55 manual, revised in 1986, the sheet flow length was limited to 300 ft. This length was changed to 150 ft for pervious surfaces and 100 ft for pavements through memoranda. The current worksheet 3 that can be filled in online does not accept a sheet flow length longer than 100 ft long.

Because of the preceding assumptions/limitations, the calculated peak runoff using the TR-55 does not closely relate with actual observations. In performing hydrologic analysis for a number of projects, the author has found that the TR-55 exaggerates peak flows. To exemplify, the author employed the TR-55 method to perform a dam breach analysis for a stream in Ho-Ho-Kus, New Jersey. The computed 100-year flood discharge was found to be nearly equal to the spillway capacity. However, no overtopping of the spillway was reported during Tropical Storm Floyd in September 1999 when the storm surpassed the 24-hour, 100-year storm by nearly 50%.

It is also to be noted that soil maps are commonly in 1 in. = 2000 ft or 1:20,000 (1 in. = 1767 ft) scale; as such, soil types cannot be accurately identified for small catchment areas. Considering this limitation, the author recommends that the TR-55 method not be employed for any area less than 10 acres (4 ha). The method also cannot be employed where soil maps are nonexistent. Also, for urbanized areas where the soil has been disturbed in the past and has been classified as urban land, the author suggests basing the calculations on the soil type(s) occurring in the neighboring areas. Recently, the NRCS has developed a website for the hydrologic soil groups and names in some states, New Jersey included. This software allows preparing user-selected soil maps, as well as soil reports for any US postal address. This resolves concerns with the soils map scale where the NRCS soil maps are available online (see Case Study 3.1).

Example 3.13

A natural watershed is composed of 40% woods in hydrologic soil group (HSG) C and 60% grass-brush also in hydrologic soil group C. Calculate the runoff depth for:

   a. 3.5 in. rainfall
   b. 90 mm rainfall

Consistent with common practice, assume soil cover is in good condition.

Solution

Using Table 3.19:

   CN = 70 for woods, HSG C
   CN = 65 for grass-brush, HSG C

Calculate composite CN:

$$CN = 70 \times 0.4 + 65 \times 0.6 = 67$$

Potential abstractions:

$$S = 1000/CN - 10 = 4.93 \text{ in.} = 125.1 \text{ mm}$$

a. $R = \dfrac{(P-0.2S)^2}{(P+0.8S)} = \dfrac{(3.5-0.2\times4.93)^2}{(3.5+0.8\times4.93)} = 0.85 \text{ in.}$

b. $R = \dfrac{(90-0.2\times125.1)^2}{(90+0.8\times125.1)} = 22.2 \text{ mm}$

Note for this 24-hour storm, nearly 25% of rainfall turns to runoff.

Example 3.14

The catchment area in the previous example is 20 acres. Calculate the peak runoff for the specified storms in that example. The storm distribution is type III and the time of concentration is calculated at 0.6 h.

Solution

For CN = 67, $I_a = 0.985$ in. (see Table 3.20); $A = 20$ acres $= 0.031$ mi^2.

a. $P = 3.5$ in.

$$I_a/P = 0.28$$

$$T_c = 0.6 \text{ h}$$

For $T_c = 0.6$ hour and $I_a/P = 0.28$, interpolate q_u using Figure 3.23:

$$q_u = 325 \text{ cms/in.}$$

Using Equation 3.67, noting that $F_p = 1$:

$$Q = q_u \times A \times R$$

$$= 325 \times 0.031 \times 0.85$$

$$Q = 8.6 \text{ cfs}$$

b. $P = 90$ mm rainfall

$$I_a = 0.985 \times 25.4 \text{ mm/in.} = 25.0 \text{ mm}$$

$$I_a/P = 25/90 = 0.28$$

From Figure 3.23, $q_u = 325$ cfs/mi^2

$$R = 22.2 \text{ mm} = 0.87 \text{ in.}$$

$$Q = 325 \times 0.031 \times 0.87 = 8.8 \text{ cfs}$$

$$Q = 8.8/35.3 = 0.25 \text{ m}^3/\text{s}$$

3.6.8 WinTR-55 Method

There has been a considerable amount of literature presenting comments and discussion on TR-55, which was revised June 1986. To address these comments, the Natural Resources Conservation Service formed a committee in 1998 to revise and modernize TR-55. WinTR-55, which is the new version of TR-55, is the outcome of work by this committee. This new version has been described in a WinTR-55 *User Guide*, issued January 2009. The NRCS has also revised and completely rewritten the TR-55 computer model. The revised software is a Windows-based program that has a significant advantage over the DOS version. The new TR-55 employs the WinTR-20 program as a driving tool to perform hydrologic analysis of small watersheds. A final version of WinTR-55 can be down-loaded free of charge from the NRCS website or simply by searching for WinTR-55. The software is labeled WinTR-55.exe; it was last updated August 5, 2009, and has incorporated a number of improvements over TR-55. Important improvement elements are

a. There is an option for performing calculations in English or metric units.
b. WinTR-55 can perform hydrologic calculations for watersheds up to 25 square miles (65 km^2), consisting of 1 to 10 subareas/reaches.
c. WinTR-55 computer model (similarly to TR-55) can calculate effective CN values for a subarea in which the pavements are partly or wholly (100%) drained to pervious area. The WinTR-55 also allows using customized CN for unconnected impervious surfaces (0% to 100%).
d. It allows using default rainfall depths or user-defined rainfall depths 0–50 in. (0–1270 mm).
e. The software accepts NRCS type I/IA/II/III or user-defined rainfall distribution (hyetograph).
f. WinTR-55 allows the use of custom dimensionless hydrographs such as Delmarva.*
g. The software incorporates two average antecedent runoff conditions.
h. Incorporating the built-in WinTR-20 computer model, the software can perform channel or storage routing based on the Muskingum-Cunge method. Also, using the WinTR-20 computer model, the software performs structure routing based on the storage-indication (level pool) method. Pipe and weir are the only structure types that are modeled by this software.

The WinTR-55 computer software limits the sheet flow to 100 ft (±30 m); however, it allows using $n = 0.8$ for calculating the sheet flow time of concentration in wood-dense underbrush soil cover. Similarly to TR-55, this software accepts a time of concentration of $0.1 \leq t_c \leq 10$ hours. Application of the method is illustrated by the following simple example.

CASE STUDY 3.1

Figure 3.25 shows a drainage area map of a 26.0 acre watershed in Monmouth County in New Jersey. This area, designated as area 2, is hydraulically separated from the downstream water-shed by a roadway which rises over 2 ft above a depression in this area. Figure 3.26 depicts a depression behind the roadway. According to the soil survey maps of Monmouth County, New

---

* The Delmarva unit hydrograph was developed by the New Jersey Department of Agriculture for runoff estimation on agricultural lands located within the coastal zone of flat topography (<5% slope). The method is characterized by a peak rate factor of 284 instead of 484 in Equation 3.70. The hydrograph was adopted by State Soil Conservation Committee on July 12, 2004 through Technical Bulletin 2004-2.0 (http://www.state.nj.us/agriculture//pdf/delmarvabulletin.pdf) Hydrograph 2004-2.0 doc. The state of Maryland has included this hydrograph in Appendix D.14 of its storm water design manual, vol. II.

FIGURE 3.25 A drainage area map of a 26.0 acre watershed in Monmouth County in New Jersey.

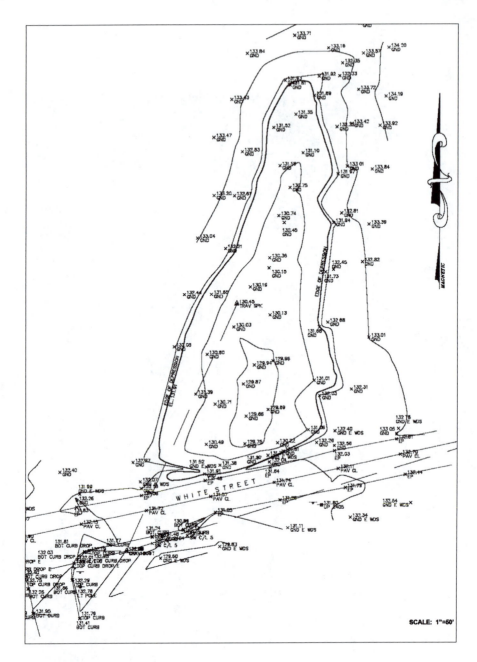

FIGURE 3.26 Topographic map of depression.

Jersey, the soil in this watershed is Lakehurst sand, which is hydrologic soil group (HSG) A. A copy of the soil survey map is provided as Figure 3.27.

Runoff calculations for the watershed are performed using WinTR-55 computer model (WinTR-55.exe). A 0.6 acre area of the watershed, which drains to and is fully retained by a depression on the northwesterly side of the watershed, is excluded from the calculations. Table 3.22 presents calculations for the runoff curve number. The paved area within this watershed, which is wholly drained to the wooded area, is inputted as 100% unconnected in a subtable that does not appear in the WinTR-55 printouts.

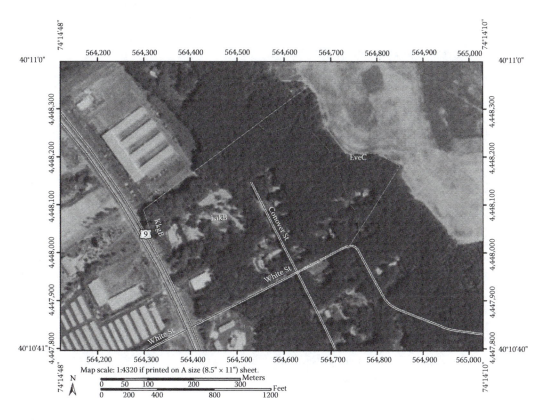

FIGURE 3.27   A copy of the soil survey map.

TABLE 3.22
CN Calculations, Area A2

Land Use	Hydrologic Soil Group	Subarea Acres	CN
Woods (good)	A	23.12	30
Paved (100% unconnected)	A	1.98	33[a]
Dirt road (100% unconnected)	A	0.30	30 (72)
		25.40	30

[a] 8% imperviousness; 100% unconnected.

Table 3.23 includes the time of concentration calculations for the watershed comprising sheet flow and shallow concentrated flow but no channel flow. Presented in Table 3.24 are the 24-hour rainfall depths for the storms of 2-, 10-, and 100-year frequency (note that the 2-year storm rainfall depth is required for T_c calculations). Also listed in this table are the runoff depths for these storm events.

Table 3.25 lists the computed peak flows for the storms of 2-, 10-, and 100-year frequency. As shown, the 2-year storm does not produce any runoff and the peak runoff during the 10-year storm is very small.

To account for the detention/retention effect of the depression and the infiltration loss thereof, routing computations are performed. It is to be noted that WinTR-55.exe is not capable of calculating infiltration or routing through a multistage outlet structure.

TABLE 3.23

Time of Concentration Calculations

Flow Type	Flow Length ft	Slope ft/ft	Manning's n	Flow Velocity ft/s	Travel Time Hour
Sheet flow	100	0.0260	0.40		0.313
Shallow conc. (unpaved)	1450	0.0186	0.05 (N/A)	2.8	0.183
					0.496

TABLE 3.24

24-Hour Rainfall and Runoff Depths

Storm Freq. (yrs.)	Rainfall Depth (in.)	Runoff Depth
2	3.4	0
10	5.2	0.005
100	8.9	0.65

TABLE 3.25

Computed Peak Flows

Storm Freq. (yrs.)	Peak Runoff (cfs)
2	0
10	0.05
100	4.79

Table 3.26 lists the stage-storage-discharge relation for the depression. The storage calculations are prepared based on topographic data shown in Figure 3.26, and the listed discharges are calculated using an estimated infiltration rate of 8 in./h through the depression. Note that the estimated infiltration rate is conservative in that for sandy soils, such as the one present at the site, the actual infiltration may be greater than 20 in./h. In fact, a percolation test in the vicinity of the depression area shows an infiltration rate of 19 in./h.

$$Q = \text{area} \times 8 \text{ in.}/(12 \times 3600) = 1.852 \times 10^{-4} \times \text{area}$$

TABLE 3.26

Stage-Storage-Discharge Infiltration Relation

Elev.	Area ft²	Avg. Area ft²	Δ Vol. ft³	Vol. ft³	Q cfs
129.75	0			0	0
		765	191[a]		
130.00	1530			191	0.28
		5370	5370		
131.00	9210			5561	1.71
		13,445	12,235		
131.57	17,680			17,796	3.27

[a] Rounded to whole number.

The inundation depth and retention volume in the depression are computed by routing the runoff hydrograph (calculated using WinTR-55) through the depression, represented by Table 3.26. For brevity, only the computation results for the 100-year storm are presented herein:

Max. water surface elev. = 131.32 ft
Max. retention storage = 9839 cf < 17,796
Max. discharge (infiltration) = 2.26 cfs < 3.27

The preceding figures indicate that the 100-year storm does not overtop the roadway. Note that the previously listed discharge represents the maximum infiltration through the inundated area, which is less than the infiltration capacity of the depression when full.

3.7 UNIVERSAL RUNOFF METHOD

The shortcomings of the rational and the SCS methods were discussed in previous sections. As noted, the rational method is based on assumptions of linearity and proportionality, which do not properly account for initial losses and which contradict actual observations—for example, see Table 3.15. The SCS method is also developed based on a number of assumptions and approximations, many of which do not agree with the physical rainfall–runoff process. The universal method proposed by the author in 2009 and further developed in the first edition of this book (2011) and Pazwash (2013) provides a realistic representation of the rainfall–runoff relation. This method is a physically based model that directly accounts for initial losses and infiltration. It requires estimating initial abstractions based on the soil cover and performing either a soil gradation analysis or a permeability test to determine infiltration.

3.7.1 Lag Time between Rainfall and Runoff

Figure 3.28 depicts a hyetograph of a uniformly distributed rainfall and initial losses, which include interception, surface depression, and infiltration for a pervious area. For a paved area, this figure may be modified by removing the interception and infiltration lines and adjusting the surface depression, also referred to as depression storage. While the rainfall intensity and infiltration rate are expressed in mm/h (in./h) in this figure, the initial abstractions including interception and depression storage have a length scale and are expressed in mm (in.). The infiltration curve on the figure is conservatively shown as having a constant rate equal to hydraulic conductivity, which is the minimum rate of infiltration and which occurs under the condition of surface saturation. For simplicity, interception and depression storage are shown as rectangular blocks in the figure.

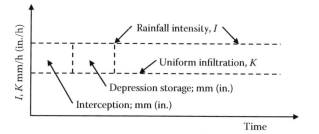

FIGURE 3.28 Representation of universal rainfall–runoff relation.

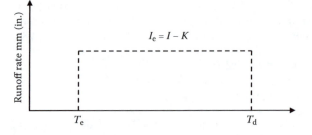

FIGURE 3.29 Excess rainfall from a pervious area.

Accounting for the initial losses, the net rainfall excess will be uniformly distributed beyond a lag time, which is shown in Figure 3.29. This figure indicates that no runoff occurs until the initial losses are satisfied. The lag time from the start of rainfall to the onset of runoff is expressed by

$$IT_e = (X_t + S_d) + KT_e \tag{3.71}$$

or

$$T_e = \frac{L}{(I - K)} \tag{3.72}$$

where

I = rainfall intensity, mm/h (in./h)
X_t = interception by trees and vegetation, mm (in.)
S_d = depression storage, mm (in.)
$L = X_t + S_d$ = initial abstractions (excluding infiltration), mm (in.)
K = permeability, mm/h (in./h)
T_e = time for runoff to begin (lag time between the beginning of rainfall and start of runoff), hours
T_d = storm duration, hours

Typical values of interception and depression storage are listed in Table 3.27. The term, L, in this table represents initial losses, which is the sum of interception and depression storage. The values in this table are based on observations and reports covered in Sections 3.2.1 and 3.2.2.

The author suggests that permeability be determined based on either the results from a permeability test or the USDA Soil Texture Permeability Rating K_0 to K_5 in Table 3.8. Considering

TABLE 3.27

Typical Values of Interception and Depression Storage (Initial Losses)

Surface Cover	X_t Interception mm (in.)	S_d = Depression Storage mm (in.) Steep Slope	Mild Slope	Initial Losses $L = X_t + S_d$
Woodlands	2.5–7.5 (0.1–0.3)	7.5 (0.30)	20 (0.5)	10–25 (0.4–1.0)
Grass–lawn	2.0–5.0 (0.08–0.2)	5 (0.20)	10 (0.4)	7.0–15 (0.3–0.6)
Landscape	2.5 (0.1)	20 (0.8)	50 (2)	25–50 (1–2)
Porous pavement, pavers[a]	0	2.0 (0.08)	3.0 (0.12)	2.0–3.0 (0.08–0.12)
Pavement	0	1.0 (0.04)	2.0 (0.08)	1.0–2.0 (0.04–0.08)

[a] Pavers with more than 20% openings may be considered as a pervious surface, having the same infiltration as grass in soils of high permeability. However, in soils of low permeability, pavers have a higher infiltration rate than lawn.

antecedent moisture condition, the initial losses for pervious areas may be adjusted, applying a factor of less than one to their potential values listed in Table 3.27. Thus, Equation 3.72 becomes

$$T_c = M_c \frac{L}{(I - K)} \tag{3.73}$$

where M_c = antecedent moisture factor is the portion of initial abstractions (interception and depression storage capacity) that is still available after a previous storm.

The author suggests $M_c = 0.5$ as a conservative estimate of the initial abstractions for pervious surfaces. When $I \leq K$, the permeability rate becomes limited to the rainfall intensity and the preceding equation implies that no runoff occurs.

3.7.2 Runoff Volume and Discharge from Pervious Surfaces

According to Figure 3.29, the depth of excess rainfall, namely runoff, from a pervious area is calculated by the equation

$$R = (T_d - T_e) \cdot (I - K) \tag{3.74}$$

or

$$R = I(T_d - T_e) - K(T_d - T_e) \tag{3.75}$$

where R = depth of runoff, mm (in.).

In Equation 3.75, the first term on the right-hand side represents the gross rainfall depth after the initial losses are satisfied and the second term accounts for infiltration losses from the pervious surface, after the runoff begins. For $I \leq K$, the infiltration becomes limited to rainfall intensity and Equation 3.74 indicates $R = 0$.

The runoff volume is calculated from the product of runoff depth times and the area of pervious surface, as follows:

$$V_p = 0.001 \, AR = 0.001 \, A \cdot (I - K) \cdot (T_d - T_e) \qquad \text{SI} \tag{3.76}$$

$$V_p = AR/12 = 0.083 \, A \cdot (I - K) \cdot (T_d - T_e) \qquad \text{CU} \tag{3.77}$$

where
V = runoff volume, m^3 (ft^3)
A = area of pervious surface, m^2 (ft^2)

Considering the uniform variation of excess rainfall, the peak discharge may be approximated by a trapezoidal geometry shown in Figure 3.30. In this figure, T_c is the time of concentration of catchment area and the area under the trapezoid equals the runoff volume given by Equations 3.76 and 3.77. According to this figure, the time of concentration T_c equals $T_e + T_t$, where T_t is travel time through the catchment area and the peak rate of runoff is given by

$$Q = \frac{V}{3600(T_d - T_e)} \tag{3.78}$$

where the factor 3600 represents seconds per hour. Combining Equations 3.76 through 3.78 results in

$$Q = 0.278 \times 10^{-6} \times A \, (I - K), \qquad \text{SI} \tag{3.79}$$

$$Q = 2.31 \times 10^{-5} \times A \, (I - K), \qquad \text{CU} \tag{3.80}$$

where Q = peak discharge m^3/s (cfs) and other parameters are as defined previously.

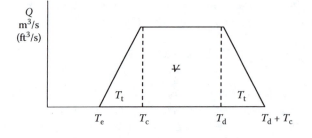

FIGURE 3.30   Peak runoff distribution.

For a catchment area composed of different types of pervious surfaces, such as lawn, woods, and landscape, Equations 3.76 through 3.78 can be amended as follows:

$$T_e = M_c \, \Sigma A \cdot L / \Sigma A \, (I - K) \tag{3.81}$$

$$V = 0.001 \, [\Sigma A \cdot (I - K)] \cdot (T_d - T_e) \qquad \text{SI} \tag{3.82}$$

$$V = 0.083 \, [\Sigma A \cdot (I - K)] \cdot (T_d - T_e) \qquad \text{CU} \tag{3.83}$$

$$Q_p = V / [3600 \, (T_d - T_e)] \tag{3.84}$$

where
  ΣA = all pervious surfaces combined
  K = soil permeability relating to each A

and other parameters were described previously.

3.7.3   LAG TIME AND RUNOFF EQUATIONS FOR IMPERVIOUS SURFACES

For an impervious surface, the rainfall–runoff relation is depicted in Figure 3.31.
  In this case the lag time is given by the following equation:

$$T_e = L / I \tag{3.85}$$

where
  $L = S_d$
  I = rainfall intensity, mm/h (in./h)

Referring to Figure 3.31, the runoff depth is given by the equation:

$$R = I \, (T_d - T_e), \text{ mm (in.)} \tag{3.86}$$

where
  T_d = rainfall duration, h
  T_e = lag time calculated from Equation 3.85, h

FIGURE 3.31   Rainfall–runoff relation for impervious surfaces.

In this case, equations for runoff volume and peak discharge are as follows:

$$V_i = 0.001 \ I \cdot A \ (T_d - T_c) \qquad \text{SI} \tag{3.87}$$

$$V_i = 0.083 \ I \cdot A \ (T_d - T_c) \qquad \text{CU} \tag{3.88}$$

$$Q_i = V/[3600(T_d - T_c)] \qquad \text{SI/CU} \tag{3.89}$$

3.7.4 EQUATIONS FOR COMPOSITE SURFACES

For a catchment area comprising pervious and paved surfaces, the lag time and runoff volume depend on whether or not pervious and impervious surfaces are connected. When the surfaces are unconnected, the lag time, runoff volume, and peak discharge are calculated separately for each surface using Equations 3.73 through 3.89. Then, the composite runoff volume and peak discharges for the entire area are obtained by adding of the volumes and discharges from pervious and impervious areas, as follows:

$$V = V_p + V_i \tag{3.90}$$

$$Q = Q_p + Q_i \tag{3.91}$$

The calculation process is illustrated by an example later in this section.

If the runoff from an impervious surface is directed to a pervious surface, that runoff may be partly or wholly retained by the pervious surface and infiltrated over time. In this case, the equation for the lag time can be derived from a graph similar to Figure 3.25, with the following result:

$$T_c = \frac{\sum M_c L \cdot A}{\sum A(I - K)} \tag{3.92}$$

In this equation $K = 0$ for impervious surfaces. Similarly to the previous case, an antecedent moisture factor M_c may be applied to pervious surfaces.

The runoff volume and peak discharge are given by the following equations:

$$V = 0.001 \ [\Sigma A \ (I - K)] \cdot (T_d - T_c) \qquad \text{SI} \tag{3.93}$$

$$V = 0.083 \ [\Sigma A \ (I - K)] \cdot (T_d - T_c) \qquad \text{CU} \tag{3.94}$$

$$Q = V/[3600 \ (T_d - T_c)] \tag{3.95}$$

or

$$Q = c_x \ [\Sigma A \ (I - K)]/3600 \qquad \text{SI/CU} \tag{3.96}$$

where

$$c_x = 0.001 \text{ SI}; \quad c_x = 0.083 \text{ CU}$$

where all parameters are as described previously. M_c for pavements may be taken as 1.0.

For a catchment composed of various types of pervious and impervious surfaces, the equations for the lag time, runoff volume, and peak discharge are organized in Section 3.7.4.1.

3.7.4.1 Universal Runoff Equations for Composite Surfaces

1. Unconnected pervious and impervious areas
   a. Pervious areas

$$T_c = \frac{\Sigma M_c L \cdot A}{\Sigma A \ (I - K)}$$

$$V_p = c_x [\Sigma A \ (I - K)] \cdot (T_d - T_c) \qquad \text{SI}$$
$$Q_p = V_p/3600(T_d - T_p) \qquad \text{SI, CU}$$

or

$$Q_p = c_x [\Sigma A \ (I - K)]/3600$$

b. Impervious areas

$$T_e = \frac{\Sigma L \cdot A}{I \Sigma A}$$

$V = c_x I (\Sigma A) \cdot (T_d - T_e)$ SI, CU
$Q_i = V_p/3600(T_d - T_p)$ SI, CU
$V = V_p + V_i$
$Q = Q_p + Q_i$

2. Impervious areas drain to pervious areas

$$T_e = \frac{\Sigma M_c L \cdot A}{\Sigma A (I - K)}$$

$V = c_x[\Sigma A (I - K)] \cdot (T_d - T_e)$ SI, CU
$Q_i = V/3600(T_d - T_e)$ SI, CU
$R = V/\Sigma A$ SI, CU

Definitions:

A = area in m^2 (ft^2)
I = rainfall intensity, mm/h (in./h)
K = permeability, mm/h (in./h)
S_d = surface retention, mm (in.)
X_i = plants interception, mm (in.)
$L_i = X_i + S_d$ initial losses for plants, mm (in.)
$L_i = S_d$ surface retention for pavements, mm (in.)
R = runoff depth, mm (in.)
V = runoff volume, m^3 (ft^3)
Q = discharge, m^3/s (ft^3/s)
M_c = antecedent moisture condition factor; $M_c = 1.0$ impervious; $M_c = 0.5$ pervious
T_e = runoff lag time, h
T_d = rainfall duration, h
$c_x = 0.001$ SI, $1/12 = 0.083$ CU

Example 3.15

A 1500 m^2 residential lot is currently wooded. Based on a soil gradation analysis, the site soil is characterized as sandy clay loam. Calculate the runoff volume and the peak runoff for the following rainfalls:

    a. 10 year – 10 minute storm of 150 mm/h intensity
    b. 10 year – 60 minute storm of 50 mm/h intensity

Solution

From Table 3.22, estimate initial losses L_p at 20 mm. According to Figure 3.14, sandy clay loam has K_2 permeability rating; $K = 15$–50 mm/h (see Table 3.8).
    Estimate $K = 30$ mm/h.

    a. 10 minute, 150 mm/h storm
       Using Equations 3.73 through 3.76 to calculate the lag time, runoff depth, and runoff volume, respectively:

$$T_e = \frac{0.5 \times 20}{(150 - 30)} = 0.0833\,h$$

$$R = (0.1667 - 0.0833)(150 - 30)$$

$$R = 10.04\ mm$$

$$V = 0.001\,R \cdot A = 15.07\ m^3$$

Peak runoff, using Equation 3.79 is

$$Q = 0.278 \times 10^{-6}\,(150 - 30) \times 1500 = 0.05 \text{ m}^3/\text{s}$$

Alternatively, use Equation 3.78:

$$Q = \frac{V}{3600\,(T_d - T_e)} = \frac{15.07}{\left[3600\,(0.1667 - 0.0833)\right]}$$

$$Q = 0.05 \text{ m}^3/\text{s}$$

b. 60 minute, 50 mm/h rainfall

$$T_e = 0.5 \times 20/(50 - 30) = 0.5 \text{ h} = 30 \text{ min}$$

Runoff begins 30 minutes past the start of rainfall. Runoff depth is

$$R = (1 - 0.5)(50 - 30) = 10.0 \text{ mm}$$

Runoff volume is

$$V = 0.001\,R \cdot A = 0.001 \times 10 \times 1500 = 15.0 \text{ m}^3$$

Peak rate of runoff is

$$Q = 0.278 \times 10^{-6} \times (50 - 30) \times 1500 = 0.0125 \text{ m}^3/\text{s}$$

Note: This example shows that, contrary to the rational method and the TR-55 method, the runoff volume may not necessarily increase with the storm duration. The example further reveals that the peak rate of runoff does not increase proportionally to the rainfall intensity.

Example 3.16

Calculate the peak and volume of runoff for the single-family home shown in the following during a 30-minute storm having 3.25 in./h (83 mm/h) intensity. Perform the calculations for the following cases:

a. Direct discharge from pavements, and
b. Roof and driveway drain to lawn and landscape

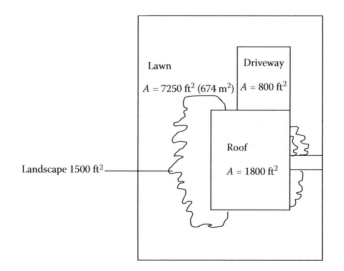

Given:

$L_i = 0.5$ in. (12.5 mm) lawn
$L_i = 1.5$ in. (38 mm) landscape
$S_d = 0.12$ in. (3 mm) driveway
$S_d = 0.04$ in. (1 mm) roof

Landscape:

$A = 1500$ ft² (139 m²)
$K = 6$ in./h (150 mm/h)

Solution

Calculations are performed using equations in Section 3.7.4.1. Spreadsheet tabulations are performed separately for cases "a" and "b."

Case A: Unconnected pervious and impervious areas

Pervious areas

A	L	K	I	M_C	$M_C LA$	$A(I - K)$
7250	0.5	3	3.25	0.5	1812.5	1812.5
1500	1.5	6	3.25	0.5	1125	−4125
				Totals	2937.5	−2312.5

$\Sigma A\ (I - K) < 0$	
No runoff from lawn and landscape	

Impervious areas

A	L	I	M_C	$M_C LA$	AI
1800	0.04	3.25	1.00	72	5850
800	0.08	3.25	1.00	64	2600
2600		Totals		136	8450

$T_i = \dfrac{136}{8450} = 0.02$ h		
$V = 340.8$ cf*	$V = 3.25 \times 2600 \times (0.5 - 0.02)/12$
$Q = 0.196$ cfs	$Q = V/[3600 \times (0.5 - 0.02)]$

Case B: Connected

A	L	K	I	M_C	$M_C LA$	$A(I - K)$
7250	0.5	3.0	3.25	0.5	1812.5	1812.5
1500	1.5	6	3.25	0.5	1125	−4125
1800	0.04	0	3.25	1.0	72	5850
800	0.08	0	3.25	1.0	64	2600
				Totals	3073.5	6137.5

$T_e = \dfrac{3073.5}{6137.5} = 0.50$ h $= 30$ min		
$V =$ 0 cf	no runoff from home
$Q =$ 0 cfs	

* Calculated using $c_x = 1/12$ (0.083).

3.7.5 Universal Method Application to Nonuniform Rainfall

For a nonuniformly distributed storm, the rainfall can be divided into discrete segments of uniform rainfall intensity. Then the universal runoff equations are applied to each segment. If initial losses are satisfied for any part of drainage area during the first segment, then no losses other than infiltration will be included in the runoff calculations for that part in other segments. The following example illustrates the application of the method for a rainfall divided into two segments.

Example 3.17

Calculate the volume and peak runoff in Example 3.17 for a 30-minute storm falling at 4.0 in./h in the first 15 minutes and 2.5 in./h in the next 15 minutes. Perform calculations for case b (pervious areas and pavements hydraulically connected).

Solution

a. The rainfall depth during the first 15 minutes is 4 in. × 15/60 = 1.0 in. Therefore, the initial losses for lawn and pavements are fully satisfied during the first part of rainfall. Also, accounting for antecedent moisture factor $M_c = 0.5$, all of the available initial losses for landscape are met. (Note: 1.5 × 0.5 = 0.75 in.)

A ft²	L in.	K in./h	I in./h	M_c	M_cLA ft² in.	A(I – K) ft² in./h
7250	0.50	3.0	4	0.5	1812.5	7250
1500	1.50	6.0	4	0.5	1125.0	–3000
1800	0.04	0	4	1.0	72.0	7200
800	0.08	0	4	1.0	64.0	3200
Total					3073.5	14,650

$$T_e = \frac{\sum M_c LA}{\sum A(I-K)} = \frac{3073.5}{14{,}650} = 0.21\text{h}$$

$$V = 0.083 \times \Sigma A(I - K) \cdot (T_d - T_e)$$

$$= 0.083 \times 14{,}650\,(0.25 - 0.21) = 48.9 \text{ ft}^3$$

$$Q = \frac{V}{3600} \cdot (T_d - T_e) = \frac{48.9}{3600\,(0.25-0.21)} = 0.34 \text{ ft}^3/\text{s}$$

b. For the second 15 minutes, $I = 2.5$ in./h.

During this period, the losses are solely due to infiltration. Spreadsheet calculations for this period are organized in the following table:

A ft²	L in.	K in./h	I in./h	M_c	M_cLA ft² in.	A(I – K) ft² in./h
7250	0	3.0	2.5	0.5	0	–3625
1500	0	6.0	2.5	0.5	0	–5250
1800	0	0	2.5	1.0	0	4500
800	0	0	2.5	1.0	0	2000
Total					0	–2375

Since $\Sigma A(I - K) < 0$, there is no runoff during the second period.

Note: The depth of rainfall in this storm is equal to 1.625 in., which is the same as that of previous example. However, this nonuniform storm produces a larger peak but smaller runoff volume than the uniform rainfall distribution of the previous problem. It is evident that neither rational method nor the TR-55 method is capable of performing such calculations.

3.8   STORM WATER MANAGEMENT MODELS

Storm water management models can be classified into two types: flow quantity models and flow quantity-runoff quality models. The former type, which provide no information on storm water quality, include the rational method, SCS method, universal method (presented in this book), and HEC-1 and HEC-HMS computer software. This type also includes a number of empirical synthetic hydrographs and regression equations such as the Snyder unit hydrograph and the USGS unit hydrograph. The latter type generally includes comprehensive computer models, which calculate not only the flow rates, but also estimate amounts of pollutants carried by runoff. The most widely used comprehensive models in the United States are Storm Water Management Model (SWMM) developed in the 1980s by the EPA, WinSLAMM (Source Loading and Management Model) also developed by the US EPA in 2007, and SWAT, developed by the US Department of Agriculture Research Services and Texas A&M AgriLife Research in 2000.

The Snyder synthetic unit hydrograph and the US Geological Survey (USGS) unit hydrograph are among the unit hydrographs widely employed in practice. These two unit hydrographs are described in Appendix 3B and Appendix 3C, respectively. The rational and TR-55 models were described in previous sections. The HEC-1 and its Windows version HEC-HMS, which were developed by the US Army Corps of Engineers (COE), are commonly employed for watershed hydrologic analysis. The HEC-1 model can also be used for detention basin routing computations. However, more user friendly software, such as Bentley's Haestad Methods Pond Pack and HydroCAD, are available for hydrologic calculations and routing computations and are commonly employed in storm water management practices. The USGS has developed and compiled regression equations for estimating the peak flows for 2- through 500-year frequency storms throughout the United States. The USGS regression equations are available for both rural areas and urban areas.

The rural equations are based on watershed and climatic characteristics within specific regions in each state. The regression equations are in the following general form:

$$Q_T = kA^aB^bC^c \tag{3.97}$$

where
Q_T = T-year rural peak flow
k = regression constant
a, b, and c = regression coefficients
A, B, and C = basin characteristics

The urban peak flow equations include seven parameters, one of which is the rural peak flow for the region calculated using Equation 3.97. These equations are described in the *Urban Drainage Design Manual* (HEC-22, 2013), a copy of which is included in Appendix 3D at the end of this chapter.

The USGS has also prepared a computer program known as StreamStats that provides not only peak discharges, but also basin characteristics, such as drainage area, length, and slope for any stream basin. The program has been fully implemented in a number of states and efforts are underway for a long-term goal of national coverage. New Jersey is among those states where the program is fully operational. The program can be downloaded from the following USGS website: http://water.usgs.gov/osw/streamstat.

The information for a watershed can be readily accessed inputting the latitude and longitude of a point of interest in a watershed or surfing for the location on the interface map. Appendix 3E presents the application of this program to a stream in New Jersey. As shown, the StreamStats program output includes a drainage area map of the stream, and basin information such as the stream length and slope, percentage of the area covered by urban use and peak discharges for the storms of 2- through 500-year frequency.

The EPA SWMM is a dynamic rainfall-runoff simulation model. This model can be used for single-event or long-term (continuous) simulation of runoff quantity and quality from urban areas. SWMM was first developed in 1971 and has since undergone several major upgrades. The latest edition, version 5, which runs under Windows, EPA SWMM 5, was produced by the Water Supply and Water Resources Division of the US EPA's National Risk Management Laboratory with assistance from the CDM, Inc. consulting firm. The SWMM 5 allows the user to perform water quality simulations and view the results in a variety of formats. These include color-coded drainage area maps, time series tables and graphs, profile plots, and statistical frequency analysis. The runoff component of the SWMM generates rainfall–runoff relation for a collection of subcatchment areas, transports the runoff through pipes and channel conveyance system, storage/treatment devices, pumps, and regulators. SWMM calculates the flow rate, flow depth, and quantity of runoff in each pipe and channel and tracks the quantity and quality of runoff generated within each subwatershed during a simulation period or multiple time intervals.

WinSLAMM was originally developed in the 1970s to investigate the relationships between sources of urban runoff pollutants and runoff quality. It has since been continually expanding. Now it includes a wide variety of source conditions and outfall control practices such as infiltration basin, wet pond, porous pavement, street cleaning, and grass swale. Unlike many other models, WinSLAMM is applicable to frequent and relatively small flows. As such, this model is most suited for water quality analysis. It can predict the sources of runoff pollutants and flows for the rainfalls that are of interest to water quality analysis.

WinSLAMM has been used in conjunction with SWMM to investigate the impact of urban runoff on receiving waters. A more refined relation between these two models is currently under development. Once the two models are merged, this application will be implemented.

SWAT (Soil and Water Assessment Tool) is a river basin scale model that quantifies the impact of land development practices in large, complex watersheds on flooding, erosion, and water quality (sediment, nutrients, and pesticides). This model, last revised in 2012, is supported by the USDA Agricultural Research Services at the Grassland, Soil and Water Research Laboratory in Temple, Texas. Unlike traditional methods that use design storms, this model allows a user to predict continuous impacts of flow. The model requires climatic (rainfall and temperature) data, topographic information, and historic and future land use patterns. The city of Austin in Texas has used this model to study Walnut Creek Watershed in Austin at the creek's confluence with the Colorado River (Lopez et al., 2012).

PROBLEMS

3.1 An undisturbed rock sample has an oven-dry weight of 425.30 g. After saturation with kerosene its weight is 476.19 g. It is then immersed in kerosene and displaces 196.07 g of kerosene. What is the porosity of the sample?

3.2 A mildly sloped, undisturbed catchment area is covered with approximately 50% woods and 50% grass. For a 60-minute storm of 50 mm/h intensity, calculate:
   a. Interception using the Brooks et al. method. Estimate $S = 4$ mm and neglect evaporation during the storm
   b. Depression storage using Equation 3.3 with $S_d = 20$ mm.

3.3 Solve Problem 3.2, using: $I = 2$ in./h, $S = 0.15$ in., and $S_d = 0.8$ in.

3.4 Calculate infiltration volume for a sandy clay loam soil, having $f_o = 2.5$ in./h, $f_c = K = 1$ in./h during a 2-hour storm of $I = 1.6$ in./h intensity. Use:
   a. Horton equation with $\alpha = 1$ hour^{-1}
   b. Horton equation and Table 3.6

3.5 Redo Problem 3.4 for the following parameters: $f_o = 62.5$ mm/h, $f_c = 25$ mm/h, and $I = 40$ mm/h.

3.6 The infiltration rate in a small area was found to be 100 mm/h at the beginning of a 6-hour storm and it decreased to an equilibrium of 20 mm/h. A total of 250 mm of water was infiltrated during the storm period. Calculate the value of α in Horton's equation.

3.7 Average rainfall intensities during each hour of a 4-hour storm over a 100 ha watershed were 30, 50, 25, and 15 mm/h, respectively. If the infiltration φ index for the storm was 20 mm/h, calculate the direct runoff from the watershed.

3.8 The rainfall intensities of a 60-minute storm falling on a 10-acre basin were as follows:

Time, min	15	30	45	60
Intensity, in./h	3.0	2.5	4.0	1.5

a. Calculate the total rainfall depth in inches.
b. Calculate the φ index if the net rainfall (overland flow) from the basin was measured to be 1.5 in.
c. Calculate the volume of runoff from the basin.

3.9 Rain falls uniformly on a sandy loam soil at a rate of 30 mm/h for 2 hours. Using the Green-Ampt method, calculate:
a. Time for the soil surface to become saturated
b. The amount of infiltration at the end of the storm period
Assume initial moisture content at 0.25 and fp = 50 mm/h.

3.10 Solve Problem 3.9 for a 2-hour storm of 1 in./h intensity; $f_p = 2.0$ in./h.

3.11 In Problem 3.9, rain falls at a rate of 15 mm/h during the first hour and 40 mm/h during the second hour. Calculate the time to surface saturation and the infiltration amount.

3.12 A falling head permeameter, similar to that shown in Figure 3.12b, is used to measure permeability of a soil sample. The head, H, is measured to drop from 400 to 150 mm in 90 seconds. Calculate the soil permeability. The permeameter dimensions are sample length, $L = 120$ mm; $r = 10$ mm; and $R = 100$ mm.

3.13 Calculate the soil permeability of Problem 3.12 for a head drop of 14 in. to 6 in. in 90 seconds. The permeater dimensions are $L = 5$ in.; $r = 0.4$ in.; and $R = 4$ in.

3.14 A 25-acre wooded area receives 2 inches of rain uniformly in 1 hour. If the time of concentration is 30 minutes, what is the peak flow rate at the watershed outlet? Base your calculations on the rational method.

3.15 Redo Problem 3.14 for 10 ha and 50 mm rainfall.

3.16 The rain falls uniformly for 1 hour at the rate of 2 in./h on a 100-acre watershed composed of: 25 acres, $C = 0.3$; 20 acres, $C = 0.4$; 40 acres, $C = 0.6$; and 15 acres, $C = 0.90$. Calculate the peak rate of runoff, assuming a time of concentration of 1 hour. What will be the peak discharge, if the time of concentration is 45 minutes?

3.17 Rain falls uniformly at a rate of 50 mm/h for 1 hour on a 40 ha watershed which comprises 10 ha having $C = 0.3$; 10 ha with $C = 0.4$; and 20 ha with $C = 0.6$. Calculate the peak rate of runoff assuming a time of concentration of 1 hour.
   If the time of concentration is 45 minutes, what will be the peak discharge? Plot a hydrograph for this storm.

3.18 A watershed contains two subareas with the following hydrologic characteristics:
a. Subarea 1: $A = 80$ acres, $C = 0.50$, $T_c = 30$ min
b. Subarea 2: $A = 100$ acres, $C = 0.35$, $T_c = 40$ min
   The 25-year rainfall intensities of the 30- and 40-minute duration storms are 3.6 and 3.0 in./h, respectively. Calculate the peak rate of runoff from each subarea and the composite discharge from the watershed.

3.19 In Problem 3.18, subareas 1 and 2 are 32 and 40 ha, respectively. Calculate the 25-year peak runoff from
   a. Each subarea
   b. The watershed
      The 25 year rainfall intensities of the 30- and 40-minute duration storms are 90 and 75 mm/h, respectively.

3.20 A 150-acre watershed includes two subareas: Subarea A, which includes 30% of the watershed, has a time of concentration of 20 min; subarea B has mild slope comprising 70% of the watershed with a time of concentration of 60 min. The abstraction can be taken as 1 in./h. Calculate the 25-year peak flow. Use the following IDF relation:

$$I = (30 \times T^{0.22})/(t_d + 18)^{0.75}$$

where I = rainfall intensity, in./h; T = return period in years; and t_d = rainfall duration in minutes. Assume a linear discharge at the watershed outlet. State any other assumptions used.

3.21 Name the factors on which the soil curve number, CN, in the TR-55 method depends.

3.22 A drainage area has a composite soil curve number of 55. Based on the SCS method, how much rain must fall before any runoff occurs?

3.23 A natural watershed has the following characteristics:
   a. Woods in good condition, soil group B, covering 40% of the area
   b. Grass-brush in good condition, soil group C, covering 60%
      Calculate the runoff depth, in centimeters, for an 11.5 cm rainfall in 24 hours. Use the SCS type III storm distribution.

3.24 Calculate runoff depth in inches in Problem 3.23 for a 24 hour-5 inch rainfall.

3.25 Using the TR-55 method, calculate the time of concentration for a drainage area having the following characteristics:
   a. Sheet flow; dense grass, length $L = 100$ ft; slope $S = 2.0\%$, 2 year-24 hour rainfall $P_2 = 3.5$ in.
   b. Shallow concentrated flow; unpaved, length $L = 250$ ft, slope $S = 1\%$
   c. Stream flow; Manning's $n = 0.06$, flow area $A = 9$ ft^2, wetted perimeter $P = 11$ ft, slope $S = 0.7\%$, length $L = 700$ ft.
      The watershed area is 10 acres.

3.26 Calculate the time of concentration of the drainage area in Problem 3.25 using the FHWA method. Base your calculation on rainfall intensity of 10-year storm frequency in Figure 3.1b.

3.27 The drainage area described in Problem 3.25 is composed of 40% wood, 30% grass, and 30% pavement. For this area, calculate the depths of runoff and the peak rates of flow for the 2-, 10-, and 100-year storms of 24-hour duration. The 24-hour storms in the area are $P_2 = 3.5$ in., $P_{10} = 5.3$ in., and $P_{100} = 8.7$ in. The storm is type III distribution and the soil is group C.

3.28 Use the rational method to calculate the peak rates of flow for the drainage area described in Problem 3.27 for the storms of 2-, 10-, and 100-year frequency. Use the New Jersey rainfall intensity–duration–frequency (IDF) curves of Figure 3.1b for this area. Base your calculations on the following runoff coefficients:
   a. $C = 0.95$ pavement
   b. $C = 0.30$ lawn
   c. $C = 0.25$ wooded area

3.29 A 10 ha watershed is composed of 60% woods, 20% lawn, and 20% pavement. The soil is uniformly hydrologic group B. Calculate the depth, volume, and peak runoff for a 24-hour, 150 mm type III storm. Assume a time of concentration of 45 minutes.

3.30 The drainage area of Problem 3.27 is composed of loamy sand soil and the pavements drain onto lawn and wooded areas. Calculate the volume and the peak runoff for the storms of 2-, 10-, and 100-year frequency and 45-minute duration based on the universal method. Use the IDF curves in Figure 3.1b.

3.31 Calculate the peak and volume of runoff in the previous problem if the pavements are not drained onto lawns and woodland.

3.32 A mildly sloped 10 ha watershed includes 4 ha of woodland, 3.5 ha of lawn, and 2.5 ha of pavement. The soil is sandy loam. Using the Universal Method, calculate the volume and peak runoff rate for a 45-minute storm of 75 mm/h intensity for the following cases:
   a. Pavements are hydraulically separate from pervious areas.
   b. Pavements are drained to lawn and woodland.

3.33 A single-family home includes a 130 m^2 dwelling, 80 m^2 driveway and paved patio, 275 m^2 lawn, and 150 m^2 landscape. Initial abstraction and hydraulic conductivity of lawn and landscape are as follows:
   a. $L = 10$ mm, $K = 50$ mm/h—lawn
   b. $L = 30$ mm, $K = 75$ mm/h—landscape
      Calculate the lag time, the runoff volume, and peak runoff for a storm of 60 mm/h, lasting 1 hour for the following cases:
   a. Roof and driveway drain directly to street
   b. Roof downspouts end at the landscape
   c. Roof and driveway drain to lawn and landscape

3.34 Solve Problem 3.33 for a rainfall intensity of 40 mm/h.

3.35 Rain falls at a rate of 50 mm/h for 30 minutes and 40 mm/h for the next 30 minutes. Calculate the lag time, runoff volume, and peak runoff from the dwelling in Problem 3.33 for cases a, b, and c.

3.36 Calculate the peak and volume of runoff for the single-family home shown below during a 30-minute storm having 3.2 in./h intensity. Perform the calculations for the following cases:
   a. Direct discharge from pavements
   b. Roof and driveway drain to lawn and landscape

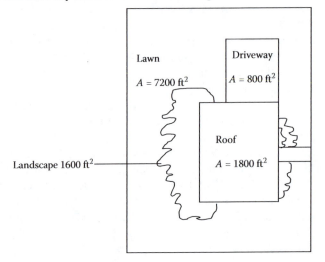

Given:

$L_i = 0.4$ in. lawn
$L_i = 1.2$ in. landscape
$S_d = 0.08$ in. driveway
$S_d = 0.04$ in. roof

Landscape:

$A = 1600$ ft^2
$K = 5$ in./h

APPENDIX 3A: THE SHEET FLOW TIME OF CONCENTRATION IN TR-55 METHOD: A COMMENTARY

The following three worksheets present calculations for the sheet flow time of concentration using the TR-55 worksheet 3. These worksheets demonstrate that segmenting the sheet flow reach, as commonly practiced and allowed in the TR-55 method, is an improper procedure. This procedure always results in an overestimation of the time of concentration and should not be practiced. Worksheet 3 in the TR-55 Manual should be revised, eliminating this erroneous procedure. Exemplified is a case where a 100 ft sheet flow reach is segmented to a 60 ft reach at 2% and a 40 ft reach at 1% grade (see the first worksheet). The second and third worksheets present time of concentration calculations for a 100 ft reach at 2% slope and 100 ft at a composite slope of 1.6%, respectively. These worksheets show that the calculated time of concentration for the segmented reach is not only longer than that of a composite slope, but also a 100 ft reach of steeper slope. The reason for this anomaly is that the time of concentration equation is not a linear function of distance; rather it is proportional to the 0.8 power of length. (Note that, e.g., $100^{0.8} = 39.8$ whereas $2 \times 50^{0.8} = 45.7$.)

U.S. Department of Agriculture
Natural Resources Conservation Service

FL-ENG-21B
06/04

TR 55 Worksheet 3: Time of Concentration (T_c) or Travel Time (T_t)

Project: _____ Designed By: _____ Date: _____

Location: _____ Checked By: _____ Date: _____

Check one: Present   Developed

Check one: T_c T_t through subarea _____

NOTES: *Space for as many as two segments per flow type can be used for each worksheet. Include a map, schematic, or description of flow segments.*

Sheet Flow (Applicable to T_c only)   Segment ID

1. Surface description (Table 3-1)					
2. Manning's roughness coeff., n (Table 3-1)	0.80	0.80			
3. Flow length, L (total L ≤ 100 ft) .... ft	60	40			
4. Two-year 24-hour rainfall, P_2 .... in	3.5	3.5			
5. Land slope, s .... ft/ft	0.020	0.010			
6. $T_t = \dfrac{0.007\,(nL)^{0.8}}{P_2^{0.5}\,s^{0.4}}$ Compute T_t .... hr	0.40	+ 0.38	=	0.77	

Shallow Concetrated Flow   Segment ID

7. Surface description (paved or unpaved)			
8. Flow length, L .... ft			
9. Watercourse slope, s .... ft/ft			
10. Average velocity, V (Figure 3-1) .... ft/s			
11. $T_t = \dfrac{L}{3600\,V}$ Compute T_t .... hr	+	=	

Channel Flow   Segment ID

12. Cross sectional flow area, a .... ft^2			
13. Wetted perimeter, P_w .... ft			
14. Hydraulic radius, $r = \dfrac{a}{P_w}$ Compute r .... ft			
15. Channel Slope, s .... ft/ft			
16. Manning's Roughness Coeff., n			
17. $V = \dfrac{1.49\,r^{2/3}\,s^{1/2}}{n}$ Compute V .... ft/s			
18. Flow length, L .... ft			
19. $T_t = \dfrac{L}{3600\,V}$ Compute T_t .... hr	+	=	
20. Watershed or subarea T_c or T_t (add T_t in steps 6, 11, and 19 .... hr		0.77	

U.S. Department of Agriculture
Natural Resources Conservation Service

FL-ENG-21B
06/04

TR 55 Worksheet 3: Time of Concentration (T$_c$) or Travel Time (T$_t$)

Project: _____   Designed By: _____   Date: _____

Location: _____   Checked By: _____   Date: _____

Check one: Present   Developed

Check one: T$_c$ T$_t$ through subarea _____

NOTES: Space for as many as two segments per flow type can be used for each worksheet.   Include a map, schematic, or description of flow segments.

Sheet Flow (Applicable to T$_c$ only)   Segment ID

1. Surface description (Table 3-1) ....

2. Manning's roughness coeff., n (Table 3-1) ....

3. Flow length, L (total L $\leq$ 100 ft) .... ft

4. Two-year 24-hour rainfall, P$_2$... in

5. Land slope, s .... ft/ft

6. T$_t$ = $\dfrac{0.007\ (nL)^{0.8}}{P_2^{0.5}\ s^{0.4}}$ Compute T$_t$ .... hr

0.80	
100	
3.5	
0.020	
0.60	

0.60 + [   ] = 0.60

Shallow Concetrated Flow   Segment ID

7. Surface description (paved or unpaved) ....

8. Flow length, L .... ft

9. Watercourse slope, s .... ft/ft

10. Average velocity, V (Figure 3-1) .... ft/s

11. T$_t$ = $\dfrac{L}{3600\ V}$ Compute T$_t$ .... hr

[   ] + [   ] = [   ]

Channel Flow   Segment ID

12. Cross sectional flow area, a .... ft^2

13. Wetted perimeter, P$_w$ .... ft

14. Hydraulic radius, r = $\dfrac{a}{P_w}$ Compute r .... ft

15. Channel Slope, s .... ft/ft

16. Manning's Roughness Coeff., n ....

17. V = $\dfrac{1.49\ r^{2/3}\ s^{1/2}}{n}$ Compute V .... ft/s

18. Flow length, L .... ft

19. T$_t$ = $\dfrac{L}{3600\ V}$ Compute T$_t$................................... hr

[   ] + [   ] = [   ]

20. Watershed or subarea T$_c$ or T$_t$ (add T$_t$ in steps 6, 11, and 19 .... hr   0.60

U.S. Department of Agriculture
Natural Resources Conservation Service

FL-ENG-21B
06/04

TR 55 Worksheet 3: Time of Concentration (T$_c$) or Travel Time (T$_t$)

Project: _____ Designed By: _____ Date: _____

Location: _____ Checked By: _____ Date: _____

Check one: Present   Developed

Check one: T$_c$ T$_t$ through subarea _____

NOTES: Space for as many as two segments per flow type can be used for each worksheet.   Include a map, schematic, or description of flow segments.

<u>Sheet Flow</u> (Applicable to T$_c$ only)   Segment ID

1. Surface description (Table 3-1) ....
2. Manning's roughness coeff., n (Table 3-1) 0.80
3. Flow length, L (total L ≤ 100 ft) .... ft   100
4. Two-year 24-hour rainfall, P$_2$.................................. in   3.5
5. Land slope, s .... ft/ft   0.016
6. $T_t = \dfrac{0.007\,(nL)^{0.8}}{P_2^{0.5}\,s^{0.4}}$ Compute T$_t$ .... hr   0.65 + ____ = 0.65

<u>Shallow Concetrated Flow</u>   Segment ID

7. Surface description (paved or unpaved) ....
8. Flow length, L .... ft
9. Watercourse slope, s .... ft/ft
10. Average velocity, V (Figure 3-1) .... ft/s
11. $T_t = \dfrac{L}{3600\,V}$ Compute T$_t$ .... hr   ____ + ____ = ____

<u>Channel Flow</u>   Segment ID

12. Cross sectional flow area, a .... ft^2
13. Wetted perimeter, P$_w$ .... ft
14. Hydraulic radius, r = $\dfrac{a}{P_w}$ Compute r .... ft
15. Channel Slope, s .... ft/ft
16. Manning's Roughness Coeff., n ....
17. $V = \dfrac{1.49\,r^{2/3}\,s^{1/2}}{n}$ Compute V .... ft/s
18. Flow length, L .... ft
19. $T_t = \dfrac{L}{3600\,V}$ Compute T$_t$................................... hr   ____ + ____ = ____
20. Watershed or subarea T$_c$ or T$_t$ (add T$_t$ in steps 6, 11, and 19) .... hr   0.65

APPENDIX 3B: SNYDER UNIT HYDROGRAPHS

The Snyder synthetic unit hydrograph was developed in 1938 for watersheds in the Appalachian highlands. However, the hydrograph has been successfully applied throughout the United States for watersheds ranging from 25 to 25,000 km² by modification of empirical constants. This hydrograph, which is also incorporated in the HEC-1 model, is characterized by five parameters that are depicted in Figure 3B.1.

These parameters are the peak discharge per unit area of watershed, q_p, specific rainfall duration T_R, basin lag time T_L, time to peak, t_p, and base time, t_b. In addition, the width of unit hydrograph at 50% and 75% of peak discharge is defined in this synthetic hydrograph. In a standard unit hydrograph, these parameters are given by the following equations:

$$T_R = t_p/5.5 \text{ hour} \tag{3B.1}$$

$$T_L = C_1 C_t \, (LL_c)^{0.3} \text{ hour} \tag{3B.2}$$

$$q_p = C_2 C_p A/T_L \text{ m}^3/\text{km}^2 \text{ (cfs/mi}^2) \tag{3B.3}$$

$$t_p = T_R/2 + T_L \tag{3B.4}$$

$$W_{50} = C_{w50} \, (q_p/A)^{-1.08} \tag{3B.5}$$

$$W_{75} = C_{w75} \, (q_p/A)^{-1.08} \tag{3B.6}$$

$$C_{w50} = 2.14 \text{ SI (770 CU)} \tag{3B.7}$$

$$C_{w75} = 1.22 \text{ SI (440 CU)} \tag{3B.8}$$

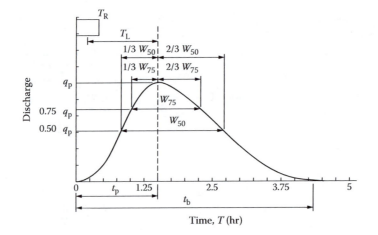

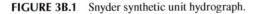

FIGURE 3B.1   Snyder synthetic unit hydrograph.

where

A = drainage area, km^2 (mi^2)

q = discharge, m^3/s (cfs)

L = length of main stream, km (m)

L_c = distance from the outlet to a point on the stream nearest to the centroid of watershed area, km (mi)

C_2 = 2.75 SI (640 CU)

The parameters C_1 and C_p in the preceding equations are derived from gauged watersheds in the same region where the Snyder hydrograph is being applied. The widths W_{50} and W_{75} each are distributed at one-third before the peak and two-thirds past it.

Assuming a triangular shape for the unit hydrograph and considering that the area under hydrograph represents 1 cm (1 in.) of direct runoff, the base time may be estimated from the following equation:

$$t_b = C_3 A / q_R, \text{ h} \tag{3B.9}$$

where C_3 = 5.56 SI (1290 CU).

For rainfall durations other than $t_b/5.5$, the lag time and the peak discharge of the unit hydrograph are adjusted as follows:

$$T_{LR} = T_L - \frac{T_R - T_R'}{4} \tag{3B.10}$$

$$q_R = \frac{q T_L}{T_{LR}} \tag{3B.11}$$

where

T_R' = rainfall duration, h.

$T_{LR'}$ = lag time for storm duration, T_R'.

T_L = obtained from Equation 3B.2.

APPENDIX 3C: USGS NATIONWIDE URBAN HYDROGRAPH

The USGS Nationwide Urban Hydrograph uses the information developed by the USGS that approximates the shape and characteristics of hydrographs. The main parameters for this hydrograph are (1) dimensionless hydrograph ordinates, (2) time lag, and (3) peak flow.

A copy of Table 3C.1 in HEC-22 (2013) that lists default values of the dimensionless hydrograph ordinates is included in this appendix. The values in this table are derived from the nationwide urban hydrograph study. These values define the shape of the hydrograph, and the lag time is given by

$$T_{L} = k_{L} L_{M}^{0.62} SL^{-0.32} (13 - BDF)^{0.47} \qquad (3C.1)$$

where

T_{L} = lag time, h
k_{L} = 0.38 SI (0.85 CU)
L_{M} = main channel length km (mi)
SL = main channel slope m/km (ft/mi)
BDF = basin development factor (see discussion in Appendix 3A.1, USGS Nationwide Urban Equations)

The peak flow can be computed using rational, SCS, or other methods (other than the universal method) described in this chapter. In applying this method, the abscissa in Table 3C.1 are multiplied by the time lag between the centroid of the rainfall and the centroid of runoff calculated using Equation 3C.1. Then the ordinates in the table are multiplied by the calculated peak flow.

TABLE 3C.1
USGS Dimensionless Hydrograph Coordinates

Abscissa	Ordinate	Abscissa	Ordinate
0.0	0.00	1.3	0.65
0.1	0.04	1.4	0.54
0.2	0.08	1.5	0.44
0.3	0.14	1.6	0.36
0.4	0.21	1.7	0.30
0.5	0.37	1.8	0.25
0.6	0.56	1.9	0.21
0.7	0.76	2.0	0.17
0.8	0.92	2.1	0.13
0.9	1.00	2.2	0.10
1.0	0.98	2.3	0.06
1.1	0.90	2.4	0.03
1.2	0.78	2.5	0.00

APPENDIX 3D: USGS REGRESSION EQUATIONS FOR URBAN PEAK DISCHARGES

The regression equations for urban peak discharges were introduced in Section 3.8 of this chapter. A copy of Table 3.4 in the HEC-22 (FHWA, 2013) is included in this appendix (called Table 3D.1 here). These equations include seven factors of which BDF (basin development factor) is of most significance. The parameters that affect the BDF are described in the following table. The equations cover peak discharges for 2-year through 500-year storm frequency. Though verified, these equations may still deviate in the order of 35% to 50% from field measurements.

The basin development factor (BDF) is a highly significant parameter in the urban equations and provides a measure of the efficiency of the drainage basin and the extent of urbanization. It can be determined from drainage maps and field inspection of the basin. The basin is first divided into upper, middle, and lower thirds. Within each third of the basin, four characteristics must be evaluated and assigned a code of 0 or 1: channel improvements, channel lining (prevalence of impervious surface lining), storm drains or storm sewers, and curb and gutter streets. With the curb and gutter characteristic, at least 50% of the partial basin must be urbanized or improved with respect to an individual characteristic to be assigned a code of 1. With four characteristics being evaluated for each third of the basin, complete development would yield a BDF of 12.

TABLE 3D.1
Nationwide Urban Equations Developed by USGS

$$UQ2 = 2.35\, A_s^{.41}\, SL^{.17}\, (R12 + 3)^{2.04}\, (ST + 8)^{-.65}\, (13 - BDF)^{-.32}\, IA_s^{.15}\, RQ2^{.47} \qquad (3\text{-}8)$$

$$UQ5 = 2.70\, A_s^{.35}\, SL^{.16}\, (R12 + 3)^{1.86}\, (ST + 8)^{-.59}\, (13 - BDF)^{-.31}\, IA_s^{.11}\, RQ5^{.54} \qquad (3\text{-}9)$$

$$UQ10 = 2.99\, A_s^{.32}\, SL^{.15}\, (R12 + 3)^{1.75}\, (ST + 8)^{-.57}\, (13 - BDF)^{-.30}\, IA_s^{.09}\, RQ10^{.58} \qquad (3\text{-}10)$$

$$UQ25 = 2.78\, A_s^{.31}\, SL^{.15}\, (R12 + 3)^{1.76}\, (ST + 8)^{-.55}\, (13 - BDF)^{-.29}\, IA_s^{.07}\, RQ25^{.60} \qquad (3\text{-}11)$$

$$UQ50 = 2.67\, A_s^{.29}\, SL^{.15}\, (R12 + 3)^{1.74}\, (ST + 8)^{-.53}\, (13 - BDF)^{-.28}\, IA_s^{.06}\, RQ50^{.62} \qquad (3\text{-}12)$$

$$UQ100 = 2.50\, A_s^{.29}\, SL^{.15}\, (R12 + 3)^{1.76}\, (ST + 8)^{-.52}\, (13 - BDF)^{-.28}\, IA_s^{.06}\, RQ100^{.63} \qquad (3\text{-}13)$$

$$UQ500 = 2.27\, A_s^{.29}\, SL^{.16}\, (R12 + 3)^{1.86}\, (ST + 8)^{-.54}\, (13 - BDF)^{-.27}\, IA_s^{.05}\, RQ500^{.63} \qquad (3\text{-}14)$$

where

UQ_T = Urban peak discharge for T-year recurrence interval, ft³/s

A_s = Contributing drainage area, sq. mi

SL = Main channel slope (measured between points which are 10% and 85% of main channel length upstream of site), ft/mi

R12 = Rainfall amount for 2-hour, 2-year recurrence, inches

ST = Basin storage (percentage of basin occupied by lakes, reservoirs, swamps, and wetlands), percent

BDF = Basin development factor (provides a measure of the hydraulic efficiency of the basin—see description of BDF HEC-22 [FHWA, 2013])

IA = Percentage of basin occupied by impervious surfaces

RQ_T = T-year rural peak flow

APPENDIX 3E: USGS STREAMSTATS PROGRAM: A CASE STUDY

This study concerns an unnamed stream along Elm Avenue in Bogota, New Jersey (see Figure 3E.1). Figure 3E.2 presents the watershed parameters of stream computed by the StreamStats software.

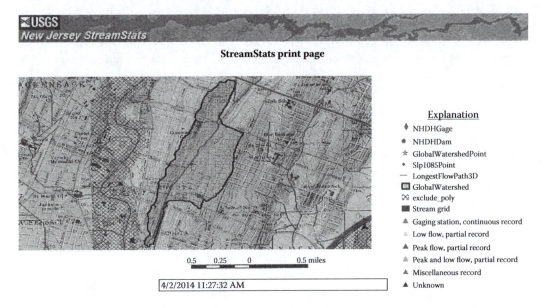

FIGURE 3E.1   StreamStats drainage area map of the stream.

StreamStats ungaged site report

Date: Wed Apr 2 2014 12:04:46 Mountain daylight time
Site location: New Jersey
NAD27 latitude: 40.8703 (40 52 13)
NAD27 longitude: −74.0331 (−74 01 59)
NAD83 latitude: 40.8704 (40 52 14)
NAD83 longitude: −74.0327 (−74 01 58)
Drainage area: 0.54 mi2

Peak flows region basin characteristics			
100% peak glaciated piedmont region 2009 5167 (0.54 mi2)			
Parameter	Value	Regression equation valid range	
		Min	Max
Drainage area (square miles)	0.54 (below min value 1.27)	1.27	56.4
Percent storage (percent)	0 (below min value 0.62)	0.62	11.6
Stream slope 10 and 85 method (feet per mi)	36	9.37	176
Basin population density (persons per square mile)	8640	645	13492

Warning: Some parameters are outside the suggested range. Estimates will be extrapolations with unknown errors.

Peak flows region streamflow statistics					
Statistic	Flow (ft³/s)	Prediction error (percent)	Equivalent years of record	90-Percent prediction interval	
				Minimum	Maximum
PK2	101		1		
PK5	162		2		
PK10	209		3		
PK25	275		4		
PK50	329		5		
PK100	385		5		
PK500	524		6		

FIGURE 3E.2   StreamStats ungaged site report.

REFERENCES

ASCE, 2006, *Standard guidelines for the design of urban storm water systems*, ASCE/EWRI 45-05, *Standard guidelines for installation of urban storm water systems*, ASCE/EWRI 46-05, *Standard guidelines for the operations and maintenance of urban storm water systems*, ASCE/EWRI 47-05.

ASCE and WEF, 1992, *Design and construction of urban storm water management systems,* American Society of Civil Engineers Manuals and Reports of Engineering Practice no. 77 and *Water Environment Federation Manual of Practice* FD-20.

Brooks, K.N., Elliott, P.R., Gregersen, H.M., and Thames, J.L., 1991, *Hydrology and the management of watershed*, Iowa State University Press, Ames, IA.

Chin, D.A., 2006, *Water resources engineering*, 2nd ed., Chapter 5, Pearson Prentice Hall, Upper Saddle River, NJ.

Chow, V.T., 1964, *Handbook of applied hydrology*, McGraw-Hill, New York.

Chow, V.T., Maiment, D.R., and Mays, L.W., 1988, *Applied hydrology*, Chapter 15, McGraw-Hill, New York.

FHWA (US Department of Transportation), October 1984, Hydrology, hydraulic engineering circular no. 19.

———— October 2002, highway hydrology, hydraulic design series no. 2, 2nd ed., publication no. FHWA-NHI02-001, Chapter 6.

———— September 2009 (revised August 2013), Urban drainage design manual, Hydraulic engineering circular no. 22, 3rd ed., publication no. FHWA-NHI-01-021, section 3.

———— August 2012, Introduction to highway hydraulics, hydraulic design series no. 4 (HDS #4).

Freeze, R.A. and Cherry, J.A., 1979, *Groundwater*, Chapter 2, Prentice Hall, Upper Saddle River, NJ.

Izzard, C.F., 1946, Hydraulics of runoff from developed surfaces, report 26, Hwy Res. Board, Washington, DC, pp. 129–146.

Linsley, R.K., Jr., Kohler, M.A., and Paulhus, J.L.H., 1982, *Hydrology for engineers*, 3rd ed., Chapters 3 and 4, McGraw-Hill, New York.

Lopez, J.M., Glick, R., and Gosselink, L., 2012, Taking a SWAT at changing urban cycles, a combined approach to evaluate changes in flooding, erosion and aquatic life, *Stormwater*, January/February, pp. 39–46.

Maidment, D.R., editor in chief, 1993, *Handbook of hydrology*, Chapter 28, McGraw-Hill Inc., New York.

Mays, L.W., editor in chief, 1996, *Water resources handbook*, Chapter 26, Urban storm water management, McGraw-Hill, New York.

McCuen, R.H., Johnson, P.A., and Regan, R.M., 2002, US Dept. of Transportation, FHWA, Highway hydrology, hydraulic design series no. 2, 2nd ed., publication no. FHWA-NHI02-001.

McCuen, R.H. and Spiess, J.M., 1995, Assessment of kinematic wave time of concentration, *ASCE Journal of Hydraulic Engineering*, 121 (3): 256–266.

New Jersey Department of Environmental Protection, Trenton, NJ, February 2004, *New Jersey storm water best management practices manual*, partly revised April 2009, http://www.state.nj.us/dep/watershedmgt/bmpmanualfeb2004.htm.

Pazwash, H., 1989, Comparison of rational and SCS-TR55 methods for urban storm water management, *Channel Flow and Catchment Runoff, Proceedings of the International Conference for Centennial of Manning's Formula and Kuichling's Rational Formula*, University of Virginia, Charlottesville, May 22–26, pp. 156–165.

———— 1992, Simplified Design of Multi-Stage Outfalls for Urban Detention Basins, Proceedings of Water Resources Sessions at Water Forum 92, August 2–6, Baltimore, MD, pp. 861–866.

———— 2009, Universal runoff model, paper R13, StormCon, Anaheim, CA, August 16–20, 2009.

———— 2013, Universal Runoff Model, Comparison with Rational and SCS Methods. *World Environmental and Water Resources, Congress 2013*, Cincinnati, OH, May 20–22, 2013.

Rawls, W.J., Brakensiek, D.L., and Miller, N., 1983, Green-Ampt infiltration parameters from soil data, *Journal of Hydraulic Division, ACSE*, 109 (1): 62–70.

Residential Site Improvement Standards, 2009, New Jersey Administrative Code, Title 5, Chapter 21 (http://www.nj.gov/dca/codes/nj-rsis), adopted 1/16/1997, last revised 6/15/2009.

Rickly Hydrological Company, 1700 Joyce Avenue, Columbus, OH 43219, 1-800-561-9677 (http://www.rickly.com/MI/Infiltrometer.htm).

Schneider, L.E. and McCuen, R.H., 2005, Statistical guidelines for curve number generation, *Journal of Irrigation and Drainage Engineering, ASCE*, 1311 (3): 282–290.

Schuller, T.R., July 1987, Controlling urban runoff: A practical manual for planning and designing urban BMPs, Metropolitan Washington Council for Governments, Washington, DC.

Standards for soil erosion and sediment control in New Jersey, July 1999, Adopted by the New Jersey State Soil Conservation Committee.

Todd, D.K., 1980, *Groundwater hydrology*, 2nd ed., Chapter 3, John Wiley & Sons, New York.

Urban Drainage and Flood Control District, revised 1991, *Urban storm drainage criteria manual*, Denver, Colorado.

USDA (US Department of Agriculture, NRCS), Soil Conservation Service, June 1986, Urban hydrology for small watersheds, technical release 55 (TR 55).

USDA (US Department of Agriculture, NRCS), January 2009, Small watershed hydrology WinTR-55 user guide.

———WinTR-55 Small watershed hydrology, Windows 7, last updated 2/7/2013.

Viessman, W., Jr., Lewis, G.L., and Knapp, J.W., 1989, *Introduction to hydrology*, 3rd ed., Harper & Row, New York.

Westphal, J.A., 2001, Hydrology for drainage system design and analysis, in L.W. Mays, ed. *Storm water collection system design handbook*, McGraw-Hill, New York.

4 Design of Storm Drainage Systems

This chapter presents procedures for the design of storm drainage elements, including inlets, storm drains, culverts, and erosion control measures at conduit outlets and lined swales. The design of inlets, culverts, and swales is primarily intended to provide a discharge capacity adequate to carry the calculated design flow.

4.1 INTRODUCTION TO ROADWAY DRAINAGE ANALYSIS

Management of storm water runoff is an important aspect of every urban development and roadway project. Traffic safety depends on surface drainage. A proper drainage design can eliminate, or at least minimize, the occurrence of the hazardous phenomenon of hydroplaning.

Roadway runoff can be managed in different ways, depending on whether the streets and roads are curbed or uncurbed. Conveyance of runoff from curbed streets and roads is discussed in this chapter. Also presented in this chapter is the design of roadside swales to carry runoff. For streets with curbs and gutters, runoff is collected and conveyed by a drainage system comprising inlets and pipes. The provision of sufficient numbers of inlets is the key to effective removal of storm water from pavements. Expensive drainage systems often flow below their capacity because of insufficient inlets.

The design of any drainage system begins with runoff calculations. These calculations were discussed in a previous chapter. For curbed streets and roads, the design is followed by gutter flow analysis, inlet calculations, and sizing of drainage pipes. The following sections discuss the flow spread in gutters, procedures for calculating the efficiency of inlets, and design of storm drains. A section also presents flow equations developed by the author for roadways with zero longitudinal profile.

4.1.1 GUTTER FLOW

The hydraulic capacity of a gutter depends on its geometry and the longitudinal grade. Applying the Manning formula to a triangular shaped gutter, which is a typical curbed gutter section, yields

$$Q = \frac{K}{n} S_x^{5/3} S^{1/2} T^{8/3} *$$

(4.1)

where
$K = 0.375$ SI
$K = 0.56$ CU
S_x = cross slope for roadway, m/m (ft/ft)
S = longitudinal slope of gutter m/m (ft/ft)
T = flow spread, m (ft)
Q = gutter flow m³/s (cfs)

---

* See Appendix 4A for derivation of this equation.

Expressing Q in terms of the flow depth, d, at the curb gives

$$Q = \left(\frac{K}{n} \right) \left(\frac{S^{1/2}}{S_x} \right) d^{8/3}$$ (4.2)

where

$$d = TS_x, \text{ m (ft)}$$ (4.3)

and K is the same as defined in Equation 4.1.

Various forms of nomographs are available for estimating the flow in triangular gutters in terms of the depth of flow. One such version is shown in Figure 4.1a in metric units and Figure 4.1b in English units. However, to improve accuracy, gutter flow calculations should be performed directly using either Equation 4.1 or 4.2.

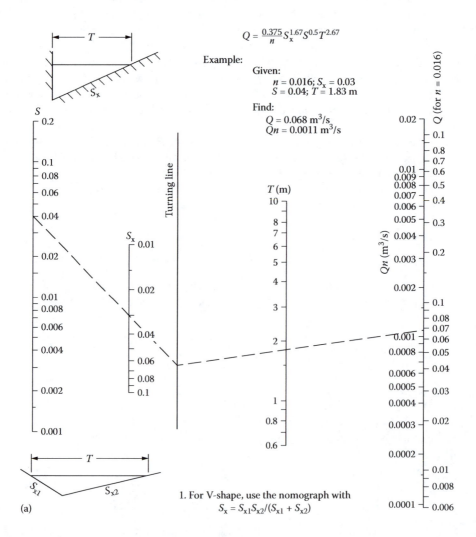

FIGURE 4.1   (a) Flow in triangular gutter sections (SI units).   (*Continued*)

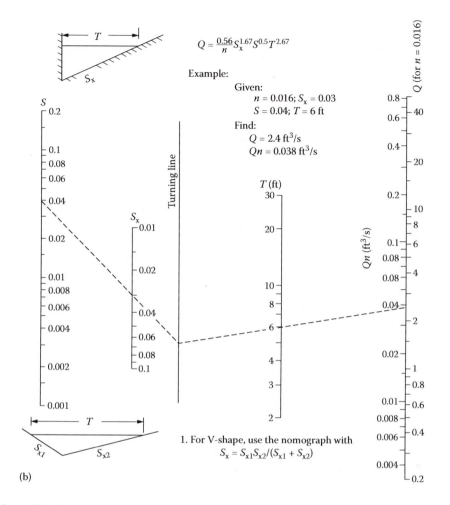

FIGURE 4.1 (CONTINUED) (b) Flow in triangular gutter sections (English units).

Rewriting Equation 4.1 in terms of the flow spread, T, gives

$$T = \left[\frac{(n/K)^{0.375}}{S_x^{0.625}} \right] \times \left(\frac{Q}{S^{0.5}} \right)^{0.375} \tag{4.4}$$

This equation is dimensionally homogeneous and can be used both in metric and customary units. The first term on the right-hand side is generally constant for a given project. Thus, substituting that constant for the first term in the preceding equation will simplify tabulation of flow spread along roadway gutters. It is to be noted that the preceding equation breaks down for roadways at zero grade. Spread flow calculations for such a condition are derived in Section 4.5 and Appendix 4B in this chapter.

The n values for various surface textures are listed in Table 4.1.

Table 4.2 summarizes the flow spread calculations for a range in flow and longitudinal slope of gutters in meters and feet. The calculations are based on a cross slope of 4%, which is a common grade for gutters, and $n = 0.013$, which represents a smooth gutter surface pavement. Highlighted in this table are the conditions in which the spread exceeds 2 m (6 ft).

TABLE 4.1

Typical Manning's *n* Values for Street and Pavement Cutters

Type of Gutter or Pavement	Manning *n*
Concrete gutter, troweled finish	0.012
Asphalt Pavement	
Smooth texture	0.013
Rough texture	0.016
Concrete Pavement	
Float finish	0.014
Broom finish	0.016

TABLE 4.2

Flow Spread in Gutter

Q L/s (m³/s)	Longitudinal Slope, S% 1	2	3	4	5	7
	in Meters					
30 (0.03)	1.35	1.18	1.10	1.04	1.0	0.93
60 (0.06)	1.75	1.53	1.42	1.35	1.29	1.21
90 (0.09)	**2.03**	1.79	1.65	1.57	1.50	1.48
120 (0.12)	**2.26**	1.99	1.84	1.75	1.67	1.65
150 (0.15)	**2.46**	**2.16**	**2.00**	1.90	1.82	1.79

Q cfs	Longitudinal Slope, S% 1	2	3	4	5	7
	in Feet					
1	4.32	3.80	3.52	3.33	3.20	3.00
2	5.61	4.92	4.56	4.32	4.15	3.89
3	**6.53**	5.73	5.31	5.03	4.83	4.53
4	**7.27**	**6.38**	5.92	5.61	5.38	5.05
5	**7.91**	**6.94**	**6.43**	**6.10**	5.85	5.49

Note: $n = 0.013$, $S_x = 4\%$.

These tables show that the smaller the longitudinal slope is, the wider is the flow spread. It is interesting to note that for longitudinal slopes of 2% or less the spread exceeds 2 m for discharges larger than 120 L/s (6 f for discharges greater than ±3.5 cfs). Thus, in a roadway having a 2 m (6 ft) wide shoulder, the spread may extend to traffic lanes on roads at small slope. To reduce the length of spread, composite gutters are sometimes used along roadways. Figure 4.2 shows composite gutter sections having a 0.6 m (2 ft) gutter section with 50 mm (2 in.) depression. A further discussion of the flow spread will be given later in this chapter.

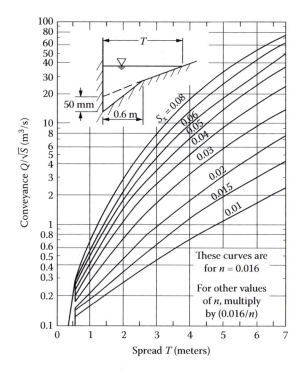

FIGURE 4.2 Discharge versus spread for a composite gutter section, metric units.

Example 4.1

A smooth asphalt roadway includes four 12 ft wide lanes, two 4 ft median shoulders and two 6 ft wide side shoulders (one at each side). The roadway is elevated and receives no off-road runoff. The road is at 1.5% longitudinal slope and the traffic lanes and the side shoulders are at 2% and 4% cross slopes, respectively. A Jersey barrier is placed at the center of the roadway. The first set of inlets are placed at side shoulders 350 ft from the road high point.

    a. Calculate the 25-year peak rate of runoff to the first inlet on each shoulder. Base your calculations on $T_c = 10$ min, $I = 6.7$ in./h.
    b. Calculate the flow spread on shoulder at the inlet for $n = 0.013$.

Solution

a. $\Sigma w = 2 \times 12 + 4 + 6 = 34$ ft
   $A = 34 \times 350 = 11,900$ sf $= 0.273$ acres
   $Q = ACI = 0.273 \times 0.95 \times 6.7 - 1.74$ cfs

b. $T = \left[\dfrac{Qn}{\left(0.56 \times S_x^{5/3} S^{1/2}\right)} \right]^{3/8}$

   $T = \left(\dfrac{1.74 \times 0.013}{0.56 \times 0.04^{5/3} \times 0.015^{1/2}} \right)^{0.375} = 70.57^{0.375} = 4.93\,\text{ft}$

Example 4.2

Redo Example 4.1 for 3.75 m wide lanes, 1.3 m center shoulder, 1.8 m side shoulder, and $I =$ 170 mm/h. The first sets of inlets are placed 100 m from the high point.

Solution

a. $\Sigma w = 2 \times 3.75 + 1.3 + 1.8 = 10.6$ m
   $A = 10.6 \times 100 = 1060$ m^2 = 0.106 ha
   $Q = 2.78 \times 10^{-3} ACI = 2.78 \times 10^{-3} \times 0.95 \times 0.106 \times 170 = 0.0476$ m^3/s = 47.6 L/s

b. $T = \left(\dfrac{Qn}{0.375 \times 0.04^{5/3} \times 0.015^{1/2}} \right)^{0.375}$

$T = \left(\dfrac{0.0476 \times 0.013}{0.375 \times 0.00468 \times 0.1225} \right)^{0.375} = 1.49$ m

4.2 TYPES OF INLETS

Figure 4.3 depicts four commonly used types of inlets, namely grate inlet, curb opening inlet, combination inlet, and slotted drain inlet. It is to be noted that curb openings and combination inlets are not common in some countries. Figures 4.4 and 4.5 show an inlet along a curb in Budapest, Hungary, and an inlet at a parking lot median in a roadside service center in Austria, respectively. The latter photo also shows a basket under an inlet grate to intercept floatables and debris.

To avoid bottles and cans from entering a curb opening, a new type of curb piece designated as an Eco-curb piece is being used on new inlets in New Jersey. The New Jersey Department of Environmental Protection (NJDEP) has also mandated that all existing curb pieces be replaced with

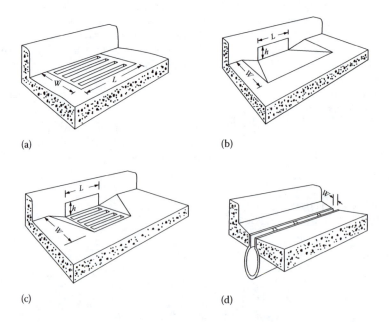

(a)   (b)

(c)   (d)

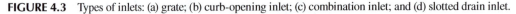

FIGURE 4.3 Types of inlets: (a) grate; (b) curb-opening inlet; (c) combination inlet; and (d) slotted drain inlet.

FIGURE 4.4   An inlet along a curb in Budapest, Hungary.

FIGURE 4.5   An inlet at a parking lot median, in a roadside service center in Austria.

the new curb pieces having openings less than 2 in. wide. Figure 4.6 shows a sketch of these curb pieces. Environmental Retrofit Solutions (ERS), a company that was headquartered in Hawthorne, New Jersey, manufactured curb pieces made of Core-Ten high strength/low-alloy weathering steel, which can be readily mounted over the existing curb inlets (see Figure 4.7). These curb pieces are now available through Campbell Foundry in Harrison, New Jersey, which acquired ERS in 2009 and also does business as Campbell-ERS. Figure 4.8a and b shows a new curb piece and traditional curb piece, respectively.

Trench drain is a modified type of slotted drain. This type of drain commonly consists of a rectangular flume and a continuous grate. A variety of trench drains are commercially available. An ACO drain is one such trench drain that comes in 0.5 m and 1 m long sections (see Figure 4.9a). ACO has recently begun manufacturing highway drains. These drains, which are made of polymer concrete, come in 4 ft long, 8 in. inside width sections with 1/16 in. per ft (0.16%) internal longitudinal slope (refer to Figure 4.9b). ACO highway drains provide significantly larger capacity than original ACO drains. ACO highway drains are also nearly four times less expensive than the older models, and as such, are most suitable and cost effective for roads of small longitudinal slope. Following the author's recommendation, these drains were installed in State Route 30 in Magnolia, Camden County, under a New Jersey DOT roadway improvement project.

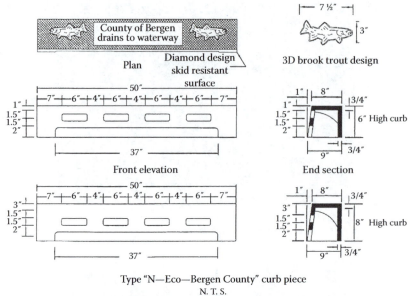

Type "N—Eco—Bergen County" curb piece
N. T. S.
(Campbell Foundry Co., pattern #2618 - BC or approved equiv.)

Note:

1. The contractor shall provide shop drawings of the inlet castings to the engineer
   for review and approval.

2. Material shall be gray cast iron ASTM A48–83, class 30B.

3. In retrofit situations this curb piece (head) will fit existing Campbell Foundry Co.
   manufactured curb inlets for NJDOT types B, B–1, B–2, D, D–1, and D–2.

4. Casting is supplied without surface coating.

FIGURE 4.6 Eco-curb piece adopted by Bergen County, New Jersey.

FIGURE 4.7 Curb piece manufactured by Campbell ERS in Harrison, New Jersey.

FIGURE 4.8 (a) A new curb piece. (b) A traditional curb piece.

(a)

ACO ROAD

Highway drain

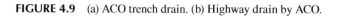

Channel

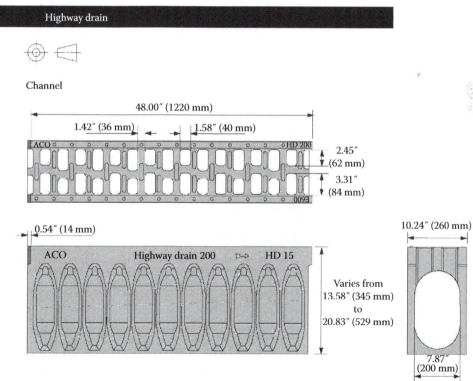

(b)

FIGURE 4.9 (a) ACO trench drain. (b) Highway drain by ACO.

(*Continued*)

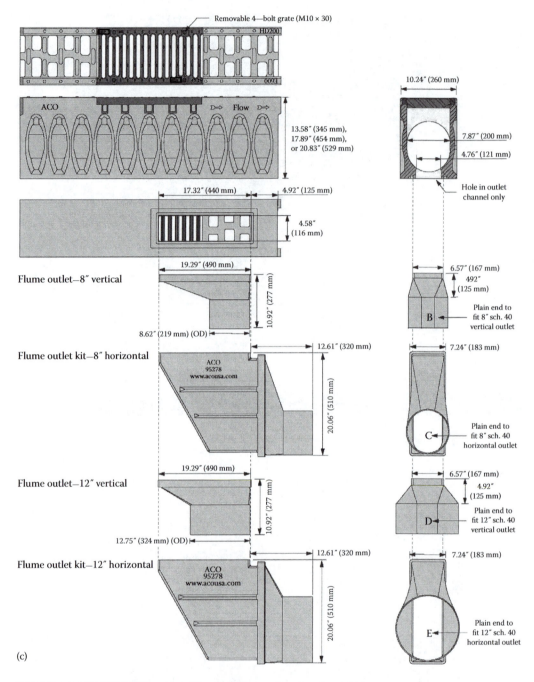

(c)

FIGURE 4.9 (CONTINUED)   (c) ACO access drain and outlet channel kits.

To provide access for maintenance, ACO highway drains are also available with ductile iron grates bolted at four corners. ACO also manufactures flume outlets for connection into a collector line or catch basin. These flumes are available in 8 or 12 in. schedule 40 vertical or horizontal outlets with iron access grates. Figure 4.9c shows an ACO access drain and outlet channel kits.

A number of foundries, including Neenah and Campbell, manufacture various sizes of grates for trench drains. Figure 4.10 exemplifies grates manufactured by Campbell Foundry in Harrison and Kearny, New Jersey (2012). This foundry makes 6 to 48 in. wide grates.

Standard heavy duty trench covers/grates are designed for A.A.S.H.T.O. HS20-44 highway loading and are suitable for use in most cargo handling areas. For any application other than standard truck traffic please contact our engineering department for design and material recommendations.

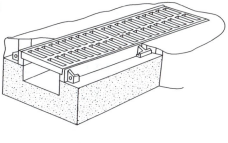

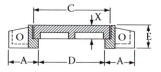

Trench frame and cover

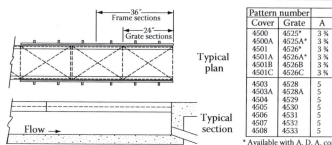

Typical plan

Typical section

Standard heavy duty							
Pattern number		Dimensions in in.					Grate
Cover	Grate	A	C	D	E	X	type
4500	4525*	3 ¾	7 ¾	6	2 ¼	2	1
4500A	4525A*	3 ¾	9 ¾	8	2 ¼	2	2
4501	4526*	3 ¾	11 ¾	10	2 ¼	2	2
4501A	4526A*	3 ¾	14	12	2 ¼	2	2
4501B	4526B	3 ¾	15 ¾	14	2 ¼	2	2
4501C	4526C	3 ¾	17 ¾	16	2 ¼	2	2
4503	4528	5	19 ¾	18	4	2 ¾	3
4503A	4528A	5	21 ¾	20	4	2 ¾	2
4504	4529	5	25 ¾	24	4	2 ¾	3
4505	4530	5	31 ¾	30	4	2 ¾	3
4506	4531	5	38	36	4	2 ¾	3
4507	4532	5	44	42	4	2 ¾	3
4508	4533	5	50	48	4	2 ¾	3

* Available with A. D. A. compliant openings (openings ½″ or less in width)

FIGURE 4.10   Campbell Foundry trench drains.

4.3   INLET DESIGN

The hydraulic capacity, namely the flow intercepted by an inlet, depends on the gutter flow, the inlet type, and the inlet location. Inlets at grades have different hydraulic characteristics from inlets at low points.

A comprehensive publication for inlet hydraulics has been prepared by the US Department of Transportation, Federal Highway Administration. This publication, which is entitled "Urban Drainage Design Manual, Hydraulic Engineering Circular No. 22 (HEC-22)," presents detailed procedures for estimating the capacity and efficiency of various types of inlets (FHWA, 2013). The following sections briefly discuss some of these procedures.

4.3.1   GRATE INLETS AT GRADE

Grates provide effective means of draining roadway and parking lot pavements. Inlet grates come in a variety of sizes and geometrics. Grates are selected considering three basic elements: hydraulic efficiency, traffic safety (vehicular, bicycle, and pedestrian), and debris clogging. Structural strength, cost, durability, and vandalism are among other elements to be considered. The grates are designated as P-1-7/8 and so on; the letter P stands for parallel bar grates and the number indicates the bar spacing in inches. There are also CV (curved vane) grates and reticuline or honeycomb patterns. Figures 4.11 and 4.12 show P-1-7/8 and reticuline grates, respectively.

Grates are effective pavement drainage elements where debris is not a problem. Grate inlets intercept gutter flow in part through their front edge and partly from their side. All the water passing over the upstream edge of the grate (called frontal flow) will be intercepted if the gutter is long

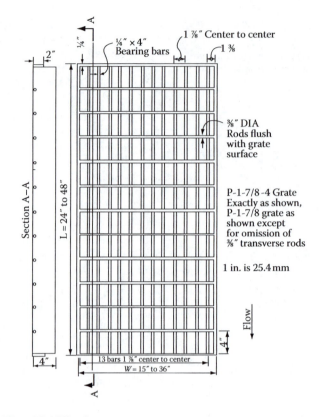

FIGURE 4.11   P-1-7/8 and P-1-7/8—four grates.

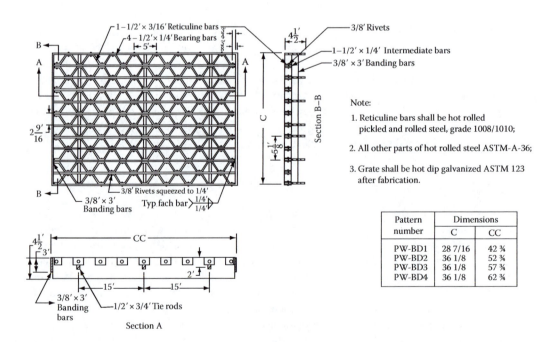

FIGURE 4.12   A reticuline grate (manufactured by Campbell Foundry in New Jersey).

enough and the flow velocity is low enough to avoid a splash over. The splash-over velocity depends on the type of grate. The ratio of the frontal flow to total gutter flow, E_o, for a uniform gutter slope, is given by

$$E_o = \frac{Q_w}{Q} = 1 - \left(1 - \frac{w}{T}\right)^{2.67}$$

(4.5)

where
Q = total gutter flow, m³/s (cfs)
w = width of grate (or depressed gutter), m (ft)
Q_w = frontal flow; flow in width w, m³/s (cfs)
T = total flow spread in gutter, m (ft)

Figure 4.13 shows a graphical solution to the preceding equation in dimensionless form for straight cross slope and depressed gutters.

The ratio of the frontal flow to gutter flow intercepted by an inlet is given by

$$R_f = 1 - 0.295(V - V_o) \quad \text{SI}$$

(4.6)

$$R_f = 1 - 0.09(V - V_o) \quad \text{CU}$$

(4.7)

where
V = velocity of flow in gutter in m/s (ft/s)
V_o = gutter velocity where splash over first occurs in m/s (ft/s)

The splash-over velocity depends on the type and length of inlet grate. Table 4.3 lists typical values of the splash-over flow velocity of various types of grates in metric and customary units. This table, for example, shows that the splash-over velocity for a 4 ft long grate varies from 7 ft/s for a

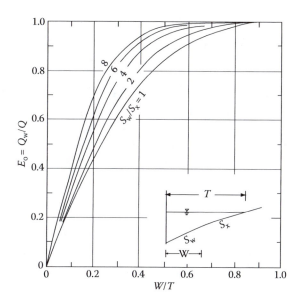

FIGURE 4.13   Ratio of frontal flow to gutter flow.

TABLE 4.3

Splash-Over Velocity of Common Inlet Grates, V_o

Grate Type	Length, ft (m)	Splash-Over Velocity, ft/s (m/s)
Reticuline	2 (0.6)	4.2 (1.25)
	4 (1.2)	7.0 (2.10)
Curved vane	2 (0.6)	5.9 (1.75)
	4 (1.2)	9.0 (2.70)
P-1-1/8	2 (0.6)	6.3 (1.90)
(P-30)	4 (1.2)	9.1 (2.75)
P-1-7/8	2 (0.6)	8.1 (2.40)
(P-50)	4 (1.2)	11.5 (3.45)

Note: P-30 and P-50 represent parallel bars spaced 30 and 50 mm on center, respectively.

---

reticuline inlet to over 11 ft/s for a P-1-7/8 inlet. Thus, if the gutter flow velocity is less than these figures, the second term in Equations 4.6 and 4.7 disappears and R_f becomes 1.0. The gutter flow velocity can be calculated by the following equation:

$$V = \frac{Q}{(S_x T^2/2)} \tag{4.8}$$

Alternatively, the gutter flow velocity can be calculated using the following equations:

$$V = \frac{0.75 S^{0.5}}{n S_x^{0.67} T^{0.67}} \quad \text{SI} \tag{4.9}$$

$$V = \frac{1.12 S^{0.5}}{n S_x^{0.67} T^{0.67}} \quad \text{CU} \tag{4.10}$$

Figure 4.14 depicts the splash-over velocity for a large number of grates in SI units.

In addition to frontal flow, water also enters a grate as side flow. This flow depends on the cross slope of the gutter, the length of the grate, and flow velocity. The ratio of side flow intercepted to the total side flow, referred to as side flow interception efficiency, can generally be calculated by

$$R_s = \left[1 + (KV^{1.8}/S_x L^{2.3})\right]^{-1} \tag{4.11}$$

where
L = length of grate, in m (ft)
V = the gutter flow velocity, in m/s (ft/s)
$K = 0.0828$ SI
$K = 0.15$ CU
S_x = shoulder cross slope, m/m (ft/ft)

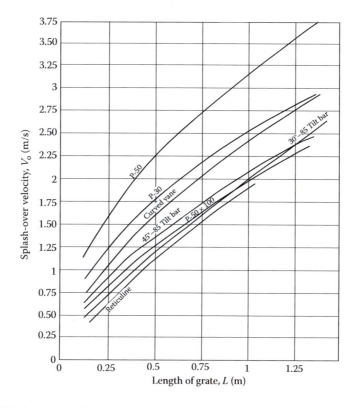

FIGURE 4.14 Splash-over velocity for various grates (SI units).

Thus, the overall efficiency and the intercepted flow of a grate are expressed by the following two equations, respectively:

$$E = R_f E_o + R_s(1 - E_o) \tag{4.12}$$

and

$$Q_i = EQ \tag{4.13}$$

It is to be noted that for a roadway at 4% slope or greater and grates of limited length, the side flow is negligible. Likewise, the intercepted flow of curb opening in combination inlets can be ignored for steep roads. In such cases, the preceding equation simplifies as

$$E - R_f E_o \tag{4.14}$$

Figures 4C.1 and 4C.2 in Appendix 4C of this chapter give the overall efficiency of 0.6 m × 0.6 m and 0.6 m × 1.2 m, 2 ft × 2 ft, and 2 ft × 4 ft grates for a 4% cross slope for a large range in gutter flow and longitudinal slope. These figures show that a 0.6 m × 0.6 m (2 ft × 2 ft) grate inlet cannot intercept 75% of a 0.09 m³/s (3 cfs) gutter flow under any roadway longitudinal slope. The previously described figures also demonstrate that, all in all, inlets may not be as effective as many engineers perceive. Therefore, inlets must be spaced close enough to effectively remove runoff from streets and roadways.

Example 4.3

Given: $T = 6$ ft, $S_x = 0.04$, $S = 0.03$, $n = 0.016$, and no bicycle traffic, calculate interception capacity for the following grates:

  a. 2 ft × 2 ft P-1-1/8 grate
  b. 2 ft × 4 ft (long) P-1-1/8 grate
  c. 2 ft × 2 ft reticuline grate

Base your calculations for the condition of no spread beyond the shoulder.

Solution

First calculate gutter flow capacity:

$$Q = \left(\frac{0.56}{n}\right) S_x^{1.667} S^{1/2} T^{8/3}$$

$$Q = \left(\frac{0.56}{0.016}\right) \times 0.04^{1.667} \times (0.03)^{1/2} \times 6^{2.667} = 3.37 \text{ cfs}$$

Calculate gutter flow velocity:

$$V = Q/A = 3.37/(1/2 \times 6^2 \times 0.04) = 4.67 \text{ ft/s}$$

Next, calculate the frontal flow to gutter flow ratio:

$$\frac{w}{T} = \frac{2}{6} = 0.33$$

$$E_o = \frac{Q_w}{Q} = 1 - \left(1 - \frac{w}{T}\right)^{2.67} = 1 - (1 - 0.33)^{2.67} = 0.66$$

Alternatively, enter Figure 4.13 with $w/T = 0.33$ and $S_x/s = 1.33$ to read $E_o = 0.66$.
The frontal flow interception efficiency of grates can be calculated using Equation 4.6 in combination with Table 4.3.

  a. For P-1-1/8, $V_o = 6.3$ ft/s $> V$ $R_f = 1.0$
  b. For P-1-1/8, four long grate, $V_o = 9.1$ ft/s $> V$ $R_f = 1.0$
  c. For reticuline, 2 ft long grate, $V_o = 4.2 < V$

$$R_f = 1 - 0.09(4.67 - 4.2) = 0.96$$

Then calculate side flow interception efficiency.
For grates (a) and (c) where $L = 2$ ft, the side flow interception can be calculated using Equation 4.11, as follows:

$$R_s = \left(1 + \frac{0.15 \times 4.67^{1.8}}{0.04 \times 2^{2.3}}\right)^{-1} = 0.076$$

Likewise, for grate (b), where $L = 4.0$ ft,

$$R_s = 0.288$$

From Equation 4.12:

a. $E = 1 \times 0.66 + 0.076(1 - 0.66) = 0.66 + 0.026 = 0.686$
   $Q_i = 3.37 \times 0.686 = 2.31$ cfs
b. $E = 1 \times 0.66 + 0.288(1 - 0.66) = 0.758$
   $Q_i = 2.55$ cfs
c. $E = 0.96 \times 0.66 + 0.076(1 - 0.66) = 0.66$
   $Q_i = 2.22$ cfs

The preceding calculations imply that less than 4% of the overall efficiency of the 2 ft long parallel grate and reticuline inlets is due to side interception and that this efficiency is approximately 13% for a 4 ft long parallel grate. It is also seen that the 2 ft parallel grate has nearly 5% more capacity than the reticuline grate and that increasing the length of grate from 2 to 4 ft increases the capture capacity by a mere 10%.

Note that the preceding results for the 2 ft × 2 ft and 2 ft × 4 ft P-1-1/8 grates can be directly derived from Figures 4C.1b and 4C.2b in Appendix 4C of this chapter.

4.3.2 Curb Opening Inlets

Curb opening inlets do not interfere with traffic and in the absence of a curb piece are also less vulnerable to clogging than grate inlets. In the past, these inlets were commonly used without any curb piece. However, as previously indicated, to address the current storm water management regulations, a curb piece is installed to prevent floatable objects from entering the drainage pipes. See Figures 4.7 and 4.8. Also as indicated previously, curb inlets offer little capacity in intercepting the runoff on steep slopes unless they are depressed.

The flow through a curb opening inlet occurs as a weir flow before the opening is fully inundated and as an orifice when the opening is submerged. For a curb inlet at level grade (no depression) the weir flow discharge is given by

$$Q = C_w L d^{1.5} \tag{4.15}$$

$$C_w = 1.66 \quad \text{SI}$$

$$C_w = 3.0 \quad \text{CU}$$

where
  L = length of curb opening in m (ft)
  d = depth of water at the curb opening in m (ft)

Under submerged conditions, the flow equation becomes

$$Q = C_o L h (2g h_o)^{1/2} \tag{4.16}$$

where
  C_o = orifice coefficient = 0.67 SI and CU
  h = height of opening, m (ft)
  $h_o = (d - h/2)$ = water depth to the center of the opening, m (ft)
  g = 9.81 m/s^2 (32.2 ft/s^2)

For an inlet at depression, the equation for orifice flow remains unchanged, while the weir flow equation modifies as

$$Q = C_w(L + 1.8W_o)d^{1.5} \qquad (4.17)$$

where
  W_o is the width of gutter depression in m (ft)
  $C_w = 1.25$ SI
  $C_w = 2.3$ CU

According to HEC-22 (FHWA, 2013) the length of curb opening inlet needed to fully intercept the gutter flow on a straight cross slope is given as

$$L_{100} = KQ^{0.42}S^{0.3}(nS_x)^{-0.6} \qquad (4.18)$$

where
  L_{100} = required curb opening length to intercept 100% of gutter flow, in m (ft)
  $K = 0.82$ SI*
  $K = 0.60$ CU
  Q = gutter flow m³/s (ft³/s)

The same publication proposes the following equation for the interception efficiency of a curb-opening inlet of length L:

$$E = 1 - \left(1 - \frac{L}{L_{100}}\right)^{1.8} \qquad (4.19)$$

Using this equation, the length of curb opening inlet required to achieve a 75% efficiency can be calculated as $L = 0.54L_{100}$.

4.3.3 SLOTTED INLETS

Slotted inlets offer little interference to traffic and can be used in curbed or uncurbed sections. They are particularly effective in airport runways, parking lots, garage entries at downhill driveways, docks, ports, and roads with small longitudinal slope where the runoff would impound on the gutter. For slotted inlets in sag locations, the inlet capacity may be calculated using the weir flow equation for a depth of up to 6 cm (0.2 ft) and the orifice flow equation for depths larger than 0.12 m (0.4 ft). Within these depths the flow is in a transient condition. The inlet capacity for the weir flow condition can be calculated using the same equation as for the curb opening (Equation 4.15). In that equation the depth at curb measured from the normal cross slope is substituted for depth of water, d, and the weir coefficient, C_w, varies with flow depth and slot length with a typical value of

$$C_w = 1.4 \quad SI$$

$$C_w = 2.5 \quad CU$$

* This factor is misprinted as 0.076 in the HEC-22 (2013).

The orifice flow condition follows the equation:

$$Q = 0.8LW(2gd)^{0.5} \tag{4.20}$$

where

L = length of slot, m (ft)
W = width of slot, m (ft)
g = acceleration of gravity, 9.81 m/s^2 (32.2 ft/s^2)
d = depth of water at slot inlet, m (ft)

This equation is valid when the depth of water is greater than the opening, which results in inundation of slot. For smaller flow depths, the weir flow equation should be used.

As was previously indicated, a number of trench drains are commercially available. One such drain is the ACO drain, which comes in sections 0.5 and 1 m long and 95 mm (3.75 in.) and upward widths and various depths. Another one is the ACO highway drain, which was previously described. Capacity and technical information on this and other trench drains can be obtained from the manufacturers' specifications.

4.3.4 COMBINATION INLETS

Combination inlets are mainly used to accept debris in order to avoid clogging of the grates. The interception capacity of a combination inlet consists of the sum of the capacities of the grate and curb opening. However, this capacity is not considerably greater than the capacity of the grate alone on roadways at 4% or greater slope. This is particularly the case for new curb pieces, which provide small openings. Therefore, the interception capacity of combination inlets is commonly calculated neglecting the curb pieces in practice.

4.3.5 NEW JERSEY INLETS

Many states have specific inlet grates. In New Jersey, for example, four types of inlets are commonly used for streets, roads, and parking areas. These are grate inlet types "A" and "E" with 2 ft × 4 ft and 4 ft × 4 ft grates, respectively, and combination inlets type "B" and "D" having 2 ft × 4 ft grates plus curb pieces. The only difference between type "B" and type "D" inlets is that the former inlet has larger inside dimensions than the latter. These inlets are depicted in Figure 4.15. Also available are 2 ft × 2 ft and smaller inlets for lawns and gardens.

The New Jersey Department of Transportation (NJDOT) has developed the following equation for the interception capacity of grate inlets (NJDOT, 2013):

$$Q_1 = \frac{Kd^{1.54}S^{0.233}}{S_x^{0.276}}$$

$$K = 2.98 \quad \text{SI} \tag{4.21}$$

$$K = 16.88 \quad \text{CU}$$

where

d = depth of water in gutter, m (ft)
S = longitudinal slope, m/m (ft/ft)
S_x = cross slope, m/m (ft/ft)
Q = flow intercepted by the grate, m^3/s (cfs)

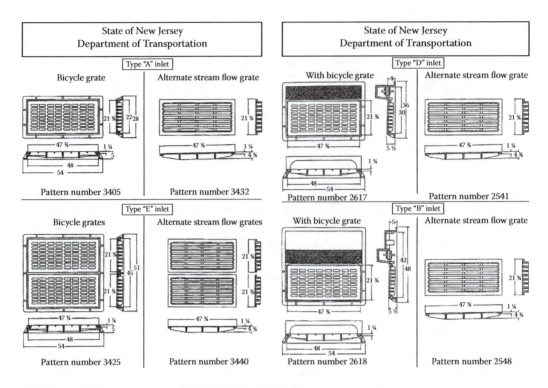

FIGURE 4.15 New Jersey types "A," "B," "D," and "E" inlets.

The type "A" inlet is similar to the FHA type P-1-7/8, 2 ft × 4 ft inlet grate. Using Equation 4.21 yields results reasonably close to, though somewhat smaller than, Figure 4C.2b (see Appendix 4C) for a 2 ft × 4 ft grate. The NJDOT suggests using Equation 4.21 for inlet grate types "A," "B," and "D" without any modification. However, combination inlets such as types "D" and "B" provide a somewhat larger interception capacity than type "A" inlets in small longitudinal slopes. In using these inlets, the author suggests applying an adjustment factor of 1.05 to the K value in Equation 4.21 only when $S \leq 2\%$.

Eliminating the depth of flow, d, between Equations 4.21 and 4.2, results in

$$Q_I = 5.23 \frac{S_x^{0.302}}{S^{0.056}} (nQ)^{0.5775} \quad \text{SI} \tag{4.22}$$

$$Q_I = 23.59 \frac{S_x^{0.302}}{S^{0.056}} (nQ)^{0.5775} \quad \text{CU} \tag{4.23}$$

where Q = gutter flow, m³/s (cfs).

These equations indicate that the flow interception capacity of an inlet increases with the gutter flow and cross slope of the gutter but decreases with the longitudinal slope. The ratio of the inlet flow to the gutter discharge is termed the inlet efficiency. This ratio represents the portion of the flow in gutter that is captured by an inlet. The remainder is the flow that bypasses an inlet and

TABLE 4.4

Efficiency of "A" Inlet at Grade[a]

Q m³/s	S = 1%	S = 2%	S = 4%	S = 6%	S = 8%	
		$S_o = 4\%, n = 0.016$				
0.04	92	88	85	83	82	
0.06	77	74	71	70	69	
0.09	65	63	60	59	58	
0.120	58	55	53	52	51	
0.150	52	50	48	47	47	

Q cfs	S = 1%	S = 2%	S = 4%	S = 6%	S = 8%	S = 10%
			$S_x = 4\%, n = 0.016$			
2	79	76	73	72	70	70
3	67	64	62	62	59	59
4	59	57	55	53	53	52
5	54	52	50	49	48	47
6	50	48	46	45	44	44

[a] For types "B" and "D" inlets, the listed efficiencies in this table may be adjusted using a factor of 1.05 for $S \leq 2\%$.

contributes to the gutter flow approaching the next downstream inlet. The efficiency of an inlet is written as

$$E = \frac{Q_1}{Q} = 5.23 \frac{S_x^{0.302}}{S^{0.056}} \frac{n^{0.5775}}{Q^{0.4225}} \quad \text{SI} \tag{4.24}$$

$$E = \frac{Q_1}{Q} = 23.6 \frac{S_x^{0.302}}{S^{0.056}} \frac{n^{0.5775}}{Q^{0.4225}} \quad \text{CU} \tag{4.25}$$

These equations demonstrate that the efficiency of an inlet decreases with increase in the gutter flow. This relation is illustrated by Table 4.4, which is prepared based on Equation 4.24 and shows the inlet efficiency versus gutter flow and roadway slope. As this table indicates, an "A" type inlet at slopes of 2% or greater provides less than 75% efficiency even for a 0.06 m³/s gutter flow. Table 4.4, relating to English units, shows that "A" inlets placed at over 2% slope capture less than 75% of gutter flows of 2 cfs or more.

4.3.6 GRATES ON SAG

The flow at an inlet at low point occurs as weir flow until the water builds up to a certain depth. Beyond that depth, the inlet functions as an orifice. For bicycle-safe grates, the NJDOT reports that inlets behave as a weir when flow depth is less than ±9 in. (230 mm) and as an orifice for greater depths. For a conservative design, the author suggests using 15 cm (6 in.) depth as the criteria for differentiating these types of flow. The flow for each type can be calculated as follows:

a. Weir flow:

$$Q_i = C_w P d^{1.5}$$ (4.26)

$$C_w = (1.66 \text{ SI}) \quad 3.0 \text{ CU} \quad d \leq 15 \text{ cm } (0.5 \text{ ft})$$

where
  C_w = weir coefficient
  P = perimeter around the open area of grate, m (ft)
  d = the depth of the approach flow, m (ft)

The perimeter for inlets at curbs and off curbs can be calculated from the following equations, respectively:

$$P = 2w + \ell$$

$$P = 2(w + \ell)$$

where w is the width and ℓ is the length of inlet grate in m (ft).
b. Orifice flow:

$$Q_i = C_o A_o (2gd)^{0.5} \quad d \geq 15 \text{ cm } (0.5 \text{ ft})$$ (4.27)

where
  C_o = orifice coefficient
  A_o = clear opening area of the inlet grate in m^2 (ft^2)
  d = depth of approach flow in m (ft)

This equation is applicable to both metric and customary units, with $C_o = 0.6$ and $g = 9.81$ m/s^2 (32.2 ft/s^2).

Types "B" and "D" inlets have curb pieces that allow water to enter the inlets when debris partly clogs the grate. For inlets without curb pieces, the area A_o in the preceding equation should be reduced by 50% to account for partial clogging. The capture efficiency of a curb inlet may be increased significantly by depressing the gutter at the inlet. Also, it is recommended to use curb opening inlets and, preferably, combination inlets on sags.

4.4 INLETS SPACING

The spacing of inlets on curbed gutters is determined based on the rate of flow in the gutter, the capacity of inlet, and the allowable spread. The first two parameters have been discussed already. The allowable spread is selected based on the degree of safety for the traffic. The NJDOT (2013) has adopted the following criteria:

- Interstate highways and freeways: full shoulder
- Land service roads: shoulder + one-third width of lane
- Ramps: one-third width of ramp, for all roadways
- Acceleration/deceleration lanes: one-half width of lane, for all roadways

A proper procedure for spacing of inlets is outlined here:

a. Calculate flow in the gutter. Include overland flow tributary to the roadway.
b. Calculate the spread in the gutter. Place the first inlet at the location where the spread approaches the allowable limit.
c. Calculate the capacity of the inlet and its efficiency. Adjust the inlet location, if necessary, to meet the required efficiency and calculate the bypassing flow.
d. Add the bypassing flow to the tributary runoff to the next inlet.
e. Repeat steps "b" through "d" to the end of the system.

The calculations can be arranged in a tabular format. Figure 4.16 shows a spreadsheet, prepared by the author. In this table, the calculations are organized into three sections. Section 1 presents gutter flow calculations in six columns. The first five columns list the inlet number, the area tributary to an inlet, the runoff coefficient, the time of concentration, and rainfall intensity. This latter parameter depends on the time of concentration and storm frequency and is determined using local rainfall intensity–duration–frequency (IDF) curves. Then the rational formula ($Q = CAI$) is applied to the listed figures in columns 2, 3, and 5 to calculate the flow in the gutter.

The next section of the table includes the spread calculations in the gutter. Inputted in columns 7 and 8 are the flow bypassing the previous inlet (nonexistent for the first inlet) and the total flow in the gutter, respectively. The next two columns list the longitudinal slope and cross slope of the gutter. Column 11 presents the calculated spread width based on the listed slopes in the previous two columns using Equation 4.4. The calculated spread width would then be compared with the allowable spread, T_a. If T is larger than T_a, then the inlet is relocated closer to the previous inlet and Q and T are recalculated in columns 2 through 11.

Allowable spread: $T_a = $ ft/m   Manning's $n =$ _____

Gutter flow						Gutter spread						Inlet efficiency			
Inlet no.	A	C	T_c	I	Q	Q_B	Q_T	S	S_x	T	$T < T_a$	Q_I	$E = Q_I/Q$	$E < E_A$	$Q_B = Q - Q_I$

FIGURE 4.16   Inlet spacing calculations (blank form).

The last section of the table examines the inlet efficiency versus its allowable value. In column 13, the inlet capacity, namely the intercepted flow, is determined using appropriate figures in Appendix 4C. Alternatively, the intercepted flow may be calculated using acceptable local or state DOT equations, such as the NJDOT's Equation 4.21. The ratio of intercepted inlet flow to the gutter flow is inputted in column 14, and this ratio is compared with the allowable efficiency. If the efficiency is acceptable, then the flow bypassing the inlet is calculated by subtracting the intercepted inlet flow from the gutter flow and is entered in column 16 and also column 7 for the next inlet. However, if the inlet efficiency is not satisfactory, the inlet spacing is reduced and calculations are repeated from column 2 through column 15. The iteration process can be significantly reduced, or even eliminated, by examining Table 4.2 for flow spread, applicable figures for inlet capacity in Appendix 4C, or Table 4.4 for inlet efficiency.

The NJDOT drainage design manual (NJDOT, 2013) specifies 400 ft (125 m) as the maximum spacing between inlets. The same manual specifies that inlets have a minimum of 75% efficiency. The Residential Site Improvements Standards (2009), enacted in New Jersey in 1997, also specify the maximum spacing between inlets in a development as 400 ft. The same standards specify 6 cfs as the maximum capacity of a "B" inlet. As shown in the previous section, this figure far exceeds the interception capacity of an inlet at grade.

Example 4.4

In Example 4.1, apply the NJDOT equation for inlet capacity to calculate:

a. The flow captured by the first inlet on each shoulder
b. The efficiency of the inlet
c. The maximum spacing to the second set of inlets, accounting for the bypass flow from the first inlets

Solution

In Example 4.1, $Q = 1.74$ cfs, $S = 1.5\%$, $S_x = 4\%$, $T = 4.93$ ft.

a. $d = TS_x = 4.93 \times 0.04 = 0.197$ ft

$$Q_i = \frac{16.88 d^{1.54} S^{0.233}}{S_x^{0.276}} = 16.88 \times 0.197^{1.54} \times \frac{0.015^{0.233}}{0.04^{0.276}} = 1.26 \text{ cfs}$$

Add 5% accounting for interception by the curb piece, $Q = 1.32$ cfs.

b. $E = \dfrac{Q_i}{Q} = \dfrac{1.32}{1.74} = 76\%$

Bypass flow $= 1.74 - 1.32 = 0.42$ cfs
c. Calculate gutter flow for 75% inlet efficiency by an iteration process.

First try $T = 5.0$ ft

$$Q = \frac{0.56}{0.013} \times 0.04^{1.667} \times 0.015^{0.5} \times 5^{2.667} = 0.02466 \times 5^{2.667} = 1.80 \text{ cfs}$$

$$d = S_x T = 0.04 \times 5 = 0.2 \text{ ft}$$

Calculate flow intercepted by the second set of inlets:

$$Q_i = \frac{16.88 \times 0.20^{1.54} \times 0.015^{0.233}}{0.04^{0.276}} = 15.425 \times 0.2^{1.54} = 1.29 \text{ cfs}$$

Add 5% for side interception

$$Q_i = 1.05 \times 1.29 = 1.35 \text{ cfs}$$

$$E = \frac{Q_i}{Q} = \frac{1.35}{1.80} = 0.75 \text{ OK}$$

Allowable roadway runoff:

$$Q = 1.80 - 0.42 = 1.38 \text{ cfs}$$

Maximum tributary area:

$$A = \frac{Q}{CI} = \frac{1.38}{0.95 \times 6.7} = 0.217 \text{ acres} = 9444 \text{ sf}$$

$$L = \frac{9444}{34} = 278 \text{ ft}$$

Place second set of inlets 275 ft from the first inlets.

Exercise: Show that for a roadway with $n = 0.016$, the spacing of a second set of inlets in the preceding example can be increased to ± 360 ft.

Example 4.5

Solve the previous example using 2 ft × 4 ft P-1-1/8 FHA grates and Figure 4C.2b ($n = 0.016$).

Solution

Enter Figure 4C.2b with $Q = 1.74$ cfs, and $S = 1.5\%$ to read:

$$Q_i = 1.5 \text{ cfs}$$

$$E = \frac{Q_i}{Q} = \frac{1.5}{1.74} = 0.86 \quad 86\%$$

Bypass runoff = 1.74 − 1.5 = 0.24 cfs

Note: Due to a higher roughness coefficient, the flow spread is wider; therefore, the inlet has a larger efficiency than the previous example. Also, the FHA P1–1-1/8 grate provides a larger opening area than the NJ type "B" bicycle-safe grates. As such, the second set of inlets can be spaced further than that calculated in the previous example.

First try $T = 6$ ft (entire side shoulder)

$$Q = \frac{0.56}{0.016} \times 0.04^{1.667} \times 0.015^{0.5} \times 6^{2.667} = 2.38 \text{ cfs}$$

Enter the same figure with $Q = 2.38$, $S_o = 0.015$ to read:

$$Q_i = 1.97 \text{ cfs}$$

$$E = \frac{1.97}{2.38} = 82.8\%$$

Contributing roadway runoff $Q = 2.38 - 0.24 = 2.14$ cfs

$$\text{Area} = 2.14/(0.95 \times 6.7) = 0.336 \text{ acre} = 14{,}645 \text{ sf}$$

$$L = \frac{14{,}645}{34} = 430 \text{ ft}$$

Set the second set of inlets at the maximum allowable spacing of 400 ft from the first set.

Example 4.6

Using the calculated discharge in Example 4.2, calculate:

   a. The flow captured by the first inlet on each shoulder
   b. The efficiency of the inlet
   c. The maximum spacing to the second inlet for 75% efficiency

Base your calculations on 0.6 m × 1.2 m FHA P-30 grate ($n = 0.016$).

Solution

Given:

$$Q = 0.0476 \text{ m}^3/\text{s, say, } 0.048 \text{ m}^3/\text{s}$$

$$S_x = 4\%, S = 1.5\%, \Sigma W = 10.6 \text{ m}$$

Enter Figure 4C.2a with $Q = 0.048$ m³/s and $S = 1.5\%$ and interpolate between $Q = 0.03$ and $Q = 0.05$ to read:

$$Q = 0.041 \text{ and } E = 86\%$$

Check efficiency:

$$E = 0.041/0.048 = 85.4\% \quad \text{OK}$$

Bypass flow:

$$Q_b = 0.048 - 0.041 = 0.007 \text{ m}^3/\text{s}$$

Try $T = 2$ m for the second set of inlets.
Calculate Q using Equation 4.1:

$$Q = (0.375/0.016) \times 0.04^{5/3} \times 0.015^{1/2} \times 2^{8/3}$$

$$Q = 0.085 \text{ m}^3/\text{s}$$

Enter the same figure with $Q = 0.085$ and $S = 0.015$ to read:

$$E = 81\% > 75\% \quad \text{OK}$$

Contributing roadway runoff:

$$Q = 0.085 - 0.007 = 0.078 \text{ m}^3/\text{s}$$

$$\text{Area} = Q/(2.78 \times 10^{-3} \times CI)$$

$$\text{Area} = 0.078/(2.78 \times 10^{-3} \times 0.95 \times 170) = 0.1737 \text{ ha} = 1737 \text{ m}^2$$

$$L = 1737/10.6 = 163.9 \text{ m}$$

Set the second inlets at a maximum allowable distance of 120 m (<400 ft).
Note that Figures 4C.1 and 4C.2 are not accurate enough for efficiency calculations.

4.5   INLETS ON ROADWAYS AT 0% GRADE

Roadways traversing through flat lands can be at zero or nearly zero longitudinal grade. A section of the New Jersey Turnpike traversing through Meadowlands is an example where the roadway is at zero or nearly zero longitudinal slope. The gutter flow in such roadways is not governed by Equation 4.1, as the longitudinal slope is zero. Rather, the water impounds on the shoulder as depicted in Figure 4.17 and flows toward the inlet by gravity. This figure presents the flow parameters for a roadway consisting of roadway lanes of total width W, shoulder width W', and inlet spacing L. It is assumed that no off-road runoff enters the shoulders. The flow enters the shoulder laterally from the traffic lanes; thus, the flow increases linearly toward inlets. However, the flow depth increases away from inlets and reaches a maximum, y_{max}, midway between two consecutive inlets.

Approximating the flow on the shoulder as a one-dimensional flow, the author (Pazwash and Boswell, 2003) has developed equations for flow spread. Appendix 4B presents the deviation for the water surface equation with the final result:

$$y^{19/3} = y_0^{19/3} + Kn^2 S_x^2 q^2 \left[\left(\frac{L}{2} \right)^3 - x^3 \right] \tag{4.28}$$

where
q is the gutter flow per unit length of roadway
y_0 is the depth of flow at the inlet
x and y are the distance from the water divide and corresponding depth of water, respectively

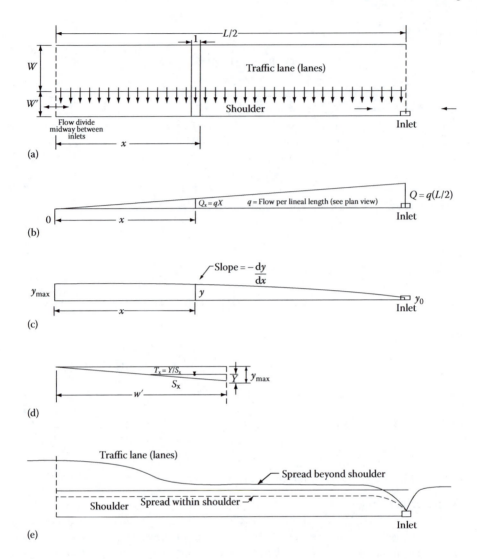

FIGURE 4.17 Gutter flow for a roadway of horizontal profile: (a) plan view; (b) cutter flow vs. distance; (c) variation of depth of flow; (d) flow cross-section; and (e) flow spread.

$$K = 21.3 \quad SI$$

$$K = 10 \quad CU$$

The depth of flow at the inlet can be calculated using the weir flow equation:

$$y_o = \left(\frac{qL}{C_w P} \right)^{2/3} \tag{4.29}$$

where
$P = L + 2w$ (L = length, w = width of inlet)
C_w = weir coefficient

At $x = 0$, where the depth of water attains its maximum, the equation reads as

$$y_{max}^{19/3} = y_0^{19/3} + KS_x^2 n^2 q^2 \left(\frac{L}{2}\right)^3 \tag{4.30}$$

Rewriting the preceding equation in terms of L:

$$L = 2\left\{\frac{[(y_{max})^{19/3} - (y_0)^{19/3}]}{KS_x^2 n^2 q^2}\right\}^{1/3} \tag{4.31}$$

The maximum spacing between inlets, L, can be calculated by setting $y_{max} = S_x W'$, where W' is the shoulder width (assumed to be the allowable spread). Since y_0 varies with L, this equation has to be solved by trial and error to determine the maximum spacing between inlets that would keep the water spread within the shoulder. To facilitate calculations, the term containing y_0 may be neglected in the preceding equation, resulting in

$$L_{max} = 2\left[\frac{(S_x W')^{19/3}}{KS_x^2 n^2 q^2}\right]^{1/3} \tag{4.32}$$

The preceding equations neglect the effects of debris and clogging on the inlets. In the case of uniformly spaced inlets, the full clogging of one inlet will increase the length of flow spread from one-half of the inlets' spacing to full spacing from the clogged inlet. Thus, the maximum spacing between inlets calculated by Equation 4.32 should be reduced by one-half. Further, to account for the depth of flow at inlets, the author recommends that the maximum spacing between inlets be additionally reduced by 20%. Thus:

$$L_{max} = 0.8\left[\frac{(S_x W')^{19/3}}{KS_x^2 n^2 q^2}\right]^{1/3} \tag{4.33}$$

Example 4.7

Calculate the maximum spacing between inlets for a flat roadway having the following parameters:

$$S_x = 3\%$$

$$W = 22 \text{ ft } (6.8 \text{ m}); \ W' = 8 \text{ ft } (2.5 \text{ m})$$

$$n = 0.016$$

$$I = 6.7 \text{ in./h } (170 \text{ mm/h}); \ 25\text{-year, } 10\text{-minute storm in New Jersey}$$

Solution

The maximum spacing for the inlets in this case is calculated in both CU and metric units as follows:

$$\Sigma W = 22 + 8 = 30 \text{ ft}; \ 6.8 + 2.5 = 9.3 \text{ m}$$

$$q = 30 \times 6.7/43{,}560 = 4.61 \times 10^{-3} \text{ cfs/ft}$$

$$q = 2.78 \times 10^{-3} \times 9.3 \times 170/10^4 = 4.40 \times 10^{-4} \text{ m}^3/\text{m}$$

$$y_{max} = S_w W' = 0.03 \times 8 = 0.24 \text{ ft}; \ 0.03 \times 2.5 = 0.075 \text{ m}$$

$$L_{max} = 107 \text{ ft}; \ 34 \text{ m}$$

This result shows that inlets must be closely spaced even in the absence of any offsite runoff. When the off-road runoff is present and significant, it is more feasible to use trench drains such as ACO highway drains or slot drains than grate inlets in flat roadways and pavements.

4.6   DESIGN OF STORM DRAINS

Storm drain pipes are designed to provide sufficient capacity to convey the calculated peak rate of runoff. Commonly, the pipes are designed based on an inlet location plan, which logically should be prepared following inlet spacing calculations. In many development projects, however, an inlet location plan is prepared before any inlet efficiency calculations are performed. The peak rate of runoff to each inlet is generally calculated using the rational formula:

$$Q = 2.78 \times 10^{-3} \, CAI \quad \text{SI} \tag{4.34}$$

$$Q = CAI \quad \text{CU} \tag{4.35}$$

The parameters in this equation as defined in Chapter 3 are

A = area, hectares (acres)
C = runoff coefficient
I = rainfall intensity, mm/h (in./h)
Q = peak runoff, m^3/s (cfs)

For the first reach of pipe, it is assumed that the pipe receives all of the runoff tributary to the most upstream inlet. This is a conservative assumption considering that inlets do not fully capture the gutter flow. Likewise, the design flow for the pipe downstream of a second inlet is calculated by adding the runoff from the tributary area to that inlet and the discharge in the preceding pipe. The calculation process is continued to the most downstream pipe, terminating at a stream, detention basin, receiving drainage system, or body of water.

The calculations may be organized in a tabular format. The US Department of Transportation and many state highway agencies, including the NJDOT, have tables for pipe sizing calculations. Figure 4.18 is a spreadsheet, prepared by the author. In this table the calculation processes are arranged into four sections presenting inlet and pipe reach description, flow calculations, pipe sizing calculations, and the depth of cover over pipes.

Type of pipes: _____   Manning's n: _____

	Pipe section			Inlet calculations						Storm drain calculations												Elevations and cover			
No	From	To	Length (ft)	Acres A (acres)	Runoff coef C	Conc. time (min)	Rain intensity (in/h)	Inlet Q (cfs)	Sum A × C	Total time (min)	Rain intensity (in/h)	Pipe discharge Q (cfs)	Slope (%)	Diameter (in)	Full flow capacity (cfs)	Normal depth (in)	Normal velocity (ft/sec)	Travel time (min)	Ground elev at upper end (ft)	Invert elev upper end (ft)	Invert elev lower end (ft)	Cover at upper elevation (ft)			

FIGURE 4.18 Storm drain design calculations (spreadsheet, prepared by the author).

The inlet calculations section, which follows the inlet and pipe description section, organizes runoff calculations using the rational method. This section, which shows the gutter flow at each inlet, is similar to the calculations presented in Figure 4.16. The third section summarizes calculations for design flow of each pipe and compares this flow with the capacity of the pipe, which is calculated using the Manning formula. The Manning's *n* value for various types of pipes can be found in Table 2.2 in Chapter 2. Also included in this section are the calculations for the flow velocity assuming full flow in each pipe.

Tables in an appendix to Chapter 2 list the flow capacity factors for round and elliptical concrete pipes and some other makes of pipes as well. Actual velocity depends on the ratio of the design discharge to the pipe capacity. Table 2.6 and Figure 2.5 in Chapter 2 facilitate the calculations for partially full flow in round pipes. A figure showing the partial full flow parameters for elliptical pipes is also included in an appendix to that chapter.

The last section in the table lists the invert elevations at the upstream and downstream ends of a pipe and the depth of cover over the pipe, accounting for the wall thickness of the pipe. The calculated cover is then compared with the minimum required cover specified by the pipe manufacturer for a given loading on the pipe to ensure that the pipe has adequate cover. The following case study illustrates the calculation process.

It is to be noted that calculations in Figure 4.18 provide a preliminary design of drainage pipes. In final design, the hydraulic grade line (HGL) calculations are also performed to determine that no surcharge occurs. These calculations include not only frictional losses, but also entrance and exit losses and local losses at structures and bends. Accounting for these losses, flow in many pipes, which according to the preliminary calculations (Figure 4.18) would be part full, may be under pressure due to the backwater effect. The calculations process is illustrated by a case study later (see Case Study 4.2).

CASE STUDY 4.1

The study relates to a drainage improvement project in the borough of New Milford in New Jersey. Figure 4.19 shows the existing drainage system in the project area. Because of inadequacy of the drainage pipes and an obstruction due to a sanitary sewer crossing through MH2, the existing manhole just to the east side of the Lake Street–New Milford intersection was experiencing occasional flooding. To mitigate flooding at this manhole, a new drainage system was designed. The proposed system is shown in Figure 4.20. As shown, the existing 24 in. pipe would be replaced by a 30 in. pipe. Also, this pipe would be dropped by over 2 ft at manhole (MH2) in order to avoid the sanitary sewer that traverses through the manhole.

The drainage area to the proposed system is delineated based on a review of the USGS Hackensack quadrangle and site visits. Figure 4.21 shows the tributary drainage areas to the new inlets. Based on a review of tax maps, Google aerial maps, and site visits, the runoff coefficient was estimated at $C = 0.46$. Figure 4.22 presents calculations for the storm drain pipes, which are designed based on a 25-year frequency storm.

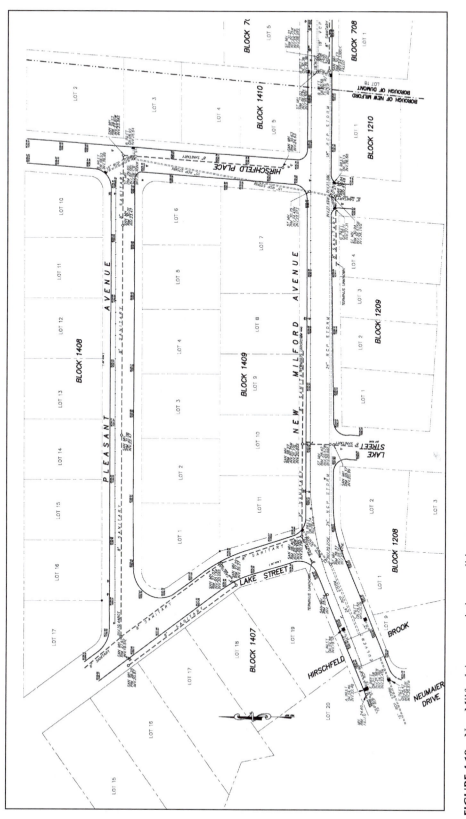

FIGURE 4.19   New Milford Avenue, existing condition map.

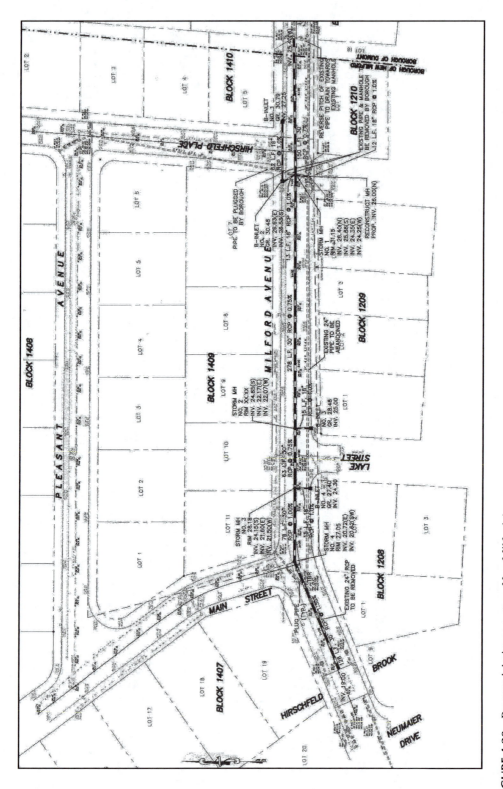

FIGURE 4.20   Proposed drainage system, New Milford Avenue.

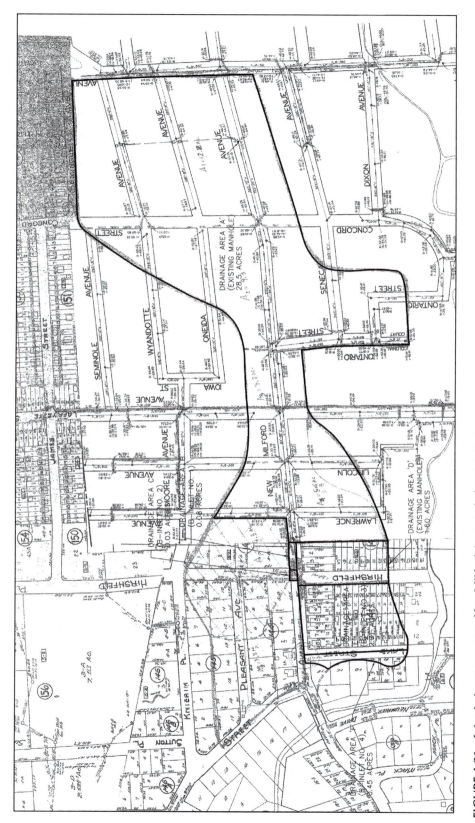

FIGURE 4.21 Inlet drainage area map, New Milford Avenue, drainage system.

Pipe section				Inlet calculations					Storm drain calculations										Elevations and cover			
No	From	To	Length (ft)	Acres *A* (acres)	Runoff coef *C*	Conc. time (min)	Rain intensity (in/h)	Inlet Q (cfs)	Sum *A* × *C*	Total time (min)	Rain intensity (in/h)	Pipe discharge Q (cfs)	Slope (%)	Diameter (in)	Full flow capacity (cfs)	Normal depth (in)	Normal velocity (ft/sec)	Travel time (min)	Ground elev at upper end (ft)	Invert elev upper end (ft)	Invert elev lower end (ft)	Cover at upper elevation (ft)
1	EX MH	MH 1	150	28.50	0.46	60	2.4	31.46	13.110	60.00	2.4	31.46	0.70%	30	37.17	20.83	6.51	0.38	31.03	25.40	24.35	3.1
2	CB 1	CB 2	40	0.08	0.9	10	6.7	0.48	0.072	10.00	6.7	0.48	1.00%	18	11.38	2.61	3.19	0.21	30.57	27.25	26.85	1.8
3	CB 2	MH 1	16	0.03	0.9	10	6.7	0.18	0.099	10.21	6.7	0.66	1.00%	18	11.38	3.11	3.48	0.08	30.48	26.75	26.59	2.2
4	EX MH	MH 1	12	1.60	0.46	20	4.6	3.39	0.736	20.00	4.6	3.39	1.00%	18	11.38	6.66	5.54	0.04	30.94	26.00	25.88	3.4
5	MH 1	MH 2	334					0.00	13.945	60.38	2.4	33.47	0.75%	30	38.48	21.60	6.74	0.83	31.15	24.25	21.74	4.4
6	CB3	CB4	38	2.71	0.46	30	3.6	4.49	1.247	30.00	3.6	4.49	2.00%	18	16.09	6.39	7.70	0.08	28.45	25.00	24.24	2.0
7	CB4	MH2	16	0.45	0.46	10	6.7	1.39	1.454	30.08	3.6	5.23	1.00%	18	11.38	8.64	5.54	0.05	27.54	24.14	23.98	1.9
8	MH2	MH3	88					0.00	15.399	61.21	2.4	36.96	1.00%	30	44.43	20.66	7.78	0.19	28.18	21.64	20.76	4.0
9	CB5	MH3	22	10.43 cfs from Lake St from Borough improvement calculations								10.43	2.00%	18	16.09	10.53	7.83	0.05	26.14	21.80	21.36	2.8
10	MH3	CULV	128						15.399	61.40	2.4	47.39	1.30%	30	50.66	25.80	8.88	0.24	27.05	20.66	19.00	3.9

FIGURE 4.22 Storm drain design (design storm = 25 years, Manning's $n = 0.012$).

CASE STUDY 4.2

Calculate the hydraulic grade line (HGL) in the proposed 30-inch pipe of Case Study 4.1.

SOLUTION

First frictional and structural losses in the pipes and structures (manholes and inlets) are calculated. The losses in structure are due to changes in velocity in the structure and also due to changes in flow direction at bends. Figure 4.23 summarizes head loss calculations for the proposed 30 in. pipes and manholes thereon.

Frictional losses in the pipes are calculated based on the Manning formula, which, as shown in Chapter 2, may be written as

$$h_f = (29n^2L/R^{1.33})V^2/2g \qquad (4.36)$$

where n is the Manning's roughness coefficient and L and R are length and hydraulic radius of each pipe reach, respectively. The losses in structures and bends are calculated applying a factor to the velocity head. The losses also include entrance and exit losses. The former loss occurs at each structure and depends on the type of connection ($K_c = 0.5$, which is for a typical square edge coefficient and is conservatively used herein). However, the exit loss only occurs where the pipe terminates at a stream, lake, or any body of water, and $K_{ex} = 1.0$.

Figure 4.24 presents calculations for hydraulic grade line for the proposed 30 in. pipe along New Milford Avenue. The water surface elevation at the MH-3 on the 30 in. pipe is calculated based on inlet control at the manhole and outlet control due to losses in the first reach of pipe (refer to Figure 4.23). Calculations for inlet–outlet control for the pipe are elaborated in an example in the next section (see Example 4.9).

Figure 4.24 indicates that flow in each pipe reach is under the backwater effect of downstream pipe and that flow in all pipes is completely full. The table also shows that the HGL rises nearly 0.1 ft (less than 1 in. as the calculations are rounded to the first decimal place) at the existing manhole. However, no surcharge would actually occur at this manhole as the 30 in. pipe has significantly more capacity than the existing 24 in. pipe which precedes this manhole and therefore will receive less flow than the calculated 47.4 cfs in the previous example.

4.7 HYDRAULIC DESIGN OF CULVERTS

A culvert is a short reach of conduit employed to convey the flow of a stream, drainage channel, or swale under a roadway or railway. The flow to a culvert can occur with its inlet and outlet faces either being fully submerged or partly open. Also, one of the faces can be partly open while the other face is submerged.

The capacity of a culvert depends on the flow that the inlet face of the culvert can accept as well as the flow that the culvert can pass under the force of gravity. Depending upon which one of these capacities limits the discharge, the culvert flow is labeled accordingly. Specifically, the culvert flow is termed as inlet control if the inlet capacity is smaller than the gravity-driven capacity. On the other hand, the outlet control condition prevails if the inlet capacity is not the limiting factor. Figure 4.25 depicts flow conditions where the inlet capacity is less than the gravity-flow capacity, and thus the flow occurs as inlet control. Figure 4.26 shows outlet control cases where water can enter the culvert at a greater rate than the culvert can convey.

Regardless of the submergence at the inlet or outlet face of a culvert, both the inlet and outlet control capacities should be calculated and the smaller of the two used in design. The inlet control

| Pipe reach | | | | | | | | Head losses | | | | | | | |
From	To	L (ft)	D (ft)	Q (cfs)	V (ft/s)	$V^2/2g$ (ft)	R	h_f	K_s	h_s (ft)	K_b	h_b (ft)	h_e (ft)	h_{xt} (ft)	H_t (ft)
MH3	CUL	128.0	2.5	47.4	9.7	1.46	0.625	1.45	0.0	0.0	0.0	0.0	0.7	1.46	3.62
MH2	MH3	88.0	2.5	37.0	7.5	0.9	0.6	0.6	0.3	0.3	0.0	0.0	0.3	0.0	1.3
MH1	MH2	334.0	2.5	33.5	6.8	0.7	0.6	1.9	0.3	0.2	0.0	0.0	0.4	0.0	2.5
EX MH	MH1	150.0	2.5	31.5	6.4	0.6	0.6	0.7	0.3	0.2	0.0	0.0	0.3	0.0	1.3

Note:

1. $h_f = (29n^2 L / R^{1.33}) V^2 / 2g$ Friction loss
2. $h_s = K_s V^2 / 2g$ Structural loss ($K_s = 0.3$ pressure flow, $K_s = 0$ open channel flow)
3. $h_b = K_b V^2 / 2g$ K_b – Bend factor = 0, no bend
4. $h_e = K V^2 / 2g$ Entrance loss ($K = 0.5$ square edge)
5. $h_{xt} = V^2 / 2g$ Exit loss
6. $H_t = h_f + h_s + h_b + h_e + h_{xt}$ Total loss

FIGURE 4.23   Head loss calculations (spreadsheet; New Milford Ave., New Milford, NJ, $n = 0.012$).

Pipe reach		Q	H_t	T_w	$T_w + T_t$	H_w	Inv. el.	I.C. elev.	HGL	T.O.S.	CL
From	To	(cfs)	Total head loss (ft)	Tailwater elev. (ft)	(ft)	Inlet control headwater (ft)	Upstream inv. elev. (ft)	Inlet control elev. (ft)	Headwater elev. (ft)	Top of structure elev. (ft)	Clearance
MH3	CUL	47.4	3.6	21.4	25.0	5.1	20.7	26.5	26.5	27.1	0.60
MH2	MH3	37.0	1.3	26.5	27.3	3.7	21.6	25.3	27.3	28.2	0.9
MH1	MH2	33.5	2.5	27.3	29.8	3.3	24.3	27.6	29.8	31.2	1.4
EX MH	MH1	31.5	1.3	29.8	31.1	3.0	25.4	28.4	31.1	31.0	−0.1

Note:
1. H_t = Total losses (See Figure 4.23)
2. T_w = Tailwater or $(D + d_c/2)$ (whichever greater for first pipe)
3. I.C. elev. = H_w + Inv. el.
4. HGL = $T_w + H_t$ or I.C. elev. (whichever is greater)
5. H_w is calculated using orifice flow equation

FIGURE 4.24 Hydraulic grade line calculations.

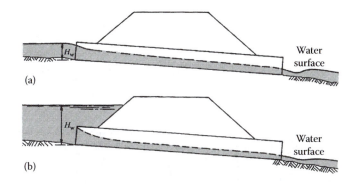

(a)

(b)

FIGURE 4.25 Inlet control schemes. (a) Inlet face open. (b) Inlet face submerged.

capacity is basically the capacity of the opening face, which acts as a weir flow when partly full and as an orifice when fully submerged. The flow equations for these cases are

$$Q = C_w LH^{1.5} \tag{4.37}$$

$$Q = CA(2gh)^{0.5} \tag{4.38}$$

where
L = span of culvert, m (ft)
H = head above culvert invert, m (ft)
h = head above center of opening, m (ft)
A = area of barrel, m² (ft²)
C_w = weir coefficient
C_o = orifice coefficient

The weir coefficient C_w and orifice coefficient C_o vary with the type of culvert, its geometry, and entrance face configurations. The US Department of Transportation, Federal Highway Administration HDS no. 5 (FHWA, 2012) includes rigorous parametric equations for various makes of culverts

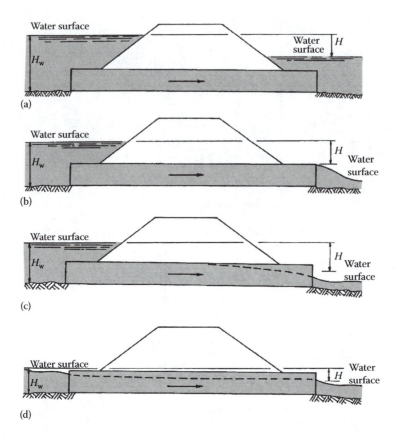

FIGURE 4.26 Outlet control schemes. (a) Both inlet and outlet faces submerged. (b) Inlet submerged, outlet full. (c) Inlet submerged, outlet partly open. (d) Inlet and outlet both partly open.

and their inlet face configurations. The same publication and the *Concrete Pipe Design Manual* (American Concrete Pipe Association, 2005) include nomographs that greatly simplify the inlet control flow calculations for circular, elliptical, arch, and box (rectangular) culverts. Figures 4.27a, 4.28a, and 4.29a, in metric units, present inlet control capacity of round, elliptical, and box culverts, respectively. Similar nomographs in customary units are shown in Figures 4.27b, 4.28b, and 4.29b.

The outlet control capacity can be calculated using the Manning formula for frictional losses and accounting for the entrance and exit losses. The energy equation in this case may be presented as follows:

$$H = \left(\sum K_\ell + \frac{19.62n^2L}{R^{4/3}} \right) \left(\frac{V^2}{2g} \right) \quad \text{SI} \tag{4.39}$$

$$H = \left(\sum K_\ell + \frac{29n^2L}{R^{4/3}} \right) \left(\frac{V^2}{2g} \right) \quad \text{CU} \tag{4.40}$$

In these equations:

V = full flow velocity in culvert, m/s (ft/s)
n = Manning's roughness coefficient
L = length of culvert, m (ft)
R = hydraulic radius, m (ft)
$\sum K_\ell$ = sum of entrance and exit loss coefficients

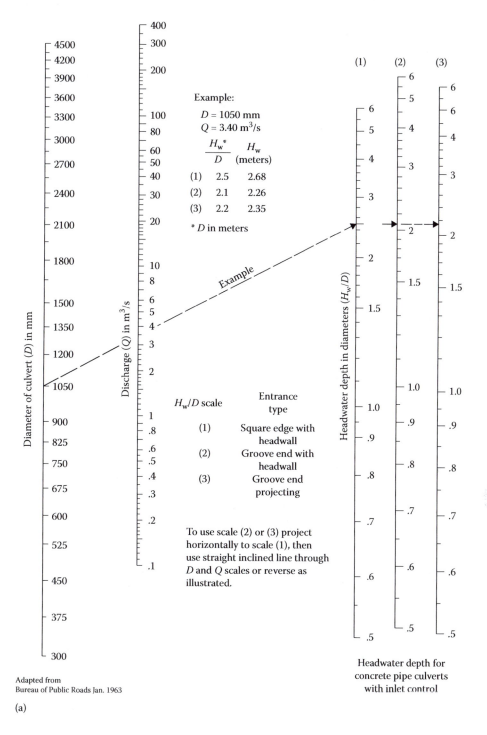

Adapted from
Bureau of Public Roads Jan. 1963

(a)

FIGURE 4.27   (a) Inlet control nomograph, round culverts (SI units).   *(Continued)*

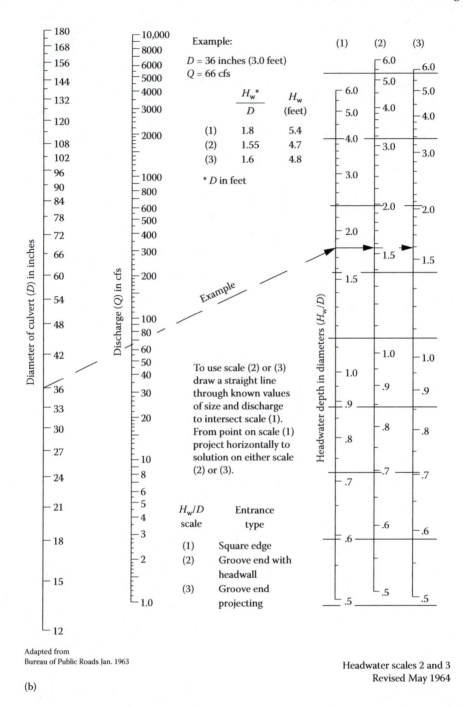

Diameter of culvert (D) in inches

180
168
156
144
132
120
108
102
96
90
84
78
72
66
60
54
48
42
36
33
30
27
24
21
18
15
12

Discharge (Q) in cfs

10,000
8000
6000
5000
4000
3000
2000
1000
800
600
500
400
300
200
100
80
60
50
40
30
20
10
8
6
5
4
3
2
1.0

Example:

D = 36 inches (3.0 feet)
Q = 66 cfs

	$\dfrac{H_w{}^*}{D}$	H_w (feet)
(1)	1.8	5.4
(2)	1.55	4.7
(3)	1.6	4.8

* D in feet

Example

To use scale (2) or (3) draw a straight line through known values of size and discharge to intersect scale (1). From point on scale (1) project horizontally to solution on either scale (2) or (3).

H_w/D scale	Entrance type
(1)	Square edge
(2)	Groove end with headwall
(3)	Groove end projecting

Headwater depth in diameters (H_w/D)

(1)
6.0
5.0
4.0
3.0
2.0
1.5
1.0
.9
.8
.7
.6
.5

(2)
6.0
5.0
4.0
3.0
2.0
1.5
1.0
.9
.8
.7
.6
.5

(3)
6.0
5.0
4.0
3.0
2.0
1.5
1.0
.9
.8
.7
.6
.5

Adapted from
Bureau of Public Roads Jan. 1963

Headwater scales 2 and 3
Revised May 1964

(b)

FIGURE 4.27 (CONTINUED)   (b) Inlet control discharge—headwater relation for circular culverts (CU).

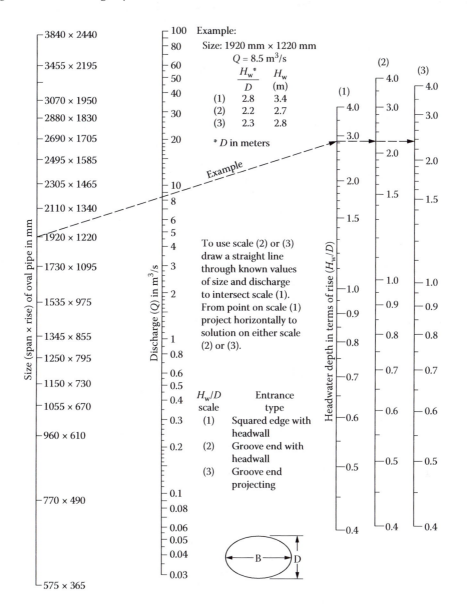

Adapted from
Bureau of Public Roads Jan. 1963

(a)

FIGURE 4.28   (a) Inlet control headwater depth for horizontal elliptical concrete pipes (SI units).   (*Continued*)

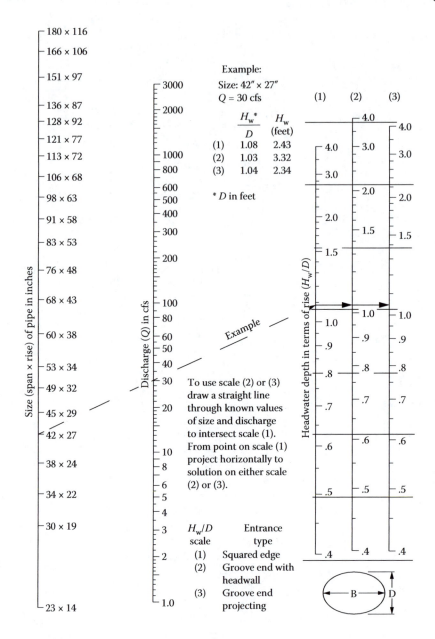

Size (span × rise) of pipe in inches

180 × 116
166 × 106
151 × 97
136 × 87
128 × 92
121 × 77
113 × 72
106 × 68
98 × 63
91 × 58
83 × 53
76 × 48
68 × 43
60 × 38
53 × 34
49 × 32
45 × 29
42 × 27
38 × 24
34 × 22
30 × 19
23 × 14

Discharge (Q) in cfs

3000
2000
1000
800
600
500
400
300
200
100
80
60
50
40
30
20
10
8
6
5
4
3
2
1.0

Example:
Size: 42″ × 27″
Q = 30 cfs

	$\dfrac{H_w{}^*}{D}$	H_w (feet)
(1)	1.08	2.43
(2)	1.03	3.32
(3)	1.04	2.34

* D in feet

Example

To use scale (2) or (3)
draw a straight line
through known values
of size and discharge
to intersect scale (1).
From point on scale (1)
project horizontally to
solution on either scale
(2) or (3).

H_w/D scale   Entrance type
(1)   Squared edge
(2)   Groove end with headwall
(3)   Groove end projecting

Headwater depth in terms of rise (H_w/D)

(1)   (2)   (3)

4.0
4.0   3.0   4.0
3.0   3.0
           2.0   3.0
                      2.0
2.0   1.5   2.0
1.5   1.5
           1.0   1.0
1.0   1.0
.9   .9   .9
.8   .8   .8
.7   .7   .7
.6   .6   .6
.5   .5   .5
.4   .4   .4

Adapted from
Bureau of Public Roads Jan. 1963

(b)

FIGURE 4.28 (CONTINUED)   (b) Inlet control discharge—headwater relation for elliptical pipes (CU).

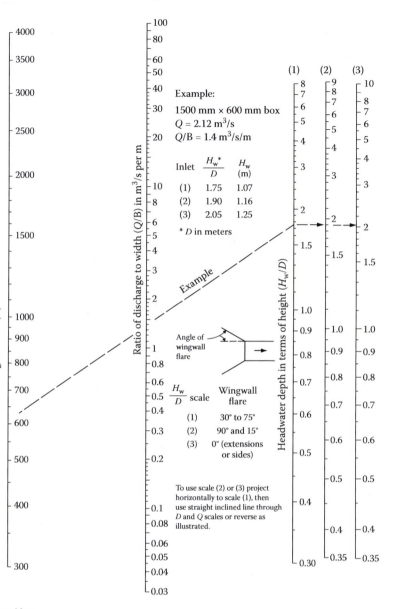

Adapted from
Bureau of Public Roads Jan. 1963

(a)

FIGURE 4.29   (a) Inlet control headwater for box culverts (SI units).   (*Continued*)

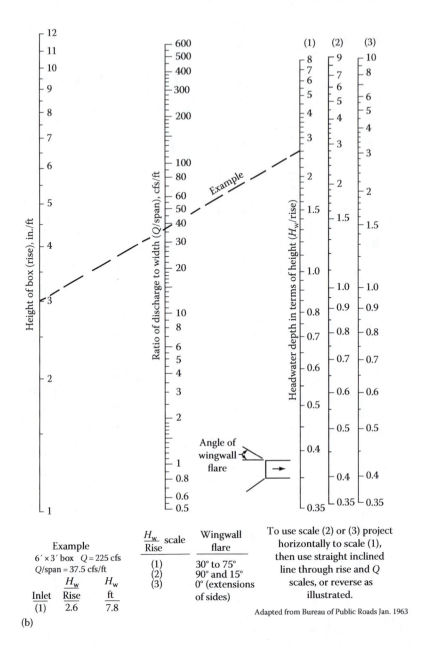

FIGURE 4.29 (CONTINUED) (b) Inlet control discharge—headwater relation for concrete box culverts (CU).

Table 4.5 lists typical values of Manning's roughness coefficient, n, for concrete, HDPE, and corrugated metal conduits, which are commonly used in practice. The entrance loss coefficient can vary from 0.2 to 0.9 depending on the entrance conditions; the former coefficient represents a tapered inlet face, and the latter a CMP culvert projecting from fill. Table 4.6 presents entrance loss coefficients for concrete and CMP culverts for various inlet face configurations. The exit loss coefficient is almost always equal to 1.0. Thus, under normal conditions, ΣK_ℓ for a single culvert may be estimated at 1.5.

The HDS no. 5 (FHWA, 2012) and the *Concrete Pipe Design Manual* (American Concrete Pipe Association, 2005) contain charts to facilitate solution of conduit capacity under outlet control.

TABLE 4.5

Manning "n" Values for Various Culverts

Conduit Type		Manning "n"
Concrete pipe	Smooth walls	0.012–0.013
Concrete box	Smooth walls	0.012–0.015
HDPE	Smooth interior	0.012
Corrugated metal pipes "n" depends on corrugation size	2–2/3 by 1/2 in.	0.022–0.027
and barrel size; smaller "n" values for larger barrels	6 by 1 in.	0.023–0.028
	6 by 2 in.	0.033–0.035
	9 by 2-1/2 in.	0.033–0.037

Note: The Manning "n" values listed in this table are recommended design values for new culverts. For concrete pipes and culverts with deteriorated walls and poor joints, "n" values may be as high as 0.018. Corrugated metal pipes, in addition to corrosion, may also experience deformation and shape change, which could significantly reduce their hydraulic capacity.

TABLE 4.6

Entrance Loss Coefficients

Type of Structure and Design of Entrance	Coefficient, K_e
Concrete Pipe	
Projecting from fill, groove end	0.2
Projecting from fill, sq. cut end	0.5
Headwall or headwall and wingwalls	
Groove end of pipe	0.2
Square-edge	0.5
Mitered to conform to fill slope	0.7
End section conforming to fill slope[a]	0.5
Beveled edges 33.7° or 45° bevels	0.2
Corrugated Metal Pipe and Arch	
Projecting from fill (no headwall)	0.9
Headwall or headwall and wingwalls	
Square-edge	0.5
Mitered to conform to fill slope	0.7
End section conforming to fill slope[a]	0.5
Concrete Box	
Headwall parallel to embankment (no wingwalls)	
Square-edged on three edges	0.5
Rounded on three edges to radius of 1/12 barrel dimension	0.2
Wingwalls at 30° to 75° to barrel	
Square-edged at crown	0.4
Crown edge rounded to radius of 1/12 barrel dimension	0.2
Wingwalls at 10° to 30° to barrel	
Square-edged at crown	0.5
Wingwalls parallel (extension of sides)	
Square-edged at crown	0.7

Source: FHWA, Hydraulic design of highway culverts, Hydraulic Design Series No. 5 (HDS5), 3rd ed., Washington, DC, 2012.

[a] "End section conforming to fill slope" made of either metal or concrete are the sections commonly available from manufacturers.

Figure 4.30 is a sample of charts for round pipes. To improve accuracy, however, Equations 4.39 and 4.40 should be used for head loss calculations.

Design calculations for a culvert to carry a given flow under a roadway or embankment are commonly performed by an iterative process. This process consists of selecting a culvert size and type and then performing calculations to determine the required head for both the inlet control and outlet control flow conditions; the larger of the two is used in design. Commonly, the required headwater for inlet control is estimated based on inlet control charts, such as Figures 4.27 through 4.29; the head losses for the outlet control are calculated using Equations 4.39 and 4.40. The calculations may be organized in a table (see Figure 4.31).

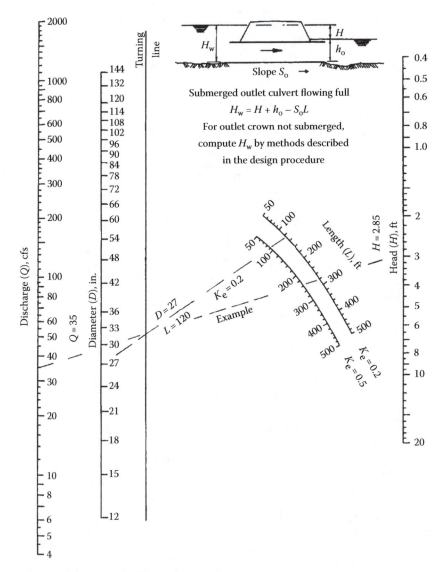

Adapted from Bureau of Public Roads Jan. 1963

FIGURE 4.30 Outlet control discharge–headwater relation for circular concrete pipe culverts flowing full, $n = 0.012$ (CU).

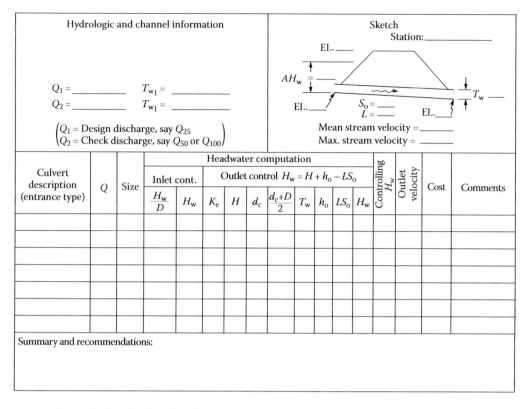

FIGURE 4.31 Hydraulic tabulation for culverts. Parameters are derived as follows: H_w/D, inlet control nomograph; K_e, Table 4.6; H, Equations 4.39 and 4.40; d_c, critical depth, Figures 4D.1 and 4D.2 in Appendix 4D or Table 2.2 in Chapter 2 (for circular culverts); h_o, T_w or $(d_c + D)/2$, whichever is greater; S_o and L are as defined on the sketch.

Example 4.8

The proposed 30 in. pipe along New Milford Avenue (Case Study 4.1) terminates at the New Milford Bridge over Hirschfeld Brook (Figure 4.19). A review of the state and FEMA (Federal Emergency Management Agency) flood maps shows the 10-year flood elevation of Hirschfeld Brook to be at approximately elev. 17.7 ft at the New Milford Avenue Bridge. Calculate the inlet control capacity of the 30 in. pipe.

Solution

The 30 in. pipe is designed based on a 25-year frequency storm. Considering that the floods in the brook lag hours behind rainfalls, the 10-year flood level of Hirschfeld Brook may conservatively be assumed as the backwater elevation at the pipe outlet. However, since the invert of the pipe is well above the 10-year flood level, the proposed pipe is not under a backwater effect of the brook. Therefore, headwater can be calculated based on the rim and invert elevations of MH#3. According to Figure 4.20, these elevations are respectively

Rim elev. = 27.05 ft
Inv. elev. = 20.66 ft

Therefore, the headwater depth is

$$H_w = 27.05 - 20.66 = 6.39 \text{ ft}$$

$$D = 30 \text{ in.} = 2.5 \text{ ft}$$

$$\frac{H_w}{D} = 2.6$$

Drawing a straight line connecting these values for pipe diameter and H_w/D on Figure 4.27b gives an inlet control capacity of $Q = 55$ cfs.

Example 4.9

Calculate the inlet capacity of the existing 30 in. pipe in the preceding example using the orifice flow equation.

Solution

Applying the orifice flow equation to MH-3 results in

$$Q = 0.6 \times (\pi \times 2.5^2/4) \times (2gH)^{0.5} \tag{4.41}$$

where H is the available head to the center of the 30 in. pipe and is given as

$$H = 27.05 - (20.66 + 1.25) = 5.14 \text{ ft}$$

where 27.05 and 20.66 are the rim elevation of MH-3 and the invert elevation of the 30 in. pipe, respectively (see Figure 4.22).

$$Q = 0.6 \times 4.91 \times (2 \times 32.2 \times 5.14)^{1/2} = 53.6 \text{ cfs}$$

This result is very close to that obtained using Figure 4.27b in the previous example.

Example 4.10

The calculated 25-year peak runoff from a 38.85 acre area tributary to the 30 in. pipe in the previous example is calculated at 47.4 cfs. Calculate the water surface elevation at manhole MH#3. The pipe is 128 ft long and, as indicated, is not under a backwater effect of Hirschfeld Brook.

Solution

Calculate headwater under inlet and outlet control conditions using Figure 4.27b and Equation 4.40, respectively.

a. Inlet control condition
   Drawing a straight line through given data for pipe diameter and discharge and extending to H_w/D line for square edge line gives a conservative value of $H_w/D = 2.15$. Therefore:

$$H_w = 2.15 \times 2.5 = 5.38 \text{ ft*}$$

$$\text{W.S. elev.} = 20.66 \text{ (inv. elev.)} + 5.38 = 26.04 \approx 26 \text{ ft}$$

---

* Note: Calculating H_w using the orifice flow equation (with $C = 0.6$) gives $H_w = 5.3$ ft.

b. Outlet control conditions

The entrance loss coefficient for this pipe square edge with headwall is $K_e = 0.5$ (see Table 4.6).

$$A = \left(\frac{\pi}{4}\right)D^2 = 4.91\,\text{ft}^2$$

$$V = \frac{47.4}{4.91} = 9.65\,\text{fts/s}$$

$$\frac{V^2}{2g} = \frac{9.65^2}{2 \times 32.2} = 1.45$$

$$R = \frac{D}{4} = 0.625\,\text{ft}$$

Use $n = 0.012$ (Table 4.9 in Section 4.9.1).

$$H = \left(1.5 + \frac{29 \times 0.012^2 \times 128}{0.625^{4/3}}\right) \times 1.45$$

$$H = (1.5 + 1.0)1.45 = 3.62\,\text{ft}$$

The preceding calculations are arranged in the following table. In this table the backwater effect is inputted as

$$T_w = \frac{(d_c + D)}{2}$$

where d_c may be derived from Table 2.2 (critical depth for circular pipes) as follows:

$$K_c = \frac{Q}{\sqrt{g}d^{2.5}}$$

$$K_c = \frac{47.4}{\left[(32.2)^{1/2} \times 2.5^{2.5}\right]} = 0.845$$

$$\frac{d_c}{D} = 0.905 \quad \text{(see Table 2.2 in Chapter 2)}$$

$$d_c = 2.26\,\text{ft}$$

Alternatively, using Figure 4D.1b in Appendix 4D gives

$$d_c = 2.25\,\text{ft}$$

$$h_o = (2.26 + 2.5)/2 = 2.38$$

$$LS_o = 128 \times 1.3\% = 1.66\,\text{ft}$$

$$H_w = H + h_o - LS_o = 3.62 + 2.38 - 1.66 = 4.34\,\text{ft}$$

Since H_w for inlet control is larger than H_w for outlet control, the 30 in. pipe is under inlet control. According to Figure 4.22, the invert of the pipe lies 6.39 ft below the rim of MH#3; therefore, the available head is sufficient and water would not surcharge from the manhole. The calculations are organized on Figure 4.32.

Project: _____   Designer: _____
                                                    Date: _____

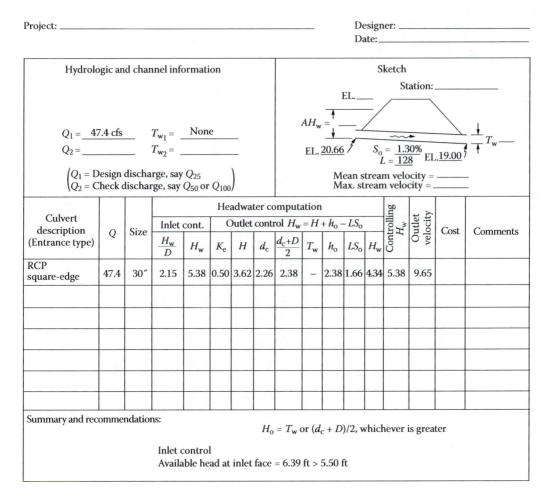

FIGURE 4.32 Table of headwater calculations, Example 4.10.

4.8 EROSION CONTROL AT OUTFALLS

An erosive flow may occur at the outlet of a pipe in a stream or channel. The potential for erosion can be determined based on either the permissible velocity or permissible shear stress for the ambient soil. The allowable velocity concept, which requires a simple analysis and is commonly used for outlet conduit protection, is discussed first. Table 4.7 includes allowable velocities for various soils. To determine soil stability these velocities are compared with the design velocity in the conduit. The design velocity is that which occurs during the erosion control design storm, which is generally the 25-year frequency storm. When design velocity exceeds the allowable velocity, the conduit outlet must be protected to avoid erosion and scour.

The protection commonly consists of a stone riprap section extending from the culvert outlet to where the channel is stable or to a reach to reduce the flow velocity below permissible values in the channel. Following a recommendation by the National Academy of Science in 1970, riprap aprons are designed based on, at a minimum, the 25-year flood. The erosion control may also be provided using a scour hole. Design calculations for riprap apron and scour hole are presented in the following sections.

TABLE 4.7

Permissible Velocities for Various Soils

	Allowable Velocity	
Soil Texture	m/s	ft/s
Sand	0.5[a]	1.75
Sandy loam	0.8	2.50
Silt loam	0.9	3.00
Sandy clay loam	1.1	3.50
Clay loam	1.2	4.00
Clay, fine gravel	1.5	5.00
Cobbles	1.7	5.50
Shale (nonweathered)	1.8	6.00

Source: NJ State Soil Conservation Committee, Standards
for soil erosion and sediment control in New
Jersey, 1999.

[a] Rounded to the nearest tenth.

4.8.1 RIPRAP APRONS

Where the outlet terminates either at a well-defined channel or a flat area such as the bed of a detention basin, a riprap apron is used to protect the soil from erosion. The length of an apron, L_a, is calculated based on empirical equations, which were developed by the US Environmental Protection Agency (EPA, 1976). In SI units, the length, L_a, and the size of stone, d_{so}, are as follows:

$$L_a = \frac{3.26q}{D_o^{1/2}} + 7D_o \quad \text{for } T_w < \frac{D_o}{2} \tag{4.42}$$

$$L_a = \frac{5.43q}{D_o^{1/2}} \quad \text{for } T_w \geq \frac{D_o}{2} \tag{4.43}$$

$$d_{so} = \frac{3.5}{T_w} q^{1.33} \tag{4.44}$$

In customary units these equations become

$$L_a = \frac{1.8q}{D_o^{1/2}} + 7D_o \quad \text{for } T_w < \frac{D_o}{2} \tag{4.45}$$

$$L_a = \frac{3q}{D_o^{1/2}} \quad \text{for } T_w \geq \frac{D_o}{2} \tag{4.46}$$

$$d_{so} = \frac{0.2}{T_w} q^{1.33} \tag{4.47}$$

where

d_{so} = median stone diameter in cm (in.)
D_o = culvert height in m (ft)
$q = Q/W_o$
Q = culvert discharge, m³/s (cfs)
W_o = maximum culvert width
T_w = tailwater depth above the invert of the culvert, m (ft)

Where T_w cannot be estimated, $T_w = 0.2\ D_o$ is used.

In a well-defined channel, the bottom width of the apron should be at least equal to the bottom width of the channel; the lining should extend at least 1 ft above the tailwater elevation, but no lower than two-thirds of the vertical conduit dimension above the culvert invert. In addition, the riprap channel section should satisfy the following conditions:

- The side slopes should be 2:1 (2 H, 1 V) or flatter.
- The bottom grade should be level (0% slope).
- There should be no drop at the end of the apron or at the terminus of the culvert.

When Reno mattress or erosion control blankets are used in lieu of riprap stone, the side slopes may be increased to 1:1.

Where there is no well-defined channel downstream of the apron, the width, W, at the end of apron should be as follows:

$$W = 3D_o + L_a \quad \text{for } T_w < \frac{D_o}{2} \tag{4.48}$$

$$W = 3D_o + 0.4L_a \quad \text{for } T_w > \frac{D_o}{2} \tag{4.49}$$

The preceding equations are valid in both SI and CU units. The width of apron at the culvert outlet should be at least three times the culvert width. Figure 4.33 shows riprap apron geometries in these cases. The size of the stone in both cases is calculated using Equations 4.44 and 4.47.

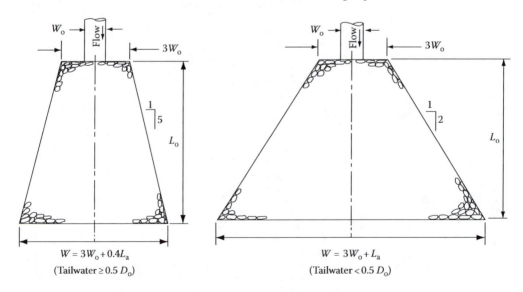

FIGURE 4.33   Riprap apron configuration at outlets.

For multiple culverts closely spaced, the length of the riprap may be sized for one culvert and the width should accommodate all culverts. When the culvert spacing equals or exceeds one-fourth of the width dimensions, the riprap length and stone size, calculated for one culvert, are increased by 25%.

The Soil Conservation Service (SCS) (presently the National Resources Conservation Service [NRCS]) has developed two charts for conduit outlet protection design: one chart for tailwater depth below the pipe center ($T_w < 0.5\ D_o$), and another for $T_w \geq 0.5\ D_o$. Figure 4.34 shows the chart for $T_w < 0.5\ D$. Using this figure, both the stone size, d_{50}, and the length of riprap apron, L_a, can be determined. The bottom grade of the apron should be constructed on a flat grade (0%). Figure 4.34 has been adopted by a number of soil conservation districts, including New York State (NY State Department of Environmental Conservation, 2005).

4.8.2 Preformed Scour Holes

Where a flat apron is either impractical or too large, existing scour holes may be modified to control erosion. In new outfalls, scour holes may be installed to achieve the same objective. Figure 4.35 shows a layout and section view of a scour hole. The depth of scour holes, y, commonly varies from $D_o/2$ to D_o, where D_o is the vertical diameter of the pipe/culvert.

The median stone diameter, d_{so}, can be calculated using the following equations:

$$d_{so} = \frac{2.8}{T_w} q^{4/3} \quad \text{when } y = \frac{D_o}{2} \tag{4.50}$$

$$d_{so} = \frac{1.8}{T_w} q^{4/3} \quad \text{when } y = D_o \tag{4.51}$$

In customary units, these equations become respectively

$$d_{so} = \frac{0.15}{T_w} q^{4/3} \quad \text{when } y = \frac{D_o}{2} \tag{4.52}$$

$$d_{so} = \frac{0.1}{T_w} q^{4/3} \quad \text{when } y = D_o \tag{4.53}$$

The parameters and units in these equations are the same as those defined for riprap apron.

The riprap stone should be fieldstone or rough unhewn quarry stone. Stone should be angular, hard, and of such quality that it does not disintegrate due to weathering or exposure to flowing water. The riprap should be composed of well-graded mixture down to 25 mm (1 in.) size particles such that 50% of the mixture by weight is larger than the calculated d_{so} size and placed on a geo-textile fabric or stone filter or a combination of both. The Standards for Soil Erosion and Sediment Control in New Jersey (1999) follow the EPA equations for riprap dimensions and specify a riprap thickness equal to three times d_{so} with no filter fabric or two times d_{so} when a filter fabric is placed beneath the riprap. Some states, Michigan among others, recommend a minimum riprap thickness of 1.5 d_{so} or 6 in., whichever is greater. The standards in New York State (2005, Section 5B) and many other states call for riprap thickness of 1.5 times maximum stone size plus thickness of filter or bedding.

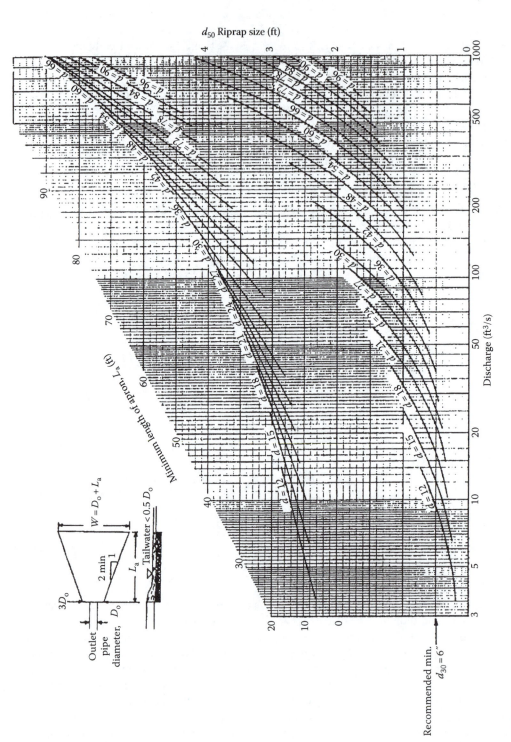

FIGURE 4.34　Conduit outlet protection design (after SCS) for round pipe and minimum tailwater ($T_w \leq 0.5\,D_o$).

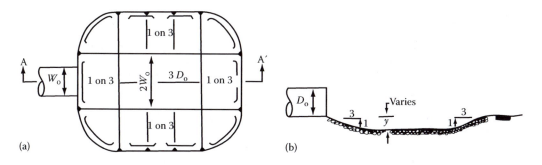

FIGURE 4.35   Preformed scour hole: (a) plan and (b) section.

Example 4.11

Runoff from a residential development is conveyed by a 750 mm pipe into an infiltration-detention basin. The 25-year discharge in the pipe is calculated at 0.5 m³/s. Calculate dimensions of the required riprap apron to protect the sand bed in the basin from erosion based on the Standards for Soil Erosion and Sediment Control in New Jersey (1999). The design tailwater depth, which in New Jersey is specified as the maximum water level at the detention basin during a 2-year storm, is calculated at 0.6 m.

Solution

$$A = \pi \times 0.75^2/4 = 0.442 \text{ m}^2$$

$$V = \frac{Q}{A} = \frac{0.5}{0.442} = 1.13 \text{ m/s}$$

Since design velocity is larger than the permissible velocity for sand, riprap apron is required.

$$T_w = 0.6 \text{ m} > \frac{D_o}{2}$$

$$q = \frac{Q}{D_o} = \frac{0.5}{0.75} = 0.667 \text{ m}^3/\text{s}$$

$$L_a = \frac{5.43q}{D_o^{1/2}} = \frac{5.43 \times 0.667}{(0.75)^{0.5}} = 4.18 \text{ m, use } 4.2 \text{ m}$$

$$W = 3D_o + 0.4L_a = 3 \times 0.75 + 0.4 \times 4.18 = 3.92 \text{ m, use } 4.0 \text{ m}$$

Using Equation 4.34:

$$d_{50} = \left(\frac{3.5}{0.6}\right) \times 0.667^{1.33} = 3.39 \text{ cm}$$

Use 7.5 cm (3 in.) min. size. Thickness of riprap 3 × 7.5 = 22.5 cm, no filter fabric.

Example 4.12

A 36 in. pipe terminates at an unlined drainage channel. The 25-year discharge from the pipe is calculated at 49 cfs. Design stone riprap to protect the channel from erosion. Base your calculations on (a) the EPA method, and (b) the SCS method (Figure 4.34). In the former method assume tailwater to be above the center of the pipe.

Solution

Given $Q = 49$ cfs and $D_o = 36$ in.:

a. Using Equations 4.42 and 4.34:

$$q = \frac{Q}{D_o} = \frac{49}{3} = 16.33$$

$$L_a = 3 \times \frac{16.33}{3^{0.5}} = 28.3 \text{ ft (use 28 ft)}$$

Conservatively assume $T_w = D_o/2 = 1.5$ ft.

$$d_{50} = \frac{0.2 \times 16.33^{1.33}}{1.5} = 5.47 \text{ in. (use 6 in.)}$$

Construct channel banks at 2:1 slope and place riprap to elevation 2 ft above the bed, which is equal to two-thirds of the pipe vertical diameter.

b. Using Figure 4.34:
Extrapolating the $d = 36$ in. line to $Q = 49$ cfs gives

$$d_{50} = 6 \text{ in.}$$

Likewise, extrapolating the upper $d = 36$ in. line to intersect $Q = 49$ cfs and extending a horizontal line to the left gives $L_a = 16$ ft.

This length is significantly shorter than that obtained based on the EPA method. The width of the riprap apron at the outfall in this method is the same as that in the previous method.

Example 4.13

The discharge pipe from an infiltration–detention basin terminates at a stream. The pipe is 450 mm in diameter, and the 100-year discharge from the basin is computed at 0.55 m³/s. To reduce the area of disturbance at the stream, a scour hole is provisioned at the outfall point. The depth of scour hole is 22.5 cm; the tailwater at the stream, which may be taken as the 2-year flood elevation, is 0.9 m deep. Size the scour hole.

Solution

Top length of scour hole, $L = 3D + \left(3\dfrac{D}{2}\right) \times 2 = 6D = 6 \times 0.450 = 2.7 \text{ m}$
Top width of scour hole, $W = 5D = 2.25 \text{ m}$
Bottom length of scour hole, $L' = 3D = 1.35 \text{ m}$
Bottom width of scour hole, $W' = 2D = 0.9 \text{ m}$

Stone size is

$$q = \frac{Q}{D} = \frac{0.55}{0.45} = 1.222\,\text{m}^2/\text{s}$$

$$d_{so} = \left(\frac{2.8}{T_w}\right) q^{1.33}$$

$$d_{so} = \left(\frac{2.8}{0.9}\right) 1.222^{1.33} = 4.06\,\text{cm}$$

Use three layers of 7.5 cm (3 in.) stone, which is the minimum size specified by the standards in many states.

4.9   DRAINAGE CHANNELS

Drainage channels, like pipes, are used to convey the runoff from developments and roadways. Channels, also referred to as swales, are also used along the toe of slopes. To avoid erosion, channels may need to be protected with lining. The lining can be rigid or flexible. Rigid linings include cast-in-place concrete, stone masonry, and grouted riprap. Gabion walls may be classified as semirigid lining. Rigid linings were widely used in the past; however, the new trend is to use flexible linings. As such, rigid linings are not discussed in this text.

Flexible linings include riprap stone, gravel mulch, grass, synthetic mats, and fiberglass roving and the like. In humid areas such as the northeast part of the United States, grass lining may serve as the most effective means of protecting swales in mildly sloped terrains. In addition to economy, grass swales are easy to maintain and aesthetically pleasing. By removing silt and suspended sediment from runoff, grass swales also improve water quality. Grass lining can be best achieved by sodding, with sods laid parallel to the flow direction and secured with pins or staples. If the runoff is diverted during the grass growth or the grass is protected until the vegetative cover is established, seeding is also satisfactory. In arid climates, in the absence of irrigation, and in steep terrains where high-flow velocities are likely to occur, other lining such as gravel, riprap stone, or gabion mattress should be considered.

Swales covered with flexible lining, such as riprap stone or grass, are generally designed based on the resistance of the cover to erosion, disregarding the properties of the native soil. An exception to this design procedure is presented in HEC-15 (FHWA, 2005), to be discussed later. In designing grass-lined swales, the ability of equipment to mow the grass is an important factor to consider. Over time, trapezoidal or V-shaped sections become parabolic and thus parabolic grass swales are more practical than other geometrics.

Drainage channels, whether lined or unlined, are designed based on two different concepts. One concept, like that of erosion control at outfalls, is the maximum permissible velocity. The other is permissible shear stress, also known as the tractive force method. The former method is empirical in nature and considers the channel to be stable against erosion as long as the actual flow velocity is less than a critical one, referred to as the maximum permissible velocity. However, in the tractive force method, the shear stress is the basis of the stability criterion. These concepts are discussed in the next two sections.

4.9.1   Permissible Velocity Concept

In a previous section, the permissible velocity of soils was introduced as a criterion for determining the need for, rather than the design of, riprap apron at a conduit outfall. The use of a definite permissible velocity for a given soil is too simplistic, as the said velocity for noncohesive material varies not only with the size of soil particle, but also with the compactness of the soil. For cohesive soils, the permissible velocity depends on a plasticity index and soil compactness. Figure 4.36 shows the USSR data for permissible velocity of noncohesive soils as a function of grain size, in millimeters and US standard sieve sizes (Chow, 1959).

The USSR data for permissible velocities for cohesive soils are presented in Figure 4.37 (Chow, 1959). The velocities in Figures 4.36 and 4.37 relate to approximately 1 m deep flow. For other flow depths a correction factor should be applied to the velocities shown in the figures. Table 4.8 lists the flow depth adjustment factor using the USSR data.

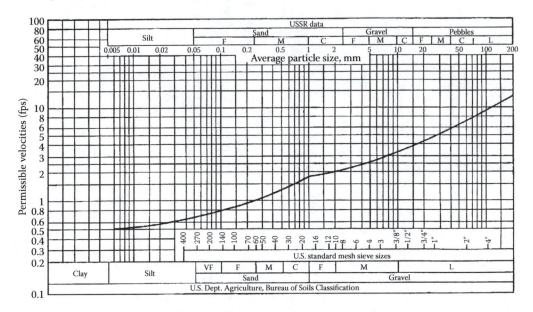

FIGURE 4.36 USSR data on permissible velocities of noncohesive soils. C—coarse; F—fine; L—large; M—medium; VF—very fine.

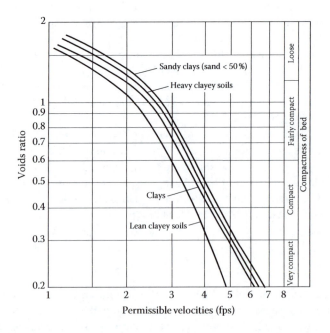

FIGURE 4.37 Curves showing USSR data on permissible velocity of cohesive soils.

TABLE 4.8

Corrections of Permissible Velocity with Depth for Both Cohesive and Noncohesive Material

Flow Depth, m (ft)	Adjustment Factor
0.3 (1)	0.8
0.6 (2)	0.92
1 (3.3)	1.0
1.5 (4.9)	1.09
2 (6.6)	1.16
3 (9.8)	1.25

TABLE 4.9

Permissible Velocity of Noncohesive Particles for Sediment-Free and Sediment-Laden Channels

Particle Size, mm	Permissible Velocity, m/s (ft/s)	
	Sediment Free	Sediment Laden[a]
2.0	0.6 (2.0)	1.0 (3.2)
5.0	0.8 (2.5)	1.3 (4.2)
10.0	1.0 (3.3)	1.6 (5.2)
50.0	1.9 (6.2)	2.6 (8.5)
100.0	2.6 (8.5)	3.2 (10.5)

[a] Sediment-laden figures apply to suspended material concentrations of over 2% by weight.

It is to be noted that clear water has a larger potential for erosion than a sediment-laden flow. This effect is depicted in a figure in the Standards for Soil Erosion and Sediment Control in New Jersey (NJ State Soil Conservation Committee, 1999). Table 4.9 is prepared based on that figure.

A correction factor for flow depth, similar to that of USSR data, applies to the permissible flow velocities in this table. The permissible velocities for sediment-free channels in Table 4.9 closely follow the USSR data. Since stable channels are expected to be sediment free, Figure 4.36 forms a more conservative basis for estimating the permissible velocity of noncohesive material.

Channel side slopes and channel bends also affect the channel stability. These parameters, however, will be treated in the tractive force method, which is more widely used than the permissible velocity concept for design of stable channels.

4.9.2 Tractive Force Method

In this method, the flow shear stress at the channel bed is compared with the permissible shear stress of the ambient soil or the flexible lining material. The flow shear stress, which reflects the hydrodynamic force of flowing water in a channel, is known as tractive force. The basis for stable channel design with flexible living is that tractive force should not exceed the critical shear stress of lining materials. In uniform flow, the average shear stress on the channel perimeter is given by

$$\tau = \gamma RS \tag{4.54}$$

where
  γ = unit weight of water, N/m³ (lb/ft³)
  R = hydraulic radius, m (ft)
  S = average bed slope, m/m (ft/ft)
  τ = shear stress, lb/ft² (CU);   Pascal = N/m² in SI units

It is to be noted that the permissible shear stress can be related to the permissible velocity. Eliminating the slope, S, between Equation 4.54 and the Manning formula results in

$$V_p = \frac{0.01 R^{1/6} \tau_p^{1/2}}{n} \quad \text{SI} \tag{4.55}$$

$$V_p = \frac{0.189 R^{1/6} \tau_p^{1/2}}{n} \quad \text{CU} \tag{4.56}$$

where τ_p and V_p are the permissible shear stress and the permissible velocity, respectively. These equations indicate that V_p is not a constant for a given channel lining; rather, it varies with the hydraulic radius. Thus, the use of permissible shear stress, which depends on hydraulic conditions, is more appropriate than the permissible velocity criteria for the selection of a lining material. For this reason, the concept of permissible velocity is losing popularity in engineering practice. In fact, many jurisdictional agencies recommend using the tractive force method, rather than the permissible velocity concept, for design of stable channels in noncohesive material and in cohesive material of low plasticity. See, for example, NJ Standards for Soil Erosion (NJ State Soil Conservation Committee, 1999).

Shear stress is distributed nonuniformly along the wetted perimeter of a channel. Distribution of shear stress in a trapezoidal channel tends toward zero at the corners with a maximum at the center bed of the channel. The maximum shear stress at the sides occurs nearly at the lower third of the slope. Figure 4.38 depicts the shear stress variation in a channel section.

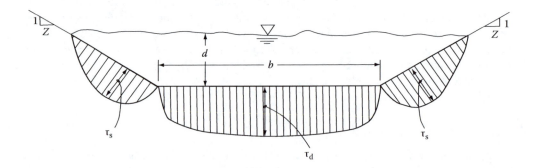

FIGURE 4.38   Variation of shear stress in trapezoidal channel.

The maximum shear stress for a straight channel is given as

$$\tau_d = \gamma dS \tag{4.57}$$

where d is the maximum depth of flow. Note that the depth of flow in rectangular and trapezoidal channels were denoted as "y" in Chapter 2. In swales, the flow depth is commonly nonuniform and the maximum depth is designated as "d" here to differentiate from "y."

In design of a stable channel, a safety factor is applied to the shear stress at the bottom of the channel, satisfying the following equation:

$$\tau_p \geq SF\tau_d \tag{4.58}$$

where

SF = safety factor, typically 1.0 to 1.5
τ_p = permissible shear stress of channel cover

4.9.3 BARE SOIL AND STONE LINING

Permissible shear stress for bare soils and stone linings is listed in Table 4.10. Figures 4.39 and 4.40 present the permissible shear stresses of noncohesive and cohesive soils as function of grain size and plasticity index, respectively.* Table 4.10 lists the permissible shear stress of noncohesive material. According to this table, the permissible shear stress of fine grained soils of d_{75} smaller than 1.3 mm (0.05 in.) is relatively constant and may be conservatively estimated at 1.0 N/m² (0.02 lb/ft²). For noncohesive soils of $1.3 < d_{75} < 15$ mm (0.05–0.6 in.), the permissible shear stress may be calculated using the following equation:

$$\tau_p = Kd_{75} \tag{4.59}$$

where

τ_p = permissible shear stress N/m² (lb/ft²)
$K = 0.75$ SI, 0.4 CU
d_{75} = 75% finer soil size, mm (in.)

The same equation may be applied to coarse and very coarse gravel and stone riprap, however, substituting d_{50} for d_{75} in the equation.

$$\tau_p = Kd_{50} \quad d_{50} > 15 \text{ m} \tag{4.60}$$

For cohesive materials, as Figure 4.40 indicates, the permissible shear stress depends on the soil plasticity index and its compactness, namely porosity or void ratio. The newest edition of HEC-15 (FHWA, 2005) presents the following formula for permissible shear stress of cohesive soils:

$$\tau_p = (C_1 PI^2 + C_2 PI + C_3)(C_4 + C_5 e)^2 C_6 \tag{4.61}$$

where

PI = plasticity index
e = void ratio (volume of void to volume of solids)
C_1 through C_6 are coefficients that depend on the soil type. (A copy of these coefficients, which are listed in Figure 4.18 in HEC-15 [FHWA, 2005], is included in Appendix 4E.)

The use of Equation 4.61 requires measurement of void ratio in addition to plasticity and soil type.

---

* Figures 4.39 and 4.40 have been removed from HEC-22 (2013). Instead, shear stresses of noncohesive and cohesive soils are presented by Equations 4.60 and 4.61 and Table 4.10.

TABLE 4.10

Permissible Shear Stresses for Lining Materials

Lining Category	Lining Type	Permissible Shear Stress	
		N/m²	lb/ft²
Bare soil cohesive (PI = 10)	Clayey sands	1.8–4.5	0.037–0.095
	Inorganic silts	1.1–4.0	0.027–0.110
	Silty sands	1.1–3.4	0.024–0.072
Bare soil cohesive (PI ≥ 20)	Clayey sands	4.5	0.094
	Inorganic silts	4.0	0.083
	Silty sands	3.5	0.072
	Inorganic clays	6.6	0.140
Bare soil noncohesive (PI < 10)	Finer than coarse sand $D_{75} < 1.3$ mm (0.05 in.)	1.0	0.02
	Fine gravel $D_{75} = 7.5$ mm (0.3 in.)	5.6	0.12
	Gravel $D_{75} = 15$ mm (0.6 in.)	11.0	0.24
Gravel mulch	Coarse gravel $D_{50} = 25$ mm (1 in.)	19.0	0.40
	Very coarse gravel $D_{50} = 50$ mm (2 in.)	38.0	0.80
Rock riprap	$D_{50} = 0.15$ m (0.5 ft)	113.0	2.40
	$D_{50} = 0.30$ m (1.0 ft)	227.0	4.80

Source: FHWA, Design of roadside channels with flexible linings, Hydraulic Engineering Circular No. 15 (HEC-15), 3rd ed., September 2005.

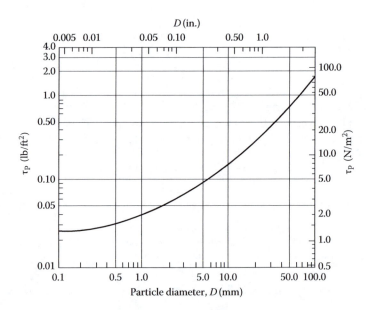

FIGURE 4.39 Permissible shear stress for noncohesive soils. (From FHWA, Urban drainage design, Hydraulic Engineering Circular No. 22 [HEC-22], 2nd ed., August 2001.)

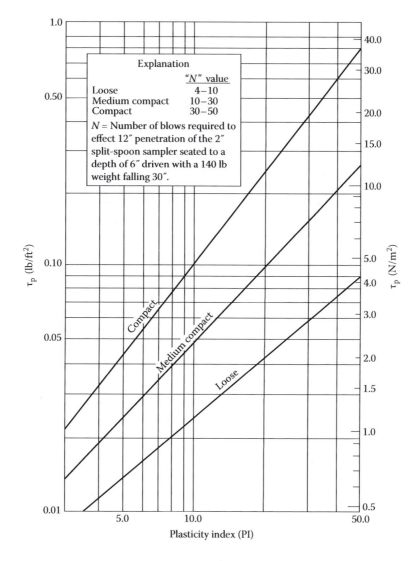

FIGURE 4.40 Permissible shear stress of cohesive soils. (From FHWA, Urban drainage design, Hydraulic Engineering Circular No. 22 [HEC-22], 2nd ed., August 2001.)

4.9.4 SIDE SLOPE STABILITY

In riprap-lined channels with side slopes steeper than 3:1, side slope stability must also be analyzed. The permissible shear stress on the side of a channel is given by

$$\tau_s = K_1 \tau_d \tag{4.62}$$

where
τ_s = side shear stress, N/m² (lb/ft²)
τ_d = bottom shear stress, N/m² (lb/ft²)
K_1 = ratio of channel side stress to bottom stress

For parabolic and triangular with rounded bottom channels, there is no sharp discontinuity along the perimeter. Thus, it can be assumed that the shear stress on the sides is equal to the bottom shear stress given by Equation 4.57, that is, $K_1 = 1$. For trapezoidal channels, the factor K_1 depends on the side slope, m, and the ratio of bottom width to depth (b/d) of the channel. The ratio of τ_s/τ_d to these parameters is shown in Figure 4.41 and can be approximated by the following equation:

$$
\begin{aligned}
K_1 &= 0.77 \quad m < 1.5 \\
K_1 &= 0.67 + 0.066\,m \quad 1.5 < m < 5 \\
K_1 &= 1 \quad m > 5
\end{aligned}
\tag{4.63}
$$

Equation 4.63 may also be applied to triangular channels with sharp angles.

For a channel lined with stone, the side stability is also affected by the angle of repose of the stone. Specifically, the ratio of tractive forces (namely, permissible shear stresses) on the side and the bottom depends on the side slope and the angle of repose of the riprap stone, and is given by

$$
K_2 = \left(1 - \frac{\sin^2 \psi}{\sin^2 \theta}\right)^{1/2}
\tag{4.64}
$$

where θ and ψ are the angle of repose and the angle of side slope, respectively. The angle of repose depends on the size and shape of the stone and is plotted in Figure 4.42. The preceding equation indicates the side angle should be smaller than the angle of repose. The median stone size for the side slopes is calculated using the following equation:

$$
(d_{50})\,\text{sides} = \frac{K_1}{K_2}(d_{50})\,\text{bottom}
\tag{4.65}
$$

where
  (d_{50}) bottom is derived from Figure 4.39
  K_1 = ratio of side shear stress to bed shear stress (Figure 4.41 or Equation 4.63)
  K_2 = ratio of tractive forces on the sides and bottom (Equation 4.64)

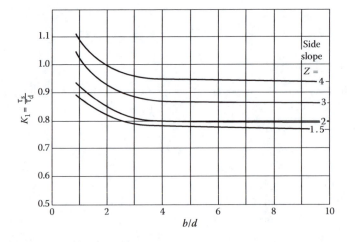

FIGURE 4.41   Channel side shear stress to bottom shear stress ratio, K_1. (From FHWA, Urban drainage design manual, Hydraulic Engineering Circular No. 22 (HEC-22), 3rd ed., August 2013.)

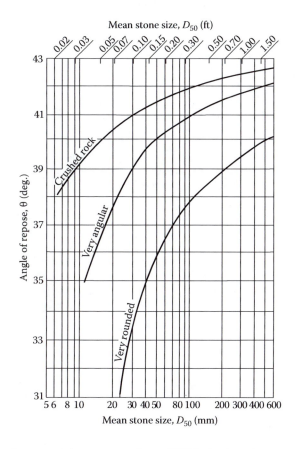

FIGURE 4.42 Angle of repose of riprap stones. (From FHWA, Design of roadside channels with flexible linings, Hydraulic Engineering Circular No. 15 (HEC-15), 3rd ed., September 2005.)

4.9.5 GRASS LINING

The permissible shear stress of grass lining depends on the density and height of grass, known as retardance degree or retardance class. Vegetal covers are grouped into five retardance classes, A through E; class A has the highest retardance and E the lowest. Table 4.11 serves as a guide in selecting the vegetal retardance and Table 4.12 provides specific information about retardance classification of various grass covers.

TABLE 4.11
Vegetal Retardance Selection Guide

Average Height of Grass, cm (in.)	Retardance Class	
	Grass Density: Good	Grass Density: Fair
75 (30)	A	B
27.5–60 (11–24)	B	C
15–25 (6–10)	C	D
5–15 (2–6)	D	D
5 (<2)	E	E

TABLE 4.12

Retardance Classification of Vegetal Covers

Retardance Class	Cover[a]	Condition
A	Weeping love grass	Excellent stand, tall, average 760 mm (30 in.)
	Yellow bluestem ischaemum	Excellent stand, tall, average 910 mm (36 in.)
B	Kudzu	Very dense growth, uncut
	Bermuda grass	Good stand, tall, average 300 mm (12 in.)
	Native grass mixture (little bluestem, bluestem, blue gamma, and other long and short Midwest grasses)	Good stand, unmowed
	Weeping lovegrass	Good stand, tall, average 610 mm (24 in.)
	Lespedeza sericea	Good stand, not woody, tall, average 480 mm (19 in.)
	Alfalfa	Good stand, uncut, average 280 mm (11 in.)
	Weeping lovegrass	Good stand, unmowed, average 330 mm (13 in.)
	Kudzu	Dense growth, uncut
	Blue gamma	Good stand, uncut, average 280 mm (11 in.)
C	Crabgrass	Fair stand, uncut 250 to 1200 mm (10 to 48 in.)
	Bermuda grass	Good stand, mowed, average 50 mm (6 in.)
	Common lespedeza	Good stand, uncut, average 280 mm (11 in.)
	Grass–legume mixture—fall, spring (orchard grass, redtop, Italian ryegrass, and common lespedeza)	Good stand, uncut, 150 to 200 mm (6 to 8 in.)
	Centipede grass	Very dense cover, average 150 mm (6 in.)
	Kentucky bluegrass	Good stand, headed, 150 to 300 mm (6 to 12 in.)
D	Bermuda grass	Good stand, cut to 60 mm (2.5 in.) height
	Common lespedeza	Excellent stand, uncut, average 110 mm (4.5 in.)
	Buffalo grass	Good stand, uncut, 80 to 150 mm (3 to 6 in.)
	Grass–legume mixture—fall, spring (orchard grass, redtop, Italian ryegrass, and common lespedeza)	Good stand, uncut, 100 to 130 mm (4 to 5 in.)
	Lespedeza sericea	After cutting to 50 mm (2 in.) height, very good stand before cutting
E	Bermuda grass	Good stand, cut to height, 40 mm (1.5 in.)

Source: US Soil Conservation Service, Handbook of channel design for soil and water conservation, SCS-61, Stillwater Outdoor Hyrdaulic Laboratory, Oklahoma, June 1954.

[a] Covers classified have been tested in experimental channels. Covers were green and generally uniform.

Permissible shear stresses of vegetative covers are listed in Table 4.13. As shown, the permissible shear stress of grass covers varies by over 10 fold from retardance A to retardance E.

As will be shown through examples, the permissible shear stresses in Table 4.13 are too conservative compared with other methods.

The HEC-15 (FHWA, 2005) presents the following method for calculating the stability of grass-lined channels. This method, unlike others, is based on the combined effect of soil shear stress and the effective shear stress of the vegetative lining, using the following equation:

$$\tau_p = \left[\frac{\tau_p, \text{soil}}{(1 - C_f)} \right] \left(\frac{n}{n_s} \right)^2 \tag{4.66}$$

TABLE 4.13

Permissible Shear Stress for Vegetal Covers

Vegetal Retardance Class	Permissible Shear Stress	
	N/m²	lb/ft²
A	177.2	3.70
B	100.6	2.10
C	47.9	1.00
D	28.7	0.60
E	16.8	0.35

Source: FHWA, Urban drainage design, Hydraulic Engineering
Circular No. 22 (HEC-22), 2nd ed., August 2001.

where

τ_p = permissible shear stress on the vegetative lining, N/m² (lb/ft²)
τ_p, soil = permissible soil shear stress, N/m² (lb/ft²)
C_f = grass cover factor (see Table 4.14)
n_s = soil grain roughness coefficient
n = overall lining roughness coefficient

The permissible shear stress, τ_p, for cohesive and noncohesive soils applicable to this method is the same as those listed in Table 4.10. Also, as indicated in a previous section, the shear stress of noncohesive and cohesive soils can be calculated using Equations 4.59 and 4.61, respectively. Soil roughness coefficient can be taken as

$$n_s = 0.016 \quad \text{for } d_{75} \le 1.3 \text{ mm (0.05 in.)}$$

For soils of larger grain size, the following equation is used:

$$n_s = \alpha_1 (d_{75})^{1/6} \tag{4.67}$$

where

n_s = soil grain roughness
α_1 = 0.015 SI; 0.026 CU
d_{75} = 75% finer stone size, mm (in.)

The overall roughness coefficient, n, for grass linings varies with both vegetative retardance class and the flow shear stress. The latter parameter affects the roughness coefficient through bending

TABLE 4.14

Grass Cover Factor, C_f

Growth Form	Excellent	Very Good	Good	Fair	Poor
Sod	0.98	0.95	0.90	0.84	0.75
Bunch	0.55	0.53	0.50	0.47	0.41
Mixed	0.82	0.79	0.75	0.70	0.62

Source: FHWA, Design of roadside channels with flexible linings, Hydraulic
Engineering Circular No. 15 (HEC-15), 3rd ed., September 2005.

TABLE 4.15

Density–Stiffness Coefficient, C_s

Grass Condition	Excellent	Very Good	Good	Fair	Poor
C_s (SI)	580	290	106	24	8.60
C_s (CU)	49	25	90	2.0	0.73

the grass stem, which in turn reduces the stem height relative to the flow depth. The coefficient n in HEC-15 (FHWA, 2005) is given as

$$n = \alpha_2 C_n \tau_o^{-0.4} \tag{4.68}$$

where

τ_o = mean boundary shear stress ($\tau_o = \gamma RS$)
C_n = grass roughness coefficient; dimensionless and the same in SI and CU
α_2 = unit conversion constant 1.0 SI (0.213 CU)

The grass roughness coefficient depends on the density–stiffness of grass and the grass cover condition and is given by

$$C_n = \beta C_s^{0.1} h^{0.528} \tag{4.69}$$

where

C_s = density–stiffness coefficient
h = stem height, m (ft)
β = unit conversion factor, 0.35 SI, 0.237 CU

Table 4.15 presents the density–stiffness coefficient for various grass cover conditions.

Depending on grass heights, which may range from 7.5 to 22.5 cm (3–9 in.) and grass cover conditions, the C_n for roadside grass channels varies from 0.1 to 0.3, with 0.2 being an average. For vegetative retardance classes C, D, and E, on which the design of grass swales is commonly based on, the coefficient C_n may be estimated as follows:

$$C_n = 0.22 \text{ (class C)}$$
$$C_n = 0.147 \text{ (class D)}$$
$$C_n = 0.1 \text{ (class E)}$$

4.9.6 Manning's Roughness Coefficient Variation with Lining

The Manning's roughness coefficient depends on the channel roughness and its relative magnitude to the flow depth. For a given lining, the channel roughness will increase with a decrease in flow depth. In urban developments and roadside channels, the flow depth typically ranges from 0.15 to 1 m (0.5 to 3.3 ft).

For riprap, cobble, and gravel lining, Blodgett (1986) has proposed an equation for the n value, which is included in HEC-15 (FHWA, 2005) and may be expressed as

$$n = \frac{\alpha D_a^{1/6}}{7.05 + 16.4 \log\left(\dfrac{D_a}{d_{50}}\right)} \tag{4.70}$$

where

D_a = average flow depth in channel, m (ft)
d_{50} = mean riprap, gravel size, m (ft)
α = unit conversion factor = 1 SI (0.82 CU)

The average flow depth is defined as the ratio of waterway area to the top width (A/T). The preceding equation gives Manning's n values, which are very conservative. For an average depth of 0.5 m (1.5 ft) and 100 mm (4 in.) stone, for example, the preceding equation gives $n = 0.048$, which is far greater than values reported in the literature (e.g., see Table 5.1 in HEC-22, 2013).

The Standards for Soil Erosion and Sediment Control in New Jersey (NJ State Soil Conservation Committee, 1999) includes a graphical relation between n and stone size, based on the following equation:

$$n = 0.0395(d_{50})^{1/6} \tag{4.71}$$

where d_{50} = stone size, in feet.

Expressing d_{50} in millimeters and inches, the preceding equation becomes

$$n = 0.015(d_{50})^{1/6} \quad \text{SI} \tag{4.72}$$

$$n = 0.026(d_{50})^{1/6} \quad \text{CU} \tag{4.73}$$

These equations, which are identical to Equation 4.67 with d_{50} substituted for d_{75}, disregard the flow depth and underestimate n values for shallow channels. For example, for a 6 in. (150 mm) riprap lining, Equation 4.73 gives $n = 0.035$, which is unrealistic for shallow flow depths. It is the author's recommendation that Equations 4.71, 4.72, and 4.73 be applied only when the depth of flow is at least five times greater than the stone size, d_{50}.

USDA-NRCS (1992) offers the following equation for the Manning's n for gravel and stone-lined channels:

$$n = \frac{\alpha y^{1/6}}{\left[14 + 21.6 \log(y/d_{50})\right]} \tag{4.74}$$

where

y = flow depth, m (ft)
d_{50} = 50% finer stone size, m (ft)
α = 1.22 SI (1 CU)

Apart from differences in coefficients between Equations 4.70 and 4.74, the latter equation is based on the average flow depth and the former on the flow depth. Since the average flow depth (area divided by the top width) is always smaller than the flow depth in trapezoidal or parabolic channels, a direct comparison between these equations cannot be made. See Example 4.20.

Both Equations 4.70 and 4.74 yield n values that are more realistic than in Equation 4.71. New York State Department of Environmental Conservation (2005), among others, has adopted the USDA-NRCS (1992) equation for riprap sizing. Table 4.16, which is prepared based on Equation 4.74, may be used for design of stone channels.

A variation of the NJ Standards Equation 4.71 is presented by the NRCS Conservation Practices Standard (2010). This equation may be expressed as

$$n = \alpha(d_{50}S)^{0.147} \tag{4.75}$$

where

S = channel slope, m/m (ft/ft)
d_{50} = median gravel/riprap size, mm (in.)
α = unit conversion constant 0.029 (SI), 0.047 (CU)

TABLE 4.16
Manning's *n* Values for Stone Linings

Lining Category	Stone Size, mm (in.)	Flow Depth, m (ft)		
		0.15/(0.5)	0.5/(1.5)	1.0/(3.3)
Gravel	25 (1)	0.029/0.029	0.026/0.026	0.025/0.025
	50 (2)	0.037/0.037	0.031/0.031	0.029/0.029
Cobbles	100 (4)	0.050/0.050	0.037/0.038	0.034/0.035
Riprap	150 (6)	–	0.043/0.044	0.038/0.039
	200 (8)	–	0.048/0.050	0.042/0.043
	300 (12)	–	–	0.048/0.049

For a 15 cm (6 in.) stone and a channel at 2% slope, the preceding equation gives $n = 0.06$, which is significantly greater than those calculated based on the NJ Standards (Equations 4.72 and 4.73).

For vegetal covers, as indicated, the Manning's *n* varies with density and height of grass, namely the retardance class. The HEC-15 (FHWA, 2005) method of calculating the *n* value was discussed in a previous section. Two other methods are discussed herein.

A relation between the Manning's roughness coefficient, *n*, with the product of flow velocity, *V*, and hydraulic radius, *R*, for various vegetal retardance classes is shown in Figure 4.43 in English units. Since the flow velocity and hydraulic radius are related to the roughness coefficient by the Manning formula, the design of grass channel using this figure requires iterative calculations. Specifically, for a trial value of *n*, the velocity or hydraulic radius is calculated using the Manning formula; and the *n* value is refined iteratively until it fits the appropriate retardance curve on this figure (see Example 4.15).

The NJ soil erosion and sediment control manual (NJ State Soil Conservation Committee, 1999) contains design charts, which eliminate any iterative calculations when *n* and *S* are given. These charts present flow velocity and Manning's *n* as a function of hydraulic radius, *R*, and channel slope, *S*, for each vegetal retardance class. The chart for retardance D, for example, is shown as Figure 4.44. The retardance D serves as the design criterion for calculating the discharge capacity of grass-lined channels in New Jersey.

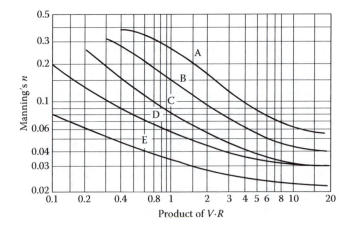

FIGURE 4.43 Relation between Manning's roughness coefficient and the product of velocity, *V*, and hydraulic radius, *R*. Curves A to E present retardance: A for very high, B for high, C for moderate, D for low, and E for very low vegetal retardance. Product of $V \cdot R$. (From Chow, V. T., *Handbook of applied hydrology*, McGraw-Hill, New York, 1964.)

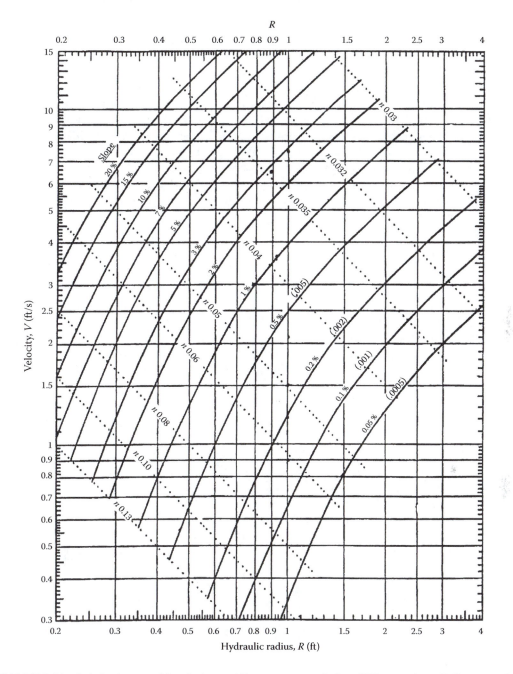

FIGURE 4.44 Relation between Manning's *n* and flow parameters, *R*, *S*, and *V* for retardance D (low vegetal retardance). (Adapted from NJ State Soil Conservation Committee, Standards for soil erosion and sediment control in New Jersey, July 1999.)

In practice, design calculations for well maintained man-made grass swales are conservatively performed assuming retardance D and/or E, which correspond to either dormant season or freshly mowed grass, respectively. To ensure stability against erosion, the flow velocity or the shear stress are calculated by any one of the previously indicated methods; namely, Equation 4.66 (HEC-15; FHWA, 2005) or Figure 4.43, is compared with permissible velocity or shear stress for vegetal cover. The permissible velocities are presented in the US Soil Conservation Service (1954) and are also specified in soil erosion standards of many states. A number of texts, referenced at the end of this chapter, include a table of the USSCS permissible velocities (see, for example, Table 7.6 in Chow, 1959, and Table 9.3 in the ASCE manual 77, 1992). In general, the velocities in grass swales should be limited to 1 m/s (3 ft/s) for common grass cover and 1.5 m/s (5 ft/s) for very sturdy grass. Also, the slope of swales should be limited to 5% for common grass cover such as crab grass and no more than 10% for exceptionally sturdy grass mixtures. The permissible shear stresses for grass linings are presented in Table 4.13 and Equation 4.66 in this chapter. As indicated, the use of Figure 4.44 eliminates any iterative calculations for the allowable velocity, which can then be used for calculating discharge capacity of a grass-lined channel.

A 10- or 25-year storm frequency is commonly specified as the design criterion for flexible lined channels. To account for the grass growth, a freeboard is provided above the design depth for grass swales. The ASCE manual (1992) suggests a freeboard equal to 15 cm (6 in.) plus the velocity head, $V^2/2g$. However, since the velocity head is commonly small for grass-lined channels, it may be ignored in practice.

Example 4.14

The flow parameters of a roadside trapezoidal channel installed on clayey sand (SC) are as follows:

$$S_o = 0.012; b = 1.0 \text{ m}; m = 3; \text{PI} = 15; \text{ and } e = 0.5 \text{ for soil}$$

The channel is lined with a good stand of mixed grass 75 mm high and is maintained in very good condition. For a design discharge of 1 m³/s calculate the depth of flow and the maximum shear stress at the channel bed and determine whether or not the grass lining is stable. Base your calculations on the HEC-15 (FHWA, 2005) method.

Solution

The calculations are performed using the following steps:

Step 1. Estimate depth of flow and calculate Manning's n from Equation 4.68.
Step 2. Calculate discharge and compare it with the desired flow. If the difference is more than 5%, go to step 3.
Step 3. Estimate a new depth and repeat calculations in steps 1 and 2.
Step 4. Compare the calculated permissible shear stress with the shear stress on the channel bed (max. in the channel).

Estimate $y = 0.40$ m

$$A = by + my^2 = 0.40 + 3 \times 0.4^2 = 0.880 \text{ m}^2$$

$$P = b + 2y\sqrt{(1 + m^2)} = 1.0 + 2 \times 0.4\sqrt{10} = 3.530 \text{ m}$$

$$R = \frac{0.88}{3.53} = 0.249 \text{ m}$$

$$\tau_o = \gamma RS = 9.81 \times 10^3 \times 0.249 \times 0.012 = 29.4 \text{ N/m}^2$$

Calculate Manning's n from Equations 4.69 and 4.68 in that order:

$$C_n = \beta C_s^{0.1} h^{0.528}$$

$$\beta = 0.35$$

$$C_s = 290 \text{ very good condition}$$

$$h = 0.075 \text{ m}$$

$$C_n = 0.35 \times 290^{0.1} \times 0.075^{0.528} = 0.157$$

$$n = \alpha C_n \tau_o^{-0.4} = 1 \times 0.157 \times 29.4^{-0.4} = 0.0406 \approx 0.041$$

$$Q = \frac{1}{n} AR^{2/3} S^{1/2} = \frac{1}{0.041} \times 0.88 \times 0.249^{2/3} \times (0.012)^{1/2}$$

$$Q = 0.931 \text{ m}^3/\text{s}$$

Extrapolate $y = 0.41$ m.
Calculate the maximum shear stress on the channel bed:

$$\tau_d = \gamma y S = 9.81 \times 10^3 \times 0.41 \times 0.012 = 48.3 \text{ N/m}^2$$

Calculate the permissible shear stress of the soil from Equation 4.61:

$$\tau_p = (C_1 PI^2 + C_2 PI + C_3)(C_4 + C_5 e)^2 \times C_6$$

The coefficients C_1 through C_6 may be obtained from Table 4E.1 in Appendix 4E for SC soil:

$$C_1 = 1.07, \; C_2 = 14.3, \; C_3 = 47.7, \; C_4 = 1.42, \; C_5 = -0.61, \; C_6 = 4.8 \times 10^{-3}$$

$$\tau_p, \text{ soil} = (1.07 \times 15^2 + 14.3 \times 15 + 47.7)(1.42 - 0.61 \times 0.5) \times 4.8 \times 10^{-3}$$

$$\tau_p, \text{ soil} = 2.69 \text{ N/m}^2 \text{ soil}$$

Next calculate the permissible shear stress on the grass using Equation 4.66:

$$\tau_p = \left[\frac{\tau_p, \text{soil}}{(1 - C_f)} \right] \left(\frac{n}{n_s} \right)^2$$

$$n = 0.041$$

$$n_s = 0.016$$

$$d_{50} < 1.3 \text{ mm}$$

$$C_f = 0.79 \quad \text{(see Table 4.14)}$$

$$\tau_p = \left[\frac{2.69}{1 - 0.79} \right] \times \left(\frac{0.041}{0.016} \right)^2 = 84.1 \text{ N/m}^2$$

$$\tau_p > \tau_d = 48.3$$

The grass is stable and the safety factor in this case is 84.1/48.3 = 1.75.

Example 4.15

A grass swale at 2% slope is to carry a discharge of 35 cfs. Design the hydraulic parameters R and A for a stable channel based on Figure 4.43.

Solution

Consider retardance degree D, relating to dormant season, and conservatively assume a permissible velocity of 4 ft/s. Determine the design R, using the following procedure:

1. Assume an n value and determine the corresponding value of VR from Figure 4.43, curve D.
2. Calculate R based on the selected maximum permissible velocity ($R = VR/V_p$)
3. Calculate the value of VR product using the Manning formula:

$$VR = \frac{1.49R^{5/3}S^{1/2}}{n} \quad CU$$

and check this value against the value of VR obtained in step 1.
4. Refine the assumed value of n and repeat steps 1 through 3 until the calculated VR value is equal to the value of VR obtained from the n versus VR curve in Figure 4.43.
5. Calculate the waterway area: $A = Q/V$.

The calculations in customary units are summarized in the following table:

Trial No.	n	VR	$R = VR/4$	$1.49 R^{5/3} S^{1/2}/n$
1	0.050	1.4	0.35	0.73
2	0.040	2.8	0.70	2.91
3	0.047	2.7	0.68	2.70

The design value for the selection of channel section is $R = 0.68$ ft and $A = Q/V = 35/4 = 8.75$ ft².

Example 4.16

Using the NJ Standards (Figure 4.44), find the hydraulic parameters for a grass-lined channel to carry a discharge of 35 cfs at 2% slope, based on

a. Permissible velocity of 4 ft/s
b. Hydraulic radius of 0.68 ft, obtained in the previous example

Solution

a. Entering Figure 4.44 with $V = 4.0$ ft/s and $S = 0.02$, the values of n and R are obtained directly, without a need for any trial, as follows:

$$n = 0.041$$

$$R = 0.69 \text{ ft}$$

The channel area $A = Q/V = 8.75$ ft².

Note: The NJ Standards method yields nearly identical results to that of Figure 4.43 without a need for any iteration.

b. Entering Figure 4.44 with $R = 0.68$ and $S = 0.02$, gives:

$$V = 3.9 \text{ ft/s}$$

$$n = 0.042$$

Example 4.17

Design a channel lined with 10 cm stone riprap to carry a discharge of 1.5 m³/s at 1% slope. Base your design on NRCS's Equation 4.74 for calculating the Manning's n. Perform your design for the following channel geometries:

a. Trapezoidal section of 3:1 side slope
b. Parabolic section

Solution

Given $Q = 1.5$ m³/s, $S = 1\%$.
First calculate hydraulic parameters of channels based on the permissible shear stress for $d_{50} = 0.1$ m using Equation 4.60:

$$\tau_p = 0.75 \times 100 \text{ mm} = 75 \text{ N/m}^2$$

The maximum shear stress occurs at the channel bed, which, according to Equation 4.57, is

$$\tau = \gamma y S$$

Conservatively apply a safety factor of 1.5:

$$\tau_p = K\tau \quad \tau = 75/1.5 = 50 \text{ N/m}^2$$

The maximum channel depth is

$$y = 50/(9.81 \times 10^3 \times 0.01) = 0.5 \text{ m}$$

All sections should be designed based on this depth.
From Equation 4.74:

$$n = 1.22 y^{1/6}/[14 + 21.6\log(y/d_{50})]$$

$$n = 1.22 \times 0.5^{1/6}/(14 + 21.6\log5) = 0.037$$

$$Q = AR^{2/3}S^{1/2}/n$$

Simplify the Manning formula for $Q = 1.5$ m³/s, $S = 0.01$, and $n = 0.037$:

$$AR^{2/3} = 1.50 \times 0.037/(0.01)^{1/2} = 0.555$$

a. Trapezoidal sections, 3:1 side slope
Express area, A, and hydraulic radius, R, in terms of flow depth, y, and bottom width, b.

$$A = by + my^2 = 0.5b + 0.75$$

$$R = A/P = (0.5b + 0.75)/(b + 2 \times 0.5\sqrt{10})$$

$$= (0.5b + 0.75)/(b + 3.16)$$

Solve the preceding equations for $AR^{2/3}$ for a trial value for b and refine it by an iterative process. The following table summarizes the calculations.

Trial No.	b m	A m²	R m	$AR^{2/3}/n$
1	1.00	1.25	0.300	0.560
2	0.90	1.20	0.296	0.532
3	0.98	1.24	0.300	0.555

Therefore:

$$A = 1.24 \ \text{m}^2$$

$$V = Q/A = 1.50/1.25 = 1.21 \ \text{m/s}$$

Add 0.15 m for free board.

$$\text{Top width } B = 0.98 + 2 \times 3 \times 0.65 = 4.88 \ \text{m}$$

b. Parabolic section (refer to Table 2.4 in Chapter 2)
For this section:

$$A = (2/3)Ty = 0.333T, \text{ where } T = \text{top width}$$

$$P = T + 8y^2/3T = T + 0.667/T$$

$$R = A/P = 0.333T/(T + 0.667/T)$$

Solve for T by a trial process (see table below):

Trial No.	T m	A m²	R m	$AR^{2/3}$
1	6.0	2.000	0.327	0.950
2	4.0	1.332	0.320	0.623
3	3.0	1.000	0.310	0.458
4	3.6	1.199	0.317	0.557

The wet section is 3.6 m wide and 0.5 m deep.

$$V = 1.5/1.199 = 1.25 \ \text{m/s}$$

Adding 0.15 m as freeboard, the center depth will be

$$y = 0.5 + 0.15 = 0.65 \ \text{m}$$

Considering that the top width in parabolic sections varies in proportion to square root of depth, the top width of section will be

$$T = 3.6(0.65/0.5)^{1/2} = 4.1 \ \text{m, top width}$$

While both sections are practical, the parabolic section is more efficient. Its top width is approximately 0.8 m narrower than the trapezoidal section.

Example 4.18

Using Equation 4.70, calculate the Manning's n value for the channels in the previous problem.

Solution

a. Trapezoidal section
    For the flow depth of 0.5, the top width and mean depths are

$$T = b + 2my = 0.98 + 2 \times 3 \times 0.5 = 3.98 \text{ m}$$

$$D = A/T = 1.24/3.98 = 0.312 \text{ m}$$

$$d_{50} = 0.1 \text{ m}$$

From Equation 4.75 which may be written as:

$$n = \frac{0.319 \times 0.312^{1/6}}{2.25 + 5.23 \log\left(\dfrac{0.312}{0.1}\right)} = 0.054$$

b. Parabolic section
    Based on the calculated top width and area for this channel, the mean depth and n values are

$$D = A/T = 1.199/3.6 = 0.333 \text{ m}$$

$$n = \frac{0.319 \times 0.333^{1/6}}{2.25 + 5.23 \log\left(\dfrac{0.333}{0.1}\right)} = 0.053$$

The n value calculated using Equation 4.70 deviates by over 45% from the $n = 0.037$ calculated based on Equation 4.74.

Example 4.19

Design the trapezoidal grass-lined channel of Example 4.14 based on the NJ standards manual (Figure 4.44).

Solution

Given $S = 1.2\%$, $Q = 1 \text{ m}^3/\text{s} = 35.3 \text{ cfs}$, and $b = 1 \text{ m} = 3.28 \text{ ft}$.
    Since both velocity and hydraulic radius are unknown, the problem requires a trial and error solution.

First try   Assume $R = 1.0$ ft

Step 1. Entering the figure with $R = 1.0$ and $S = 0.012$ gives

$$V = 4.3 \text{ ft/s}$$

$$n = 0.037$$

Step 2. Calculate area:

$$A = Q/V = 35.3/4.3 = 8.20 \text{ ft}^2$$

Calculate flow depth for the above area

$$3.28y + 3y^2 = 8.20$$

$$y = 1.20 \text{ ft (rounded to second decimal place)}$$

$$P = 3.28 + 2\sqrt{10} \times 1.20 = 10.87 \text{ ft}$$

$$R = A/P = 8.20/10.87 = 0.75 \text{ ft}$$

Second try $R = 0.85$ ft

Figure 4.44 gives $V = 3.5$ ft/s; $n = 0.040$
Calculate $A = Q/V = 35.3/3.5 = 10.09 \text{ ft}^2$

Calculate flow depth:

$$A = 3.28y + 3y^2 = 10.09$$

$$y = 1.37 \text{ ft}$$

$$P = 3.28 + 2\sqrt{10} \times 1.37 = 11.94 \text{ ft}$$

$$R = A/P = 0.845 \text{ ft} \approx 0.85 \text{ ft}$$

Therefore:

$$R = 0.85 \text{ ft} = 0.26 \text{ m}$$

$$V = 3.5 \text{ ft/s} = 1.07 \text{ m}$$

$$n = 0.040$$

$$A = 10.09 \text{ ft}^2 = 0.94 \text{ m}^2$$

$$y = 1.37 \text{ ft} = 0.418 \text{ m}$$

Note: The calculated depth of channel in this method differs nearly 1.9% from that calculated 0.35 m using the more laborious HEC-15 (FHWA, 2005).

4.9.7   CHANNEL BENDS

Because of a change in flow direction, the flow induces centrifugal forces in channel bends. The result is a rise in the water surface at the outer bends and a drop at the inner bend. This rise at the outer bend, termed as superelevation, can be estimated from the following equation:

$$\Delta y = \frac{V^2 T}{g R_c} \tag{4.76}$$

where
  V = mean velocity, m/s (ft/s)
  T = water surface width of the channel, m (ft)
  g = gravitational acceleration, 9.81 m/s^2 (32.2 ft/s^2)
  R_c = mean radius of the bend, m (ft)

To account for superelevation, an extra free board must be provided in a channel bend. Also, because of centrifugal forces, a larger shear stress is exerted on the outer bend. To account for this increase, a factor greater than one is applied to the shear stress in the straight channel section, as follows:

$$\tau_b = K_b \tau_d \tag{4.77}$$

where K_b is the bend factor and τ_b and τ_d are the shear stresses at the bend and straight section, respectively. The bend factor depends on the ratio of the bend radius, R_c, to the top width of the water surface, T. For R_c/T ratios between 2 and 10, K_b is given by

$$K_b = 2.38 - 0.206\left(\frac{R_c}{T}\right) + 0.0073\left(\frac{R_c}{T}\right)^2 \tag{4.78}$$

$$K_b = 2.0 \quad \text{for} \quad \frac{R_c}{T} \le 2.0$$

$$K_b = 1.05 \quad \text{for} \quad \frac{R_c}{T} \ge 10$$

Table 4.17 lists K_b values for R_c/T ratios between 2 and 10.

The increased shear stress persists to a distance downstream of the bend. This distance, L_b, can be calculated using the following equation:

$$L_b = \frac{K_u R^{7/6}}{n_b} \tag{4.79}$$

where
  n_b = Manning's roughness in the channel bend
  R = hydraulic radius of channel, m (ft)
  K_u = a constant, 0.74 SI, 0.60 CU

Contrary to intuition, the length L_b in the preceding equation does not depend on the bend curvature.

TABLE 4.17
Bend Factor in Channels

R_c/T	10	9	8	7	6	5	4	3	2
K_b	1.05	1.13	1.20	1.30	1.41	1.53	1.670	1.83	2.0

Example 4.20

A trapezoidal channel of 1 m bottom width and 3:1 side slope is lined with 15 cm (6 in.) riprap. The channel is at 1% slope and includes a straight section and a bend with a centerline radius of 18 m (59 ft). For a discharge of 1.0 m³/s:

  a. Calculate the maximum shear stress in the straight reach and in the bend.
  b. Determine if the lining is stable.
  c. Calculate the minimum distance past the bend where the channel protection must extend.

Solution

The solution will be performed in SI units; the solution in customary units is left as an exercise to interested readers.

Step 1. Estimate the n value.
   From Table 4.16 estimate $n = 0.043$ (assuming $y \approx 0.5$ m)
   Using Manning's formula:

$$AR^{2/3} = \frac{Qn}{S^{1/2}} = \frac{1 \times 0.043}{(0.01)^{1/2}} = 0.43$$

$$b = 1 \text{ m}$$

$$A = y + 3y^2 ; R = \frac{(y + 3y^2)}{1 + 2\sqrt{10}y}$$

Substitute A and R in the Manning formula:

$$\frac{(y \pm 3y^2)^{5/3}}{(1 + 2\sqrt{10}y)^{2/3}} = 0.43$$

Simplify:

$$f(y) = \frac{(y + 3y^2)^{5/2}}{(1 + 2\sqrt{10}y)} = (0.43)^{1.5} = 0.282$$

Solve for y by trial and error. The following table summarizes calculations:

Trial No.	y	$f(y)$
1	0.500	0.4197
2	0.450[a]	0.299
3	0.445[a]	0.289

[a] From Equation 4.74, $n = 0.044 \dfrac{(y + 3y)^{5/2}}{1 + 2\sqrt{10}y} = 0.292$.

Interpolate between $y = 0.450$ and 0.445 to obtain $y = 0.446$ m.
Check n, using Equation 4.74.

$$n = \left[1.22 \times (0.446)^{1/6}\right] / \left[14 + 21.6\log(0.446/0.15)\right] = 0.044$$

Therefore:

$$y = 0.446 \text{ m}$$

$$R = \frac{(0.446 + 3 \times 0.446^2)}{(1 + 2 \times 0.446 \times 10^{1/2})} = 0.273 \text{ m}$$

a. Shear stress on the channel bed in straight section:

$$\tau_d = \gamma y S = 9.81 \text{ kN/m}^3 \times 0.446 \times 0.01 = 0.044 \text{ kPa} = 44 \text{ Pa}$$

Shear stress at bend:

$$\text{Top width, } T = 1 + 2 \times 3 \times 0.446 = 3.68 \text{ m}$$

$$R_c/T = 18/3.68 = 4.9$$

From Table 4.18, by interpolation (or Equation 4.78):

$$K_b = 1.55$$

$$\tau_b = K_b \tau_d = 1.55 \times 44 = 68 \text{ Pa}$$

b. Estimate the permissible shear stress from Equation 4.58:

$$d_{50} = 150 \text{ mm}$$

$$\tau_p = 0.75 \times 150 = 112.5 \text{ Pa}$$

The $d_{50} = 15$ cm (6 in.) stone is stable.

c. The influence distance of bend can be calculated using Equation 4.79:

$$L_b = \frac{K_u R^{7/6}}{n}$$

$$L_b = \frac{0.74 \times 0.273^{7/6}}{0.044} = 3.7 \text{ m } (12.1 \text{ ft})$$

4.9.8 Composite Lining

In practice, sometimes two different linings are employed in a single channel. An example is a channel that experiences low flows where concrete, riprap, or gravel is used in the bottom and a more cost-effective lining, such as grass, is used in the upper section. Figure 4.45 shows a grass-lined

FIGURE 4.45   Grass-lined swale with gravel bed in Saint-Sauveur, Canada. (Photo by the author, 2014.)

swale with gravel bed in Saint-Sauveur, Canada. Flow calculations in a composite channel involve the use of an equivalent Manning's *n* for the entire perimeter of the channel. The equivalent, also referred to as effective roughness coefficient, n_e, is calculated using the following equation:

$$n_\mathrm{e} = \left[\frac{P_\mathrm{L}}{P} + \left(1 - \frac{P_\mathrm{L}}{P} \right) \left(\frac{n_\mathrm{s}}{n_\mathrm{L}} \right)^{3/2} \right]^{2/3} n_\mathrm{L} \qquad (4.80)$$

where
   P_L = perimeter of low flow lining
   P = total flow perimeter
   n_s = Manning's *n* for the side slope lining
   n_L = Manning's *n* for the low flow lining

   When vegetation is used as side slope lining, a transition lining should be used adjacent to the low flow channel to avoid erosion until the vegetative lining is established.
   The stability calculations for composite channel are similar to those described for grass lining with additional steps as follows:

   a. Calculating the effective *n* (n_e) using Equation 4.80.
   b. Calculating the shear stress at the maximum depth τ_d and the sear stress on the side slope (Equations 4.57 and 4.62).
   c. Comparing the shear stresses τ_d and τ_s to the permissible shear stresses τ_p for each lining. If τ_d or τ_s is greater than the τ_p for the respective lining, a different type of lining should be investigated.

4.10   OTHER LININGS

This section discusses gabion boxes and Reno mattresses (which are among the oldest methods of channel protection) and some of the newest methods such as turf reinforcement mats and erosion control blankets.

4.10.1 GABION BASKETS AND MATTRESSES

Gabion is a riprap enclosed tightly in wire baskets. The wire basket is made of zinc-coated (galvanized) or PVC-coated steel wire woven in the form of a rectangular container and reinforced on corners with heavier gage wire (see Figure 4.46). The baskets are installed along the channel sides, connected together and filled with well graded, durable stone. Figure 4.47 shows a gabion basket under installation and Figure 4.48 depicts a channel protected by a Reno mattress. The baskets vary from 15 cm (6 in.) thick mattresses to 1 m (3 ft) thick box-like gabions. The wire mesh binds the

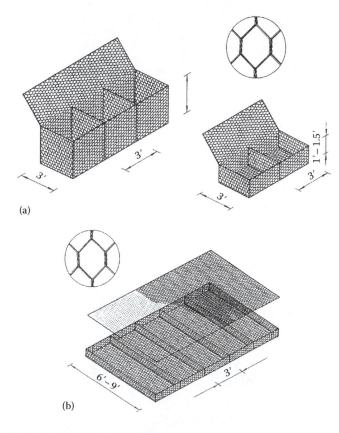

(a)

(b)

FIGURE 4.46   Gabion baskets (a) and mattress (b).

FIGURE 4.47   Gabion wall being installed. (Photo by the author.)

FIGURE 4.48   Reno mattress covering stream banks. (Photo by the author.)

stones together, preventing their movement. Gabion baskets provide significantly more resistance to erosion from flowing water than riprap. Figure 4.49 shows gabion walls at drainage outfalls. The outfalls (designed by the author) are equipped with Tideflex tide gates to prevent water in the tidal channel from backing up into the drainage system during high tides.

To simplify calculations, gabion baskets and gabion mattresses are commonly designed based on permissible velocity, rather than permissible shear stress. Table 4.18 gives the maximum permissible velocity of gabion baskets and mattresses.

During the past three decades, prefabricated concrete blocks and interlocked concrete mats have been introduced to the market. In addition to the ease of installation, these blocks and mats have a more pleasing appearance than gabions. Keystone was one of the first companies that manufactured

FIGURE 4.49   Gabion walls at drainage outfall at a tidal channel. (Photo by the author.)

TABLE 4.18
Maximum Flow Velocity for Gabion Mattress

Basket/Mattress Thickness, cm (in.)	Permissible Velocity, m/s (ft/s)
15 (6)	2.0 (6)
23 (9)	3.5 (11)
30 (12)	4.3 (14)

Source: NJ State Soil Conservation Committee, Standards for soil erosion and sediment control in New Jersey, July 1999.

FIGURE 4.50 Typical view of eroded Glenwood Brook in Millburn, New Jersey. (Photo by the author.)

FIGURE 4.51 Keystone walls under construction in 1990. (Photo by the author.)

concrete blocks in the United States. Now many companies make various types of blocks and mats. ArmorLoc, ArmorFlex, PetraFlex, Allan Block, Versa-Lok, Verdura, and Drivable Grass are some of the blocks and mats available on the market.

Figure 4.50 depicts a typical view of eroded Glenwood Brook in Millburn, New Jersey. To restore the brook channel, Keystone walls were erected along the banks of the brook. Figure 4.51 depicts Keystone walls under construction in 1990. This bank stabilization project, designed by the author, was the largest of its kind in New Jersey. With only minor repairs (mostly along the streambed), the walls are still standing tall.

4.10.2 Turf Reinforcement Mats (TRMs)

Turf reinforcement refers to a means of providing a structure to the soil/vegetation matrix that helps the establishment of vegetation and supports the vegetation once established. Turf may also be reinforced by using a gravel mulch. This latter method involves adding coarse to very coarse gravel into soil and seeding the soil–gravel layer. The gravel–soil mixture provides a nondegradable lining. Gravel mulches are designed based on their permissible shear stress (see Table 4.10). The density, size, and gradation of the gravel are the main properties that affect the erosion control performance of gravel mulch. The gravel should be applied to the soil at a rate of 25% of the mixture. The thickness of mixture for fine graded soils should be 95–100 mm (3–4 in.) and the gravel should be applied at 6 kg/m^2 per centimeter depth (3 lb/ft^2 per inch).

A turf reinforcement mat (TRM) consists of a long-lasting synthetic fiber, filament, or netting of sufficient thickness. The mats provide strength and void space to retain soil and to establish grass roots within this matrix. The mats are placed along the flow direction and secured with staples at intervals specified by their manufacturer. Mats can be installed first and covered with top soil and then seeded. Alternatively, the area can be covered with top soil and seeded before the mat is installed. In the former method, the plant roots grow within the mat; in the latter, the grass stems grow through the mat. To provide immediate protection, a growth medium may be added to the TRM to form an intimate bond with the TRM matrix, seed, and soil.

A large number of companies manufacture TRMs. One of the commercially available TRMs is Green Armor, manufactured by Profile. This TRM consists of a three-dimensional matrix composed of thermally fused nylon filaments. Enkamat Turf Reinforcement, which is a widely used TRM and was also made by Profile, is now available at Coldbond (http://www.coldbond-geosynthesics.com). The bond medium in this system, which is called Flexterra, is hydraulically filled into the TRM. Since germination and growth occur within the cavernous matrix, the system maximizes the root entanglement and helps improve long-term performance. Depending on application, Enkamat TRM rolls come in groups: Enkamat, Enkamat J, and Enkamat Flatback. Enkamat is available in Enkamat 7010 and 7020. The latter Enkamat is 17 mm thick and has up to 1800 m of artificial root structure filament per square meter of mat. Enkamat J is 10 mm thick and Enkamat Flatback comes in Enkamat 7210, 7220, and 7225, up to 19 mm thick and 2700 m of filament per square meter. Enkamat products are especially suited for dry slopes. Futerra is another TRM from Profile. This TRM, which was developed in 1972, is one of the most specified TRMs. Futerra R45 is a high-performance TRM (HP-TRM), suitable for steep slopes and channel stabilization, and is also manufactured by Profile.

Landlok, manufactured by Propex, is another brand-name TRM made of geosynthetics. This TRM is available in Landlok 300, which is a second-generation TRM, and Landloks 435, 450, and 1051, which are first-generation TRMs. Landlok 300 comes in 8.5 ft wide by 106 ft long (2.6 m × 32.3 m) rolls. Landlok makes stitch-bonded and woven-bonded TRMs, which last up to 10 years and 25 years, respectively. Propex also makes Pyramat, which is a high-performance HP-TRM suitable for high-velocity flows and lasts up to 50 years.

ArmorMax is another HP-TRM, also manufactured by Propex, which is expected to last 50 years or longer. The US Army Corps of Engineers used this product to control erosion at their Rapid Repair Levee Break Laboratory in Vicksburg, Mississippi. More than 4500 m² (5000 yd²) of ArmorMax were installed at this facility by January 2012; and less than 2 months after its application, the entire eroded area was covered with thick grass vegetation. Appendix 4F provides pictures and brief descriptions of Landlok TRMs by Propex, Inc. Also included in this appendix are the physical properties of ArmorMax, also manufactured by Propex.

4.10.3 Erosion Control Blankets (ECBs)

An erosion control blanket (ECB) is a degradable product composed of natural or polymer fibers that are physically or chemically bonded together to form a uniform continuous mat. ECBs are stiffer, thicker, and denser than open-weave textile linings such as jute net, woven paper net, and straw with net, which were in common use in the past. Similar to TRMs, ECBs are placed on the channel banks parallel to the flow direction and secured with staples at specified spacings. A 30 to 60 cm (1 to 2 ft) overlap is provided to ensure full coverage. Depending on type, ECBs degrade from 1 to 3 years; but long-term protection is provided by the established vegetation. Curled wood mat, reinforced polypropylene fiber, and coconut fiber are among erosion control blankets. Propex manufactures a number of ECBs under the brand name Landlok. These ECBs have longevity of 1 year for Landlok 407 to 3 years for Landlok C2. Among these ECBs is Landlok SuperGro, made of lightweight geo-composite material with a rapidly degradable polypropylene screen and a thin web of green soil reinforcement polypropylene fiber. SuperGro comes in 8 ft × 1125 ft (2.44 m × 342.9 m) rolls covering 1000 square yards (836 m²) per roll. This ECB, which degrades in 1.5 years, weighs 0.7 ounce per square yard (24 g/m^2) and has a specific gravity of 0.9. Appendix 4G includes information on all Propex ECBs in general, and SuperGro in particular.

4.10.4 Properties of ECBs and TRMs

TRMs and ECBs are collectively known as rolled erosion control products, RECPs. The density, stiffness, and thickness are the main properties of RECPs in terms of their erosion control performance. These physical properties are measured by a series of standard tests that are referred to as index tests (see Table 4.19).

The Manning's *n* values of RECPs vary from one product to another and are also a function of the shear stress. The roughness factor of each product is commonly measured by a full-scale laboratory test and is included in the manufacturer's specification. The shear stresses of RECPs are also specified by their manufacturers.

Table 4.20 exemplifies specified erosion control properties of Landlok ECBs manufactured by Propex. The listed permissible shear stress and velocity for Landlok ECBs in this table represent

TABLE 4.19
Index Tests for RECPs (ECBs and TRMs)

Property	Index Test	Description
Density	ASTM D 6475	Standard test method for mass per unit area for erosion control blankets
	ASTM D 6566	Standard test method for measuring mass per unit area of turf reinforcement mats
	ASTM D 6567	Standard test method for measuring the light penetration of turf reinforcement mats
Stiffness	ASTM D 4595	Test method for tensile properties of geotextile by the wide-width strip method
Thickness	ASTM D 6525	Standard test method for measuring nominal thickness of erosion control products

TABLE 4.20
Landlok Erosion Control Blankets' Performance Values (CU and SI Units) by Propex

Material	Functional Longevity	Maximum Short-Term Shear Stress and Velocity (Unvegetated)[a]		Manning's *n*	
		Shear Stress	Velocity	0.6 in. (0–150 mm)	C-Factor
Landlok® 407	Short-term degradable (1 year)	–	–	–	–
Landlok S1	Short-term degradable (1 year)	2.0 lb/ft^2 96 N/m^2	n/r	n/r	0.14
Landlok S2	Short-term degradable (1 year)	2.0 lb/ft^2 96 N/m^2	5.0–6.0 ft/s 1.5–1.8 m/s	0.027	0.21
Landlok SuperGro®	Extended-term degradable (1.5 years)	2.0 lb/ft^2 96 N/m^2	–	–	–
Landlok CS2	Extended-term degradable (2 years)	2.0 lb/ft^2 96 N/m^2	5.0–6.0 ft/s 1.5–1.8 m/s	0.021	0.09
Landlok C2	Long-term degradable (3 years)	2.0 lb/ft^2 96 N/m^2	5.0–6.0 ft/s 1.5–1.8 m/s	0.018	0.06

Note: "n/r" not recommended for use in swales and low-flow channels.

[a] Typical design limits for natural vegetation are a maximum shear stress of 2.0 lb/ft^2 (96 N/m^2) and a velocity limit of 5.0 to 6.0 ft/s (1.5 to 1.8 m/s).

the condition of no vegetation growth. Also listed in this table are the longevity, Manning's n, and a soil loss factor, C, of Landlok ECBs.

ECBs and TRMs reduce tractive force of flowing water before it reaches the underlying soil. Therefore, to control erosion, the applied shear stress at the soil surface should be less than the permissible shear stress of the soil. As the shear stress on the erosion control mats or blankets increases beyond their limits, the lining is detached from the soil, and the flow directly contacts the soil surface. This limit is defined as the shear stress that results in 12.5 mm (0.5 in.) soil loss. To avoid this condition, a safety factor is applied to permissible shear stresses of TRMs and ECBs as specified by the manufacturer. Depending on the application, the safety factor varies from 1.5 to 2, in that the specified shear stresses are reduced by this factor.

4.10.5 DESIGN OF RECP LINED CHANNELS

The Manning's n and the permissible shear stress of reinforced erosion control products (TRMs and ECBs) vary from one product to another. Therefore, a single equation cannot be established for these products. To design a channel lined with any RECP, a table of n values versus applied shear stresses should be obtained from the manufacturer. This information typically includes an upper shear stress, the middle shear stress, the lower shear stress, and the n value corresponding to each. The upper and lower shear stress values must be equal to twice and one-half of the middle shear stresses, respectively.

The manufacturer's information is used to calculate the n values from the following equation:

$$n = a\tau_o^b \tag{4.81}$$

where
τ_o = mean boundary shear stress N/m^2 (lb/ft^2)
coefficient "a" and exponent "b" depend on the n value at the mid range of applied shear and the n values corresponding to the range of shear stresses, respectively.

These coefficients are calculated from the following equations:

$$a = n_m/\tau_m^b \tag{4.82}$$

$$b = -\sqrt{[\ln(n_m/n_\ell)\ln(n_u/n_m)]}/0.693 \tag{4.83}$$

where
τ_m = middle shear stress
n_m = n value corresponding to τ_m
n_ℓ = n value corresponding to lower τ (τ_ℓ)
n_u = n value corresponding to upper τ (τ_u)

The permissible shear stress on an RECP lining is calculated both for the underlying soil and the RECP. In the case of TRMs, the presence of vegetation also affects erosion resistance properties.

RECPs dissipate shear stress before it reaches the soil surface. To control erosion, the shear stress at the soil surface should be less than the permissible shear of the soil. As the shear stress on the surface of RECPs increases, the lining is detached from the soil and the current may erode the soil. The effective shear stress on the soil surface is related to the RECP shear stress and the design shear stress in the channel by the following equation:

$$\tau_e = (\tau_d - \tau_\lambda/4.3)(\alpha/\tau_\lambda) \tag{4.84}$$

where

τ_c = effective shear stress on the soil N/m² (lb/ft²)

τ_λ = shear stress on the RECP that results in 12.5 mm (0.5 in.) of erosion

τ_d = design shear stress N/m² (lb/ft²)

α = conversion constant: 6.5 (SI), 0.14 (CU)

For stability, the permissible shear stress for the RECP should be at least equal to the design shear stress, namely the shear stress on the bed. Likewise, the effective shear stress should be no more than the effective shear stress on the soil, τ_λ. Substituting τ_p for τ_d and τ_p, soil for τ_c in the preceding equation results in the following equation for the permissible shear stress of the RECP:

$$\tau_p = (\tau_\lambda/\alpha)(\tau_p,\text{soil} + \alpha/4.3) \qquad (4.85)$$

The following example illustrates the design procedure for an RECP lined channel.

Example 4.21

To control erosion in a trapezoidal earthen channel an RECP of the following performance data is selected:

Roughness Rating

Applied Shear, N/m²	n Value
45	0.039
90	0.035
180	0.030

τ_λ = 80 N/m² (shear on lining at 12.5 mm soil loss)

The channel is at 2.0% slope; bottom width and side slopes are 1.0 m and 3:1, respectively, and the soil is clayey sand (SC) of PI = 16, e = 0.45. Determine if the RECP lining is satisfactory as a temporary lining for a discharge of 0.70 m³/s.

Solution

Estimate depth at 0.35 m.

Calculate hydraulic radius, shear stress, and the Manning's n as follows:

$$A = by + my^2 = 1 \times 0.35 + 3 \times 0.35^2 = 0.7175 \text{ m}^2$$

$$P = b + 2y\sqrt{(m^2 + 1)} = 1 + 2 \times 0.35 \sqrt{10} = 3.21$$

$$R = 0.223 \text{ m}$$

$$\tau_o = \gamma RS = 9.81 \times 10^3 \times 0.223 \times 0.020 = 43.8 \text{ N/m}^2$$

$$b = -\sqrt{[\ln(0.035/0.039)\ln(0.030/0.035)]}/0.693$$

$$b = -0.024$$

$$a = n_m/\tau_m^b = 0.035/(90)^{-0.024} = 0.039$$

$$n = a\tau_o^b = 0.039 \times 43.8^{-0.024} = 0.036$$

Then calculate the discharge:

$$Q = 1/0.036 \times 0.7175 \times (0.223)^{2/3} \times (0.020)^{1/2} = 1.047 \text{ m}^3/\text{s}$$

The calculated discharge is nearly 50% greater than the design discharge. Try a smaller flow depth.

$$y = 0.285 \text{ m}$$

$$A = 0.285 + 3 \times 0.285^2 = 0.529 \text{ m}^2$$

$$P = 1 + 2 \times 0.285\sqrt{10} = 2.802 \text{ m}$$

$$R = 0.189 \text{ m}$$

$$\tau_o = \gamma R S = 9.81 \times 10^3 \times 0.189 \times 0.02 = 37.0 \text{ N/m}^2$$

$$n = 0.039 \times 37.0^{-0.024} = 0.036$$

$$Q = 1/0.036 \times 0.529 \times 0.189^{2/3} \times (0.02)^{1/2} = 0.689 \text{ m}^3/\text{s}$$

The calculated flow differs by 1.6% (less than 5%) of the design flow. Proceed with calculating the soil and the RECP permissible shear stresses using Equations 4.61 and 4.85, respectively.

$$\tau_p, \text{soil} = (C_1 PI^2 + C_2 PI + C_3)(C_4 + C_5 e)^2 C_6$$

See Appendix 4E for C values.

$$= (1.07 \times 16^2 + 14.3 \times 16 + 47.7)(1.42 - 0.61 \times 0.45)^2 \times 0.0048 = 3.47 \text{ N/m}^2$$

$$\tau_p = (\tau_\lambda/\alpha)(\tau_p, \text{soil} + \alpha/4.3) = (80/6.5)(3.47 + 6.5/4.3) = 61.3 \text{ N/m}^2$$

Calculate the maximum shear stress on the channel bottom.

$$\tau_d = \gamma y S = 9.81 \times 10^3 \times 0.285 \times 0.02 = 55.9 \text{ N/m}^2$$

The selected RECP is acceptable as a temporary lining until the vegetation is established. Note that the permanent channel stability for the established grass has to be evaluated separately.

PROBLEMS

4.1 Calculate the flow spread in a 2 m wide asphalt gutter for a flow of 0.08 m³/s. The gutter is at 2% longitudinal and 4% cross slopes, respectively.

4.2 Calculate the spread in a 6 ft wide asphalt gutter for a flow of 2.5 cfs. The gutter is at 1.5% longitudinal slope and 4% cross slope.

4.3 A 2 m wide smooth asphalt shoulder is at 4% cross slope and 0.5% longitudinal slope. Does the spread extend beyond the shoulder for a flow of 0.09 m³/s?

4.4 A 6 ft wide smooth asphalt shoulder is at 4% cross slope. For a flow of 2.5 cfs, does the spread extend beyond the shoulder if the longitudinal slope is 0.5%?

4.5 Calculate the flow spread on a 2 m wide shoulder at 4% cross slope for gutter flows of 0.03 m³/s, 0.06 m³/s, and 0.09 m³/s. The roadway and shoulder have an asphalt cover ($n = 0.016$). Perform calculations for a longitudinal slope of 3%.

4.6 Redo Problem 4.5 for a 6 ft wide shoulder and discharges of 1, 2, and 4 cfs.

4.7 In Problem 4.6, calculate the flow capture and efficiency of a "B" inlet using the NJDOT method.

4.8 Recalculate flow capture and efficiency in Problem 4.7 using the HEC-22 charts.

4.9 A grate inlet is to be placed at a roadway sag flush with a 15 cm curb to capture the gutter flow, calculated at 0.15 m³/s. Calculate the minimum required length of a 60 cm wide grate. For grate in sags assume that 50% of the grate opening and 25% of the grate perimeter are clogged by debris.

4.10 Calculate the required grate opening area in Problem 4.9 for a discharge of 6 cfs and 2 ft wide grate. The curb is 6 in. high. Consider weir flow through grate, in this case.

4.11 In Example 4.2 of this chapter, calculate the maximum spacing of the first set of inlets from the high point for a 75% efficiency.

4.12 A pond discharges through a 1.2 m × 1.2 m square box culvert at a rate of 2.5 m³/s. The culvert is very long, Manning's n is 0.014, and $S = 0.005$. Calculate the elevation of water level in the pond at the inlet face of the culvert. The exit loss in the pond and the entrance loss can be neglected. Assume that the outlet is unsubmerged.

4.13 Redo Problem 4.12 if a 1200 mm circular culvert is used in lieu of the box culvert. Use $n = 0.013$.

4.14 Solve Problem 4.12 for a discharge of 85 cfs and 48 in. culvert.

4.15 Solve Problem 4.12 if the culvert is 300 ft long and the culvert outlet is submerged up to its crown. Account for inlet and exit losses in this case.

4.16 A 60 in. RCP culvert carries a stream under a roadway. The culvert is 100 ft long, at 2% slope, and its upstream invert lies 10 ft below the edge of road. Calculate the capacity of the culvert for the following conditions:
   a. The culvert outlet is unsubmerged.
   b. The outlet is submerged 2 ft above its crown.

4.17 A box culvert is to carry a discharge of 10 m³/s. The culvert is 20 m long and has 0.4% slope. The depth of water at the upstream face of the culvert is not to rise more than 1 m above its crown. Select a suitable size culvert. The downstream face of the culvert is unsubmerged.

4.18 For a channel in alluvial silt, the Manning's n value and the maximum permissible velocity are 0.02 and 2 ft/s, respectively. Calculate the corresponding permissible tractive force if the channel slope is 0.9%.

4.19 Calculate the discharge and the cross-sectional area of a channel excavated in a non-cohesive soil having permissible shear stress of 5 Pa, angle of repose of 32°, and $n = 0.025$. The channel slope is 0.4%.

4.20 Design a parabolic swale lined with Bermuda grass to carry a discharge of 35 cfs, at 2% slope. Base your design on retardance class D and the NJ Soil Erosion Standards.

4.21 Design the grass swale of Problem 4.20 for a discharge of 1 m³/s, using Figure 4.43 (n vs. VR relation).

4.22 Design a swale, lined with riprap stone of $d_{50} = 15$ cm, to carry a discharge of 1.5 m³/s. The channel has an average slope of 1.5%. Base your design on the permissible shear stress method and Equation 4.70 for n, and the following channel geometry:
   Trapezoidal channel of 3:1 side slope

4.23 Redo Problem 4.22 for $d_{50} = 6$ in. and 50 cfs discharge and
   a. Equation 4.70
   b. NRCS Equation 4.74

4.24 A trapezoidal grass-lined swale is to carry a discharge of 1.5 m³/s at 1.5% slope. The soil is clayey sand of PI = 20 and $e = 0.5$. The swale is 1.25 m wide at the bottom and has 3:1 side slope. For a mixed-grass lining in very good condition, 75 mm thick, design the swale using the HEC-15 (FHWA, 2005) method.

4.25 A trapezoidal channel of bottom width of 2 m and side slopes of 2:1 has a bend with a radius of 15 m. For a discharge of 8 m³/s, water depth at the inner wall of the bend is 1.0 m. Calculate water depth at the outer bank around this bend. Slope of this channel is mild.

4.26 Calculate the water depth at the outer bank of a trapezoidal channel of 2:1 side slope and 6 ft bottom width at a 50 ft radius bend. The channel discharge is 250 cfs and the depth of water at the inner bend is 3 ft.

4.27 Calculate the length of the channel past the bend in Problem 4.26, which requires protection. Base your calculations on a Manning's n of 0.040.

4.28 The trapezoidal channel of Problem 4.25 is lined with $d_{50} = 15$ cm stone and is at 1.3% slope. Determine whether or not the stone is stable. Base your analysis on permissible shear stresses listed in Table 4.10.

4.29 Calculate the effective Manning's n for the trapezoidal channel in Problem 4.24 if it is lined with concrete at the bottom and 75 mm high grass at the sides. Assume grass cover to be in good condition. Base your calculations for $n = 0.016$ for concrete.

4.30 In Problem 4.24, if the slope of the channel is 3%, will the grass be stable?

4.31 An RECP having the following listed roughness rating is used to control erosion in a trapezoidal earthen channel of 3:1 side slope:

Roughness Rating	
Applied Shear (N/m²)	n Value
40	0.042
80	0.039
160	0.035

$$\tau_\ell = 75 \text{ N/m}^2$$

The bottom width and the channel slope are 0.9 m and 1.5%, respectively. The soil is clayey sand (SC) having PI = 16 and $e = 0.5$. Determine if the RECP lining is satisfactory for a discharge of 0.5 m³/s.

APPENDIX 4A: DERIVATION OF GUTTER FLOW EQUATION

In a gutter cross section, the top width exceeds the flow depth at the curb by several fold (25 times for a 4% cross slope). As such, a direct application of the Manning formula to gutter cross section does not accurately reflect the flow in the gutter. To be accurate, the flow should be calculated by the sum of elemental flows through the section.

Figure 4A.1 illustrates flow in a triangular curb gutter. The partial flow through an element of width, dx, can be calculated from the following equation:

$$dQ = Vy dx \tag{4A.1}$$

where V is the average flow velocity and y is the depth of flow within the segment. Using the Manning formula in SI units, the average flow velocity within the segment can be calculated by

$$V = \frac{y^{2/3} S^{1/2}}{n} \tag{4A.2}$$

where n is the Manning's roughness coefficient of the gutter section and S is the longitudinal slope of the gutter. According to Figure 4A.1:

$$dx = \frac{dy}{S_x} \tag{4A.3}$$

Substituting V and dx from Equations 4A.2 and 4A.3 into Equation 4A.1 yields

$$dQ = \left(\frac{1}{n}\right)\left(\frac{S^{1/2}}{S_x}\right) y^{5/3} dy \tag{4A.4}$$

Integrating the preceding equation within the limits of the flow depth from 0 to d across gutter, the total flow can be calculated as

$$Q = \int_0^d \frac{1}{n} y^{5/3} \frac{S^{1/2}}{S_x} dy$$

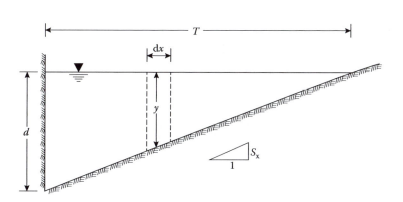

FIGURE 4A.1   Triangular gutter flow.

$$Q = 0.375 \left(\frac{S^{1/2}}{S_x} \right) \frac{d^{8/3}}{n} \tag{4A.5}$$

Substituting for $d = TS_x$ in the preceding equation gives

$$Q = 0.375 S^{1/2} S_x^{5/3} \frac{T^{8/3}}{n} \tag{4A.6}$$

In customary units, the gutter flow equation can be derived by simply applying a 1.49 factor instead of 1 in Equation 4A.2, resulting in

$$Q = 0.56 \left(\frac{S^{1/2}}{S_x} \right) \frac{d^{8/3}}{n} \tag{4A.7}$$

and

$$Q = 0.56 S^{1/2} S_x^{5/3} \frac{T^{8/3}}{n} \tag{4A.8}$$

APPENDIX 4B: DERIVATION OF FLOW EQUATIONS
FOR INLETS ON ROADWAYS AT 0% GRADE

The flow at a distance of x from the midway point between any two inlets can be calculated from the following equation:

$$Q_x = qx \tag{4B.1}$$

where q represents the flow per lineal length of roadway (refer to Figure 4B.1). Using the rational method, the unit discharge is related to the rainfall intensity, I, in./h (mm/h) by the following equations:

$$q = \frac{2.78 \times 10^{-3}(W + W')}{10,000} \times I \quad \text{SI} \tag{4B.2}$$

$$q = \frac{W + W'}{43,200} \times I \quad \text{CU} \tag{4B.3}$$

where W and W' are the width of roadway and gutter, respectively.

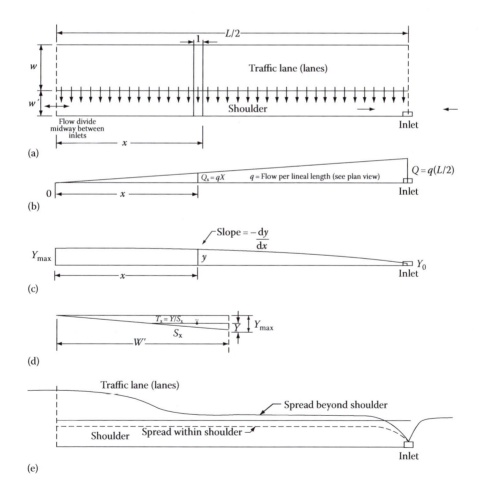

FIGURE 4B.1 Gutter flow for a roadway of horizontal profile: (a) plan view; (b) cutter flow vs. distance; (c) variation of depth of flow; (d) flow cross-section; and (e) flow spread.

A runoff coefficient of 1.0 is used for paved roadway in these equations. Using Manning's formula, Q_x can be related to the gutter flow parameters shown in Figure 4B.1 as follows:

$$A = \frac{y^2}{2S_x}; \quad R = \frac{y}{2} \quad \text{(refer to flow cross section)}$$

$$Q = AR^{2/3}\frac{S^{1/2}}{n} = 0.315 y^{8/3}\frac{S^{1/2}}{S_x n} \quad \text{SI} \tag{4B.4}$$

where S is the energy slope.

Equating Equations 4B.1 and 4B.4 and raising to the second power yield

$$q^2 x^2 = 0.1 y^{16/3}\frac{S}{S_x^2 n^2}$$

Approximating the energy slope S by $-dy/dx$ in the preceding equation,

$$y^{16/3}dy = -10.08 S_x^2 n^2 q^2 x^2 dx$$

Integrating:

$$y^{19/3} = -21.3 S_x^2 n^2 q^2 x^3 + C \tag{4B.5}$$

The constant of integration, C, can be calculated from the condition $y = y_o$ at $x = L/2$ where y_o is the depth of water at the inlet and L is the spacing between two consecutive inlets. The resulting equation is

$$y^{19/3} = y_o^{19/3} + 21.3 S_x^2 n^2 q^2 \left[\left(\frac{L}{2}\right)^3 - x^3\right] \tag{4B.6}$$

In customary units, the preceding equation reads as

$$y^{19/3} = y_o^{19/3} + 10.0 S_x^2 n^2 q^2 \left[\left(\frac{L}{2}\right)^3 - x^3\right] \tag{4B.7}$$

APPENDIX 4C: HYDRAULIC DESIGN CHARTS FOR INLETS

Note: Figures 4C.1 and 4C.2 give the overall efficiency of a 0.6 m × 0.6 m and 0.6 m × 1.2 m, 2 ft × 2 ft, and 2 ft × 4 ft grates for a 4% cross slope for a large range in gutter flow and longitudinal slope.

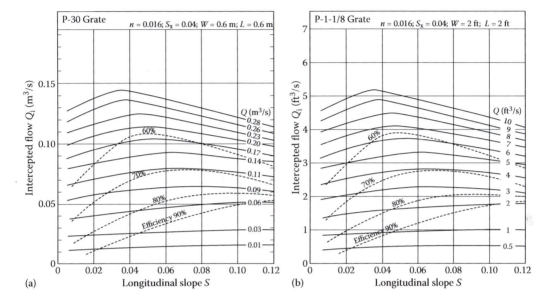

FIGURE 4C.1 (a) Interception capacity of a 0.6 m × 0.6 m, P-30 grate (SI units). (b) Interception capacity of a 2 ft × 2 ft, P1-1/8 grate (CU).

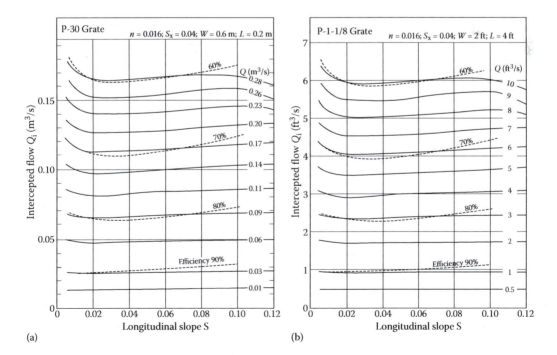

FIGURE 4C.2 (a) Interception capacity of a 0.6 m × 1.2 m P-30 grate (SI units). (b) Interception capacity of a 2 ft × 4 ft, P1-1/8 grate (CU).

APPENDIX 4D: CRITICAL FLOW CHARTS FOR ROUND AND ELLIPTICAL PIPES

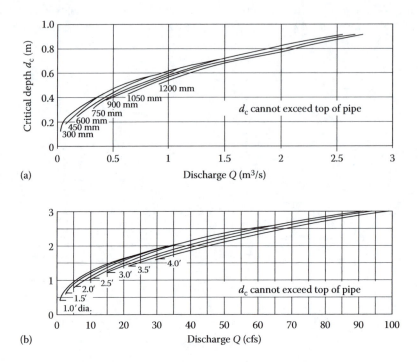

(a)

(b)

FIGURE 4D.1   (a) Critical depth, circular pipe (SI units). (b) Critical depth, circular pipe (CU).

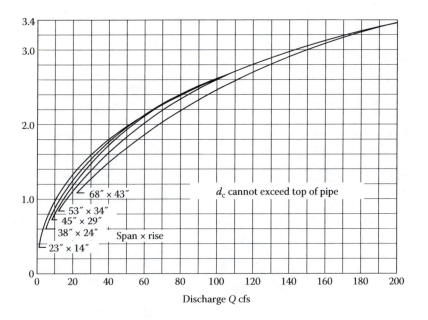

FIGURE 4D.2   Critical depth, horizontal elliptical pipe (CU).

APPENDIX 4E: PERMISSIBLE SHEAR STRESS OF COHESIVE MATERIAL IN HEC-15

TABLE 4E.1
Coefficients for Permissible Soil Shear Stress

ASTM Soil Classification	Applicable Range	C_1	C_2	C_3	C_4	C_5	C_6 (SI)	C_6 (CU)
GM	$10 \leq PI \leq 20$	1.07	14.3	47.7	1.42	−0.61	4.8×10^{-3}	10^{-4}
	$20 \leq PI$			0.076	1.42	−0.61	48	1.0
GC	$10 \leq PI \leq 20$	0.0477	2.86	42.9	1.42	−0.61	4.8×10^{-3}	10^{-3}
	$20 \leq PI$			0.119	1.42	−0.61	48	1.0
SM	$10 \leq PI \leq 20$	1.07	7.15	11.9	1.42	−0.61	4.8×10^{-3}	10^{-4}
	$20 \leq PI$			0.058	1.42	−0.61	48	1.0
SC	$10 \leq PI \leq 20$	1.07	14.3	47.7	1.42	−0.61	4.8×10^{-3}	10^{-4}
	$20 \leq PI$			0.076	1.42	−0.61	48	1.0
ML	$10 \leq PI \leq 20$	1.07	7.15	11.9	1.48	−0.57	4.8×10^{-3}	10^{-4}
	$20 \leq PI$			0.058	1.48	−0.57	48	1.0
CL	$10 \leq PI \leq 20$	1.07	14.3	47.7	1.48	−0.57	4.8×10^{-3}	10^{-4}
	$20 \leq PI$			0.076	1.48	−0.57	48	1.0
MH	$10 \leq PI \leq 20$	0.0477	1.43	10.7	1.38	−0.373	4.8×10^{-3}	10^{-3}
	$20 \leq PI$			0.058	1.38	−0.373	48	1.0
CH	$20 \leq PI$			0.097	1.38	−0.373	48	1.0

Source: FHWA, Design of roadside channels with flexible linings, Hydraulic Engineering Circular No. 15 (HEC-15), 3rd ed., September 2005.

Note: CH = inorganic clays of high plasticity, fat clays; CL = inorganic clays of low to medium plasticity, gravelly clays, sandy clays, silty clays, lean clays; GC = clayey gravels, gravel–sand–clay mixtures; GM = silty gravels, gravel–sand silt mixtures; MH = inorganic silts, micaceous or diatomaceous fine sands or silts, elastic silts; ML = inorganic silts, very fine sands, rock flour, silty or clayey fine sands; SC = clayey sands, sand–clay mixtures; SM = silty sands, sand–silt mixtures.

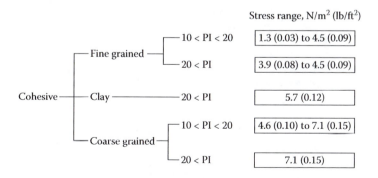

FIGURE 4E.1   Cohesive soil permissible shear stress.

APPENDIX 4F: PROPEX TURF REINFORCEMENT
MATS AND ARMORMAX PROPERTIES

PROPEX EROSION CONTROL PRODUCT GUIDE
PERMANENT SOLUTIONS

MODERATE			SEVERE
LANDLOK® STITCH-BONDED TRMS	LANDLOK® WOVEN TRMS	PYRAMAT® WOVEN HPTRMS	ARMORMAX™ SYSTEM
▶ 1st generation turf reinforcement mats (TRMs)	▶ 2nd generation turf reinforcement mats (TRMs)	▶ High performance turf reinforcement mat (HPTRM)	▶ Anchored reinforced vegetation system consisting of HPTRM and earth percussion anchors
▶ Moderate-flow channels, bank protection and steep soil slopes	▶ Moderate-flow channels, bank protection, and steep soil slopes where greater loading and/or survivability is required	▶ High-flow channels, extreme slopes, pipe inlets & outlets and other arid/semi-arid applications	▶ Earthen levees and stream, river and canal banks
▶ Up to 10 years*		▶ Up to 50 years*	▶ Storm water channels in arid and semi/arid environments
	▶ Up to 25 years*		▶ Surficial slope stabilization
			▶ Up to 50 years or greater*

*Design life performance may vary depending upon field conditions and applications.

For downloadable documents like construction specifications, installation guidelines, case studies and other technical information, please visit our web site at **geotextile.com**. These documents are available in easy-to-use Microsoft® Word format.

PROPEX® | THE ADVANTAGE CREATORS™

GEOSYNTHETICS

Propex Inc.
6025 Lee Highway, Suite 425
PO Box 22788
Chattanooga, TN 37422

PH: 423 899 0444
PH: 800 621 1273
FAX: 423 899 7619
www.geotextile.com

ARMORMAX™
Anchored Reinforced Vegetation System

ArmorMax™ Anchored Reinforced Vegetation System is the most advanced flexible armoring technology available for severe erosion challenges. The ArmorMax system can be used in **non-structural applications** where additional factors of safety are required, including protecting earthen levees from storm surge and wave overtopping and stream, river and canal banks from scour and erosion. In addition, this system is ideally suited to protect storm water channels in arid and semi-arid environments where vegetation densities of less than 30% coverage are anticipated. For **structural applications**, the system can be engineered to provide surficial slope stabilization to resist shallow plane failures. Consisting of our woven three-dimensional High Performance Turf Reinforcement Mat (HPTRM) with X3® fiber technology and earth percussion anchors, you can count on the ArmorMax system to hold its ground.

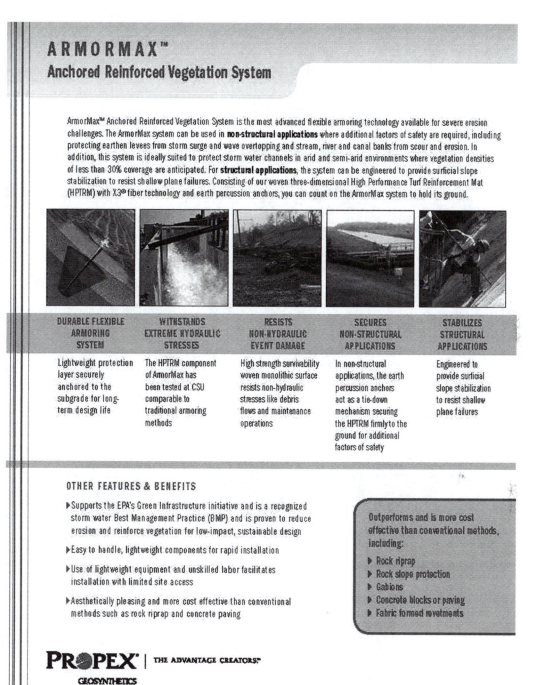

DURABLE FLEXIBLE ARMORING SYSTEM	WITHSTANDS EXTREME HYDRAULIC STRESSES	RESISTS NON-HYDRAULIC EVENT DAMAGE	SECURES NON-STRUCTURAL APPLICATIONS	STABILIZES STRUCTURAL APPLICATIONS
Lightweight protection layer securely anchored to the subgrade for long-term design life	The HPTRM component of ArmorMax has been tested at CSU comparable to traditional armoring methods	High strength survivability woven monolithic surface resists non-hydraulic stresses like debris flows and maintenance operations	In non-structural applications, the earth percussion anchors act as a tie-down mechanism securing the HPTRM firmly to the ground for additional factors of safety	Engineered to provide surficial slope stabilization to resist shallow plane failures

OTHER FEATURES & BENEFITS

▶ Supports the EPA's Green Infrastructure initiative and is a recognized storm water Best Management Practice (BMP) and is proven to reduce erosion and reinforce vegetation for low-impact, sustainable design

▶ Easy to handle, lightweight components for rapid installation

▶ Use of lightweight equipment and unskilled labor facilitates installation with limited site access

▶ Aesthetically pleasing and more cost effective than conventional methods such as rock riprap and concrete paving

Outperforms and is more cost effective than conventional methods, including:

▶ Rock riprap
▶ Rock slope protection
▶ Gabions
▶ Concrete blocks or paving
▶ Fabric formed revetments

PROPEX | THE ADVANTAGE CREATORS™
GEOSYNTHETICS

ARMORMAX™
Anchored Reinforced Vegetation System

WOVEN THREE-DIMENSIONAL HPTRM PROTECTION LAYER FEATURING X3® FIBER TECHNOLOGY

▶ Unique X3 fiber shape provides over 40% more surface area than conventional fibers to capture the moisture, soil and water required for rapid vegetation growth

▶ Exhibits extremely high tensile strength as well as superior interlock and reinforcement capacity with both soil and root systems

▶ Maximum ultraviolet protection for long-term design life

▶ Netless, rugged material construction stands up to the toughest erosion applications where high loading and/or high survivability conditions are required

EARTH PERCUSSION ANCHORS TO SECURE THE MAT TO THE GROUND

▶ Made of corrosion resistant aluminum alloy, gravity die cast and heat treated to give considerable increase in mechanical strength and durability both during installation and in service

▶ Connected to a threaded rod or stainless tendon to fully enhance corrosion resistance particularly at the soil/air interface

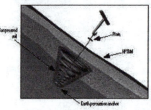

▶ As the load exerted on the soil by the ArmorMax system increases, a body of soil above the anchor is compressed and provides resistance to any further anchor movement — permanently securing the mat to the ground

ARMORMAX NON-STRUCTURAL APPLICATIONS

The figures below illustrate the ArmorMax system for non-structural applications. The system is comprised of the HPTRM and typically Type 2 earth percussion anchors.

LEVEE ARMORING   ARID/SEMI-ARID STORM WATER CHANNELS   CANAL, STREAM AND RIVER BANK PROTECTION

ARMORMAX STRUCTURAL APPLICATION

The figures below illustrate the use of ArmorMax in a structural application for surficial slope stabilization. The system is comprised of the HPTRM and Type 1A or 1B earth percussion anchors as specified by the project engineer.

SHALLOW PLANE FAILURE   APPLY ARMORMAX SYSTEM   VEGETATION GROWTH

KEY PHYSICAL PROPERTIES OF ARMORMAX™

▶ Material Composition: Patented ultraviolet protection package in HPTRM, stainless steel tendons and galvanized threaded rods provide long-term design assurance.

▶ Tensile Strength: HPTRM boasts 4000 x 3000 lb/ft (58.4 x 43.8 kN/m) of tensile strength, which exceeds the U.S. EPA's definition of a High Performance Turf Reinforcement Mat.

▶ Seedling Emergence: HPTRM features X3® fiber technology, which offers 40% more fiber surface area to capture the critical sediment and moisture needed to increase seed germination within the first 21 days.

▶ Flexibility: Allows the system to conform and maintain intimate contact with the prepared subgrade.

▶ Holding Strength: Based on anchor size, tendon rod length and on-site soil parameters the anchor foot provides up to an ultimate of 500 to 5000 lbs of pullout resistance per earth percussion anchor. Actual holding strengths depend upon soil characteristics, anchor type and installation techniques.

ARMORMAX PROPERTY TABLES[1] ENGLISH & METRIC VALUES

	PROPERTY	TEST METHOD	VALUE[2]	HPTRM
	HIGH PERFORMANCE TURF REINFORCEMENT MAT			
PHYSICAL	MASS/UNIT AREA	ASTM D-6566	MARV	13.5 oz/yd² 455 g/m²
	THICKNESS	ASTM D-6525	MARV	0.4 in 10.2 mm
	LIGHT PENETRATION (% Passing)	ASTM D-6567	TYPICAL	10%
	COLOR	VISUAL	–	GREEN, TAN
MECHANICAL	TENSILE STRENGTH (Grab)	ASTM D-6818	MARV	4000 x 3000 lb/ft 58.4 x 43.8 kN/m
	TENSILE ELONGATION	ASTM D-6818	MARV	25%
	RESILIENCY	ASTM D-6524	MARV	80%
	FLEXIBILITY/STIFFNESS	ASTM D-6575	TYPICAL	0.534 in-lbs 615,000 mg-cm
DURABILITY	UV RESISTANCE @ 6000 HOURS	ASTM D-4355	MINIMUM	90%
	ROLL SIZES	MEASURED	TYPICAL	8.5 ft x 90 ft 2.6 m x 27.4 m

	PROPERTY	ANCHOR LENGTH (ft) (Minimum Installation Depth)	MAXIMUM PULL-OUT (Field Tested)
NON-STRUCTURAL	**EARTH PERCUSSION ANCHORS**		
	TYPE 2	2.0 ft 0.6 m	500 lbs 226.8 kg
STRUCTURAL	TYPE 1A[3]	3.5 ft 1.1 m	2,000 lbs 907.2 kg
	TYPE 1B[3]	3.5 ft 1.1 m	5,000 lbs 2268 kg

NOTES: 1. The property values listed are effective 12/2006 and are subject to change without notice.
2. MARV indicates minimum average roll value calculated as the typical minus two standard deviations. Statistically, it yields a 97.7% degree of confidence that any sample taken during quality assurance testing will exceed the value reported.
3. Maximum tendon/wedge grip strength capacity is 2000 lbs. Threaded rods with bolted steel plates up to 5000 lbs.

APPENDIX 4G: LANDLOK EROSION CONTROL BLANKETS
AND SUPERGRO PROPERTIES BY PROPEX, INC.

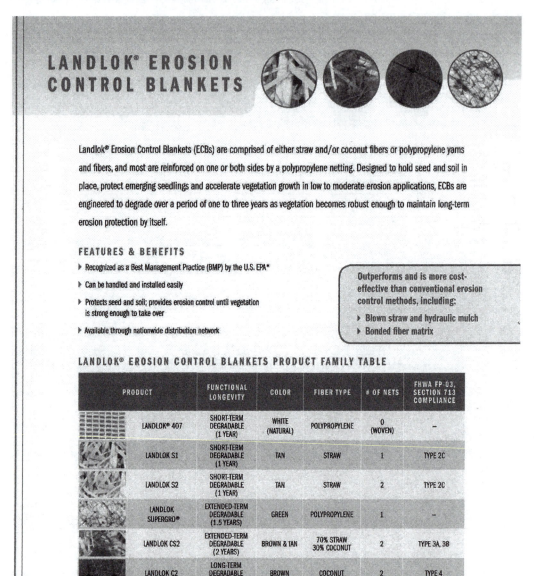

LANDLOK® EROSION CONTROL BLANKETS

Landlok® Erosion Control Blankets (ECBs) are comprised of either straw and/or coconut fibers or polypropylene yarns and fibers, and most are reinforced on one or both sides by a polypropylene netting. Designed to hold seed and soil in place, protect emerging seedlings and accelerate vegetation growth in low to moderate erosion applications, ECBs are engineered to degrade over a period of one to three years as vegetation becomes robust enough to maintain long-term erosion protection by itself.

FEATURES & BENEFITS

▶ Recognized as a Best Management Practice (BMP) by the U.S. EPA*

▶ Can be handled and installed easily

▶ Protects seed and soil; provides erosion control until vegetation is strong enough to take over

▶ Available through nationwide distribution network

> Outperforms and is more cost-effective than conventional erosion control methods, including:
> ▶ Blown straw and hydraulic mulch
> ▶ Bonded fiber matrix

LANDLOK® EROSION CONTROL BLANKETS PRODUCT FAMILY TABLE

PRODUCT	FUNCTIONAL LONGEVITY	COLOR	FIBER TYPE	# OF NETS	FHWA FP-03, SECTION 713 COMPLIANCE
LANDLOK® 407	SHORT-TERM DEGRADABLE (1 YEAR)	WHITE (NATURAL)	POLYPROPYLENE	0 (WOVEN)	–
LANDLOK S1	SHORT-TERM DEGRADABLE (1 YEAR)	TAN	STRAW	1	TYPE 2C
LANDLOK S2	SHORT-TERM DEGRADABLE (1 YEAR)	TAN	STRAW	2	TYPE 2C
LANDLOK SUPERGRO®	EXTENDED-TERM DEGRADABLE (1.5 YEARS)	GREEN	POLYPROPYLENE	1	–
LANDLOK CS2	EXTENDED-TERM DEGRADABLE (2 YEARS)	BROWN & TAN	70% STRAW 30% COCONUT	2	TYPE 3A, 3B
LANDLOK C2	LONG-TERM DEGRADABLE (3 YEARS)	BROWN	COCONUT	2	TYPE 4

*U.S. EPA: United States Environmental Protection Agency

PROPEX | THE ADVANTAGE CREATORS.™

GEOSYNTHETICS

LANDLOK® EROSION CONTROL BLANKETS PROPERTY TABLE[1] ENGLISH & METRIC UNITS

	PROPERTY	TEST METHOD	VALUE	LANDLOK® 407	LANDLOK S1	LANDLOK S2	LANDLOK SUPERGRO®	LANDLOK CS2	LANDLOK C2
MECHANICAL	TENSILE STRENGTH	ASTM D-6818	TYPICAL	460 x 250 lb/ft 6.7 x 3.7 kN/m	50 x 65 lb/ft 0.73 x 0.95 kN/m	75 x 75 lb/ft 1.1 x 1.1 kN/m	N/A	100 x 100 lb/ft 1.5 x 1.5 kN/m	150 x 150 lb/ft 2.2 x 2.2 kN/m
MECHANICAL	TENSILE ELONGATION	ASTM D-6818	TYPICAL	30% x 40	20%	25%	N/A	30%	25%
PHYSICAL	MASS PER UNIT AREA	ASTM D-6475	TYPICAL	2.2 oz/yd² 75 g/m²	8.5 oz/yd² 288 g/m²	8.8 oz/yd² 298 g/m²	0.70 oz/yd² 23.7 g/m²	8.8 oz/yd² 298 g/m²	8.8 oz/yd² 298 g/m²
PHYSICAL	THICKNESS	ASTM D-6525	TYPICAL	–	0.11 in 2.8 mm	0.25 in 6.35 mm	–	0.40 in 10.2 mm	0.30 in 7.6 mm
ENDURANCE	FUNCTIONAL LONGEVITY	OBSERVED	TYPICAL	SHORT-TERM DEGRADABLE (1 YEAR)	SHORT-TERM DEGRADABLE (1 YEAR)	SHORT-TERM DEGRADABLE (1 YEAR)	EXTENDED-TERM DEGRADABLE (1.5 YEARS)	EXTENDED-TERM DEGRADABLE (2 YEARS)	LONG-TERM DEGRADABLE (3 YEARS)
PACKAGING	ROLL WIDTH	MEASURED	TYPICAL	12.5 ft 3.81 m	8 ft 2.45 m	8 ft 2.45 m	7.5 ft 2.3 m	8 ft 2.45 m	8 ft 2.45 m
PACKAGING	ROLL LENGTH	MEASURED	TYPICAL	432 ft 131.7 m	112.5 ft 34.3 m	112.5 ft 34.3 m	1200 ft 365.8 m	112.5 ft 34.3 m	112.5 ft 34.3 m
PACKAGING	ROLL WEIGHT	CALCULATED	TYPICAL	98 lb 44 kg	53 lb 24 kg	53 lb 24 kg	44 lb 20 kg	55 lb 25 kg	55 lb 25 kg
PACKAGING	ROLL AREA	MEASURED	TYPICAL	600 yd² 502 m²	100 yd² 84 m²	100 yd² 84 m²	1000 yd² 836 m²	100 yd² 84 m²	100 yd² 84 m²

NOTES: 1. The property values listed are effective 08/2006 and are subject to change without notice.

LANDLOK® EROSION CONTROL BLANKET PERFORMANCE VALUES ENGLISH & METRIC UNITS

MATERIAL	FUNCTIONAL LONGEVITY	MAXIMUM SHORT-TERM SHEAR STRESS AND VELOCITY (UNVEGETATED)[2]		MANNING'S "n"	C-FACTOR
		SHEAR STRESS	VELOCITY	0-6 in (0-150 mm)	
LANDLOK® 407	SHORT-TERM DEGRADABLE (1 YEAR)	–	–	–	–
LANDLOK S1	SHORT-TERM DEGRADABLE (1 YEAR)	2.0 lb/ft² 96 N/m²	n/r	n/r	0.14
LANDLOK S2	SHORT-TERM DEGRADABLE (1 YEAR)	2.0 lb/ft² 96 N/m²	5.0 - 6.0 ft/sec 1.5 - 1.8 m/sec	0.027	0.21
LANDLOK SUPERGRO®	EXTENDED-TERM DEGRADABLE (1.5 YEARS)	2.0 lb/ft² 96 N/m²	–	–	–
LANDLOK CS2	EXTENDED-TERM DEGRADABLE (2 YEARS)	2.0 lb/ft² 96 N/m²	5.0 - 6.0 ft/sec 1.5 - 1.8 m/sec	0.021	0.09
LANDLOK C2	LONG-TERM DEGRADABLE (3 YEARS)	2.0 lb/ft² 96 N/m²	5.0 - 6.0 ft/sec 1.5 - 1.8 m/sec	0.018	0.06

NOTES: 1. "n/r" not recommended for use in swales and low-flow channels.
2. Typical design limits for natural vegetation are a maximum shear stress of 2.0 lb/ft² (96 N/m²) and a velocity limit of 5.0 to 6.0 ft/sec (1.5 to 1.8 m/sec).

LANDLOK® EROSION CONTROL BLANKETS

APPLICATION RECOMMENDATIONS FOR LANDLOK® EROSION CONTROL BLANKETS

	APPLICATION	FUNCTIONAL LONGEVITY	PRODUCT STYLE	INSTALLED COST[1]	ANCHOR RECOMMENDATIONS
SLOPES[2]	1.5H:1V	LONG-TERM DEGRADABLE (3 YEARS)	LANDLOK® C2	\$2.00 - 2.75/yd^2 \$2.39 - 3.29/m^2	2 ANCHORS/yd^2 2.5 ANCHORS/m^2
	2H:1V	EXTENDED-TERM DEGRADABLE (2 YEARS)	LANDLOK CS2	\$1.75 - 2.25/yd^2 \$2.09 - 2.69/m^2	2 ANCHORS/yd^2 2.5 ANCHORS/m^2
	3H:1V	SHORT-TERM DEGRADABLE (1 YEAR)	LANDLOK S2	\$1.25 - 1.75/yd^2 \$1.50 - 2.09/m^2	1.5 ANCHORS/yd^2 1.8 ANCHORS/m^2
	3H:1V	SHORT-TERM DEGRADABLE (1.5 YEARS)	LANDLOK SUPERGRO®	–	1.5 ANCHORS/yd^2 1.8 ANCHORS/m^2
	4H:1V OR FLATTER	SHORT-TERM DEGRADABLE (1 YEAR)	LANDLOK S1	\$1.00 - 1.50/yd^2 \$1.20 - 1.79/m^2	1 ANCHOR/yd^2 1.2 ANCHORS/m^2
	4H:1V OR FLATTER	SHORT-TERM DEGRADABLE (1 YEAR)	LANDLOK 407	\$1.00 - 1.50/yd^2 \$1.20 - 1.79/m^2	1 ANCHOR/yd^2 1.2 ANCHORS/m^2
SWALES & LOW-FLOW CHANNELS[3]	SHEAR STRESS UP TO 2.0 lbs/ft^2 (96 N/m^2) VELOCITY UP TO 5.0 to 6.0 ft/sec (1.5 to 1.8 m/sec)	SHORT-TERM DEGRADABLE (1 YEAR)	LANDLOK S2	\$1.25 - 1.75/yd^2 \$1.50 - 2.09/m^2	2.5 ANCHORS/yd^2 3 ANCHORS/m^2
		EXTENDED-TERM DEGRADABLE (1.5 YEARS)	LANDLOK SUPERGRO®	–	1.5 ANCHORS/yd^2 1.8 ANCHORS/m^2
		EXTENDED-TERM DEGRADABLE (2 YEARS)	LANDLOK CS2	\$1.75 - 2.25/yd^2 \$2.09 - 2.69/m^2	2.5 ANCHORS/yd^2 3 ANCHORS/m^2
		LONG-TERM DEGRADABLE (3 YEARS)	LANDLOK C2	\$2.00 - 2.75/yd^2 \$2.39 - 3.29/m^2	2.5 ANCHORS/yd^2 3 ANCHORS/m^2

NOTES: 1. Installed cost estimates range from large to small projects according to material quantity. The estimates include material, seed, labor and equipment. Costs vary greatly in different regions of the country. 2. For slopes steeper than 1.5 H:1V, please see our Landlok® TRM and Pyramat® HPTRM product brochure. 3. For channels with shear stress greater than 2.0 lbs/ft^2 (96 N/m^2) and a velocity greater than 5.0 to 6.0 ft/sec (1.5 to 1.8 m/sec), please see our Landlok TRM and Pyramat® HPTRM product brochure.

KEY PROPERTIES OF LANDLOK® EROSION CONTROL BLANKETS

- Mass Per Unit Area: Ensures a consistent distribution of fibers within the matrix, which leads to improved erosion protection.
- Functional Longevity: Product range allows selection of the best product for the application.

SUGGESTED SPECIFICATION FOR LANDLOK® SUPERGRO®

DESCRIPTION

Project shall consist of ground surface preparation, seeding, fertilizing, and furnishing and installing Landlok® SuperGro® erosion control fabric or approved equal.

MATERIAL DESCRIPTION

SuperGro is a flexible composite of a uniform blanket of polypropylene fibers reinforced with polypropylene netting, green in color. Specifically designed to prevent surface erosion of freshly landscaped areas.

MATERIAL SPECIFICATIONS

Weight: 0.7 ounce per square yard
Specific gravity: 0.9
Ultraviolet degradable
Fire retardant (meets flammability test CS191-53)
Chemically inert
Roll size: 8.0 ft x 1,125 ft (2.4 m x 342.9 m)
          1,000 sy per roll

MATERIAL INSTALLATION

Material shall be installed in accordance with the following procedure:

1. Prepare the ground surface by mechanical means and/or rake.

2. Seed and fertilize the area.

3. Unroll SuperGro mat with netting side up.

   NOTE: Do not stretch. Make sure net is relaxed on the surface to allow conformance with ground surface.

4. Anchor the mat by placing the pins at 4-ft. ± 1-ft. intervals in adjacent panel overlap areas.

   NOTE: Be sure the pins are well secured in the ground. Size of the pins should be 4- to 6-inch U-shaped type, depending on the ground condition. Wood pegs may be used if preferred.

5. Overlap end of the rolls and adjacent panel sides from 2 to 6 inches.

6. Anchor the top, bottom and any end of roll overlaps.

7. Water lightly after installation if possible; this will enhance grass growth and interlock the fibers into the soil.

MAINTENANCE

The contractor shall be required to perform all maintenance necessary to keep the treated area in a satisfactory condition until the work is finally accepted.

If any staples become loosened or raised or if any fabric comes loose, torn or undermined, satisfactory repairs shall be made immediately without additional compensation. If seed is washed out before germination, the area shall be fertilized, reseeded or restored without additional compensation.

METHOD OF MEASUREMENT

Erosion control fabric shall be measured by the square yard complete-in-place in accordance with plan dimensions, not including additional SuperGro for overlaps.

Landlok® SuperGro® easily conforms to the ground on uneven surfaces.

For downloadable documents like construction specifications, installation guidelines, case studies and other technical information, please visit our web site at **geotextile.com**. These documents are available in easy-to-use Microsoft® Word format.

PROPEX | THE ADVANTAGE CREATORS™

GEOSYNTHETICS

Propex Inc.
6025 Lee Highway, Suite 425
PO Box 22788
Chattanooga, TN 37422

PH: 423 899 0444
PH: 800 621 1273
FAX: 423 899 7619
www.geotextile.com

REFERENCES

American Concrete Pipe Association, 2005, *Concrete pipe design manual*, 17th printing.

ASCE and WEF, 1992, Design and construction of urban storm water management systems, American Society of Civil Engineers Manuals and Reports of Engineering Practice no. 77 and Water Environment Federation manual of practice FD-20.

Blodgett, J.C., 1986, Rock riprap design for protection of stream channels near highway structures, vol. 1— Hydraulic characteristics of open channels, USGS Water Resources Investigation Report 86-4127.

Campbell Foundry Company Catalogue, 2012, 22nd ed., Harrison, Kearny, New Jersey, http://www.campbell .com.

Chow, V. T., 1959, *Open channel hydraulics*, McGraw-Hill, New York.

―――― 1964, *Handbook of applied hydrology*, McGraw-Hill, New York.

EPA (US Environmental Protection Agency), 1976, Erosion and sediment control, surface mining in the eastern U.S., EPA-62515-76-006, Washington, DC.

FHWA (US Dept. of Transportation), 1967, Use of riprap for bank protection, Washington, DC, June.

―――― 1984, Drainage of highway pavements, Hydraulic Engineering Circular No. 12, March.

―――― 1995, Best management practices for erosion and sediment control, Report No. FHWA-FLP-94-005, June.

―――― 2001a, Introduction to highway hydraulics, Hydraulic Design Series (HDS) No. 4, Washington, DC.

―――― 2001b, Urban drainage design manual, Hydraulic Engineering Circular No. 22, 2nd ed.

―――― 2005, Design of roadside channels with flexible linings, Hydraulic Engineering Circular No. 15 (HEC-15), 3rd ed., September.

―――― 2012, Hydraulic design of highway culverts, Hydraulic Design Series (HDS), No. 5, Washington, DC, 3rd ed., April.

―――― 2013, Urban drainage design manual, Hydraulic Engineering Circular No. 22, 3rd ed., August.

Futerra Blankets, Profile Products LLC, 750 Lake Look Road, Suite 440, Buffalo Grove, IL 60089, Ph. 866-325-6262 (http://www.netlessblanket.com).

Green Armor Systems Enkamat TRM by Profile Products LLC, 750 Lake Road, Suite 440, Buffalo Grove, IL 60089, Ph. 800-508-8680 (http://www.profileproducts.com).

LandLok, Erosion Control Blankets and TRMs, Propex Inc., 6025 Lee Highway, Suite 425, P.O. Box 22788, Chattanooga, TN 37422, Ph. 423-899-0444/800-621-1273 (http://www.geotextile.com).

Maccaferri Inc., 10303 Governor Lane Boulevard, Williamsport, MD 21795-3116, Ph. 301-223-6910, (hdqtrs @maccaferri-usa.com; http://www.maccaferri-usa.com).

Mays, L.W., ed., 1996, *Water resources handbook*, Chapter 10, Storm water management, McGraw-Hill, New York.

Neenah Foundry Catalog "R," 14th ed., Box 729, Neenah, WI (http://www.neenahfoundry.com).

NJDOT (New Jersey Department of Transportation), 2013, Drainage design, section 10, in Roadway design manual, April.

NJ State Soil Conservation Committee, 1999, Standards for soil erosion and sediment control in New Jersey, July.

NRCS, Conservation Practices Standards, 2010, Lined waterway or outlet, code 468, September.

NY State Department of Environmental Conservation, 2005, New York State standards and specifications for erosion and sediment control, August 2005, Albany, NY.

NY State Soil Erosion and Sediment Control Committee, 1997, Guidelines for urban erosion and sediment control, April 1997.

Pazwash, H. and Boswell, S.T., 2003, Proper design of inlets and drains for roadways and urban developments, *ASCE EWRI Conference, World Water and Environmental Resources Congress*, June 22–26, Philadelphia, PA.

Residential Site Improvements Standards, 2009, New Jersey administrative code, title 5, Chapter 21, adopted January 6, 1977, last revised June 15, 2009.

USDA, Natural Resources Conservation Service, 1992, *Engineering field handbook*, Washington, DC.

US Soil Conservation Service, 1954, Handbook of channel design for soil and water conservation, March 1947, revised June 1954, Stillwater Outdoor Hydraulic Laboratory, Oklahoma.

5 Storm Water Management Regulations

5.1 INTRODUCTION, FEDERAL REGULATIONS

Prior to the enactment of the Clean Water Act, sanitary sewage, combined sewage, and industrial wastewater were directly discharged into open waters without receiving any treatment. To restore and maintain chemical and biological integrity of the nation's waters, the US Environmental Protection Agency (EPA) passed the Clean Water Act (CWA) of 1972. The enactment of this Act changed the traditional discharge of point source pollutants to rivers, lakes, estuaries, and wetlands. The CWA prohibited discharge of dredged or fill material as well as untreated wastewater from municipal and industrial sources into streams, lakes, wetlands, and other water bodies of the United States unless a National Pollutant Discharge Elimination System (NPDES) permit was obtained. NPDES permits were issued under CWA section 404.

5.1.1 NPDES, Phase I Program

The CWA somewhat improved the quality of our nation's waters. This Act, however, did not address the nonpoint source pollutants carried with storm water runoff. Following an extensive study, known as the National Urban Runoff Program (NURP), which was conducted between 1979 and 1983, the US EPA amended the 1972 Clean Water Act. The amendment is known as the Water Quality Act of 1987 (Pub. L 100-4), also known as the Pollution Prevention Act, phase I of the US EPA Storm Water Program. To accelerate compliance with the 1987 Water Quality Act, EPA started action in 1990 to promulgate numeric water quality criteria for those states that had not adopted sufficient water quality standards for toxic pollutants (http://www.epa.gov/waterscience/standards /about/history.html). The phase I program established the Nonpoint Source Management Program, which as indicated in Chapter 1, covered any construction activity disturbing 5 acres or more of land. The program, which was implemented in 1992, also affected medium and large municipal separate storm sewer systems (MS4s) generally serving populations of 100,000 or more. In addition, the permit coverage under NPDES phase I also included 10 categories of industrial activities.

For wetlands, regulatory guidance requiring compensatory mitigation has been set by various agencies, including the Corps of Engineers, the EPA, the US Fish and Wildlife Services, and the National Marine Fisheries Service since 1990. Because of a general deficiency in effectiveness of compensatory wetlands to mimic the natural ones, the US EPA and the Corps on March 2, 2006, proposed revisions to compensatory mitigation regulations. On April 10, 2008, US EPA and the Corps published a final rule that improved and consolidated existing regulations and established equivalent standards for all types of mitigation under the CWA section 404. This rule, which came into effect on June 9, 2008, was intended to provide greater consistency and ecological effectiveness of mitigation projects. The rule made changes to where and how the compensatory mitigation is required.

Despite the NPDES—phase I, degraded water bodies still existed. According to the 1998 National Water Quality Inventory, a biennial summary of state surveys of water quality, approximately 40% of the surveyed water bodies in the United States were still impaired by pollution and did not meet water quality standards (EPA, 2000a). A leading source of the impairment was polluted runoff. Based on the inventory, nearly 35% of assessed river miles, 45% of lakes, and 44% of estuaries have been impaired by urban/suburban storm water runoff (EPA, 2000b). As noted in Chapter 1, the National Rivers and Streams Assessment Report (EPA, 2013a) indicates that of the 1.19 million miles (1.92 × 10⁶ km) of the nation's surveyed rivers and streams, 46% are in poor biological condition, 23% in fair condition, and only 21% in good condition. It is estimated that up to 60% of the nation's existing water pollution problems are attributed to storm water nonpoint sources.

A 2000 inventory report by states, territories, and interstate commissions was even more grim. Of the 33% of the US water assessed for this national inventory, 40% of streams, 45% of lakes, and 50% of estuaries were not clean enough for fishing and swimming (EPA, 2000b). As of 2007, approximately half of the rivers, lakes, and bays under EPA oversight were still unsafe for fishing and swimming. These are very significant numbers given that urban/suburban areas cover a small percentage of US land. Nearly one-half of the lake and river impairment was due to storm water runoff from construction activities.

5.1.2 NPDES, Phase II Program

To lessen adverse impacts on water quality, the EPA published phase II of the storm water program in the *Federal Register* on December 8, 1999 (vol. 64, no. 235, NPDES). This program, which implemented under Section 402 (p. 6) of the CWA, became effective on March 10, 2003, and extended the phase I coverage to include the following two additional classes of storm water discharges nationwide:

1. Operators of small MS4s located in "urbanized areas." A "small" MS4 is any MS4 not already covered by the NPDES, phase I program. MS4s in phase II include those with 50,000 or more population and at least 1000 people per square mile population density.
2. Operators of a small construction site that disturb 1 acre or more of land.

The preceding classes were required to obtain a phase II MS4 permit and construction general permit (CGP), respectively. The CGP regulates the discharge of storm water runoff from construction sites. Under the new program, the number of permitees in the NPDES system was increased from 100,000 to more than 500,000. MS4s also cover 11 industrial categories that discharge to an MS4 or to the waters of the United States. Any one of the 11 industrial categories (except construction) included in MS4 may certify for a condition of no exposure if its industrial material and operations are not exposed to storm water. Such industries are exempt from storm water general permits. Phase II MS4s are covered by general permit. All regulated MS4s are required to

- Develop or implement a storm water management program (SWMP) or storm water pollution prevention plan (SWPPP) to reduce the contamination of storm water runoff and to prohibit illicit discharges
- Provide adequate long-term operation and maintenance measures
- Train staff to protect storm water when maintaining MS4 infrastructure and performing daily municipal activities such as parks and open-space maintenance, land disturbances, and new construction and storm water systems maintenance

To address the phase II MS4 requirements, some municipalities developed programs for data collection and reporting while others retained consultants to do the job for them. In the state of Maine, for example, all 28 regulated small municipalities, after ironing out the details, deployed a standardized program. This partnership eased the job of each municipality and reduced its cost (Brzozowski, 2005). More information on the MS4s may be obtained from the following EPA website: http://cfpub.epa.gov/npdes/storm water/munic.cfs.

The NPDES phase II program required operators of construction sites (disturbing at least 1 acre of land) to prepare a SWPPP. An EPA publication titled, "Developing Your Storm Water Pollution

Prevention Plan: A Guide for Construction Sites," is available on the EPA website (www.epa .gov/npdes/swpppguide). The EPA has also developed a user friendly template to help in writing an effective SWPPP. This template, which is in Word format, is available at http://cfpub.epa.gov/npdes /storm water/swppp.cfm. The template, dated October 2, 2007, can be used for construction site operators in five unauthorized states, which include Alaska, Idaho, Massachusetts, New Hampshire, and New Mexico, as well as the District of Columbia and Indian country lands where the EPA is the NPDES permitting agency. The SWPPP template is also available for other states that are authorized by the EPA to implement their storm water NPDES permitting program. In addition, a sample inspection report template (customizable non-PDF version) in Word format was issued by the EPA on October 2, 2007.

Short courses and training sessions have also been offered around the country. Basically, there are two parts to an SWPPP: the narrative and the sediment and erosion control drawings to address eight factors:

- Site evaluation, assessment, and planning
- Erosion and sediment control best management practices (BMPs)
- Good housekeeping BMPs
- Post construction BMPs
- Inspections
- Record keeping and training
- Final stabilization
- Certification and notification

A two-part report by Brzozowski (2008a,b) presents more information on SWPPP.

The EPA's construction general permit (CGP), which went into effect in 2003, expired on July 1, 2008. The EPA reissued a final 2008 construction general permit (CGP), which was valid for a period of 2 years and applied only to new sites disturbing 1 acre or more or smaller sites that are part of a larger development plan.

The EPA developed a national regulation, called the Effluent Limitations Guidelines for the Construction and Development Industry. This guideline, known as the C&D rule, was issued in December 2009. The C&D rule placed numeric limitation on the turbidity and pollutants.

On February 16, 2012, the EPA reissued the previously mentioned CGP, which had expired in 2011. In some states the effective date came 2 to 3 months later: April 9 in Idaho, April 13 in Washington state, and May 9 for some areas in Minnesota and Wisconsin. During the process of developing the current CGP, discussion among the regulatory community centered on the numeric limits outlined in the C&D rule. The draft 2012 CGP contained language on the effluent limitation on turbidity. However, because of a series of court actions, the EPA removed all numeric turbidity limits from the final CGP. While collecting data to evaluate and support the numeric limits in the future, EPA has indicated the 2012 CGP will not reopen before it expires at midnight on February 16, 2017. In a presentation, EPA representatives have given a fairly good idea of what the future rule will contain (EPA, 2013b).

The most significant changes in the effective CGP are

- The review period increased from 7 to 14 days.
- Eligibility for emergency-related construction. This does not cover situations where cationic treatment chemicals will be used unless specific authorization is received from the EPA.
- Conditional emergency eligibility is offered for sites discharging to sensitive waters such as impaired waters or waters with high water quality (EPA's tiers 2, 2.5, or 3).
- Sites must maintain a 50 ft buffer adjacent to surface waters. If a 50 ft buffer cannot be maintained, sediment removal effectiveness for the natural 50 ft buffer must be calculated and sediment control measures that will have at least the same effectiveness must be provided.

- Poststorm inspection must be made after events of 0.25 in. or greater.
- Timelines for maintenance and repair activities are aggressive. If a minor repair is identified, it must be corrected by the end of the next day. Sediment up gradient of inlet protection must be removed by the end of the same day it is found, if feasible, but no later than the end of the next day.
- All exposed portions of a site where earth-moving activities will not resume for 14 days or more must be immediately stabilized. The period limit is 7 days for steep slope (15% or greater), within the 50 ft buffer zone or if the receiving waters are identified as impaired for sediment or sediment related parameters such as nutrients.

Following the enactment of phase II NPDES, authorized states (namely 45 states where the EPA is not the NPDES permitting authority) developed regulations that are as strict as or stricter than the NPDES phase II requirements. The New Jersey Department of Environmental Protection (NJDEP), for example, developed its municipal storm water regulation program. The program addresses pollutants entering the state waters from any MS4s systems owned or operated by federal, state, county, and local agencies. Under this program, New Jersey Pollutants Discharge Elimination System (NJPDES) permits are issued to the municipalities throughout the state, highway systems, and public complexes that include large public colleges and hospitals, as well as parks. By March 3, 2004, nearly all New Jersey's municipalities and county, state, and interstate transportation entities and large public complexes were required to apply for NJDEPS permits.

On February 2, 2004, the NJDEP adopted the storm water rules (N.J.A.C.7 and 8) and issued the four final general permits as follows: the tier A storm water permit; the tier B storm water permit; the public complex storm water permit; and the highway storm water permit (NJDEP, 2004b). Tiers A and B are generally located in the more densely populated regions in the state or along or near the coast and rural areas and noncoastal regions, respectively. The permits address storm water quality issues of new and existing development and redevelopment by requiring the preparation of storm water program and implementation of specific permit requirements known as statewide basic requirements (SBRs). SBRs may also require the permitee to implement related BMPs. Tiers A and B permits, public complex permits and highway permits may require the implementation of additional measures.

To improve the quality of state open water, the NJDEP also requires no disturbance to existing vegetation within a buffer along streams and lakes. This buffer depends on stream classification and ranges from 50 ft (15 m) for non-trout-producing streams to 300 ft (45 m) for C class—trout production streams and lakes. By filtering the silt and pollutants, the vegetation improves the quality of runoff entering open waters. In the following sections, an overview of the current storm water management regulations in the United States is presented. This will be followed by a brief discussion of specific regulations in New Jersey, where the author resides, and the Maryland and New York. Maryland has one of the most stringent, though rational, storm water management regulations in the United States and the State of New York closely follows the Maryland's regulations.

5.2   AN OVERVIEW OF CURRENT STORM WATER MANAGEMENT REGULATIONS

The current storm water management regulations in various states in the United States include runoff quantity and water quality standards. In general, runoff quantity regulations are intended to avoid channel erosion and to prevent increased flooding downstream of projects disturbing 1 acre or more. These regulations call for maintaining the existing peak rates of runoff for given storm events. The storm events vary from state to state but generally include 1- or 2-year storm frequency for channel protection, 10- or 25-year storm frequency to control overbank flooding, and 100-year storm frequency for extreme floods. Some states, Maryland and New Jersey included, also have adopted regulations on groundwater recharge to offset the impact of developments on inhibiting infiltration.

The current trend in runoff quantity control aims at maintaining the volume of runoff. In this regard, the city of Atlanta passed a storm water ordinance in February 2013 that covers both residential and commercial properties. The ordinance applies to new commercial projects that disturb more than 1 acre (±4000 m²) of land and commercial redevelopment projects that add or replace more than 500 ft² (45 m²) of impervious surface. Such developments are required to infiltrate the first inch of runoff through infrastructure measures. In addition, the owners or developers of residential projects are required to infiltrate the first inch of runoff. This ordinance, however, disregards the soil type. While infiltrating 1 in. of rainfall in a sandy soil is readily achievable, it may not be practical in rocky areas. In 2013, the Department of Energy & Environment of Washington D.C. modified its stormwater management regulation. The rule, which went into effect on July 14, 2015, requires that major substantial improvement projects (meaning a new structure of over 5000 ft² (450 m²) footprint or reconstruction exceeding 50% of the pre-project assessed value) must retain 0.8 in. (2 cm) of stormwater runoff.

Maryland's Stormwater Design Manual (2009) requires that environmental site design (ESD) be used to the maximum extent practicable to reduce the volume of runoff from the 1-year storm design to the levels equivalent to woods in good condition. This requirement will also address groundwater recharge, water quality, and channel protection volumes, to be described later in this chapter.

With regard to water quality, the design criterion is a frequent storm, commonly under 1-year frequency. The philosophy is that a storm washes the bulk of pollutants from the ground and, in particular, paved surfaces, in the beginning of a storm and there is much less pollutant loading when the rains are sustained.

In earlier practices, the water quality storm was specified as 0.5 in. of rainfall, based on the concept that capturing and treating early runoff having a higher concentration and/or mass of contaminants is more effective than treating the later portion of the runoff. The concept of first flush can be associated with a rainfall season. In parts of the United States and many other parts of the world, the rainfall is seasonal. In Southern California, the bulk of the rainfall occurs from approximately November to March with the months of January and February having the greatest rainfall. The first large or first few storms of the season transport larger masses of pollutants than later in the season. To quantify this effect, concentration first flush and mass first flush have been identified (Kayhanian and Stenstrom, 2008). The mass first flush may be defined as the first flush in which 80% of the pollutant mass is emitted in the first 30% of the runoff. The concept of mass first flush is not so relevant to the northeastern United States, where the rain falls fairly uniformly throughout the year.

The concept of first flush has been fading away and giving way to water quality storm. In New Jersey, for example, the water quality storm is defined as 1.25 in. of rainfall in a 2-hour period. This storm represents the 90th percentile of rainfalls in 1 year. In New York State, the water quality storm is also the 90% annual rainfall. The NPDES phase II and some states define the water quality storm as the 85% annual rain event.

The pollutants removal criteria selected by different states are typically tied to the quality issues of receiving waters. Many states use total suspended solids (TSS) as a surrogate pollutant parameter. It is presumed that effective TSS removal (typically 70% to 80%) will provide for the control of other pollutants such as total phosphorus and heavy metals. In a number of states, including New Jersey and Washington state, the treatment goal is 80% total suspended removal for new pavements. The treatment goal in other states, such as California, Maryland, Georgia, and New York, is the percent of the water quality runoff volume. To achieve the intended goal, many states specify capturing a certain percentage (typically 85% to 90%) of the storm water depth over a 1- or 2-hour period, which amounts to 1 in. (2.5 cm) or more of rainfall.

New Jersey relates TSS removal rate with the retention time for extended detention basins. In some states, including California, New Jersey, and New York, the criterion for wet ponds is the ratio of the pond storage volume to the volume of runoff generated from the water quality storm. However, the treatment strategy is rapidly evolving. There is a recent shift from the TSS removal rate to the removal of other pollutants and, in particular, total phosphorus, total nitrogen, and dissolved metals.

For example, the state of Washington's Department of Ecology regulates multiple parameters. This regulatory agency requires 80% removal of TSS; for developments tributary to fish-bearing streams, the treatment of soluble zinc is required as well. In Maryland and North Carolina, total nitrogen (TN) is generally a pollutant of concern because of the effects of nitrogen on river systems. In some other states or municipalities, such as Hillsboro, Oregon, the total phosphorus (TP) is a regulated pollutant. Total phosphorus is also regulated in Virginia and, since April 2008, New York has added phosphorus to the water quality criteria. In Washington state, both phosphorus and dissolved metals are regulated pollutants. Carver County, Minnesota, has standards for TSS and TP removal as follows:

TSS removal = 90% for 1.25 in. storm
Total phosphorus (TP) = 50% removal for 2.5 in. storm

These regulations are in addition to a 2002 rule requiring the first 1/3 in. (8.5 mm) of storm runoff from impervious areas to be treated for water quality. Acceptable practices for water quality treatment are infiltration, filtration, and bio-infiltration. Carver County also requires maintaining the predevelopment peak rates of runoff for the 2-, 10-, and 100-year frequency storms.

It is to be noted that the percentage of removal criteria is not a good measure of BMP performance and is misleading. Also, the pollutants in storm water are a mix of various chemical and physical substances bound together. Chemicals are absorbed onto solid particles and particles stick to larger objects. Therefore, 80% or any TSS removal specified by a regulatory agency may not reflect the actual performance of a BMP. To define the actual percentage of TSS removal, the amount of materials entering a BMP and leaving it has to be measured. Though it appears simple, it is not. The TSS removal of a wet pond, for example, not only varies from site to site, but also is dependent on rainfall duration and intensity.

The impact on the environment does not solely depend on the percentage of removal (output loading) of given pollutants; it is also a function of the input loading. A simplified example shown in Figure 5.1 indicates that removing 50% of a lower pollutant loading (20 mg/L) is twice as effective as removing 80% of a higher pollutant concentration (100 mg/L). This example, though an oversimplification, indicates that reducing the pollutant production (loading) is as important as, if not more important than, the effectiveness of a BMP. While performance of a BMP varies with location, rainfall pattern and land uses, the average outflow concentration is useful in comparing different types of BMPs.

Equally, if not more, important is the volume of runoff. Reducing runoff volume plays a significant role in overall pollutant load reduction. Practices such as pervious pavements, disconnection of

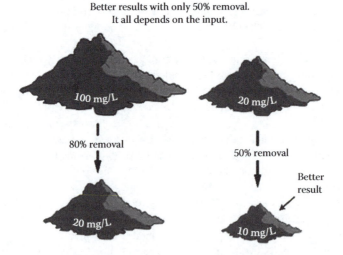

FIGURE 5.1 Input–output pollutant concentration relation. (From EPA.)

roof and pavement runoff, infiltration basins, bio-retention and the like, which capture a portion of runoff and infiltrate it into soil, reduce the total outflow pollutant load. Thus, inflow concentration, runoff loading, and outflow concentration—all three—are key elements in pollutant removal criteria.

5.2.1 EISA Section 438

To address storm water runoff from federal facilities, Congress enacted Section 438 of the Energy Independence and Security Act (EISA) of 2007. Section 438 of EISA of 2007 relates to storm water runoff requirements for federal development projects and reads as follows:

> The sponsor of any development or redevelopment project involving a Federal facility with a footprint that exceeds 5000 square feet [465 m²] shall use site planning, design, construction, and maintenance strategies for the property to maintain or restore, to the maximum extent technically feasible, the predevelopment hydrology of the property with regard to the temperature, rate, volume, and duration of flow.

The intension of EISA Section 438 is to preserve or restore the hydrology of a federally owned facility during construction or reconstruction. More specifically, EISA Section 438 is intended to ensure that aquatic biota, stream channel stability, and natural aquifer recharge of receiving water are not adversely impacted by changes in runoff temperature, volumes, durations, and rates due to federal projects (EWRI Currents, 2009).

Subsequent to EISA Section 438, President Obama signed Executive Order 13514 on "federal leadership in environmental, energy and economic performance" calling upon all federal agencies to "lead by example" in addressing various environmental issues including storm water runoff. Pursuant to this executive order, the US Environmental Protection Agency prepared a technical guidance for federal facilities to meet with Section 438 of EISA (2009). EISA Section 438 raises a new bar on storm water management and imposes a far more stringent storm water requirement on federal facilities than any requirement governing private projects in any state in the nation.

5.3 NJDEP STORM WATER MANAGEMENT REGULATIONS

The current storm water management regulations in the state of New Jersey were adopted on February 2, 2004 (NJDEP, 2004b). The regulations expired in February 2009 and were extended for 1 year through February 2010. On April 19, 2010, the NJDEP issued amended storm water management rules (NJDEP, 2010). The amended rules basically maintained the 2004 storm water management standards and included new sections and revisions. The rules also redefined the storm water management goals and included measures to

- Reduce soil erosion from any development or construction project
- Prevent, to the greatest extent feasible, an increase in nonpoint source pollution
- Minimize pollutants in storm water runoff from new and existing development
- Protect public safety through the proper design and operation of storm water management basins

It also designated the following entities as storm water management planning agencies:

- A municipality
- A county
- A county water resources agency or association
- A designated planning agency under N.J.A.C. 7:15 (NJGS, 2008)
- A soil conservation district, in coordination with the state soil conservation committee
- The Delaware River Basin Commission

TABLE 5.1

New Jersey Storm Water Management Regulations

SWM Criteria	Requirements
Water quality	1.25 in. of rainfall in 2 hours
	80% TSS removal for new pavement
	50% TSS removal for reconstruction
	0% TSS removal for roofs and pervious areas
Peak flow rate	Reductions below predevelopment values for 2-, 10-, and 100-year storms
	50% for 2-year storm
	25% for 10-year storm
	20% for 100-year storm
Groundwater recharge	Maintaining predevelopment annual recharge or infiltrate the increased runoff volume for the 2-year storm

- The Pinelands Commission
- The Delaware and Raritan Canal Commission
- The New Jersey Meadowlands Commission
- The NJDEP
- Other regional, state, or interstate agencies

The NJDEP storm water management regulations, referred to as New Jersey Pollutant Discharge Elimination System (NJPDES), govern any "major development," defined as the one that either disturbs 1 acre or more of land or that creates at least 0.25 acres of "new" impervious area. Land disturbance is defined as placement of impervious surface, exposure or movement of soil or bedrock, and clearing or removal of vegetation. A new impervious area means

- Any net increase in impervious surfaces on site
- Any replacement of an existing drainage system by a system of larger capacity
- Any proposed collection and discharge of runoff into a regulated area from an existing impervious surface, the runoff from which currently sheets flows into vegetation

The regulated area in New Jersey is any water course with a drainage area of 50 acres (20 ha) or larger. While the 1 acre land disturbance criteria follow the EPA's phase II storm water management rule, the 0.25 acre increase as defined previously is above and beyond the federal regulations.

In the NJDEP regulations, the uncompacted gravel and porous pavements are not considered as an impervious surface in terms of water quality. Likewise, brick pavers having 25% or more opening count as pervious areas. The storm water management regulations in New Jersey require provisions for measures for runoff quantity, storm water quality, and groundwater recharge. These requirements are summarized in Table 5.1 and are described in more detail in the next few sections.

5.3.1 Runoff Quantity Requirement

Because of extensive annual flood damages in New Jersey, the runoff quantity regulations in this state are more stringent than those of many, if not all, states in the nation. The regulations require performing hydrologic and hydraulic analyses to demonstrate no adverse increase in both runoff rates and volumes or design of measures to create certain reductions in the peak rates of runoff as described next:

1. Demonstrate that the post construction runoff hydrographs for the 2-, 10-, and 100-year frequency storms do not exceed at any point and time respective hydrographs for preconstruction runoff, or

2. Demonstrate that the post construction peak runoff rates are no greater than the preconstruction peak runoff rates for the storms of 2-, 10-, and 100-year frequency. And, in addition, any increased volume or change in the peak timing of the runoff will not increase flood damage at or downstream of site, or

3. Design storm water management measures so that the post development runoff rates for the 2-, 10-, and 100-year storm events are 50%, 75%, and 80% of the respective preconstruction peak runoff rates. The reductions apply only to the portion of the site that is disturbed by the project. Specifically, any offsite runoff entering onto the site and portions of the site which remain intact are not subject to the previously indicated reductions.

In practice, major developments provide measures to address condition 3. Tidal areas are exempt from the runoff quantity requirements 1, 2, or 3, unless the increased volume of storm water would aggravate flood-related damages downstream of the point of discharge.

Agricultural projects that meet the definition of major development in N.J. A.C. 7:8-1.2 need to submit an application to the Soil Conservation District for review and approval.

5.3.2 STORM WATER QUALITY STANDARDS

The water quality standards only apply when the project creates 0.25 acres or more of "new" impervious area, as defined previously. The water quality standard requires provision of measures to reduce the post construction TSS load in storm water runoff generated from the water quality storm as follows:

- Eighty percent for any new impervious area*
- Fifty percent for any impervious surface to be reconstructed

Building roofs, gravel surfaces, porous pavements, pavers with 25% or more in opening area, and lawn and landscapes are not subject to any TSS removal. Also, any repavement that does not require land disturbance, such as milling and repaving, is considered maintenance and does not require any TSS removal. The water quality design storm in New Jersey is 1.25 in. of rainfall nonuniformly distributed in 2 hours. This rainfall represents the 90th percentile of average annual storms. Figure 5.2 shows the rainfall intensity curve and Table 5.2 reflects the cumulative rainfall depth and the rainfall intensity of the water quality storm.

Table 5.3 lists the presumed TSS removal rates for certain BMPs when designed in accordance with the New Jersey Storm Water Best Management Practices Manual (NJDEP, 2004a). Appendix 5A includes figures for TSS removal of vegetated filter strips for hydrologic soil groups (HSGs) A through D as approved by NJDEP (2004a). If more than one BMP is used in series, the composite TSS removal is calculated by

$$R = A + B - (A \times B)/100 \tag{5.1}$$

where
R = combined TSS removal for two systems
A and B = TSS removal rate for each BMP

If a site contains two or more drainage areas, the required TSS removal for each area should be met. However, if some of the drainage areas converge to a common discharge point, the TSS calculations are based on the weighted TSS removal. The NJDEP does not accept two similarly

---

* The required TSS removal is 95% where the runoff is discharged to a special water resources protection area. Such areas include category C1 streams and their 300 ft buffer.

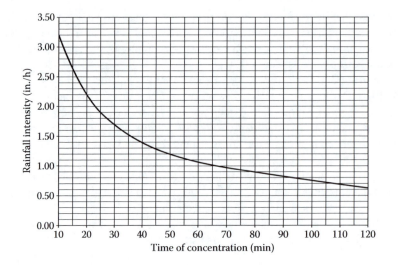

FIGURE 5.2   NJDEP 1.25 in./2 h water quality storm rainfall intensity curve.

TABLE 5.2
NJDEP Water Quality Storm, Temporal Distribution

Time (min)	Cumulative Rainfall (in.)	Incremental Rainfall (in.)	Intensity (in./h)
0	0.0000	0.0000	0.00
5	0.0083	0.0083	0.10
10	0.0166	0.0083	0.10
15	0.0250	0.0084	0.10
20	0.0500	0.0250	0.30
25	0.0750	0.0250	0.30
30	0.1000	0.0250	0.30
35	0.1330	0.0330	0.40
40	0.1660	0.0330	0.40
45	0.2000	0.0340	0.41
50	0.2583	0.0583	0.70
55	0.3583	0.1000	1.20
60	0.6250	0.2667	3.20
65	0.8917	0.2667	3.20
70	0.9917	0.1000	1.20
75	1.0500	0.0583	0.70
80	1.0840	0.0340	0.41
85	1.1170	0.0330	0.40
90	1.1500	0.0330	0.40
95	1.1750	0.0250	0.30
100	1.2000	0.0250	0.30
105	1.2250	0.0250	0.30
110	1.2334	0.0084	0.10
115	1.2417	0.0083	0.10
120	1.2500	0.0083	0.10

Note: Intensity = (incremental rainfall/5 min) × 60 min/h.

TABLE 5.3
TSS Removal Rate for BMPs

BMP	TSS Removal Rate%
Bio-retention systems	90
Constructed wetlands	90
Infiltration basins	80
Wet ponds	50–60
Extended detention basins	40–60
Vegetative filter strips	60–80
Sand filler	80
Manufactured treatment devices	50 or 80[a]

[a] TSS removal rates as verified by the New Jersey Corporation for Advanced Technology (NJCAT) and certified by the NJDEP.

manufactured treatment devices used in series. Also, when using an infiltration basin, NJDEP requires that the MTD be placed before the basin. The regulations also call for the minimal use of nutrients through the application of nonstructural measures and the reduction in phosphorous pollutants to the maximum extent practicable.

5.3.3 GROUNDWATER RECHARGE STANDARDS

All major developments must address groundwater recharge requirements as follows:

1. The infiltration from the site and storm water management features maintain the average annual predevelopment groundwater recharge or
2. The storm water management measures infiltrate the difference between the post- and preconstruction runoff volumes for the 2-year storm*

The New Jersey Geological Survey has developed a method for performing pre- and post annual groundwater recharge volume calculations. This method is described in "Geological Survey Report, No. 32 (1993), A Method of Evaluating Groundwater Recharge Areas in New Jersey" and is known as the GSR-32 method. The GSR-32 is a spreadsheet that calculates the pre- and post construction recharge depths and volumes for any municipality in New Jersey, based on soil cover and storm water infiltration systems. The spreadsheet can be downloaded free of charge from the NJ Geological Survey website (http://www.state.nj.us/dep/njgs/). Figure 5.3 exemplifies the application of this spreadsheet for a project involving improvements to Garden State Parkway in New Jersey. Sheet 1 of 2 of the table shows a reduction (deficit) of 36,900 cubic feet (1044 m³) due to proposed impervious surfaces. Sheet 2 of the table presents information about the retention system (which consisted of Cultec chambers in stone trench) and indicates a recharge volume of 91,254 cubic feet.

Sites with contaminated soil are exempt from groundwater recharge requirements. Also, the state discourages placing groundwater recharge systems in sites with limestone substrata. Groundwater recharge requirements also do not apply to urban redevelopment areas and the previously developed portions of a site, namely any area that had been graded or filled or paved or occupied by structures. The most feasible means of addressing groundwater recharge is to infiltrate the roof runoff, which is considered pure. However, for roof-top garages and industrial buildings where contaminants can deposit on the roofs, a pretreatment is required.

---

* The second method of calculation is not clearly defined. Specifically, the method does not specify the use of the rational method or the SCS TR-55 method.

(Continued)

New Jersey Groundwater Recharge Spreadsheet Version 2.0 November 2003

Annual Groundwater Recharge Analysis (based on GSR-32)

	Select Township ↓	Average Annual P (in)	Climatic Factor
	MIDDLESEX CO., PERTH AMBOY CITY	47.8	1.53

Pre-Developed Conditions

Land Segment	Area (acres)	TR-55 Land Cover	Soil	Annual Recharge (in)	Annual Recharge (cu.ft)
1	0.64	Open space	Downer	15.9	36.902
2	0.13	Woods	Woodstown	12.9	6.080
3					
4					
5					
6					
7	0				
8	0				
9	0				
10	0				
11	0				
12	0				
13	0				
14	0				
15	0				
Total =	0.8			Total Annual Recharge (in) 15.4	Total Annual Recharge (cu-ft) 42.982

Procedure to fill the Pre-Development and Post-Development Conditions Tables

For each land segment, first enter the area, then select TR-55 Land Cover, then select Soil. Start from the top of the table and proceed downward. Don't leave blank rows (with A=0) in between your segment entries. Rows with A=0 will not be displayed or used in calculations. For impervious areas outside of standard lcs select "Impervious Areas" as the Land Cover. Soil type for impervious area are only required if an infiltration facility will be built within these areas

---

Project Name:	GSP INTERCHANGE 88/89
Description:	This is a test application
Analysis Date:	09/01/03

Post-Developed Conditions

Land Segment	Area (acres)	TR-55 Land Cover	Soil	Annual Recharge (in)	Annual Recharge (cu.ft)
1	0.64	Impervious areas	Keyport	0.0	-
2	0.13	Open space	Woodstown	12.9	6.082
3	0				
4	0				
5	0				
6	0				
7	0				
8	0				
9	0				
10	0				
11	0				
12	0				
13	0				
14	0				
15	0				
Total =	0.8			Total Annual Recharge (in) 2.2	Total Annual Recharge (cu.ft) 6.082

Annual Recharge Requirements Calculation

% of Pre-Developed Annual Recharge to Preserve =	100%		Total Impervious Area (sq. ft)	27,678

Post-Development Annual Recharge Deficit=	**36,900**	(cubic feet)

Recharge Efficiency Parameters Calculations (area averages)

RWC=	2.75	(in)	DRWC=	0.00	(in)
ERWC=	0.65	(in)	EDRWC=	0.00	(in)

FIGURE 5.3 Groundwater recharge calculations: sheet 1 of 2.

Project Name	Description		Analysis Date	BMP or LID Type
GSP INTERCHANGE 88/89	This is a test application		09/04/03	

Recharge BMP Input Parameters

Parameter	Symbol	Value	Unit
BMP Area	ABMP	2112.0	sq.ft
BMP Effective Depth, this is the design variable	dBMP	8.4	in
Upper level of the BMP surface (negative if above ground)	dBMPu	48.0	in
Depth of lower surface of BMP, must be>=dBMPu	dEXC	56.4	in
Post-development Land Segment Location of BMP, Input Zero if Location is distributed or undetermined	SegBMP	2	unitless

Root Zone Water capacity Calculated Parameters

Parameter	Symbol	Value	Unit
Empty Portion of RWC under Post-D Natural Recharge	ERWC	1.08	in
ERWC Modified to consider dEXC	EDRWC	0.00	in
Empty Portion of RWC under Infilt. BMP	RERWC	0.00	in

Recharge Design Parameters

Parameter	Symbol	Value	Unit
Inches of Runoff to capture	Cdesign	0.26	in
Inches of Rainfall to capture	Pdesign	0.34	in
Recharge Provided Avg. over imp. Area		16.0	in
Runoff Captured Avg. over imp. Area		16.0	in

Parameters from Annual Recharge Worksheet

Parameter	Symbol	Value	Unit
Post-D Deficit Recharge (or desired recharge volume)	Vdef	36,900	cu.ft
Post-D Impervious Area (or target Impervious Area)	Aimp	68,607	sq.ft
Root Zone Water Capacity	RWC	4.61	in
RWC Modified to consider dEXC	DRWC	0.00	in
Climatic Factor	C-factor	1.50	no units
Average Annual P	Pavg	47.8	in
Recharge Requirement over Imp. Area	dr	15.9	in

BMP Calculated Size Parameters

Parameter	Symbol	Value	Unit
ABMP/Aimp	Aratio	0.03	unitless
BMP Volume	VBMP	1,478	cu.ft

System Performance Calculated Parameters

Parameter	Symbol	Value	Unit
Annual BMP Recharge Volume		91,264	cu.ft
Avg BMP Recharge Efficiency		100.0%	Represents % Infiltration Recharged
%Rainfall became Runoff		78.3%	%
%Runoff Infiltrated		42.7%	%
%Runoff Recharged		105.0%	%
%Rainfall Recharged		82.2%	%

CALCULATION CHECK MESSAGES

Volume Balance--> Solve Problem to satisfy Annual Recharge
dBMP Check--> OK
dEXC Check--> OK
BMP Location---> OK

OTHER NOTES

Pdesign is accurate only after BMP dimensions are updated to make each volume = deficit volume. The portion of BMP infiltration prior to filling and the area occupied by BMP are ignored in these calculations. Results are sensitive to dBMP, make sure dBMP selected is small enough for BMP to empty in less than 3 days. For land Segment Location of BMP if you select "impervious areas" RWC will be minimal but not zero as determined by the soil type and a shallow root zone for this Land Cover allowing consideration of lateral flow and other losses.

How to solve for different recharge volumes: By default the spreadsheet assigns the values of total deficit recharge volume "Vdef" and total proposed impervious area "Aimp" from the "Annual Recharge" sheet to "Vdef" and "Aimp" on this page. This allows solution for a single BMP to handle the entire recharge requirement assuming the runoff from entire impervious area is available to the BMP. To solve for a smaller BMP or a LID-IMP to recharge only part of the recharge requirement, set Vdef to your target value and Aimp to impervious area directly connected to your infiltration facility and then solve for ABMP or dBMP. To go back to the default configuration clik the "Default Vdef & Aimp" button.

FIGURE 5.3 (CONTINUED) Groundwater recharge calculations: sheet 2 of 2.

Aboveground infiltration systems serve as effective and feasible means of addressing not only groundwater recharge but also water quality requirements. If designed as a detention-infiltration basin, the basin can also regulate the peak rates of runoff, addressing runoff quantity requirements.

5.3.4 RUNOFF CALCULATION METHODS

The NJ storm water management regulations specify the use of the following methods in performing runoff quantity, water quality, and groundwater recharge calculations.

1. Rational/modified rational method: The rational method is employed for calculating peak rates of runoff and the modified rational method for calculating volume of runoff and performing routing computations. This method is acceptable for up to 20 acres (8 ha).
2. TR-55 method: This method is used for SCS 24-hour TYPE III storms.
3. The USDA Natural Resources Conservation Service (NRCS) methodology as described in Section 4, *National Engineering Handbook* (NEH-4).*

Table 3.5 in Chapter 3 included the SCS 24-hour rainfall depths in various counties in New Jersey. The IDF curves for shorter duration storm were also shown in Figure 3.1a and b in metric and English units, respectively.

In calculating the runoff coefficients (including the soil curve number) and groundwater recharge, the pervious and wooded portions of a site should be assumed to be in good cover calculations. Existing improvements may be accounted for in the calculations, provided that they have existed for at least 5 years without interruption. Also, in computing preconstruction storm water runoff, land features significant to runoff such as ponds, depressions, wetlands, and culverts, which may result in ponding, should be accounted for.

5.3.5 STANDARDS FOR STORM WATER MANAGEMENT STRUCTURES

Structural storm water management measures shall be designed to

- Minimize maintenance
- Take into account the existing site conditions such as wetlands, flood plains, slopes, depth to seasonal high water table, drainage pattern, and presence of limestone
- Be readily accessible to facilitate maintenance and repairs. Trash racks should be installed at the inlet and outlet structures. Racks shall be parallel bars with on-center spacing of no greater than one-third the diameter of the orifice or the width of the weir, but no less than 1 in. and no more than 6 in.
- Have an orifice no smaller than 2.5 in. in diameter at the intake of the outlet structure
- Have a minimum of 2 ft (60 cm) separation between the bottom of an infiltration basin and spring high water table

5.3.6 NONSTRUCTURAL STORM WATER STRATEGIES

The NJ storm water management regulations emphasize, as a primary consideration, the use of nonstructural storm water strategies to the maximum extent practicable in meeting the soil erosion, groundwater recharge, runoff quantity, and runoff quality standards. To ensure that this requirement is met, the state has developed a form called Nonstructural Strategies Point System (NSPS). This form, which was adopted in 2006, used to be completed for every major project; however due

---

* Note: Section 4 is designated as Part 630 Hydrology in the updated *National Engineering Handbook*, in part revised through November 2010.

to court cases, its use has been abandoned. The storm water management rule N.J.A.C. 7:8-5.3(c) requires that the land used as a nonstructural storm water measure be preserved by deed or through other means.

5.3.7 Municipal Storm Water Management Review

To assist municipalities in the development of municipal control ordinances relating to review of subdivisions and site plans, the NJDEP prepared a model ordinance, in April 2004. This ordinance, entitled "Model Storm Water Control Ordinance for Municipalities," was included as Appendix D to the New Jersey Best Management Practices Manual. All municipalities were required to adopt a municipal storm water management plan (MSWMP) within 12 months from the date of tier A and tier B municipal storm water general permits, which were issued by the NJDEP to address the municipal storm water requirements mandated by the US EPA.*

In addition, within 12 months after the adoption of MSWMP, municipalities were required to adopt storm water control ordinances and submit both the MSWMP and ordinance(s) to their counties for review. Within 60 days of receipt, the counties were required to approve, conditionally approve, or disapprove the submitted MSWMP and ordinance(s). To assist the municipalities and counties with review, the NJDEP prepared a checklist titled "Municipal Regulation Checklist and Municipal Storm Water Management Plans and Storm Water Control Ordinances," in May 2005; sample ordinances titled "Sample Municipal Storm Water Management Plan" and "Model Storm Water Control Ordinance for Municipalities" were also prepared. These are included as appendices in the New Jersey Storm Water Best Management Practices Manual (2004).

5.3.8 Suggestions for Improving the NJDEP Regulations

The regulations regarding the TSS removal are generic in that they disregard the rate of generation of pollutants. One site can produce twice as much pollutant as another. As such, removing 50% of TSS from the latter results in a higher water quality improvement than removing 80% from the former. A truck stop generates many times more pollutants, and particularly oil and grease, than driveways and streets in an influential community. As such, the treatment target should be based on the pollution generation rate. The regulations also assume that lawn and landscape areas are undisturbed virgin land and pollutant free. As indicated in Chapter 1, fertilizers, pesticides, and other chemicals can degrade water quality more than a paved area such as a driveway.

The practicality of maintenance is also ignored. The regulations imply acceptance of a 2.5 in. (63 mm) orifice for an underground detention basin. A 2.5 or even 3 in. (75 mm) orifice is highly susceptible to clogging due to silt, leaves, small debris, and even a tennis ball. While in an open detention basin, the clogging can be noticed; in an underground system where the outlet structure is hidden, the clogging will be left undetected, leaving the detention basin partly filled with water before a storm event occurs. Considering practicality and maintenance, the author recommends using a 6 in. (150 mm) orifice as the smallest opening in any underground detention basin. Following this recommendation, hundreds of underground detention and/or infiltration basins were thus constructed in New Jersey.

In regard to water quality, it is more prudent to use a device that is inexpensive and can be readily maintained than another of superior quality—however costly to maintain and hidden from view. The author has observed that many underground sand filters approved by the NJDEP were rendered totally ineffective due to lack of maintenance. There are cases where the sand filter was ignored years after installation. Once the sand is clogged, the runoff enters the system and is discharged from the overflow weir without any treatment. Figure 5.4 shows a sand filter that remained

---

* Tier A municipalities are generally located within the more densely populated regions or along or near the coast. Tier B municipalities are located in more rural areas and noncoastal regions.

FIGURE 5.4   Grates on a sand filter, unmaintained. (Photo by the author.)

unmaintained for 5 years after its installation and was totally clogged. Installing inlet filters such as Flo-Gard+Plus or similar devices, which are easy to maintain, will be far more effective than a sophisticated water treatment device that is left idle. To improve the water quality of storm water runoff, the regulations, like in some other states, should include the removal of total phosphorus and nitrogen and dissolved metals in certain situations.

For extremely flat sites, the 100 ft sheet flow limit is too small. The sheet flow lengths of 250 feet for pervious and 150 feet for impervious surfaces (post development) are reasonable for performing T_c calculations.

The regulations totally exclude single-family homes from any water quality, and more importantly, any runoff quantity control. A single-family residence in an affluent community in New Jersey can disturb 40,000 ft² (3720 m²) of virgin land and create 10,000 ft² (1020 m²) of pavement, yet be free of any storm water management regulations. It should be realized that the majority of residential units in any suburban community comprise single-family homes. While each home has seemingly little impact on storm water runoff, collectively they create an enormous increase in the peak and volume of runoff and degradation of storm-water quality, as well. It is time that the regulations not only be concerned with the size of a project, but also based on its per capita impact. To be successful, it is imperative to focus on not just large projects, but rather on the urban life as a whole.

Maryland is an exceptional state where the jurisdictional criteria for storm water management is 5000 ft² (465 m²) of disturbance. As such, the regulations cover many more land disturbances, including large and midsize single-family homes.

In his profession, the author has had the opportunity to review/consult on hundreds of development projects in a number of municipalities, mostly in northern New Jersey. To control runoff, he has been requiring a net zero increase in runoff rate/volume from single family homes, which are exempt from the NJDEP regulations. Depending on site and soil conditions, his recommendations to prevent the increased runoff have included dry wells or underground retention-infiltration chambers for collection of roof runoff, landscape depressions, and rain gardens for driveways and building roofs. The use of concrete paver blocks (pavers) for driveways is also among his recommendations for reducing the runoff volume and improving water quality.

5.4   STATE OF MARYLAND STORM WATER MANAGEMENT REGULATIONS

The Maryland Department of the Environment developed a manual titled "The 2000 Maryland Storm Water Design Manual, in two volumes, in April 2000. Volume I contained five chapters and four appendices. In addition, the Maryland Department of the Environment prepared an ordinance titled "Model Storm Water Management Ordinance" in July 2005 that was supplemented in 2007. The state amended the storm water design manual on May 4, 2009, and the new regulations went

TABLE 5.4

Maryland's Unified Storm Water Sizing Criteria

Sizing Criteria	Description of Storm Water Sizing Criteria
Water quality volume (WQ_v) (acre-feet) or (cu-ft)	$WQ_v = [(P)(R_v)(A)]/12$ P = rainfall depth in inches and is equal to 1.0 in. in the eastern rainfall zone and 0.9 in. in the western rainfall zone (Figure 2.1 in the manual) R_v = volumetric runoff coefficient, and A = area in acres (or ft²)
Recharge volume (Re_v) (acre-feet)	Fraction of WQ_v, depending on predevelopment soil hydrologic group $Re_v = [(S)(R_v)(A)]/12$ S = soil-specific recharge factor in inches
Channel protection storage volume (Cp_v)	Cp_v = 24 hour (12 hour in USE III and IV watersheds) extended detention of post developed 1-year, 24-hour storm event Not required for direct discharges to tidal waters and the Eastern Shore of Maryland (see Figure 5.5)
Overbank flood protection volume (Q_p)	Controlling the peak discharge rate from the 10-year storm event to the predevelopment rate (Q_{p10}) is optional; consult the appropriate review authority For Eastern Shore: Provide peak discharge control for the 2-year storm event (Q_{p2}). Control of the 10-year storm event (Q_{p10}) is not required
Extreme flood volume (Q_f)	Consult with the appropriate reviewing authority. Normally, no control is needed if development is excluded from 100-year floodplain and downstream conveyance is adequate

Source: Table 2.1, 2009 Maryland Storm Water Design Manual.

into effect in May of 2010. Volume I of the 2009 manual presents storm water management criteria in the state. It includes five chapters and five appendices, presenting storm water design standards, storm water management systems, and NPDES permit requirements. The new manual and 2010 regulation made significant changes in storm water management design and implementation.

In Maryland, the storm water management rules apply to any construction activity disturbing 5000 ft² (465 m²) or more of land. Chapter 1 in the manual is introductory and presents performance standards (14 in all) for storm water management practices. The standards basically include minimizing the generation of storm water, pretreatment, meeting groundwater recharge, removing 80% of TSS and 40% of the total annual post development phosphorus load, control of peak flows, preparation of a storm water pollution plan (SPP) for certain industrial areas, and provisions for storm water discharges from hotspots and redevelopment areas. One of the new requirements in the manual is to implement environmental site design (ESD) to the maximum extent practicable to mimic predevelopment conditions. Chapter 2 in volume I contains storm water management rules, titled "Unified Storm water Sizing Criteria." These rules, which include runoff quantity control, water quality control, and groundwater recharge, are summarized in Table 5.4 and briefly discussed in this section.

5.4.1 WATER QUALITY VOLUME, WQ_v

The water quality volume, WQ_v, is the storage needed to capture and treat the runoff from 90% of the average annual rainfall. This 90% rainfall event is 1 in. in the eastern zone and 0.9 in. in the western zone. The eastern zone covers over two-thirds of the counties in the easterly half of the state. The water quality volume is calculated by

$$WQ_v = (P)(R_v)(A)/12 \qquad (5.2)$$

where

$R_v = 0.05 + 0.009 \, (I)$
A = area in acres
P = 90% of average annual rainfall (0.9 or 1 in.)
I = percent impervious cover on site
WQ_v = water quality volume, acre-feet

In SI units, the preceding equation reads as

$$WQ_v = 10 \cdot (P)(R_v)(A) \tag{5.3}$$

where

P = 22.9 or 25.4 mm
A = area in hectares
WQ_v = water quality volume, m^3

A minimum of WQ_v = 0.2 in./acre (0.5 mm/ha) shall be met at sites or drainage areas with less than 15% impervious cover. The water quality requirement can be met by providing 24-hour draw-down time of WQ_v in an extended detention basin or 24-hour drawdown of a portion of WQ_v in conjunction with a storm water pond or wetland system (refer to Chapter 3 in the manual).

5.4.2 RECHARGE VOLUME CRITERIA, RE_v

The criteria for groundwater recharge is to maintain the predevelopment average annual recharge of the hydrologic soil present at the site. The recharge volume, Re_v, is calculated from the following equation:

$$Re_v = (S)(R_v)(A)/12 \tag{5.4}$$

where
S = soil-specific recharge factor (see Table 5.5)
R_v and A are as defined for water quality criteria (Table 5.4)

In metric units, the preceding equation reads as

$$Re_v = 10 \cdot (S)(R_v)(A) \tag{5.5}$$

where
A = area in hectares
R_v = recharge volume, m^3

TABLE 5.5
Soil-Specific Recharge Factor, S

Hydrologic Soil Group	in.	(mm)
A	0.38	(9.65)
B	0.26	(6.60)
C	0.13	(3.30)
D	0.07	(1.78)

The recharge volume, which is smaller than WQ_v, is inclusive in it. Specifically, satisfying WQ_v by a structural practice such as infiltration basin, bioretention, or nonstructural practice (e.g., buffers, disconnection of roof runoff, or combination thereof), the recharge requirement is also met. However, when utilizing separate measures for water quality and groundwater recharge, Re_v may be subtracted from WQ_v in sizing the water quality BMP.

If more than one hydrologic soil group (HSG) is present at the site, the composite S value is calculated based on the proportion of the total site area within each HSG. The composite S calculations will be applied to percent volume runoff when structural measures are used and to the percent impervious area when nonstructural measures are employed. Acceptable nonstructural measures include filter strips to treat parking lot and rooftop runoff, sheet flow discharge to stream buffers, and grass channels to treat roadway runoff.

Groundwater recharge requirement does not apply to redevelopment projects or any portion of a site designated as a hotspot. In addition, the local review agencies may relax the recharge volume requirements if a site is situated on unsuitable soil such as marine clay, karst, or an urban redevelopment area. In such situations nonstructural practices should be used for a portion of the site (percent area method) to the maximum extent practicable. The use of percent volume and percent area method is clarified by the following example:

Example 5.1

A 30-acre residential development site comprises 60% hydrologic group B and 40% hydrologic group C soil and has 35% impervious cover. Calculate the required recharge volume Re_v by the percent volume and percent area methods.

Solution

First, calculate composite S based on Table 5.5:

$$S = 0.26 \times 0.60 + 0.13 \times 0.4 = 0.208$$

a. Percent volume method:

$$R_v = 0.05 + 0.09(I) = 0.05 + 0.009 \times 35 = 0.365$$

$$Re_v = (S)(R_v)(A)/12$$

$$Re_v = 0.208 \times 0.365 \times 30/12 = 0.19 \text{ ac-ft}$$

b. Percent area method:

$$Re_v = (A_i) \times \text{imp. percentage}$$

$$A_i = 30 \times 0.35 = 10.5 \text{ ac}$$

$$Re_v = 0.208 \times 10.5 = 2.18 \text{ acres}$$

The recharge requirement may be met by either retaining 0.19 ac-ft using the structural method or disconnecting 2.18 acres of impervious area and directing them to a nonstructural system such as vegetated buffer.*

---

* The manual refers to recharge requirement as "treatment," which is unclear.

TABLE 5.6
Rainfall Depths of the 1-, 2-, 10-, and 100-year, 24-hour Storm Events in Maryland

County	1 Year, 24 h (in.)	2 Year, 24 h (in.)	10 Year, 24 h (in.)	100 Year, 24 h (in.)
Allegany	2.4	2.9	4.5	6.2
Anne Arundel	2.7	3.3	5.2	7.4
Baltimore	2.6	3.2	5.1	7.1
Calvert	2.8	3.4	5.3	7.6
Caroline	2.8	3.4	5.3	7.6
Carroll	2.5	3.1	5.0	7.1
Cecil	2.7	3.3	5.1	7.3
Charles	2.7	3.3	5.3	7.5
Dorchester	2.8	3.4	5.4	7.8
Frederick	2.5	3.1	5.0	7.0
Garrett	2.4	2.8	4.3	5.9
Harford	2.6	3.2	5.1	7.2
Howard	2.6	3.2	5.1	7.2
Kent	2.7	3.3	5.2	7.4
Montgomery	2.6	3.2	5.1	7.2
Prince George's	2.7	3.3	5.3	7.4
Queen Anne's	2.7	3.3	5.3	7.5
St. Mary's	2.8	3.4	5.4	7.7
Somerset	2.9	3.5	5.6	8.1
Talbot	2.8	3.4	5.3	7.6
Washington	2.5	3.0	4.8	6.7
Wicomico	2.9	3.5	5.6	7.9
Worcester	3.0	3.6	5.6	8.1

Source: Maryland Stormwater Design Manual, 2009.

5.4.3 CHANNEL PROTECTION STORAGE VOLUME CRITERIA, Cp_v

To protect downstream channels from erosion, 24-hour extended detention of the 1-year, 24-hour storm event shall be provided. An exception is the Eastern Shore where C_p is not required. Also, in use III and IV watersheds, only 12 hours of extended detention time shall be required. Table 5.6 lists 24-hour rainfall depths of 1-, 2-, 10-, and 100-year frequency in all counties in Maryland.* See Figure 5.5 for counties map. To meet the Cp_v requirement, detention ponds and underground vaults are commonly used. The rationale for this requirement is that runoff will be stored and released in such a gradual manner that erosive velocities will not be created in downstream channels. Because of large storage requirements, infiltration is not recommended for Cp_v. The Cp_v requirement does not apply to direct discharges to the tidal waters and Eastern Shore of Maryland. It is also not required when the 1-year post development peak discharge is less than 2 cfs. Included in the manual is a simplified solution for addressing the required detention time.

Off-site areas should be modeled as present land use in good condition for the Cp_v calculations. If a site consists of multiple drainage areas, Cp_v may be distributed proportionally to each drainage area and the Cp_v for the entire drainage area is addressed.

* Eastern Shore includes Kent, Queen Anne's, Caroline, Tabbot, Dorchester, Wicomico, Worcester, and Somerset Counties (see Figure 5.5).

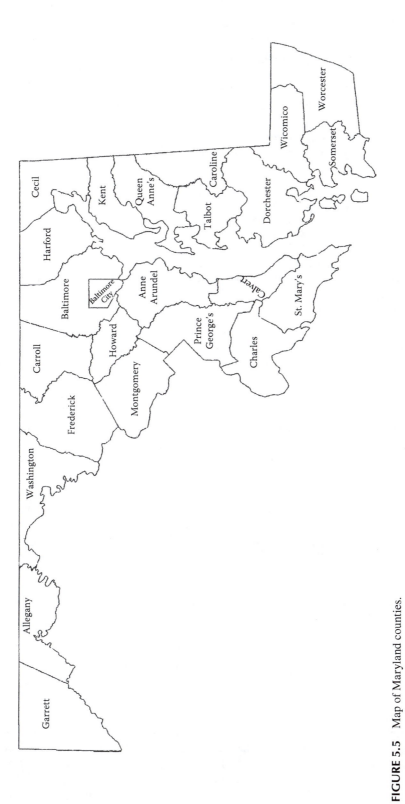

FIGURE 5.5 Map of Maryland counties.

5.4.4   Overbank Protection Volume Criteria, Q_p

The criterion for overbank protection is a 10-year, 24-hour storm event, Q_{p10}, except for the Eastern Shore where the criterion is a 2-year storm event, Q_{p2}. Calculations for the Q_{p2} and Q_{p10} are performed based on TR-55 or TR-20, using rainfall depths shown in Table 5.6. In computing the predevelopment runoff, the land shall be assumed meadow in good cover condition. Offsite areas, too, should be assumed as "present land use condition" with good vegetative cover. The length of sheet flow, in T_c calculations, is limited to 150 ft for predevelopment conditions and 100 ft for post development conditions.

Detention basins/ponds and underground chambers/vaults are commonly employed to address Q_{p2} and Q_{p10}. Similarly to Cp_v, overbank protection does not apply to direct discharges to tidal waters.

5.4.5   Extreme Flood Volume Criteria, Q_F

The objectives of the extreme flood criteria are to

a. Prevent flood damage from large storm events
b. Avoid expanding the boundaries of predevelopment 100-year Flood Emergency Management Agency (FEMA)
c. Protect the BMP control structures

The criteria for extreme flood Q_f is either a 100- or 10-year storm event depending on whether or not the area downstream of the site is situated within or out of the 100-year floodplain, respectively. In the latter case, hydraulic/hydrologic analyses may be required to demonstrate no adverse impact on downstream structures. The analyses typically extend to the first downstream tributary of equal or greater drainage area than that of the development or to any downstream dam, bridge, highway culvert, or point of restricted stream flow. In performing the 100-year flood analysis, offsite areas should be modeled as ultimate (full development) condition.

5.4.6   BMP Design

Chapter 3 of the manual covers performance criteria for urban BMP design such as general feasibility, conveyance, pretreatment, treatment, environment, and maintenance. An introductory note explains that, in small watersheds, increases in temperature due to development have a primary impact on the quality of receiving waters. To minimize this effect, BMP designs in such watersheds are required to

*   Minimize permanent pools
*   Limit extended detention times for C_p to 12 hours
*   Maintain existing forested buffer
*   Bypass available base flow

The Maryland storm water management manual identifies the following BMPs:

a. Five types of ponds, P-1 through P-5: P-1 is a micropool extended detention pond, P-2 a wet pond, P-3 a wet extended detention pond, P-4 a multiple pond system, and P-5 a pocket pond. Each type is defined with descriptive illustration and its features, such as riser, valves, drains, buffers, and setbacks, landscaping, and maintenance, specified.
b. Four types of storm water wetlands, W-1 through W-4: these represent shallow wetland, extended detention shallow wetland, pond/wetland system, and pocket wetland, respectively. Each type is illustrated by a figure and its application is defined. The manual also

requires that 25% of the total WQ_v be retained in deep water zones with a minimum depth of 4 ft (1.2 m), that at least 35% of the surface area of wetlands be less than 6 in. (15 cm) deep, and that 65% of the total surface area be shallower than 18 in. (45 cm).

c. Infiltration trenches/basins: Infiltration systems are required to exfiltrate the entire WQ_v, less the pretreatment volume, through the bed of the systems. A porosity, $n = V_v/V_t$, of 0.40 is allowed for the design of stone trenches (reservoirs) for infiltration practices.

d. Filtration systems: The manual classifies filtering systems into six groups: F-1 through F-6. These are surface sand filter, underground sand filter, perimeter sand filter, organic filter, pocket sand filter, and bioretention, respectively. These systems are not to be designed to meet Cp_v and Q_p. Rather, filter practices are generally combined with a separate facility to provide those controls.

The filtration surface area is designed based on the following listed permeabilities, K, and using the following equation:

$$A = WQ_v/KT \tag{5.6}$$

where K for
sand = 3.5 ft/day (1.1 m/day)
peat = 2.0 ft/day (0.6 m/day)
leaf compost = 8.7 ft/day (2.65 m/day)
bioretention soil = 0.5 ft/day (0.15 m/day)
WQ_v = water quality volume, ft³ (m³)
T = drain time, days

A maximum 40-hour infiltration time is recommended for sand bed and 2 days for bioretention soils. Dry or wet pretreatment equivalent to at least 25% of the calculated WQ_v is required prior to the filtration system. A typical pretreatment system consists of a sedimentation basin that has a length-to-width ratio of 2:1 and a minimum surface area calculated by

$$A_s = E'(Q_o/V_f) \tag{5.7}$$

where
A_s = sedimentation basin area, ft² (m²)
Q_o = discharge rate from basin = $WQ_v/24$ hour, ft³/day (m³/day)
V_f = fall velocity of particles, ft/s (m/s)

and, for $I \le 75\%$, use:
V_f = 0.0004 ft/s (0.00013 m/s) for particle size = 20 μm

and, for $I > 75\%$, use:
V_f = 0.033 ft/s (0.01 m/s) for particle size = 40 μm
I = percent impervious
E' = sediment trapping efficiency constant

The reason for using larger settling velocity for $I > 75\%$ is that sites with greater than 75% imperviousness have a higher percentage of coarse grained sediments.

The sediment trapping efficiency constant, E', is related to the sediment trap efficiency, E, by the following equation:

$$E' = -\ln[1 - (E/100)] \tag{5.8}$$

For $E = 90\%$, $E' = 2.30$.

Substituting for E' and V_f in Equation 5.7 gives

$$I \le 75\%$$

$$A_s = 0.066 \ WQ_v, \ \text{ft}^2 \tag{5.9}$$

$$A_s = 0.217 \ WQ_v, \ \text{m}^2 \tag{5.10}$$

$$I > 75\%$$

$$A_s = 0.0081 \ WQ_v, \ \text{ft}^2 \tag{5.11}$$

$$A_s = 0.027 \ WQ_v, \ \text{m}^3 \tag{5.12}$$

The criteria for a filtering system is that the entire treatment system (including pretreatment) hold at least 75% of WQ_v prior to infiltration. Minimum filter bed thickness is typically 18 in. (45 cm) for infiltration basins and 12 in. (30 cm) for sand filters.

e. Swales: Swales are designed to treat the full WQ_v and may be dry swale or wet swale, designated as O-1 and O-2, respectively. Dry swale is basically a vegetated open channel, and wet swale has an expanded basin with wetland vegetation and constricted outlet. Figure 5.6 shows a schematic plan view of a wet swale.

Design criteria for swales (open channel) area:
1. Swales shall be designed for the 10-year storm.
2. The peak flow velocity for the 10-year storm shall be nonerosive.
3. Channels shall have moderate side slopes (flatter than 3:1)—in no case steeper than 2:1.
4. A minimum ponding time of 30 minutes is recommended for WQ_v treatment. The maximum allowable ponding time shall be less than 48 hours. An underdrain system shall be provided in dry swales to meet the maximum ponding time requirement.

Pretreatment storage of 0.1 in. (2.5 mm) of runoff per impervious acre is required. This storage is usually provided by installing check dams at pipe inlet and/or driveway crossings. For lateral inflow, a pea gravel diaphragm of gentle side slopes shall be provided along the top of the channel. Direct discharge of concentrated flow shall be prevented.

Chapter 4 of the manual discusses selection of the BMP at a site. The selection is made based on the following factors:

- Watershed factors
- Terrain factors
- Storm water treatment suitability
- Physical feasibility factors
- Community and environmental factors
- Location and permitting factors

One of the watershed factors is development in the intensely developed area of the Maryland critical area (a zone extending 1000 ft from mean high tide and the landward edge of tidal wetlands). BMPs in these areas shall comply with the "10% rule." This rule requires that post development phosphorus load must be reduced to 10% below predevelopment loads. Other items in watershed factors are coldwater streams, sensitive streams, wetlands and reservoir protection, and shellfish/beach.

Terrain factors include low relief areas containing karst and carbonaceous rock. Storm water treatment suitability includes meeting Re_v, C_{pv}, Q_{p2}, and Q_{p10}, as well as the ability to accept hotspot runoff. Storm water treatment suitability is summarized in Table 5.7, which is adapted from Table 4.3 in the manual.

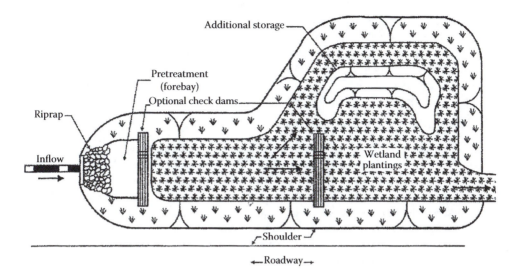

FIGURE 5.6 Schematic plan view of wet swale.

TABLE 5.7
BMP Selection Based on Storm Water Treatment Suitability

Code	BMP List	Re_v Ability	C_{pv} Control	Q_p Control	Additional Safety Concerns	Space	Accept Hotspot Runoff
P-1	Micropool ED	No[a]	Yes	Yes	No	Yes	Yes[c]
P-2	Wet pond	No[a]	Yes	Yes	Yes	Varies	Yes[c]
P-3	Wet ED pond	No[a]	Yes	Yes	Yes	Yes	Yes[c]
P-4	Multiple pond	No[a]	Yes	Yes	Yes	No	Yes[c]
P-5	Pocket pond	No[a]	Yes	Yes	Varies	Yes	Yes[c]
W-1	Shallow wetland	Varies[b]	Yes	Yes	No	No	Yes[c]
W-2	ED wetland	Varies[b]	Yes	Yes	Varies	Varies	Yes[c]
W-3	Pond/wetland	Varies[b]	Yes	Yes	Yes	No	Yes[c]
W-4	Pocket wetland	No	Varies	Varies	No	Varies	Yes[c]
I-1	Infiltration trench	Yes	Varies	Varies	No	Yes	No[c]
I-2	Infiltration basin	Yes	Varies	Varies	No	Varies	No[c]
F-1	Surface sand filter	Varies[b]	Varies	Varies	No	Yes	Yes[d]
F-2	Underground SF	No	No	No	Varies	Yes	Yes
F-3	Perimeter SF	No	No	No	No	Yes	Yes
F-4	Organic filter	Varies[b]	Varies	Varies	No	Yes	Yes[d]
F-5	Pocket sand filter	Varies[b]	Varies	Varies	No	Yes	Yes[d]
F-6	Bioretention	Yes	Varies	Varies	No	Varies	Yes[d]
O-1	Dry swale	Yes	No	No	No	Varies	Yes[d]
O-2	Wet swale	No	No	No	No	Varies	No

Source: Table 4.3 of the Maryland Storm Water Design Manual, Updated through May 4, 2009, vol. I and II, prepared by Center for Watershed Protection, Ellicott City, and the Maryland Department of the Environment, Baltimore, MD.

[a] Structures that require impermeable liners or that intercept groundwater may not be used for groundwater recharge.

[b] Re_v may be provided by exfiltration (see Section 3.4 in the manual).

[c] Not allowed unless pretreatment to remove hydrocarbons, trace metals, and toxicants is provided.

[d] Yes, but only if bottom of facility is lined with impermeable filter fabric that prevents leachate infiltration.

Physical feasibility factors include soils, water table, drainage area, slope restriction, and ultra-urban sites. Community and environmental factors entail ease of maintenance, community acceptance, construction cost, habitat quality, and others. The manual includes a table each of relevant BMP selection for physical feasibility factors, community and environmental factors, and location and permitting factors checklists.

5.4.7 Environmental Site Design (ESD)

Chapter 5 in the current manual (adopted May 4, 2009) has been expanded considerably following the state's "Storm Water Management Act of 2007." The Act requires establishing a comprehensive plan for storm water management approval, implementing environmental site design (ESD) to the maximum extent practicable (MEP), and ensuring that structural practices (discussed in Chapter 3 of the manual) are used only where absolutely necessary. There are many storm water management strategies that seek to replicate natural hydrology; these are known as better site design, low-impact development, green infrastructure, or sustainable site design. For consistency ESD was adopted as a more generic classification for use in Maryland. The Act defines ESD as "using small scale storm water management practices, nonstructural techniques or better site planning to mimic natural hydrologic, runoff characteristics and minimize the impact of land development on natural resources." Under this definition, ESD includes

- Optimizing conservation of natural features (e.g., drainage patterns, soil, vegetation)
- Minimizing impervious surfaces (e.g., pavement, concrete channels, roofs)
- Slowing down runoff to maintain discharge timing and to increase infiltration and evapotranspiration
- Using other nonstructural practices or innovative technologies approved by MDE

Studies indicate that, generally, stream quality and watershed health diminish when impervious cover exceeds 10% and become severely degraded beyond 25%. Thus, fundamental principles of ESD during the planning process help minimize adverse impacts of imperviousness. Table 5.8 provides a summary of site development strategies.

The manual also covers procedures for the site development phase, review of site development plans, and final plan design and review. To achieve the goal of maintaining predevelopment runoff characteristics, the Act has established the following performance standards that must be met:

- The standard for characterizing predevelopment runoff characteristics for new development projects shall be woods in good hydrologic condition.
- ESD shall be implemented to the MEP to mimic predevelopment conditions.
- As a minimum, ESD shall be used to address both Re_v and WQ_v requirements.
- Channel protection obligations are met when ESD practices are designed according to the reduced runoff curve number method described in Section 5.4.8.

The required ESD runoff depth and storage volume are calculated using the following equations:

$$R_E = (P_E)(R_v) \qquad\qquad\qquad (5.13)$$

$$ESD_v = (P_E)(R_v)(A)/12 \qquad\qquad\qquad (5.14)$$

where
  R_E = required runoff depth
  P_E = rainfall target for reduced runoff curve numbers (tabulated for hydrologic soil groups A through D) in inches. Appendix 5B includes tables for reduced curve numbers

TABLE 5.8
Summary of Site Development Strategies

Better Site Design Technique	Recommendations
Using narrower, shorter streets, rights-of-way, and sidewalks	Streets may be as narrow as 22 ft in neighborhoods serving low traffic volumes; open space designs and clustering will reduce street lengths; rights-of-way can be reduced by minimizing sidewalk width, providing sidewalks on one side of the road, and reducing the border width between the street and sidewalks
Cul-de-sacs	Allow smaller radii for turnarounds as low as 33 ft; use a landscaped island in the center of the cul-de-sac and design these areas to treat storm water runoff
Open vegetated channels	Allow grass channels or biofilters for residential street drainage and storm water treatment
Parking ratios, parking codes, parking lots, and structured parking	Parking ratios should be interpreted as maximum number of spaces; use shared parking arrangements; minimum parking stall width should be less than 9 ft and stall length less than 18 ft; parking garages are encouraged rather than surface lots
Parking lot runoff	Parking lots are required to be landscaped and setbacks are relaxed to allow for bioretention islands or other storm water practices in landscaped areas
Open space	Flexible design criteria should be provided to developers who wish to use clustered development and open space designs
Setbacks and frontages	Relax setbacks and allow narrower frontages to reduce total road length; eliminate long driveways
Driveways	Allow for shared driveways and alternative impervious surfaces
Rooftop runoff	Direct to pervious surfaces
Buffer systems	Designate a minimum buffer width and provide mechanisms for long-term protection
Clearing and grading	Clearing, grading, and earth disturbance should be limited to that required to develop the lot
Tree conservation	Provide long-term protection of large tracts of contiguous forested areas; promote the use of native plantings
Conservation incentives	Provide incentives for conserving natural areas through density compensation, property tax reduction, and flexibility in the design process

Source: Center for Watershed Protection (1998).

$R_v = 0.05 + 0.009(I)$, dimensionless runoff coefficient
ESD_v = required storage volume, ft^3
I = impervious cover percent
A = the drainage area in square feet

In SI units the preceding equations become

$$R_E = (P_E)(R_v) \tag{5.15}$$

$$ESD_v = 0.001\ (P_E)(R_v)(A) = 0.001\ R_E \cdot A \tag{5.16}$$

where
R_E = required runoff depth, mm
P_E = rainfall target, mm
A = the drainage area, m^2
R_v = dimensionless runoff coefficient
ESD_v = required storage volume, m^3

5.4.7.1 ESD Storm Water Management Requirements

Storm water treatment requirements in the 2009 manual are as follows:

Treatment: ESD practices shall be used to treat the runoff from 1 in. of rainfall (P_E = 1 in., 25 mm) on all new developments where storm water management is required.

- Cp_v: ESD practices shall be used to the MEP to address Cp_v for all sites where the peak discharge, q_i, is more than 2 cfs.
- Cp_v shall be based on 1-year, 24-hour design storm runoff calculated using reduced RCN (see Tables in Appendix 5B). If the RCN for a drainage area reflects "woods in good condition," then Cp_v is met for that area.

If the targeted rainfall is not met, any remaining Cp_v requirement shall be treated using structural practices in Chapter 3 of the manual (Section 5.4.6 in this chapter).

The runoff stored in ESP practices may be subtracted from the overbank flood protection and extreme flood volumes (Q_{p2}, Q_{p10}, Q_f), where these are required.

The following example elaborates the Cp_v calculations to address ESD.

Example 5.2

Compute the ESD storm water design criteria for a residential development with the following characteristics:

Site area = 35 acres
Drainage area = 35 acres
Soils: 50% B, 50% C
Impervious coverage = 33% uniformly distributed
1-year, 24-hour storm = 2.6 in. (Baltimore County)

Solution

Step 1. Determine predevelopment condition:

a. Determine the composite soil curve number for predevelopments assuming woods in good condition:

$$B = 55, 17.5 \text{ acres}$$

$$C = 70, 17.5 \text{ acres}$$

$$\text{RCN} = (55 \times 17.5 + 70 \times 17.5)/35 = 62.5$$

The target RCN = 62.5.

b. Determine target P_E using Table 5B.1 (see Appendix 5B):

$$I = 33\%; \text{ check RCN for both } I = 30\% \text{ and } I = 35\%$$

$$I = 30\% \ P_E = 1.6 \text{ in. soil group B} \quad \text{see Table 5.9}$$

$$I = 30\% \ P_E = 1.6 \text{ in. soil group C} \quad \text{see Table 5.9}$$

$$I = 35\% \ P_E = 1.8 \text{ in. soil group B} \quad \text{see Table 5.9}$$

$$I = 35\% \ P_E = 1.6 \text{ in. soil group C} \quad \text{see Table 5.9}$$

TABLE 5.9

Target Rainfall for Group B and C Soils (for Example 5.2)

%I	RCN	P_E = 1 in.	1.2 in.	1.4 in.	1.6 in.	1.8 in.
		Hydrologic Soil Group B				
15%	67	55				
20%	68	60	55	55		
25%	70	64	61	58		
30%	72	65	62	59	55	
35%	74	66	63	60	56	
		Hydrologic Soil Group C				
15%	78	70				
20%	79	70				
25%	80	72	70	70		
30%	81	73	72	71		
35%	82	74	73	72	70	
40%	84	77	75	73	71	

Since 33% is closer to 35% than 30%,

Use P_E = 1.8 conservatively for group B; PE = 1.6 for group C

$$\text{Composite } P_E = (1.8 \times 17.5 + 1.6 \times 17.5)/35 = 1.7 \text{ in.}$$

c. Calculate runoff depth:

$$R_v = 0.05 + 0.009 \times I = 0.35$$

$$R_E = (P_E)(R_v) = 1.7 \times 0.35 = 0.59 \text{ in.}$$

ESD targets for the project are

$$P_E = 1.7 \text{ in.}$$

$$R_E = 0.59 \text{ in.}$$

$$ESD_v = R_E \cdot A/12$$

$$ESD_v = 0.59 \times 35/12 = 1.72 \text{ ac-ft} = 74,960 \text{ ft}^3$$

Step 2. Determine storm water management requirements after using ESD:

a. Assume that ESD practices were implemented to treat 1.4 in. of rainfall (P_E = 1.4 in.) over the entire project. Calculate reduced RCN: Enter Table 5.9 and read:
   - B soils: I = 30% CN = 59; I = 35% CN = 60
   - C soils: I = 30% CN = 71; I = 35% CN = 72

   Use the larger CN values to calculate RCN:

$$RCN = \frac{60 \times 17.5 + 72 \times 17.5}{35} = 66$$

b. Calculate Cp_v Requirements:

Since RCN = 66 is greater than the target RCN (62.5, woods in good condition), Cp_v must be addressed.

Also since $P_E \geq 1$ in., Cp_v is the runoff volume from the 1-year 24-hour storm ($P = 2.6$ inch).

Calculate runoff depth, R, for CN = 66 and $P = 2.6$ in.

$$S = 1000/66 - 10 = 5.15$$

$$R = (2.6 - 0.2 \times 5.15)^2/(2.6 + 0.8 \times 5.15) = 0.37 \text{ in.}$$

$$Cp_v = 0.37 \text{ in.} \times 35 \text{ acres} = 1.29 \text{ ac-ft} = 56{,}190 \text{ ft}^3$$

Therefore, structural practices such as detention basin or shallow wetland to provide an additional Cp_v of 56,190 ft³ must be employed.

5.4.8 ADDRESSING ESD

The required runoff depth, R_E, and storage volume, ESD, may be addressed using

- Alternative surfaces
- Nonstructural practices
- Microscale practices

5.4.8.1 Alternative Surfaces

Alternative surfaces include

- Green roofs
- Permeable pavements
- Reinforced turf

5.4.8.1.1 Green Roofs

The green roof's contribution to ESD sizing criteria is calculated based on the reduced curve numbers (RCNs) listed in Table 5.10. The manual presents a detailed description of green roof installation, landscaping, inspection, and maintenance.

5.4.8.1.2 Permeable Pavements

Permeable pavements, similarly to green roofs, reduce the soil curve number based on the thickness of the sub-base. Effective RCNs for pervious pavements are listed in Table 5.11. Design requirements and soil conditions for pervious pavements are as follows:

- **Treatment:** All permeable pavement systems shall meet the following conditions:
  - Applications that exceed 10,000 ft² (930 m²) shall be designed as infiltration practices using the design methods for infiltration trenches. A porosity (η) of 30% and an effective area of the trench (A_t) equal to 30% of the pavement surface area shall be used.
  - A sub-base layer of clean, open graded, washed aggregate with a porosity (η) of 30% (1.5 to 3 in. stone is preferred) shall be used below the pavement surface. The sub-base may be 6, 9, or 12 in. (15, 22.5, or 30 cm) thick.
  - *Filter cloth shall not be used between the sub-base and soil subgrade.* If needed, a 12 in. (30 cm) layer of washed concrete sand or pea gravel (1/8 to 3/8 in. stone) may be used to act as a bridging layer between the sub-base reservoir and subsurface soils.

TABLE 5.10
Reduced CNs for Green Roofs

Roof Thickness in. (mm)	Effective RCN[a]
2 (50)	94
3 (75)	92
4 (100)	88
6 (150)	85
8 (200)	77

[a] Green roofs do not provide any percolation, act as a small retention basin, and discharge all the rainfall they receive once saturated. See the author's discussion on green roofs in Chapter 8. The RCN in Table 5.10 reflects initial losses (retention depths) varying from 0.13 in. (35 mm) for a 2 in. green roof to 0.6 in. (15 mm) for 8 in. roof.

TABLE 5.11
Effective RCNs for Permeable Pavements

Subbase	Hydrologic Soil Group			
	A	B	C	D
6 in. (15 cm)	76[a]	84[a]	93[b]	–
9 in. (22.5 cm)	62[c]	65[c]	77[c]	–
12 in. (30 cm)	40	55	70	–

[a] Design shall include 1–2 in. minimum overdrain (inv. 2 in. below pavement base) per 750 s.f. of pavement area.
[b] Design shall include 1–2 in. minimum overdrain (inv. 2 in. below pavement base) per 600 s.f. of pavement area.
[c] Design shall include 1–3 in. minimum overdrain (inv. 2 in. below pavement base) and a 0.50 in. underdrain at subbase invert.

- **Soils:**
  - Permeable pavements shall not be installed in HSG D or on areas of compacted fill. Underlying soil types and condition shall be field verified prior to final design.
  - For applications that exceed 10,000 ft² (930 m²), underlying soils have an infiltration rate (f) of 0.50 in. per hour or greater. This rate may be initially determined from NRCS soil textural classification and subsequently confirmed by geotechnical tests in the field.
  - The invert of the sub-base reservoir shall be at least 4 ft (1.2 m) (2 ft or 0.6 m on the lower Eastern Shore) above the seasonal high water table.

5.4.8.1.3 Reinforced Turf

Reinforced turf consists of interlocking modular units with interstitial areas for planting gravel or grass. These systems are suitable for light traffic areas and can be installed in all soil groups, but work best in sandy soils. Reinforced turf should not be used in hotspots that generate higher concentrations of hydrocarbons, trace metals, or toxicants.

5.4.8.2 Nonstructural Practices

Disconnecting impervious cover and treating runoff closer to its source is the next step in implementing ESD. These nonstructural practices include

TABLE 5.12

ESD Sizing Factors for Rooftop Disconnection

Disconnection Flow Path Length, ft (m)		P_E
Eastern Shore	Western Shore	in. (mm[a])
15 (4.9)	12 (3.0)	0.2 (5)
30 (9.8)	24 (7.9)	0.4 (10)
45 (14.8)	36 (11.8)	0.6 (15)
60 (19.7)	48 (15.7)	0.8 (20)
75 (24.6)	60 (19.7)	1.0 (25)

[a] Rounded to whole numbers.

- N-1 Disconnection of rooftop runoff
- N-2 Disconnection of nonrooftop runoff
- N-3 Sheet flow to conservation areas

Design parameters for each of these practices are outlined in this section.

5.4.8.2.1 Disconnection of Rooftop Runoff

The ESD sizing factor for rooftop runoff disconnection depends on the length of pervious area downsteam of downspouts and is specified in Table 5.12.

5.4.8.2.2 Nonrooftop Runoff Disconnection

The minimum length of flow path through vegetated areas shall be 10 ft, and the maximum contributing impervious flow path 75 ft. ESD sizing for nonrooftop disconnection depends on the ratios of impervious and pervious flow paths to the vegetated buffer areas and varies from 0.2 to 1.0 in. (5 to 25 mm).

5.4.8.2.3 Sheetflow to Conservation Areas

Conservation areas (called vegetated buffers in many jurisdictions, including New Jersey) are an effective means of treating storm water runoff. The ESD, P_E, depends on the width of the conservation area and ranges from 0.6 in. for 50 ft buffer to 1.0 in. for 100 ft buffer. The conservation area shall be 20,000 square feet or larger and its minimum effective width is 50 ft.

5.4.8.3 Microscale Practices

Microscale practices are small water-quality treatment or flow-control measures to capture runoff from discrete impervious areas. Microscale practices include

- M-1: rainwater harvesting
- M-2: submerged gravel wetlands
- M-3: landscape infiltration
- M-4: infiltration berms
- M-5: dry wells
- M-6: microbioretention
- M-7: rain gardens
- M-8: swales
- M-9: enhanced filters

Microscale practices should meet the following performance standards:

- Microscale practices used for new development shall promote runoff reduction and water quality treatment through infiltration, filtration, evapotranspiration, rainwater harvesting, or a combination of these techniques.
- Microscale filters used for new development shall be designed to promote recharge (e.g., enhanced filter) and be planted as part of the landscaping plans.

5.4.8.3.1   Rainwater Harvesting (Cisterns and Rain Barrels)

Rainwater harvesting systems shall meet the following conditions:

- Screens and filters shall be used to remove sediment, leaves, and other debris from runoff for pretreatment and can be installed in the gutter or downspout prior to storage.
- Rain barrels and cisterns shall be designed to capture at least 0.2 in. (5 mm) of rainfall from the contributing rooftop area. A P_E value based on the ESD_v captured and treated shall be applied to the contributing rooftop area.
- Where rainwater harvesting systems are connected to indoor plumbing, the Re_v requirement shall be addressed separately.
- The design shall plan for dewatering to vegetated areas.
- The design of large commercial and industrial storage systems shall be based on water supply and demand calculations. Storm water management calculations shall include the discharge rate for distribution and demonstrate that captured rainwater will be used prior to the next storm event.
- Large capacity systems shall provide dead storage below the outlet and an air gap at the top of the tank. Gravity-fed systems should provide a minimum of six inches of dead storage. For systems using a pump, the dead storage depth will be based on the pump specifications.

5.4.8.3.2   Submerged Gravel Wetlands

A submerged gravel wetland is a small-scale filter using wetland plants in a rock media to provide water quality treatment. Runoff drains into the lowest elevation of the wetland are distributed throughout the system and discharge at the surface. Pollutant removal is achieved in a submerged gravel wetland through biological uptake from algae and bacteria growing within the filter media. Wetland plants provide additional nutrient uptake and physical and chemical treatment processes allow filtering and absorption of organic matter. Roadway median is a suitable location where submerged gravel wetland can be used.

Submerged gravel wetlands shall meet the following requirements:

- Pretreatment shall be provided for 10% of the total ESD_v. An aboveground forebay area or below ground pretreatment chamber may be used.
- Storage for 75% of ESD_v for the entire drainage area contributing to the wetland shall be provided. A P_E value based on the ESD_v captured and treated shall be applied to the contributing drainage area. Temporary ponding depth shall not be greater than the tolerance levels of the wetland vegetation. Temporary storage of ESD_v may be provided above the gravel bed.
- Storage calculations shall account for the porosity of the gravel media.
- The gravel substrate shall be no deeper than 4 ft (1.2 m).

5.4.8.3.3   Landscape Infiltration

Landscape infiltration utilizes on-site vegetative planting areas to capture, store, and treat storm water runoff. Rainwater is stored initially, filters through the planting soil and gravel media below, and then infiltrates into native soils. These practices can be integrated within the overall site design by utilizing a variety of landscape features for storage and treatment of storm water runoff. Storage may be provided in constructed planters made of stone, brick, or concrete, or in natural areas excavated and backfilled with stone and topsoil.

Landscape infiltration can be best implemented in residential and commercial land uses. Residential areas with compact housing such as clustered homes and townhouses can utilize small green spaces for landscape infiltration. Because space in these instances prevents structural pretreatment, the drainage area to these practices should be limited to less than 10,000 ft^2 (930 m^2). Larger drainage areas may be allowed where soil testing is performed and pretreatment forebays can be implemented. Successful application is dependent upon soil type and groundwater elevation. Landscape infiltration is not recommended in HSG C and D soils.

If landscape infiltration systems are designed in accordance with the guidance in the manual, their ESD sizing criteria are given by

$$P_E = (20 \text{ in.})(A_f)/DA \text{ (in.)} \tag{5.17}$$

$$P_E = (50 \text{ cm})(A_f)/DA \text{ (mm)} \tag{5.18}$$

where A_f and DA are the surface area of landscape infiltration and the drainage area tributary thereof. The infiltration area should be at least 2% of the contributing drainage area. Landscape infiltration design criteria are as follows:

- The drainage area to any individual practice shall be 10,000 ft^2 or less.
- Landscape infiltration facilities located in HSG B (i.e., loams, silt loams) shall not exceed 5 ft (1.5 m) in depth. Facilities located in HSG A (i.e., sand, loamy sand, sandy loam) shall not exceed 12 ft (3.6 m) in depth.
- Landscape infiltration facilities shall be designed to fully dewater the entire ESD$_v$ within 48 hours. Temporary storage of the ESD$_v$ may be provided above the facility.
- A 12 to 18-in. (30–45 cm) layer of planting soil shall be provided as a filtering media at the top of the facility.
- A minimum 12-in. (30 cm) layer of gravel is required below the planting soil.
- A 12-in. (30 cm) layer of clean sand shall be provided at the bottom to allow for a bridging medium between the existing soils and stone within the bed.
- The storage volume for the ESD$_v$ shall be determined for the entire system including the temporary ponding area, the soil, and the sand and gravel layers in the bottom of the facility. Storage calculations shall account for the porosity ($n = 0.40$) of the gravel and soil media.
- Pretreatment measures shall be implemented along the main storm water runoff collection system where feasible. These include installing gutter screens, a removable filter screen on rooftop downspout pipes, a sand layer or pea gravel diaphragm at the inflow, or a two to three-inch surface mulch layer.

Figure 5.7 depicts plan view of a landscape infiltration which the author has referred to as landscape depression.

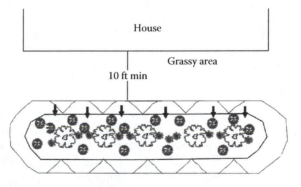

FIGURE 5.7 Plan view of landscape infiltration (depression).

5.4.8.3.4 Infiltration Berms

An infiltration berm is a mound of earth composed of soil and stone that is placed along the contour of a relatively gentle slope. This practice may be constructed by excavating upslope material to create a depression and storage area above a berm or earth dike. Storm water runoff flowing downslope to the depressed area filters through the berm and maintains sheetflow. Infiltration berms should be used in conjunction with practices that require sheetflow (e.g., sheetflow to buffers) or in a series of steeper slopes to prevent flow concentration.

Infiltration berms shall have the following design criteria:

- Berms shall be installed along the contour at a constant elevation and be level.
- When used in a series along a slope, the elevation at the downstream toe of each berm shall be the same elevation as the crest of the next berm downslope.
- The berm shall be asymmetric in shape. The crest should be 2 ft wide.
- The berm shall be graded so that a concave shape is provided at the up-gradient toe.
- The design shall consider soils suitable to resist slope failure and slumping. Side slopes should be very shallow and a ratio of 3:1 is recommended for mowed berms.
- A berm shall consist of a 6 in. layer of compacted topsoil with a gravel or aggregate interior.

The storage volume created behind and up to the crest of the berm may be used to address pretreatment, or Re_v, or contribute to ESD_v requirements.

5.4.8.3.5 Dry Wells

A dry well is an excavated pit or structural chamber filled with gravel or stone that provides temporary storage of storm water runoff from rooftops. The storage area may be constructed as a shallow trench or a deep well. Rooftop runoff is directed to these storage areas and infiltrates into the surrounding soils prior to the next storm event. The pollutant removal capability of dry wells is directly proportional to the amount of runoff that is stored and allowed to infiltrate.

Dry wells can be used in both residential and commercial sites and are best suited for treating runoff from small drainage areas such as a single rooftop or downspout. Dry wells are not appropriate for treating runoff from large impervious areas such as a parking lot. Successful application is dependent upon soil type and groundwater elevation.

Dry wells shall meet the following conditions:

- Pretreatment measures shall be installed to allow filtering of sediment, leaves, or other debris. This may be done by providing gutter screens and a removable filter screen installed within the downspout pipe or other locally approved methods. The removable filter screen should be installed below the overflow outlet and easily removable so that homeowners can clean the filter.
- A dry well shall be designed to capture and store the ESD_v. A P_E value based on the ESD_v captured and treated shall be applied to the contributing drainage area. The storage area for the ESD_v includes the sand and gravel layers in the bottom of the facility. Storage calculations shall account for the porosity of the gravel and sand media.
- The drainage area to each dry well shall not exceed 1000 ft^2 (93 m^2). Drainage areas should be small enough to allow infiltration into the ground within 48 hours (e.g., 500 ft^2 to each downspout). Infiltration trenches may be used to treat runoff from larger drainage areas (see Section 3.3 in the manual).
- Dry wells located in HSG B (i.e., loams, silt loams) shall not exceed 5 ft (1.5 m) in depth. Dry wells located in HSG A (i.e., sand, loamy sand, sandy loam) shall not exceed 12 ft (3.6 m) in depth.

- The length of a dry well should be longer than the width to ensure proper water distribution and to maximize infiltration.
- A 1 ft (30 cm) layer of clean sand shall be provided in the bottom of a dry well to allow for bridging between the existing soils and trench gravel.

Dry wells shall be installed in HSG A or B. The depth from the bottom of a dry well to the seasonal high water table, bedrock, hard pan, or other confining layer shall be greater than or equal to 4 ft (2 ft on the lower Eastern Shore).

5.4.8.3.6 Microbioretention

Microbioretention practices capture and treat runoff from discrete impervious areas by passing it through a filter bed mixture of sand, soil, and organic matter. Filtered storm water is either returned to the conveyance system or partially infiltrated into the soil. Microbioretention practices are versatile and may be adapted for use anywhere there is landscaping.

Microbioretention is a multifunctional practice that can be easily adapted for new and redevelopment applications in commercial and industrial projects. Storm water runoff is stored temporarily and filtered in landscaped facilities shaped to take runoff from various sized impervious areas. Microbioretention provides water quality treatment and aesthetic value, and it can be applied as concave parking lot islands, linear roadway or median filters, terraced slope facilities, residential cul-de-sac islands, and ultra-urban planter boxes.

The P_E values determined by Equation 5.19 may be applied to the ESD sizing criteria when microbioretention systems are designed according to the guidance that follows. Re_v requirements are also met when the P_E meets or exceeds the soil-specific recharge factor (see Table 5.5).

$$P_E = 15 \text{ in. } (A_f/DA) \tag{5.19}$$

$$P_E = 38 \text{ cm } (A_f/DA) \tag{5.20}$$

Microbioretention practices shall meet the following conditions:

- The drainage area to any individual practice shall be 20,000 ft^2 (1860 m^2) or less.
- Microbioretention practices shall capture and store at least 75% of the ESD$_v$.
- The surface area (A_f) of microbioretention practices shall be at least 2% of the contributing drainage area. A P_E value based on Equation 5.19 shall be applied to the contributing drainage area. Temporary storage of the ESD$_v$ may be provided above the facility with a surface ponding depth of 12 in. (30 cm) or less.
- Filter beds shall be between 24 and 48 in. (50 cm and 1 m) deep.
- Filter beds shall not intercept groundwater. If designed as infiltration practices, filter bed inverts shall be separated at least four ft vertically (2 ft on the lower Eastern Shore) from the seasonal high water table.
- A surface mulch layer (maximum 2 to 3 in. thick) should be provided to enhance plant survival and inhibit weed growth.
- The filtering media or planting soil, mulch, and underdrain systems shall conform to the specifications found in Appendix B.4 of the manual.

Setbacks:

- Microbioretention practices should be located down gradient and set back at least 10 ft (3 m) from structures. Microbioretention variants (e.g., planter boxes) that must be located adjacent to structures should include an impermeable liner.
- Microbioretention practices shall be located at least 30 ft (10 m) from water supply wells and 25 ft (7.5 m) from septic systems. If designed to infiltrate, then the practice shall be

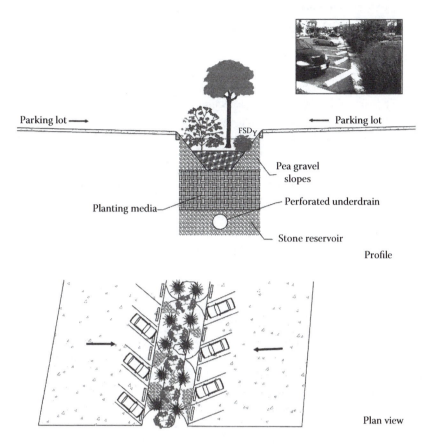

FIGURE 5.8 Microbioretention in a parking lot median.

located at least 50 ft (30 m) from confined water supply wells and 100 ft from unconfined water supply wells.

- Microbioretention practices shall be sized and located to meet minimum local requirements for clearance from underground utilities.
- Any trees planted in microbioretention practices shall be located to avoid future problems with overhead electrical and telecommunication lines.

Figure 5.8 shows a variation of microbioretention employed at a parking lot median.

5.4.8.3.7 Rain Gardens

A rain garden is a shallow, excavated landscape feature or a saucer-shaped depression that temporarily holds runoff for a short period of time. Rain gardens typically consist of an absorbent planted soil bed, a mulch layer, and planting materials such as shrubs, grasses, and flowers. An overflow conveyance system is included to pass larger storms. Captured runoff from downspouts, roof drains, pipes, swales, or curb openings temporarily ponds and slowly filters into the soil over 24 to 48 hours.

The ECS sizing criterion for rain gardens is the P_E calculated by the following equation:

$$P_E = (10 \text{ in.})(A_f/DA), \text{ in.} \tag{5.21}$$

$$P_E = (25 \text{ cm})(A_f/DA), \text{ cm} \tag{5.22}$$

Rain gardens are designed based on the following conditions:

- The drainage area to a rain garden serving a single lot in a residential subdivision shall be 2000 ft² (186 m²) or less. The maximum drainage area to a rain garden for all other applications shall be 10,000 ft² (930 m²). Microbioretention or bioretention should be considered when these requirements are exceeded.
- The surface area (A_f) of rain gardens shall be at least 2% of the contributing drainage area. A P_E value based on Equation 5.21 shall be applied to the contributing drainage area. Temporary storage of the ESD$_v$ may be provided above the facility with a surface ponding depth of 6 in. or less.
- Excavated rain gardens work best where HSG A and B are prevalent. In areas of HSG C and D, at-grade applications or soil amendments should be considered.
- A minimum 6–12 in. (15–30 cm) layer of planting soil shall be provided.
- A mulch layer 2–3 in. (5–7.5 cm) deep shall be applied to the planting soil to maintain soil moisture and to prevent premature clogging.
- The planting soil and mulch shall conform to the specifications found in Appendix B.4 of the manual.

5.4.8.3.8   Swales

Swales are channels that provide conveyance, water quality treatment, and flow attenuation of storm water runoff. Swales remove pollutants through vegetative filtering, sedimentation, biological uptake, and infiltration into the underlying soil media. Three design variants covered in this section include grass swales, wet swales, and bioswales. Implementation of each is dependent upon site soils, topography, and drainage characteristics.

Swales can be used for primary or secondary treatment on residential, commercial, industrial, or institutional sites. Swales can also be used for retrofitting and redevelopment. The linear structure allows use in place of curb and gutter along highways, residential roadways, and along property boundaries. Wet swales are ideal for treating highway runoff in low-lying or flat terrain with high groundwater. Bioswales can be used in all soil types due to the use of an underdrain. Grass swales are best suited along highway and roadway projects. Swales shall meet the following criteria:

- Swales shall have a bottom width between 2 and 8 ft (0.6 and 2.4 m).
- The channel slope shall be less than or equal to 4.0%.
- The maximum flow velocity for the ESD$_v$ shall be less than or equal to 1.0 fps (0.3 m/s).
- Swales shall be designed to safely convey the 10-year, 24-hour storm at a nonerosive velocity with at least 6 in. of freeboard.
- Channel side slopes shall be 3:1 or flatter.
- A thick vegetative cover shall be provided for proper functioning.

The following criteria apply to each specific design variant:

- **Grass swales:** Grass swales shall be used for linear applications (e.g., roadways) only, and shall be as long as the treated surface. The surface area (A_f) of the swale bottom shall be at least 2% of the contributing drainage area. The maximum flow depth for ESD$_v$ treatment should be 4 in. (10 cm), and the channel should have a roughness coefficient (Manning's n) value of 0.15. This can be accomplished by either maintaining vegetation height equal to the flow depth or using energy dissipaters like check dams, infiltration berms, or riffle/pool combinations. Equation 5.21 for rain gardens ($P_E = 10$ in. × A_f/DA) is used for ESD sizing of grass swales.
- **Bioswales:** The surface area (A_f) of the bioswale bottom shall be at least 2% of the contributing impervious area and a P_E value based on Equation 5.19 ($P_E = 15$ in. × A_f/DA) shall be

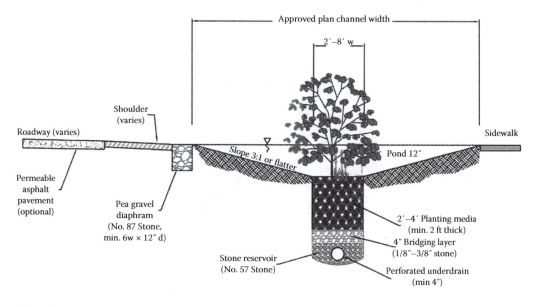

FIGURE 5.9 Section view of bioswale.

applied to the contributing drainage area. Bioswales shall be designed to temporarily store at least 75% of the ESD$_v$. A 2–4 ft (0.6–1.2 m) deep layer of filter media shall be provided in the swale bottom. The use of underdrains is recommended for all applications, and is essential in HGS C or D soils (ref. Appendix B.4 of the manual). Figure 5.9 is a schematic cross section of a bioswale.

- **Wet swales:** Wet swales shall be designed to store at least 75% of the ESD$_v$. A P_E value equivalent to the volume captured and treated shall be applied to the contributing drainage area. Wet swales should be installed in areas with a high groundwater table and check dams or weirs may be used to enhance storage.

5.4.8.3.9 Enhanced Filters

An enhanced filter is a modification applied to specific practices (e.g., microbioretention) to provide water quality treatment and groundwater recharge in a single facility. This design variant uses a stone reservoir under a conventional filtering device to collect runoff, remove nutrients, and allow infiltration into the surrounding soil.

The structural storm water filtering systems or the microfiltering structures can be modified easily for most development projects. Depending on soil conditions, a stone reservoir can be sized appropriately to provide Re_v for the drainage area to the system. These practices are subject to the same constraints and design requirements as conventional and microscale filters.

Enhanced filters shall meet the following conditions:

- Enhanced filters shall be coupled with properly designed filters to address both ESD and Re_v requirements.
- At a minimum, enhanced filter reservoirs shall be designed to store the Re_v. The stone reservoir volume is equal to the surface area multiplied by depth and by the porosity (n) of the stone [volume = surface area (ft^2) × depth (ft) × 0.4].*

---

* In metric units, the stone volume in m^3 is calculated as follows: volume = surface area (m^2) × depth (m) × 0.4.

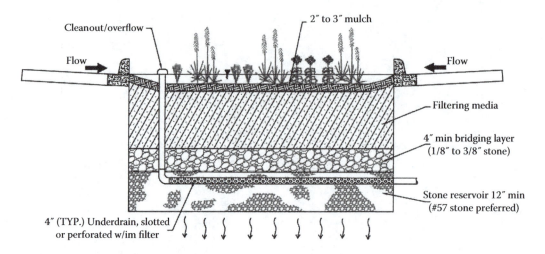

FIGURE 5.10   An enhanced filter.

- When using variation A, the stone reservoir (#57 stone preferred) shall be at least 12 in. thick below the underdrain.
- A 12 in. (30 cm) layer of sand or pea gravel (1/8 to 3/8 in. stone) may be used to act as a bridging layer between the stone reservoir and subsurface soils.
- The invert of the stone reservoir shall be separated at least 4 ft (2 ft on the lower Eastern Shore) from the seasonal high water table.
- Enhanced filters shall be located at least 25 ft (7.5 m) from septic systems, 100 ft (30 m) from unconfined water supply wells, and 50 ft (15 m) from confined water supply wells.
- Enhanced filters shall be sized and located to meet minimum local requirements for clearance (both vertical and horizontal) from sewer and water lines. Designs may need to include special protection if underground utilities cross through enhanced filters. Figure 5.10 depicts a variant of enhanced filters.

5.4.9   Redevelopment

5.4.9.1   Introduction

Redevelopment is defined as any construction, alteration, or improvement performed on sites where the existing land use is commercial, industrial, institutional, or multifamily residential and existing site impervious area exceeds 40%. The term "site" is defined as a single tract, lot, or parcel of land, or a combination of tracts, lots, or parcels of land that are in one ownership or are contiguous and in diverse ownership where development is to be performed as part of a unit, subdivision, or project. When calculating site imperviousness, the local approving agency may allow lands protected by forest preservation, conservation easements, or other mechanisms to be subtracted from the total site area. This will create incentive to preserve and protect natural resources in redevelopment projects.

5.4.9.2   Redevelopment Policy

The 40% site impervious area threshold will determine whether a project will be regulated as new development or redevelopment. When redevelopment requirements apply, all existing impervious areas located within a project's limit of disturbance (LOD) are required for management. Because redevelopment projects present a wide range of constraints and limitations, the following policy allows for flexibility and an evaluation of options that can work in conjunction with broader watershed goals and local initiatives:

1. Storm water management shall be addressed for redevelopment according to the following criteria:
    a. Reduce existing impervious area within the LOD by at least 50%
    b. Implement ESD practices to the MEP to provide water quality treatment for at least 50% of existing impervious area within the LOD
    c. Use a combination of impervious area reduction and ESD implementation for at least 50% of existing impervious areas
2. Alternative storm water management measures may be used provided that the developer satisfactorily demonstrates to the approving authority that impervious area reduction and ESD have been implemented to the MEP. Alternative storm water management measures include but are not limited to
    a. An on-site structural BMP
    b. An off-site structural BMP to provide water quality treatment for an area equal to or greater than 50% of existing impervious areas
    c. A combination of impervious area reduction, ESD implementation, and on-site or off-site structural BMP for an area equal to or greater than 50% of existing impervious area within the LOD
3. An approving agency may develop separate programmatic policies for providing water quality treatment for redevelopment projects when the preceding requirements cannot be met. These policies shall be reviewed and approved by the Maryland Department of Environment (MDE) and may include but are not limited to
    a. Retrofitting existing structural BMPs
    b. Stream restoration
    c. Trading policies that involve other pollution control programs
    d. Watershed management plans
        i. Storm water management shall be addressed according to new development requirements for any net increase in impervious area
        ii. Other criteria

5.4.10 SPECIAL CRITERIA

5.4.10.1 Sensitive Waters

Chapter 5 of the manual also includes "special criteria" for sensitive waters, such as nontidal cold water and recreation trout waters. Also discussed in the chapter are "at-source techniques for mitigating thermal impacts." Lighter colored materials like white or gray concrete reflect solar radiation, resulting in less elevated temperatures. A material's ability to reflect solar heat is measured as its solar reflectance index or "SRI" and varies from 0 (a black surface) to 100 (a white surface) and above. In thermally sensitive watersheds, designers should consider using materials with SRI values greater than 29 for paving and steep-sloped (≥2:12) roofing, and materials with SRI values greater than 78 for low-sloped (≤2:12) roofing. SRI values for selected paving and roofing material are listed here:

Material	SRI
Asphalt	0
Gray concrete (new)	35
White concrete (new)	86
Gray asphalt shingles	22

In addition to selecting the type of cover, the following techniques help reduce thermal impacts of storm water management practices:

- Maximize the infiltration capacity of each practice. Increasing infiltration reduces the amount of surface runoff and lowers the thermal energy flowing into coldwater streams.
- Design filtering practices (e.g., microbioretention) so that underdrains are at least 4 ft below the surface. Soil temperatures at this depth are cooler and fluctuate little in response to surface weather conditions. As runoff flows through, thermal energy is dissipated and effluent temperatures are decreased.
- Use shade-producing plants in landscaped practices. Trees, shrubs, and noninvasive vines on trellises can be used to screen impervious areas from the sun.

5.4.10.2   Wetlands, Waterways, and Critical Areas

Construction of storm water management facilities within 100-year floodplain wetlands and their buffers and tidal wetlands requires state and federal permits. In addition to these restrictions, runoff from new developments and redevelopments must be treated prior to directly discharging into jurisdictional wetlands and waters of the state.

Maryland's Critical Area Act has designated all land within 1000 ft of tidal waters and adjacent tidal wetlands as the "critical area." All development located within the critical area must address additional criteria. These criteria include those relating to the protection of habitat that are applied uniformly throughout the critical area and those that relate to water quality and imperviousness and are specific to land classifications as follows.

Within the critical area, land is designated as either intensely developed area, limited development area, or resource conservation area (IDA, LDA, and RCA, respectively) based on uses that existed at the time the local programs were adopted. The IDAs are those areas of concentrated development where there is little natural habitat. Any new development and redevelopment projects within the IDA must include storm water management practices to reduce post development phosphorus loads to at least 10% below predeveloped levels, commonly known as the 10% rule.

LDAs are those regions where development density is low to moderate and wildlife habitat is not dominated by agriculture, wetlands, forests, or other natural areas. Similarly, RCAs are characterized by the dominance of agriculture or protected resources like forests or wetlands. Within these areas, any new development or redevelopment must address standard water quality requirements, conserve natural areas, and incorporate corridors to connect wildlife and plant habitat. To accomplish these goals, imperviousness, alternative surfaces, or "lot coverage" is generally limited to 15% of the property or project area. There are also strict limits on clearing of existing woodland or forests. All clearing of these areas requires at least a 1:1 replacement.

To protect habitat, a forested buffer is required on all new development in all three land designations. Extending a minimum of 100 ft (30 m) from the mean high water line of tidal waters or the landward edge of tidal wetlands and tributary streams, this buffer acts as a water quality filter and protects important riparian habitat within the critical area. This distance may be expanded to include adjacent sensitive areas like hydric or highly erodible soils or steep slopes. If a sensitive area exists within a subdivision in the RCA, the minimum width of the buffer is 200 ft. Disturbance associated with new development is generally prohibited within the buffer, and, accordingly, storm water practices (e.g., microscale practices, structural facilities) cannot be located within it.

5.5   STATE OF NEW YORK STORM WATER REGULATIONS

5.5.1   Introduction

The state of New York storm water management regulations are contained in a storm water management design manual, dated August 2010. This manual was prepared by the Center for Watershed Protection in Elliot City, Maryland, and closely follows Maryland's storm water design manual, 2009. This manual supersedes the previous manual dated 2003.

TABLE 5.13

New York State Storm Water Sizing Criteria

Water Quality (WQ_v)	Standards
	90% rule:
	$WQ_v = [(P)(R_v)(A)]/12$
	$R_v = 0.05 + 0.009(I)$
	I = impervious cover (percent)
	Minimum $R_v = 0.2$
	P = 90% rainfall event number
	A = site area in acres
Runoff reduction volume (RR_v)	RR_v (ac-ft) = reduction of the total WQ_v by application of green infrastructure techniques and SWPs to replicate predevelopment hydrology. The minimum RR_v is defined as the specified reduction faster (S) provided that objective technical justification is determined.
Channel protection (Cp_v)	Default criterion:
	Cp_v = 24-hour extended detention of post developed 1-year, 24-hour storm event, remaining after runoff reduction. Where conditions allow, runoff reduction of total Cp_v is encouraged.
	Option for sites larger than 50 acres:
	Distributed runoff control—geomorphic assessment to determine the bankfull channel characteristics and thresholds for channel stability and bedload movement.
Overbank flood (Q_p)	Control the peak discharge from the 10-year storm to 10-year predevelopment rates.
Extreme storm (Q_f)	Control the peak discharge from the 100-year storm to 100-year predevelopment rates. Safely pass the 100-year storm event.

Note: Channel protection, overbank flood, and extreme storm requirements may be waived in some instances if the conditions specified in this chapter are met.

Two new chapters have been added to the 2009 manual. These include Chapters 9 and 10, as will be discussed later. The New York storm water management regulations are summarized in Table 5.13 and discussed in more detail in this section.

5.5.2 WATER QUALITY VOLUME

The water quality volume, abbreviated as WQ_v, represents 90% of the average annual storm water runoff and is the design criteria for sizing storm water management systems to improve water quality. The WQ_v is directly related to the amount of impervious cover and is calculated by the following equation:

$$WQ_v = (P)(R_v)(A)/12 \text{ (CU)} \qquad (5.23)$$

$$WQ_v = 10 \, PR_vA \text{ (SI)} \qquad (5.24)$$

where
WQ_v = water quality volume, in ac-ft (m³)
P = 90% rainfall event number, in. (mm)
$R_v = 0.05 + 0.009(I)$, where I is percent impervious cover
A = site area (contributing area), acres (ha)

A minimum R_v of 0.2 will be applied to regulated sites. Figure 5.11 shows the 90% rainfall event contours in the state.

The impervious cover in the preceding equation includes: paved and gravel roads, parking lots, driveways and sidewalks, buildings, and other impermeable surfaces such as pools, patios, and

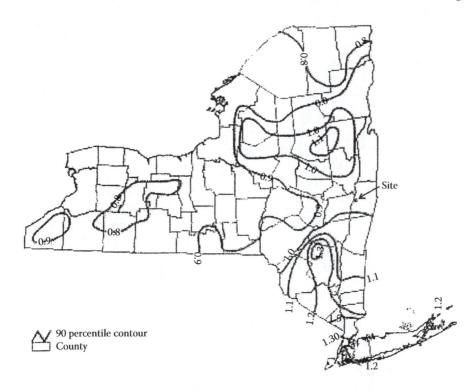

FIGURE 5.11   Ninety percent rainfall event in New York State.

sheds. Porous pavements and modular pavers are considered 50% impervious (C = 0.5). Where direct measurement of impervious cover is impractical, the land use–impervious cover relationship may be based on Table 4.2 in the manual. In that table, the impervious coverage for urban development varies from 9% for open urban land, such as golf courses and parks, to 14% for 1-acre lot residential, to 41% for town houses, and to 72% on average for commercial sites.

Runoff reduction shall be achieved by infiltration, groundwater recharge, reuse, recycle, and evapotranspiration of 100% of predevelopment water quality volume.

5.5.3 WQ_v TREATMENT PRACTICES

The WQ_v will be treated by an acceptable practice that meets the following criteria:

1. It can capture and treat the full water quality volume (WQ_v).
2. It is capable of 80% TSS removal and 40% TP removal.
3. It has acceptable longevity in the field.
4. It has a pretreatment mechanism.

Acceptable water quality treatment practices are listed next:

1. Storm water ponds: Practices have either a permanent pool of water or a combination of permanent pool and extended detention capable of treating the WQ_v. The manual presents description for five different types of ponds denoted as P-1 through P-5 (see Table 5.4 in the manual).
2. Storm water wetlands: Practices include significant shallow marsh areas and may also incorporate small permanent pools and extended detention storage to achieve the full WQ_v.

3. Infiltration practices: Practices capture and temporarily store the WQ_v before allowing it to infiltrate into the soil.
4. Filtering practices: Practices capture and temporarily store the WQ_v and pass it through a filter bed of sand, organic matter soil, or other acceptable treatment media.
5. Open channel practices: Practices are explicitly designed to capture and treat the full WQ_v within dry or wet cells formed by check dams or other means.

Table 5.14 presents more detailed information about the preceding acceptable practices. In addition, extended detention basins are acceptable for water quality treatment if the basin provides 24 hours of extended detention time for the WQ_v and includes a micropool. A local jurisdiction may reduce the detention time to as little as 12 hours in trout water to prevent stream warming.

For any off-site area entering an on-site storm water management facility, treatment should be provided based on its current condition. If water quality treatment is provided off-site, the facility needs only to treat on-site runoff.

5.5.4   Stream Channel Protection Volume Requirement (Cp_v)

Stream channel protection volume requirement (Cp_v) is intended to protect stream channels from erosion. This goal is accomplished by providing 24-hour extended detention of the 1-year, 24-hour storm event. For sites that discharge to a trout water, only 12 hours of extended detention are required to meet this criterion. Figure 5.12 shows the 1-year, 24-hour storm events in New York State.

For developments greater than 50 acres and with impervious cover greater than 25%, it is recommended that a detailed geomorphic assessment be performed to determine the appropriate level of control. Appendix J in the manual provides guidance on how to conduct this assessment.

The Cp_v requirement does not apply in certain conditions, including the following:

* The entire Cp_v volume is recharged at a site.
* The site discharges directly into tidal waters or fourth-order or larger streams. Within New York State, streams are classified using: New York State Codes Rules and Regulations (NYCRR), Volumes B–F, Parts 800-941, West Publishing, Eagan, MN.

However, this classification system does not provide a numeric stream order. The methodology identified in this manual is consistent with Strahler-Horton methodology. Specifically, a small branch is designated as a first-order stream. When two branches merge, a second order is created downstream of the confluence. Third-order stream refers to those downstream of the confluence of the second-order streams and so on.

Detention ponds or underground vaults are methods to meet the Cp_v requirement (and subsequent Q_{p10} and Q_f criteria). Note that, although these practices meet water quantity goals, they are unacceptable for water quality because of poor pollutant removal, and they need to be coupled with a practice listed in Table 5.14. The Cp_v requirement may also be provided above the water quality (WQ_v) storage in a wet pond or storm water wetland.

Bases for determining channel protection storage volume are the following:

* TR-55 and TR-20 (or approved equivalent) shall be used to determine peak discharge rates for the 1-year, 24-hour storm event.
* Off-site areas should be modeled as "present condition" for the 1-year, 24-hour storm event.
* The length of overland flow used in time of concentration (T_c) calculations is limited to no more than 100 ft for post development conditions.

TABLE 5.14

Acceptable Storm Water Management Practices for Water Quality

Group	Practice	Description
Pond	Micropool extended detention pond (P-1)	Pond that treats the majority of the water quality volume through extended detention, and incorporates a micropool at the outlet of the pond to prevent sediment resuspension
	Wet pond (P-2)	Pond that provides storage for the entire water quality volume in the permanent pool
	Wet extended detention pond (P-3)	Pond that treats a portion of the water quality volume by detaining storm flows above a permanent pool for a specified minimum detention time
	Multiple pond system (P-4)	A group of ponds that collectively treat the water quality volume
	Pocket pond (P-5)	A storm water wetland design adapted for the treatment of runoff from small drainage areas that has little or no base flow available to maintain water elevations and relies on groundwater to maintain a permanent pool
Wetland	Shallow wetland (W-1)	A wetland that provides water quality treatment entirely in a wet shallow marsh
	Extended detention wetland (W-2)	A wetland system that provides some fraction of the water quality volume by detaining storm flows above the marsh surface
	Pond/wetland system (W-3)	A wetland system that provides a portion of the water quality volume in the permanent pool of a wet pond that precedes the marsh for a specified minimum detention time
	Pocket wetland (W-4)	A shallow wetland design adapted for the treatment of runoff from small drainage areas that has variable water levels and relies on groundwater for its permanent pool
Infiltration	Infiltration trench (I-1)	An infiltration practice that stores the water quality volume in the void spaces of a gravel trench before it is infiltrated into the ground
	Infiltration basin (I-2)	An infiltration practice that stores the water quality volume in a shallow depression, before it is infiltrated into the ground
	Dry well (I-3)	An infiltration practice similar in design to the infiltration trench and best suited for treatment of rooftop runoff
Filtering practices	Surface sand filter (F-1)	A filter practice that treats storm water by settling out larger particles in a sediment chamber, and then filtering storm water through a sand matrix
	Underground sand filter (F-2)	A filtering practice that treats storm water as it flows through underground settling and filtering chambers
	Perimeter sand filter (F-3)	A filter that incorporates a sediment chamber and filter bed as parallel vaults adjacent to a parking lot
	Organic filter (F-4)	A filtering practice that uses an organic medium such as compost in the filter, in the place of sand
	Bioretention (F-5)	A shallow depression that treats storm water as it flows through a soil matrix and is returned to the storm drain system
Open channels	Dry swale (O-1)	An open drainage channel or depression explicitly designed to detain and promote the filtration of storm water runoff into the soil media
	Wet swale (O-2)	An open drainage channel or depression designed to retain water or intercept groundwater for water quality treatment

Source: Table 3.3, New York State storm water management manual prepared by Center for Watershed Protection, Ellicot City, MD, for New York State Department of Environmental Conservation, Albany, NY, August, 2010.

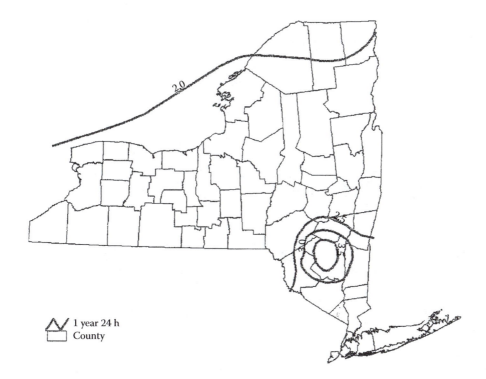

FIGURE 5.12 One-year, 24-hour design storm.

- Twenty-four-hour detention time for Cp_v is not required at sites where the resulting diameter of the extended detention orifice is too small. A minimum 3 in. orifice with a trash rack or 1 in. if the orifice is protected by a standpipe, with slots with an area less than the internal orifice, is recommended to prevent clogging.
- Extended detention storage provided for the channel protection (Cp_v-ED) does not meet the WQ_v requirement. Both water quality and channel protection storage may be provided in the same BMP, however.
- The Cp_v detention time for the 1-year storm is defined as the time difference between the center of mass of the inflow hydrograph (entering the BMP) and the center of mass of the outflow hydrograph (leaving the BMP).

5.5.5 Overbank Flow Control Criteria (Q_p)

The primary purpose of the overbank flooding design criteria is to prevent an increase in the frequency and magnitude of out-of-bank flooding generated by urban development. Overbank control requires storage to attenuate the post development 10-year, 24-hour peak discharge rate (Q_p) to predevelopment rates. The overbank flood control requirement (Q_p) does not apply in certain conditions, including when the site discharges directly into tidal waters or fourth-order (fourth downstream) or larger streams.

Design calculations for overbank flood control are performed on the following basis:

- TR-55 and TR-20 (or approved equivalent) will be used to determine peak discharge rates.
- When the predevelopment land use is agriculture, the curve number for the predeveloped condition shall be "taken as meadow."
- Off-site areas should be modeled as "present condition" for the 10-year storm event.

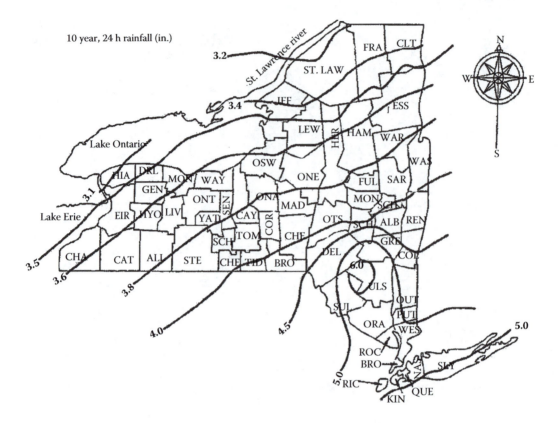

FIGURE 5.13   Ten-year, 24-hour design storm.

- Figure 5.13 indicates the depth of rainfall (24 h) associated with the 10-year storm event throughout the State of New York.
- The length of overland flow used in T_c calculations is limited to no more than 150 ft for pre-development conditions and 100 ft for post development conditions. On areas of extremely flat terrain (<1% average slope), this maximum distance is extended to 250 ft for predevelopment conditions and 150 ft for post development conditions.

5.5.6   EXTREME FLOOD CONTROL CRITERIA (Q_f)

The intent of the extreme flood criteria is to (a) prevent the increased risk of flood damage from large storm events, (b) maintain the boundaries of the predevelopment 100-year floodplain, and (c) protect the physical integrity of storm water management practices. One hundred-year control requires storage to attenuate the post development 100-year, 24-hour peak discharge rate (Q_f) to predevelopment rates.

The 100-year storm control requirement can be waived if

- The site discharges directly to tidal waters or fourth-order (fourth downstream) or larger streams.
- Development is prohibited within the ultimate 100-year floodplain.
- A downstream analysis reveals that 100-year control is not needed.

Detention structures involving dams must provide safe overflow of the design flood. The flow rates and floodplain extents referred to herein should not be confused with those developed by FEMA for use in the NFIP.

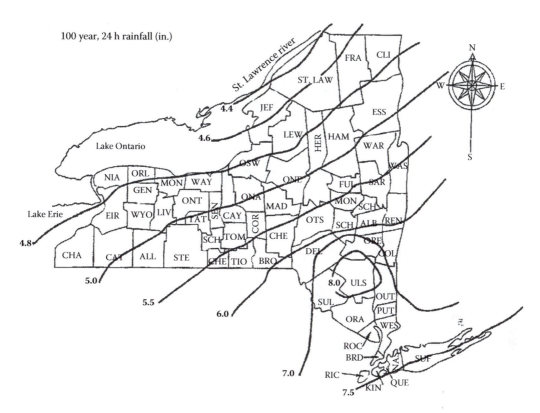

FIGURE 5.14 One hundred-year, 24-hour design storm for Q_f.

The extreme flood discharge design criterion is the 100-year, 24-hour storm event throughout the state of New York. Figure 5.14 indicates the depth of this storm event. The design criteria for this storm event are the following:

- The Q_f is calculated using the same methods discussed for overbank flood control, Q_p.
- When determining the storage required to reduce 100-year flood peaks, off-site areas are modeled under current conditions.
- When calculating the storage required to safely pass the 100-year flood, off-site areas are modeled assuming ultimate (fully developed) conditions.

5.5.7 DOWNSTREAM ANALYSIS

A downstream analysis for overbank and extreme flood control is recommended for sites greater than 50 acres to size facilities in the context of a larger watershed. The analysis will help ensure that storage provided at a site is appropriate when combined with upstream and downstream flows. For example, detention at a site may in some instances exacerbate flooding problems within a watershed. This section provides brief guidance for conducting this analysis, including the specific points along the downstream channel to be evaluated and minimum elements to be included in the analysis.

Downstream analysis can be conducted using the 10% rule. That is, the analysis should extend from the point of discharge downstream to the point on the stream where the site represents 10% of the total drainage area. For example, the analysis limits for a 50-acre site would include points on the stream from the point of discharge to the nearest downstream point with a drainage area of 500 acres. The required elements of the downstream analysis are described here:

- Compute predevelopment and post development peak flow and velocities for design storms (e.g., 10 year and 100 year) at all downstream confluences with first-order or higher streams up to and including the point where the 10% rule is met. These analyses should include scenarios both with and without storm water quantity control practices in place, where applicable.
- Evaluate hydrologic and hydraulic effects of all culverts and/or obstructions within the downstream channel.
- Assess water surface elevations to determine if an increase in water surface elevations will impact existing buildings and other structures.

Overbank and extreme flood requirements may be waived if both of the following conditions are met:

a. Peak flow rates increase by less than 5% of the predeveloped condition for the design storm (e.g., 10 year or 100 year).
b. No downstream structures or buildings are impacted.

5.5.8   CONVEYANCE SYSTEM DESIGN CRITERIA

The manual recommends the targeted storm frequencies for conveyance as the 2- and 10-year events. The 2-year event is used to ensure nonerosive flows through roadside swales, overflow channels, and pond pilot channels and over berms within practices. Figure 5.15 presents rainfall depths for the

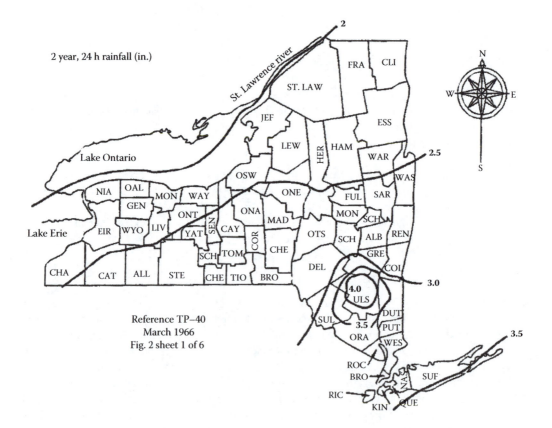

FIGURE 5.15   Two-year, 24-hour rainfall (in.).

2-year, 24-hour storm event throughout New York State. The 10-year storm is typically used as a target sizing for outfalls, and as a safe conveyance criterion for open channel practices and overflow channels. Note that some agencies or municipalities may use a different design storm for this purpose.

5.5.9   STORM WATER HOTSPOTS

A storm water hotspot is defined as a land use or activity that generates higher concentrations of hydrocarbons, trace metals, or toxicants than are found in typical storm water runoff, based on monitoring studies. If a site is designated as a hotspot, it has important implications for how storm water is managed. First and foremost, storm water runoff from hotspots cannot be allowed to infiltrate into groundwater, where it may contaminate water supplies. Second, a greater level of storm water treatment is needed at hotspot sites to prevent pollutant wash-off after construction. This treatment plan typically involves preparing and implementing a SWPPP that involves a series of operational practices at the site that reduce the generation of pollutants from a site or prevent contact of rainfall with the pollutants. Table 5.15 provides a list of designated hotspots for the state of New York.

The following land uses and activities are not normally considered hotspots:

- Residential streets and rural highways
- Residential development
- Institutional development
- Office developments
- Nonindustrial rooftops
- Pervious areas, except golf courses and nurseries, which may need an integrated pest management (IPM) plan

While large highways (average daily traffic volume [ADT] greater than 30,000) are not designated as a storm water hotspot, it is important to ensure that highway storm water management plans adequately protect groundwater.

TABLE 5.15
List of Storm Water Hotspots

The following land uses and activities are deemed *storm water hotspots*:

- Vehicle salvage yards and recycling facilities[a]
- Vehicle fueling stations
- Vehicle service and maintenance facilities
- Vehicle and equipment cleaning facilities[a]
- Fleet storage areas (bus, truck, etc.)[a]
- Discharge due to industrial activity
- Marinas (service and maintenance)[a]
- Outdoor liquid container storage
- Outdoor loading/unloading facilities
- Public works storage areas
- Facilities that generate or store hazardous materials[a]
- Commercial container nursery
- Other land uses and activities as designated by an appropriate review authority

[a] The land use or activity is required to prepare a storm water pollution prevention plan under the SPDES storm water program.

5.5.10   REDEVELOPMENT PROJECTS

Chapter 9 in the 2010 New York State storm water management manual concerns redevelopment projects. This chapter and Chapter 10, titled "Enhanced Phosphorus Removal Standards," were absent from the previous storm water management manual, dated August 2003. These were issued in April 2008 and included in the 2010 manual. A brief overview of these chapters is presented as follows.

Redevelopments, like new projects, are subject to four criteria: water quality treatment, channel protection, overbank flooding, and control of extreme storms. If a project proposes no increase in impervious area or if it does not change hydrology that would increase the discharge rate, the over-bank flooding and control of extreme flows—namely, the 10- and 100-year criteria—do not apply. The water quality criteria may be met using the following options:

1. The project reduces the existing impervious coverage by a minimum of 25%.
2. If the 25% reduction in impervious area is not met, the project captures and treats a minimum of 25% of the water quality volume (WQ_v) from the redevelopment area through standard practices. The plan should target areas with the greatest pollution generation potential, such as parking lots and service stations.
3. The project proposes the use of alternative practices to treat 75% of WQ_v from the disturbed area as well as additional runoff, if any, beyond the disturbed area.
4. The plan proposes a combination of impervious area reduction and standard or alternative practices that provide a weighted average equal to or more than options 2 or 3. When the level of water quality treatment must be acceptable for redevelopment applications, proprietary practices include wet vaults, underground infiltration systems, and manufactured water quality devices (hydrodynamic gravity or water separators and media filters).

5.5.11   ENHANCED PHOSPHORUS REMOVAL STANDARDS

Chapter 10 in the 2010 manual presents a detailed description of the sources, transport, and treatment processes of phosphorus. The chapter includes a discussion on

a. Prevention or reduction of runoff as a highly effective means for reducing the total loads of phosphorus generated and the size and cost of downstream treatment systems
b. Various storm water management systems for phosphorus control
c. Effectiveness and ability of storm water management practices to reduce the discharge and remove phosphorus

An alternate procedure for calculating the water quality volume, WQ_v, in this chapter is specified as using the TR-55 method and 1-year, 24-hour storm for phosphorus treatment. The 1-year, 24-hour storm in New York State varies from 1.8 to 3.2 in. (45 to 80 mm).

PROBLEMS

5.1   Which agency passed the Clean Water Act (CWA) of 1972? What did the Act cover?
5.2   What are MS4s and what type of permit do they require?
5.3   What is a construction general permit (CGP)? When did the current CGP go into effect?
5.4   What does NPDES stand for? When did the NSPES phase II become effective and what does it require?
5.5   What does TSS mean? What is the required TSS removal in New Jersey? What is the requirement in your state?
5.6   What does EISA Section 438 cover?
5.7   What is the new trend in storm water quantity control?

5.8   What is defined as a major development in New Jersey?

5.9   Are major developments required to provide peak flow reductions in New Jersey? What are those reductions?

5.10 What is the water quality requirement in New Jersey? Are all development projects required to provide storm water treatment?

5.11 Calculate the volume of water quality storm for a 10-acre development in New Jersey. What is the volume for a 4.0 ha site? The impervious coverage is 30%. State any assumptions made.

5.12 Calculate the water quality runoff volume for a 10-acre development, having the following impervious/pervious coverages:
   a. 15% building
   b. 25% parking
   c. 60% lawn/landscape
   How much of the runoff is subject to water quality treatment? Hint: Lawn and landscape generate little runoff during a water quality storm.

5.13 The runoff from the building and parking area of Problem 5.12 is routed through an extended detention basin that provides 40% TSS removal. Is this adequate to address the state of New Jersey storm water management regulations? If not, what is the required TSS removal deficiency? What can be done to address the TSS removal for this site?

5.14 Calculate the required groundwater recharge in Problem 5.12 using option 2 in the NJDEP regulation. The site soil is hydrologic group B and the 2-year, 24-hour storm is 3.4 in.

5.15 What is the water quantity sizing criteria in the state of Maryland? Calculate the required water quality volume for an 8-acre residential site in the eastern zone. The impervious coverage on the site is 2.4 acres.

5.16 Redo Problem 5.15 for a 2 ha site with 30% impervious coverage.

5.17 Calculate the water quality volume WQ_v of the residential project in Problem 5.15 for a site in the state of New York. The site is marked "site" on Figure 5.11.

5.18 Calculate the recharge volume for the residential site of Problem 5.15. The site soil is hydrologic group B.

5.19 Calculate the channel protection volume, Cp_v, for the site of Problem 5.18. The site is located in Frederick County. Base your calculations on a time of concentration of 30 minutes for the predevelopment conditions and 15 minutes for post development.

5.20 Calculate overbank flood protection volume, Q_p, for Problem 5.15. The site is situated in Frederick County.

5.21 Calculate the required ESD runoff depth and storage volume for Example 5.2 in this chapter in metric units. For simplicity, round the site area to 14 ha.

APPENDIX 5A: NJDEP-APPROVED TSS REMOVAL RATE FOR VEGETATED FILTER STRIPS

5A.1 Required Filter Strip Length

The maximum slope to achieve the adopted TSS removal rates shown in Table 5A.1. The required filter strip length can be determined from Figures 5A.1 to 5A.5, based upon the filter strip's slope, vegetated cover, and the soil within its drainage area. As shown in the figures, the minimum length for all vegetated filter strips is 25 ft.

TABLE 5A.1
Maximum Filter Strip Slope

		Maximum Filter Strip Slope (%)	
Filter Strip Soil Type	Hydrologic Soil Group	Turf Grass, Native Grasses, and Meadows	Planted and Indigenous Woods
Sand	A	7	5
Sandy loam	B	8	7
Loam, silt loam	B	8	8
Sandy clay loam	C	8	8
Clay loam, silty clay, clay	D	8	8

Source: NJDEP's storm water best management practices manual, adopted February 2, 2004.

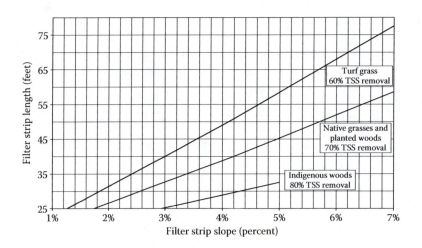

FIGURE 5A.1   Vegetated filter strip length; drainage area soil: sand HSG: A.

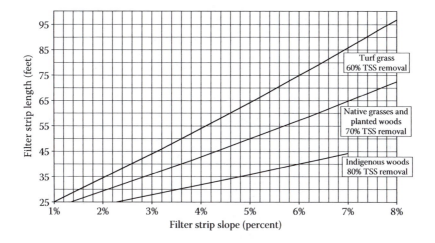

FIGURE 5A.2   Vegetated filter strip length; drainage area soil: sandy loam HSG: B.

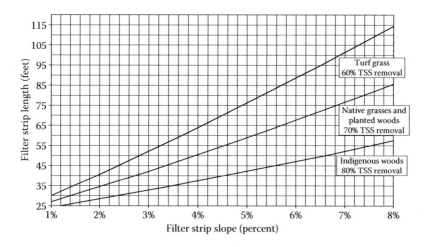

FIGURE 5A.3   Vegetated filter strip length; drainage area soil: loam, silt loam HSG: B.

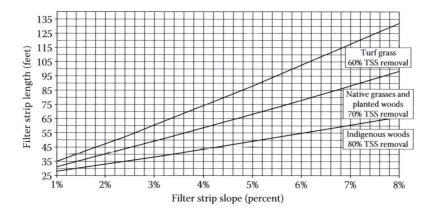

FIGURE 5A.4   Vegetated filter strip length; drainage area soil: sandy clay loam HSG: C.

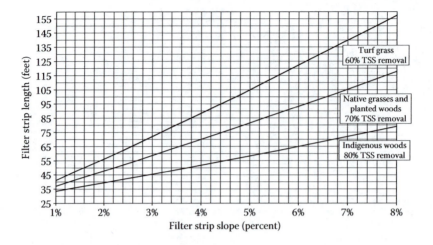

FIGURE 5A.5 Vegetated filter strip length; drainage area soil: clay loam, silty clay, Clay HSG: D.

APPENDIX 5B: MARYLAND'S REDUCED CURVE NUMBERS FOR ESD SIZING REQUIREMENT

Table 5B.1 shows rainfall targets/runoff curve number reductions used for ESD for different hydrolic soil groups.

TABLE 5B.1
Rainfall Targets/Runoff Curve Number Reductions Used for ESD

%I	RCN[a]	PE = 1 in.	1.2 in.	1.4 in.	1.6 in.	1.8 in.	2.0 in.	2.2 in.	2.4 in.	2.6 in.
					Hydrologic Soil Group A					
0%	40									
5%	43									
10%	46									
15%	48	38				*Cp*ᵥ addressed				
20%	51	40	38	38						
25%	54	41	40	39						
30%	57	42	41	39	38					
35%	60	44	42	40	39					
40%	61	44	42	40	39					
45%	66	48	46	41	40					
50%	69	51	48	42	41	38				
55%	72	54	50	42	41	39				
60%	74	57	52	44	42	40	38			
65%	77	61	55	47	44	42	40			
70%	80	66	61	55	50	45	40			
75%	84	71	67	62	56	48	40	38		
80%	86	73	70	65	60	52	44	40		
85%	89	77	74	70	65	58	49	42	38	
90%	92	81	78	74	70	65	58	48	42	38
95%	95	85	82	78	75	70	65	57	50	39
100%	98	89	86	83	80	76	72	66	59	40
					Hydrologic Soil Group B					
0%	61									
5%	63									
10%	65									
15%	67	55								
20%	68	60	55	55		*Cp*ᵥ addressed				
25%	70	64	61	58						
30%	72	65	62	59	55					
35%	74	66	63	60	56					
40%	75	66	63	60	56					
45%	78	68	66	62	58					
50%	80	70	67	64	60					
55%	81	71	68	65	61	55				
60%	83	73	70	67	63	58				
65%	85	75	72	69	65	60	55			
70%	87	77	74	71	67	62	57			
75%	89	79	76	73	69	65	59			

(Continued)

TABLE 5B.1 (CONTINUED)
Rainfall Targets/Runoff Curve Number Reductions Used for ESD

%I	RCN[a]	PE = 1 in.	1.2 in.	1.4 in.	1.6 in.	1.8 in.	2.0 in.	2.2 in.	2.4 in.	2.6 in.
80%	91	81	78	75	71	66	61			
85%	92	82	79	76	72	67	62	55		
90%	94	84	81	78	74	70	65	59	55	
95%	96	87	84	81	77	73	69	63	57	
100%	98	89	86	83	80	76	72	66	59	55

Hydrologic Soil Group C

%I	RCN[a]	PE = 1 in.	1.2 in.	1.4 in.	1.6 in.	1.8 in.	2.0 in.	2.2 in.	2.4 in.	2.6 in.
0%	74									
5%	75									
10%	76									
15%	78									
20%	79	70								
25%	80	72	70	70						
30%	81	73	72	71			Cp_v addressed			
35%	82	74	73	72	70					
40%	84	77	75	73	71					
45%	85	78	76	74	71					
50%	86	78	76	74	71					
55%	86	78	76	74	71	70				
60%	88	80	78	76	73	71				
65%	90	82	80	77	75	72				
70%	91	82	80	78	75	72				
75%	92	83	81	79	75	72				
80%	93	84	82	79	76	72				
85%	94	85	82	79	76	72				
90%	95	86	83	80	77	73	70			
95%	97	88	85	82	79	75	71			
100%	98	89	86	83	80	76	72	70		

Hydrologic Soil Group D

%I	RCN[a]	PE = 1 in.	1.2 in.	1.4 in.	1.6 in.	1.8 in.	2.0 in.	2.2 in.	2.4 in.	2.6 in.
0%	80									
5%	81									
10%	82									
15%	83									
20%	84	77								
25%	85	78				Cp_v addressed				
30%	85	78	77	77						
35%	86	79	78	78						
40%	87	82	81	79	77					
45%	88	82	81	79	78					
50%	89	83	82	80	78					
55%	90	84	82	80	78					
60%	91	85	83	84	78					
65%	92	85	83	84	78					
70%	93	86	84	84	78					
75%	94	86	84	84	78					
80%	94	86	84	82	79					

(Continued)

TABLE 5B.1 (CONTINUED)
Rainfall Targets/Runoff Curve Number Reductions Used for ESD

%I	RCN[a]	PE = 1 in.	1.2 in.	1.4 in.	1.6 in.	1.8 in.	2.0 in.	2.2 in.	2.4 in.	2.6 in.
85%	95	86	84	82	79					
90%	96	87	84	82	79	**77**				
95%	97	88	85	82	80	78				
100%	98	89	86	83	80	78	**77**			

Source: Maryland Storm Water Design Manual, updated through May 4, 2009, vol I. and II, prepared by Center for Watershed Protection, Ellicott City, and the Maryland Department of the Environment, MD (http://www.mde.state.md.us).

Note: Cp_v addressed (RCN = woods in good condition) (marked in bold).

[a] RCN applied to Cp_v calculations.

REFERENCES

Brzozowski, C., 2005, NPDES reporting requirements storm water managers share how they cope with the tremendous amount of data a storm water program generates, *Storm Water*, May/June, 52–59.
——— 2008a, Getting to know the SWPPP, *Storm Water*, July/August, 20–39.
——— 2008b, Getting to know the SWPPP, part 2 Adventures in erosion and sediment control, *Storm Water*, September, 72–81.
EPA (Environmental Protection Agency), 2000a, The quality of our nation's waters, a summary of the national water quality inventory: 1998 Report to Congress, EPA-841-S-00-001, Office of Water, Washington, DC, June.
——— 2000b, National water quality inventory report, EPA-841-R-02-001 (http://www.epa.gov/305b).
——— 2009, Technical Guidance on Implementing the Stormwater Runoff Requirements for Federal Projects under Section 438 of the Energy Independence and Security Act, http://www.epa.gov/owow/nps/lid/section438.
——— 2012, Construction general permit, February (http://cfpub.epa.gov/npdes/storm water/cgp.cfm).
——— 2013a, National rivers and streams assessment 2008–2009, a collaborative survey, draft, EPA/841/D-13/001, February 28.
——— 2013b, http://www.epa.gov/npdes/sw_rule_presentation_July2013.pdf.
EWRI Currents, 2009, Federal facilities face a new storm water hurdle, *Newsletter of the Environmental and Water Resources Institute of the ASCE*, 11 (1): 1–2.
Kayhanian, M. and Stenstrom, M. K., 2008, First flush characterization for storm water treatment, *Storm Water*, March/April, 32–45.
Maryland Model Storm Water Management Ordinance, July 2005, and supplement, Jan. 2007, Maryland Department of the Environment, Baltimore, MD (http://www.mde.state.md.us).
Maryland Storm Water Design Manual, Updated through May 4, 2009, vol. I and II, prepared by Center for Watershed Protection, Ellicott City, and the Maryland Department of the Environment, MD (http://www.mde.state.md.us).
New York State storm water management design manual, 2010, prepared by Center for Watershed Protection, Ellicott City, MD, for New York State Department of Environmental Conservation, Albany, NY, August.
NJDEP (New Jersey Department of Environmental Protection), 2004a, New Jersey Storm Water Best Management Practices Manual, NJDEP, Trenton, NJ, February, partly revised and amended September 2014.
——— 2004b, DEP storm water management rules, N.J.A.C.7:8-5 and 7:8-6, February 2.
——— 2010, Amended storm water management rules, N.JA.C.7:8, April 19.
NJGS (New Jersey Geological Survey), 1993, GSR-32 (geological survey report No. 32) (http://www.state.nj.us/dep/njgs/N.J.A.C.7:15).
US Dept. of Agriculture, Natural Resources Conservation Service, 2007, *National engineering handbook* (NEH), part 630 hydrology, Sept. 2007, revised through November 2010, Chapter 16, hydrographs, dated March 2007.

6 Manufactured Water Treatment Devices

6.1 AN OVERVIEW

In practice, a variety of structural storm water management systems are employed to improve water quality. These systems in part include vegetated swales, extended detention basins, wet ponds, infiltration basins, bioretention basins, constructed wetlands, and sand filters. In addition, there exists a variety of nonstructural storm water management measures, such as porous pavements, rain gardens, and vegetative buffers. In recent years, various water quality devices have been manufactured to address the federal and states water quality regulations. This chapter discusses manufactured water quality devices (also referred to as water treatment devices) and presents a brief comparison of these devices with structural water quality systems. Structural and nonstructural storm water management systems will be discussed later in this book.

Among structural storm water management systems, detention basins and ponds, if properly designed, can serve as effective measures for runoff quantity control; they can also improve water quality, though not as effectively as they control flow rate. Infiltration basins and bioretention basins are most suited for water quality. While some structural best management practices (BMPs) are reported to remove up to 95% of the total suspended sediment, their actual effectiveness is site dependent and can also vary from storm to storm.

The effectiveness of structural BMPs in removing total phosphorus and nitrate is far smaller than their total suspended solids (TSSs) removal ability. Some chemicals, such as chlorine derived from road de-icing, cannot be removed by any structural BMP. Also, absorption of some pollutants, such as metals, by soil may cause soil toxicity, turning infiltration basins into contaminated sites over a period of a few decades. Apart from inadequate efficacy, structural BMPs are expensive. The cost includes not only the construction, but also the land in the case of open detention/retention/infiltration basins, and maintenance as well.

A manufactured device is a prefabricated or cast-in-place storm water treatment structure utilizing vortex separation or filtration processes, absorption/adsorption materials, vegetative media, or other technology to remove pollutants from runoff. Manufactured treatment devices are intended to capture sediments, floatables, hydrocarbons, and other pollutants in storm water runoff before discharging to a drainage conveyance system or another storm water management facility. A manufactured treatment device is adequate for small drainage areas and particularly for small pavement areas, such as a parking lot or gas station. For larger sites, manufactured water quality devices may be used for pretreatment of runoff before discharging to other larger capacity water quality structures, such as detention basins and ponds. These devices are particularly suited for pretreatment before underground detention and/or retention/infiltration basins.

In general, water quality devices are designed based on the peak discharge. However, if a device is placed past a detention system, it can be sized based on the attenuated flow from the system. Such application, which reduces the size and cost of the device, is only suited for filter media devices.

Manufactured devices are commonly referred to by their trade names and generally require more maintenance than structural BMPs. Some manufactured water quality devices are equipped with filter media to improve water quality. These devices require more frequent maintenance and are more expensive to maintain than other devices and many structural systems. In terms of relative maintenance, media filters require more maintenance and grass swales the least. The cost of maintenance also varies from one practice to another; vegetated swales require the least budget while media filters are more costly, though not as costly to maintain as sand filters.

Sand and media filters have been accepted as BMPs for water quality and both have been approved by some jurisdictional agencies including the New Jersey DEP for 80% TSS removal. Sand filters comprise a horizontal bed of sand underlain by filter cloth, aggregate, and a perforated drain pipe. The silt deposit eventually occludes the sand and inhibits percolation. To maintain the filter, the water over the sand has to be vacuumed and the top layer, if not the entire sand bed, removed and replaced by clean sand. On the other hand, media filter maintenance is straightforward and simply involves replacing media filters when they are found to be losing draining capacity. Also, the cost of maintaining sand filters is nearly twice as much as for media filters. According to data reported by the Virginia Planning Board Commission, the average annual cost of maintaining sand filters is estimated at $2000 per acre of impervious area compared with $1000 for media filters (Doerfer, 2008). A study by Weiss et al. (2005) provides valuable information on effectiveness and cost of total suspended solids and phosphorus removal for extended detention basins, ponds, sand filters, constructed wetlands, bioretention basins, and infiltration trenches.

A single BMP may not remove all pollutants of concern. Therefore, rather than employing an individual BMP, various treatment options should be evaluated in selecting a set of BMPs that addresses the storm water management requirements (WERF, 2005). First, the extent to which a BMP reduces the amount of runoff should be considered. The storage of runoff for reuse should also be considered. Next, the selection should carefully consider the rate of runoff that can be treated by a BMP. Many manufactured devices have a bypass system to divert flows higher than their capacity.

An important consideration in selecting a manufactured device (and any BMP) is its overall cost compared with other devices. In analyzing the overall cost, not only the initial installation but also the long-term maintenance cost should be evaluated for the project; keeping in mind that a given BMP does not cost the same at different sites. It is also to be noted that storm water treatment technology is rapidly changing. Today's leading technology may not be the best approach a few years from now.

There are a large number of national or state publications on the performance of storm water management BMPs. An excellent publication on the subject is "National Management Measures to Control Nonpoint Source Pollution from Urban Areas," released by the Environmental Protection Agency (EPA) in November 2005, publication no. EPA-841-B-05-004, 518 pp. This publication includes such subjects as watershed assessment and protection, site development, construction site erosion and sediment control, pollution prevention, operation and maintenance, and evaluation of program effectiveness. The following references are but a few of other good national and state publications:

- "Managing Stormwater Runoff to Prevent Contamination of Drinking Water," EPA, "Source Water Protection Practices Manual," 2008
- "Critical Assessment of Stormwater Treatment Control and Selection Issues," published by WERF, 2005
- "Stormwater Best Management Practice Handbook Portal: Construction," published by California Stormwater Quality Association, November 2009
- "Stormwater Best Management Practices Manual," published by the New Jersey Department of Environmental Protection, February 2004, partly revised 2009, Chapter 6: "Standard for Manufactured Treatment Devices"
- "Construction Site Monitoring Program Guidance Manual," published by Caltrans Division of Environmental Analysis, CTSW-RT-11-255-20.1, August 2013

Another major resource is the International Stormwater BMP Database, which provides useful information on storm water management practices. This database site, which was established by the US Environmental Protection Agency and the American Society of Civil Engineers (ASCEs) in 1996, has since grown considerably and includes other partners, such as the Water Environment Research Foundation (WERF), Federal Highway Administration (FHWA), American Public Works Association, Environmental and Water Resources Institute (EWRI) of the ASCE, and services that provide performance data and BMP evaluation. The database can be accessed online (http://www.bmpdatabase.org). In 2010 WERF, FHWA, and EWRI cosponsored a comprehensive storm water BMP performance analysis technical paper series relying on data contained in the International Stormwater BMP database. The technical reports can be downloaded (http://www .bmpdatabase.org/BMPperformance.htm). The paper series published in 2010 and 2011 contained some 400 studies and by December 2012 over 100 more new BMP performance studies had been added to the BMP database. By 2013, over 800 BMP-related literature sources were catalogued and reviewed.

6.2   CERTIFICATION OF WATER QUALITY DEVICES

A large number of manufactured water quality devices have been introduced to the market following the EPA phase II storm water management rule. Stormceptor, CDS® (Continuous Deflective Separation), and Vortechemics were among the first manufactured water quality devices that came to the market in mid-1990s. The CDS water quality device, for example, was introduced to the United States in 1996. The number of manufactured devices has since grown to over three dozen and is still growing. Manufactured devices were faced with objections from institutions initially. Entities like ASCE and EPA began to consider them in their third-party independent evaluation during the beginning of the 2000s. Many jurisdictional agencies only accepted manufactured devices as a supplement to structural or nonstructural BMPs and as a last resort. This attitude is gradually changing as the experiments with the manufactured devices prove their effectiveness and efficiency.

In selecting a water quality device (and any water quality BMP) for a project, the practitioner needs to have reliable information about its performance, maintenance requirements, and life cycle. In addition, because of the site-specific nature of storm water management practice, BMPs must be tailored to suit a given site. Ideally, the selection should be made based on test data collected within the same state or a region of the same or similar hydrologic nature. However, dependable data specific to locality are not always available for manufactured water treatment devices.

Also, reported data by manufacturers are generally based on test protocols that may vary from one manufacturer to another. The claimed total suspended sediment removal for a manufactured water quality device may not specify the type and size of sediment tested.

Particle size is important in pollutant removal. As indicated in Chapter 1, large particles settle much faster than fine particles; as such, tests that utilize sand show a far larger TSS removal rate than silt. However, fine particles, which transport pollutants such as metals, constitute the sediments of concern, and these particles take a long time to settle. Also, due to variations, the natural environment does not lend itself to a standardized test for soil removal. Different regions having varying geologic conditions generate different size particles in runoff. In addition, sediment size can vary from storm to storm. An intense rainfall can produce and transport larger particles and also significantly larger sediment load than a drizzle.

Another issue on evaluating test data is whether the study was conducted in a laboratory or in the field. Laboratory tests show how a device performs under a set of strictly controlled testing protocols. As such, the effectiveness of different devices can be closely compared on that basis. However, since the size, type, and concentration of particles entering a manufactured treatment device (MTD) vary from one site to another, the results of laboratory testing may or may not reflect the actual field performance of a device.

Field tests have the benefit of representing the device's performance under actual circumstances. To be conclusive, however, field tests should be conducted within a large range of storm conditions and varying sites. Only a few states have undertaken the task of preparing some kind of document that local storm water practitioners can use. New Jersey is one of the states that have developed a rigorous testing and certification procedure for manufactured water quality devices. This procedure is briefly described in the following section.

6.2.1   NJCAT Certification

In New Jersey, the NJDEP Division of Science, Research and Technology (DSRT) used to be responsible for certifying final pollutant removal rates for all manufactured treatment devices. The certification was issued based upon one of the following:

1. Verification of the device's pollutant removal rates by the New Jersey Corporation for Advanced Technology (NJCAT) in accordance with the New Jersey Energy and Environmental Technology Verification Program at N.J.S.A. 13:D-134 et seq.
2. Verification of the device's pollutant removal rates by a technology assessment protocol-ecology (TAPE) program of another state or government agency that is recognized by New Jersey through a formal reciprocity agreement, provided that such verification is conducted in accordance with the protocol, "Stormwater Best Management Practices Demonstration Tier II Protocol for Interstate Reciprocity."
3. Verification of the device's pollutant removal rates by other third-party testing organizations (such as NSF), provided that such verification is conducted in accordance with the previously indicated protocol. Other testing protocols may be considered if they are determined by the NJDEP to be equivalent to the tier II protocol. Appendix 6A exemplifies the environmental verification of Terre Kleen™ (a water treatment device manufactured by Terre Hill Concrete Products of Terre Hill, Pennsylvania) by the EPA and NSF.

On January 25, 2013, the New Jersey Department of Environmental Protection (NJDEP) issued a new process of verification for MTDs. Prior to approval by the NJDEP, an MTD must obtain verification from the New Jersey Corporation of Advanced Technology (NJCAT). The process for NJCAT verification is included in a manual titled "Procedure of Obtaining Verification of a Stormwater Manufactured Treatment Device from New Jersey Corporation of Advanced Technology for Use in Accordance with the Stormwater Management Rules, N.J.A.C.7:8." This procedure is available at http://www.njcat.org.

Under the 2013 procedure, any MTD that had previously received verification from the NJCAT must be reverified in accordance with the new procedure prior to its expiration date of certification from the NJDEP in order to maintain its verified status. The 2013 protocol indicates that verification of an MTD shall be based on the results of a series of laboratory and analytical tests performed in strict accordance with this document. No field testing is required at this time.

The laboratory testing must satisfy one of the following applicable testing protocols:

- "New Jersey Department of Environmental Protection Laboratory Protocol to Assess Total Suspended Solids Removal by a Hydrodynamic Sedimentation Manufactured Treatment Device," dated January 25, 2013
- "New Jersey Department of Environmental Protection Laboratory Protocol to Assess Total Suspended Solids Removal by a Filtration Manufactured Treatment Device," dated January 25, 2013

Laboratory testing will evaluate the MTD's treatment process, determine performance, and assess expected life span. These tests shall be conducted either by an independent test facility or by the manufacturer. Laboratory testing conducted by the manufacturer must be performed under the direct supervision of an independent third-party observer.

Analytical testing is defined as the evaluation of total suspended solids in accordance with ASTM D3977-97. This test shall be conducted by the manufacturer or an independent analytical laboratory or independent test facility. Analytical testing conducted by the manufacturer must be performed under the direct supervision of an independent third-party observer.

If the manufacturer is using its own laboratory for either laboratory or analytical testing and a third-party observer is being used:

1. The observer shall verify compliance with the laboratory test plan.
2. The observer shall observe the testing for its full duration.
3. The observer shall have no personal conflict of interest regarding the test results.
4. The qualifications of the laboratory and the independent observer must be approved by the NJCAT.

The qualification of an independent laboratory, independent test facility and third-party observer must be approved by the NJCAT prior to testing. Appendix 6B describes the testing process. Upon completion of the tests, documents are submitted to NJCAT for review and verification. Required documents for review include: description of technology, laboratory setup, performance claims, supporting documents, design limitations, and maintenance plans. The performance claims provide information on the size and capacities of the MTD as follows:

1. For hydrodynamic separation MTDs, referred to as hydrodynamic sedimentation (HDS) MTDs by NJCAT, the performance data include
   a. Verified TSS removal rates
   b. Maximum treatment flow rate (MTFR)
   c. Maximum sediment storage depth and volume
   d. Effective treatment area
   e. Detention time and volume
   f. Effective sedimentation area
   g. Online or offline installation
   h. The basis for determining all of the above, including all pertinent calculations

Note: The TSS removal efficiency will be determined by NJCAT; if the TSS removal efficiency is greater than 50% for HDS MTDs, the TSS removal efficiency shall be rounded down to 50%. For HDS MTDs with TSS removal efficiencies that are less than 50%, NJCAT will not grant verification.

2. For filtration MTDs, the following information must be provided
   a. Verified TSS removal rates
   b. MTFR and maximum draindown cartridge flow rate (if applicable)
   c. Maximum sediment storage depth and volume
   d. Effective treatment area
   e. Detention time and wet volume
   f. Effective sedimentation area
   g. Effective filtration area
   h. Sediment mass loading capacity
   i. Maximum allowable inflow drainage area

    j.   Online or offline installation with maximum online flow rate (if applicable)
    k.   The basis for determining all of the above, including all pertinent calculations

Note: The TSS removal efficiency will be determined by NJCAT; if the TSS removal efficiency is greater than 80% for filtration MTDs, the TSS removal efficiency shall be rounded down to 80%. For filtration MTDs with TSS removal efficiencies that are less than 80%, NJCAT will not grant verification.

Following the review, NJCAT prepares a verification report, which confirms that the tested MTD has met the technical and regulatory standards. NJCAT then updates its website (http://www.njcat.org) with an electronic version of the report and provides NJDEP's Stormwater Management Unit a link to the website. Formal certification of the MTD is established on the NJDEP Stormwater website (http://www.njstormwater.org). The website is updated from time to time as new MTDs are approved.

The NJDEP, Bureau of Nonpoint Pollution Control, did not extend the testing deadlines, which expired on January 25, 2015. Table 6.1 lists the names, TSS removal rate (80% or 50%), and laboratory and/or field test certification of MTDs. Appendix 6C includes certification for Filterra® Bioretention Systems, a division of Americast (now a part of Contech Stormwater Solutions), which is the first bioretention cell certified for 80% TSS removal by the NJDEP.

It is to be noted that the NJDEP has no certification for catch basin inserts, which are practical and cost effective for small areas. However, some other jurisdictional agencies accept inlet filters for water treatment. Greenville, South Carolina, for example, has classified MTDs into three types as follows:

MTD type 1. Separation devices (standard storm water MTD); contains a sump for sediment deposition with a series of chambers, baffles, or weirs to trap trash, oil, grease, and other contaminants.
MTD type 2. Filtration devices (impaired water bodies, total maximum daily load requirements); contains a sedimentation chamber and a filtering chamber. MTD type 2 contains filter materials or vegetation to remove specific pollutants such as nitrogen, phosphorus, copper, lead, zinc, and bacteria.
MTD type 3. Catch basin inserts (limited space); may contain filter media including polypropylene, porous polymers, treated cellulose, and activated carbon designed to absorb specific pollutants such as oil, grease, hydrocarbons, and heavy metals. MTD type 3 must provide overflow features that do not reduce the original hydraulic capacity of the catch basin.

MTDs are applicable for a maximum drainage area of 3.0 acres. Size all MTDs to treat, at a minimum, the entire water quality event (WQE) with no bypass. The water quality storm in the Greenville standards is specified as 1.8 in., 24-hour type II storm or rainfall intensities, which range from 2.16 in./h for a 5-minute storm to 1.34 in./h for a 30-minute storm. In New Jersey, the water quality storm as indicated in Chapter 5 is a 1.25 in., 2-hour storm, which is a significantly more intense and voluminous storm than the specified water quality storm in Greenville. For a 10-minute storm, for example, the New Jersey water quality storm has 3.2 in./h intensity.

The Greenville County technical specifications also correctly indicate that a manufactured device rate for 90% or greater TSS removal may only remove 2% of clay particles, but it can remove 100% of silt, sand, and small and large aggregates.

TABLE 6.1
Water Treatment Devices Certified by the NJDEP

Storm Water Management Manufactured Treatment Devices Certified by NJDEP	MTD Laboratory Test Certifications	Field Test Certifications	Superseded Certifications	Certified TSS Removal Rate	Maintenance Plan
AquaFilter Filtration Chamber by AquaShield, Inc.		Certification	Superseded	80%	Plan
Aqua-Swirl Concentrator by Aqua-Shield, Inc.		Certification	Superseded	50%	Plan
Continuous Deflective Separator (CDS) Unit by CONTECH Stormwater Solutions, Inc.	Certification	Certification	Superseded	50%	Plan
Downstream Defender by Hydro International, Inc.	Certification		Superseded	50%	Plan
Dual Vortex Separator by Oldcastle® Stormwater Solutions	Certification			50%	Plan
Filterra Bioretention System by CONTECH Engineered Solutions	Certification		Superseded	80%	Plan
Jellyfish® Filter by Imbrium Systems Corporation		Certification	Superseded	80%	Plan
Media Filtration Systems by CONTECH Stormwater Solutions, Inc.		Certification	Superseded	80%	Appendix A
StormPro Stormwater Treatment Device by Environment 21, LLC	Certification			50%	
StormVault by Jensen Precast, Inc.		Certification	Superseded	80%	Appendix A
Stormwater Management StormFilter by CONTECH Stormwater Solutions, Inc.		Certification	Superseded	80%	Plan
Up-Flo Filter by Hydro International		Certification	Superseded	80%	Plan
Vortechs Stormwater Treatment System by CONTECH Stormwater Solutions, Inc.		Certification	Superseded	50%	Plan

6.3 TYPES OF MANUFACTURED DEVICES

Manufactured treatment devices come in various types providing different levels of water quality improvements. MTDs may be classified into three main groups:

1. Devices or filters that are installed in or over catch basins.
2. Hydrodynamic separation water quality structures.
3. Filter media water quality devices. These devices provide the greatest total suspended solids removal and can also partially remove other pollutants such as phosphorus and hydrocarbons. Filter media devices may be more effective than some structural water treatment systems such as grass swales.

6.3.1 Catch Basin Inserts

Catch basin inserts come in different varieties. One type consists of a series of trays with the top tray serving as initial sediment trap and the underlying trays composed of media filters. Another type employs filter fabric to remove pollutants. Yet another type includes a plastic or metal basket that fits directly into the catch basin. Most popular devices are the basket type and include inserts and hoods. Table 6.2 lists some of the inlet filters manufactured in the United States.

Among the filters listed in Table 6.2, Ultra-Urban Filter, manufactured by AbTech Industries of Scottsdale, Arizona, and FloGard+Plus® Filter, produced by Kristar Enterprises of Santa Rosa, California, are two of the more widely used catch basin inserts. Kristar also makes filters for trench drains. Clearwater BMP is another widely used inlet filter, which is manufactured by Clearwater Solutions of Vista California.

Ultra-Urban Filter was one of the first catch basin inserts and was introduced to the market by AbTech Industries, which was founded in 1996. FloGard+Plus by Kristar (now a part of Oldcastle® Stormwater Solutions) is another product that has been on the market for a long time. Ultra-Urban and FloGard+Plus filters each consists of a screen, a filter liner, and a nonleaching oil-absorbent material contained in a pouch or similar removable strainer. Each of these filters has an overflow

TABLE 6.2
List of Catch Basin Inserts

Product Name	Company	Filter Type/Name	Website Address
Ultra-Urban Filter	AbTech Industries	Smart Sponge	http://www.abtechindustries.com
FloGard+Plus	Kristar Enterprises/ Oldcastle® Stormwater Solutions	Perk Filter/stainless steel	http://www.kristar.com
Stormdrain Solutions	Stormdrain Solutions	PolyDak	http://www.stormdrains.com
Aqua-Guardian Catch Basin Insert	Aqua Shield	Hydrophobic Cellulose	http://www.aquashieldinc.com
Hydro-Kleen Stormwater Filteration System	ACF Environmental	[a]	http://www.acfenvironmental.com
Triton	Contech	Media Pak	http://www.Contechstormwater.com
Blocksom Filters[b]	Blocksom & Co.	Natural Filter	http://www.blocksom.com
REM GeoTrap Filter Insert	SWIMS	Filter Media Cartridge	http://www.SwimsClean.com
Fabco Catch Basin Insert	Fabco Industries	[a]	http://www.fabco-industries.com
Grate Inlet Skimmer Box	Suntree Technologies Inc.	[a]	http://info@suntreetech.com

[a] Unspecified.

[b] Blocksom filter is a mat placed over inlet grate. This mat is used as a sediment and erosion control measure during construction (see Chapter 9).

weir to pass flows beyond the capacity of the filter to a downstream storm drain. These filters offer simple means of removing trash, floatables, and a large fraction of sediment; leaves and grass needles; and, as indicated, oil and grease from storm water runoff. Inlet filters have not received certification from the NJDEP; nonetheless, they are most practical for developed areas where there is no room for a water quality BMP and it is unfeasible to make major modifications of existing drainage structures. These filters can be readily inserted into existing inlets on parking lots and streets and roads. The author recommends the use of these filters on inlets that drain to dry wells and other types of underground retention–infiltration basins. By removing silt, leaves, trash, and oil and grease from runoff, inlet filters prolong the useful life of underground retention–infiltration basins. The author accepts the inlet filters for 40% TSS removal in the nonmajor projects that he reviews.

Sponges in the previously indicated inlet filters are very effective in removing hydrocarbons. For example, Smart-Sponge used in Ultra-Urban Filter is capable of (a) removing three times its own weight in hydrocarbons, (b) inhibiting growth of mildew and mold, and (c) transforming hydrocarbons into a stable solid per the EPA's toxicity characteristic leach procedure (TCLP). A modified type of Smart Sponge called Smart Sponge Plus is further capable of destroying bacteria on contact. Catch basin filters not only serve as low-cost BMPs (cost from $1200 to $2500 depending on the inlet size) but also are easy to maintain. Figure 6.1 shows the components of a FloGard+Plus filter and Figure 6.2 shows an Aqua-Guardian catch basin insert. Unlike some other inlet filters, upon entering Aqua-Guardian insert, storm water accumulates in the sediment chamber. As the insert fills, water flows through the locked filter screen standpipe and is dispersed over the filter media, where the sediment, petroleum hydrocarbons, phosphates, and heavy metals (such as zinc) are removed, before exiting the base plate.

Both AbTech Industries and Oldcastle® Stormwater Solutions manufacture other types of water quality devices. For example, AbTech makes a water quality device named Stormwater Antimicrobial Treatment Unit and Oldcastle® manufactures a water quality device commercially known as Dual-Vortex Hydrodynamic Separator, which is certified for 50% TSS removal by the NJDEP. Figure 6.3 shows various water quality devices manufactured by Oldcastle® Stormwater Solutions. Included in these devices is a FloGard+Plus insert for trench drains. For more information on the water quality devices manufactured by these companies, the reader is referred to their websites (http://www.abtechindustries.com) and (http://contactstormwater@oldcastle.com). BioClean Environmental Services of Oceanside, California, is another company that makes filters for trench drains. These filters are especially designed for high levels of hydrocarbons, oils, and grease. They also capture sediment and organics. Figure 6.4 shows a simple screen filter installed in a lawn inlet to capture leaves, grass cut debris, and coarse sediment. The filter is intended to prolong the useful life of an underground Cultec chamber placed under the lawn inlet.

The REM Geo-Trap™ catch basin filter insert by Revel Environmental Manufacturing and SWIMS (Storm Water Inspection and Maintenance Services, Inc.) of Discovery Bay, California, is a more recent product introduced to the market. This insert includes a nonreactive high-density polyethylene (HDPE) plastic construction with UV inhibitors of round, square, rectangular, or custom shape; media filter cartridges are available for the removal of sand, silt, litter, hydrocarbons, metals, and antifreeze. Filter media cartridges can be interchanged with the REM Triton Series. REM Geo-Trap inserts have standard inner and outer diameters of 4 in. (100 mm) and 12 in. (300 mm), respectively. The insert can be placed in 17 in. by 17 in. to 52 in. by 52 in. catch basins. Each filter weighs 1 lb, can trap 25 lb of oil, and costs $1000 on the average. Swim's REM Geo-Trap filter inserts were installed in 12 existing catch basins on the parking lot of South Philadelphia Sports Complex in 2010 (Aird, 2012).

In Pacifica, California (population 37,000), FloGard+Plus catch basin inserts were installed in 40 inlets in 2011. Because of proximity to the ocean, stainless steel mesh was installed in the inlets (Goldberg, 2012). Fabco Industries has installed over 3000 catch basin filter inserts in Nassau County, New York, alone. Inlet filters manufactured by Blocksom & Co. are made of natural fibers and can be used on grates and curb openings. These filters can be readily attached to inlets without

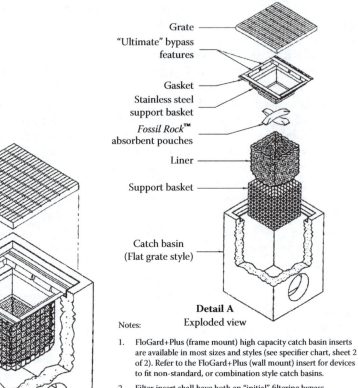

Grate
"Ultimate" bypass features
Gasket
Stainless steel support basket
Fossil Rock™ absorbent pouches
Liner
Support basket

Catch basin (Flat grate style)

Detail A
Exploded view

Notes:

1. FloGard+Plus (frame mount) high capacity catch basin inserts are available in most sizes and styles (see specifier chart, sheet 2 of 2). Refer to the FloGard+Plus (wall mount) insert for devices to fit non-standard, or combination style catch basins.

2. Filter insert shall have both an "initial" filtering bypass and "ultimate" high flow bypass feature.

3. Filter support frame shall be constructed from stainless steel Type 304.

4. Allow a minimum of 2.0 ft, of clearance between the bottom of the grate and top of outlet pipe(s), or refer to the FloGard insert for "shallow" installations.

5. Filter medium shall be *Fossil Rock*, installed and maintained in accordance with manufacturer specifications.

6. Storage capacity reflects 80% of maximum solids collection prior to impeding filtering bypass.

7. Filtered flow rate includes a safety factor of two.

FloGard+Plus® Filter
-Installed into catch basin-

U.S. patent # 6,00,023 & 6,877,029

FIGURE 6.1 FloGard+Plus inlet filter components.

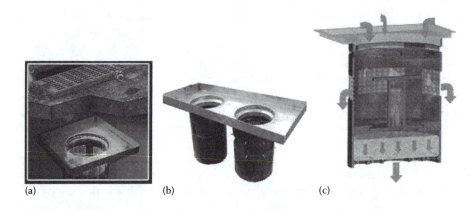

(a)   (b)   (c)

FIGURE 6.2 Aqua-Guardian™ catch basin insert. (a) Single insert, (b) double insert, and (c) cross section.

Oldcastle® Stormwater Solutions products for use from "course of construction through post construction"

Inlet Filtration & Trash Capture Products

FLOGARD+PLUS™

FloGard+PLUS Catch Basin Inserts and Trench Drain Filters remove sediment and debris, as well as petroleum hydrocarbons prior to runoff entering waterways.

Catch Basin Insert Filter

Trench Drain Filter

FLOGARD™

FloGard Downspout Filter is installed in commercial or industrial applications to remove pollutants normally found on building roofs.

Downspout Filter

FloGard Trash and Debris Guard is used to filter stormwater leaving sites with little fall from a parking surface, through a parkway culvert to curb and gutter outfall. Hydrocarbon removal is optional.

Trash & Debris Guard

FloGard T-Series Inlet Filter is an economical alternative catch basin insert for the collection of sediment and debris from stormwater runoff and other sources both during and after construction.

T-Series Catch Basin Insert Filter

NET TECH

Net Tech Gross Pollutant Trap for stormwater outlets to open channels.

Hydrodynamic Separation Products

DUAL VORTEX

Dual Vortex Hydrodynamic Separator, with integral "flow through" high-flow bypass, is an effective system for the removal/retention of sediment, debris and floatable pollutants from stormwater runoff.

Media Filtration Products

PERK FILTER™

PerkFilter is a media filtration device that captures and retains sediment, oils, metals and other target constituents from urban runoff and reduces the total discharge load. It is available in multiple configurations, including catch basins, vaults, and manholes, allowing the specifier maximum design flexibility.

Bioretention/Biofiltration & LID Products

BIOMOD™

BioMod Modular Bioretention System is a modular precast concrete biofiltration cell system which adds consistency to design and construction with features to enhance filter performance, structural integrity and reduce construction and on-going maintenance costs.

Oldcastle Stormwater Solutions
7921 Southpark Plaza • Suite 200
Littleton, CO 80120
Phone: (800) 579-8819 • Fax: (707) 524-8188
www.oldcastlestormwater.com
contactstormwater@oldcastle.com

TREEPOD™

TreePod Biofilter is a tree box filter design that is proven to be effective at the removal of ultra-fine and dissolved pollutants normally found in stormwater runoff. The TreePod offers flexible and economical designs which simplify the design and construction of a storm drain system, and its one piece precast construction assures ease of installation and a long service life.

SWALEGARD™

SwaleGard Grassy Swale Pre-filter improves filtration performance and service life of all vegetated drainage systems, filtering sediment, debris and free oils from runoff prior to entering a "green" BMP, such as a swale or pond.

Pre-Filter

SwaleGard Overflow Filter is a simple, effective and economical device that is designed to retain gross pollutants, such as trash, debris and coarse sediment, within a bioretention cell without impeding peak flow bypass needs.

Overflow Filter

FIGURE 6.3 Various Oldcastle® Stormwater Solutions water quality devices.

FIGURE 6.4 Lawn inlet screen filter.

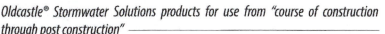

a need to remove grates. Blocksom filters come in 27 in. × 30 in. mats (69 cm × 76 cm), 27 in. × 21 ft (0.69 × 6.4 m) rolls, and 27 in. × 75 ft (22.9 m) rolls and these are uniformly 1.5 in. (3.8 cm) thick. Figure 6.5 shows a 27 in. × 30 in. mat. Blocksom inlet filters offer an economical means of removing silt, floatables, and debris and can be cleaned by sweeping the surface and sides. They are also durable, are not damaged by vehicular traffic, and are especially suited during construction. For more information, visit http://www.blocksom.com.

SNOUT is one of the oil–water separator hoods most widely employed in the United States. These hoods are manufactured by Best Management Practices Inc. of Lyme, Connecticut. Over 20,000 SNOUT hoods have been installed in various types of projects during the past decade to separate oil and other debris from water. The SNOUT is attached to a catch basin wall over any type of discharge pipe to trap floatables, trash, sediment and oil and grease. To improve hydraulics and to prevent contaminants from being drawn downstream, SNOUT is equipped with an antisiphon flow vent. A clean-out port is also provided for easy access to the pipe. SNOUT is made of marine-grade fiberglass, which is a strong, yet lightweight, plastic composite. These hoods are easy to install and less expensive to maintain than inlet filters. However, since they are hidden underground, SNOUTs are more vulnerable to maintenance neglect than catch basin inserts and are less effective in removing suspended solids and oil and grease. Figure 6.6 shows a SNOUT and its functioning. More information on SNOUT can be found at its website (http://www.bmpinc.com), which allows the user to size a SNOUT that suits his or her needs.

FIGURE 6.5   Blocksom filter 27 in. × 30 in. mat.

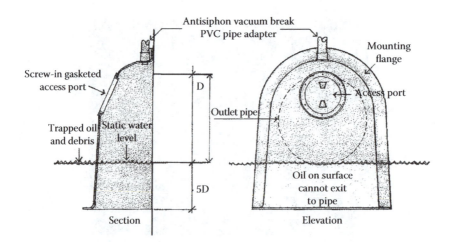

FIGURE 6.6   SNOUT water quality inlet hood.

6.3.2 HYDRODYNAMIC SEPARATION WATER QUALITY DEVICES

As indicated, a large number of water quality devices have been introduced to the market over the past 25 years. Table 6.3 is a partial list of the hydrodynamic separation water treatment devices, also referred to as hydrodynamic sedimentation devices, some of which are approved by NJDEP. These devices are most effective in removing heavy particles, floatables, and oils (in some devices) from runoff. However, they do not have media filter, which removes fine particulate and in part phosphorus and heavy metals. There are other devices recently introduced to the market that have not received certification from the NJDEP. Among these are StormSafe by Fabco Industries of Farmingdale, New York, and CrystalClean Separator by Crystal Springs Technologies of Lawrenceville, Georgia. The latter devices come in a single vault (model 1056) and twin vaults (model 2466). The single vault resembles Vortechs' system shown in Figure 6.7 and comes in seven sizes, ranging from 6 ft long by 4 ft wide (1.83 m × 1.22 m) to 12 ft long by 6 ft wide (3.66 m × 1.83 m). The height of all these units is uniformly 6 ft (compared to 7 ft for Vortechs'). Their capacities as specified by the manufacturer range from 6 cfs (0.17 m³/s) to 24 cfs (0.68 m³/s). The twin vault comes in four sizes, all 6 ft high with a maximum flow rate of 12 to 36 cfs (0.34–1.02 m³/s).

A number of water quality devices listed in Table 6.3. Some of these devices, including Aqua-Swirl and Stormceptor, separate suspended solids from inflowing flow by the process of swirl action and sedimentation. In Aqua-Swirl and Stormceptor, for example, the water enters the device through a tangential inlet pipe, which produces a swirl flow pattern that causes suspended solids to settle. The settlement occurs during each storm event and between successive storms. A combination of gravitation and hydrodynamic drag forces encourages the solids to drop out of the flow and migrate to the center of the chamber where the velocities are the lowest. During high flows, runoff entering Vort Sentry is directed into a treatment chamber through a secondary inlet allowing the capture of

TABLE 6.3
Partial List of Hydrodynamic Sedimentation Devices[a]

Product Name	Design Flow/Size	Manufacturer Name (Website Address)
Aqua-Swirl[a]	0.45–12.2	AquaShield, Inc. (http://www.aquashieldinc.com)
BaySeparator[b]		BaySaver Technologies, Inc.[b] (http://www.BaySaver.com)
CDS[c,d]	0.7–6.3 cfs	Contech Stormwater Solutions[d] (http://www.ContechStormwater.com)
Downstream Defender[a,c]	1.12–10.08	Hydro International/Water Quality Rocha (http://www.hydro-international.biz)
Dual Vortex Separator (DVS)		Oldcastle® Stormwater Solutions
HydroGard		Hydroworks, LLC
Nutrient Separator Baffle Box		Suntree Technologies
Stormceptor OSR		Imbrium Systems Corporation
Stormcepter STS[c]		Imbrium Systems Corporation
Terre Kleen[c]		Terre Hill Stormwater Systems (http://www.terrekleen.com)
StormPro	0.51–8.00	Environmental 21, LLC
Vortechs System[e]	0.63–10.1 cfs	Contech Stormwater Solutions (http://www.ContechStormwater.com)

Note: Only Aqua-Swirl, CDS, Downstream Defender, DVS, StormPro, and Vortechs System are approved by the NJDEP (as of October 2015).

[a] Approved flows by NJDEP as of October 2015.

[b] Partnered with Advanced Drainage Systems (ADS) in 2008.

[c] Approved for online (inline) or offline use by NJDEP (see Appendix 6C).

[d] CDS was acquired by Contech in 2008.

[e] Vortechs was acquired by Contech in 2005.

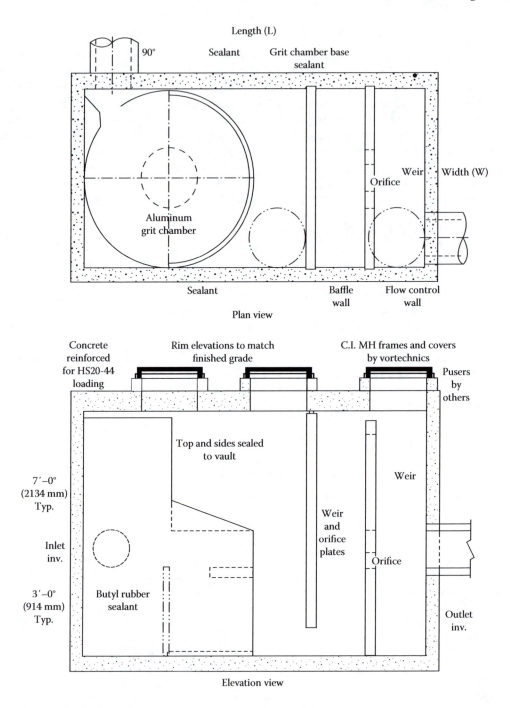

FIGURE 6.7   Plan and elevation views of Vortechs® from Contech Engineered Solutions.

floatables and debris. The bypass avoids a high velocity or turbulence in the treatment chamber, which helps prevent the washout of previously captured pollutants.

CDS water quality units include a stainless screen with 2400 µm mesh openings for effective removal of floatable and any sediment particle larger than 2.4 mm. Also, CDS units and some other devices can be equipped with a certain type of sponge for removing phosphate and metals.

The depth of CDS units and other swirl action water treatment devices varies with their treatment capacity and can be as deep as 20 ft. However, Vortechs® water quality structures as indicated are invariably 7 ft (2.13 m) deep, which is advantageous for large flow applications. Figure 6.7 provides plan and elevation views of Vortechs' systems and Table 6.4 lists dimensions of Vortechs' models, as approved by the NJDEP. The treatment capacity of Vortechs System, as reported by Contech, ranges from 1.6 cfs (45 L/s) for model 1000 to 25 cfs (0.72 m³/s) for model 16000. Tables 6.5 and 6.6 provide treatment capacity and dimensions of Dual Vortex Separator (DVS) devices manufactured by Oldcastle® Stormwater Solutions.

Figure 6.8 depicts dimensions and plan/section views of a CDS 4045 unit, which is rated for 7.5 cfs (0.21 m³/s). Because of a large range in treatment capacity, CDS units are used more than

TABLE 6.4
Vortechs' Models and Sizes

Model	Size (L × W) ft (m)[a]	Treatment Capacity, cfs (L/s)[b]	Grit Chamber Capacity ft³ (m³)
1000	9 × 3 (2.74 × 9.14)	0.63 (17.8)	7.1 (0.20)
2000	10 × 4 (3.05 × 1.22)	1.12 (31.7)	12.6 (0.36)
3000	11 × 5 (3.35 × 1.52)	1.75 (49.6)	19.6 (0.56)
4000	12 × 6 (3.66 × 1.83)	2.50 (70.8)	28.3 (0.80)
5000	13 × 7 (3.96 × 2.13)	3.40 (96.3)	38.5 (1.09)
7000	14 × 8 (4.27 × 2.44)	4.50 (127.4)	50.3 (1.42)
9000	15 × 9 (4.57 × 2.74)	5.70 (161.4)	63.6 (1.80)
11000	16 × 10 (4.88 × 3.05)	7.00 (198.2)	78.5 (2.22)
16000	18 × 12 (5.49 × 3.66)	10.1 (286.0)	113.1 (3.20)

[a] Rounded to second decimal place.
[b] Capacities as approved by the NJDEP.

TABLE 6.5
MTFRs and Required Sediment Removal Intervals for DVS Models

DVS Model	Manhole Diameter (ft)	Maximum Treatment Flow Rate (cfs)	Effective Treatment Area (sf)	Hydraulic Loading Rate (gpm/sf)	50% Max. Sediment Volume (cf)	Sediment Removal Interval (months)
DVS-36	3	0.56	7.07	35.7	5.30	67
DVS-48	4	1.00	12.57	35.7	9.42	67
DVS-60	5	1.56	19.63	35.7	14.73	67
DVS-72	6	2.25	28.27	35.7	21.21	67
DVS-84	7	3.06	38.48	35.7	28.86	67
DVS-96	8	4.00	50.27	35.7	37.70	67
DVS-120	10	6.25	78.54	35.7	58.90	67
DVS-144	12	9.00	113.10	35.7	84.82	67

Note: Sediment removal interval calculated using the *monthly* calculation in Section B, Appendix A of the NJDEP HDS protocol. In certain areas, DVS units are available in other diameters. Units not listed here are sized not to exceed a hydraulic loading rate of 35.7 gpm/sf and maintain an acceptable aspect ratio. Fifty percent sediment storage volume is equal to the effective treatment area × 9" of sediment. The maximum sediment storage volume occurs at 18" of sediment depths.

TABLE 6.6

Dimensional Overview for DVS Models

DVS Model	Manhole Diameter (ft)	Maximum Treatment Flow Rate (cfs)	Treatment Chamber Depth (ft)	Sediment Sump Depth (ft)	Total Depth below Inverts (ft)	Aspect Ratio (Dia/Depth)	50% Max. Sediment Volume (cf)	Oil Storage Capacity (cf)
DVS-36	3	0.56	3.00	1.50	4.50	1.00	5.30	6.07
DVS-48	4	1.00	3.50	1.50	5.00	0.88	9.42	15.08
DVS-60	5	1.56	4.50	1.50	6.00	0.90	14.73	28.63
DVS-72	6	2.25	5.50	1.50	7.00	0.92	21.21	48.54
DVS-84	7	3.06	6.50	1.50	8.00	0.93	28.86	79.21
DVS-96	8	4.00	7.50	1.50	9.00	0.94	37.70	116.45
DVS-120	10	6.25	9.00	1.50	10.50	0.90	58.90	225.80
DVS-144	12	9.00	10.50	1.50	12.00	0.88	84.82	388.30

Note: Treatment chamber depth is defined as the depth below the invert to the top of the sediment storage area (18" above the bottom of the unit). The aspect ratio is the unit's diameter/treatment chamber depth. The aspect ratio for the tested unit is 0.88. An aspect ratio of 0.88 or greater indicates that the treatment depth of the unit is proportional to or deeper than required based on the diameter-to-depth relationship in the tested model. An aspect ratio less than 0.88 would indicate insufficient treatment chamber depth. The detention time is the treatment chamber wet volume/JMTFR. The total wet volume includes the volume of the sediment sump.

many other devices for treatment of runoff in urban storm water management practices. Another reason for their high application is that CDS units are one of the least expensive treatment devices for a given discharge. The city of Laguna Beach in California, for example, selected CDS hydrodynamic separator as a most cost-effective BMP method for removing gross pollutants from runoff (Wieske and Penna, 2002). The city installed 3 cfs capacity CDS units offline to pretreat runoff before diverting it into the city combined sewer or returning it into the mainline storm drain. The author has used CDS water quality units on several projects, some of which are discussed in Chapter 7. Appendix 6D includes a table of various sizes of CDS water quality units. Figure 6.9 provides details of an Aqua-Swirl Model AS-5 CFD PCS (polymer coated steel) storm water treatment system.

Nonfilter media manufactured water quality devices are generally approved for 50% TSS removal by the NJDEP. StormVault is the only nonmedia filtration system that has been approved for 80% TSS removal in New Jersey. This device, which used to be offered by Contech Engineered Solutions, is now manufactured by Jensen Precast. Some manufacturers, such as BaySaver Technologies, Inc., claim 50% to 80% TSS removal for BaySeparator Stormwater Treatment System; however, this device is not approved by the NJDEP.

The majority of hydrodynamic separation water treatment devices settle sediment due to a swirling action. The Terre Kleen Stormwater Device by Stormwater Systems, a subsidiary of Terre Hill Concrete Products in Terre Hill, Pennsylvania, functions on the basis of settling of particles in a laminar flow over an inclined plate. This process is similar to that which has been practiced for many decades in the water treatment industry. To increase the area of contact, a number of parallel plates are placed in a chamber. The water enters the chamber from one or two pipes; as it fills the chamber and flows slowly upward through the plates, suspended matter slides downward and settles to the bottom. The clean water that rises to the top of the chamber exits from an outlet pipe. Figure 6.10 shows the filtration process of a Terre Kleen storm water treatment device. To remove hydrocarbons, oil absorption booms are added to the primary chamber. The booms float among the litter and settle when they become saturated with oil.

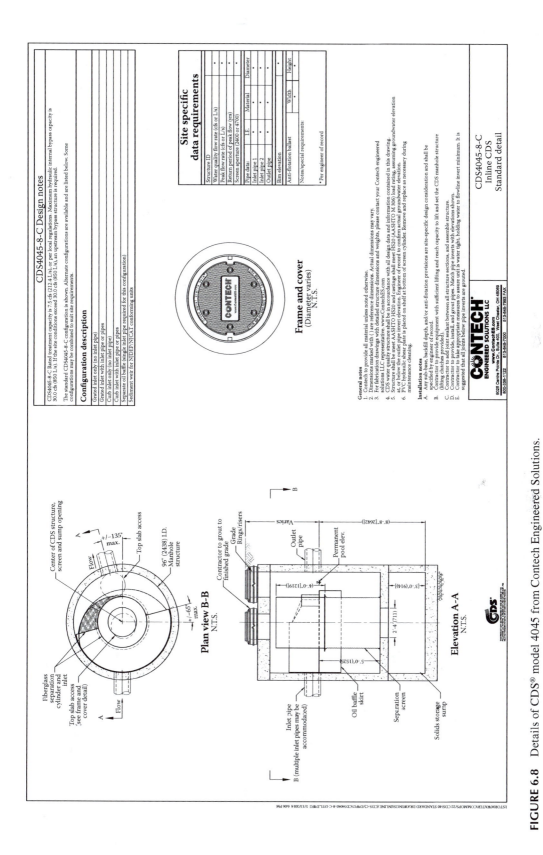

FIGURE 6.8 Details of CDS® model 4045 from Contech Engineered Solutions.

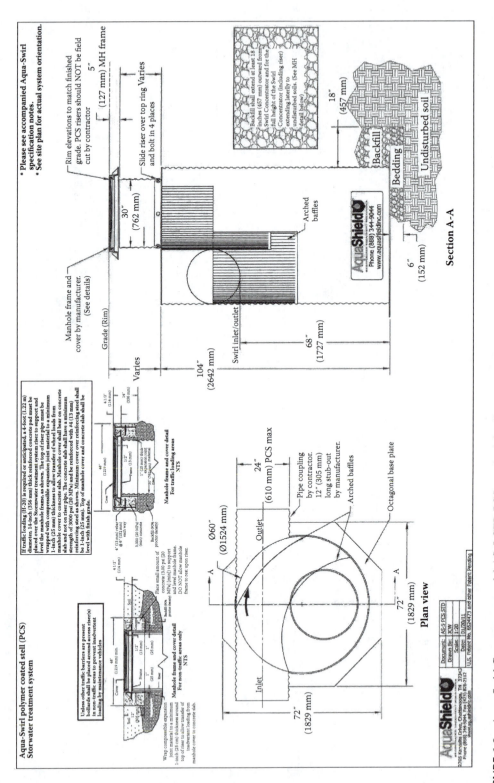

FIGURE 6.9 Aqua-Swirl Concentrator model AS-5 CFD PCS standard detail.

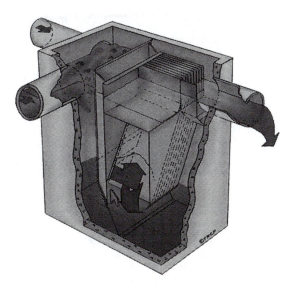

FIGURE 6.10 Terre Kleen™ stormwater treatment device.

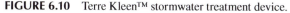

NJDEP used to accept water quality devices placed online (also referred to as inline). Each device was rated for certain water treatment capacity. Flows in excess of the capacity would be discharged through an internal overflow system. In early 2008, NJDEP modified the storm water device design requirement to avoid any pollutant washout—"scour"—during high flows. Specifically, NJDEP mandated that all storm water manufactured treatment devices were only approved as offline water quality devices until the department received new test data verifying that a device would not wash out silt-sized pollutants during intense rain events. Based on the test results, some of the hydrodynamic sedimentation devices listed in Table 6.3 are approved for online applications. CDS (high-efficiency continuous deflective separator) has been approved for online or offline application for sizes ranging from 4 to 12 ft in diameter. This device, as indicated, is rated for 50% TSS removal. Downstream Defender is also approved for online application. StormFilter (rated for 80% TSS removal) by Contech Construction Products Inc. was also accepted for peak diversion configuration by the NJDEP on September 1, 2010. The use of internal diversion in these indicated devices eliminates the need for a diversion structure, offering a compact design. It is expected that some other devices follow suit in being accepted for online use. Generally, the hydrodynamic sedimentation units are placed before detention basins, wet ponds, and, particularly, aboveground or underground infiltration and retention–infiltration basins. However, for underground detention systems (such as solid pipes), they may be placed past the detention system, so that they are sized for attenuated flow.

6.3.3 MEDIA FILTRATION WATER QUALITY DEVICES

To be approved for 80% TSS removal and to address the phosphorous removal requirement, a number of storm water treatment devices containing media filtration cartridges have been introduced to the market since the turn of the century. Media filtration water treatment devices are generally far more expensive than the catch basin inserts and hydrodynamic separation systems. They also require more maintenance and have significantly smaller flow capacity than other aforementioned devices. To be cost effective, media filter systems are generally placed after a detention basin or pond. Thus, they would be utilized as a post-treatment device for an attenuated flow rate. This type of application will be illustrated by a case study later in this chapter.

Calculations for the filter surface of a manufactured media filter (and sand filter, as well) are performed using Darcy's equation, as follows:

$$A_f = \frac{3600QD}{K(D+d/2)}$$ (6.1)

where
A_f = filter surface area m^2 (ft^2)
Q = average flow rate through the filter associated with the water quality storm, m^3/s (cfs)
K = hydraulic conductivity of the filter, m/h (ft/h)
D = thickness of filter medium, m (ft)
d = maximum water depth, m (ft)

The term $(D + d/2)/D$ in the preceding equation represents the average hydraulic gradient. The average flow rate, Q, is related to the design water quality volume, WQ_v, by

$$Q = \frac{WQ_v}{T}$$

where T denotes the drawdown time (or dewatering time) in the filter.

A number of the filter media devices are listed in Table 6.7. Also listed on this table are all the devices approved by the NJDEP for 80% TSS removal as of October 2015. A brief description of some of these devices is presented as follows.

Media filter devices are mostly approved for treatment of runoff for small flows and small pavement areas: 0.13 acre for Media Filtration System, 0.11 to 0.255 acres for Stormwater Management StormFilter (depending on cartridge height), 0.3 acre for Up-Flow Filter, and 0.7 acre for BayFilter.

TABLE 6.7
Media Filtration Water Treatment Devices

Product Name	NJDEP Approved Capacity	Manufacturer Website Address
BayFilter Stormwater Treatment[a] by BaySaver Technologies, Inc.	0.067 cfs (30 gpm) per cartridge	Advanced Drainage Systems (ADS) http://www.ads-pipe.com
AquaFilter™	0.037 cfs/ft^2 (16.5 gpm/ft^2) of filter area	AquaShield, Inc. http://www.aquashieldinc.com
Filterra Bioretention Systems	See Table 6.10	
Jellyfish® Filter	80 gpm Hi-Flow Cartridge (54 in. high)	Imbrium Systems Corp. http://www.imbriumsystems.com (now a part of Contech)
Media Filtration System (MFS)	0.04 cfs (18 gpm) per 22 in. cartridge	Contech Stormwater Solutions http://www.Contechstormwater.com
Storm Vault (offline)		Jensen Precast Inc. http://www.jensenprecast.com
Stormwater Management StormFilter	2.05 gpm/ft^2 3 models 12–27 in. high filters	Contech Stormwater Solutions http://www.Contechstormwater.com
Up-Flow Filter	0.0557 cfs (25 gpm) per filter; Max. area ≈ 0.66 acres	Hydro International Stormwater

[a] Not approved by the NJDEP.

An exception is Jellyfish, which is approved up to a 4.63 cfs flow rate. This implies that Jellyfish filter can treat runoff from up to nearly 2 acres of impervious surface.

StormFilter, manufactured by Stormwater Management, headquartered in Portland, Oregon, was the first filter media water treatment device introduced in the United States. This company changed its name to Stormwater 360 and was later acquired by Contech Stormwater Solutions, Inc., in 2008. Stormwater Management StormFilter is manufactured as a single unit or a linear grate storm filter unit. In a single unit, a number of cartridges (filters) are installed in a concrete chamber. The water enters the chamber, passes through the media filter, and starts filling the cartridge center tube. Air below the hood is purged through a one-way check valve as the water rises. When water reaches the top of the float, buoyant forces pull the float free and allow filtered water to drain. After the storm, the water level starts to drop until it reaches the scrubbing regulators. Air then rushes through the regulator, releases water, and creates air bubbles that agitate the surface of the filter media, causing accumulated sediment to drop to the vault floor. This surface cleaning mechanism helps restore the filter's permeability between storm events. StormFilter is manufactured in three sizes (heights): 27, 18, and 12 in., requiring nearly 3, 2.0, and 1.5 ft drop between the inflow and outflow, respectively. In comparison, the Media Filtration (formerly CDS Filter Media) system comes in 22 and 12 in. filters, requiring 2.3 and 1.5 ft drop. Jellyfish, however, requires less than an 18 in. head (drop between inflow and outflow) pipes regardless of its size (model). StormVault, manufactured by Jensen Precast, Inc., is the only storm water treatment device that requires practically no drop between the inflow and outflow pipes. However, this device is several times larger and far more expensive than nonmedia filter water quality devices.

The linear grate StormFilter consists of a multichamber catch basin unit that can hold up to 29 filter cartridges. This system receives sheet flow runoff through surface grates. It provides treatment with a shallow configuration, allowing treatment of runoff from a small drainage area where the available drop between the inflow and outflow is limited. Unlike this filter, the flow in many other filters, such as Media Filtration by Oldcastle® Stormwater Solutions, Up-Flo Filter by Hydro International, and Jellyfish filter cartridges from Imbrium Systems occurs in an upward direction.

Contech also manufactures Infiltration StormFilter structure. This structure includes a perforated manhole (like a dry well) at the bottom and an upper compartment that includes StormFilter cartridges. The water entering the structure is filtered by StormFilter units, drains down to a perforated manhole, and infiltrates into the ground. Figure 6.11 shows an Infiltration StormFilter unit. Contech also makes an oil–water separator device known as VortClarex.

Since 2005, Contech has acquired companies such as CDS and Vortechnics Stormwater Management 360. In 2013, Contech acquired Jellyfish® filter from Imbrium Systems, which also makes Stormceptor. StormVault, which as indicated is the only device with no filter media that is approved for 80% TSS removal by the NJDEP, was initially introduced by Contech. However, because of its large size and high cost, Contech stopped making this device a few years ago and now it is manufactured by Jensen Precast, Inc.

BaySaver Treatment Technologies manufactures BayFilter. This filter is rated for 30 gpm (115 L/min) per cartridge by the NJDEP and appears to be a cost-effective device both in terms of initial cost and maintenance. BayFilter has been acquired by Advance Drainage Systems.

Jellyfish, manufactured by Imbrium Systems Corporation of Rockville, Maryland, is another storm water treatment device that has received final certification for 80% TSS removal by the NJDEP. In this system, an insert deck divides the structure into an upper and a lower chamber. Water is treated in the lower chamber, which contains a permanent pool of water. Flow enters the lower chamber tangentially and directed around and under the cartridges in this chamber. Each 12 in. diameter cartridge consists of ninety-one 54 in. long filtration tentacles. Water is infiltrated in the filtration tentacles, flows upward, and enters into a backwash pool created by a 6 in. weir. During rainfall, filtered water overflows the weir and enters the outlet pipe. After the runoff

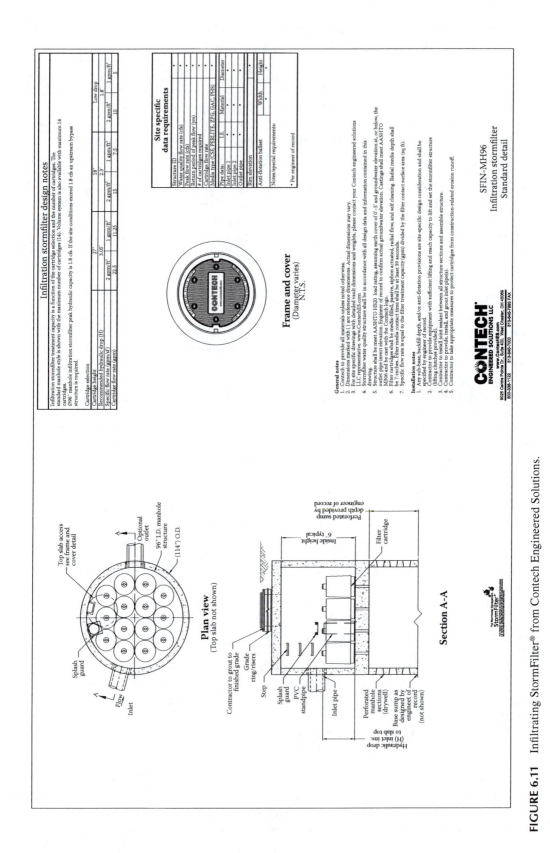

FIGURE 6.11 Infiltrating StormFilter® from Contech Engineered Solutions.

subsides, filtered water in the backwash pool drains down through the Jellyfish cartridge tentacles and brings the sediment that has accumulated on the tentacles to the sump. Water draining from the backwash pool displaces water from the lower chamber and the displaced water exits through the Jellyfish drawdown cartridge that is outside the backwash pool weir. Figure 6.12 depicts these three functions of a Jellyfish filter. Due to self-cleansing capability, Jellyfish filters are fairly easy to clean and less expensive to maintain than some other storm water filtering devices currently on the market.

The Jellyfish filter cartridges have the highest flow capacity of all currently available filter media water treatment devices. Each 54 in. (1.37 m) high cartridge is approved for 80 gpm treatment capacity by the NJDEP. As previously indicated, Jellyfish can treat direct runoff from up to approximately 2 acres of pavement. If, however, a Jellyfish filter is placed after a detention basin, it has the capacity to treat significantly larger impervious areas. Table 6.8 provides the models and the globally accepted treatment capacities of Jellyfish as specified by its manufacturer, Imbrium Systems. It is to be noted that the NJDEP-approved capacities of Jellyfish are the same as those shown in the table (visit http://www.nj.gov/dep/stormwater/treatment/html for the NJDEP approval). A comparison of the Jellyfish filter with other filter media water quality devices is presented in Table 6.9. In September 2009, Imbrium introduced a second-generation, more efficient storm water treatment device named Sorbtive Filter. This filter incorporates Imbrium's oxide-coated Sorbtive Media, which has a high phosphorus removal capability, including dissolved phosphorus—a major cause of the algae blooms—and reduces dissolved oxygen in ponds and bays. Sorbtive Filter has received approval from the Maryland Department of the Environment (MDE) for use in sand filters, bioretention cells and other microfilters throughout the state (Imbrium News Release, September 21, 2009). Imbrium Systems manufactures Stormceptor and Jellyfish.

Because of their high costs and small capacities, media filter manufactured treatment devices are installed after a detention system. As such, they need to be sized for attenuated outflow from the system, which is a fraction of the peak flow from the water quality storm. Such an application is exemplified in a case study in Chapter 7.

Examples of the filter media applications are as follows:

- San Diego International Airport. To address the storm water requirements for the rapid expansion of runways, terminal, and overnight parking for planes, the airport installed a storm water management StormFilter unit by Contech in 2010. This unit contains 179 cartridges in a 48 ft long × 24 ft wide (14.63 m × 7.32 m) concrete chamber. The StormFilter controls discharge of heavy metals, petroleum products that leak from airplanes and service vehicles, and any trash, litter, and debris into San Diego Bay (Aird, 2012).
- To address the NPDES permit, the city of Mount Dora, Florida, installed two Nutrient Separating Baffle Boxes and Skin Boss Upflow Filtration System and Sungate Damper by Suntree Technologies, Inc. in 2012. These two units treat storm water from 261 acres (105.6 ha). Each Nutrient Separating Baffle Box is 19 ft long × 13 ft wide (5.79 m × 3.96 m) and 12 ft (3.66 m) high. The Nutrient Separating Baffle Box has three components: a steel mesh basket, a baffle box, and the Skin Boss Filtration System. The steel mesh basket collects leaves and trash above the waterline in the water treatment device. After passing through the basket, silt and sediment drop into the baffle box and the water enters the filtration system, which captures hydrocarbons and nutrients. The system's hydro variant technology design enables the system to adjust automatically to changing water levels during storm events. The filtration media, which are called "Bold and Gold" and have been developed by the University of Central Florida, include recycled material. This material effectively removes phosphorus and nitrogen. The water that passes through the filter is discharged from the damper at the end of the unit (Aird, 2012).

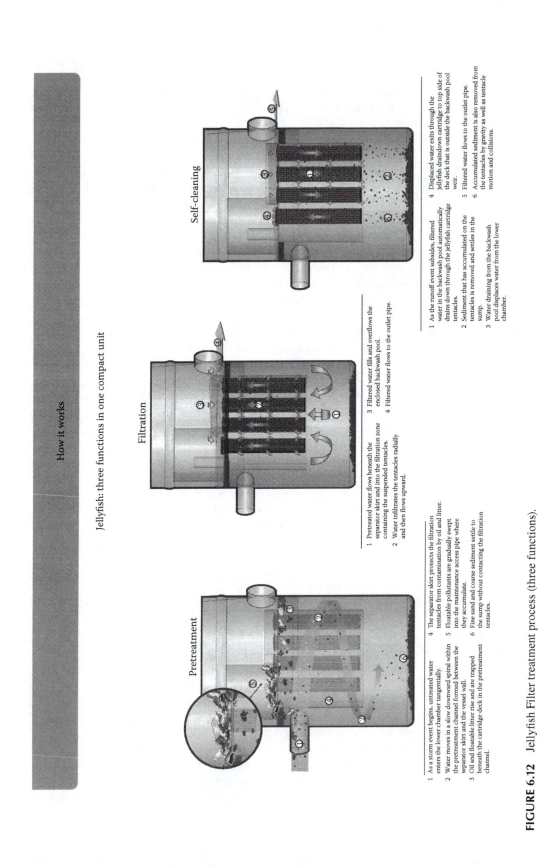

How it works

Jellyfish: three functions in one compact unit

Pretreatment

1. As a storm event begins, untreated water enters the lower chamber tangentially.
2. Water moves in a slow downward spiral within the pretreatment channel formed between the separator skirt and the vessel wall.
3. Oil and floatable litter rise and are trapped beneath the cartridge deck in the pretreatment channel.
4. The separator skirt protects the filtration tentacles from contamination by oil and litter.
5. Floatable pollutants are gradually swept into the maintenance access pipe where they accumulate.
6. Fine sand and coarse sediment settle to the sump without contacting the filtration tentacles.

Filtration

1. Pretreated water flows beneath the separator skirt and into the filtration zone containing the suspended tentacles.
2. Water infiltrates the tentacles radially and then flows upward.
3. Filtered water fills and overflows the enclosed backwash pool.
4. Filtered water flows to the outlet pipe.

Self-cleaning

1. As the runoff event subsides, filtered water in the backwash pool automatically drains down through the jellyfish cartridge tentacles.
2. Sediment that has accumulated on the tentacles is removed and settles in the sump.
3. Water draining from the backwash pool displaces water from the lower chamber.
4. Displaced water exits through the jellyfish draindown cartridge to top side of the deck that is outside the backwash pool weir.
5. Filtered water flows to the outlet pipe.
6. Accumulated sediment is also removed from the tentacles by gravity as well as tentacle motion and collisions.

FIGURE 6.12   Jellyfish Filter treatment process (three functions).

TABLE 6.8

Maximum Treatment Flow Rates for Standard (54 in. Cartridge Length) Jellyfish Filter Models

Manhole Diameter (ft/m)[a]	Model No.	Hi-Flo Cartridges (54 in. Length)	Draindown Cartridges (54 in. Length)	Maximum Treatment Flow Rate (gpm/cfs)
Catch Basin		Varies	Varies	Varies
4	JF4-2-1	2	1	200/0.45
6	JF6-3-1	3	1	280/0.62
	JF6-4-1	4	1	360/0.80
	JF6-5-1	5	1	440/0.98
	JF6-6-1	6	1	520/1.16
8	JF8-6-2	6	2	560/1.25
	JF8-7-2	7	2	640/1.43
	JF8-8-2	8	2	720/1.60
	JF8-9-2	9	2	800/1.78
	JF8-10	10	2	880/1.96
10[a]	JF10-11-3	11	3	1000/2.23
	JF10-12-3	12	3	1080/2.41
	JF10-13-3	13	3	1160/2.58
	JF10-14-3	14	3	1240/2.76
	JF10-15-3	15	3	1320/2.94
	JF10-16-3	16	3	1400/3.12
12[b]	JF12-17-4	17	4	1520/3.39
	JF12-18-4	18	4	1600/3.57
	JF12-19-4	19	4	1680/3.74
	JF12-20-4	20	4	1760/3.92
	JF12-21-4	21	4	1840/4.10
	JF12-22-4	22	4	1920/4.28
	JF12-23-4	23	4	2000/4.46
	JF12-24-4	24	4	2080/4.63
Vault		Varies	Varies	Varies

[a] The MTFR for a 10 ft diameter unit occurs with model JF-10-16-3. Since this leaves four unoccupied cartridge receptacles in the 10 ft diameter deck, the design engineer has the option to add up to four additional cartridges to increase the sediment capacity of the system. However, the MTFR may not be increased above that of the JF10-16-3.

[b] The MTFR for a 12 ft diameter unit occurs with model JF12-24-4. Since this leaves four unoccupied cartridge receptacles in the 12 ft diameter deck, the design engineer has the option to add up to four additional cartridges to increase the sediment capacity of the system. However, the MTFR may not be increased above that of the JF12-24-4.

- In Lake Tahoe at the border between the states of California and Nevada, the loss of clarity of this blue water is a major issue. The sediment, which is finer than 16 μm and is generated from heavily used roads and developments in the Lake Tahoe Basin, is the primary cause of turbidity. An area along Route 28 in Incline Village was one of the areas selected for runoff treatment. A Jellyfish filter from Imbrium was installed in this area. The selection of this device was based on its smaller footprint and lighter cartridges than other water treatment devices. The use of this device has reportedly stabilized the loss of clarity of water, which was continually a problem over the past 30 years (Goldberg, 2012).

TABLE 6.9
Proprietary Filtration System Comparison[a]

System Parameters

Verified Filter Technology	Cartridge Surface Area (ft²)	Cartridge Rated Flow (gpm)	Flow Rate per Surface Area (gpm/ft²)	Cartridge Diameter (in.)	"Footprint" Cartridge Plan View (ft²)	Wet Cartridge Weight (lb)	Min. Head Required (in.)	Full Flow Heal Required (in.)
StormFilter	11.25	22.5	2.0	19	2.0	75–250	18–33	18–33
Jellyfish filter	381	80	0.21	12	0.8	50	6	18
BayFilter	43	15–30	0.52	26	3.7	400	28	40
UpFlo Filter	≈1.1	25	22.7	Pie wedge	1.1	80	20	31
AquaFilter	4	20	16.5	2 ft × 2 ft	4.0	50	18	24

[a] NJDEP certified storm water manufactured treatment devices.

A number of other water treatment devices, which have yet to achieve a use designation through TAPE program, have been introduced to the market in the past year or two. One such device is BioStorm® storm water treatment system by Bio-Microbics of Shawnee, Kansas. This device comes in six different sizes ranging from 0.5 to 10 cfs (14–280 L/s) capacity. Figure 6.13 shows BioStorm 1.5, which is rated for 1.5 cfs (42 L/s).

6.4 BIORETENTION CELLS

Bioretention devices, which have been introduced to the market recently, not only function as water treatment devices, but also serve as small retention–infiltration structures. Because of such difference, this type of manufactured device is discussed separately from other devices. A bioretention structure (cell) typically includes a 2.3–3.0 ft (0.8–1.0 m) thick engineered soil mix, a filter element, and plants. The engineered soil mix generally consists of 50% construction sand, 30% top soil, and 20% organic matter by volume.

Bioretention cells can be coupled with catch basins to collect runoff from street gutters and parking lots. In this application, bioretention cells may be planted with salt-tolerant grasses and plants to survive road salt. Experiments at the University of Maryland indicate that bioretention cells are effective in reducing the flow, trapping trash and floatables from runoff, and lowering suspended solids, phosphorus, and metal concentrations.

Filterra Bioretention System appears to be the first such cell manufactured in the United States. This system was developed by its parent company, Americast, in 2000. In 2008, the city of Virginia Beach in Virginia installed the Bacterra™ System from Filterra Bioretention Systems at Mount Trashmore to reduce the amount of bacteria and pollutants discharging into the Lynnhaven River watershed. Filterra Bioretention System, as indicated previously, was the only bioretention cell that has been approved for 80% TSS removal by the NJDEP as of July 2014. Filterra is now a part of Contech Engineered Solutions. Figure 6.14 depicts a Filterra bioretention system and Figure 6.15 shows its cross section. Table 6.10 lists the treatment capacities of Filterra boxes as approved by the NJCAT on August 23, 2013. The listed capacities in the table are expressed in terms of the pavement area being treated.

Contech used to manufacture a bioretention cell, named Urban Green BioFilter. This system includes a biofiltration bay and storm water management StormFilter (a media filter) to collect and treat the runoff from streets and parking areas. Urban Green BioFilter used to be employed to catch curb and gutter flow or as an area drain in parking lots. Since acquiring Filterra, which is approved for 80% TSS removal in New Jersey, Contech has stopped making urban Green Biofilter.

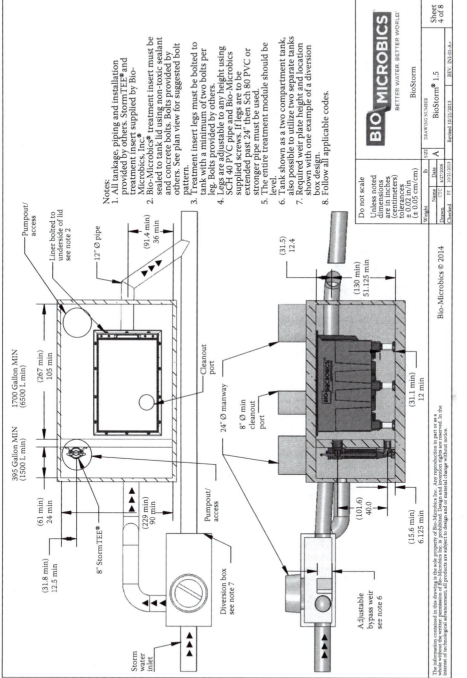

FIGURE 6.13   BioStorm 1.5 by Bio-Microbics bioretention cells.

(Continued)

BioStorm® specifications

I. Stormwater treatment system

The stormwater treatment system shall be a BioStorm® stormwater treatment system. The BioStorm® treatment system shall consist of a Diversion Structure that collect the first flush of stormwater and the remaining flow up to the design flow capacity of the stormwater treatment system. Following the Diversion Structure there shall be a settling tank with an effluent screening device referred to as the StormTEE®. This settling tank shall be used to collect the large settling debris and the floating objects that are present in the first flush of the stormwater. Following the first cell there shall be a second cell that includes the BioStorm® solids and floatable oil recovery system. The entire system of the Diversion Structure and the two cells make up the stormwater treatment system.

II. Diversion structure

The Diversion Structure shall be a concrete vault designed to be installed in the stormwater outfall pipe. This concrete vault shall contain a specially designed concrete weir that directs the first flush into the first cell of the BioStorm® treatment system. The weir shall be specifically designed so that the height of the weir directs the flow to the first cell and provides the correct hydraulic flow to the rest of the BioStorm® treatment system. The excess flow passes over the weir and then passes through the Diversion Structure into the outfall piping which takes the excess flow to the receiving stream or discharge point.

III. BioStorm® first cell

The first cell of the BioStorm® shall be a concrete vault used to settle the large debris and floatables generally present in the first flush. The material captured in the first cell is generally the street litter and trash, such as cans, bottles, paper, plastic cups, leaves and lawn waste, normally present in the first flush of stormwater. Included at the discharge end of the first cell shall be a StormTEE® screen which is used to filter the effluent from the first cell through the 9.5 mm (3/8 inch) slots. The hydraulic capacity of the StormTEE® shall limit the flow through the first cell into the second cell. The StormTEE® shall cause the excess flow to be backed up into the Diversion Box where it overflows the weir into the outfall pipe.

The StormTEE® shall have a manual plunger which can be used in between storm events to clean the surface of the StormTEE®. The StormTEE® shall be constructed from all noncorrosive plastic materials designed to withstand the forces from the high flow rates and the debris that is customary in stormwater. There shall be no moving parts or electrical requirements needed for this screening device.

The solids which settle out in the first cell shall be easily removed on a periodic basis by means of a vacuum truck. The floatables which also are retained in the first cell shall also be easily removed with the vacuum truck on a periodic basis as required.

IV. BioStorm® second cell

The second cell of the BioStorm® system shall receive the filtered effluent from the first cell. The second cell shall be a concrete vault that contains the BioStorm® solids and floatable hydrocarbon recovery device. The BioStorm® recovery device shall include a plastic liner which contains a honeycomb media that is used to settle fine soil particles. The BioStorm® recovery device shall also be designed so that the honeycomb media will act as an oil coalescer which allows hydrocarbon particles to attach to the media and grow in size so that they will float to the surface within the upper zone of liner. The BioStorm® recovery device liner shall extend above the operating water level so that the hydrocarbons that float to the surface are contained within the liner.

The effluent from the first cell shall enter the second cell so that the flow is directed around the BioStorm® recovery liner to reduce the velocity of the incoming flow. The flow shall then enter the BioStorm® recovery system at the front of the liner and flow through the honeycomb media to the opposite end of the liner. The honeycomb media shall be constructed of polypropylene, oleophilic sheets which are corrugated and angled at 60 degrees from the vertical. The projected surface shall be a minimum of 15 m²/m³. The effluent from the recovery system shall be collected from a point below the static water level to allow the floatable hydrocarbons to remain on the surface inside the liner. The honeycomb media shall be designed with vertical height adjustable legs that allow the solids which settle off the honeycomb media to settle out under the bottom of the liner in an undisturbed zone in the second cell. There shall be no moving parts or electrical requirements needed for the solids and floatable hydrocarbon recovery device.

The effluent from the recovery liner shall exit the second cell of the BioStorm® system through the effluent pipe which extends outside of the concrete vault. The effluent pipe shall then tie into the stormwater outfall piping to the receiving stream or discharge point.

The solids which settle out in the second cell and under the recovery liner shall be easily removed on a periodic basis by means of a vacuum truck. The floatable hydrocarbons which are retained in the recovery liner shall be easily removed with the vacuum truck on a periodic basis as required.

BIO MICROBICS
BETTER WATER. BETTER WORLD™
BIOSTORM

Do not scale		Drawing number	Sheet 8 of 8
Unless noted dimensions are in inches [centimeters] tolerances ± 0.02 in/in [± 0.05 cm/cm]		Specifications	

Weight		lb	Size A
	Name	Date	
Drawn	PF	4/27/2006	
Checked	PF	10/21/2013	Revised 10/21/2013 Rev.

BIO-MICROBICS © 2014

FIGURE 6.13 (CONTINUED) BioStorm 1.5 by Bio-Microbics bioretention cells.

FIGURE 6.14 Filterra® stormwater bioretention filtration system.

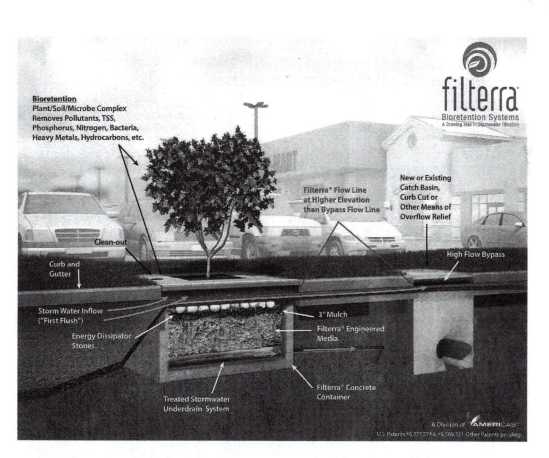

FIGURE 6.15 Cross section of Filterra® bioretention.

TABLE 6.10

Filterra New Jersey Sizing Table

Available Filterra Box Sizes (ft)	Total Contributing Drainage Area (Acres)	Outlet Pipe
4 × 4	<0.09	4 in. PVC
4 × 6 or 6 × 4	>0.09 to 0.13	4 in. PVC
4 × 8 or 8 × 4	>0.13 to 0.17	4 in. PVC
6 × 6	>0.17 to 0.19	4 in. PVC
6 × 8 or 8 × 6	>0.19 to 0.26	4 in. PVC
6 × 10 or 10 × 6	>0.26 to 0.32	6 in. PVC
6 × 12 or 12 × 6	>0.32 to 0.39	6 in. PVC
7 × 13 or 13 × 7	>0.39 to 0.49	6 in. PVC

Note: For approximate sizing only. All boxes are a standard 3.5 ft depth (INV to TC). A standard PVC pipe coupling is cast into the wall for easy connection to discharge piping. Dimensions shown are internal. Please add 1 ft to each for external (using 6: walls). For C = 0.95/CN = 98, where lower values are used, please contact Filterra. This sizing table is valid for New Jersey following NJDEP water quality design storm event of 1.25 in. in 2 h (NJAC 7:8-5.5(a)). Filterra infiltration rate 140 in./h. This sizing is scalable and equates to a ratio of filter surface area/drainage area = 0.0042 (0.42). EPA = SWMM 5 model used to create this sizing table. Please contact Filterra for sizing tables for other large treatment goals.

[a] C = 1.0.

CASE STUDY 6.1

This case represents storm water management provision for a parking lot expansion in the borough of Paramus, Bergen County, New Jersey.

PROJECT DESCRIPTION

Figure 6.16 depicts the pre- and postimprovement layouts of the parking lot. Shown on this map are the limits of parking lot expansion, a pre-existing paved basketball court that was to be removed, and the drainage facilities within the immediate vicinity of the site. The parking lot drains to an 18-in. RCP to the west and an 18-in. CMP to the east of the lot. According to the map, the net increase in impervious area measures to be 3020 ft^2 (280 m^2).

RETENTION SYSTEM CALCULATIONS

To offset the increased runoff, a retention system was provisioned to fully retain the runoff from a large portion of the expanded parking area. This system receives runoff from 4150 ft^2 (386 m^2) of pavement, which is larger than the net increase in impervious area. The provisioned system did not alter the pre-existing drainage pattern in that the runoff from the site would continue to be discharged by the 18-in. pipes to the wooded area north of the parking lot.

The retention system was designed for a 10-year, 60-minute storm event. In New Jersey, this storm has 2 in./h intensity and amounts to 2 in. of rainfall. The system comprises 60 linear feet (18.3 ft) of 24-in. perforated HDPE pipe in a 6 ft by 4.58 ft (1.83 × 1.4 m) stone trench. Figure 6.17 shows the cross section of the system. A 6-in. HDPE overflow pipe is provisioned

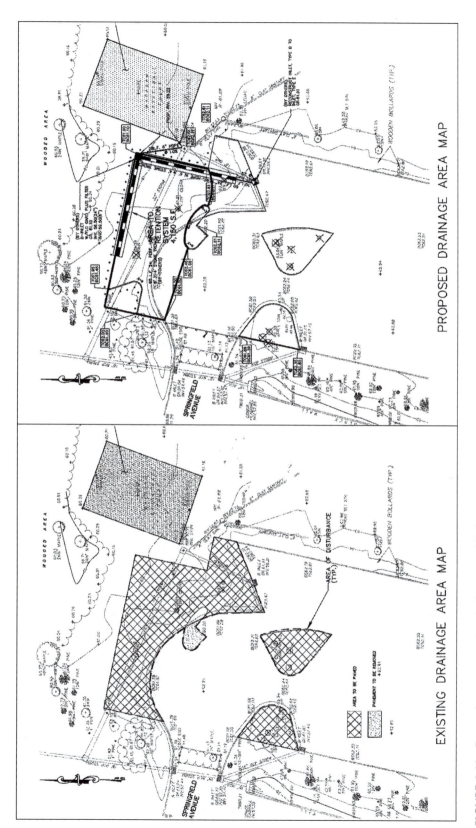

FIGURE 6.16 Pre- and postconstruction site layouts and D.A. maps.

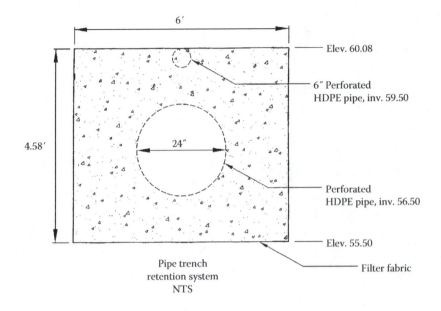

FIGURE 6.17   Pipe trench retention system.

for discharge of storms in excess of the retention capacity of the system. Calculations for the runoff volume and storage volume of the retention system are presented as follows:

1. Runoff volume = area × C × rainfall depth = 4150 × 0.95 × 2/12 in./ft = 657 ft³ (18.60 m³), where 0.95 represents the runoff coefficient (ratio of runoff to rainfall volume)
2. Retention storage volume:

$$\text{Pipe vol.} = \left(\pi \times \frac{2^2}{4} \right) \times 60 = 185 \ \text{ft}^3 \ (5.24 \ \text{m}^3)$$

$$\text{Void vol.} = \left(6 \times 4 - \pi \times \frac{2.25^2}{4} \right) \times 60 \times 0.4 = 481 \ \text{ft}^3 \ (13.62 \ \text{m}^3)$$

Total storage = 188 + 481 = 669 ft³ (18.94 m³) > 657 ft³ (18.60 m³)

In these calculations 2.25 ft and 4 ft represent the outer diameter of the pipe and the effective depth of retention system below the 6-in. overflow pipe respectively.

3. Water quality and maintenance provisions:

The project was subject to no water quality criteria. However, to avoid silt, leaves, oils, and debris causing premature clogging of the retention trench, the new catch basin was equipped with a FloGard+Plus Filter. In addition to prolonging the useful life of the retention system, the filter would also improve the quality of storm water runoff.

During a site visit, we observed that silt had accumulated in front of the headwall and within the existing 18 in. RCP on the westerly side of the project area. To improve drainage conditions, we recommended that the drainage system be cleaned. We also recommended that the inlet filter be cleaned four times annually.

CASE STUDY 6.2

This case relates to a commercial development project in a ±1.45 acres parcel of property in Starbuck Island, which is a part of Village of Green Island in Albany County, New York. A large portion of the property lies within the floodplain of the Hudson River.

PROJECT DESCRIPTION

The predevelopment site layout is shown in Figure 6.18. There existed a one-story building and concrete pavement on the northwest corner of the site, and the remainder was covered by woods and grass. The site had no drainage system and, due to topography, the runoff from the site used to flow in all directions, entering into the private properties to the north and south, the Hudson River to the east, and Osgood Avenue to the west. A portion of the site was also draining to a depression at the center of the site.

The proposed project included the construction of an office building, a self-service car wash, and associated parking stalls and driveways. Figure 6.19 shows the proposed site layout and the limits of the area to be disturbed by the project. According to the FEMA flood insurance rate map of Green Island, dated June 4, 1980, the 100-year flood lies at elevation 27.0 ft at the site. The site of the office building, the car wash, and all parking stalls was raised above the 100-year flood elevation by constructing a retaining wall along portions of northerly and easterly property lines and placing fill behind walls.

DRAINAGE DESIGN

The runoff from the driveway and the parking areas is collected by a drainage system consisting of four inlets and two manholes and is routed through an underground detention basin comprising 290 ft (88.4 m) of 30 in. (±750 mm) HDPE pipe. The roof runoff from the office building is discharged through roof leaders onto the undisturbed woodland behind the building.

It is to be noted that the New York Department of Environmental Commission (DEC) does not require a detention system to regulate the peak rates of runoff to the Hudson River. However, the objective of the proposed detention system is to attenuate the peak flows to reduce the size of StormFilter, a manufactured water quality device, which is designed to meet with the NYDEC water quality standards.

Consistent with the New York State drainage standards, the drainage system is designed based on the 10-year frequency storm. Flow calculations are performed based on the rational formula using runoff coefficients of 0.3 for pervious areas and 0.95 for impervious surfaces. The system collects runoff from the entire 0.85 acre (3440 m^2) paved area, routes it through a detention system and discharges on the overbanks of the Hudson River. Calculations for outlet control protection are performed based on the "New York Guidelines for Urban Erosion and Sediment Control" (1999), conservatively neglecting any attenuation effect of the detention basin. Though calculations indicate no need for a riprap, a 5 ft × 5 ft (1.5 m × 1.5 m) stone spreader is provisioned at the outfall to disburse the runoff.

WATER QUALITY DESIGN

Water quality calculations are performed based on the NY state water quality standards, which, as described in a previous chapter, call for treatment of the runoff volume generated during the water quality storm, which is 90% of the average annual rainfall (NY State Stormwater Management Design Manual, 2003, effective at the time of design). Based on the water quality map of New York State (Figure 5.1 in Chapter 5), the 90% rainfall event at the site is approximately 1 in. (25 mm). For this rainfall event, the water quality runoff volume, denoted as WQ_v, is calculated based on 1.025 acres (±4148 m^2), namely the entire area of disturbance due to the project. This area includes

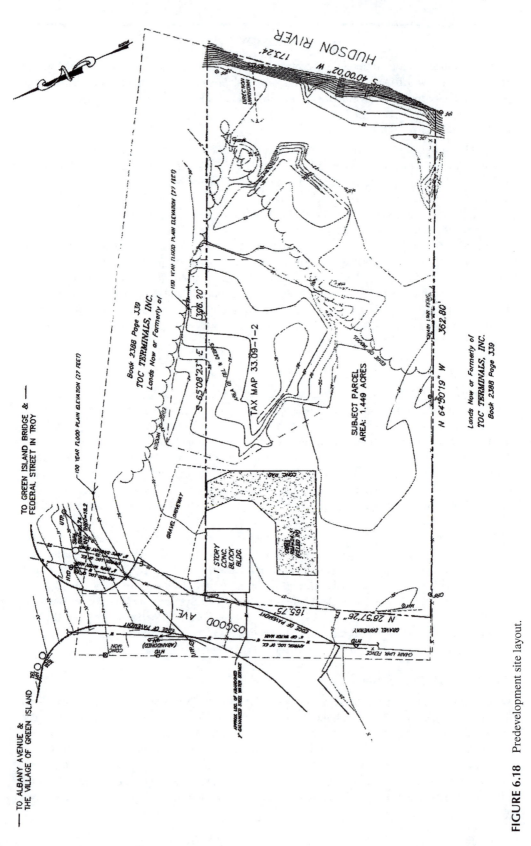

FIGURE 6.18   Predevelopment site layout.

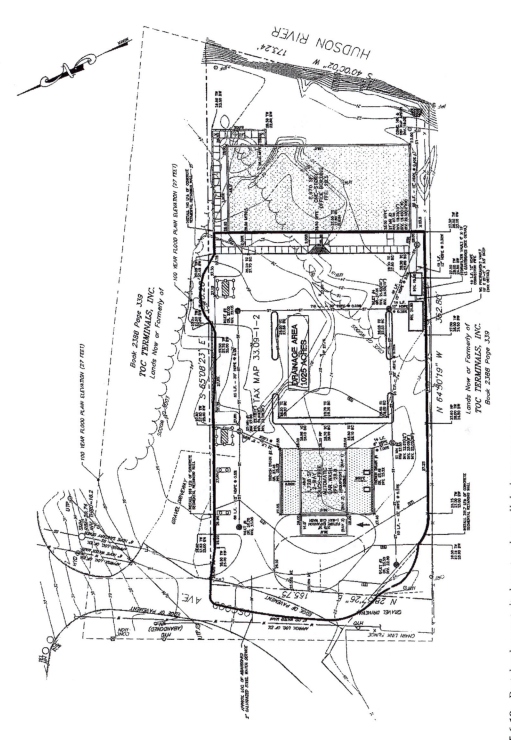

FIGURE 6.19 Post development site layout and overall drainage area map.

0.85 acre (±3440 m²) of impervious cover, including pavements and building roof and 0.175 acre (±708 m²) of lawn. The calculations are presented as follows:

$$WQ_v = \frac{(PR_v A)}{12}$$

$$A = 1.025 \text{ acres}$$

$$P = 1 \text{ in.}$$

$$I = \frac{(0.95 \times 0.85 + 0.3 \times 0.175)}{1.025} = 0.82 \quad \text{impervious ratio}$$

$$R_v = 0.05 + 0.009 \times 1 = 0.788$$

$$WQ_v = 0.0673 \text{ ac-ft}$$

$$2930 \text{ ft}^3$$

To treat this volume of runoff, a water quality system comprising a water quality pretreatment chamber and a StormFilter is provisioned. The pretreatment structure is shown in Figure 6.20. This structure consists of a 21 ft, 2 in. (9.94 m) long reach of a 10 ft wide by 3.5 ft high (3.28 m ×

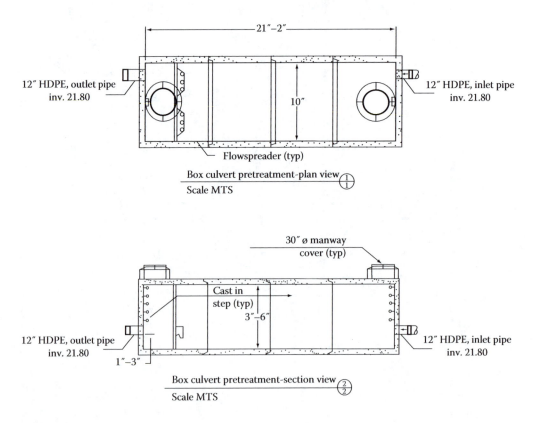

FIGURE 6.20 Pretreatment chamber detail.

1.15 m) box culvert providing 740 cf (21 m^3) of storage capacity and as such it meets with the NYDEC Stormwater Management Design Manual (2003), which required a minimum pretreatment storage of 25% of WQ_v. Calculations for the storage volumes of the pretreatment chamber and the detention system, which comprised 290 lineal feet (98.4 m) of 30 in. (750 mm) pipe, are presented as follows:

- Pretreatment chamber $V = 22.167 \times 10 \times 3.5 = 740$ ft^3 (21 m^3)
- Detention pipe $V = 290 \times (\pi \times 2.5^2/4) = 1424$ ft^3 (40.3 m^3)
- Four inlets/manholes = $4 \times 4 \times 2.5$ ft (deep); each $V = 160$ ft^3 (4.53 m^3)
- Total volume = 1424 + 160 = 1584 ft^3 (44.85 m^3)
- 25% of $WQ_v = 733$ ft^3 (20.76 m^3)

The water quality storm is routed, in its entirety, through the water treatment system, namely the pretreatment chamber and the StormFilter. Larger storms are bypassed through a 15 in. (375 mm) overflow pipe located above the crown of the detention pipes. The design of StormFilter was coordinated with Stormwater Management, Inc. (which has since been acquired by Contech). Figure 6.21 presents design calculations provided by the manufacturer. This table indicates that a StormFilter with 13 cartridges is required to treat the water quality storm; each cartridge has 7.5 gpm treatment capacity.

It is to be noted that StormFilter had not been approved by the NYDEC in February 2005, when the project was designed. However, since the StormFilter was approved for 80% TSS removal in New Jersey, the NYDEC accepted its use through a reciprocity program.

StormFilter sizing based on the NYDEC design methodology

| Project Name: | Center Island South, LLC | | Inpute | XXX |
| Date: | 2/10/2005 | | Result | XXX |

Site characteristic input

		System design	
Design storm, P (inches)	1.00		
Total area, A_T (acres)	1.03	Water surface elevation (ft)	4.50
Impervious area, A_I (acres)	0.84		
Percent of WQv to be temporarily stored in system	75%	**Volume StormFilter**	
		Vault size	8 × 16
WQ$_v$ calculations		Live storage volume provided (cu ft)	375
Percent impervious cover, I	82%		
Volumetric runoff coefficient, RV	0.79	**Pretreatment structure**	
Water quality volume, WQ$_v$ (ac-ft)	0.067	Vault size	10 × 12^{-2}
Water quality volume, WQ$_v$ (cu-ft)	2930	Total storage provided (cu ft)	740
Required live storage volume (75% of WQ$_v$) (cu ft)	2198	Live storage volume (cu ft)	425
Required pretreatment volume (25% of WQ$_v$) (cu ft)	733		
		Storage summary	
StormFilter design constants (per MDE manual)		Total live storage provided in system (cu ft)	798
Filter bed depth, D$_f$ (ft)	0.58	Additional live storage required (cu ft)	1574
Coeff. of perm. of filter media, k (ft/day)	8.7	Diameter of storage pipe (in)	30
Avg. height of water above filter bed, H$_f$ (ft)	0.75	Length of storage pipe required (ft)	290
Design filter bed drain time, T$_f$ (days)	1.67		
Surface area of stormfilter cartridge (sq ft)	7.1		
		**All volumes are based upon 4.5' water surface elevation	
Sizing calculations			
Surface area of equivalent filter bed (sq ft)	88.0		
Number of filter cartridges required	13		
Maximum filtration rate (cfs)	0.22		

FIGURE 6.21 StormFilter sizing calculations. (Provided by Stormwater Management, Inc.)

PROBLEMS

6.1 When would you use a manufactured treatment device?

6.2 A project in New Jersey is required to treat a water quality flow of 1.7 cfs for 50% TSS removal. What manufactured treatment device(s) satisfy this requirement?

6.3 What would be the answer to Question 6.2 for a discharge of 0.05 m³/s?

6.4 Select an MTD that can provide 80% TSS removal for 1.5 cfs water quality runoff.

6.5 Select a Jellyfish filter rated for 0.03 m³/s treatment.

6.6 Select an MTD to provide 80% TSS removal for 0.5 cfs water quality runoff.

6.7 Calculate the required storage of a detention system to attenuate the peak water quality from 1.5 cfs to 0.5 cfs. Base your calculations on modified rational method for a storm of 30-minute duration and a time of concentration of 15 minutes. If 48 in. solid pipes are used to provide the storage, what would be the required length of the pipe?

6.8 A commercial project results in 0.5 acre of new pavement and 0.4 acre of milled and resurfaced existing pavement. Calculate the required TSS removal and peak water quality runoff for this project based on the NJDEP Stormwater Management Rules. Base your calculations for a 10-minute time of concentration and select an MTD to address the water quality requirements for the project.

6.9 Solve Problem 6.8 for 2000 m² of new pavement and 1500 m² of repaved surfaces.

6.10 If the existing pavement in Problem 6.8 has to be reconstructed, what will be the required TSS removal? What MTD(s) would you choose to address the TSS removal requirement?

6.11 Solve Problem 6.9 if the existing pavement has to be reconstructed. Base your selection using the manufacturer's data.

6.12 To reduce the size of the MTD in Problem 6.10, the runoff is routed through an underground detention system that comprises solid pipes that would reduce the peak water quality runoff by 70%. What will be the peak water quality flow to be treated for TSS removal and which device(s) can you use in this case?

APPENDIX 6A: VERIFICATION OF TERRE KLEEN BY EPA AND NSF

THE ENVIRONMENTAL TECHNOLOGY VERIFICATION PROGRAM

U.S. Environmental Protection Agency   NSF International

ETV JOINT VERIFICATION STATEMENT

TECHNOLOGY TYPE:	**STORMWATER TREATMENT TECHNOLOGY**
APPLICATION:	**SUSPENDED SOLIDS TREATMENT**
TECHNOLOGY NAME:	**TERRE KLEEN™ 09**
TEST LOCATION:	**HARRISBURG, PENNSYLVANIA**
COMPANY:	**TERRE HILL CONCRETE PRODUCTS**
ADDRESS:	**485 Weaverland Valley Road** **PHONE: (800)242-1509**
	Terre Hill, Pennsylvania 17581 **FAX: (717)445-3108**
WEB SITE:	**http://www.terrehill.com**
EMAIL:	**precastsales@terrehill.com**

NSF International (NSF), in cooperation with the U.S. Environmental Protection Agency (EPA), operates the Water Quality Protection Center (WQPC), one of five active centers under the Environmental Technology Verification (ETV) Program. The WQPC recently evaluated the performance of the Terre Kleen™ 09 (Terre Kleen), manufactured by Terre Hill Silo Company, Inc. T/D/B/A Terre Hill Concrete Products (THCP). The Terre Kleen device was installed at the Department of Public Works (DPW) facility in Harrisburg, Pennsylvania. The testing organization (TO) for the evaluation was headed by a faculty member from the Environmental Engineering Department of The Pennsylvania State University—Harrisburg (PSH) in Middletown, Pennsylvania.

EPA created ETV to facilitate the deployment of innovative or improved environmental technologies through performance verification and dissemination of information. The ETV Program's goal is to further environmental protection by accelerating the acceptance and use of improved and more cost-effective technologies. ETV seeks to achieve this goal by providing high-quality, peer-reviewed data on technology performance to those involved in the design, distribution, permitting, purchase, and use of environmental technologies.

ETV works in partnership with recognized standards and testing organizations, and with stakeholder groups, which consist of buyers, vendor organizations, and permitters; and with the full participation of individual technology developers. ETV evaluates the performance of innovative technologies by developing test plans that are responsive to the needs of stakeholders, conducting field or laboratory tests (as appropriate), collecting and analyzing data, and preparing peer-reviewed reports. All evaluations are conducted in accordance with rigorous quality assurance protocols to ensure that data of known and adequate quality are generated and that the results are defensible.

6A.1 TECHNOLOGY DESCRIPTION

The following description of the Terre Kleen was provided by the vendor and does not represent verified information.

The Terre Kleen device combines primary and secondary chambers, baffles, a screen, and inclined sedimentation, as well as oil, litter, and debris/sediment storage chambers, into a self-contained concrete structure. The primary benefit of the Terre Kleen device is its ability to efficiently settle solids in the inclined cells (lamella plates) located in the secondary chamber using hydrodynamic principles. The design of the unit provides for underground installation as an inline treatment device, where it may be applied at a critical source area, or a larger unit may be installed in a storm sewer main to provide treatment for larger flows. Installation can be performed using conventional construction techniques. Terre Kleen units can be designed to provide specific removal efficiencies based on the size characteristics of the suspended solids and flow rate of storm water to the device.

The Terre Kleen device addresses the concern of being space-effective, providing high particle removal efficiency given the device's relatively small footprint. The ability to install the device below grade allows for the use of the above-ground space and makes it easier for the device to be retrofitted into a preexisting storm sewer system. The design allows for some treatment of all water that enters the primary settling chamber of the device, even if the flows exceed the capacity of the secondary (lamella inclined plate) chamber. The treated and bypassed water recombine prior to discharge from the device. Resuspension of captured material below the inclined plates is minimized because the stormwater enters the inclined cells sideways instead of scouring the top of the sediment.

The vendor claims that the Terre Kleen device installed for the verification test will remove 100% of particles 200 μm and larger in stormwater when the device is operating at the design storm flow of 3.49 ft³/s (cfs), which is based on the 25-year storm for Harrisburg. THCP also claims that at lower flows, removals of particles smaller than 200 μm will also be achieved.

6A.2   VERIFICATION TESTING DESCRIPTION

6A.2.1   Methods and Procedures

The test methods and procedures used during the evaluation are described in the *Environmental Technology Verification Test Plan for Terre Hill Concrete Products: The Terre Kleen, City of Harrisburg, Pennsylvania* (November 2004). The Terre Kleen device was installed at the downstream end of the stormwater collection system at the City of Harrisburg Department of Public Works facility. The drainage area is part of the city's maintenance yard occupied by the Bureau of Sanitation and includes runoff from buildings and paved and unpaved parking areas having a 90–95% impervious drainage area initially estimated at approximately 1.27 acres, but was later estimated to be approximately 2.5–3 acres after topographic maps with finer contours were made available.

Verification testing consisted of collecting data during a minimum of 15 qualified events that met the following criteria:

- The total rainfall depth for the event, measured at the site, was 0.2 in. (5 mm) or greater.
- Flow through the treatment device was successfully measured and recorded over the duration of the runoff period.
- A flow-proportional composite sample was successfully collected for both the inlet and the outlet over the duration of the runoff event.
- Each composite sample was composed of a minimum of five aliquots, including at least two aliquots on the rising limb of the runoff hydrograph, at least one aliquot near the peak, and at least two aliquots on the falling limb of the runoff hydrograph.
- There was a minimum of 6 hours between qualified sampling events.

Automated samplers and flow monitoring devices were installed and programmed to collect composite samples from the inlet and outlet, and to measure the stormwater flow into and out of the device. In addition to the flow and analytical data, operation and maintenance data were recorded.

Samples were analyzed for total suspended solids (TSS) and suspended sediment concentration (SSC). The samples were also analyzed to quantify the mass of particles greater than 250 urn in size and to determine the particle size distribution for particles ranging in size from 0.8 to 240 µm.

6A.3 Verification of Performance

The performance verification of the Terre Kleen device consisted of an evaluation of flow, sediment reduction, and operations and maintenance data collected during 15 qualified storm events over a period of approximately 11 months.

6A.3.1 Test Results

The precipitation data for the rain events are summarized in Table 6A.1.

The flow monitoring and analytical results were evaluated using event mean concentration (EMC) and sum of loads (SOL) comparisons. The EMC evaluates treatment efficiency on a percentage basis, with the calculation being made by dividing the outlet concentration by the inlet concentration and multiplying the quotient by 100. The EMC was calculated for each analytical parameter and each individual storm event. The SOL comparison evaluates the treatment efficiency on a percentage basis by comparing the sum of the inlet and outlet loads (the parameter concentration multiplied by the runoff volume) for all storm events. The calculation is made by subtracting the quotient of the total outlet load divided by the total inlet load from 1, and multiplying the difference by 100. SOL results can be summarized on an overall basis since the load calculation takes into account both the concentration and volume of runoff from each event. The SOL calculation was also conducted for TSS and SSC samples with sediment particles greater than 250 µm. The analytical data ranges, EMC range, and SOL reduction values are shown in Table 6A.2.

TABLE 6A.1
Rainfall Data Summary

Event Number	Date	Start Time	Rainfall Amount (in.)	Rainfall Duration (hr:min)	Peak Flow Rate (cfs)[a]	Runoff Volume (ft³)[a]
1	6/29/05	12:00	0.31	2:00	0.83	750
2	7/7/05	18:40	1.68	15:00	0.82	7900
3	8/16/05	09:35	0.43	11:10	0.029	210
4	8/27/05	19:05	0.68	14:00	0.76	1800
5	9/16/05	18:55	1.22	5:40	2.0	4900
6	10/13/05	05:20	0.63	21:55	0.50	960
7	10/21/05	22:45	1.17	24:15	0.80	3800
8	11/16/05	10:30	0.20	14:40	0.013	110
9	11/22/05	23:20	0.52	9:45	0.37	1300
10	11/29/05	04:55	1.04	19:05	1.2	6500
11	12/25/05	11:50	0.45	8:40	0.26	580
12	1/2/06	10:45	0.99	25:40	0.14	940
13	1/11/06	12:50	0.42	11:05	0.20	480
14	4/3/06	14:40	0.75	7:50	0.36	1500
15	5/13/06	16:20	0.71	54:10	0.089	660

[a] Runoff volume and peak discharge rate measured at the outlet monitoring point, with the exception of event 14, which was measured at the inlet monitoring point. See the verification report for further details.

TABLE 6A.2
Analytical Data, EMC Range, and SOL Reduction Results

Parameter	Inlet Range (mg/L)	Outlet Range (mg/L)	EMC Range (%)	SOL Reduction (%)	SOL Reduction Particle Size >250 μm (%)	SOL Reduction Particle Size <250 μm (%)
TSS	58–6900	35–980	−88–86	44	85	35
SSC	75–7000	35–1500	−11–87	63	98	32

Both the TSS and SSC analytical parameters measure sediment concentrations in water. However, the TSS analysis uses an aliquot drawn by the analyst from the sample container, while the SSC analysis uses the entire contents of the sample container. Heavier solids may not be picked up in the drawn aliquot for the TSS analysis, such that the TSS will tend to be more representative of the lighter solids concentrations.

The particle size distribution data showed that the Terre Kleen was approximately 98% effective in removing particles 200 μm or larger. When the particle size distribution data are combined with the hydrologic data, it shows that the performance of the device generally removed all of the particles 200 μm or larger when treating flows of 2.0 cfs or lower. The rated flow capacity (3.49 cfs) of the Terre Kleen was not exceeded during any of the 15 storm events. This device is designed to treat the entire entering flow (bypass over the plates was monitored after the primary chamber, and at no time during the testing were the plates bypassed).

6A.3.2   System Operation

The Terre Kleen was installed in February 2005, with no major issues noted. The Terre Kleen™ device was cleaned prior to the start of testing in March 2005 and was inspected frequently during verification. A review of the storm event records in January 2006 showed that two late January storms had substantial negative removals. Therefore, the decision was made to clean the device at the end of January 2006. Sediment depths prior to pump-out were between 50% and 75% of the maximum design sediment depth, measured at several points in the device. This maintenance activity consisted of using a sewer vactor truck from the City of Harrisburg to dewater and remove sediment from the device. A sample of the sediment was analyzed for Toxicity Characteristic Leachate Procedure (TCLP) metals, and the concentrations were lower than the hazardous waste limits of 40 CFR Section 261.42.

6A.3.3   Quality Assurance/Quality Control

NSF personnel completed a technical systems audit during testing to ensure that the testing was in compliance with the test plan. NSF also completed a data quality audit of at least 10% of the test data to ensure that the reported data represented the data generated during testing. In addition to quality assurance (QA) and quality control audits performed by NSF, EPA personnel conducted an audit of NSF's QA Management Program.

6A.3.4   Note for This Revision

The original verification statement was signed in September 2006 but revised in July 2008 to reflect a change in the method the drainage area size and the runoff volume and peak runoff intensity were calculated. See Sections 3.2 and 5.1.1 of the verification report for information on the revised drainage area size and runoff calculations, respectively.

Original signed by   Original signed by

Sally Gutierrez   October 14, 2007   Robert Ferguson   October 3, 2007

Sally Gutierrez   Date   Robert Ferguson   Date

Director   Vice President

National Risk Management Research Laboratory   Water Systems

Office of Research and Development   NSF International

United States Environmental Protection Agency

NOTICE: Verifications are based on an evaluation of technology performance under specific, predetermined criteria and the appropriate quality assurance procedures. EPA and NSF make no expressed or implied warranties as to the performance of the technology and do not certify that a technology will always operate as verified. The end user is solely responsible for complying with any and all applicable federal, state, and local requirements. Mention of corporate names, trade names, or commercial products does not constitute endorsement or recommendation for use of specific products. This report is not an NSF Certification of the specific product mentioned herein.

6A.3.5   Availability of Supporting Documents

Copies of the *ETV Verification Protocol Stormwater Source Area Treatment Technologies Draft 4.1, March 2002*, the test plan, the verification statement, and the verification report (NSF Report Number 06/29/WQPC-WWF) are available from

> ETV Water Quality Protection Center Program Manager (hard copy)
> NSF International
> P.O. Box 130140
> Ann Arbor, Michigan 48113-0140

NSF website: http://www.nsf.org/etv (electronic copy)

EPA website: http://www.epa.gov/etv (electronic copy)

Appendices are not included in the verification report but are available from NSF upon request.

APPENDIX 6B: NJCAT 2013 PROCEDURE:
APPENDIX A—MTD VERIFICATION PROCESS

1. A manufacturer shall file a MTD verification application with NJCAT in a form approved by NJCAT. See Appendix B.*
2. NJCAT and the manufacturer will meet in person or by telephone to review the verification application for:
   a. Administrative and clerical accuracy and completeness;
   b. Compliance with the applicable laboratory testing protocol;
   c. Prior approval from NJCAT of any necessary items such as laboratory certification; third party independent observer; testing entity; etc.
3. If the MTD is a new technology for which there is no approved protocol, NJCAT and the Manufacturers Working Group (MWG) shall meet with the manufacturer to create a laboratory testing protocol, which will be approved by NJDEP. Those interested in being part of the MWG shall contact NJCAT.
4. Upon completion of the initial review and attainment of all necessary prior approvals, the manufacturer shall commence the laboratory testing in strict accordance with the applicable laboratory testing protocol.
5. Upon completion of the laboratory testing, the manufacturer shall submit to NJCAT a complete laboratory test report with all data collected and analyzed, including the information listed in Section 5, with the exception of 5.G.6 and 5.G.7.
6. Within 30 days of receipt of the laboratory test report, NJCAT shall meet in person or by telephone with the manufacturer to discuss the report and issue a preliminary opinion letter regarding the manufacturer's compliance with the applicable laboratory protocol and, if not, specifying in detail the areas of noncompliance with the protocol.
7. If outstanding issues exist, NJCAT and the manufacturer shall meet within 10 days of the issuance of the preliminary opinion letter to discuss possible resolution of the outstanding issues.
8. If the outstanding issues are resolved or there were no issues identified in the preliminary review report, NJCAT will issue a final verification report within 90 days of issuing the preliminary report or if necessary, within 90 days of resolving the outstanding issues. If outstanding issues are not resolved, see Appendix C. The final verification report shall be posted on the NJCAT website and available for written public comment for 30 days. Anyone intending to comment must provide written notification to the manufacturer and NJCAT within 14 days of the verification report being posted on the website. Written public comments with supporting documentation must be submitted to the manufacturer and NJCAT no later than 30 days after the initial posting of the verification report on the website. If NJCAT is able to resolve written comments during this time, those comments will be addressed by NJCAT, if NJCAT is unable to resolve the comment's concern then the commenter will be given the opportunity by NJCAT to request the submission of those comments including all supporting documentation, to the Review Panel (Appendix C) for resolution.
9. If no comments were referred to the Review Panel, NJCAT shall issue a final verification report within 10 days of the end of the public comment period. If comments are referred to the Review Panel, NJCAT shall issue a final verification report within 30 days of the Review Panel issuing their resolution (see Appendix C for the resolution process).
10. Once a final verification report is issued, NJCAT shall add the new MTD to the list of verified MTDs at http://www.njcat.org. NJCAT shall include on its website the final verification report.

* Any mention of appendices in this appendix refers to appendices in the NJCAT 2013 Procedure, not this book.

11. NJCAT shall notify NJDEP once their website has been updated and a MTD has been verified. NJCAT shall send NJDEP the name of the manufacturer, name of MTD and the respective TSS removal efficiency as notification.

12. NJDEP certification has been awarded once the name of the manufacturer, name of MTD and the respective TSS removal efficiency have been placed at http://www.njstormwater .org. NJDEP will not update their website until the list of verified MTDs has been updated by NJCAT at http://www.njcat.org.

APPENDIX 6C: NJDEP CERTIFICATION LETTER FOR FILTERRA BIORETENTION SYSTEMS, FORMERLY A DIVISION OF AMERICAST, INC. NOW A PART OF CONTECH, AND SIZES AND APPROVED TREATMENT FLOW RATES OF DOWNSTREAM DEFENDER BY HYDRO INTERNATIONAL

State of New Jersey

DEPARTMENT OF ENVIRONMENTAL PROTECTION
Bureau of Nonpoint Pollution Control
Division of Water Quality
Mail Code 401-02B
Post Office Box 420
Trenton, New Jersey 08625-0420
609-633-7021 Fax: 609-777-0432
http://www.state.nj.us/dep/dwq/bnpc_home.htm

CHRIS CHRISTIE
Governor

KIM GUADAGNO
Lt. Governor

BOB MARTIN
Commissioner

May 19, 2015

Derek M. Berg
CONTECH Engineered Solutions, LLC
71 US Route 1, Suite F
Scarborough, ME 04074

Re: MTD Lab Certification for the
Filterra Bioretention System
By CONTECH Engineered Solutions, LLC

TSS Removal Rate: 80%

Dear Mr. Berg:

This certification letter is being written to update the Filterra Bioretention System lab certification to reflect an ownership change from Filterra Bioretention System, A Division of Americast, Inc. to Contech Engineered Solutions, LLC.

The Stormwater Management rules under N.J.A.C. 7:8-5.5(b) and 5.7(c) allow the use of manufactured treatment devices (MTDs) for compliance with the design and performance standards at N.J.A.C. 7:8-5 if the pollutant removal rates have been verified by the New Jersey Corporation for Advanced Technology (NJCAT) and have been certified by the New Jersey Department of Environmental Protection (NJDEP). Filterra® Bioretention Systems has requested a Laboratory Certification for the Filterra Bioretention System.

This project falls under the "Procedure for Obtaining Verification of a Stormwater Manufactured Treatment Device from New Jersey Corporation for Advanced Technology" dated January 25, 2013. The applicable protocol is the "New Jersey Department of Environmental Protection Laboratory Protocol to Assess Total Suspended Solids Removal by a Filtration Manufactured Treatment Device" dated January 25, 2013.

NJCAT verification documents submitted to the NJDEP indicate that the requirements of the afore-mentioned protocol have been met or exceeded. The NJCAT letter also included a recommended certification TSS removal rate and the required maintenance plan. The NJCAT Verification Report with the Verification Appendix for this device is published online at http://www.njcat.org/verification-process/technology-verification-database.html.

The NJDEP certifies the use of the Filterra Bioretention System by Contech Engineered Solutions, LLC at a TSS removal rate of 80%, when designed, operated and maintained in accordance with the information provided in the Verification Appendix.

Be advised a detailed maintenance plan is mandatory for any project with a Stormwater BMP subject to the Stormwater Management Rules, N.J.A.C. 7:8. The plan must include all of the items identified in Stormwater Management Rules, N.J.A.C. 7:8-5.8. Such items include, but are not limited to, the list of inspection and maintenance equipment and tools, specific corrective and preventative maintenance tasks, indication of problems in the system, and training of maintenance personnel. Additional information can be found in Chapter 8: Maintenance of the New Jersey Stormwater Best Management Manual.

If you have any questions regarding the above information, please contact Titus Magnanao, of my office at (609) 633-7021.

Sincerely,

James J. Murphy, Chief
Bureau of Nonpoint Pollution Control

C: Chron File
    Richard Magee, NJCAT
    Madhu Guru, DLUR
    Elizabeth Dragon, BNPC
    Lisa Schaefer, BNPC
    Titus Magnanao, BNPC
    Ravi Patraju, NJDEP

APPENDIX 6D: DIMENSION AND CAPACITY OF CDS MODELS

TABLE 6D.1

Dimension and Capacity of CDS Models

CDS Model	Treatment Capacity (cfs/L/s)	Minimum Sump Storage Capacity (yd³/m³)	Minimum Oil Storage Capacity (gal/L)
CDS2015-G	0.7 (19.8)	0.5 (0.4)	70 (265)
CDS2015-4	0.7 (19.8)	0.5 (1.4)	70 (265)
CDS2015	0.7(19.8)	1.3 (1.0)	92 (348)
CDS2020	1.1 (31.2)	1.3 (1.0)	131 (496)
CDS2025	1.6 (45.3)	1.3 (1.0)	143 (541)
CDS3020	2.0 (56.6)	2.1 (1.6)	146 (552)
CDS3030	3.0 (85.0)	2.1 (1.6)	205 (776)
CDS3035	3.8 (106.2)	2.1 (1.6)	234 (885)
CDS4030	4.5 (127.4)	5.6 (4.3)	407 (1540)
CDS4040	6.0 (169.9)	5.6 (4.3)	492 (1862)
CDS4045	7.5 (212.4)	5.6 (4.3)	534 (2012)
CDS2020-D	1.1 (31.2)	1.3 (1.0)	131 (495)
CDS3020-D	2.0 (56.6)	2.1 (1.6)	146 (552)
CDS3030-D	3.0 (85.0)	2.1 (1.6)	205 (776)
CDS3035-D	3.8 (106.2)	2.1 (1.6)	234 (885)
CDS4030-D	4.5 (127.4)	4.3 (3.3)	328 (1241)
CDS4040-D	6.0 (169.9)	4.3 (3.3)	396 (1499)
CDS4045-D	7.5 (212.4)	4.3 (3.3)	430 (1627)
CDS5640-D	9.0 (254.9)	5.6 (4.3)	490 (1854)
CDS5653-D	14.0 (396.5)	5.6 (4.3)	599 (2267)
CDS5668-D	19.0 (538.1)	5.6 (4.3)	733 (2774)
CDS5678-D	25.0 (708.0)	5.6 (4.3)	814 (3081)
CDS3030-DV	3.0 (85.0)	2.1 (1.6)	205 (776)
CDS5042-DV	9.0 (254.9)	1.9 (1.5)	294 (1112)
CDS5050-DV	11.0 (311.5)	1.9 (1.5)	367 (1389)
CDS7070-DV	26.0 (736.3)	3.3 (2.5)	914 (3459)
CDS10060-DV	30.0 (849.6)	5.0 (3.8)	792 (2997)
CDS10080-DV	50.0 (1416.0)	5.0 (3.8)	1057 (4000)
CDS100100-DV	64.0 (1812.5)	5.0 (3.8)	1320 (4996)

Note: The NJDEP approved treatment capacities are significantly smaller than those listed in this table. See Table 6D.2.

TABLE 6D.2

NJDEP Approved Rates for CDS Models

New Jersey Treatment Rates for CDS Models Based on a Surface Area
Specific Loading Rate of 33.2 gpm/ft²

CDS Model	Manhole Diameter (ft)	Treatment Flow Rate (cfs)	CDS Model No.
CDS-4	4	0.93	CDS2015-4
CDS-5	5	1.5	CDS2025-5
CDS-6	6	2.1	CDS3030-6
CDS-8	8	3.7	CDS4040-8
CDS-10	10	5.8	CDS5653-10
CDS-12	12	8.4	Offline

REFERENCES

AbTech Industries, 4110 North Scottsdale Road, Suite 235, Scottsdale, AZ 85251; Ph. 800-545-8999 (http://www.abtechindustries.com).

ACF Environmental, 2831 Cardwell Drive, Richmond, VA 23234, Ph. 800-448-3636 (http://www.acfenvironmental.com).

Advanced Drainage Systems, Inc., 4640 Trueman Blvd., Hillard, OH 43026, Ph. 800-821-6710 (http://www.ads-pipe.com).

Aird, J., 2012, Separation devices for storm water runoff, *Stormwater*, September, pp. 48–53.

Aqua Shield, Inc., 2705 Kanasita Drive, Chattanooga, TN 37343, Ph. 888-344-9044 (http://www.aquashieldinc.com).

BaySaver Technologies, Inc., 1302 Rising Ridge Rd., Mount Airy, MD 21771, Ph. 800-BAYSAVER (http://www.baysaver.com).

Best Management Products, Inc., 53 Mt. Archer Road, Lyme, CT 06371, Ph. 800-504-8008 (http://www.bmpinc.com).

BioClean Environmental Services, Inc., P.O. Box 869, Oceanside, CA 92049, Ph. 760-433-7640 (http://www.biocleanenvironmental.net).

Bio-Microbics, Inc., Shawnee, KS, Ph: 913-422-0770 (http://www.biomicrobics.com).

Brzozowski, C., 2004, Options for urban storm water treatment, *Stormwater*, January/February, pp. 58–68.

ClearWater Solutions, 2259 Lone Oak Lane, Vista, CA 92084, Ph. 800-758-8817 (http://www.ClearWater.BMP.com).

Contech Stormwater Solutions, West Chester, OH, Ph. 800-925-5240 (http://www.Contechstormwater.com).

Crystal Stream Technologies, 2090 Sugarloaf Parkway, Suite 245, Lawrenceville, GA 30045, Ph. 770-979-6516 (http://engineering@crystalstream.com).

Doerfer, J., 2008, Filter maintenance sand filters vs. media filters, *Stormwater Solutions*, July/August, p. 8.

EPA, 2005, National management measures to control non-point source pollution from urban runoff, publication no. EPA-841-B-05-004, November (http://www.epa.gov/owow/nps/urbanmm/index.html).

———2008, National management measures to control non-point source pollution from urban areas, publication no. EPA-841-B-05-004.

EWRA Currents, 2009, Federal facilities face a new storm water hurdle, *Newsletter of the Environmental and Water Resources Institute of the ASCE*, 11 (1): 1–2.

Fabco Industries, Inc., 170 Wilbur Pl., Ste. 2, Bohemia, NY 11716, Ph. 631-244-3536 (http://www.fabco-industries.com).

Filterra Bioretention Systems, a division of Americast, 11352 Virginia Precast Rd., Ashland, VA 23005, Ph. 866-349-3458 (support@filterra.com).

Goldberg, S., 2012, Separation devices for storm water runoff, *Stormwater*, June, pp. 52–58.

Hydro International PLC, 94 Hutchins Drive, Portland, ME 04102, Ph. 207-756-6200 (http://www.hydro-international.biz).

Imbrium Systems Corporation, 9420 Key West Avenue, Suite 140, Rockville, MD 20850 (http://www.imbrium systems.com).

Imbrium Systems, Inc., 2 St. Claire Ave. W., Ste. 2100, Toronto, ON M4V 1L5, Canada, Ph. 800-565-4801.
Jensen Precast, 825 Steneri Way, Sparks, NV 89431, Ph. 775-352-2700 (http://www.jensenprecast.com).
Kristar Enterprises, Inc., 360 Sutton Place, P.O. Box 6419, Santa Rosa, CA 95406, Ph. 800-579-8819 (http://
    www.kristar.com).
New Jersey Department of Environmental Protection, 2004, Stormwater Best Management Practices Manual,
    February.
New York Guidelines for Urban Erosion and Sediment Control, 4th printing, April 1999, prepared by
    Urban Soil Erosion and Sediment Control Printing, printed by Empire State Chapter Soil and Water
    Conservation Society, c/o Cayuga County SWED, 7413 County House Road, Auburn, NY 13021.
New York State Stormwater Management Design Manual, August 2003, prepared by Center for Watershed
    Protection, 8390 Main Street, Ellicott City, MD 21043, for New York State Department of Environmental
    Conservation, 625 Broadway, Albany, NY 12233. (This manual was updated in August 2010.)
Revel Environmental Manufacturing, Corp. office, Ph. 925-676-4736 (http://www.remfilters.com).
Stormdrain Solutions, Inc., Ph. 877-687-7473 (http://www.stormdrains.com).
StormTrap LLC, 2495 W. Bungalow Rd., Morris, IL 60450, Ph. 877-867-6872 (http://www.Stormtrap.com).
Suntree Technologies, Inc., 798 Clearlake Road, Suite 2, Cocoa, FL 32922, Ph. 321-637-7552 (http://info
    @suntreetech.com).
SWIMS (Storm Water Inspection and Maintenance Services, Inc.), P.O. Box 1627, Discovery Bay, CA 94514,
    Ph. 925-516-8966 (http://www.SwimsClean.com), e-mail (swimsclean@aol.com).
TerreKleen, Terre Hill Concrete Products, 485 Weaverland Valley Road, PO Box 10, Terre Hill, PA 17581, Ph.
    800-242-15009 (http://www.terrekleen.com).
Weiss, P.T., Gulliver, J.S., and Erickson, A.J., 2005, The cost and effectiveness of stormwater management
    practices, Minnesota Department of Transportation report 2005-23 (http://www.cts.umn.edu/publications
    /researchreports/reportdetail.html?id. = 1023).
WERF, 2005, Critical assessment of storm water treatment and control selection issues (http://werf.org).
Wieske, D. and Penna L.M., 2002, Stormwater strategy, *ASCE Civil Engineering*, 72 (2): 62–67.

7 Structural Storm Water Management Systems

Detention, retention, and infiltration basins are conventional storm water structures that are employed to control the peak rate and/or the volume of runoff, and to improve the water quality. Constructed wetlands, bioretention basins, and vegetated swales are also used in storm water management practices—more so for water quality than peak flow attenuation. Also employed are sand filters and vegetated buffers, primarily for improving water quality. This chapter presents simplified design procedures, examples, and case studies for various types of detention, retention, and infiltration basins. Also included in this chapter are recommended design criteria for storm water management systems.

7.1 DETENTION BASINS/WET PONDS

Detention basins are storage facilities that impound the storm water runoff temporarily and release it slowly through their outlet structures. Figure 7.1 depicts a grass-lined detention basin with center fountain, designed by the author. Figure 7.2 shows a typical section of a detention basin–outlet structure system. Detention basins are intended to fully drain out following a rainfall event and may be constructed aboveground or underground. Open detention basins not only regulate the peak rate of runoff, but also serve as a means of enhancing the runoff quality.

Wet ponds are differentiated from detention basins in that the former are impounded at all times, while the latter remain dry between storm events. Both of these facilities are widely employed to control peak rates of runoff and can be quite effective if properly designed and well maintained.

By prolonging the discharge, detention basins and wet ponds also improve water quality, through partial settling of suspended sediment from runoff. This effect depends on retention time, namely the time it takes the stored water to be released. A detention basin that prolongs the release time is known as an extended detention basin. The longer the retention time is, the greater is the treatment efficiency of a detention basin. The efficiency can be increased by incorporating a presettling chamber in an underground basin or a forebay in an aboveground basin to trap coarse sediment.

The effectiveness of a detention basin for water quality depends on site conditions and the types of developments. In general, detention basins can be more effective in commercial developments than residential ones since the runoff can have a higher concentration of sediments. In wet ponds, the storage above the normal pool level creates a detention effect and the stagnant water helps improve water quality. In fact, a wet pond can serve as a highly effective best management practice. In addition to improving the water quality and controlling the peak runoff, wet ponds provide aesthetic amenity, aquatic habitat for fish, and, occasionally, recreation. Figure 7.3 shows a wet pond. In coastal areas where the water table is high, ponds are the most common practice to address storm water management requirements. In South Carolina's coastal counties, wet ponds are not only a practical option but also the most popular system to meet the storm water runoff requirements. Thousands of ponds have been constructed since 1987 under the NPDES phase II program (Drescher et al., 2011).

Removal of pollutants in wet ponds occurs through various factors. The main factor is the gravity settling of suspended matter and pollutants attached thereto. As discussed in Chapter 1, the settling velocity of particles in a stagnant water body decreases exponentially with the particle size. Therefore, in a wet pond, the water quality effect for a given sediment type depends on the residence time, namely the time that takes water equal in volume to that of runoff to be discharged

FIGURE 7.1 Grass-lined detention basin with center fountain. (Photo by the author.)

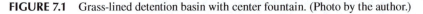

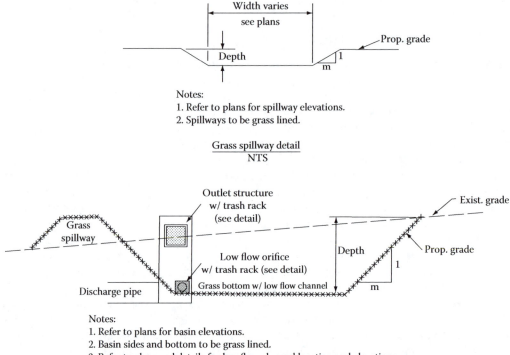

FIGURE 7.2 Typical detention basin cross section.

from the pond. Because of mixing effect with the stagnant water, a wet pond is more effective than a detention basin of equal retention time. Aquatic plants, if present, and microorganisms also improve water quality by uptake of nutrients and degradation of organic contaminants. Aquatic vegetation also removes metals. In areas of high water table where aboveground detention basins and underground detention–infiltration basins are not practical, wet ponds may serve as one of the best management practices.

Detention basins and wet ponds are designed through an iterative process. The process begins using a trial storage volume and preparation of grading plan of the basin. This is followed by preliminary sizing of the outlet openings and developing a storage–stage–discharge relation for the

FIGURE 7.3   Wet pond. (Photo by the author.)

selected detention outlet structure system. Routing computations are then performed to evaluate functioning of the system. Finally, outlet openings are refined until the desired discharges from the basin/pond are attained. This iterative process can be accelerated by first estimating the required detention storage through approximate methods. Example and case studies in this chapter illustrate the estimation procedure and the design process.

7.1.1   Flow Routing through Detention Basins

The runoff which enters into a detention basin is partly discharged through an outlet structure and partly impounds in the basin. This routing process is sketched in Figure 7.4.

Figure 7.5 shows typical inflow–outflow hydrographs through a detention basin. This figure indicates that a detention basin attenuates the runoff hydrograph, but prolongs discharge from its catchment area.

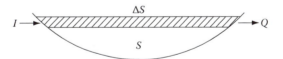

FIGURE 7.4   Routing process through a detention basin.

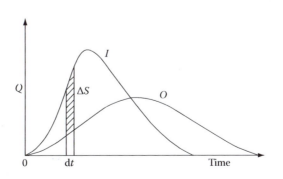

FIGURE 7.5   Flow routing through a detention basin.

According to Figure 7.5, the continuity equation during a time interval dt is given by

$$I(t) - O(t) = \frac{\Delta S}{dt} \tag{7.1}$$

where I and O represent the inflow and outflow discharge at time, t, and ΔS is the change in the volume of water stored in the detention basin within the time period, dt. In finite difference form, the preceding equation becomes

$$\left[\frac{(I_1 + I_2)}{2}\right] \cdot \Delta t - \left[\frac{(O_1 + O_2)}{2}\right] \cdot \Delta t = \Delta S = S_2 - S_1 \tag{7.2}$$

where the subscripts 1 and 2 represent the quantity of variables I, O, and S at the beginning and at the end of time period Δt, respectively. Separating the known and unknown values of variables, the preceding equation can be written as

$$\left(\frac{2S_2}{\Delta t} + O_2\right) = (I_1 + I_2) + \left(\frac{2S_1}{\Delta t} - O_1\right) \tag{7.3}$$

Since the inflow hydrograph is known (obtained from runoff calculations) and the outflow discharge and storage volume are also known at the beginning of a time period, the preceding equation gives the value of the term on the left side. To calculate the unknowns O_2 and S_2, a storage–outflow function for the detention basin–outlet structure needs to be developed.

This function can be developed by preparing a relation between the storage and elevation in the detention basin and also a relation between the discharge and elevation for the outlet structure. Specifically, for a given water surface elevation, the amounts of storage and discharge are calculated using the grading plan of the detention basin and the geometry of the outlet structure, respectively. Based on these calculations, a plot of outflow discharge, O, versus $2S/\Delta t + O$ is prepared. This plot, sketched in Figure 7.6, together with Equation 7.3 can then be used to determine the outflow discharge at the end of the time period.

In practice, Equation 7.3 is commonly solved using the so-called reservoir routing, also known as level pool or hydrologic routing method. In this method, discharge and storage are related to water surface elevation or stage and the variable $2S/\Delta t + O$ is also related to stage. Thus by solving Equation 7.3, the stage and, in turn, discharge and storage can be determined. Equation 7.3 is based on an implicit assumption that variables I, O, and S are linearly related with time. Therefore, to arrive at a fairly accurate result, Δt must be small and properly selected. A number of computer software programs are available to perform computations. One such software is PondPack, available through Haestad Methods, which has been acquired by Bently; another is StormCAD. Computations may also be performed using the HEC-1 computer software.

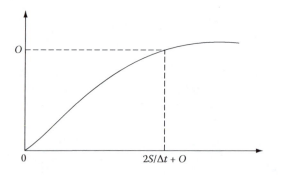

FIGURE 7.6 Outflow–storage function relation.

7.1.2 Outlet Structure Design

An outlet structure forms an integral part of detention basin design. A detention basin is effectively utilized when the openings and stages of its outlet structures are properly selected. An outlet structure may be single stage or multistage. A single-stage outlet comprises one opening, and multistage outlets include two or more openings commonly of varying size and geometry and at different levels. To attenuate runoff from storms of various frequencies, a multistage outlet structure should be employed. A simplified procedure for design of detention basins with multistage outlet is presented by Pazwash (1992). Example 7.1 will illustrate the design process.

Orifices, weirs, and grates are the most commonly used openings in an outlet structure. Discharges from these openings are discussed as follows.

7.1.2.1 Orifice

An orifice refers to an opening that is submerged at its inlet face, outlet face, or both. The flow through an orifice is calculated from the following equation:

$$Q = C_o A (2gh)^{1/2} \tag{7.4}$$

where

C_o = orifice coefficient
A = cross sectional area of orifice, m^2 (ft^2)
h = head above orifice center for free flow outlet or the difference in water surface elevation for submerged outlet, m, (ft)
g = gravitation acceleration, 9.81 m/s^2 (32.2 ft/s^2)
Q = discharge, m^3/s (cfs)

The preceding equation is dimensionally homogeneous and therefore can be used in System International (SI) and customary units (CU). The orifice coefficient is nearly independent of the depth of submergence and is approximated as $C = 0.6$ in practice.

Orifice openings are commonly employed as low-level opening for the control of more frequently occurring storms, such as 1- to 2-year frequency. A low-level orifice is also employed to prolong discharge from a detention basin in order to improve water quality.

7.1.2.2 Rectangular Weir

For a rectangular weir, the ideal discharge can be calculated from the following equation:

$$Q = \sqrt{2g} \int_0^H h^{1/2} \, dh$$

which integrates to

$$Q = \frac{2}{3} \sqrt{2g} \, L H^{3/2} \tag{7.5}$$

However, because of flow contractions at the sides and drawdown of water surface on the weir, the effective area is smaller than LH. To account for this effect, a coefficient of discharge, C_d, is applied to the ideal discharge in the preceding equation. Thus, the actual discharge is given by

$$Q = C_d \left(\frac{2}{3} \right) \sqrt{2g} \, L H^{1.5} \tag{7.6}$$

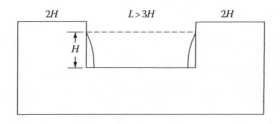

FIGURE 7.7   Side contractions in a rectangular weir.

For convenience, the term C_d (2/3) $\sqrt{2g}$ in the preceding equation is replaced by a coefficient, C_w, resulting in

$$Q = C_w LH^{1.5} \tag{7.7}$$

where
  C_w = weir coefficient
  L = length of weir, m (ft)
  H = depth of water above the weir crest, m (ft)

Based on experiments by Rehbock in Karlsruhe Hydraulic Laboratory in Germany, C_d and C_w are found to be somewhat dependent on H/P, where P is the depth of water below the weir crest (Pazwash, 2007). In practice, however, this dependence is neglected and Equation 7.7 is expressed as

$$Q = 1.8\ LH^{1.5}\ \text{(SI)} \tag{7.8}$$

$$Q = 3.3\ LH^{1.5}\ \text{(CU)} \tag{7.9}$$

The preceding equations are accurate for $H/L < 0.4$, which is not far from the usual operating range of outlet structures.

For rectangular weirs with side contractions where the weir is narrower than the width of the structure, there will be a lateral contraction of water nappe. Experiments by Francis have shown that under the conditions shown in Figure 7.7, the effect of side contraction is to reduce the width of nappe by 0.1 H on each side. Accounting for contractions on both sides, the effective weir length will be $L - 0.2\ H$.

It is to be noted that a rectangular opening in an outlet structure functions as a weir as long as it is not submerged. Once submerged, the opening behaves as an orifice. Under that condition, discharge should be calculated using Equation 7.4.

7.1.2.3 Triangular Weir

Triangular weirs, also known as V-notch weirs, serve as effective means of flow control. These weirs have the advantage that they can function for very small heads and also provide a very wide range in flow. The vertex angle of V-notch weirs commonly ranges from 15° to 90°, in practice.

In Figure 7.8, the ideal discharge through an elemental area dA is given as

$$dQ = \sqrt{2gh} \times dA \tag{7.10}$$

where

$$dA = 2x \times dh$$

$$x = h \tan\left(\frac{\theta}{2}\right)$$

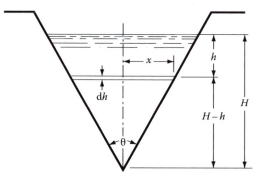

FIGURE 7.8 Triangular weir.

Substituting for h and dA in Equation 7.10 and integrating from $h = 0$ to $h = H$ gives ideal discharge through a V-notch weir:

$$Q = \left(\frac{8}{15}\right)\sqrt{2g}\,\tan\left(\frac{\theta}{2}\right)H^{5/2} \tag{7.11}$$

Introducing a coefficient of discharge C_d to account for side contractions, the preceding equation becomes

$$Q = C_d\left(\frac{8}{15}\right)\sqrt{2g}\,\tan\left(\frac{\theta}{2}\right)H^{5/2} \tag{7.12}$$

This equation is dimensionally homogeneous; as such, it applies to both SI and CU. The coefficient C_d ranges from 0.58 to over 0.62 depending on the angle of vertex and the head over the weir; the smaller the head is, the larger is the C_d. Conservatively, using $C_d = 0.585$, the preceding equation simplifies as

$$Q = 1.38\,\tan\left(\frac{\theta}{2}\right)H^{2.5}\,(\text{SI}) \tag{7.13}$$

$$Q = 2.5\,\tan\left(\frac{\theta}{2}\right)H^{2.5}\,(\text{CU}) \tag{7.14}$$

For 90° and 30° weirs, the preceding equations become

$$Q = KH^{2.5} \tag{7.15}$$

$$K = 1.38\,(\text{SI}),\ 2.50\,(\text{CU}) \quad 90°$$

$$K = 0.37\,(\text{SI}),\ 0.67\,(\text{CU}) \quad 30°$$

For heads smaller than 15 cm (0.5 ft), the discharge coefficient increases due to incomplete contraction. However, this variation may be ignored, still using Equations 7.13 and 7.14 in detention basin design.

7.1.2.4 Cipolleti Weir

A Cipolleti weir is a trapezoidal weir with 1:4 (1H, 4V) side slopes. This weir, shown in Figure 7.9, has been developed to compensate for side contractions in a rectangular weir. Thus, the weir Equations 7.8 and 7.9 may be used disregarding side contractions.

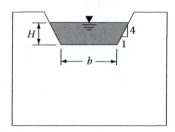

FIGURE 7.9 Cipolleti weir.

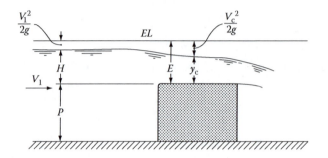

FIGURE 7.10 Broad-crested weir.

7.1.2.5 Broad-Crested Weir

A broad-crested weir is defined as a weir with its width along the flow at least three times the depth of water over the weir. If the weir is high enough to create a backwater effect, critical depth occurs on the crest of the weir. As shown in Chapter 2, the flow equations for a critical flow in a rectangular channel are

$$y_c = \frac{2}{3} E$$

$$V_c = \sqrt{g y_c}$$

$$Q = L y_c V_c = \left(\frac{2}{3}\right)^{1.5} \sqrt{g}\, L E^{1.5} \tag{7.16}$$

For high weirs (that is, large P/H; refer to Figure 7.10), the velocity of approach becomes small and E can be approximated by H, the depth of approach flow. In this case, the preceding equation becomes

$$Q = 1.7\, LH^{1.5}\ \text{(SI)} \tag{7.17}$$

$$Q = 3.09\, LH^{1.5}\ \text{(CU)} \tag{7.18}$$

7.1.2.6 Overflow Grates

Overflow grates are installed atop outlet structures to serve as an emergency measure to release water beyond the design storm. Flow-through grates were discussed in Chapter 4. As indicated, discharge from a grate occurs as weir flow at small depths of water and as orifice flow when the grate becomes fully submerged. Flow equations in these cases are given by the following equations respectively:

$$Q = C_w\, LH^{1.5} \tag{7.19}$$

$$Q = C_o f_c\, A (2gH)^{1/2} \tag{7.20}$$

where

    C_w = weir coefficient = 1.8 SI, 3.3 CU
    L = overall permeater length of grate, m (ft)
    C_o = orifice flow coefficient = 0.6 (square edge)
    f_c = clogging factor to account for opening covered with leaves, usually taken as $f_c = 0.67$
    A = opening area of grate, m² (ft²)
    H = depth of water approaching the grate, m (ft)

7.1.2.7 Stand Pipes

Stand pipes are commonly employed for discharge from sediment basins during construction. Flow into a stand pipe occurs as spillway flow for shallow depths (see Figure 7.11). In this case, the discharge is given by

$$Q = C_L \pi D h^{1.5} \tag{7.21}$$

where C_L is the spillway coefficient (C_L = 1.8 SI; 3.3 CU) and h and D are the depth of water over the stand pipe and pipe diameter, respectively. When the depth of water over the pipe exceeds 0.5 D, the inlet to standby pipe becomes fully submerged and the discharge follows the orifice equation:

$$Q = C_o \left(\frac{\pi D^2}{4} \right) \times (2gh)^{1/2} \tag{7.22}$$

where C_o is the orifice coefficient $C_o = 0.6$.

7.1.2.8 Hydro-Brake, Fluidic-Cone

Hydro-Brake is a device that functions similarly to an orifice at low head. However, when submerged, the device creates a rotational flow in the conic section, which forms an air-cored vortex in the outlet section. The core fills the center of the outlet and obstructs the flow. As a result, a Hydro-Brake with a relatively large opening performs like a small orifice. Hydro-Brake Fluidic-Cones come in 9 to 24 in. (23–60 cm) conic diameters and 3 to 10 in. (7.5–25 cm) outlet diameters. Figure 7.12 shows a Fluidic-Cone and discharge-head variation for 9-, 12-, and 15-in. cones. For more information, visit www.contechstormwater.com.

7.1.2.9 Thirsty Duck

Thirsty Duck is a buoyant flow control device that controls outflow from a detention basin. It uses the principle of buoyancy to suspend a discharge control device, such as orifice or weir, at a constant depth below the water surface. By maintaining a constant headwater over the control mechanism, this device

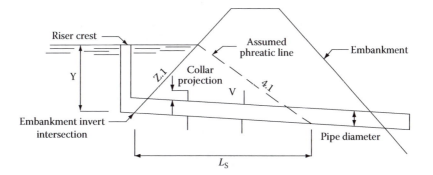

FIGURE 7.11 Stand pipe in a detention basin/pond (temporary during construction; permanent in a wet pond).

Fluidic-Cone™

Sizes and configurations

Length (*L*) refers to the point of attachment of the sleeve, plate, or flange.

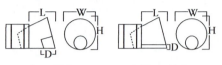

Boxed inlet   Slotted inlet

Model	Width (W) in.	mm	Height (H) in.	mm	Depth (D) in.	mm	Length in.	mm
9″	9	229	9	229	3–5	76–127	6	152
12″	12	305	11.75	298	4–6	102–152	8	203
15″	15	381	14.75	375	5–8	127–203	10	254
18″	18	457	17.75	451	3	76	12	305
21″	21	533	20.5	521	3.5	89	14	356
24″	24	610	23.5	597	4	102	16	406

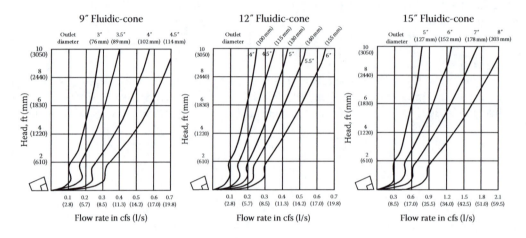

FIGURE 7.12 Fluidic-cone dimensions and discharge rates.

provides a constant discharge. Thirsty Duck comes in five models: ER100 series, ER200 series, and TD100, 200 and 300 series. Using Thirsty Duck, the discharge from a detention basin can be kept at the maximum allowable limit; thus the storage volume of detention basins and ponds can be minimized.

The device does not have any active mechanism or component. As with any other storm water control device, clogging due to debris can cause the structure to malfunction. The device can be accessed for clean-out during routine maintenance. The structure housing the Thirsty Duck device may be equipped with a locking hatch cover to prevent vandalism. Thirsty Duck can be contacted at www.Thirsty-Duck.com.

Example 7.1

Runoff calculations for a 3.29-acre residential development are performed based on the rational method. The existing peak rates of runoff for the storms of 2-, 10-, and 100-year frequency are

$$Q_2 = 3.82 \text{ cfs}$$

$$Q_{10} = 5.40 \text{ cfs}$$

$$Q_{100} = 7.63 \text{ cfs}$$

The local storm water management regulations require that the post development peak rates of runoff from the site be 50%, 25%, and 20% smaller than those of predevelopment for the 2-, 10-, and 100-year frequency storms, respectively. To meet these regulations and to provide a prolonged retention time to improve water quality, an outlet structure is provisioned comprising:

3 in. orifice at elevation 341.50 ft
12 × 12 in. opening at elevation 343.50 ft
Inlet grate (Campbell no. 3220) at elevation 344.75 ft

Calculate the elevation–discharge rating for this structure.

Solution

Discharge through openings is calculated as follows:

a. 3 in. orifice

$$Q_o = C_o A(2gh)^{1/2} = 0.6 \times 0.049 \, (64.4)^{1/2} \, (El - El_o) = 0.236 \, (El - 341.63)^{1/2}$$

b. 12 in. × 12 in. primary opening. Flow through this opening occurs as weir flow until the opening is submerged and as orifice flow when the water level rises above the opening.

$$Q_p = C_o LH^{1/5} = 3.3 \times 1 \times H^{1.5} = 3.3 \, (EL - 343.50)^{1.5} \quad \text{weir flow}$$

$$Q_p = 0.6 \times A(2gH)^{1/2} = 0.6 \times 1 \times 1 \times (64.4)^{1/2} \, (EL - 344.0)^{0.5} \quad \text{orifice flow}$$

$$Q_p = 4.81 \, (EL - 344)^{0.5} \quad \text{orifice flow}$$

c. Overflow grate. This grate has a perimeter length of 9.50 ft and 1 in. wide by 5-1/8 in. long slots with a total opening area of 345 in.^{2} (2.396 ft^2). Because of the small width of openings, the flow over the grate may be calculated using an orifice flow equation for depths of flow over 3 in.:

$$Q_g = CA(2gh)^{1/2} f_c$$

where $C = 0.6$ and $f = 0.66$ is the suggested clogging factor:

$$Q_g = 7.7 \, h^{1/2}$$

Table 7.1 shows the discharge–elevation rating table for the structure. This table implies that the 100-year water level in the detention basin will not rise above elevation ±344.85 ft. At this elevation, the discharge is nearly equal to the allowable discharge for the 100-year storm, which is 7.63 × 0.8 = 6.10 cfs.

TABLE 7.1

Stage-Discharge Relation for the Outlet Structure

EL (ft)	Q_o[a]	Q_p	Q_g	Q_t
341.50	0			0
342.00	0.14			0.14
343.00	0.28			0.28
343.50	0.32	0		0.32
344.00	0.36	1.17		1.53
344.50	0.40	3.40		3.80
344.75	0.42	4.17	0	4.59
345.00	0.43	4.81	3.81[a]	9.05

[a] A clogging factor equal to 0.66 is applied.

7.2   PRELIMINARY SIZING OF DETENTION BASINS

This section presents simple procedures for estimating the required storage volume of detention basins. The calculations cover the rational and modified rational methods and the SCS TR-55 method, as well as the universal runoff method developed by the author.

7.2.1   RATIONAL AND MODIFIED RATIONAL METHODS ESTIMATION

Assuming that the discharge from a detention basin varies linearly with time, the inflow–outflow hydrographs may be shown by Figure 7.13. In this figure Q_p and O_o are peak inflow and peak discharge from the detention basin, respectively. The area between inflow–outflow hydrographs to the left of Q_o represents the water stored in the basin. Based on this figure, the required detention storage to attenuate the peak flow from Q_p to Q_o, namely the allowable discharge from the basin, is given as

$$S = \left(\frac{1}{2}\right) \times (Q_p - Q_o)T_b \times 60 \tag{7.23}$$

where T_b is the base time of inflow hydrographs, in minutes, and the factor 60 represents the number of seconds per minute.

Inversely, the outflow discharge for a given detention storage may be calculated as

$$Q_o = Q_p - \frac{2S}{(60T_b)} \tag{7.24}$$

Since Equations 7.23 and 7.24 are dimensionally homogeneous, they are applicable to both CU and SI units. As indicated by these equations, the storage is related to the time base of hydrograph, which is dependent on the time of concentration. For storms lasting longer than the time of concentration, assuming that discharge from a detention basin varies linearly with time, the inflow–outflow hydrographs can be presented by Figure 7.14. According to this figure, which represents the case

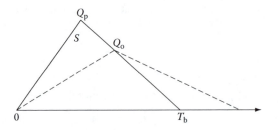

FIGURE 7.13   Detention storage estimation, rational hydrograph.

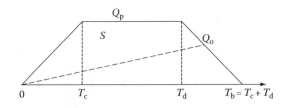

FIGURE 7.14   Detention storage estimation, modified rational hydrograph. Q_o = allowable discharge.

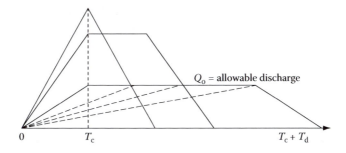

FIGURE 7.15 Required detention storage estimation, modified rational hydrograph.

of modified rational hydrograph, the required detention storage to attenuate the peak inflow to the allowable discharge, Q_o, is calculated as follows:

$$S = \left(Q_p T_d - \frac{1}{2} Q_o T_b \right) 60 \tag{7.25}$$

where $T_b = T_d + T_c$.

Using the modified rational method, the required storage should be calculated for storm durations ranging from the time of concentration to the duration at which the peak runoff equals the allowable discharge from the detention basin. Figure 7.15 shows the calculation process, in graphical form.

7.2.2 SCS TR-55 Method Estimation

The TR-55 method (1986) includes a figure for estimating the required storage volume. Figure 7.16 shows the nondimensional relation between runoff volume and discharge ratios for type I through type III storms: one curve for types I and IA and another for types II and III. In this figure the ratio of storage volume to runoff volume is plotted against the ratio of the peak outflow to peak inflow discharge. This figure indicates that for a given outflow to inflow ratio, types II and III storms require a significantly larger storage volume than types I and IA. Using this figure allows for estimating

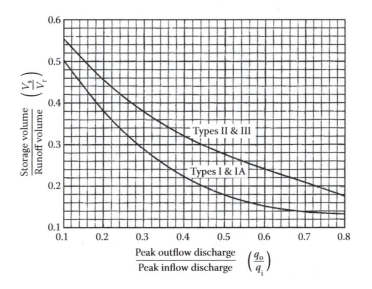

FIGURE 7.16 Approximate detention storage volume for types I, IA, II, and III storms.

storage for a given discharge or, inversely, estimating outflow discharge from a detention basin of a given storage volume (see Example 7.2). The required storage volume can be calculated as follows:

$$V_s = V_r \left(\frac{q_o}{q_i} \right)$$ (7.26)

where

V_s = detention storage volume, m³ (ft³)
V_r = runoff volume, m³ (ft³)
q_o = outflow (allowable) discharge, m³/s (cfs)
q_i = inflow discharge, m³/s (cfs)

7.2.3 UNIVERSAL METHOD OF STORAGE VOLUME ESTIMATION

In the universal method (Pazwash, 2009, 2011), approximating the discharge variation with time by a straight line, the inflow–outflow hydrographs may be represented by Figure 7.17. It can be readily shown that the required detention storage in this method is given as

$$S = \left[(T_d - T_e)Q_i - \left(\frac{1}{2} \right)(T_d + T_c - 2T_e)Q_o \right] 60$$ (7.27)

For a storm duration $T_d = T_c$, the preceding equation simplifies as follows:

$$S = (Q_i - Q_o)(T_c - T_e)60$$ (7.28)

The parameters T_e, T_t, and T_d in this method are the lag time between the rainfall and the onset of runoff, travel time, and storm duration, respectively. Similarly to the modified rational method, the calculations may be performed for various rainfall durations. However, unlike the modified rational method, the runoff volume does not continually increase with the storm duration. In fact, for sustained storms, the entire rainfall volume may be dissipated through infiltration.

7.2.4 ADJUSTING DETENTION STORAGE VOLUME ESTIMATION

All of the previously described methods of storage volume estimation are based on the assumption that outflow from a detention basin varies linearly with time. Actual outflows even for a single outlet opening deviate from this assumption. The deviation increases with the number of outlet openings and, particularly, when a small, low-flow orifice is incorporated for water quality improvement. A method of estimating the required detention storage for a multistage outlet is presented by Pazwash (1992). For preliminary design of detention basins, the author suggests

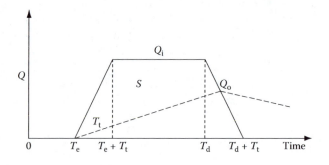

FIGURE 7.17 Routed hydrograph in the universal method.

that the storage calculated by any one of the methods be increased by 50% to 75% depending on the type of outlet structure. It is to be noted, however, that the previously described estimation methods are intended for preliminary sizing of detention basins and that they cannot serve as a substitute for routing computations.

Example 7.2

The post development peak discharges in Example 7.1, which are calculated based on the rational and modified rational method, are summarized in Table 7.2.
   Calculate the required detention storage using

a. Rational hydrograph ($T_d = T_c = 15$ min)
b. Modified rational method

The post development discharges are as noted in Example 7.1.

Solution

a. Rational method
   The runoff coefficient is calculated at 0.45. As such, use $T_d = 2.3 \, T_c$; see Table 3.14 in Chapter 3 (suggested by the author).
   Allowable discharges are calculated based on the existing peak runoff noted in Example 7.1.

$$Q_o = 3.82 \times 50\% = 1.91 \text{ cfs (2 years)}$$

$$Q_o = 5.40 \times 75\% = 4.05 \text{ cfs (10 years)}$$

$$Q_o = 7.63 \times 80\% = 6.10 \text{ cfs (100 years)}$$

Applying Equation 7.23 to 2-, 10-, and 100-year storms:

$$S = \frac{1}{2}(4.74 - 1.91) \times (2.3 \times 15 \times 60) = 2929 \text{ ft}^3 \text{ (2 years)}$$

TABLE 7.2
Peak Runoff for Various Storm Durations

T_d (min)	Q_2	Q_{10}	Q_{100}
$T_d = T_c = 15$	4.74	7.25	10.21
20	4.29	6.07	8.58
30	3.26	4.88	6.96
45	2.66	3.70[a]	5.33[a]
60	2.07	2.96[a]	4.44[a]

[a] Peak inflow less than allowable discharge (see Example 7.1). No further reduction is required.

TABLE 7.3

Required Storage Volume Estimation

	Two-Year Storm, Q_a = 1.91 ft^3				Ten-Year Storm, Q_a = 4.05		One Hundred-Year Storm, Q_a = 6.10	
T_d (min)	20	30	45	60	20	30	20	30
Q_i (cfs)	4.29	3.26	2.66	2.07	6.07	4.88	8.58	6.96
S (ft^3)	3143	3290	3744	3155	3032	3317	3891	4293

$$S = \frac{1}{2}(7.25 - 4.05) \times (2.3 \times 15 \times 60) = 3312 \text{ ft}^3 \text{ (10 years)}$$

$$S = \frac{1}{2}(10.21 - 6.10) \times (2.3 \times 15 \times 60) = 4255 \text{ ft}^3 \text{ (100 years)}$$

b. Modified rational method

Storage is calculated utilizing Equation 7.25:

$$S = [Q_i T_d - 0.5 \, Q_a(T_d + T_e)]60$$

Calculations for the storms of 2-, 10-, and 100-year frequency are organized in Table 7.3.

The largest amount of calculated storage volume is 4293 ft^3 (30 min. storm). Accounting for nonlinear discharge variation, the ponding effect of 3 in. orifices, and free board, add 50% to the calculated required storage. Therefore, the detention basin storage volume should be approximately 7000 ft^3.

Example 7.3

To attenuate the peak runoff from the site of the development described in Examples 7.1 and 7.2, an aboveground detention basin is provisioned. The storage–stage relation for this detention basin is calculated based on the grading plan of the basin and is tabulated in Table 7.4. The outlet structure of this basin is described in Example 7.1. Compute the peak discharges for this detention basin.

TABLE 7.4

Storage-Elevation Rating

Elevation (ft)	Area (ft^2)	Avg. Area (ft^2)	Δ Volume[a] (ft^3)	Volume (ft^3)
341.5	1360			0
		1530	760	
342.0	1680			760
		2035	2035	
343.0	2390			2795
		2720	2720	
344.0	3050			5515
		3500	3500	
345.0	3950			9015

Note: The calculated ΔV deviates less than 0.25% from the prismatic calculation, given by the equation:

$\Delta_t = [A_1 + A_2 + (A_1 A_2)^{1/2}]/3$.

[a] Δ Vol. = average area × ΔH.

Solution

Haestad Method PondPack computer software is utilized to perform the computations. The routing computations for the 100-year, 30-minute storm, which according to Example 7.2 requires the largest detention storage, are included in Table 7.5. The computations for other storm durations and frequencies are summarized in Table 7.6. A comparison of the existing allowable and proposed peak runoff rates from the site is presented in Table 7.7.

In this project, the detention basin is intentionally oversized to reduce downstream flooding, which is caused by an inadequate drainage system. As a result, the detention basin creates peak flow reductions far beyond the requirements.

The purpose of a 3 in. orifice at the outlet structure is to create a prolonged retention time in the detention basin in order to allow coarse and medium sized suspended sediment to settle. In this project the 3 in. orifice alone could not meet the applicable water quality standards. Therefore, a water treatment device was incorporated downstream of the outlet structure to supplement the dual-purpose detention basin.

Note: The required detention storage volume calculations in Example 7.2 yield critical storm durations of 45 minutes for the 2-year frequency storm and 30 minutes for the 100-year storms (based on the modified rational method). These approximate results agree with the more accurate routing computations. Thus, in practice, the critical storm duration is determined using the

---

TABLE 7.5
One Hundred-Year, 30 Minute Storm Routing Computations Summary

Inflow hydrograph: A:PR100-30.HYD

Rating table file: A:SPAR-DET.PND

INITIAL CONDITIONS

Elevation = 341.50 ft

Outflow = 0.00 cfs

Storage = 0 cu-ft

	Given Pond Data		Intermediate Routing Computations	
Elevation (ft)	Outflow (cfs)	Storage (cu-ft)	$2S/t$ (cfs)	$2S/t + 0$ (cfs)
341.50	0.0	0	0.0	0.0
342.00	0.1	760	5.1	5.2
343.00	0.3	2795	18.6	18.9
343.50	0.3	4145	27.6	28.0
344.00	1.5	5515	36.8	38.3
344.50	3.7	7250	48.3	52.0
344.75	4.6	8130	54.2	58.8
345.00	8.2	9015	60.1	68.3

Note: Time increment (t) = 5.0 min.

Starting pond W.S. elevation = 341.50 ft

    Summary of peak outflow and peak elevation

        Peak inflow = 6.96 cfs

        Peak outflow = 4.46 cfs[a]

        Peak elevation = 344.71 ft

    Summary of approximate peak storage

        Initial storage = 0 cu-ft

        Peak storage from storm = 8005 cu-ft

        Total storage in pond = 8005 cu-ft

[a] The computed peak discharge for 100-year, 15 minute storm is 4.43 cfs, which is nearly the same as that of a 100-year, 30 minute storm.

TABLE 7.6

Routing Computation Summary

Storm Duration (min)	2-Year Storm		10-Year Storm		100-Year Storm	
	Inflow	Outflow	Inflow	Outflow	Inflow	Outflow
15	4.74	0.94	7.25	2.58	10.21	4.42
20	4.29	0.68	6.07	2.03	8.58	3.84
30	3.26	1.02	4.88	2.59	6.96	4.46
45	2.66	1.45	3.70	2.76	5.33	4.33
60	2.07	1.39	2.96	2.52	–	–

TABLE 7.7

Comparison of Pre- and Post Development Peak Flows

Storm Freq. (years)	Q_E = Ex. Peak (cfs)	Q_P = Prop. Peak (cfs)	Reduction % $(1 - Q_P/Q_E)$
2	3.82	1.45	62
10	5.40	2.76	49
100	7.63	4.46	42

graphical solution and the routing computations are performed for that duration to determine discharge. This procedure shortens the routing computations.

7.3   EXTENDED DETENTION BASINS

Extended detention basins, also referred to as dual-purpose detention basins, are those basins that not only attenuate the peak rates of runoff, but also improve the water quality. The latter effect is achieved by using a small, low-level opening in the outlet structure to prolong the release of runoff from the detention basin. The opening is commonly sized based on the water quality storm, which as indicated in a previous chapter, varies from state to state. Required storage volume of a detention basin can be calculated based on the methods previously discussed in this chapter. As indicated, a multiplier should be applied to the calculated storage volume to account for the nonlinear discharge from the detention basin and particularly the discharge from the low level opening that is to impound the water quality storm, in whole or in part. The sizes of openings in the outlet structure, which produce an efficient use of detention basins, are commonly determined through an iterative process. A simple procedure for sizing the openings is presented in a paper by the author (Pazwash, 1992). Design procedures for dual-purpose detention basins are presented by case studies in this chapter. To avoid standing water, which creates mosquito breeding, the bottom of detention basins should lie above the water table. A shallow ponding also promotes algae growth, which is unsightly.

CASE STUDY 7.1

This study relates to the design of storm water management elements for a three-story self-storage commercial building on a 4-acre lot that is subdivided from a 29.84-acre parcel of land. The project site, which is located in the northeast section of Hillsborough Township, Somerset County, New Jersey, is identified as lot 12.01, block 65 in the township tax map (see Figure 7.18) and is occupied by a nursery and garden center. The site is bounded by the Conrail to the south, Route 206 to the north, and the remainder of lot 12.01 to the east. Figure 7.19 shows

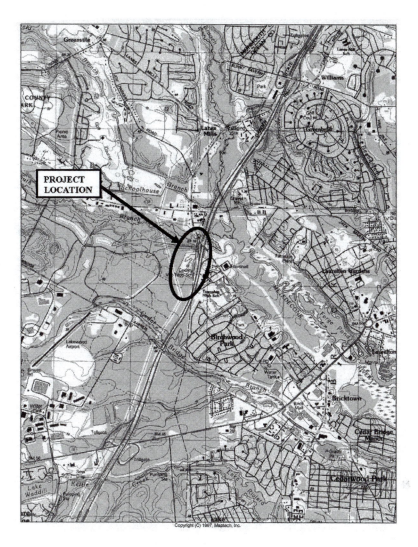

FIGURE 7.18 Project location map portion of lot 12.01, in block G5 tax map sheets 6 and 7, township of Hillsborough.

the predevelopment site layout. This plan indicates that the site is largely disturbed, covered by buildings, gravel parking, and storage sheds. The site is relatively flat, sloping from 71 ft (21.6 m) elevation along the railroad to the south to an elevation range of 67 to 69 ft (20.4–21 m) along Route 206 to the north.

Figure 7.20 depicts the post development site layout and the footprint of the building, which is 40,110 ft^2 (3726 m^2) on plan area. Since the project increases imperviousness, an extended detention basin is provisioned to reduce the peak runoff rates from the site in compliance with the state's storm water management regulations and to improve the water quality. The detention basin is proposed at the northwesterly corner of the site that is depressed and is currently discharged by a pipe to an existing inlet on Route 206.

To supplement the water quality effect of the extended detention basin, manufactured water treatment devices are incorporated in the storm water management system. In addition, underground infiltration chambers are provisioned to address the groundwater recharge requirement for the project. Design calculations for the proposed storm water management elements are presented as follows.

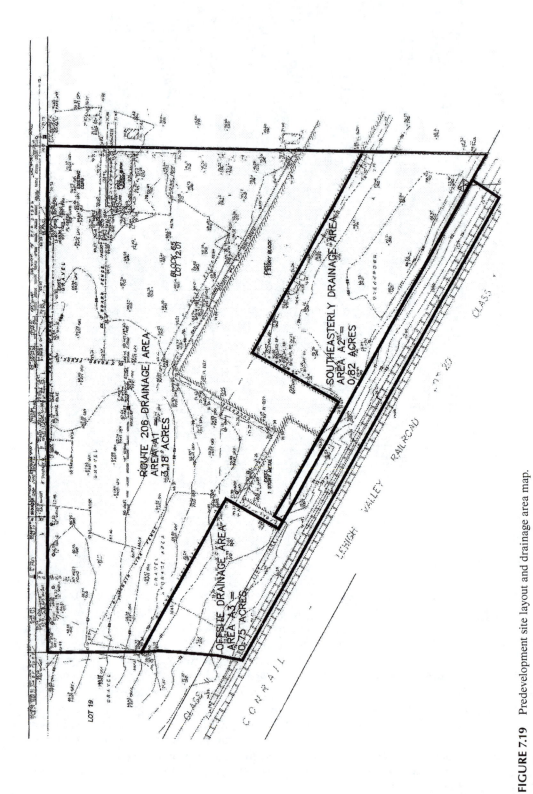

FIGURE 7.19 Predevelopment site layout and drainage area map.

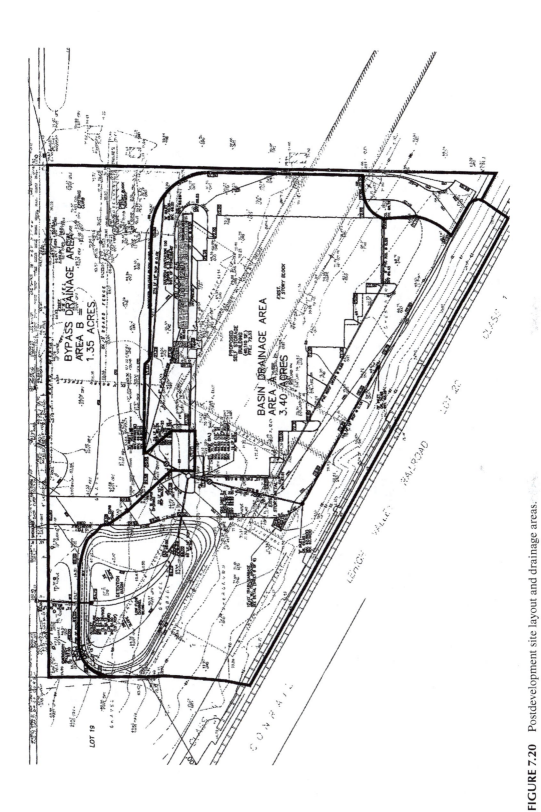

FIGURE 7.20 Postdevelopment site layout and drainage areas.

RUNOFF CALCULATIONS

Based on topography, the site is divided into three subareas. These subareas are designated A1, A2, and A3 in Figure 7.19. Area A1, containing 3.18 acres, flows into an existing storm drain system connected to Route 206 storm drainage; area A2, containing 0.82 acres, is a self-contained area. Area A3 is a 0.75 acre off-site drainage area, the runoff from which flows onto the project site. Since area A2 does not contribute to any runoff from the site, it is excluded from the predevelopment runoff calculations.

The postdevelopment drainage subareas are delineated in Figure 7.20. As seen, the detention basin receives runoff from the offsite area, designated as area A3 in Figure 7.19. Runoff calculations for the project are prepared using rational and modified rational methods. Specifically, the peak rates of runoff are calculated using the former method and the routing computations are prepared based on the latter method. Runoff coefficients of 0.95, 0.60, 0.45, and 0.35 are selected for the impervious, gravel pavement, bare soil, and lawn/landscaped areas, respectively. Consistent with the applicable storm water management regulations, runoff calculations are performed for the storms of 2-, 10-, and 100-year frequency. The calculations also include the 25-year storm frequency in compliance with drainage standards for the county. Table 7.8 summarizes calculated values of the runoff coefficients and the peak rates of runoff from the site. Tables 7.9 and 7.10 list the calculated values of the runoff coefficients and peak rates of runoff and allowable discharges from areas A1 and A3, respectively. The allowable discharges are calculated by applying the required reduction factors to the peak rates of runoff to the onsite area (area A1), noting that offsite runoff is not subject to a reduction.

Runoff calculations for the postdevelopment conditions are performed by separating the tributary area to the detention basin, area A, from the bypass area, area B. Calculations are arranged in Table 7.11, which includes the tabulated runoff coefficients for areas A and B, the peak rates of runoff from these areas, and the allowable discharges from the detention basin. The latter discharges are conservatively calculated algebraically by deducting the bypass runoff from the allowable discharge from the site (shown in Table 7.8) for each storm event.

TABLE 7.8
Predevelopment Runoff Calculations and Allowable Discharges from the Site

Total lot area = 4.00 acres

Total area of disturbance (areas A1 and A2) = 4.00 acres

Drainage area A1 to Route 206 drainage system = 3.18 acres

Drainage areas A2 to southeasterly property corner = 0.82 acre

(This area is self-contained on site and is excluded from the predevelopment calculations.)

Drainage area A3 (offsite tributary to basin) = 0.75 acre

TABLE 7.9
Composite Runoff Coefficients for Areas A1 and A3

| | Imp., C = 0.95 | | Bare Soil, C = 0.45 | | Gravel, C = 0.60 | | Total | |
	(Acre)	A × C	(Acre)	A × C	(Acre)	A × C	Area	$C_w = (\Sigma A*C)/A$
Area A1	0.73	0.69	1.10	0.50	1.35	0.81	3.18	0.63
Area A3	–	–	0.75	0.34	–	–	0.75	0.45

T_c = 15 min.

$Q_{A1} = A \times C \times I = 2.00 \times I.$

$Q_{A3} = A \times C \times I = 0.34 \times I.$

TABLE 7.10
Peak Rates of Runoff

Storm Frequency (years)	I (in./h)	Q_{A1} (cfs)	Reduction	Q_{A3} (cfs)	Q_S (cfs)
2	3.5	7.00	50%	1.19	4.69
10	4.8	9.60	25%	1.63	8.83
25	5.6	11.20	–	1.90	13.10
100	6.8	13.60	20%	2.31	13.19

Note: Allowable from site, $Q_S = \% \, (Q_{A1}) + Q_{A3}$.

TABLE 7.11
Post Development Runoff Calculations and Allowable Discharges from the Detention Basin

Drainage area A to basin = 3.40 acres
Drainage area B bypassing basin = 1.35 acres

	Imp., C = 0.95		Landscape, C = 0.35		Offsite, C = 0.45		Total	
	(Acre)	$A \times C$	(Acre)	$A \times C$	(Acre)	$A \times C$	Area	$C_w = (\Sigma A * C)/A$
Area A	1.93	1.83	0.72	0.32	0.75	0.34	3.40	0.73
Area B	0.19	0.18	1.16	0.41	–	–	1.35	0.43

$T_c = 15$ min.
$Q_A = A \times C \times I = 2.48 \times I$
$Q_B = A \times C \times I = 0.58 \times I$

Storm Frequency (years)	I (in./h)	Q_A (cfs)	Q_B (cfs)	Q_O (cfs)
2	3.5	8.68	2.03	2.66
10	4.8	11.90	2.78	6.05
25	5.6	13.89	3.25	9.86
100	6.8	16.86	3.94	9.25

Note: Allowable from basin, $Q_O = Q_S{}^a - Q_B$.
[a] See Table 7.33.

DETENTION BASIN DESIGN

To simplify routing computations, first the critical storm duration, namely the one that requires the largest detention storage, is calculated based on the modified rational method. Calculations are organized in Table 7.12 for storms of 2-, 10-, and 100-year frequency. This table indicates a critical storm duration of 60 minutes for a 2-year storm and 30 minutes for the 10- and 100-year storms. The calculations also indicate a required detention storage of approximately 11,650 ft³. However, considering that the detention basin is dual purpose and, more importantly, that there is a large room available to place a basin, a detention basin approximately 2.4 times larger than required is provisioned. This will lessen surcharges experienced from an inlet on Route 206 to which the outlet pipe from the detention basin will be connected. To prolong the retention time in the detention basin, a 3 in. orifice is proposed for the discharge of the water quality storm. Tables 7.13 and 7.14 provide storage and discharge calculations for the detention basin, respectively.

Table 7.15 summarizes routing computations for the 2-, 10-, and 100-year frequency storms. Also listed in this table are the calculated allowable discharges from the detention basin (refer

TABLE 7.12
Critical Storm Calculations, Modified Rational Method

$V = [T_b \times 60s \times Q_c] - [0.5 \times T_b \times 60s \times Q_O] \, T_b \, (\text{base time}) = T_d + T_c, \, T_c = 15 \, \text{min}$

Q_O = allowable discharge

Refer to Table 7.11 for calculations of Q_O

Storm Duration (min)	I (in./h)	Q_A (cfs)	Required Vol. (ft³)
2-Year Storm			$Q_O = 2.66$
15	3.50	8.68	5418
20	3.00	7.44	6135
30	2.40	5.95	7123
45	1.80	4.46	7265
60	1.50	3.72	7407
90	1.10	2.73	6352
10-Year Storm			$Q_O = 6.05$
15	4.80	11.90	5269
20	4.00	9.92	5552
30	3.20	7.94	6117
45	2.50	6.20	5850
60	2.00	4.96	$Q_A < Q_O$
100-Year Storm			$Q_O = 9.25$
15	6.80	16.86	6853
20	5.60	13.89	10,103
30	4.50	11.16	11,651
45	3.60	8.93	$Q_A < Q_O$
60	3.00	7.44	$Q_A < Q_O$

TABLE 7.13
Storage–Elevation Relation

Elevation	Area (ft²)	Ave. Area	Storage (ft³)	Σ Volume (ft³)
64.75	0		0	0
		50		
65.0	100		13	13
		700		
65.5	1300		350	363
		3580		
66.0	5860		1790	2153
		8900		
66.5	11,940		4450	6603
		12,270		
66.75	12,600		3068	9670
		12,950		
67.0	13,300		3238	12,908
		13,980		
67.5	14,660		6990	19,898
		15,450		
68.0	16,240		7725	27,623

TABLE 7.14

Discharge–Elevation Relation

			Q (cfs)		
Elevation	3 in.	36 in. × 6″	Cover	Emergency Spillway	Total
64.75	0				0
65.0	0.06				0.06
65.5	0.12				0.12
66.0	0.17				0.17
66.5	0.20				0.20
66.75	0.21	0.00			0.21
67.0	0.23	1.13			1.35
67.5	0.25	5.11	0	0	5.36
68.0	0.28	7.22	0	15.91	23.41

3 in. orifice—inv. 64.75 ft

36 in. × 6 in. weir—inv. 66.75 ft

Top of concrete outlet structure—elevation 67.25 ft

Solid cover on outlet structure—elevation 67.50 ft

Emergency spillway—inv. 67.50 ft

Equations:

Orifice flow

$Q_o = f[CA\ (2gh)^{1/2}]$, $C = 0.6$

Clogging factor, $f = 0.66$

Weir flow

$Q_w = CLH^{1.5}$, $C = 3.0$

3 in. Orifice

$A = 0.049$ ft^2

H = elevation—64.875 ft

36 in. × 6 in. Weir

$L = 3$ ft

H = elevation—66.75

Above elevation 67.25 ft use orifice equation

$A = 1.50$ ft^2

H = elevation—67.00

Emergency spillway, 15 ft. weir

$L = 15$ ft

H = elevation—67.50

Velocity over spillway:

$V = Q/A = 15.91/5.75 = 2.8$ ft/s

Permissible velocity for 6 in. riprap is 6.0 ft/s

TABLE 7.15

Routing Computations Summary

Storm Frequency (years)	Critical Storm Duration (min.)	Q_i, Inflow	Q_o, Outflow	Allowable from Detention
2	60	3.72	1.01	2.66
10	30	7.94	1.39	6.05
100	30	11.16	3.95	9.25

TABLE 7.16
Comparison of Predevelopment, Allowable and Post Development Peak Rates of Runoff for the Site

Storm Frequency (years)	$Q_E{}^a$ (cfs)	$Q_A{}^b$ (cfs)	Q_P (cfs)
2	8.19	4.69	3.04
10	11.23	8.83	4.17
100	15.91	13.19	7.89

[a] Predevelopment flow includes offsite area.
[b] See Table 7.8 for allowable peak flow from site, Q_A. $Q_P = Q_o + Q_B$.

to Table 7.11). As shown, the proposed extended detention basin reduces the peak flows far in excess of the regulations.

Table 7.16 presents a comparison of the pre- and post development discharges from the project site.

WATER QUALITY PROVISION

The NJDEP storm water management regulations call for removing 80% TSS from the increased impervious areas and 50% for the pavements to be reconstructed. Calculations for the required TSS removal rate are arranged in Table 7.17. These calculations indicate a required TSS removal rate equal to 25% for the entire site.

To determine the TSS removal provided by the extended detention basin, the New Jersey water quality storm, namely 1.25 in. of runoff in 2 hours, is routed through the detention basin. It is to be noted that the water quality storm has a nonuniform distribution; however, according to the author's experience, routing computations can be simplified based on the average rainfall intensity (0.625 in./h) without a sacrifice in accuracy. The routing computation, thus

TABLE 7.17
Required TSS Removal Rate for the Project Area

Ground Cover	Area (ac.)	Percent TSS Required Removal Rate	A × Percent TSS
Ex. impervious (excludes ex. roof)	0.17	50%	0.09
Proposed roof	0.92	0%	0.00
Increase impervious	1.03	80%	0.82
Grass/landscape (to basin)[a]	0.72	0%	0.00
Offsite area	0.75	0%	0.00
	$\Sigma A = 3.59$		$\Sigma A \times \% \text{TSS} = 0.91$

Required TSS removal rate $= \dfrac{0.91}{3.59} = 25\%$

Provided TSS removal rate for the project area
CDS units = 50%
Detention = 42% (see Figure 7.21)
Composite = 71%

[a] Grass area bypassing detention basin (1.16 acres) is excluded from the calculations. This area will not be treated for water quality and does not require any treatment for water quality.

performed, shows a retention time (defined as the time from the peak discharge from the basin until 90% of the maximum water quality volume stored in the detention basin is drained) of 13 hours. According to Figure 7.21, the proposed detention basin provides 42% TSS removal for the runoff from the pavement area tributary of which is 1.01 acres, excluding the roof. However, since the new impervious area (excluding the roof), which is 1.03 acres and requires 80% TSS removal, the detention basin alone is insufficient. To address the required TSS removal, two manufactured water treatment devices are placed en route of drainage lines before the detention basin. The location of these devices, which are CDS (continuous deflective system) units, are shown in Figure 7.20. The composite TSS removal rate provisioned by the detention basin and the water quality units is calculated considering that the area tributary to the CDS units, namely the area tributary to the proposed four inlets, receives treatment by not only the CDS units but also the detention basin. However, the remainder of the paved area, which is routed directly through the detention basin, is treated by the basin alone. Table 7.18 summarizes the calculations of the TSS removal rate provided by the system. In this table the composite TSS removal rate of the water quality storm that passes through the CDS units is calculated as follows:

$$TSS = TSS_1 + TSS_2 - \frac{TSS_1 \times TSS_2}{100}$$

$$TSS = 50 + 42 - \frac{50 \times 42}{100} = 71\%$$

where TSS_1 and TSS_2 are the TSS removal rates of each CDS unit and the detention basin, respectively.

Flow calculations for the CDS units are prepared based on the drainage areas to the proposed inlets. Based on an inlet drainage area map (not included herein), the CDS units 1 and 2 receive runoff from 0.52 and 1.04 acres, respectively. Tables 7.19 and 7.20 present flow calculations for the CDS units. Based on the calculated flows, CDS model number PMSU20-25 (now named

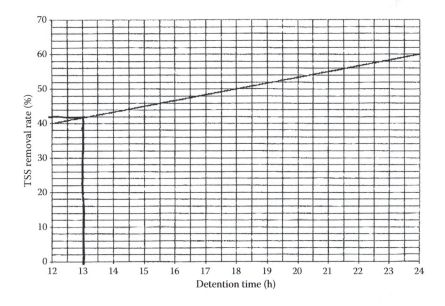

FIGURE 7.21   TSS removal rate versus detention time. (Adapted from New Jersey Srormwater BMP Manual, Chapter 9.4: "Standard for Extended Detention Basins," February 2004, pp. 9.4–3.)

TABLE 7.18
Provided TSS Removal Rate for Paved Areas

Ground Cover	Area (ac.)	Percent TSS Required Removal Rate	A × Percent TSS
To CDS units and basin	1.56	71%	1.108
To basin only	0.37[a]	42%	0.155
Bypass	0.10	0%	0.000
	ΣA = 3.59		$\Sigma A \times \%$ TSS = 1.263

Note: Provided TSS removal rate = 1.263/3.59 = 0.35%.

[a] 1.93 acres tributary area to the basin; 1.56 acres treated by CDS units.

TABLE 7.19
Calculations for CDS Unit Sizing: Composite Runoff Coefficient

	Imp., C = 0.95		Pervious, C = 0.35			
	(ac)	A × C	(ac)	A × C	Total Area	$C_w = (\Sigma \times AC)/A$
CDS unit 1	0.41	0.39	0.11	0.05	0.52	0.84
CDS unit 2	0.38	0.36	0.66	0.30	1.04	0.63

1.25 in. in 2 h
T_C = 15 min
I = 2.6 in./h

TABLE 7.20
Flow Calculations and CDS Sizing

CDS Unit	Q = CIA	Model No.	Treatment Rate
1	1.14	PMSU 20-25	1.7
2	1.70	PMSU 20-25	1.7

CDS 2025-W), capable of treating 1.7 cfs, is selected for unit 1 and unit 2. (This device is designated as CDS-6 in the NJDEP verification letter and has a treatment capacity of 1.6 cfs.)

GROUNDWATER RECHARGE CALCULATIONS

The site is located where the NJDEP groundwater recharge requirement is to be met. The criteria for meeting this requirement were presented in a previous chapter. In this project, the annual groundwater recharge criterion is selected for the design of a groundwater recharge system. The groundwater recharge calculations are prepared using the New Jersey Geological Survey (NJGS) spreadsheet GSR-32 (1993). This spreadsheet performs calculations of the annual groundwater deficit due to a project and also calculates the groundwater recharge provided by a selected infiltration system.

The spreadsheet calculations for the groundwater recharge deficit are presented on sheet 1 of 2 of Figure 7.22. The calculations in this table, which are based on the 4.0-acre area of disturbance, indicate a deficit of 29,034 ft³ (822 m³). To offset this deficit, three rows of 15 units each Cultec Contractor HD 100 in a 115 ft by 11 ft (35 × 3.35 m) stone trench are provisioned.

Annual groundwater recharge analysis (based on GSR-32)

New Jersey Groundwater Recharge Spreadsheet Version 2.0 November 2003	Select Township↓	Average Annual P (in)	Climatic factor
	Somerset Co., Hillsborough TWP	45.7	1.50

Land Segment	Area (acres)	TR-55 Land cover	Soil	Annual Recharge (in)	Annual Recharge (cu.ft)
1	0.73	Impervious areas	Penn	0.0	–
2	1.35	Gravel, dirt	Penn	5.6	27,198
3	1.92	Open space	Penn	12.6	87,678
4					
5					
6					
7					
8					
9					
10					
11					
12					
13					
14					
15					
Total =	4.0			Total Annual Recharge (in) 7.9	Total Annual Recharge (cu.ft) 114,875

Procedure to fill the pre-developemtn and post-development conditions tables

For each land segment, first enter the erea, the select TR-55 Land Cover, then select Soil. Start from the top of the table and proceed downward. Don't leave blank rows (with A–C) in between your segment entries. Rows with A=D will not be displayed or used in calculations. For impervious areas outside of standard lots select "Impervious Areas" as the Land Cover. Soil type for impervious areas are only required if an infiltration facility will be built within these areas.

Project name:	RAIA Properties Corp
Description:	Block 65, Lot 12.01
Analysis state:	04/02/07

Land Segment	Area (acres)	TR-55 land cover	Soil	Annual Recharge (in)	Annual Recharge (cu.ft)
1	2.12	Impervious areas	Penn	0.0	–
2	1.88	Open space	Penn	12.6	85,851
3					
4					
5					
6					
7					
8					
9					
10					
11					
12					
13					
14					
15					
Total =	4.0			Total Annual Recharge (in) 5.9	Total Annual Recharge (cu.ft) 85,851

Annual recharge requirements calculations

% of Pre-developed annual recharge to preserve	100%	Total Impervious Area (sq.ft)	92,347
Post-development annual recharge deficit =	29,034	(Cubic feet)	

Recharge efficiency parameters calculations (area averages)

RWC =	391	(in)		DRWC =	0.57	(in)
ERWC =	0.98	(in)		EDRWC =	0.14	(in)

(*Continued*)

FIGURE 7.22 Annual groundwater recharge analysis based on GSR-32.

Project Name	Description		Analysis Date		BMP or LID Type		
RAIA Properties Corp	Block 65, Lot 12.01		04/02/07		CULTEC Contractor 100HD (3 Rows X 15 Units)		

Recharge BMP input parameters				Root zone water capacity calculated parameters				Recharge design parameters			
Parameter	Symbol	Value	Unit	Parameter	Symbol	Value	Unit	Parameter	Symbol	Value	Unit
BMP Area	ABMP	1265.0	sq.ft	Empty portion of RWC under post-D natural recharge	ERWC	1.02	in	Inches of Runoff to capture	Qdesign	1.22	in
BMP effective depth, this is the design variable	dBMP	10.9	in	ERWC modified to consider dEXC	EDRWC	0.40	in	Inches of Rainfall to capture	Pdesign	1.44	in
Upper level of the BMP surface (negative if above ground)	dBMPu	9.0	in	Empty portion of RWC under infilt. BMP	RERWC	0.31	in	Recharge provided avg. over imp. area		29.3	in
Depth of lower surface of BMP, must be >=dBMPu	dEXC	27.5	in					Runoff captured avg. over imp. area		31.9	in
Post-development land segment location of BMP, Input Zero if location is distributed or undetermined	SegBMP	2	unitless								

Parameters from annual recharge worksheet				BMP calculated size parameters				Calculation check messages			
Parameter	Symbol	Value	Unit	ABMP/Aimp	Aratio	0.11	unitless	Volume balance -> Solve problem to satisfy annual recharge			
Post-D deficit recharge (or desired recharge volume)	Vdef	29,024	cu.ft	BMP volume	VBMP	1,149	cu.ft	dBMP check -> OK			
Post-D impervious area (or target impervious area)	Aimp	12,000	sq.ft	System performance calculated parameters				dEXC check -> OK			
Root zone water capacity	RWC	4.07	in	Annual BMP recharge volume		29,294		BMP location -> OK			
RWC modified to consider dEXC	DRWC	1.59	in	Avg BMP recharge efficiency		91.8%	Represents % infiltration recharged	Other notes			
Climatic factor	C-factor	1.50	no units	% Rainfall became runoff		77.9%		Pdesign is accurate only after BMP dimension are updated to make rech volume = deficit volume. The portion of BMP infiltration prior to filling and the area occupied by BMP are ignored in these calculations. Results are sensitive to dBMP, make sure dBMP selected is small enough for BMP to empty in less than 3 days. For land segment location of BMP if you select "Impervious areas" RWC will be minimal but not zero as determined by the soil type and a shallow root zone for this land cover allowing consideration of lateral flow and other losses.			
Average annual P	Pavg	45.7	in	% Runoff infiltrated		89.6%					
Recharge requirement over imp. area	dr	3.8	in	% Runoff recharged		10.7%					
				% Rainfall recharged		8.3%					

How to solve for different recharge volumes: By default the spreadsheet assigns the values of total deficit recharge volume "Vdef" and total proposed impervious area "Aimp" from the "Annual recharge" sheet to "Vdef" and "Aimp" on this page. This allows solution for a single BMP to handle the entire recharge requirement assuming the runoff from entire impervious area is available to the BMP. To solve for a smaller BMP or a LID-IMP to recharge only part of the recharge requirement, set Vdef to your target value and Aimp to impervious area directly connected to your infiltration facility and then solve for ABMP or dBMP. To go back to the default configuration click the "Default Vdef and Aimp" button.

FIGURE 7.22 (CONTINUED) Annual groundwater recharge analysis based on GSR-32.

This recharge system, which is shown in Figure 7.20, receives roof runoff from a portion of the self-storage building.

Sheet 2 of 2 of the spreadsheet in Figure 7.22 shows that by draining 12,000 ft² (1115 m²) of the roof area into the Cultec chambers, 29,294 ft³ (829.5 m³) of annual groundwater recharge can be attained. The parameters in this table are

$$\text{ABMP} = \text{recharge area} = 115 \times 11 = 1265 \text{ ft}^2$$

$$\Psi = 45 \text{ units at } 22.28 \text{ ft}^3/\text{unit} + 0.4 \ (18.5 \text{ in.}/12)[2 \times 11 + 2 \times 115.0'] = 1158 \text{ ft}^3$$

where
2 ft × 11 reflects the area of stone at the sides of chambers
dBMP = effective depth of recharge system = 1158/1265 = 0.92 ft = 10.9 in.
dBMP_u = depth of top of infiltration system = 9.0 in.
dEXC = depth of infiltration system bottom = 9 + 18.5 = 27.5 in.

Other parameters in Figure 7.22 Sheet 2 are filled automatically by the spreadsheet when inputting the name of a municipality in New Jersey.

To determine that the bottom of the infiltration chambers and the extended detention basin lie above the ambient water table two test pits were conducted at the site. Soil samples were also sent to a laboratory for gradation tests and soil permeability ratings. Groundwater was encountered at 5 ft, 8 in. (1.7 m) below the existing grade and motting (which is an indicator of seasonal high water table) was found at 3 to 5 ft (1–1.5 m) depth. The soil gradation indicated a soil permeability rating of K3 in one pit and K2 in another. These ratings reflect a permeability range of 2–6 in./h (50–150 mm/h) and 0.6–2 in./h (15–50 mm/h), respectively. The tests verify that the designed bottom of the detention basin and the infiltration chambers lie above the water table. As such, the chambers will function satisfactorily.

CASE STUDY 7.2

CASE DESCRIPTION

This case relates to a residential project in the town of North Greenbush in Rensselaer County in New York. The project named Westview Estates includes the construction of 38 single-family homes in a 44.5-acre (18.0 ha) parcel of property along County Route 74, also known as Winter Street Extension. The project disturbs approximately 23.7 acres (9.59 ha) while the remainder of the property remains intact and serves as conservation land. Figure 7.23 depicts the predevelopment layout of the vicinity of the development area. This plan, which is prepared based on 2 ft contour lines of an aerial map of the site, shows that a single building and its driveway, covering 0.5 acres (2024 m²), are the sole impervious areas on the site. Figure 7.24 shows the postdevelopment layout and two extended detention-water quality ponds. Design calculations for these ponds, which are provisioned to address the storm management requirements for the project, are discussed herein.

DESCRIPTION OF STORM WATER MANAGEMENT PROVISIONS

Figure 7.23 shows the limits of the area to be disturbed by the project, which, as indicated, measures 23.7 acres. Of this, approximately 18.9 acres (7.65 ha) are tributary to northerly properties and an existing drainage ditch along Winter Street Extension and 4.8 acres (1.94 ha) flow in southerly and easterly directions to neighboring properties. The remainder of the site area discharges to a northerly property.

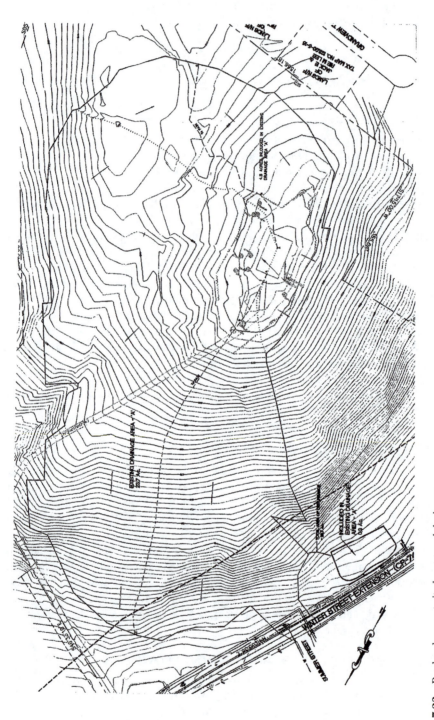

FIGURE 7.23 Predevelopment site layout and drainage.

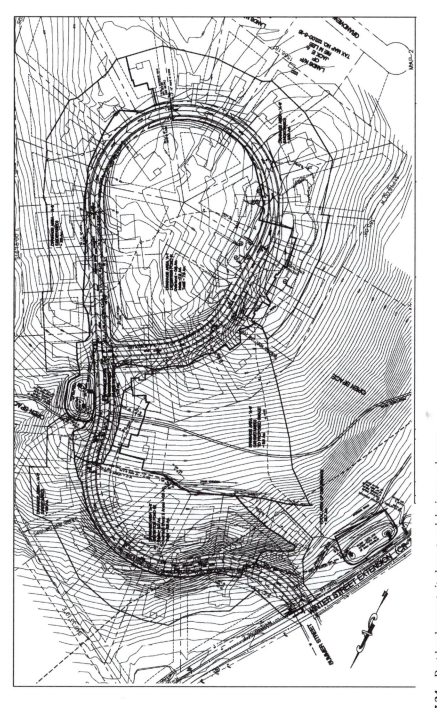

FIGURE 7.24 Postdevelopment site layout and drainage subareas.

Delineated on Figure 7.24 are the drainage areas tributary to the detention ponds and subareas bypassing them. To reduce the runoff to the northerly property, where the soil is reported to be soggy, the discharge from detention pond 1 is redirected by a proposed swale to the southerly side of the site, where it will sheet flow to the existing drainage ditch. A close review of Figures 7.23 and 7.24 indicates that the proposed detention pond 1 collects runoff from a large portion of the area that would otherwise flow to the northerly property and, as such, it reduces the runoff to this property. The plans also indicate that the project will reduce the tributary area to the easterly and southerly properties from 4.8 acres (1.94 ha) to 3.3 acres (1.34 ha) and thereby reduce the runoff to these properties as well.

RUNOFF CALCULATIONS

Runoff calculations are prepared using the SCS TR-55 method. According to the Rensselaer County soil survey maps, the site soil is mostly hydrologic group C. The predevelopment soil curve number and time of concentration are calculated to be 70 and 0.4 hours, respectively. Consistent with the New York State storm water management regulations, the calculations are performed for the storms of 1-, 10-, and 100-year frequency. For the ease of comparison with the postdevelopment conditions, the entire 23.7 acres of the disturbance area, which is designated as area A on Figure 7.23, is lumped as one area in performing the predevelopment runoff calculations. Table 7.8 lists the computed predevelopment peak rates of runoff for various storm events.

Runoff calculations for the postdevelopment conditions include the areas tributary to the detention ponds, designed as areas A-1 and A-2 on Figure 7.24. Also included in the calculations are the undetained areas, which comprise 12.8 acres (5.18 ha) in total and are designated as area A-3 on this plan.

DETENTION PONDS DESIGN

Figure 7.25 shows schematic details of the proposed detention ponds. This type of pond is referred to as pocket pond P-5 in the NY State Stormwater Management Design Manual (2003 and 2010). In accordance with this manual, each pond is provided with a forebay and a permanent pool. The forebay has a minimum capacity equal to 10% of the water quality volume, WQ_v, and the permanent pool a capacity at least equal to 50% WQ. The water quality volume in New York State is the 90% rule, defined in a previous chapter. Calculations for the water quality volumes in ponds 1 and 2 are presented next.

WQ_v CALCULATIONS

Detention pond 1:

$$WQ_v = PR_v \frac{A}{12}$$

$$A = 7.5 \text{ acres}$$

$$P = 0.95 \text{ in. (see Figure 7.26)}$$

$$I = \text{percent pavement} = 26\%$$

$$R_v = 0.05 + 0.009(I) = 0.284$$

$$WQ_v = 0.95 \times 0.284 \times \frac{7.5}{12} = 0.284 \text{ ac-ft} = 7357 \text{ ft}^3 (208.3 \text{ m}^3)$$

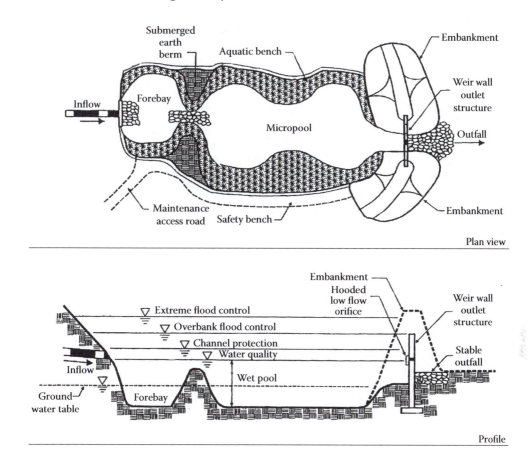

FIGURE 7.25 Pocket pond (P-5).

Detention pond 2:

$$A = 3.3 \text{ acres}$$

$$P = 0.95$$

$$I = 24\%$$

$$R_v = 0.05 + 0.009 \times 0.24 = 0.266$$

$$WQ_v = 0.07 \text{ ac-ft} = 3050 \text{ ft}^3 \text{ } (86.4 \text{ m}^3)$$

Detention pond 1 has been designed with a forebay and permanent pool, each extending from elevation 469.5 ft to 473.0 ft with a storage volume as follows*:

$$\text{Forebay } 750 \text{ ft}^3 > 10\% \text{ } WQ_v$$

$$\text{Permanent pool } 3700 \text{ ft}^3 > 50\% \text{ } WQ_v$$

---

* Calculations for the storage volume of forebays and permanent pools of the detention ponds, which are calculated from the elevation–area relation for each, are not included herein, for brevity.

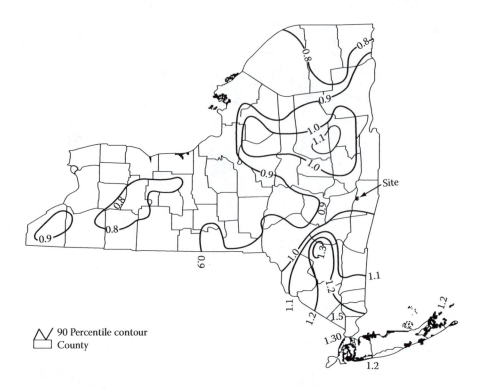

FIGURE 7.26   Water quality storm in New York State.

Likewise, the forebay and permanent pool of detention pond 2 are designed as follows:

Forebay elev. 394 ft to 397.25 ft, storage = 310 ft³

Permanent pool elev. 394 ft to 397.25 ft, storage = 1550 ft³

CP_v, Q_p, AND Q_f DESIGN CALCULATIONS

The detention ponds 1 and 2 are provided with an active storage above their permanent pools. Also, the lowest opening in the outlet structure of each pond is set just above its permanent pool water level. Figure 7.27 depicts the openings and elevations of outlet structures of detention ponds 1 and 2. Storage and discharge relations with elevation for detention pond 1 are tabulated in Tables 7.21 and 7.22, respectively. Tables 7.23 and 7.24 organize the storage and discharge calculations for detention pond 2. It is to be noted that the storage volumes of forebay and permanent pool, which are noneffective in flow attenuation, are not included in the active detention storage tabulations.

Routing computations for detention ponds are performed for storms of 1-, 10-, and 100-year frequency. These storms represent channel protection (CP_v), overbank flood (Q_p), and extreme flood (Q_f) in New York State, respectively. The computations show that both of the detention ponds have adequate capacity to control the 24-hour SCS storms of up to and including the 100-year frequency. Tables 7.25 and 7.26 summarize the routing computations for the detention ponds 1 and 2, respectively. These tables indicate that the proposed detention ponds create significant reductions in the peak rates of runoff. The attenuation effect is most profound for the 1-year frequency storm, which occurs more frequently than others.

Calculations for the postdevelopment composite discharges from the site are performed by adding routed hydrographs from detention ponds 1 and 2 and direct runoff hydrograph from the undetained area A-3. The calculations include storms of 1-, 10-, and 100-year frequency and are summarized in Table 7.27.

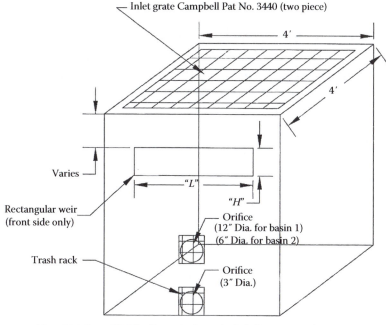

Inlet grate Campbell Pat No. 3440 (two piece)

4′

4′

Varies

"L"

"H"

Rectangular weir
(front side only)

Orifice
(12″ Dia. for basin 1)
(6″ Dia. for basin 2)

Trash rack

Orifice
(3″ Dia.)

Note: Detail provided for dimensioning only. Reinforcement and
structural detailing to be completed by structure
manufacturer and will be responsibility of the contractor.
Detail shop drawings signed and sealed by a licensed
professional engineer to be submitted for approval prior
to construction.

Outlet structure schedule

Pond	Bottom of pond	3 in. orifice invert	6 in. orifice invert	15 in. orifice invert	Weir			Top of structure[a]
					L	H	Invert	
1	473.00	473.00	–	475.00	18 in.	12 in.	476.50	478.10
2	397.25	397.25	398.00	–	12 in.	6 in.	399.00	400.00

[a] Set at the computed 100-year water surface elevation.

FIGURE 7.27 Outlet structure detail.

Table 7.28 presents a comparison of the predeveloped and postdevelopment peak rates of runoff discharged from the area to be distributed by the project. This table indicates that the proposed storm water management system for the project not only meets, but also exceeds the NYS storm water management standards.

7.4 UNDERGROUND DETENTION BASINS

Underground detention basins are commonly installed under pavements such as parking lots and driveways. Where open space is sparse, underground detention systems are far more practical than aboveground detention basins. In highly developed areas where no open space is available, underground detention systems form a most feasible means of attenuating the peak rates of runoff.

TABLE 7.21
Elevation–Storage Chart, Detention Pond 1 (Upper)

Elevation	Area (ft²)	Δ Volume (ft³)	Cumulative Volume (ft³)
473.0	3940		
		4340	
474.0	4740		4340
		5170	
475.0	5600		9510
		6060	
476.0	6520		15,570
		7010	
477.0	7500		22,580
		8015	
478.0	8530		30,595
		2205	
478.25	8800		32,800[a]
		2240	
478.5	9100		35,040
		2310	
478.75	9350		37,350
		Total:	37,350 ft³

[a] Emergency spillway of pond at elevation 478.25 ft.

However, it is to be noted that many jurisdictional agencies do not accept underground detention basins for water quality.

Solid and perforated pipes, chambers, vaults, and modular structures are employed in constructing underground detention basins. Solid pipes and vaults provide only detention storage. However, perforated pipes and chambers in stone trench also provide storage in stone void and, in addition, dissipate runoff through infiltration. As such, these systems are more cost effective than solid pipes. To avoid contaminating groundwater, the bottom of the stone trench should be at least 2 ft (0.6 m) above the water table. The soil should have a minimum permeability rate of 1 in./h (25 mm/h) to allow the system to fully drain within 3 but preferably 2 days.

7.4.1 SOLID AND PERFORATED PIPES

Solid reinforced concrete, high-density polyethylene (HDPE), and corrugated metal pipes are widely employed in practice. Among these, corrugated metal pipes (CMPs) are far more vulnerable to deformation and deterioration than other types of pipe. The author has witnessed many CMPs that were rusted, deformed, and/or collapsed in less than 20 years of service. Thus, the use of this type of pipe in any underground detention system is discouraged.

Using solid pipes in parallel, a minimum spacing should be kept between pipes to allow rock placement and adequate compaction. For advanced drainage system (ADS) N-12 (dual wall pipes), the minimum spacing is 12 in. (30 cm) or one-half of nominal pipe size, whichever is greater. The same spacing should be allowed for concrete pipes (American Concrete Pipe Association, 2005). Perforated RCP and HDPE pipes are installed in stone trench with a minimum 4 in. (10 cm) bedding for 12 to 24 in. (300–600 mm) pipes and 6 in. (15 cm) for 30 to 60 in. (750–1500 mm) pipes. Also, a 6 in. stone cover is commonly placed over the pipes.

TABLE 7.22
Elevation–Outflow Chart, Detention Pond 1 (Upper Pond)

Elevation	Flow Rate (cfs)					
	Q_o (3 in.)	Q_o (15 in.)	Q_w	Q_s	Q_{grate}	Q_{total}
473.0	0					0
473.5	0.14					0.14
474.0	0.22					0.22
474.5	0.28					0.28
475.0	0.32	0				0.32
475.5	0.36	1.10				1.46
476.0	0.40	3.62				4.02
476.5	0.43	5.53	0			5.96
477.0	0.46	6.93	1.59			8.98
477.5	0.49	8.09	4.50			13.08
478.0	0.52	9.10	7.22			16.85
478.1	0.53	9.29	7.58		0	17.40
478.25	0.53	9.57	8.08	0	9.25	27.43
478.5	0.55	10.02	8.85	7.00	15.00	41.41

Equations used:

3 in. Orifice—Inv. 473.0 ft

$Q_o = CA \sqrt{(2gH)} = 0.6 \times 0.049 \times \sqrt{(64.4)} \times H^{0.5}$ $H = $ Elev. $- 473.125$

15 in. Orifice—Inv. 475.0 ft

$Q_o = CA\sqrt{(2gH)} = 0.6 \times 1.227 \times \sqrt{(64.4)} \times H^{0.5}$ $H = $ Elev. $- 475.625$

18 in. $\times$ 12 in. Weir—Inv. 476.5 ft

$Q_w = CLH^{1.5}$ (Elev. 476.5 – 477.5) $= 3.0 \times 1.5 \times H^{1.5}$ $H = $ Elev. $- 476.5$

$Q_w = CA\sqrt{(2gH)}$ (Elev. > 477.5) $= 0.6 \times 1.5 \times \sqrt{(64.4)} \times H^{0.5}$ $H = $ Elev. $- 477.00$

20 ft wide emergency spillway—Crest 478.25 ft

$Q_w = CLH^{1.5}$ (Elev. > 478.25) $= 2.8 \times 20 \times H^{1.5}$ $H = $ Elev. $- 478.25$

Outlet structure grate opening (Elev. 478.10 ft)

$Q_{grate} = $ refer to Campbell Foundry flow charts for flat drop inlets with stream flow grates (pattern no. 3440)

TABLE 7.23
Elevation–Storage Chart, Detention Pond 2 (Lower Pond)

Elevation	Area (ft²)	Δ Volume (ft³)	Cumulative Volume (ft³)
397.25	4940		
		1260	
397.5	4200		1260
		2730	
398	5710		3990
		6250	
399	6785		10,240
		7350	
400	7910		17,590[a]
		2020	
400.25	8200		19,610
		2090	
400.5	8500		21,700
		Total:	21,700 ft³

[a] Emergency spillway of pond at elevation 400.00 ft.

TABLE 7.24

Elevation–Outflow Chart, Detention Pond 2 (Lower Pond)

| | Flow Rate (cfs) | | | | | |
Elevation	Q_o (3 in.)	Q_o (6 in.)	Q_w	Q_s	Q_{grate}	Q_{total}
397.25	0					0
397.5	0.08					0.08
398.0	0.19	0				0.19
398.5	0.25	0.47				0.72
399.0	0.30	0.82	0			1.12
399.5	0.35	1.06	1.06			2.46
400.0	0.38	1.25	2.08	0	0	3.72
400.25	0.40	1.34	2.41	7.00	12.00	23.15
400.5	0.42	1.42	2.69	19.80	17.00	41.33

TABLE 7.25

Routing Computations Summary, Area A.1, Detention Pond 1

Storm Freq. (years)	Q_{In} (cfs)	Q_{Out} (cfs)
1	5.0	0.35
10	18.0	5.75
100	31.0	14.46

TABLE 7.26

Routing Computations Summary, Area A.2, Detention Pond 2

Storm Freq. (years)	Q_{In} (cfs)	Q_{Out} (cfs)
1	2.0	0.01
10	6.0	1.00
100	10.0	3.53

TABLE 7.27

Composite Peak Rates of Runoff Postdevelopment Conditions

Storm Freq. (years)	Pond 1 Q (cfs)	Pond 2 Q (cfs)	Area A-3 Q (cfs)	Site Total Q^a (cfs)
1	0.35	0.01	6.0	6.0
10	5.75	1.0	24.0	29.4
100	14.46	3.53	41.0	56.7

[a] Calculated by adding hydrographs.

TABLE 7.28

Comparison of Existing and Proposed Peak Rates of Runoff

Entire Area of Disturbance (Area A)

Storm Freq. (years)	EX. Q (cfs)	PR. Q (cfs)	Overall Reduction (%)
1	8	6	25%
10	34	29.4	14%
100	60	56.7	5%

Table 7.29 presents ADS's recommended spacing for perforated HDPE pipes, which come in 12 to 60 in. (300–1500 mm) size. This table also lists the storage volumes in pipe, the void volume in stone trench, and the total retention storage. The table serves as an aid in design and is applicable for cases where a large number of pipes are laid in parallel in a wide stone trench. Because of having thicker walls, perforated concrete pipes require a slightly larger stone trench than that of HDPE pipes for a given pipe size. Accurate calculations of storage volume for any perforated pipe in stone trench can be performed by adding the inside volume of pipes to the void volume in stone trench. The latter volume is calculated by deducting the outer volume of pipes from the trench volume and multiplying the difference by the stone porosity, namely the ratio of void to the overall stone volume. A 40% ratio is used in common practice.

Because of ease of construction and lower initial cost, both solid and perforated HDPE pipes are more widely used than solid or perforated reinforced concrete pipes for underground detention basins. It is to be noted, however, that the structural integrity of the HDPE pipes is more sensitive to proper installation, including backfill material and compaction, than that of reinforced concrete pipes. It is advisable to perform a long-term cost analysis before selecting a make of pipe.

7.4.2 CHAMBERS

In recent years a variety of chambers, mostly made of plastic material, have been introduced to the market. Plastic chambers, which are arch shaped, can be stacked up to save storage and transportation cost and are rapidly becoming more popular than perforated pipes. At present, ADS, StormTech™ (now a division of Advanced Drainage System [ADS]), Cultec Inc., StormChamber, and Triton are major manufacturers of plastic chambers in the United States. The StormTech™ chambers, which are made of polypropylene (PP), came in two different sizes, designated as SC-740 and SC310 for over two decades. SC 740 is 51 in. (1295 mm) wide by 30 in. (762 mm) high and SC-310 is 34 in. (864 mm) wide by 16 in. (406 m) high. Each chamber is 90.7 in. (2300 mm) long, providing an effective (installed) length of 85.4 in. (2170 mm). Figure 7.28 shows SC-740 and SC-310 chambers and their dimensions. Also noted in this figure are weights and storage volumes of these chambers. Tables 7.30 and 7.31 present tabulation of gradual storage of the SC-740 and SC-310 chambers, respectively. These tables are handy for detention basin design. A single StormTech SC-740 chamber installed in stone trench with 6 in. (15 cm) stone base and 6 in. stone cover provides 2.2 ft³/ft² (0.67 m³/m²) storage. The storage per unit area of SC-310 is 1.3 ft³/ft² (0.4 m³/m²). This implies that, in terms of storage volume, these chambers are equivalent to 2.2 ft deep and 1.3 ft deep aboveground detention basins, respectively. More recently, StormTech started to make a larger chamber, designated as MC-3500, which is one of the largest chambers currently available on the market. This chamber is 90 in. (2286 mm) long by 75 in. (1905 mm) wide and 45 in. (1143 mm) high. MC-3500 weighs 124 lb and has a chamber storage of 110 ft³ (3.11 m³) and minimum installed storage of 162.8 ft³ (4.61 m³). StormTech™ now makes a chamber for use under parking areas. This chamber is known as DC-780, is 90.7″ × 51″ × 30″ high (2304 × 1296 × 762 mm), which is the same size as SC-740, has a net storage capacity of 46.2 ft³ (1.3 m³). It also makes MC-4500 chamber, which is another chamber for use under parking lots. This chamber is 52″ (1321 mm) long × 100″

TABLE 7.29
Storage Capacities of N-12®, N-12® ST, and N-12® WT Pipes

Nominal Inside Diameter, in. (mm)	Average Outside Diameter, in. (mm)	"X" Spacing, in. (mm)	"S" Spacing, in. (mm)	"C" Spacing, in. (mm)	Pipe Volume, ft³/ft (m³/m)	Stone Void Volume, ft³/ft (m³/m)	Total Retention Storage, ft³/ft (m³/m)	Retention Surface Area Required, ft²/ft³ (m²/m³)	Detention Surface Area Required, ft²/ft³ (m²/m³)
12 (300)	14.5 (368)	8 (210)	10.9 (280)	25.4 (650)	0.81 (0.07)	0.84 (0.08)	1.65 (0.15)	1.3 (4.2)	2.7 (8.6)
15 (375)	18 (457)	8 (210)	10.9 (280)	28.9 (750)	1.2 (0.11)	1.1 (0.10)	2.3 (0.21)	1.1 (3.5)	1.97 (6.4)
18 (450)	21 (533)	9 (230)	14.3 (360)	35.3 (900)	1.8 (0.16)	1.4 (0.13)	3.2 (0.29)	0.93 (3.0)	1.6 (5.4)
24 (600)	28 (711)	10 (260)	13.4 (340)	41.4 (1050)	3.1 (0.29)	2.0 (0.18)	5.1 (0.47)	0.68 (2.2)	1.1 (3.6)
30 (750)	36 (914)	18 (460)	17.1 (430)	53.1 (1350)	4.9 (0.46)	3.1 (0.28)	8.0 (0.74)	0.55 (1.8)	0.90 (3.0)
36 (900)	42 (1067)	18 (460)	21 (530)	63.0 (1600)	7.1 (0.66)	4.2 (0.39)	11.3 (1.05)	0.47 (1.5)	0.74 (2.4)
42 (1050)	48 (1219)	18 (460)	24 (610)	72 (1830)	9.2 (0.87)	5.8 (0.53)	15.0 (1.40)	0.40 (1.3)	0.65 (2.1)
48 (1200)	54 (1372)	18 (460)	24.5 (620)	78.5 (2000)	12.4 (1.15)	6.7 (0.62)	19.1 (1.77)	0.34 (1.1)	0.53 (1.7)
60 (1500)	67 (1702)	18 (460)	23 (580)	90 (2290)	19.3 (1.79)	8.5 (0.78)	27.8 (2.57)	0.27 (0.89)	0.39 (1.3)

Note: See figure below for typical cross section used in volume calculations. Bedding depth assumed 4 in. for 12–24 in. pipe and 6 in. for 30–60 in. pipe.

1. Based on A-profile pipe.
2. Actual ID values used in calculation.
3. Stone porosity assumed 40%.
4. Stone height above crown of pipe is not included in void volume calculations.
5. Calculation is based on the average OD of the pipe.

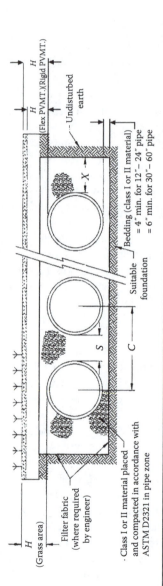

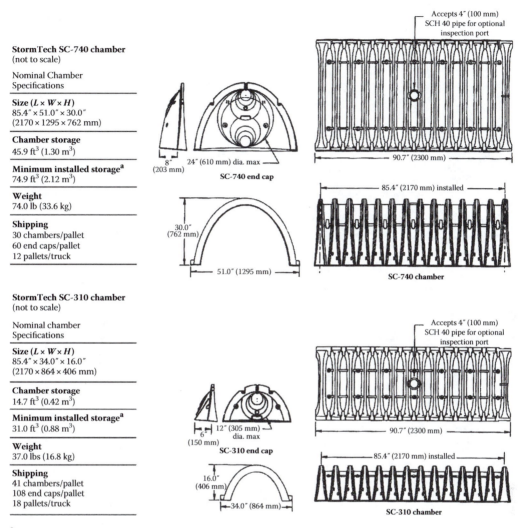

StormTech SC-740 chamber
(not to scale)

Nominal Chamber
Specifications

Size ($L \times W \times H$)
85.4″ × 51.0″ × 30.0″
(2170 × 1295 × 762 mm)

Chamber storage
45.9 ft³ (1.30 m³)

Minimum installed storage[a]
74.9 ft³ (2.12 m³)

Weight
74.0 lb (33.6 kg)

Shipping
30 chambers/pallet
60 end caps/pallet
12 pallets/truck

StormTech SC-310 chamber
(not to scale)

Nominal chamber
Specifications

Size ($L \times W \times H$)
85.4″ × 34.0″ × 16.0″
(2170 × 864 × 406 mm)

Chamber storage
14.7 ft³ (0.42 m³)

Minimum installed storage[a]
31.0 ft³ (0.88 m³)

Weight
37.0 lbs (16.8 kg)

Shipping
41 chambers/pallet
108 end caps/pallet
18 pallets/truck

[a]This assumes a minimum of 6 inches (150 mm) of stone below, above and between chamber rows.

FIGURE 7.28 StormTech chambers, dimensions, and details.

(2540 mm) wide × 60″ (1524 mm) high, weights 120 lbs. (54.4 kg) and provides a chamber storage of 106.5 ft³ (3.01 m³) and minimum installed storage of 162.6 ft³ (4.60 m³).

On October 1, 2008, Contech introduced a large chamber called Chamber Maxx in West Chester, Ohio. This chamber, which is made of polypropylene resin, is 98.4 in. (2.50 m) long, 51.4 in. (1306 mm) wide, and 30.3 in. (770 mm) high and has 49 cf of storage capacity. The installed lengths of starter, intermediate, and end chambers are 96.2, 85.4, and 88.5 in., respectively.

The Cultec chambers are made of high molecular weight HDPE and come in a wide range of sizes varying from 12.5 to 48 in. (317–1200 mm) high and 16 to 54 in. (406–1372 mm) wide. The largest chamber, which is now manufactured by Cultec, is called Recharger 900 HD. Dimensions and storage capacity of this chamber, which is the largest available chamber, are

Length: 9.25 ft (2.82 m)
Width: 78 in. (1.98 m)
Height: 48 in. (1.22 m)

TABLE 7.30

Storage-Stage Relation for SC-740 Chamber

Depth of Water in System (in.)		Cumulative Chamber Storage (ft^3)	Total System Cumulative Storage (ft^3)
	StormTech SC-740 Chamber		
42		45.90	74.90
41	↑	45.90	73.77
40		45.90	72.64
39	Stone	45.90	71.52
38	cover	45.90	70.39
37		45.90	69.26
36	↓	45.90	68.14
35		45.85	66.98
34		15.69	65.75
33		45.41	64.46
32		44.81	62.97
31		44.01	61.36
30		43.06	59.66
29		41.98	57.89
28		40.80	56.05
27		39.54	54.17
26		38.18	52.23
25		36.74	50.23
24		35.22	48.19
23		33.64	46.11
22		31.99	44.00
21		30.29	41.85
20		38.54	39.57
19		26.74	37.47
18		24.89	35.23
17		23.00	32.96
16		21.06	30.68
15		19.09	28.36
14		17.08	26.03
13		15.04	23.68
12		12.97	21.31
11		10.87	18.92
10		8.74	16.51
9		6.58	14.09
8		4.41	11.66
7		2.21	9.21
6	↑	0	6.76
5		0	5.63
4	Stone	0	4.51
3	foundation	0	3.38
2		0	2.25
1	↓	0	1.13

Note: For StormTech SC-740 chamber add 1.13 ft^3 of storage for each additional inch of stone foundation.

TABLE 7.31

Storage-Stage Relation for SC-310 Chamber

Depth of Water in System (in.)		Cumulative Chamber Storage (ft³)	Total System Cumulative Storage (ft³)
	StormTech SC-310 Chamber		
28		14.70	31.00
27		14.70	30.21
26	↑	14.70	29.42
25		14.70	28.63
24	Stone	14.70	27.84
23	cover	14.70	27.05
22		14.70	26.26
21	↓	14.64	25.43
20		14.49	24.54
19		14.22	23.58
18		13.68	22.47
17		12.99	21.25
16		12.17	19.97
15		11.25	18.62
14		10.23	17.22
13		9.15	15.78
12		7.99	14.29
11		6.78	12.77
10		5.51	11.22
9		4.19	9.64
8		2.83	8.03
7		1.43	6.40
6	↑	0	4.74
5		0	3.95
4	Stone	0	3.16
3	foundation	0	2.37
2		0	1.58
1	↓	0	0.79

Note: For StormTech SC-310 chamber add 0.79 ft³ of storage for each additional inch of stone foundation.

Installed length: 7 ft (2.13 m)

Length adjustment per run: 2.25 in. (0.69 m)

Chamber storage: 17.62 ft³/ft; 162.99 ft³/unit; 1.637 m³/m

Minimum installed storage: 27.25 ft³/ft; 190.73 ft³/unit; 2.53 m³/m; 5.40 m³/unit

Minimum center to center spacing: 7.25 ft (2.21 m)

Maximum allowable cover: 6 ft (1.83 m)

This listed storage does not include any void volume in stone trench.

StormChamber™, manufactured by Hydrologic Solutions Inc., is one of the larger chambers currently available on the market. This chamber, made of HDPE, is 8.5 ft (2.59 m) long, 5 ft (1.52 m) wide, and 34 in. (86.4 cm) high. The chamber alone provides 77 ft³ (2.18 m³) of storage volume and has a design storage capacity of 115 to 161 ft³ (3.26 m³) per chamber installed in stone trench. This chamber is strong enough for stacked installation. Figure 7.29 shows a StormChamber withstanding the weight of a jeep.

FIGURE 7.29   StormChamber withstanding the weight of a jeep.

Triton Stormwater Solution of Brighton, Michigan, introduced a chamber made of high-strength resin in 2005. This chamber is nearly 45% lighter than HDPE chambers per cubic foot of storage. Triton chambers come with a side portal feed, which serves as a lateral to which the chambers connect. Chambers are available in six sizes, ranging from 59 × 36 × 35 in. (W, H, L) to 59 × 36 × 102 in. (1.5 × 0.91 × 2.59 m). Triton chambers are soy-oil based and a carbon-neutral product. They are lighter but stronger than other plastic chambers at this time. These chambers are independently tested and approved for double-stack installation. Triton, like StormChamber, is strong enough to carry heavy traffic load in such configurations. They have been also safely installed at a depth of over 20 ft (6 m).

7.4.3 PLASTIC AND CONCRETE VAULTS

More recently a number of underground storm water storage vaults and modules have become commercially available. These modules are made of either reinforced concrete or plastic material. They can be used as a detention system by using either a concrete slab base or impermeable PVC liner or as a detention/retention–infiltration basin using a geotextile fabric. Depending on their type, these modules provide anywhere from 90% to 97% void space; as such, they need less room than both solid or perforated pipes or chambers in stone trench. Some of the widely employed modules are described next.

    a. StormTrap: This trademark module is manufactured by a company of the same name with headquarters in Morris, Illinois. The modules are single or multicells made of reinforced concrete. StormTraps may be placed either on a concrete foundation or stone base, serving as detention basin and detention–infiltration basin, respectively. These cells come 1 ft, 2 in. (35.6 cm) to 5 ft, 8 in. (1.73 m) high in 1 in. intervals and can be stacked up with the bottom traps upside down to double the depth of the system—2 ft, 4 in. to 11 ft, 4 in. high. Such an arrangement has been used in a soccer field in the city of Downey, California. Figure 7.30 shows single and double StormTrap schemes. StormTraps come in type II and type IV, as mid and side units, respectively. Each unit is 15.33 ft (4.67 m) long; type II is 8.42 ft (2.57 m) wide and type IV is 6.63 ft (2.03 m) wide. Each 5 ft (1.52 m) high, type II and type IV units provide 590 cf (16.71 m³) and 448 cf (12.69 m³) storage volume respectively.

    b. Atlantis Water Management for Life, established in 1986 and headquartered in Bellingham, Washington, manufactures plastic cells and tanks for different applications. These products are commercially available under the names of Turf Cell, Drainage Cell, and D-Raintank. Turf Cell can be used under permeable pavements, Drainage Cell under sidewalks, and D-Raintank as a retention–infiltration basin for the roof runoff or under parking lots. D-Raintanks come in six different sizes (see Table 7.32). These include a minimodule and a single module. They also include double, triple, quad, and penta modules which are formed by stacking two, three, four, and five single modules, respectively. Figure 7.31 shows a single tank module (Atlantis).

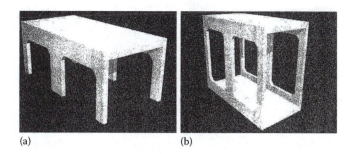

FIGURE 7.30 (a) Single and (b) double cell StormTrap.

TABLE 7.32
Model Number and Sizes of D-Raintank

Part No.	Product	Size (L × H × W)	Boxes/ft³	Ft³/box
70000	**Mini** module	26.72 in. × 9.36 in. × 15.91 in.	0.422	2.37
70003	**Single** module	26.72 in. × 17.55 in. × 15.91 in.	0.225	4.44
70004	**Double** module	26.72 in. × 34.32 in. × 15.91 in.	0.115	8.69
70005	**Triple** module	26.72 in. × 51.09 in. × 15.91 in.	0.077	12.98
70006	**Quad** module	26.72 in. × 67.86 in. × 15.91 in.	0.0588	17.17
70007	**Penta** module	26.72 in. × 84.63 in. × 15.91 in.	0.047	21.42

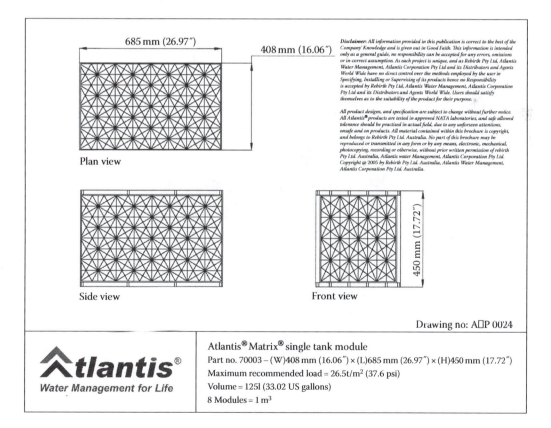

FIGURE 7.31 Atlantis® Matrix® single tank module.

c. StormTank® is another trademark modular tank, manufactured by Brentwood Industries. These modules are constructed of rugged lightweight polypropylene sides, top and bottom panels, and rigid PVC columns. StormTanks come in five different modules ranging from 18 to 36 in. high, invariably 36 in. long by 18 in. wide (914 × 456 m). Dimensions and installed storages of these modules are listed in Table 7.33.

Figure 7.32 depicts a StormTank module and its dimensions. The system components are shipped separately packed to save shipping cost and assembled onsite without a need for any tools or bonding agents. These modules provide 97% void volume, which is one of the most effective space saver tanks currently on the market. Similar to other modular storage tanks and cells, StormTanks can be stacked up on each other to reduce the footprint of the detention system. In a turf field project in North Bergen, New Jersey, the author used 36 in. high StormTanks as an underground detention system. Brentwood also makes a plastic chamber, named "The Arch."

d. Rainstore[3] is another plastic module manufactured by Invisible Structures, Inc. of Golden, Colorado. Each Rainstore[3] module is a 1 m × 1 m (40 in. × 40 in.) × 10 cm (4 in.) high with 36 columns and weighs 6.3 kg (14 lb). Figure 7.33 depicts geometry and dimensions of a Rainstore unit. Columns are thin-walled cylinders made by injecting molded recycle resins of either high-impact polypropylene (HIPP) or HDPE. Cylinders are 10 cm (4 in.) tall, 10 cm

TABLE 7.33
Dimensions and Volumes of StormTanks

Module	Height, in. (mm)	Nominal Void Space (%)	Storage, ft³ (m³)	Installed Storage, ft³ (m³)	Weight, lb (kg)
ST-18	18 (457)	95.5	6.44 (0.18)	9.14 (0.26)	27.7 (10.0)
ST-24	24 (610)	96.0	8.66 (0.25)	11.36 (0.32)	26.3 (11.9)
ST-30	30 (762)	96.5	10.88 (0.31)	13.58 (0.38)	29.5 (13.3)
ST-33	33 (838)	96.9	11.99 (0.34)	14.69 (0.42)	29.82 (13.5)
ST-36	36 (914)	97.0	13.10 (0.37)	15.80 (0.45)	33.1 (15.0)

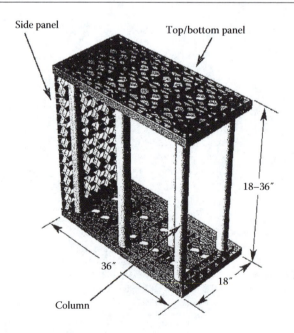

FIGURE 7.32 StormTank module by Brentwood Industries.

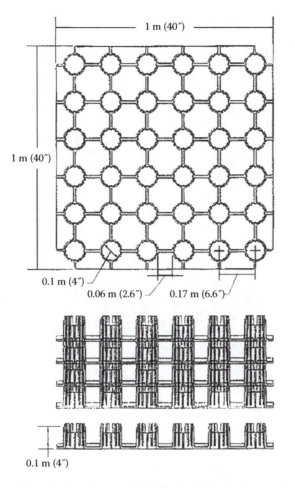

FIGURE 7.33 Rainstore, detention/retention module. (By Invisible Structures, Inc.; www.invisiblestruc tures.com.)

in diameter, 5 mm (0.2 in.) thick, and spaced 16.7 cm (4.6 in.) apart. T-shaped beams connect cylinders and resist external lateral soil/water pressure. A stack of 10 units occupies 1 m³ (35.3 ft³) and holds approximately 250 gal (950 L) of water. Rainstore units can be placed against each other on geotextile fabrics and can be stacked up to 2.4 m (7.9 ft) high.

e. Arrow Concrete Products of Granby, Connecticut makes concrete box structures for detention-retention applications. The boxes (know as Retain-it®) come in three models including end, meddle and side pieces. Each piece is 8 ft × 8 ft (2.44 × 2.44 m) and is available in 2, 3, 4 and 5 ft (0.61, 0.91, 1.22 and 1.52 m) standard heights. Intermediate heights can be made by special order. The boxes may be placed on a concrete platform to serve as a detention basin or a stone base to form a retention-infiltration basin. A 3 ft (0.91 m) high model alone can store approximately 130 ft³ (3.7 m³) of water.

f. Terre Arch™/Terre Box™. Terre Arch is a modular multichambered precast concrete storm water storage structure manufactured by Terre Hill Stormwater Systems of Terre Hill Concrete Products in Terre Hill, Pennsylvania. This open bottom, Roman arch shaped structure comes in two different sizes: Terre Arch 26 and Terre Arch 48. The former is an 8 ft wide by 19 ft long by 34 in. high (2.44 m × 5.79 m × 0.86 m) structure and includes four 52 in. wide by 26 in. high chambers. The latter is an 8 ft × 20 ft × 55 in. (2.44 × 6.1 × 1.4 m) structure with three 48 in. high arches. Figure 7.34 shows Terre Arch 26 detention–infiltration structure. Each Terre Arch 26 section weighs 13,500 lb (6100 kg), covers 152 ft²

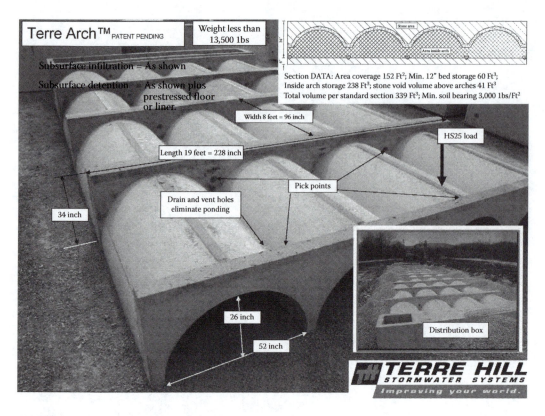

FIGURE 7.34   Terre Arch 26 modular arch structure.

(14.12 m²), and provides 238 ft³ (6.74 m³) of net storage. Filling valleys between arches with stone, the Terre Arch 26 and Terre Arch 48 provide 277 ft³ (7.84 m³) and 541 ft³ (15.32 m³) of store volume, respectively.

Terre Arch fiber-reinforced design is lightweight but has an HS-25 load bearing rating. Thus, unlike the HDPE chamber, it can withstand heavy trucks and machinery with no cover. Also, because of large size and minimal handling, compaction, and backfilling requirements, Terre Arch 26 and Terre Arch 48 structures save installation time and cost. According to the manufacturer, the finished cost per unit storage volume of Terre Arch competes with that of HDPE chambers.

g. Terre Hill Stormwater Systems also manufactures Terre Box. This is a watertight joint seal concrete box system for the detention/retention of storm water. Designed for underground installation, Terre Box is also rated for HS-25 load bearing and can be installed below parking lots or traveled roadways. This system is a preferred choice for below-grade detention systems when recharge is not desired.

h. OldCastle Precast, one of the larger manufacturers of precast concrete in the United States, makes a concrete vault named StormCapture. This chamber is an open-bottom box with side openings and can be custom made to suit a specific application. It can be used as an underground detention system, retention–infiltration basin, and rainwater harvesting cistern. OldCastle can be reached by phone at 800-579-8819 or online (www.oldcastleprecast.com).

i. Oldcastle® Stormwater Solutions also makes a cube-like modular hollow box called CUDO® Water Storage System. This box, which is made of high-strength injection molded polypropylene plastic, is designed to support HS-20 traffic loads. CUDO® measures 24″ × 24″ × 24″ (61 × 61 × 61 cm) and can be installed as a single layer or stacked up to four units to increase storage volume.

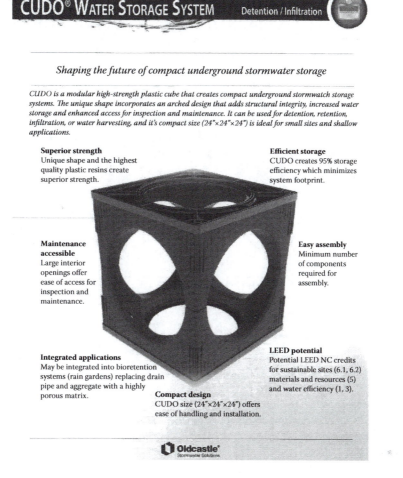

FIGURE 7.35   CUDO® Water Storage System.

Each unit provides 95% storage efficiency and comes in two halves and two end caps, which save transportation cost and can be readily assembled in the field. Figure 7.35 shows a single CUDO® configuration. More information on CUDO® can be obtained from its manufacturer's website (http://www.oldcastlestormwater.com) or by phone: (800) 579-8819.

CASE STUDY 7.3

A roadway improvement project in Woodcliff Lake, New Jersey, results in a 0.37 acre increase in impervious area and 1.02 acres of land disturbance. To address applicable storm water management regulations, a detention basin is provisioned to reduce the predeveloped peak rates of runoff from the disturbed area by 50%, 25%, and 20% for the 2-, 10-, and 100-year frequency storms, respectively. Due to offsite runoff entering the roadway, the overall drainage area at the point of discharge downstream of the detention basin is 9.14 acres (37 ha). Because of limited open space, an underground detention system comprising HDPE pipes is selected in this project. This case study presents step-by-step calculations for the detention basin design and water treatment device.

RUNOFF CALCULATIONS

Runoff calculations in this project are performed based on the rational method. Since a large percentage of the area is covered by paved surfaces and the roadway is drained by a drainage system, a 10-minute time of concentration is used in calculating the peak rates of runoff. Calculations for the composite runoff coefficient for the area of disturbance, E1, and the undisturbed and/or offsite area, E2 (shown on a drainage area map that is not included here), the peak rates of runoff from these areas and the allowable discharges, Q_A, from the site are summarized in Tables 7.34 through 7.36, respectively. It is to be noted that the portion of the roadway to be milled and repaved is not considered as area of disturbance.

Calculations for the postconstruction condition are prepared by differentiating the tributary area to the detention basin, P_1, from the area bypassing it, P_2. Composite runoff coefficient and peak runoff calculations for these areas are summarized in Tables 7.37 and 7.38.

DETENTION BASIN DESIGN

Calculations for the approximate detention storage volume and the critical storm durations are performed using the rational and modified rational method (Equations 7.23 and 7.25).

TABLE 7.34
Runoff Coefficient Preconstruction

	Reconstruct Ex. Pav. C = 0.95	Grass to Be Paved C = 0.35	Pav. to Remain[a] C = 0.95	Grass[b] C = 0.35	Woods[c] C = 0.25	Total Area (ac)	$C_W = \Sigma(A \times C)/A$
E1	0.65	0.37				1.02	0.73
E2			2.59	4.43	1.10	8.12	0.53
Total	0.65	0.37	2.59	4.43	1.10	9.14	0.55

Note: Peak rates of runoff, existing conditions:
$Q_{E1} = 1.02 \times 0.73 \times I = 0.745 \times I.$
$Q_{E2} = 8.12 \times 0.53 \times I = 4.304 \times I.$
$Q_{E\text{-TOTAL}} = 9.14 \times 0.55 \times I = 5.027 \times I.$
[a] This area includes 2.06 acres of road to be milled/repaved and 0.53 offsite impervious area.
[b] This area includes 1.21 acres of grass within project limits and 3.22 offsite grass area.
[c] All woods is offsite area.

TABLE 7.35
Peak Rates of Runoff, Preconstruction

Storm Frequency	I (in./h)	Q_{E1} (cfs)	Q_{E2} (cfs)	$Q_{E\text{-TOTAL}}$ (cfs)
2	4.2	3.13	18.08	21.11
10	5.8	4.32	24.96	29.16
100	8.0	5.96	34.43	40.22

Note: $T_c = 10$ min.

TABLE 7.36
Allowable Peak Flows from Site

Storm Frequency	Q_{E1} (cfs)	Reduction	Q_{E2} (cfs)	Q_A (cfs)
2	3.13	50%	18.08	19.64
10	4.3.2	25%	24.96	28.20
100	5.96	20%	34.43	39.20

Note: $Q_A = \%Q_{E1} + Q_{E2}$. Calculations for the offsite area runoff coefficient are estimated based on aerial photography.

TABLE 7.37
Runoff Coefficients, Post Construction

	Imp. C = 0.95	Grass C = 0.35	Woods C = 0.25	Total Area (ac)	$C_w = \Sigma(A \times C)/A$
P_1	1.59	2.53	1.10	5.22	0.51
P_2	2.02	1.90	0.00	3.92	0.66
Total	3.61	4.43	1.10	9.14	0.57

Note: Peak rates of runoff, proposed conditions:
$Q_{P1} = 5.22 \times 0.51 \times I = 2.662 \times I.$
$Q_{P2} = 3.92 \times 0.66 \times I = 2.587 \times I.$
$Q_{P\text{-TOTAL}} = 9.14 \times 0.57 \times I = 5.210 \times I.$

TABLE 7.38
Peak Runoff, Post Construction

Storm Frequency	I (in./h)	Q_{P1} (cfs)	Q_{P2} (cfs)	$Q_{P\text{-TOTAL}}$ (cfs)
2	4.2	11.18	10.87	22.05
10	5.8	15.44	15.00	30.44
100	8.0	21.30	20.70	41.99

Note: $T_c = 10$ min.

Calculations are performed for storms of 2-, 10-, and 100-year frequency and are organized in Table 7.39. This table indicates a minimum detention basin volume of 2100 ft³. It also shows the critical storm duration of 10 minutes for storms of 2-, 10-, and 100-year frequency. Considering nonlinear variation of discharge and to further reduce discharges downstream of the site, the estimated storage volume is increased by nearly 60% in the design of detention pipes. Figures 7.36 and 7.37 show typical cross section and layout of detention pipes,

TABLE 7.39

Required Detention Storage Estimation

Modified rational method, critical storm calculations

$V_R = [T_D \times 60s \times Q_{P1}] - [0.5 \times T_b \times 60s \times Q_O] T_b$ (base time) $= T_D + T_c, T_c = 10$ min.

$Q_{P1} = 5.22 \times 0.51 \times I = 2.662 \times I$

$Q_{P2} = 3.92 \times 0.66 \times I = 2.587 \times I$

$Q_O =$ Allowable peak outflow from basin $= Q_A - Q_{P2}$

See Table 7.36 for Q_A calculations.

Storm Duration (min)	I (in./h)	Q_{P1} (cfs)	Q_{P2} (cfs)	Q_O (cfs)	Required Vol., V_R (ft³)
2-Year Storm					$Q_A = 19.64$
10	4.20	11.18	10.87	8.78	1800[a]
15	3.60	9.58	9.31	10.33	$Q_{P1} < Q_O$
20	3.00	7.99	7.76	11.88	$Q_{P1} < Q_O$
30	2.40	6.39	6.21	13.43	$Q_{P1} < Q_O$
45	1.80	4.79	4.66	14.98	$Q_{P1} < Q_O$
10-Year Storm					$Q_A = 28.20$
10	5.80	15.44	15.00	13.20	1680
15	4.20	11.18	10.87	17.34	$Q_{P1} < Q_O$
20	4.00	10.65	10.35	17.86	$Q_{P1} < Q_O$
100-Year Storm					$Q_A = 39.20$
10	8.00	21.30	20.70	18.50	2100
15	6.90	18.37	17.85	21.35	$Q_{P1} < Q_O$
20	5.80	15.44	15.00	24.20	$Q_{P1} < Q_O$
30	4.60	12.25	11.90	27.30	$Q_{P1} < Q_O$

[a] $V = 60 \times T_b (Q_{P1} - Q_O)/2; T_b = 2.5 T_c$.

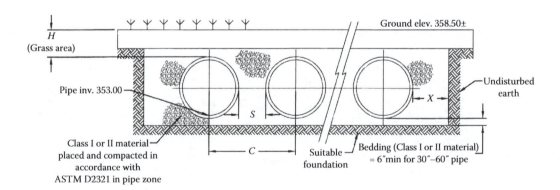

FIGURE 7.36 Typical detention basin cross section detail.

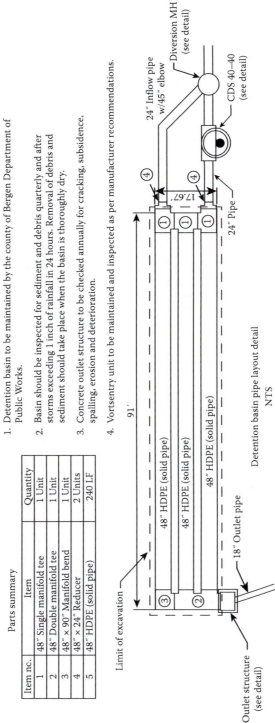

Detention basin maintenance notes:

1. Detention basin to be maintained by the county of Bergen Department of Public Works.

2. Basin should be inspected for sediment and debris quarterly and after storms exceeding 1 inch of rainfall in 24 hours. Removal of debris and sediment should take place when the basin is thoroughly dry.

3. Concrete outlet structure to be checked annually for cracking, subsidence, spalling, erosion and deterioration.

4. Vortsentry unit to be maintained and inspected as per manufacturer recommendations.

Parts summary

Item no.	Item	Quantity
1	48″ Single manifold tee	1 Unit
2	48″ Double manifold tee	1 Unit
3	48″ × 90″ Manifold bend	1 Unit
4	48″ × 24″ Reducer	2 Units
5	48″ HDPE (solid pipe)	240 LF

FIGURE 7.37 Detention basin pipe layout detail.

TABLE 7.40

Storage–Elevation Relation

Pipe dia.: 4 ft.

Three rows of 80 L.F. of 48 in. HDPE; headers on either end

(See Figures 7.36 and 7.37.)

Elev. (ft)	Depth (ft)	Depth/Dia.[a]	Ratio to Total Volume[a]	Volume (ft³)
353	0	0.00	0	0
353.5	0.5	0.13	0.075	255
354	1	0.25	0.2	680
354.25	1.25	0.31	0.27	918
354.5	1.5	0.38	0.33	1122
355	2	0.50	0.5	1700
355.5	2.5	0.63	0.65	2210
356	3	0.75	0.82	2788
356.5	3.5	0.88	0.93	3162
357	4	1	1	3400

[a] Part full volumes are calculated based on Figure 2.5 in Chapter 2.

respectively. As shown, the proposed detention system comprises three 80 ft (24.4 m) long rows of 48 in. (1.22 m) HDPE pipes connected together with 48 in. headers. Table 7.40 lists the storage-elevation relation for this underground detention basin, which provides 3400 ft³ (96.3 m³) of storage volume.

OUTLET STRUCTURE DESIGN

The maximum allowable discharge from the detention basin is calculated at 18.50 cfs (see Table 7.39). To attain this allowable discharge, the outlet structure includes a 6-in. low-flow orifice designed to discharge rainfalls up to the water quality storm and a 1.25 ft wide weir above the water quality level. Table 7.41 lists the discharge calculations for each of these openings and the composite discharge from the outlet structure.

ROUTING COMPUTATIONS

Routing computations are performed for the critical storm durations determined in the preliminary sizing of the detention basin. Routing computations for the 100-year, 10-minute duration storm are shown in Table 7.42.

The computations indicate that the detention system is effectively (97%) utilized and that by providing a larger detention basin than that calculated by the preliminary design, the outflow is reduced below the allowable discharge of 18.50 cfs (see Table 7.39). Table 7.43 summarizes the routing computations for the storms of 2-, 10-, and 100-year frequency. The composite discharges at the outfall point are conservatively calculated by adding the peak discharge from the detention basin to the peak runoff from bypass areas. The calculations are shown in Table 7.44.

Table 7.45 presents a comparison of the calculated allowable and postconstruction peak runoff at the point of discharge. As can be seen, the proposed oversized detention basin system reduces the peak runoff below the allowable values. If composite discharges were calculated adding hydrographs, the postdevelopment peak runoff would be smaller than those shown in Tables 7.44 and 7.45.

WATER QUALITY PROVISION

The project increases impervious coverage by 0.37 acres. Since this is larger than the 0.25 acre threshold, the project is subject to the state of New Jersey water quality requirement. To address

TABLE 7.41

Discharge Rating of Outlet Structure

Orifice			Weir		
Inv.: 353			Inv.: 354.25		
Dia.(in): 6			L (ft): 1.25		
Area(sf): 0.196					

Elev.	H (ft.)	Q (cfs)	H (ft.)	Q (cfs)	Total Q (cfs)
353	0	0			0.00
353.5	0.25	0.47			0.47
354.0	0.75	0.82			0.82
354.25	1.00	0.95	0.00	0.00	0.95
354.5	1.25	1.06	0.25	0.47	1.53
354.75	1.50	1.16	0.50	1.33	2.48
355.0	1.75	1.25	0.75	2.44	3.69
355.5	2.25	1.42	1.25	5.24	6.66
356.0	2.75	1.57	1.75	8.68	10.25
356.5	3.25	1.70	2.25	12.66	14.36
357.0	3.75	1.83	2.75	17.10	18.93

Equations used:

Orifice flow: $Q_o = [CA (2gh)^{1/2}]$ $C = 0.6$

Orifice (partially submerged): $Q_O = (\pi/4)(CLH^{1.5})$ $C = 3.0$

Weir flow: $Q_w = CLH^{1.5}$ $C = 3.0$

Submerged weir: $Q_{Sw} = [C(L \times H) (2gh)^{1/2}]$ $C = 0.6$

Head (orifice): H = Elev. − Invert + 0.5 dia.

Head (weir): H = Elev. − Invert

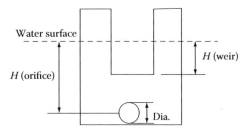

this requirement a CDS unit is provisioned for the project. The unit is placed before the detention basin and offline from the inflow pipe such that storms larger than the water quality storm bypass the unit.*

Calculations for the required TSS removal for the project and the TSS removal provided by the CDS water quality device are presented in Tables 7.46 and 7.47, respectively. The TSS removal rate criteria are defined under the former table. Calculations in the latter table are simply based on the paved area, which is treated by the water quality device before entering the detention system.

The size of the CDS unit provisioned in this project is determined based on the impervious area tributary to the unit and the rainfall intensity of the water quality storm.

---

* In 2008, the New Jersey DEP mandated placing manufactured water quality devices offline. Michigan appears to be the only other state in the nation that requires the same. The NJDEP has modified this requirement based on additional test data (see Chapter 5) and has approved some MTDs including CDS for online application.

TABLE 7.42

Routing Computations, 100-Year, 10-Minute Storm

Inflow Hydrograph		Routing Computations				
Time (min)	Inflow (cfs)	I1 + I2 (cfs)	2S/t – O (cfs)	2S/t + O (cfs)	Outflow (cfs)	Elevation (ft)
0.0	0.00	–	0.0	0.0	0.00	353.00
5.0	10.65	10.7	6.4	10.7	2.12	354.64
10.0	21.30	32.0	5.3	38.4	16.53	356.74
15.0	10.65	32.0	5.8	37.3	15.71	356.65
20.0	0.00	10.7	7.7	16.5	4.37	355.12
25.0	0.00	0.0	5.4	7.7	1.15	354.34
30.0	0.00	0.0	3.8	5.4	0.83	354.01
35.0	0.00	0.0	2.5	3.8	0.65	353.75
40.0	0.00	0.0	1.5	2.5	0.51	353.55
45.0	0.00	0.0	0.8	1.5	0.32	353.34
50.0	0.00	0.0	0.5	0.8	0.18	353.19
55.0	0.00	0.0	0.3	0.5	0.10	353.11
60.0	0.00	0.0	0.2	0.3	0.06	353.06
65.0	0.00	0.0	0.1	0.2	0.03	353.04
70.0	0.00	0.0	0.0	0.1	0.02	353.02
75.0	0.00	0.0	0.0	0.0	0.01	353.01

Summary of peak outflow and peak elevation:

Peak inflow = 21.30 cfs.

Peak outflow = 16.53 cfs.

Peak elevation = 356.74 ft.

Summary of approximate peak storage:

Initial storage = 0 cu-ft.

Peak storage from storm = 3275 cu-ft.

Total storage in pond = 3275 cu-ft.

TABLE 7.43

Routing Computations Summary 2-, 10-, and 100-Year Storm Events

Storm Freq. (years)	Rainfall Intensity (in./h)	Storm Duration (min)[a]	Q_{IN} (cfs)	Q_{OUT} (cfs)
2	4.20	10	11.18	8.30
10	5.80	10	15.44	11.90
100	8.00	10	21.30	16.53

[a] Critical storm duration; see Table 7.39.

TABLE 7.44

Composite Peak Flows from Site 2-, 10-, and 100-Year Storms

Storm Freq. (years)	Q_{P2}[a] (cfs)	Q_{OUT}[b] (cfs)	Total Q (cfs)
2	10.87	8.30	19.17
10	15.00	11.90	26.90
100	20.70	16.53	37.23

[a] Q_{P2} is peak flow of the bypass area. See Table 7.39.

[b] Q_{OUT} is peak outflow from the detention basin. See Table 7.43.

TABLE 7.45

Comparison of Calculated Allowable and Proposed Peak Rates of Runoff from Site

Storm Freq. (years)	$Q_A{}^a$ (cfs)	$Q_P{}^b$ (cfs)
2	19.64	19.17
10	28.20	26.90
100	39.20	37.23

Note: Area = 9.14 acres.

[a] Q_A = allowable from site (see Table 7.36).

[b] Q_P = total proposed flow (calculated by adding peak flows; see Table 7.44).

TABLE 7.46

Required TSS (Total Suspended Solids) Removal Rate for the Project Area

Ground Cover	Area (ac.)	% TSS Required Removal Rate	$A \times$ % TSS
Increase in impervious area[a]	0.37	80%	0.30
Ex. pavement to be fully reconstructed[b]	0.65	50%	0.33
	$\Sigma A = 1.02$		$\Sigma A \times$ % TSS = 0.62

Note: Pavement being milled/repaved does not count as disturbance and does not trigger the storm water management rules.

[a] New pavement is subject to 80% TSS removal.

[b] Pavement being fully reconstructed is considered disturbed and pursuant to the storm water management rules requires a TSS removal rate of 50%.

TABLE 7.47

Provided TSS Removal Rate for the Project Area

Ground Cover	Area (ac.)	% TSS Provided Removal Rate	$A \times$ % TSS
Impervious to water quality unit	1.59	50%	0.80
	$\Sigma A = 1.59$		$\Sigma A \times$ % TSS = 0.80

Note: 0.80 > 0.62. Proposed water quality treatment is adequate.

Water Quality Unit Sizing

Water quality peak flow rate to basin

A (acres) = 1.59[a]
C_W = 0.90
T_C = 10 min
I = 3.2 in./h[b]
Q (cfs) = 4.58 cfs

Note: Use CDS Model CDS 40-40, which is rated for 6 cfs treatment rate by the manufacturer.
[a] Excludes grass and woods areas; they retain runoff during WQ storm.
[b] See Chapter 3.

CASE STUDY 7.4

This case exemplifies the application of underground chambers in stone trench. The project involves widening of New Jersey Garden State Parkway in the vicinity of Interchange 88/89. The project location is shown in Figure 7.38, which is prepared based on the US Geological Survey Lakewood Quadrangle. Figure 7.39 depicts the limits of road widening, which extends nearly one-half mile (800 m) along north- and south-bound lanes of the parkway. The project is located entirely within a CAFRA (Coastal Area Facility Review Act) zone and disturbs approximately 0.77 acre (0.31 ha) of land. Of this, 0.642 acre (0.26 ha) will be new pavements due to the road widening and the remainder includes clearing some wooded areas associated with grading along the pavements. The existing pavements within the project limits, which will be milled or resurfaced, are not considered as a land disturbance and therefore not subject to the NJDEP's storm water treatment requirements.

STORM WATER MANAGEMENT PROVISIONS

To control the peak rates of runoff, an underground detention–infiltration system is provisioned along the right shoulder of southbound lanes. The system comprises two rows of Cultec Recharger V8 chambers placed in an 11.5 ft (3.77 m) wide by 183.7 ft (60.25 m) long stone trench. This trench is 46 in. (125.6 cm) thick and includes a 6 in. (15 cm) stone base and 6 in. stone cover. The system provides 5186 ft³ (147 m³) of storage. Figure 7.40 shows a typical section of the Recharger V8 detention system. Before entering the chambers, the entire runoff is routed through a CDS PMSU 30-30 (redesignated as CDS30-30) water quality device that has 3 cfs treatment capacity. This device removes floatable objects, silt, debris, oil, and grease, thereby avoiding premature clogging of the chambers and lessening their maintenance. The CDS unit is approved for 50% total suspended solids (TSS) removal, is easy to maintain, and partially addresses the water quality requirements for the project. To supplement this device, existing woodlands along the northbound shoulder are utilized as a vegetative strip, which provides 80% TSS removal for the runoff from the area tributary thereof.

RUNOFF CALCULATIONS

Figure 7.41 shows the preconstruction drainage areas of the north- and south-bound lanes of the parkway. This figure is prepared using the 1 ft contour lines generated from the topographic survey conducted in connection with the design. The drainage areas shown in this figure cover the tributary areas to the project site. Figure 7.42 shows the postproject drainage areas. As shown in this figure, the proposed chambers collect 2.754 acres of the drainage area in the south-bound lanes with the remainder of 0.286 acres of the south-bound drainage area and 1.152 acres of the north-bound lanes remaining undetained.

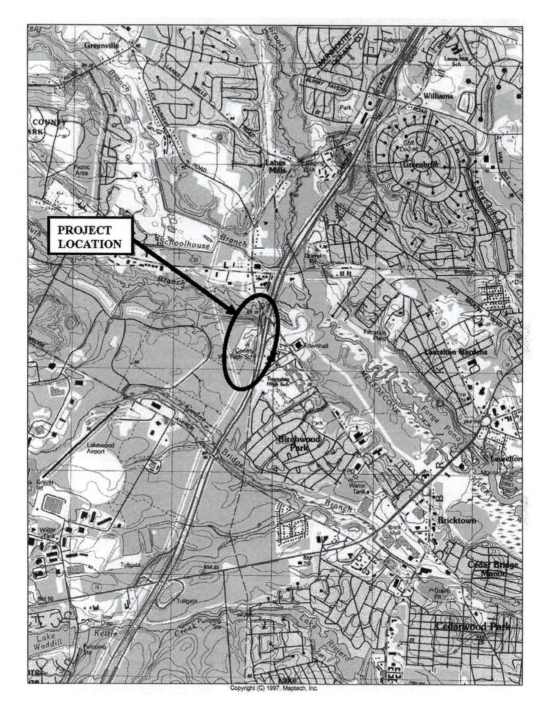

FIGURE 7.38   Location map.

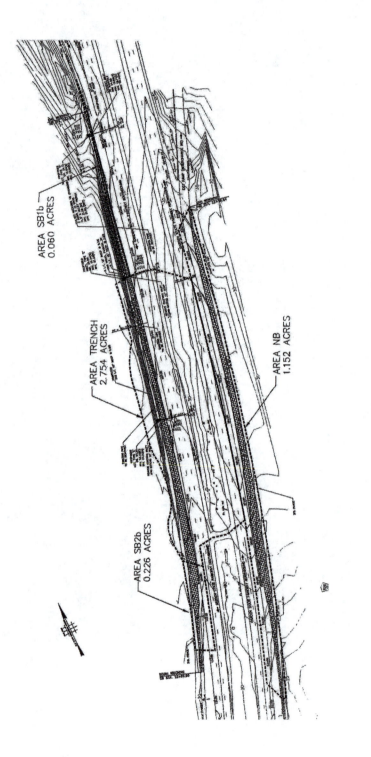

FIGURE 7.39 Overall map of altered area.

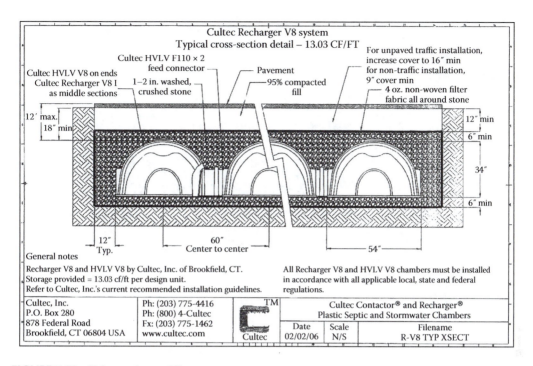

FIGURE 7.40   Cultec recharger V8 system.

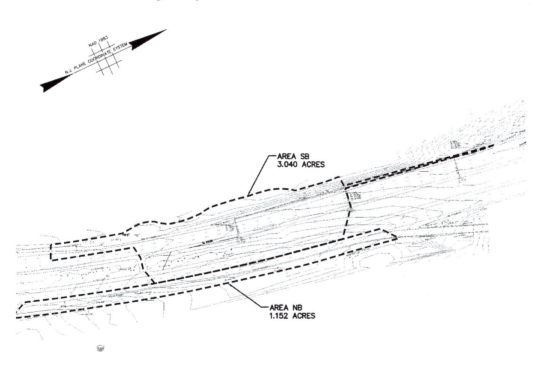

FIGURE 7.41   Preconstruction drainage areas.

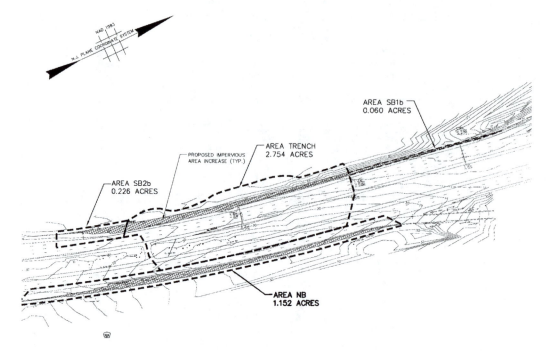

FIGURE 7.42   Postproject drainage areas.

   Since the area of disturbance is significantly smaller than the drainage area tributary to the detention basin, calculations are first prepared to establish the required flow reductions from the project area. Runoff calculations for the project area are prepared based on the rational method using a time of concentration of 10 minutes, which is commensurate with the area of disturbance. In compliance with the NJDEP's storm water management regulations, the calculations include the storms of 2-, 10-, and 100-year frequency. Tables 7.48 and 7.49 present the pre- and postproject runoff coefficient calculations for the disturbed area and the required reductions in the peak rates of runoff, respectively.

   Runoff calculations for the entire 4.192 acres (1.697 ha) drainage area are performed differentiating the tributary area to the proposed underground chambers and the bypass areas. While the calculations show a time concentration of 15 minutes, runoff calculations for the detention basin are conservatively performed using $T_c = 10$ minutes, which is the same as that of the bypass areas including the tributary area to the northbound lanes. Table 7.50 presents runoff

---

TABLE 7.48

Calculations for Existing and Proposed Peak Flows

Area being altered: 0.770 acres

Open space being paved: 0.642 acres

Woods to be open space: 0.128 acres

	Imp. $C = 0.95$	Grass $C = 0.35$	Woods $C = 0.25$	Total Area (ac)	$C_W = \Sigma(A \times C)/A$
Pre	0	0.642	0.128	0.770	0.33
Post	0.642	0.128	0	0.770	0.85

$Q_E = 0.770 \times 0.33 \times I.$

$Q_P = 0.770 \times 0.85 \times I.$

TABLE 7.49

Required Flow Reductions

Required reduction, Q_R

$Q_R = Q_P - \%Q_E$

Storm Frequency	I (in./h)	Q_E (cfs)	Reduction	Q_P (cfs)	Q_R (cfs)
2	4.2	1.07	50%	2.75	2.22
10	5.8	1.47	25%	3.80	2.69
100	8	2.03	20%	5.24	4.83

Note: Storm duration and time of concentration = 10 minutes.

TABLE 7.50

Calculations for Postdevelopment Peak Flows for Detained and Bypass Areas and Allowable Discharges from the Detention-Infiltration System

	Imp. $C = 0.95$	Grass $C = 0.35$	Woods $C = 0.25$	Total Area (ac)	$C_W = \Sigma(A \times C)/A$	
			Bypass Areas			
SB1b	0.06	0	0	0.060	0.95	
SB2b	0.182	0.044	0	0.226	0.83	
NB	0.905	0.247	0.000	1.152	0.82	
Total	1.147	0.291	0	1.438	0.83	
			Detained Areas			
Trench	1.575	0.7	0.479	2.754	0.68	
Storm Frequency	I (in./h)[a]	Q_B (cfs)[b]	Q_D (cfs)[c]	Q_P (cfs)[d]	Q_R (cfs)[e]	Q_A (cfs)[f]
2	4.2	5.01	7.87	12.88	2.22	5.65
10	5.8	6.92	10.86	17.78	2.69	8.17
100	8.0	9.55	14.98	24.53	4.83	10.15

[a] Storm duration and time of concentration = 10 minutes.
[b] Q_B = bypass flow = $1.438 \times 0.83 \times I$.
[c] Q_D = inflow to underground chambers = $2.754 \times 0.68 \times I$.
[d] Q_P = total postdevelopment flow.
[e] See Table 7.49 for calculations of Q_R (required flow reduction).
[f] $Q_A = Q_D - Q_R$, allowable max. discharge from chambers.

calculations together with support data for the postproject peak rates of runoff from the entire site area. Also listed in this table are the maximum allowable discharges from the proposed detention system, accounting for the required flow reductions shown in Table 7.49.

DETENTION–INFILTRATION SYSTEM DESIGN

To avoid iterative routing computations, simple calculations are first performed to determine the critical storm duration, namely the one that will require the largest detention storage volume for each storm event. Table 7.51 presents the critical storm duration calculations performed using the modified rational method. Conservatively, the percolation losses are neglected in these preliminary calculations. The critical storm duration is found to be 15 minutes for the 2- and 100-year storm events and 10 minutes for the 10-year storm event; the required detention storage is estimated to be 2934 ft³ (83.1 m³).

TABLE 7.51

Calculations for Critical Storm Duration and Preliminary Sizing of Detention Basins

Modified rational method

$V = [T_D \times 60s \times Q_D] - [0.5 \times T_b \times 60s \times Q_A]^a$

T_b (base time) $= T_D + T_c$, $T_c = 15$ min

Q_D = Detention inflow $= 2.754 \times 0.68 \times I$

Q_A = Max. allowable discharge from detention infiltration basin

Storm Duration (min)	I (in./h)	Q_D (cfs)	Required Volume (ft³)
2-Year Storm		$Q_{ALLOWABLE} = 5.65$	
10	4.20	7.87	1662
15	3.60	6.74	1830
20	3.00	5.62 < 5.65	1657
10-Year Storm		$Q_{ALLOWABLE} = 8.17$	
10	5.80	10.86	2019
15	4.80	8.99	1963
20	4.00	7.49 < 8.17	1636
100-Year Storm		$Q_{ALLOWABLE} = 8.17$	
10	8.00	14.98	2709
15	6.80	12.73	2934
20	5.60	10.49	1931

[a] Preliminary sizing calculations conservatively neglect infiltration.

Tables 7.52 and 7.53 list elevation–storage and elevation–discharge relations for the detention–infiltration system, respectively. The former table shows that the proposed system is nearly 77% larger than the preliminary calculations of detention basin sizing indicate. The listed discharges on the latter table account for the seepage loss through the bottom of the stone trench housing the chambers. The seepage loss is calculated to be 0.49 cfs (13.9 L/s) based on a soil permeability equal to 10 in./h (250 mm/h), which is one-half of the results from three percolation tests conducted at the site of the chambers; factor 0.5 represents a design safety factor equal to 2.0.

Routing computations are performed using hydrographs for the critical storm durations, which are highlighted in Table 7.51 and the storage–discharge elevation-rating (Tables 7.52 and 7.53). The computations are summarized in Table 7.54. Listed in this table are the outflows from the system, adjusted for the rate of seepage loss. The table also presents a comparison of the outflows and the allowable discharges from the Cultec chambers. The table shows that the oversized detention system creates flow reductions 16% to 36% more than required.

7.5   WATER TREATMENT STRUCTURES

Manufactured water treatment devices were discussed in a previous chapter. Vegetative swales and sand filters, which are among the most common water quality structures, are discussed herein. Filter strip, which is a non-structural water treatment device, will be discussed in the next chapter.

7.5.1   VEGETATIVE SWALES

Grassed waterways, also referred to as swales, are one of the earliest measures taken to treat the storm water runoff. Swales are used alongside roadways or in medians of divided highways. They

TABLE 7.52

Stage–Storage Calculations

Cultec recharge V8 chambers

No. of units: 50

2 rows × 25 units

11.5′ × 183.67′ stone bed area, 6″ stone box, 6″ stone cover

| Elev. (ft) | Storage (ft³) | | |
	Depth	Incremental	Cumulative
23.83	3.83	70.41	5186.1
23.75	3.75	211.22	5115.7
23.50	3.5	213.18	4904.4
23.25	3.25	253.51	4691.3
23.00	3	323.49	4437.8
22.75	2.75	357.93	4114.3
22.50	2.5	381.26	3756.3
22.25	2.25	397.61	3375.1
22.00	2	409.16	2977.5
21.75	1.75	415.92	2568.3
21.50	1.5	421.59	2152.4
21.25	1.25	432.7	1730.8
21.00	1	434.23	1298.1
20.75	0.75	441.42	863.9
20.50	0.5	211.22	422.4
20.25	0.25	211.22	211.2
20.00	0	0	0.0

are also employed in urban development as a substitute for inlets and pipes. Grass swales can provide moderate improvements in urban runoff quality. To effectively treat the runoff and to be stable, the swale should be properly designed and regularly maintained. The swale should also be a shallow trench no more than 1.5 ft (0.45 m) deep and have side slopes less than 3:1 (3H, 1V) to facilitate mowing. Figure 7.43 shows a typical section of a grass swale.

Swales are practical in areas of moderate longitudinal slope where the flow velocity will be less than that which would cause erosion during the design storm, which is specified as a 10-year event by many jurisdictional agencies. In the Northeast and other parts of the United States, where there is sufficient rain during the growing season, swales may not need any watering. However, when constructed in semiarid climates, swales require irrigation to maintain the grass growth.

To be effective, the water depth during the water quality storm should not rise significantly above the grass. Once the flow rises over the grass, the grass begins to bend down in the direction of flow and the water passes over the grass without receiving much treatment.

Manning's n value for grass swales may be as high as 0.3 or 0.4 before submergence begins. However, when the grass becomes deeply submerged, the n value may be dropped by 10-fold to 0.04. The effect is a faster velocity, shorter resident time, and low pollutant removal. Swale tests by Yu et al. (2001) with controlled flows at detention times ranging from 5.5 to 18 minutes showed pollutant removal efficiencies of 48% to 86%. Tests by others, referenced in Minton (2005), indicate similar results. However, particles smaller than 15 microns require prolonged detention time. Also, due to erosion, swales lose their effectiveness. The New Jersey Department of Environmental Protection used to accept grass swales as a water treatment structure. However, in the current regulations, readopted in April 2010, grass swales are no longer considered an effective means of water quality improvement.

TABLE 7.53

Outlet Structure Elevation–Discharge Relation

15 in. dia. outlet

Invert = bottom of chamber + 6 in.

Inv. = 21.0 ft

Elev.	Depth	H	Q (cfs)	Inf. (cfs)	Total Q
23.83	3.83	2.21	8.78	0.49	9.27
23.75	3.75	2.13	8.62	0.49	9.11
23.50	3.5	1.88	8.09	0.49	8.58
23.25	3.25	1.63	7.53	0.49	8.02
23.00	3	1.38	6.93	0.49	7.42
22.75	2.75	1.13	6.27	0.49	6.76
22.50	2.5	0.88	5.53	0.49	6.02
22.25	2.25	0.63	4.67	0.49	5.16
22.00	2	1.00	2.95	0.49	3.44
21.75	1.75	0.75	1.92	0.49	2.41
21.50	1.5	0.50	1.04	0.49	1.53
21.25	1.25	0.25	0.37	0.49	0.86
21.00	1	0	0	0.49	0.49
20.00	0	0	0	0	0.00

Fully Submerged	Partly Submerged	Infiltration
$Q = CA \, (2gH)^{0.5}$ dia. = 15 in.	$Q = (\pi/4)(CLH^{1.5})^a$	$Q = A \times I \times (FT/12 \text{ in.})$ $(h/3600 \text{ s})$
$A = 1.227 \text{ ft}^2$	$L = 1.25 \text{ ft}$	Trench area = 11.5 × 183.67 ft
$C = 0.6$	$C = 3.0$	$A = 2112 \text{ ft}^2$
$H = \text{Elev.} - 21.625$	$H = \text{Elev.} - 21.0$	$I = 10 \text{ in./h}$
$Q = 5.91(H^{0.5})$	$Q = 2.95(H^{1.5})$	$Q = 0.49 \text{ cfs}$

[a] This equation is proposed for estimating weir flow discharge from circular openings.

TABLE 7.54

Peak Inflows, Outflows, and Allowable Flows for Detention–Infiltration System

Storm Freq. (years)	Q_{IN}	Q_{DIS}	Q_O	Q_A
2	6.74	4.66	4.17	5.65
10	8.99	6.48	5.99	8.17
100	12.73	9.22	8.73	10.15

Note:

Q_A = allowable discharge. See Table 7.50.

Q_{DIS} = peak discharge from chambers.

Q_O = outflow = peak discharge – infiltration rate (0.49 cfs).

7.5.2 SAND FILTERS

Sand filters are intended primarily for water quality improvement; they provide no flow rate control. A typical sand filter system consists of two or three chambers. The three-chamber system includes a sedimentation chamber, filtration chamber, and discharge chamber. Many sand filters have only two chambers: one for sedimentation and the other for filtration and discharge. The sedimentation

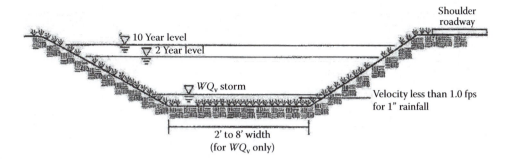

FIGURE 7.43 TSS removal for vegetated filter strips drainage area soil: loam, silt loam HGS:B. (Adapted from NJDEP, 2004, Chapter 9.)

chamber removes coarse sediment and floatables and the filtration chamber contains a sand bed, typically 18 to 24 in. (45 to 60 cm) thick, for removing finer sediments and other pollutants. Figure 7.44 shows a two-chamber sand filter. Water enters the first chamber through grates, overflows the weir partitioning the two chambers, and enters the second chamber filled with sand. The infiltrated water is discharged directly through an outlet pipe or indirectly through a 6 in. perforated pipe, which may terminate to a drainage system or detention basin.

A variety of above- or underground sand filter designs are used in different parts of the country (EPA, 1999). Regardless of deviations in shape, sand filters are generally designed to treat the water quality storm and may be sized based on the following equation:

$$Q = A_s K \frac{\left[(D + d/2)/D\right]}{3600} \tag{7.29}$$

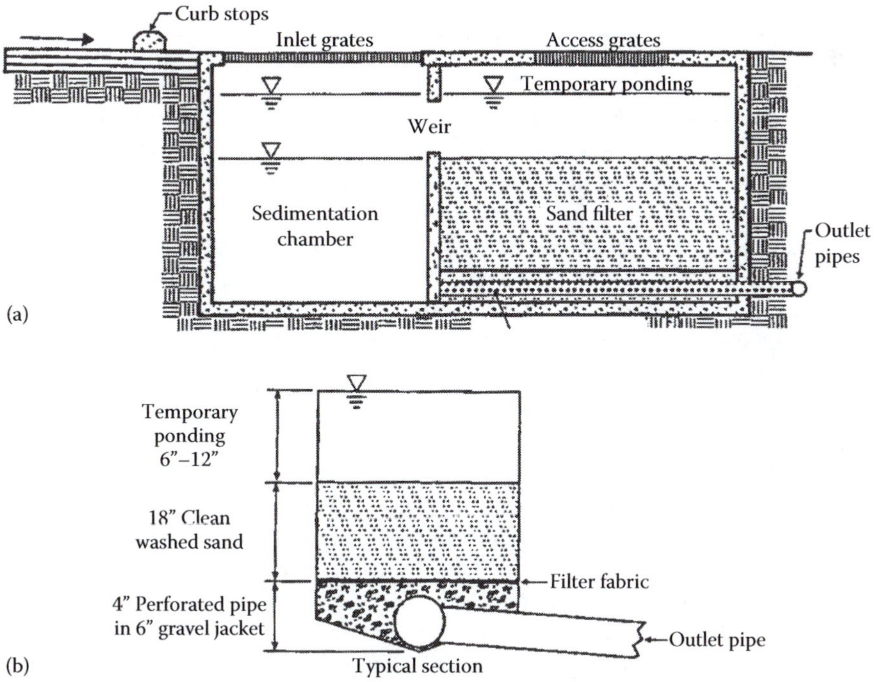

(a)

(b)

FIGURE 7.44 A two-chamber sand filter. (a) Profile and (b) section.

where

Q = average infiltrated flow rate through the sand bed in m³/s (cfs)

D = thickness of sand filter, m (ft)

d = maximum water depth (depth of water on sand bed at the end of storm event, m (ft))

K = sand permeability, m/h (ft/h)

A_s = surface area of sand bed, m² (ft²)

The average discharge from the sand filter, Q, is related to the water quality volume, WQ_v, by

$$Q = \frac{WQ_v}{T} \tag{7.30}$$

where T is dewatering time of the sand filter in hours.

The Washington, DC sand filter design consists of three chambers, including a discharge chamber, and is sized for 0.5 in. (1.3 cm) of runoff. For such water quality rainfall, sand filters are commonly designed to handle runoff from up to 1 acre (0.4 ha) impervious drainage areas. In New Jersey and Maryland, where the water quality storms are 1.25 in. and 0.9 and/or 1 in. respectively, sand filters should be used for smaller areas. By placing sand filters past detention basins, they may be sized for the attenuated runoff from a larger area. In general, sand filters are preferred over infiltration trenches and infiltration basins, where contamination of groundwater with common pollutants such as BOD, suspended solids, or coliform is of concern and the water table is high.

Sand filters are accepted for 80% TSS removal by many jurisdictional agencies. They are also effective in removing some other pollutants if properly maintained. However, they are totally useless if neglected. The author has witnessed sand filters that were fully clogged and discharged the inflow runoff from their overflow weir without any treatment. To prolong their useful life, sand filters are provided with a pretreatment stilling basin. Figure 7.45 shows an organic filter in which peat/sand mixture is substituted for sand bed.

CASE STUDY 7.5

This study represents the combined use of a detention basin and use of a vegetative filter strip to treat the runoff from a parking lot in the borough of Franklin Lakes in New Jersey. Figure 7.46 depicts the existing parking lot layout and drainage pipes that traverse this lot. Figure 7.47 shows the layout of the parking lot expansion. Also shown on the latter plan is a vegetative filter strip and a shallow detention basin provisioned to address the water quality and peak flow reduction requirements for the project pursuant to the NJDEP's storm water management regulations adopted in February 2004. Design calculations for the detention basin are discussed here and the application of filter strip to address the water quality requirement for the project is presented in Chapter 8.

DETENTION BASIN DESIGN

Delineated on Figure 7.47 are the limits of the drainage area tributary to the detention basin and the area where the runoff bypasses the basin. To allow for a meaningful comparison between the pre- and postdevelopment peak rates of runoff, the same overall drainage area as that of postdevelopment conditions is drawn on the predevelopment plan.

Runoff calculations for the project site are prepared based on the rational method using a runoff coefficient of 0.95 for pavements and 0.35 for lawns. Tables 7.55 and 7.56 summarize calculations of pre- and postdevelopment runoff coefficients and peak runoff rates for the subareas shown on Figures 7.46 and 7.47, respectively. Listed in the latter table are the maximum allowable discharges from the site, all calculated based on a 10-minute storm. In designing a

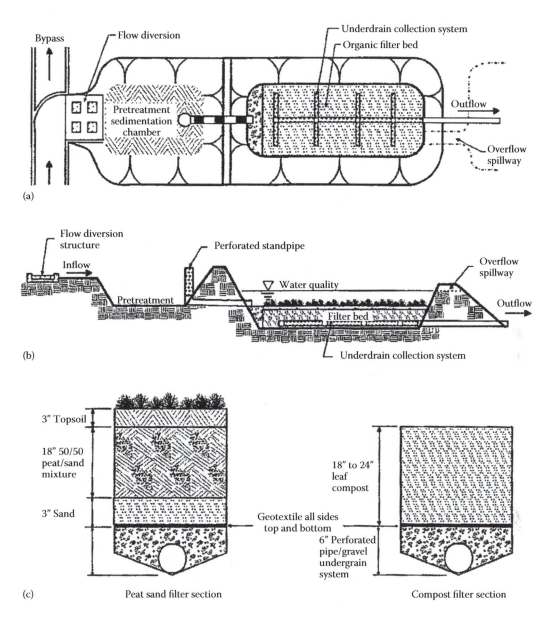

FIGURE 7.45 Organic filter (modified sand filter). (a) Plan, (b) profile, and (c) section. (From NY State Stormwater Management Design Manual, Figure 6.18, 2010.)

detention basin-outlet structure system, the critical storm duration, namely the one that requires the largest detention storage to address the required flow reductions from the site, is first determined. Table 7.57 presents calculations of the critical storm duration for the storms of 2-, 10-, and 100-year frequency. According to this table a detention basin with minimum storage volume of 2550 ft³ is required. Also, critical storm durations are found to be 30 minutes for the 2-year storm event and 15 minutes for the 10- and 100-year storm events. Considering that the preceding calculations are based on a linear outflow hydrograph assumption and accounting for a freeboard above the maximum water surface elevation, a detention basin with a capacity nearly twice as large as the calculated storage volume is designed.

To prolong the retention time for the water quality storm, the outlet structure is equipped with a 2.5 in. low-flow orifice, which is the smallest permissible orifice. Also, an 18 in. wide by 6 in.

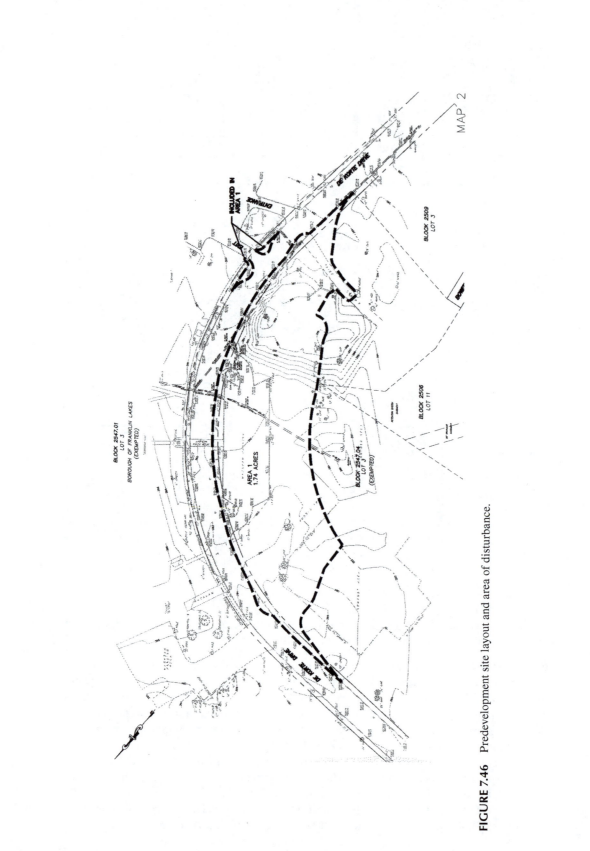

FIGURE 7.46 Predevelopment site layout and area of disturbance.

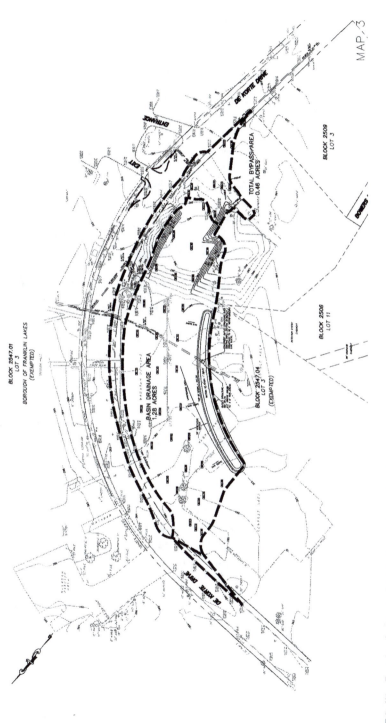

FIGURE 7.47 Postdevelopment site layout and drainage subareas.

TABLE 7.55
Runoff Coefficient Calculations

	Predeveloped Conditions		
	Area (ft²)	C	A × C
Gravel	1565	0.65	1017
Paved	16,630	0.95	15,799
Grass	57,575	0.35	20,151
Total	75,770		36,967

Area 1: area of disturbance = 1.74 acres (75,770 ft²)
$C_W = (\sum A \times C)/\text{Area} = 0.49$

	Postdevelopment Conditions				
	Paved C = 0.95	Grass C = 0.35	Total Area (ft²)	Total (A × C)	C_w
Basin	30,575	24,975	55,550	1.28	0.68
Bypass	5820	14,400	20,220	0.46	0.52

TABLE 7.56
Calculated Peak Rates of Runoff Pre- and Postdevelopment Conditions

Storm Freq. (years)	I (in./h)	Q_E[a]	Q_{DET}[b]	Q_{BYPASS}[c]	Q_A[d]
2	4.2	3.58	3.66	1.00	1.79
10	5.8	4.95	5.05	1.39	3.71
100	8.0	6.82	6.96	1.91	5.46

[a] $Q_E = 1.74 \times 0.49 \times I$.
[b] $Q_{DET} = 1.28 \times 0.68 \times I = 0.87 \times I$.
[c] $Q_{BYPASS} = 0.46 \times 0.52 \times I = 0.24 \times I$.
[d] $Q_A = Q_E \times \%$ reduction = allowable discharge from site.

high weir is provided to release storms of 2- through 100-year frequency. Figure 7.48 depicts the proposed outlet structure and Table 7.58 summarizes calculations of storage and discharge relation with stage. Routing computations for the storms of 2-, 10-, and 100-year frequency are performed based on the previously indicated critical storm durations and are summarized in Table 7.59. Composite peak rates of runoff for the postdevelopment conditions are calculated by adding the hydrographs of direct runoff and the routed flow from the detention basin. Table 7.60 lists composite peak runoff rates for the postdevelopment conditions and presents a comparison of these discharges with the predevelopment peak runoff rates from the project site, namely the area of disturbance. Since the detention basin is oversized, it creates greater reductions in the peak rates of runoff than the regulations require.

7.6   INFILTRATION BASINS

Infiltration basins are primarily intended for water quality and/or groundwater recharge; however, they may be designed to serve as a multipurpose infiltration–detention system. To improve water quality, infiltration basins are sized to fully contain the runoff from a given storm event. This storm,

TABLE 7.57

Required Detention Storage Calculations

$T_c = 10$ min

$V = [T_D \times 60s \times Q_{DET}] - [0.5 \times T_b \times 60s \times Q_o]$

   T_b (base time) $= T_D + T_c$

   $Q_o =$ max. allowable outflow from basin, $Q_A - Q_{BYPASS}$

   $Q_{BYPASS} = 0.24 \times I$ (see Table 7.56)

   $Q_{DET} = 0.87 \times I$ (see Table 7.56)

   $Q_A =$ Allowable from site (see Table 7.56)

Storm Duration (min)	I (in./h)	Q_{DET} (cfs)	Q_{BYPASS} (cfs)	Q_o (cfs)	V (ft³)
		2-Year Storm			
10	4.20	3.66	1.01	0.78	1724
15	3.60	3.13	0.86	0.93	2126
20	3.00	2.61	0.72	1.07	2170
30	2.40	2.09	0.58	1.21	2303
45	1.80	1.57	0.43	1.36	1989
60	1.45	1.26	0.35	1.44	n/a[a]
		10-Year Storm			
10	5.80	5.05	1.39	2.32	1638
15	4.90	4.26	1.18	2.53	1938
20	4.00	3.48	0.96	2.75	1703
30	3.30	2.87	0.79	2.92	n/a
		100-Year Storm			
10	8.00	6.96	1.92	3.54	2054
15	6.90	6.01	1.66	3.80	2552
20	5.80	5.05	1.39	4.07	2397
30	4.70	4.09	1.13	4.33	n/a

[a] $Q_o > Q_{DET}$.

which is generally known as the water quality storm, is specified by local or state jurisdictional agencies. An infiltration basin may be placed online or offline. In the former case it receives the entire runoff; in the latter, it infiltrates only the water quality storm and diverts the excess runoff through an overflow weir. Discharge from the weir is then routed through an aboveground or underground detention basin. Figure 7.49 depicts an online infiltration basin with a pretreatment stilling basin.

An infiltration basin may or may not be lined with vegetation. However, to improve the pollutant removal and to ease maintenance, the bottom of an infiltration basin is normally covered with a 6 to 12 in. (15–30 cm) thick sand. The sand traps sediments from runoff and can be replaced when the infiltration rate is found to be diminished.

The proper design of an infiltration basin is essential in its effective functioning. A soil log and percolation test must be taken at the location of the infiltration basin to measure the ambient water table level and soil permeability. Available soil survey maps are not accurate enough to serve as the basis of design. A minimum percolation rate of 0.5 in./h (13 mm/h) is accepted by a number of jurisdictional agencies. However, since the infiltration rate may decrease during extended wet weather conditions, the author recommends 1 in./h (25 mm/h) as a minimum acceptable percolation rate. The bottoms of basins should be located at least 2 ft (60 cm), but more preferably 3.3 ft (1 m) above the high water table or the bedrock. It is unfeasible to construct an infiltration basin in poor soils. Likewise, it is not advisable to place good fill over poor soil unless the fill consists of a high-permeability material such as sand and is 1 to 2 ft (30–60 cm) thick.

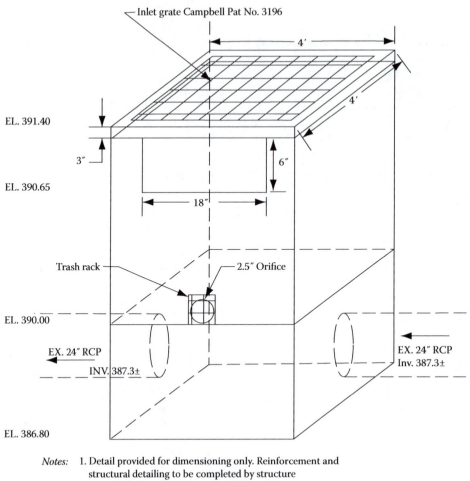

Inlet grate Campbell Pat No. 3196

4'

4'

EL. 391.40

3"

6"

EL. 390.65

18"

Trash rack   2.5" Orifice

EL. 390.00

EX. 24" RCP
INV. 387.3±

EX. 24" RCP
Inv. 387.3±

EL. 386.80

Notes: 1. Detail provided for dimensioning only. Reinforcement and
           structural detailing to be completed by structure
           manufacturer and will be responsibility of the contractor.
           Detail shop drawings signed and sealed by a licensed
           professional engineer to be submitted for approval prior
           to construction.
        2. Outlet structure to be constructed over existing 24" RCP
           to be utilized as an outlet pipe from the basin.

FIGURE 7.48   Pond outlet structure detail (N.T.S.).

TABLE 7.58
Elevation–Storage–Discharge Relation

Elevation (ft)	Storage (cf)	Discharge (cfs)
390.00	0	0
390.65	1795	0.12
391.00	2995	1.09
391.15	3580	1.76
391.40	4630	2.74
391.50	5050	5.78

TABLE 7.59

Routing Computation Summary

Storm Frequency (years)	Storm Duration (min)	Q_{DET}	Q_{OUT}
2	30	2.09	0.99
10	15	4.26	1.18
100	15	6.01	2.23

TABLE 7.60

Comparison of Existing and Proposed Peak Rates of Runoff

Storm Frequency (years)	Q_E	Q_{PR}[a]	Reduction %[b]
2	3.58	1.57	56
10	4.95	2.36	52
100	6.82	3.89	43

[a] Q_{PR} = composite postdevelopment discharge, calculated by adding hydrographs of Q_{out} and Q_{BYPASS}.

[b] Reduction = $(1 - Q_{PR}/Q_E) \times 100$.

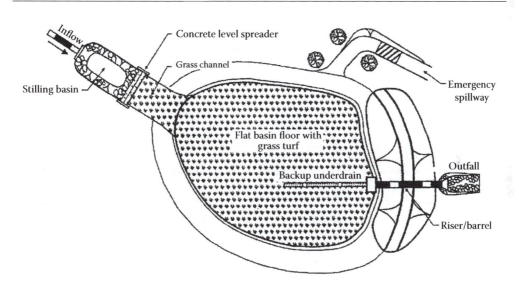

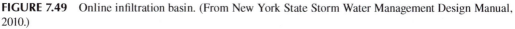

FIGURE 7.49 Online infiltration basin. (From New York State Storm Water Management Design Manual, 2010.)

An important criterion in the design of infiltration basins is the water quality storm. The water quality storm definition, as discussed in a previous chapter, varies from state to state. In New Jersey, for example, the water quality storm is specified as a storm of 2-hour duration, having nonuniform temporal distribution and amounting to 1.25 in. (32 mm) of rainfall. As far as the sizing of infiltration basins is concerned, the rainfall intensity has no significance. Preferably, the basin should be sized to fully retain the entire volume of design storm and drain it over time. Drawdown time (also known as dewatering time) and the infiltration bed area are related by the following equation:

$$T = \frac{3600(WQ_v/A_f K)}{[(D + d/2)/D]} \tag{7.31}$$

where

T is drawdown time in hours
A_f is the area of infiltration bed in m^2 (ft^2)
other terms are the same as defined for sand filters

For an infiltration basin, the drawdown time is commonly limited to 24 or 48 hours, depending on whether the basin is vegetated or not. In practice, infiltration basins are commonly sized to store the water quality storm in full or in part and Equation 7.31 is used to calculate the drawdown time.

It is to be noted that clean sand has a permeability of 2 to 5 ft per hour (not per day). However, a design criterion of 2 to 4 ft/day (0.6 to 1.3 m/day) is used to represent relatively clogged sand, the condition when the infiltration basin/sand filter should be maintained. Climate plays an important role in drawdown time. In cold climates, the soil media supporting vegetation take longer to become anaerobic and a longer drawdown time may be allowed. In humid climates, sand beds of infiltration basins (and sand filters) should have a higher permeability to be completely dry before the next storm. To discharge large storm events, a spillway may be incorporated in the infiltration basin design. Alternatively, the tops of infiltration basins may be constructed as a level spreader or a diversion system is provisioned.

In areas of high soil permeability, the infiltration basin may be expanded to serve as an infiltration–detention basin. In such a case, the basin will be provided with an outlet structure to attenuate the peak rates of runoff beyond the water quality storm. The design of a dual purpose infiltration–detention basin follows the same procedure as a detention basin: namely, the development of stage–storage–discharge relation and routing computations. The infiltration loss through the bottom of the basin during the design storm helps reduce the required storage volume. It is to be noted, however, that the infiltration during the storm event is generally small and is conservatively neglected in design. The following case study exemplifies design of a detention–infiltration basin.

CASE STUDY 7.6

This case study relates to an infiltration–detention basin to address storm water quality and quantity requirements for a residential project in a municipality in New Jersey. The site of development is situated partly within the flood plain of the Third River, which abuts the easterly property boundary. The project disturbs more than 1 acre (0.4 ha) of land; as such, it is subject to the storm water management regulations adopted by the New Jersey Department of Environmental Protection (NJDEP) on February 2, 2004, and the state residential site improvement standards (RSIS) as well.

PREPARATION OF DRAINAGE AREA MAPS

Figure 7.50 shows an existing drainage area map. This map, which is prepared using a 1 in. = 80 ft scale, 1 ft contour lines, and aerial map of the site, indicates that the runoff from a major portion of the site travels overland to the Third River, while only a small portion of the site drains to the existing access road west of the site. Since runoff from the latter area flows to a tributary of the Third River that merges with the river shortly downstream of the site, the entire area of disturbance is considered as a single drainage area.

Figure 7.51 shows the proposed site layout, including the detention–infiltration basin and drainage conveyance system. Also shown on this map are the main drainage areas, designated as areas A1 and B2. Area A1 includes most of the development site, which is a tributary to the proposed detention basin. This basin, which is located in the southeasterly corner of the property, discharges to the Third River and is designed as a dual-purpose detention–infiltration basin to address both flow reductions and water treatment requirements. Area B2 is the balance of the site where the runoff occurs as overland flow. This area includes the stream buffer and a

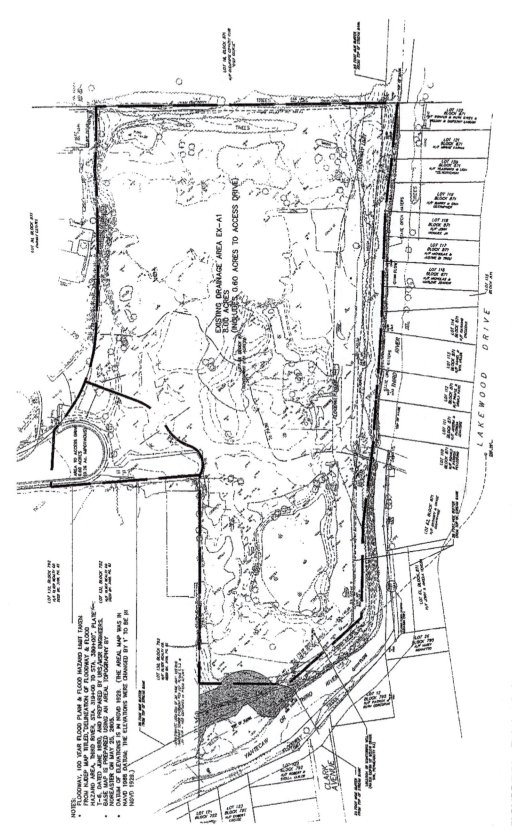

FIGURE 7.50 Existing drainage area map.

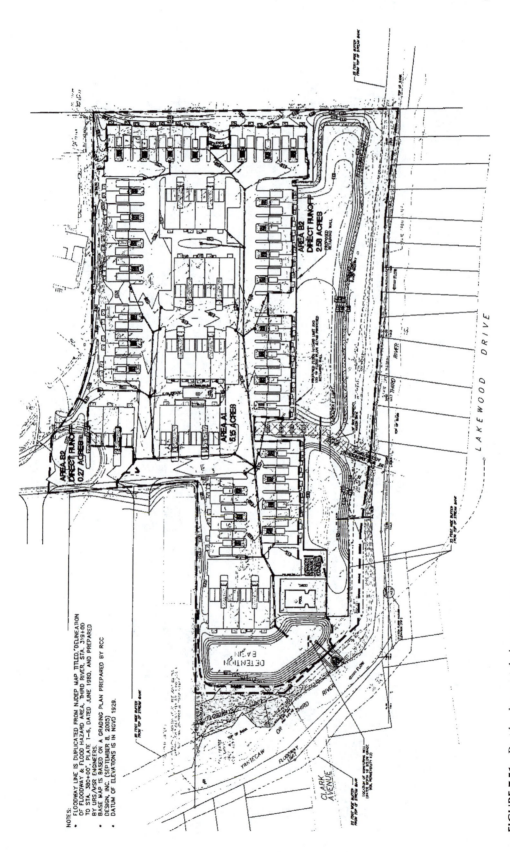

FIGURE 7.51   Postdevelopment drainage area map.

TABLE 7.61

Calculated Peak Rates of Runoff Existing Conditions

Storm Frequency (years)	Rainfall Amount (in.)	Q (cfs)
2	2.4	14
10	5.2	24
25	6.4	31
100	8.7	43

vegetated depression east of the development and south of the detention basin. Also included in area B2 is a small portion of the development that, due to topography, drains to the access road west of the site.

RUNOFF CALCULATIONS

Runoff calculations for the site are prepared using the SCS TR-55 method. Based on the soil survey map of the county in which the site is located, the site soil is urban land, Boonton–Wethersfield; the latter two soils are classified as hydrologic group C. In compliance with the RSIS and the NJDEP's storm water management regulations, the runoff calculations include the storms of 2-, 10-, and 100-year frequency. Additionally, calculations are performed for the 25-year frequency, which accords with the local soil conservation district standards. Table 7.61 lists the computed existing peak rates of runoff from the site.

Runoff calculations for the proposed conditions are performed separating the area tributary to the detention–infiltration basin, namely, area A1, and the area bypassing it, area B2. Similar to the existing conditions, the runoff calculations for the proposed conditions are prepared for the storm of 2-, 10-, 25-, and 100-year frequency. Consistent with the RSIS, a minimum time of concentration of 10 minutes is used for the area tributary to the detention basin; the Haestad methods software, which is employed for runoff computations, rounds this time to 0.2 hour.

INFILTRATION BASIN DESIGN

The proposed storm water management system for the project includes a detention–infiltration basin. Also, an underground rain cistern is provisioned to collect the roof runoff from two of the interior buildings. The water collected by the rain cistern is intended to satisfy irrigation demands of lawns and landscaped areas. This will not only reduce the volume of runoff discharged to detention–infiltration, but will also conserve the water demands of the development. To arrive at a conservative detention basin design, however, it is assumed that the roof runoff from these two buildings will drain directly to the detention basin in the event that the storm occurs when rain cistern is full.

To meet with the NJDEP's 80% total suspended solids removal, the proposed dual purpose basin has been designed to fully retain and infiltrate the water quality storm. This type of basin, which has been accepted for a TSS removal rate effectiveness of 80%, is chosen for this site considering practicality and ease of maintenance. Figure 7.52 is a schematic detail of the outlet structure. This figure indicates that the crest of the weir is located 2 ft (0.6 m) above the bottom of the basin. Table 7.62 presents discharge rating for the outlet structure and Table 7.63 shows the elevation–storage relation for the basin, accounting for the void volume in the 1 ft (30 cm) thick sand bed.

Calculations for the volume of runoff entering into the basin during the water quality storm are included in Table 7.64. As shown, storms up to and including the water quality storm are fully contained below the weir in the outlet structure and infiltrated into the sand bed.

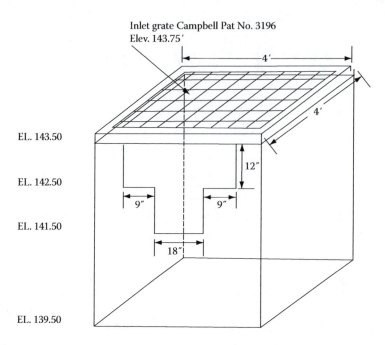

FIGURE 7.52   Outlet structure detail.

TABLE 7.62
Discharge–Elevation Relation[a]

Elev.	Q (cfs)				
	18 in. × 24 in.[b]	18 in. × 12 in.	Grate	20 ft spillway	Total
141.5	0.00				0.00
142.0	1.59				1.59
142.5	4.50	0.00			4.50
143.0	8.27	1.59			9.86
143.5	12.73	4.50			17.23
143.75	16.14	6.25	0	0	22.40
144.0	17.69	7.22	4.48	7.00	36.39

[a]   Refer to Figure 7.52.
[b]   Center weir.

Conservatively, the infiltration losses during the water quality storm are neglected in the calculations. The maximum depth of standing water within the infiltration basin is less than 2 ft during the 2-hour water quality storm and, as such, the infiltration basin is in compliance with the NJDEP Stormwater Best Management Practices Manual (2004).

Routing computations for the detention–infiltration basin are performed for the storms of 2-, 10-, 25-, and 100-year frequency. Table 7.65, summarizing the computations, indicates that the proposed detention basin produces a significant attenuation in the peak rates of runoff. Calculations for the composite discharges from the site are prepared by adding the outflow hydrographs from the basin to the runoff hydrographs for the bypass area. The composite

TABLE 7.63
Storage–Stage Relation

Elev.	Area (ft²)	Ave. Area (ft²)	Volume (ft³)	Σ Volume (ft³)
138.5	6900		0	0
	(12 in. sand; 30% void)	6900		
139.5	6900		2070	2070
		7240		
140.0	7580		3620	5690
		8240		
141.0	8900		8240	13,930
		9240		
141.5	9580		4620	18,550
		9940		
142.0	10,300		4970	23,520
		10,650		
142.5	11,000		5325	28,845
		11,370		
143.0	11,740		5685	34,530
		12,110		
143.5	12,480		6055	40,585
		12,860		
144.0	13,240		6430	47,015

TABLE 7.64
Water Quality–Area A1—Detention-Infiltration Basin

Cover	C	Area (ac.)	$C \times A$
Impervious	0.95	3.5	3.33
Grass	0.35	1.38	0.48
Brick paver	0.6	0.27	0.16
		5.15	
			$C_w = 0.77$

Water quality storm = 0.625 in./h for 2 h
$Q = 0.625$ in./h $\times C_w \times A = 2.48$ cfs
$WQ_v = Q \times 2$ h $\times 3600$s/h = 17,865 ft³
Storage provided below outlet = 18,550 ft³
See Table 7.63

TABLE 7.65
Detention Basin Routing Computations Summary 24 h.
SCS Type III Storms

Storm Frequency (years)	Q_{IN} (cfs)	Q_{OUT} (cfs)
2	11.0	2.10
10	17.0	6.94
25	22.0	11.33
100	31.0	19.90

TABLE 7.66

Calculated Composite Peak Rates of Runoff Proposed Conditions

Storm Frequency (years)	Q^a (cfs)
2	3.0
10	9.7
25	15.5
100	29.4

[a] Calculated by adding hydrographs.

See table below for 100 years storm calculations.

Time (h)	100 Y—Out (cfs)	B2—100 Y (cfs)	100 Y—Total (cfs)
	Calculations for 100-Year Storm		
12.2	2.2	12.0	14.1
12.3	8.1	14.0	22.0
12.4	15.3	12.0	27.3
12.5	19.4	10.0	29.4
12.6	19.9	7.0	26.9
12.7	18.4	5.0	23.4

TABLE 7.67

Comparison of Existing and Proposed Peak Rates of Runoff from the Site

Storm Frequency (years)	Q_E (cfs)	Q_P (cfs)	Reductions (%)
2	14	3.0	79
10	24	9.7	60
25	31	15.5	50
100	43	29.4	32

discharges are listed in Table 7.66 and a comparison of the existing and proposed peak rates of runoff is presented in Table 7.67. The latter table indicates that the selected storm water management system for the project reduces the peak rates of runoff far in excess of the applicable requirements.

DESIGN OF DRAINAGE SYSTEM

Consistent with the RSIS, the internal drainage pipes are designed based on a 25-year storm. Flow calculations are based on the rational formula using a runoff coefficient of 0.35 for lawn and landscaped areas and 0.95 for impervious areas. Discharge capacity of the proposed RCP drainage pipes is based on a Manning's roughness coefficient of 0.013. An 18 in. RCP is used as the outlet pipe from the detention–infiltration basin. This pipe has ample capacity to convey the 25-year storm discharge based on the slope of the pipe. The pipe can also adequately discharge the 100-year storm under pressure due to the standing water in the basin. To avoid erosion of the river, a scour hole is provisioned at the terminus of the outlet pipe at the river. The design of the scour hole is conservatively based on the 100-year storm discharge from the basin.

7.7 RETENTION–INFILTRATION BASINS

A retention–infiltration basin is similar to a detention–infiltration basin in that it stores the runoff and infiltrates it over time. However, these basins, unlike detention–infiltration basins, have zero discharge while filling but discharge without any attenuation once full. This major difference is illustrated by Figure 7.53. This figure shows that for a given storage volume, a detention basin attenuates the inflow runoff, but a retention basin may surcharge at the same rate as the runoff that enters it. Thus, contrary to a common design procedure by many engineers, the required detention storage calculations are not relevant to retention basin design. To avoid frequent surcharge, retention–infiltration basins should be designed to fully retain the runoff they receive during a design storm. This storm should be selected based on the type of basin, nature of development, and site climatic conditions.

Retention–infiltration basins may be constructed aboveground or underground. In urban development projects the latter type of basin is more common than the former. To discharge runoff in excess of the design storm, retention basins may be provided with an emergency outlet or spillway. A flat spreader on the top of aboveground detention basins and an overflow pipe in underground basins serve this purpose. Perforated pipes or chambers in a stone trench, storage modules, and dry wells (commonly referred to as seepage pits) are the most widely employed retention basins for small pavement areas and, particularly, dwelling roofs of single-family homes. Retention systems comprising chambers and storage modules were described previously. Dry wells and their design procedure are discussed herein.

7.7.1 Dry Wells

A dry well (seepage pit) is a perforated hollow cylinder installed in stone trench. Traditionally, cylinders are made of concrete; more recently these have also been fabricated of plastic materials. Figure 7.54 shows a typical 7.0 ft (2.1 m) O.D., 5.0 ft (1.5 m) high concrete seepage pit, manufactured by Peerless Concrete Products Company in Butler, New Jersey. This company makes seepage pits ranging from 6 ft, 6 in. O.D., 6.0 ft I.D. to 10 ft (3.0 m) O.D., and 3 ft to 9 ft, 10 in. (1 to 3.2 m) high. The top slab thickness of pits varies from 4 in. (10 cm) in nontraffic areas such as backyards to 8.0 in. (20 cm) for HS-20 load rating. Soil erosion and DEP/DEC (Department of Environmental Protection/Conservation) manuals in many states include details of seepage pits filled with gravel (see Figure 9.25 in Chapter 9). This arrangement, however, reduces the effective retention volume of a dry well cylinder by over 60%.

To avoid premature clogging due to grass clippings, leaves, silt, and debris, seepage pits are not recommended for retainage of runoff from lawn areas and supermarket parking lots. They are best suited for roof runoff, where few impurities are present. Even for this application, roof leaders/gutters should have a screen to remove leaves and roofing grit. In New Jersey, where the author practices, and many parts of the country as well, seepage pits are widely used for the retainage of runoff from dwelling roofs to maintain/reduce runoff from residential lots. To prevent frequent surcharges, the author recommends that seepage pits be designed at a minimum based on a 10-year, 60-minutes duration storm. In New Jersey, this storm has approximately 2 in./h (50 mm/h) intensity

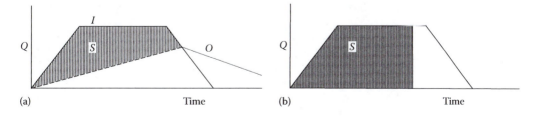

FIGURE 7.53 Flow routing through (a) detention basin and (b) retention basin.

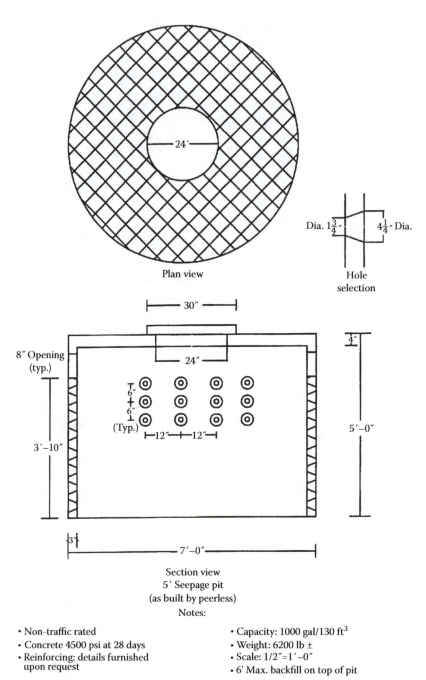

Dia. $1\frac{3}{4}''$ $4\frac{1}{4}''$ Dia.

Hole
selection

Plan view

Section view
5′ Seepage pit
(as built by peerless)
Notes:

- Non-traffic rated
- Concrete 4500 psi at 28 days
- Reinforcing: details furnished
  upon request

- Capacity: 1000 gal/130 ft^3
- Weight: 6200 lb ±
- Scale: 1/2″=1′–0″
- 6′ Max. backfill on top of pit

FIGURE 7.54 A 7 ft O.D., 5.0 ft high seepage pit. (Manufactured by Peerless Concrete Products Company, 2014, in Butler, New Jersey.)

and amounts to 2 in. (50 mm) of rainfall. Universally, the required storage volume of a roof runoff seepage pit can be calculated as follows:

$$V = A \times P \text{ (SI)} \tag{7.32}$$

$$V = 0.083\, A \times P \text{ (CU)} \tag{7.33}$$

where
   V = rainfall volume = retention storage, L (ft^3)
   A = roof area, m^2 (ft^2)
   P = rainfall depth, mm (in.)

Alternatively, the required retention volume may be calculated using the following equation, which is based on rainfall intensity:

$$V = A \times I \times T_d \text{ (SI)} \tag{7.34}$$

$$V = 0.083\, A \times I \times T_d \text{ (CU)} \tag{7.35}$$

where
   I = rainfall intensity, mm/h (in./h)
   T_d = storm duration, hours
   A and V are as defined before

The preceding equations conservatively neglect percolation losses through the bottom of the seepage pit during the design storm. For storms longer than the design (suggested 1 hour), the percolation losses may be included in the seepage pit calculation. This is illustrated by Example 7.4. The example shows that for soils of high and moderate permeability, the suggested 10-year, 60-minute storm provides a satisfactory design. To discharge larger storms, roof drains can be equipped with an overflow pipe (see Figure 7.55).

Figure 7.56 depicts the increase in runoff from a vacant lot due to construction of a single-family home. In this figure, the lot size, dwelling roof, and driveway areas are labeled as A, R, and D, respectively. For simplicity, the changes in runoff are expressed in terms of the product of runoff coefficient and area, CA. Also, for numerical illustration, runoff coefficients of 0.3 and 0.95 are employed for the pervious and paved areas, respectively; pervious areas commonly consist of grass–woods combination for open lots and lawn–landscape area for the improved sites. If the roof runoff is retained by seepage pit(s) or underground chambers in stone trench, the pre- and postdevelopment peak rates of runoff are proportional to the following (Pazwash, 2012):

$$Q_{\text{Pre}} \approx 0.3A \tag{7.36}$$

$$Q_{\text{Post}} \approx 0.3(A - R - D) + 0.95D \tag{7.37}$$

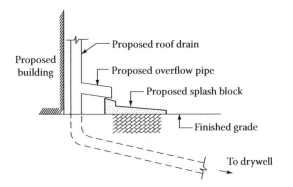

FIGURE 7.55   Drywell overflow.

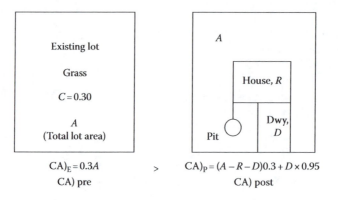

$$CA)_E = 0.3A \qquad > \qquad CA)_P = (A - R - D)0.3 + D \times 0.95$$
$$CA)\ pre \qquad\qquad\qquad\qquad CA)\ post$$

FIGURE 7.56 Schematic changes in peak runoff from a residential lot.

respectively, where:

    $A =$ lot area, m^2 (ft^2)
    $R =$ roof area, m^2 (ft^2)
    $D =$ driveway, m^2 (ft^2)

The rainfall intensity is the proportionality factor in these equations. The condition that the post development peak runoff would not exceed that of predevelopment is determined from the following equation:

$$0.3(A - R - D) + 0.95D \leq 0.3A \tag{7.38}$$

This equation simplifies as

$$0.3R \geq 0.65D \tag{7.39}$$

Thus, if the roof area is approximately 2.2 times larger than the driveway area and the roof runoff is fully retained, the postdevelopment peak runoff rate will be smaller than that of the predevelopment condition. Equations 7.38 and 7.39 are based on an assumption that the time of concentration is the same for pre- and postdevelopment conditions, which is a reasonable assumption for single-family homes.

The preceding principle can be applied to any runoff coefficient. For example, using a runoff coefficient of 0.25 for pervious areas, Equation 7.39 changes to

$$R \geq 2.8D \tag{7.40}$$

If the preceding condition (the driveway being 2.8 times smaller than the roof area) cannot be met, then the driveway may be covered with paver blocks rather than asphalt pavement. This will reduce the runoff coefficient for the driveway to 0.5 or less, and therefore the condition of no increase in peak runoff rate (Equation 7.40) becomes

$$R \geq D \tag{7.41}$$

Figure 7.57 depicts the changes in the peak runoff from a reconstructed single-family home in which the runoff from the new dwelling roof is retained by a seepage pit. In this case, the condition of no increase in peak runoff from the lot is expressed by the following equation:

$$[C_p(A - R - D) + C_I D]_{Post} \leq [C_p(A - R - D) + C_I(R + D)]_{pre} \tag{7.42}$$

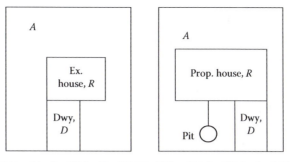

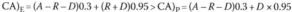

$$CA)_E = (A - R - D)0.3 + (R + D)0.95 > CA)_P = (A - R - D)0.3 + D \times 0.95$$

FIGURE 7.57 Changes in peak runoff.

In the preceding equation, C_p and C_I are the runoff coefficients for pervious and paved areas, respectively, and the other parameters are as defined previously. Except for unusual circumstances, the preceding equation holds true; therefore, retaining the roof runoff will satisfy the condition of no increase in the peak runoff from the reconstructed dwelling.

Example 7.4

Design a seepage pit to retain the runoff from a 2250 sf (209 m²) dwelling roof for a 10-year, 60-minute storm having 2 in./h (50 mm/h) intensity. Also, determine if a surcharge occurs during storms of 25-year frequency. The soil permeability is measured at 4 in./h at the site and the local rainfall intensity–duration relation for the 25-year storm is tabulated in Table 7.68.

Solution

Required storage volume is calculated using Equation 7.35:

$$\mathcal{V} = 0.083\, A \times I \times T_d = 0.083 \times 2250 \times 2 \times 1$$

$$\mathcal{V} = 373.5\ \text{ft}^3 (10.6\text{m}^3)$$

Select a 7.0 ft O.D., 5.0 ft deep pit (Figure 7.56) in 11 ft × 11 ft × 6.5 ft stone trench (min. 24 in. stone envelope and 18 in. stone base).

$$\text{Volume of pit} = \left(\frac{\pi}{4}\right) \times 6.5^2 \times 5 = 165.9\ \text{cf}$$

TABLE 7.68

Twenty-Five Year Storm

Duration-Intensity Relation

Duration (h)	Intensity (in./h)
1.5	1.50
2.0	1.20
3.0	0.90
4.0	0.73
6.0	0.53

TABLE 7.69

Seepage Pit Behavior during 25-Year Storms

T_d (h)	I (in./h)	V_{rain} (ft^3)	V_{per} (ft^3)	$V_{pit} + V_{per}$ (ft^3)	Storage Filled (%)
1.5	1.50	420.2	60.5	464.0	90.6
2.0	1.20	448.2	80.7	484.2	92.7
3.0	0.90	504.3	121.0	524.5	96.1
4.0	0.73	545.3	161.3	564.8	96.5
6.0	0.53	593.9	242.0	645.5	92.0

Note: The drywell fully retains storms of 25-year frequency.

Calculate void volume in stone trench, estimating porosity at 40%.

$$\text{Void volume} = [11 \times 11 \times 6.5 - (\pi/4) \times 7^2 \times 5]0.4 = 237.6 \text{ ft}^3$$

$$\text{Total volume} = 165.9 + 237.6 = 403.5 \text{ ft}^3 > 373.5$$

During the 25-year frequency storms, the rainfall amount is compared with the seepage pit volume (403.5 ft^3) plus the percolation from the bottom of the seepage pit, which is calculated as follows:

$$\text{Percolation} = (11 \text{ ft} \times 11 \text{ ft} \times 4 \text{ in./h}/12) \times T_d$$

$$\text{Percolation volume} = 40.33 \times T_d$$

where T_d is the storm duration in hours. In the preceding equation, seepage losses from the sides of stone trench are ignored. Table 7.69 lists the calculated rainfall volume, the percolation volume, the sum of percolation and seepage pit volumes, and the percent of the seepage pit volume utilized during each storm. In this table:

$$V_{rain} 0.083 \times 2250 \times I \times T_d = 186.75 \times I \times T_d$$

$$V_{per} = 40.33 \, T_d$$

CASE STUDY 7.7

This case involves a residential–commercial development project in a 5.4-acre (21,850 m^2) parcel of property in the town of Secaucus, in Hudson County, New Jersey. The site is situated within the jurisdiction of the New Jersey Meadowlands Commission and is partly located within the tidal floodplain of the Hackensack River. Figures 7.58 and 7.59 depict the site conditions. Figure 7.60 shows the predevelopment site layout. The site is covered by a building, asphalt pavement, gravel parking, and strips of grass.

The development consists of a retail building, three residential apartment buildings, and associated parking and driveways. The postdevelopment site layout is shown on Figure 7.61.

STORM WATER MANAGEMENT PLAN

The predevelopment drainage subareas are delineated on Figure 7.60. Shown on this plan are areas E1 and E2; area E1 drains to the existing drainage system in Seaview Drive extension and New County Road. The runoff from area E2 drains to the rear properties, namely lots 9.01 and 10, and mostly enters into an inlet in lot 9.01. Because of the inadequacy of the drainage system in these lots and drainage system downstream thereof, this inlet experiences frequent

FIGURE 7.58   Predevelopment gravel pavement.

FIGURE 7.59   Predevelopment paved parking.

flooding. To alleviate flooding in the northerly properties, the storm water management system is designed to divert the runoff from a portion of area E-2 to the drainage system on Seaview Drive extension. However, to offset the impact of this diversion on the roadway drainage system, an aboveground detention basin is incorporated in the storm water management system. To further reduce flooding in the rear properties, the runoff from the roofs of residential buildings 1, 2, and 3 will be directed to underground retention–infiltration basins and fully detained for a selected design storm. The layout of the drainage system is shown on Figure 7.61. Also delineated on this plan are the postdevelopment drainage subareas, labeled as area P1, area P1-D, and area P2. Area P1, covering 1.46 acres, drains directly to the roadway drainage system; area P1-D, consisting of 1.75 acres, is routed through the detention basin and is also discharged into the roadway drainage system. Area P2 includes 0.9 acre of green area in the back of the residential buildings and is the only area tributary to the rear properties. An underground retention infiltration basin is also provisioned for building 4 (office building) to reduce the runoff to Seaview Drive extension.

RUNOFF CALCULATIONS

Runoff calculations for the project are performed based on the SCS TR-55 method. According to the General soils map of Essex and Hudson Counties (1993), the site soil is classified as

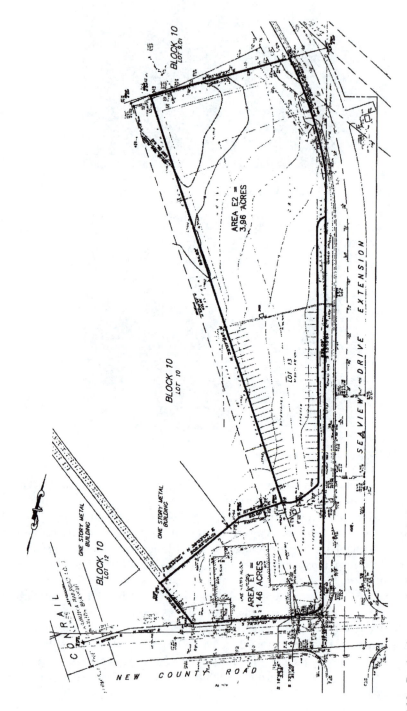

FIGURE 7.60 Predevelopment site layout and drainage area.

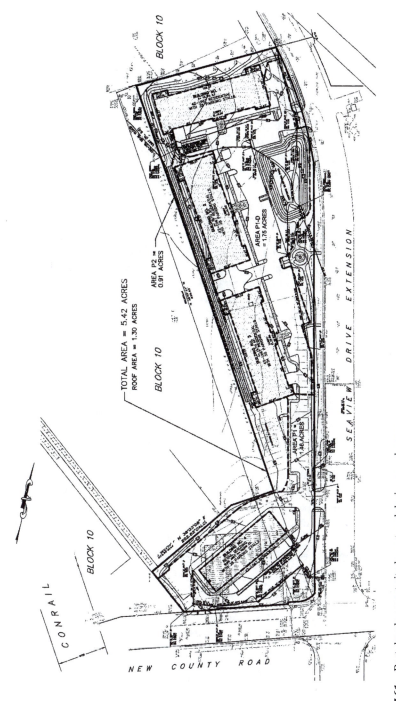

FIGURE 7.61   Post development site layout and drainage subareas.

TABLE 7.70

Computed Peak Runoff, CFS, Pre- and Postdevelopment, Subareas

Storm Freq. (years)	24-h Rainfall (in.)	Predeveloped		Postdeveloped		
		E1	E2	P1	P2	P1-D
2	3.3	4	8	3	1	3
10	5.0	6	14	5	2	6
100	8.3	10	25	9	4	11

hydrologic group C. Considering that the site is mostly impervious, runoff calculations are prepared using a time of concentration of 10 minutes for both pre- and postdevelopment conditions. In compliance with the NJDEP's storm water management regulations (2004), the runoff calculations include the storms of 2-, 10-, and 100-year frequency. Design calculations for internal drainage pipes are performed for the 25-year frequency storm, which accords with the RSIS's design criteria and the Meadowlands Commission drainage standards. Table 7.70 summarizes the computed peak runoff rates from the pre- and postdevelopment drainage subareas shown in Figures 7.60 and 7.61, respectively. Also listed in this table are the SCS 24-hour rainfall depths in Hudson County. It is to be noted that the roof areas of all four buildings, which are fully detained by the underground retention–infiltration systems are excluded from the computations. A review of Table 7.70 indicates that the proposed storm water management system reduces the peak rates of runoff to the rear properties by sixfold or more for storms of 2- through 100-year frequency but it will increase the runoff to the roadway drainage system. This impact is offset by the proposed detention basin, described next.

DETENTION BASIN DESIGN

The detention basin shown on Figure 7.61 is approximately 4 ft (1.2 m) deep, 10,340 ft² (961 m²) on plan area, and provides over 26,700 ft³ (756 m³) of storage volume. The basin is provided with an outlet structure consisting of a low-flow 4 in. (100 mm) orifice and an overflow grate just 2 in. below the top of the detention basin. The intent of this design is to attenuate the runoff to the maximum extent feasible. Figure 7.62 depicts the outlet structure and Table 7.71 presents elevation–storage–discharge rating for the detention system.

Runoff hydrographs of area PD-1 are routed through the proposed detention basin and the computed discharges are summarized in Table 7.72.

The composite discharges to the roadway drainage system are computed by adding the outflow hydrograph from the detention basin with the runoff hydrograph from area P-1 for each storm event. Table 7.73 lists the composite discharges to the Seaview Drive extension from the site. Also listed in this table are the predevelopment peak discharges to the roadway. A review of this table demonstrates that the provisioned storm water management system reduces the runoff to the roadway drainage system.

ROOF RUNOFF RETENTION SYSTEM DESIGN

The roof runoff retention–infiltration systems are designed to fully retain the roof runoff generated by the 10-year, 60-minute storm event. This storm amounts to 2 in. of rainfall in New Jersey. Overflow pipes emanating from the systems are designed to discharge runoff from longer duration storms. While considered as an extra conservative design criterion by some practitioners, the selected storm avoids frequent surcharge from the retention system.

In this project a shallow retention–infiltration system, consisting of Cultec chambers, is selected to retain roof runoff. As shown on Figure 7.61, the roofs of buildings 2 and 3 are drained to a common retention system in the back of the buildings and the roof; buildings 1 and 4 are drained to two separate systems. Design calculations for the retention system of building 4 (commercial/

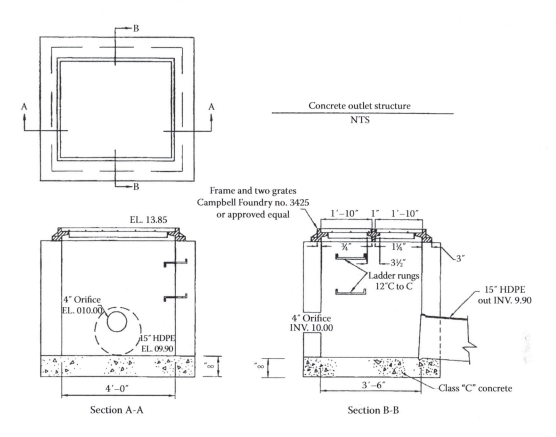

FIGURE 7.62 Outlet structure detail.

TABLE 7.71
Detention Basin Elev.–Storage–Discharge Relation

Elev., ft (m)	Storage, ft³ (m³)	Discharge, cfs (L/s)
10.0 (3.28)	0	0
10.5 (3.44)	625 (17.7)	0.24 (6.8)
11.0 (3.61)	2695 (76.3)	0.38 (10.8)
11.5 (3.77)	5800 (164.3)	0.48 (13.6)
12.0 (3.94)	9315 (263.9)	0.57 (16.1)
12.5 (4.10)	13,140 (372.2)	0.64 (18.1)
13.0 (4.26)	17,200 (487.3)	0.71 (20.1)
13.5 (4.43)	21,760 (616.4)	0.77 (21.8)
14.0 (4.59)	26,730 (757.2)	0.82 (23.2)

TABLE 7.72
Routing Computation Summary

Storm Freq. (years)	Inflow Peak, cfs (L/s)	Outflow Peak, cfs (L/s)
2	3.0 (84.6)	0.49 (13.9)
10	6.0 (169.9)	0.63 (17.8)
100	11.0 (311.5)	0.80 (22.7)

TABLE 7.73

Comparison of Existing and Proposed Discharges to New County Road and Seaview Drive Extension (Area 1)

Storm Freq. (years)	Q_{E1} (cfs)	Q_{P1} (cfs)	Q_O (cfs)	Total Q_P (cfs)
2	4	3	0349	3.4
10	6	5	0.63	5.5
100	10	9	0.80	9.6

Note:

Q_{P1} = direct runoff to roadway.

Q_O = peak outflow from detention basin.

Total Q_P = composite flow to roadway; computed adding Q_{P1} and Q_O hydrographs.

office building), with 10,000 ft² (929 m²) of plan area, are exemplified herein. The retention system includes one row of 22 units of Cultec Recharger 280 HD chambers in a 157 ft by 5.92 ft (47.85 × 1.8 m) stone trench. According to the manufacturer, Recharger 280 HD chamber alone provides 6.079 ft³ of storage volume per lineal foot (0.565 m³/m). This chamber is 8 ft long by 47 in. wide by 26.5 in. high (2.43 m × 119.4 cm × 67.3 cm). Calculations for the required storage volume and the retention volume of the retention–infiltration system are presented as follows.

ROOF RUNOFF VOLUME

a. Roof area = 10,000 ft²
b. Design storm = 2 in. (50 mm) rainfall
c. Required storage volume = (roof area) × (rainfall) = 10,000 × 2 × 0.98/12 = 1633 ft³ (46.2 m³)

where the coefficient 0.98 represents the percentage of rainfall that generates runoff; surface retention is estimated at 2%.

STORAGE VOLUME OF RETENTION SYSTEM

Trench dimensions are 157 ft × 5.92 ft

These dimensions are calculated as follows:

Length = 20 × 7.0 (inner pieces) + 2 × 7.5 (end pieces) + 2 × 1 (stone extension) = 157 ft

where 7.5 ft is the effective width of end pieces

Width = 47 in. (chamber) + 2 × 12 in. (side stone) = 71 in. = 5.92 ft

Depth = 6 + 26.5 + 6 = 38.5 in. = 3.21 ft (0.98 m)

Bed area = 157 (71/12) = 928.92 ft² (86.30 m²)

Volume in chambers:

$$(157 - 2) \times 6.079 \text{ cf/ft} = 942.2 \text{ ft}^3 \ (26.7 \text{ m}^3)$$

Void volume in stone trench:

$$(928.92 \times 38.5/12 - 942.2) \times 0.4 = 815.2 \text{ ft}^3 \ (23.1 \text{ m}^3),$$

where 0.4 represents porosity (void volume/trench volume)

$$\text{Total volume} = 942.2 + 815.2 = 1757.4 \text{ ft}^3 \text{ (49.8 m}^3)$$

The calculations show that the retention–infiltration system can fully retain the roof runoff. It is to be noted that the calculations conservatively neglect percolation through the bottom of stone trench during the storm period; however, the calculations also ignore the volume occupied by the chamber walls.

To ensure proper functioning of the roof runoff retention–infiltration systems, two test pits were conducted at the site. Water tables were also measured at these pits. One of the test pits was dug at the location of the detention basin and the other behind the proposed building 2, where a retention–infiltration system was to be placed. The following is a copy of the May 2, 2007, letter report from Johnson Soils Company (a geotechnical engineering firm in northern New Jersey) presenting the test results from two soil logs (see Figure 7.63).

Letter report from Johnson Soils Company presenting the test results from two soil logs

752 Grand Avenue
Ridgefield, NJ 07657
Telephone: 201-943-1793
Fax: 201-943-0951
Email: johnsonsoils@gmail.com

May 2, 2007

Via fax and mail
201-641-1831

Attn: Margita Batistic

Boswell Engineering
330 Phillips Avenue
P.O. Box 3152
South Hackensack, NJ 07606

Re: Hudson County Motors
   TP-1 & TP-3
   Secaucus, NJ
   #07-159

Dear Margita:

Following are the test results for the samples submitted on 4/27/2007. The hydrometer analysis was done accordingly to ASTM-D-422.

Sample #	SAND	SILT	CLAY	Permeability Rating
TP-1	64	30	6	K-2**
TP-3	65	26	9	K-3

TP-1: 28.0% of the sample was coarse fragments.
57.2% of the total sand sample was fine and very fine SAND.
TP-3: 29.3% of the sample was coarse fragments.
40.5% of the total sand sample was fine and very fine SAND.

SOIL PERMEABILITY RATING IS FROM THE SOIL PERMEABILITY / TEXTURAL TRIANGLE. Based on the average textural analysis, the soil permeability rating for TP-1 is K-2 (0.6-2 in./hr.), and the soil permeability rating for TP-3 is K-3 (2-6 in./hr.).

**Please note the permeability rating has been adjusted to the next lowest class due to the high percentage of fine and very fine SAND as per Chapter 9A Standards for Individual Subsurface Sewage Disposal Systems 7:9A-6.3 (h) 2.

Very truly yours,

Matthew Glennon
Project Supervisor   Enc. Graphs

• Subsurface Investigation • Geotechnical Engineering • Construction Testing •

The tests show a permeability rating of K2 at test pit 1 (TP-1, behind building 2) and K3 at test pit 2 (TP-3, at detention basin site). Depths of water table were also measured at these pits and are noted here:

TP-1: approximate ground elevation 10.2 ft
    Water at 3 ft, 6 in. depth
TP-3: approximate ground elevation 12.3 ft
    Water at 4 ft, 0 in. depth

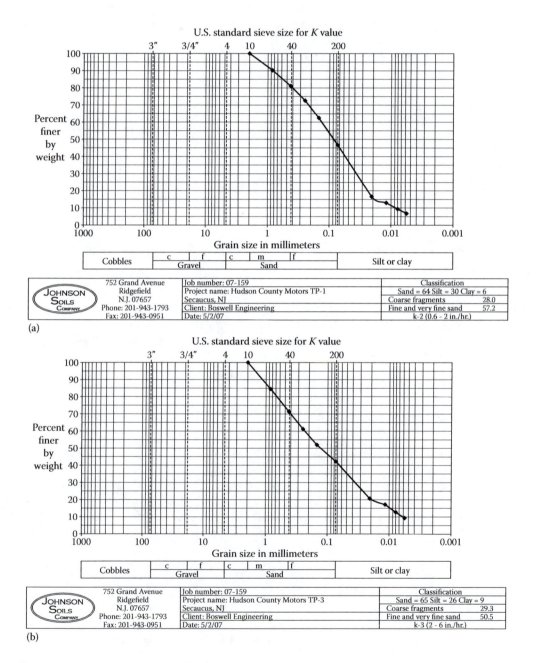

(a)

(b)

FIGURE 7.63 Sieve analysis of soil samples.

These data indicate that the water table lies approximately 2 ft (0.6 m) below the bottom of the retention system and 1.7 ft (0.5 m) below the bottom of the detention basin.

Time to drain system: Based on soil permeability results, the drain time of the infiltration basin behind building 4 is estimated as follows:

$$\text{Average depth of system} = \frac{1757.4}{928.92} = 1.89 \text{ ft} = 22.7 \text{ in.}$$

Soil rating = K2, 0.6–2 in. per hour. (See soil tests results.)

Average permeability = 1.3 in./h

$$\text{Time to drain} = \left[\frac{(\text{depth})}{(1.3\,\text{in./h})} \right] \times 2 \text{ safety factor}$$

$$= 34.9\,\text{h}$$

$$= 1 \text{ day, } 10.9\,\text{h}$$

WATER QUALITY PROVISIONS

The project replaces gravel pavement (which is not considered as an impervious surface in terms of water quality) by bituminous pavement. As such it produces a net increase of 0.15 acre in impervious area. Since the increase in impervious area is under 0.25 acre, the project is exempt from the state water quality requirement. However, three CDS water quality units are provisioned to treat the storm water discharges from the site. One of these units is placed in area P1-D (past outlet structure of detention basin) and two units are in area P1. Design calculations for these CDS units are provided in Table 7.74.

TABLE 7.74

Water Quality Calculations: TSS Removal Rates and CDS Sizing

Required TSS Removal Rate

Cover	Area (ac.)
Existing impervious	1.96
Increase impervious (does not include bldg.)	0.15

Increase in impervious is less than 0.25 acres; therefore, water quality is not required. However, CDS units are provided to treat most of the site's impervious area.

Proposed TSS Removal Rate

Area	Area (ac.)	TSS Removal Rate	$A \times \%$
Total impervious area to CDS units	1.96	50%	0.98

Weighted C Value for CDS Unit No. 1

Cover	Area (ac.)	C Value	$C \times A$[a]
Impervious	0.66	0.95	0.63
Grass	0.09	0.35	0.03
Total area and C value	0.75		0.66

Weighted C Value for CDS Unit No. 3

Cover	Area (ac.)	C Value	$C \times A$
Impervious	0.152	0.95	0.14
Grass	0.038	0.35	0.01
Total area and C value	0.190		0.16

CDS Unit Sizing

	$C \times A$	I (in./h)	Q (cfs)	CDS Unit Model No.[b]
CDS unit no. 1	0.66	3.2	2.11	CDS 30-30
CDS unit no. 2			0.41[c]	CDS 20-15-4
CDS unit no. 3	0.16	3.2	0.50	CDS 20-15-G

[a] Rounded to second decimal place.

[b] CDS model no. based on NJDEP and NJCAT treatment rates (conditionally approved at the time of design).

[c] Peak flow discharge from detention basin during water quality storm event.

PROBLEMS

7.1   Storage versus outflow relation for a detention basin is given below. Calculate the storage–outflow function, $2S/\Delta t + O$, versus O for each of the tabulated values, using $\Delta t = 5$ minutes. Plot a graph of the storage-outflow function versus O.

Storage (10^3ft^3)	75	81	87.5	100	110.2
Outflow (cfs)	2	6	12.0	22	34.0

7.2   Redo Problem 7.1 for the storage versus outflow discharge relation shown below:

Storage (10^3m^3)	2	3	3.5	4	5
Outflow (m^3/s)	0.06	0.15	0.35	0.70	1.00

7.3   Consider a 0.5-acre detention basin with vertical walls. The inflow discharge increases linearly from zero to 60 cfs at 30 minutes and then decreases linearly to zero at 75 minutes. The outlet from the detention basin is a 24 in. RCP culvert, located at the bottom of the basin. The basin is 5 ft deep and is initially empty. Use the level pool routing procedure with a 5-minute time interval to calculate the maximum discharge and the maximum water depth in the detention basin. The outlet pipe is 50 ft long at 2% slope and has an unsubmerged outlet.

7.4   Solve Problem 7.3 for the following conditions:
Detention basin size = 2050 m^2, 1.5 m deep
Inflow peak = 1.75 m^3/s
Diameter of pipe = 60 cm
Use a time interval of 10 minutes.

7.5   Using the TR-55 method, the pre- and postdevelopment 100-year peak rates of runoff for a 20-acre site are calculated at 25 and 50 cfs, respectively. Estimate the required detention storage to regulate the 100-year discharge from the site to 80% of the predevelopment peak rate of runoff. The soil curve number for the proposed condition is calculated at CN = 80 and depth of the 100-year, 24-hour storm is 8.0 in. Use type III storm.

7.6   Solve Problem 7.5 for a 5 ha site and 100-year peak runoff discharge at 1 m^3/s and 2 m^3/s, respectively. The 100-year, 24-hour storm is 200 mm.

7.7   Using the rational method, the 100-year pre- and postdevelopment peak rates of runoff from a 50-acre residential development site are calculated at 20 and 50 cfs, respectively. Estimate the required storage to regulate the 100-year discharge from the site to 80% of the predevelopment peak rate of runoff. Assume that the runoff from the entire site is drained to a detention basin and the postdevelopment time of concentration is 30 minutes.

7.8   In a 10 ha residential development, the pre- and postdevelopment peak rates of runoff are calculated at 1 m^3/s and 2 m^3/s, respectively. Estimate the required detention storage to attenuate the postdevelopment peak runoff to 80% of the predevelopment condition. Use the rational method and assume T_c = 30 minutes for a postdevelopment hydrograph.

7.9   The postdevelopment conditions of an urban watershed are as follows:
A = 20 ha
T_c = 30 min
C = 0.3 for lawn/landscaped
C = 0.95 for impervious areas
Impervious coverage = 25%

Calculate the required detention storage volume to reduce the 2-year and 100-year postdevelopment discharges from the watershed to 50% and 80% of the respective pre-developed conditions where T_c = 60 minutes and the watershed is woodland, C = 0.25.

Base your calculations on the rational/modified rational method and assume a linear discharge from the detention basin. Use Figure 3.1a (Chapter 3) rainfall curves.

7.10 Solve Problem 7.7 for a 50-acre watershed; use Figure 3.1b rainfall curves.

7.11 Repeat Problem 7.9 for a 50-acre watershed using the TR-55 method, assuming hydrologic group B, and 100% wood cover for existing conditions and 20% wood cover for the postdevelopment conditions. The 2- and 100-year SCS 24 storms are 3.3 and 8.0 in., respectively. The storm is type III.

7.12 In Problem 7.11, design a single outlet, using the following geometrics:
   a. Circular orifice
   b. Rectangular opening
   c. V notch weir
   The detention basin is 4.5 ft deep. Allow 6 in. ± freeboard.

7.13 A 75-acre development will consist of 35% pavement, 35% lawn, and 30% woods. Using the rational method, calculate the 100-year peak rate of runoff if the time of concentration is 30 minutes. The entire site runoff is drained to a detention basin. Using the IDF curves of Figure 3.1b in Chapter 3, estimate the required detention storage to create a 20% reduction in the 100-year peak rate of runoff below that of the existing conditions. The site is entirely wooded under the predeveloped conditions and has a time of concentration of 60 minutes.

7.14 Redo Problem 7.13 using the SCS method. The site soil is hydrologic group C. The 100-year, 24-hour storm is 8.3 in. The storm is type III.

7.15 Solve Problem 7.13 under the following conditions:
        Area = 30 ha
        T_c = 60 minutes, predevelopment
        T_c = 30 minutes, postdevelopment
   Use IDF curves of Figure 3.1a.

7.16 Calculate the required detention storage in Problem 7.13 using the universal runoff model. The soil is silty sand and permeability is measured at 2.5 in./h. The runoff from all pavements is directly connected to the detention basin.

7.17 Calculate the required detention storage in Problem 7.15 using the universal runoff model and 50 mm/h permeability.

7.18 A 20 ha development site is wooded and has a permeability rate of 35 mm/h. The development creates 5 acres of pavement and turns another 5 acres into lawn, all hydraulically connected. The time of concentration is calculated at 30 minutes and 20 minutes for the pre- and postdevelopment conditions, respectively. Using the universal runoff model, developed by the author, estimate the required detention storage to produce 50% and 20% reductions in the peak rates of runoff for the 2- and 100-year frequency storms, respectively. The local IDF curves show the following rainfall intensities:

Storm Frequency (years)	Storm Duration	Rainfall Intensity (mm/h)
2	20	50
	30	40
100	20	120
	30	80

7.19 Calculate the required detention storage in Problem 7.18, if the pavements and pervious areas separately drain to the detention basin.

7.20 A detention basin is designed to attenuate the runoff from 5.16 acres comprising 4.65 acres of roadway/pavement, 0.28 acres of lawn, and 0.23 acre of brick pavers. The allowable discharges from the detention basin are calculated at 7.94, 10.90, and 15.88 cfs for the storms of 2-, 10-, and 100-year frequency, respectively. Elevation–discharge–storage relations for the detention basin are listed in the following table:

Elevation (ft)	Discharge (cfs)	Storage (ft³)
869.5	0	0
870.0	0.8	1,150
870.5	2.0	3,400
871.0	2.7	5,670
871.5	4.1	7,950
872.0	6.0	10,200
872.5	8.4	12,600
873.0	11.0	15,200
873.5	13.8	17,950
874.0	17.4	20,980

   a. Perform calculations to determine the critical storm duration, namely the storm that requires the largest detention storage. Use the IDF curves in Figure 3.1b in Chapter 3.
   b. Perform routing computations for the critical storm.
   c. Compare the computed peak outflows from the detention basin with the allowable discharge from it.

7.21 Design a seepage pit to fully retain the runoff from the 150 m² roof of a residential dwelling for rainfalls up to 50 mm. The pit is to be placed in a 50 cm thick stone envelope and 25 cm stone base.

7.22 A seepage pit is to be used to fully retain the runoff from a 1500 square foot roof of a residential dwelling during the 10-year, 60-minute storm. This storm has a 2 in./h intensity and amounts to 2 in. of rainfall. Calculate the required storage volume of the seepage pit.

   A 6.5 ft O.D., 6 ft I.D., 4 ft high seepage pit, enveloped in a 2 ft thick stone and a 1 ft thick stone base is proposed for this dwelling. Is this pit adequate? Use a void ratio of 40%.

7.23 In lieu of the seepage pit in Problem 7.22, 36 in. perforated HDPE pipes in stone trench are employed. Size this retention system and sketch its layout.

7.24 In Problem 7.21, StormTech MC-3500 chambers in stone trench are to be used in lieu of the seepage pit. Calculate the number of chambers required and define the footprint dimensions of the stone trench.

7.25 A 5000 m² commercial building roof is drained into Terre Arch 26 structures. Calculate how many Terre Arch units are required to retain 30 mm of rainfall. The structures are to be placed on 150 mm thick stone base and their valleys also filled with stone. If the soil permeability is 50 mm/h, how many hours will it take for the water to drain?

REFERENCES

Advanced Drainage Systems (ADS), Inc., 4640 Trueman Blvd., Hillard, OH 43026-2438, Ph. 800-733-7473 (info@ads-pipe.com, www.ads-pipe.com).

American Concrete Pipe Association, 2005, Concrete pipe design manual, 17th printing.

ASCE and WEF, 1992, Design and construction of urban stormwater management systems, American Society of Civil Engineers Manuals and Reports of Engineering Practice no. 77 and Water Environment Federation manual of practice FD-20.

Atlantis Water Management for Life, Home: 3/19-21 Gibbes Street, Chatswood, NSW 2067, Australia, Ph. 612-9417-8344 (info@atlantiscorp.com).

Atlantis Water Management for Life, US Office: 702 Kentucky Street #194, Bellingham, WA 98225, Ph. 888-734-6533 (www.atlantiswatermanagement.com).

Brentwood Industries, Global Headquarters, 610 Morgantown Rd., Reading, PA 19611, Ph: 610-374-5109 (www.brentwoodindustries.com).

Campbell Foundry Company, 2014, Catalogue.

Contech Construction Products, 9025 Center Pointe Drive, Suite 400, West Chester, OH 45069, Ph. 513-645-7993 (info@contech-cpi.com, www.contech-cpi.com).

Cultec Inc., 878 Federal Road, P.O. Box 280, Brookfield, CT 06841, Ph. 203-775-4416 (www.cultec.com).

Drescher, S.R., Sanger, D.M. and Davis, B.C., 2011, Stormwater ponds and water quality, potential for impacts on natural receiving water bodies, *Stormwater*, Nov./Dec., 14–23.

EPA (US Environmental Protection Agency), 1999, Stormwater technology fact sheet sand filters, EPA-832-F-99-007, September.

Essex-Hudson County Soil Survey Map, 1993.

Minton, G.R., 2005, Revisiting design criteria for stormwater treatment systems, *Stormwater*, pp. 28–43.

Neenah Foundry, 2008, Catalog "R," 14th ed., Box 729, Neenah, WI (www.neenahfoundry.com, www.nfco.com).

New Jersey Geological Survey, 1993, Geological survey report, GSR-32.

New Jersey State Soil Conservation Committee, 1999, Standards for soil erosion and sediment control in New Jersey, July.

New York State Department of Environmental Conservation, 2005, New York State standards and specifications for erosion and sediment control, August.

New York State Soil Erosion and Sediment Control Committee, 1997, Guidelines for urban sediment control, April.

New York State Stormwater Management Design Manual, 2010, prepared by Center for Watershed Protection, 8390 Main Street, Ellicott City, MD 21043, for New York State Department of Environmental Conservation, 625 Broadway, Albany, NY 12233.

NJDEP (New Jersey Department of Environmental Protection) N.J.A.C.7:8, last amended April 19, 2010, Stormwater management.

———, 2004, Stormwater best management practices manual.

NRCS (US Department of Agriculture), Soil Conservation Service, 1986, Technical release no. 55, Urban hydrology for small watershed, June.

——— 2009, Small watershed hydrology, Win TR-55 user guide, January.

Old Castle Precast, Ph. 888-965-3227 (888-oldcast) (www.oldcastleprecast.com).

Pazwash, H., 1992, Simplified design of multi-stage outfalls for urban detention basins, *Proceedings of the ASCE Water Resources Sessions of the Water Forum 92*, Baltimore, August 2–6, pp. 861–866.

——— 2007, Fluid mechanics and hydraulic engineering, Tehran University Press.

——— 2009, Universal runoff model, paper R13, *StormCon*, Anaheim, CA, August 16–20.

——— 2011, *Urban storm water management*, CRC Press, Boca Raton, FL.

——— 2012, Storm water management techniques for single family homes, presented at the *8th Annual NJAFM Conference*, Somerset, NJ, October 2–3, 2012.

Peerless Concrete Products Co., 2014, 246 Main Street, Butler, NJ 07405, Ph. 973-838-3060 (www.peerlessconcrete.com).

Residential site improvements standards, 2009, New Jersey Administrative Code Title 5, Chapter 21, adopted January 6, 1977, last revised June 15, 2009.

StormChamber™, Hydrologic Solutions, Inc., P.O. Box 672, Occoquan, VA 22125, Ph. 877-426-9128/703-492-0686 (http://www.hydrologicsolutions.com, www.stormchambers.com, info@stormchambers.com).

StormTech, a Division of ADS, 70 Inwood Rd., Suite 3, Rocky Hill, CT 06067, Ph. 888-892-2694 (www.stormtech.com).

StormTrap LLC, 2495 W. Bungalow Road, Morris, IL 60450, Ph. 877-867-6872/815-941-4549 (info@stormtrap.com).

Terre Hill Stormwater Systems, P.O. Box 10, 485, Weaverland Valley Road, Terre Hill, PA 17581, Ph. 800-242-1509 (www.terrestorm.com).

Triton Stormwater Solutions, 9864 E. Grand Drive, Suite 110, Brighton, MI 48116, Ph. 810-222-7652 (www.tritonsws.com).

Yu, S.I., Kuo, J., Fassman, E., and Pan, H., 2001, Field test of a grass swale performance removing runoff pollution, *Journal of Water Resources Management*, 137 (3): 168–171.

8 Newer Trends in Storm Water Management (Green Infrastructure)

A new trend in storm water management is to control the runoff at its source rather than collecting and conveying it by a drainage system into a detention or retention basin. Source control practices include, but are not limited to, low-impact development (LID); use of permeable pavements, rain gardens, rain barrels, and cisterns; and disconnection of paved areas.

8.1 INTRODUCTION/SOURCE REDUCTION/CONTROL

8.1.1 INTRODUCTION

Source reduction, also known as source control, refers to measures to reduce the volume of runoff and its peak flow before the runoff is discharged from its source into storm drains or a combined sewer system.

Effectiveness of structural best management practices (BMPs) in general and storm water treatment devices in particular in removing pollutants was discussed in previous chapters. As noted, structural BMPs and manufactured devices may not remove all pollutants effectively. Some pollutants, such as salt, cannot be removed by any structural BMP. Apart from inefficacy, structural BMPs are expensive. The cost includes not only the construction and maintenance, but also the land in the case of open detention, retention, or infiltration basins. Although structural BMPs may sometimes be necessary, they should be considered as a supplement rather than the prime water quality measure in any storm water management design. A sound practice should be based on prevention, realizing that source reduction and/or control is a far better option than treatment. In fact, source reduction is an Environmental Protection Agency (EPA)-mandated component of the National Pollutant Discharge Elimination System (NPDES).

8.1.2 SOURCE REDUCTION

Source reduction can be effectively achieved through reducing impervious surfaces, including narrower streets, sidewalks, and driveways; small cul-de-sacs; and shared driveways. Taller residential and office buildings also reduce the building footprints and impervious surfaces. Reducing impervious surfaces not only generates less runoff and pollutants but also reduces cost and maintenance. While local codes may not permit narrower streets or smaller cul-de-sacs, waivers or alternatives such as "tee" turn-a-rounds or looping lanes can be discussed with local officials. Although the public may view narrower streets as unsafe, narrower roads in fact reduce the speeds at which vehicles can travel and make the roads safer. Sweden, which has narrow streets and reduced city speed limits, has the lowest number of pedestrian fatalities in the world.

Using pavers or pervious concrete in lieu of concrete walkways and patios and grass pavers for emergency and maintenance vehicles also provides source reduction. For an in-depth discussion on source reduction, the reader is referred to Chapter 5 in the New York State Storm Water Management Design Manual (2010).

A genuine source reduction, as will be discussed later in this chapter, should be based not on a development project or site basis, but rather on a per-capita basis. On this basis, multifamily residential developments, high-rise buildings, and, above all, city living reduce the per-capita impervious coverage and the rate and volume of runoff on a per-capita basis.

8.1.3 SOURCE CONTROL

A variety of measures can be employed to achieve source control. These measures are intended to reduce runoff at its source and include rain gardens, bioretention basins, roadside drainage swales, subsurface gravel wetlands, roof leaders' disconnection from drainage systems, pervious pavements, blue roofs, green roofs, rainwater harvesting, and tree pods along sidewalks. If these measures, which are also known as green infrastructures, are properly implemented, it is possible to avoid any increase in runoff from a development site during frequent storms such as 1-, 2-, and even 5-year storms and to create significant reduction in both volume and peak rate of runoff for storms of 10-, 25-, and even 100-year frequency.

Green infrastructure techniques also utilize the natural features of the site to reduce runoff. Figure 8.1 exemplifies layout of a subdivision site plan in which natural conservation areas are preserved.

The Federal Clean Water Needs Survey estimates that over $100 billion of infrastructure investment is needed in the next 20 years to address storm water and sewer overflows. Many jurisdictional agencies in the United States recommend the use of green infrastructure measures as a primary method of storm water management. Cities are increasingly considering green infrastructure as a cost effective solution to address the storm water and sewer overflow problems. A number of cities in the United States are also experimenting with and/or implementing some of these measures. The city of Portland, Oregon, and New York City are among the cities with storm and sewer overflows that are exemplified here.

With a population of 2.2 million, Portland is the third most populated city in the northwestern United States. The city has a Green Streets Program, which combines rain gardens and such LID features as permeable pavement, green roofs, and roadside/sidewalk swales and planters that allow water to infiltrate (Rogers and Faha, 2007). With this program, the city estimates to reduce the peak

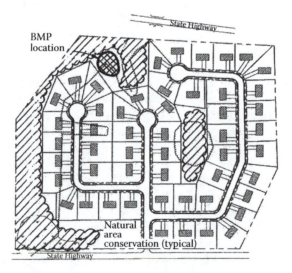

FIGURE 8.1   Schematic layout plan of a residential subdivision with natural conservation areas preserved. (From New York State Stormwater Management Design Manual, Figure 5.32, 2010.)

flows by as much as 85%, storm water volume by 60%, and water pollution by up to 90%. To further reduce the combined sewer overflows (CSOs) during heavy storm events that pollute the Willamette River and Columbia Slough Watershed, the city constructed two massive tunnels along the east and west banks of the river. The east tunnel is 22 ft (6.7 m) in diameter and 6 miles (9.7 km) long and the west tunnel is 14 ft (4.27 m) in diameter and 3.5 miles (5.6 km) long. The latter tunnel crosses under the river and merges with the former tunnel. The tunnels were bored using a microtunneling process, which had also been used in the Big Dig project in Boston.

New York City, like many other older cities, is served by combined sewer systems that convey both storm water and sewer to treatment plants. There are approximately 450 CSOs in the five boroughs of New York City and, during sustained or heavy rainfall, overflow occurs to rivers and bays. To reduce CSOs the city intends to manage runoff from 10% of impervious surfaces through green infrastructure by 2020. To evaluate the effectiveness of various measures, the city began pilot studies in a number of sites throughout the borough of Queens in 2011. Included in the measures were enhanced tree pots in 20 ft long × 5 ft wide (6 × 1.5 m) and 2 ft (0.6 m) deep sandy soils, which receive runoff from 2000 to 6000 ft^2 (200–600 m^2) of area; bioswales along open curb cuts; and installation of 6400 ft^2 (595 m^2) of porous asphalt and 4200 ft^2 (390 m^2) of crushed glass in a section of Queens. The pilot study has indicated that green infrastructure can be effective in one of the most densely developed urban areas in the United States (McLaughlin et al., 2012).

8.1.4 Source Reduction Benefits

Source reduction, which may also be termed a decentralized solution, has the following advantages compared to structural BMPs:

- *Effectiveness*: Source reduction can be more effective than structural BMPs for a number of pollutants, including salt, phosphorus, and metal. Some pollutants, such as salt and soluble phosphorus, cannot be removed by detention basins and ponds.
- *Cost reduction*: Source reduction reduces the size of required BMPs to control runoff quantity and water quality, thereby saving the overall cost of storm water management provisions.
- *Total elimination of water quality devices*: Source reduction may eliminate the need for any water quality device. Pursuant to the author's recommendation, the need for any water quality devices was eliminated in some large single-family homes in Franklin Lakes, New Jersey, simply by using pavers in lieu of paved driveways.
- *Damage prevention*: Source control can avoid flood and flood-related damage to downstream properties. This is particularly true during construction where a large amount of sediment is produced on site. Soil erosion control measures greatly reduce sediment discharge to drainage systems and detention basins. Serving as an expert witness, the author found that due to inadequate soil erosion control measures and lack of maintenance, the runoff from a residential project under construction caused millions of dollars of damage to a new-car dealership downstream thereof. Specifically, during a heavy storm, the runoff from the development surcharged from the drainage system, which was half filled with silt, and the muddy water flowed onto the dealership lot, ruining many new cars. The legal and punitive costs to the developer were much more than the cost of source control: namely, installing and maintaining adequate soil erosion measures.

Source boundaries may be defined up to the storm water conveyance system (Baker, 2007). However, accounting for pollutants entering catch basins due to road de-icing and oil leaks from cars and trucks, the conveyance system may be included in the source boundary. It is to be noted that contrary to a common perception, lawns produce more pollutants than roads. Lawn runoff is laden with nutrients and phosphorus. In fact, the total concentration of nitrogen and phosphorus in lawns is higher than effluent from an advanced secondary treatment plant. Therefore, it is prudent to begin source control from lawns. To reduce pollution, soil should not be compacted during construction to avoid reducing infiltration. Also, fertilizer and pesticides should be used at the minimum amount necessary and not be applied before rainfall.

To reduce lawn pollution, some states have passed laws that restrict the use of lawn fertilizers above a low limit of phosphorus. A widespread enactment of such laws will eventually lower the concentration of phosphorus in lawn runoff due to fertilizer, which is the largest source of this urban pollutant. Homeowners are generally unaware of the ill effects of lawn chemicals on the environment. They may participate in any lawn pollution reduction plan if they are informed. Public outreach is the key in educating homeowners and in the success of any public planning. A list of measures for source reduction is included in Section 8.15 in this chapter.

8.2 LOW-IMPACT DEVELOPMENT

Urban development severely disrupts natural hydrology. Low-impact development is an environmental friendly development approach with minimal impacts on hydrologic regime and water quality. Specifically, the objectives of LID, which is a site-based process, are

- Minimizing increase in the amount of runoff by reducing impervious covers, such as roads, driveways and parking areas
- Maximizing on-site infiltration and retention by conserving natural vegetation, maintaining natural drainage courses, and minimal use of pipes and inlets
- Reducing clearing and grading to minimize erosion and sedimentation
- Providing measures to store runoff in small depressions, and retention and detention basins spread out throughout the site
- Maintaining predevelopment time of concentration by routing runoff and disconnecting impervious surfaces such as roofs and driveways from roads
- Minimizing or eliminating storm water treatment systems, which require high maintenance and fail when neglected
- Promoting public awareness of, knowledge of, and participation in pollution prevention measures to protect the environment

The public is not fully aware of the extent to which traditional development involving clear cutting large strips of land harms the environment. Compacted lawns create more runoff than natural open space and woodland. LID is intended to eliminate or reduce these problems through an alternate, comprehensive approach to storm water management including retention and storage of rainfall. The LID storage techniques include green roofs, rain gardens, rain barrels, porous pavements, pavers, constructed wetlands, conservation of existing woodland and open space, landscape bioretention cells and basins, green rain basins (lawn and landscape depressions), grass swales, infiltration trenches, at or below grade stone basin, and landscaping. A review of literature on LID can be found in an EPA publication (2008). The following case exemplifies low-impact development.

The Ipswich River in northeastern Massachusetts, which is the source of drinking water for more than 330,000 residents and businesses, had been classified as a highly stressed stream because of low or no flow along its upper reach and chronic fecal coliform bacteria. To improve water quality and increase low flow, the town of Wilmington implemented a LID storm water management project within the Silver Lake Watershed. Silver Lake is a 28.5 acre (11.5 ha) kettle lake, has a drainage area of 132 acres (53 ha), and is situated in the headwaters of the Ipswich River. The project included the installation of a number of LID storm water management measures in a parking lot in the vicinity of the Silver Lake town beach and a residential area on the opposite side of the lake, as follows:

- Concrete pavers to replace asphalt pavement along one side of a street
- Porous asphalt and interlocking concrete pavers on a parking lot

- Residential roof rain harvesting systems, 200 to 850 gal. (0.76 to 3.2 m^3) capacity
- An underground 8000 gal. (30 m^3) tank to collect roof runoff from an elementary school
- Block pavers in a parking lot
- Gravel Pave (stone in a geoweb cell) in an overflow parking lot
- Flexi-Pave (poured-in-place rubber paving surface) in another overflow parking lot
- Rain gardens (12 in all) along streets to collect runoff
- Bioretention basins (depressed landscape and engineered soil mixture) along the perimeter and a median island in the paver parking area
- Two vegetated swales to replace drainage pipes terminating at the lake

In situ permeability tests were conducted using single and/or double ring infiltrometers at rain gardens, bioretention basins, porous asphalt, concrete pavers, and Gravel Pave. The measured permeability ranged from 12 in./h (300 mm/h) in a rain garden to 22 in./h (560 mm/h) in bioretention basins to 57 in./h (1450 mm/h) in pavers, to 78 in./h (1980 mm/h) in porous asphalt, and over 9 ft/min (2.75 m/min) for Gravel Pave. Overall, the porous surfaces performed better than the design infiltration rates. To maintain its permeability, the porous asphalt has been vacuum swept twice a year. The project was completed in the summer of 2006 for a construction cost of $340,000 in total (Roy and Braga, 2009).

8.3 SMART GROWTH

Smart growth is an urban planning designed to achieve environmental, community, and economic improvements. In the early 1970s, transportation and community planners introduced the idea of compact cities and communities. Peter Calthorpe, an architect, was among those who promoted the idea of urban villages that relied on public transportation, bicycling, and walking, instead of cars. Architect Andres Duany recommended changing design codes to promote a sense of community and discourage driving (http://en.wikipedia.org/wiki/Smart-Growth). Smart growth is an alternate to urban sprawl, traffic congestion, and unconnected neighborhoods. The principles of this planning contradict the ongoing concepts of suburban planning based on detached dwellings and long commutes that benefited oil companies and the automobile industry.

According to the EPA (http://www.epa.gov/smartgrowth/about_sg.htm; Nisenson, 2004), the environmental goals of smart growth include water savings achieved through development strategies that include

- Compact development pattern
- Mixed use development
- Preservation of open space and critical environmental areas
- Providing walkable neighborhoods and a variety of transportation choices
- Making better use of existing infrastructure
- Offering a range of housing opportunities
- Involving community and stakeholders in decision making

The best tool for achieving smart growth is local zoning ordinances. Considering smart growth's inherent benefits, communities across the country are adopting smart growth strategies as the BMPs to manage storm water runoff. Examples of smart growth techniques as BMPs are outlined here:

- Regional planning
- Infill development
- Redevelopment policies
- Special development districts (e.g., brownfields redevelopment)
- Tree and canopy programs

- Parking policies to reduce the number of spaces needed
- "Fix it first" infrastructure policies
- Smart growth street designs
- Storm water utilities

Smart growth reduces the cost of infrastructure significantly. At an average, it costs $50,000 to $60,000 to service a new structure in an undeveloped, also known as greenfield area, compared with $5000 to $10,000 for extending and maintaining the energy delivery system in a brownfield—namely, abandoned industrial and commercial sites. According to the Department of Energy, the loss of energy in transmission is 9%. Redevelopment reduces/eliminates this loss.

Smart growth, like many other types of urban planning, has opponents. They argue that the phrase "smart growth" implies that other growth and development strategies are not smart. There is also debate about whether transit-proximate development is actually smart growth when it is not transit oriented. The National Motorist Association strongly objects to some elements of smart growth including any tactics intended to reduce personal automobile use. Some libertarian groups, such as the Cato Institute,* criticize smart growth due to a potential increase in land values that would prohibit average income people from buying detached houses.

On June 16, 2009, the EPA joined the Partnership for Sustainable Communities with the US Department of Housing and Urban Development and the US Department of Transportation. The objective of sustainable partnership is to help improve access to affordable housing and to offer more transportation options and lower transportation costs while protecting the environment in communities.

For more detailed information on smart growth, one can visit the EPA website (http://www.epa.gov/smartgrowth/), which also links to many publications, and smart growth online (http://www.smartgrowth.org).

8.4   GREEN INFRASTRUCTURE

Green infrastructure as a concept originated in the White House during then President Clinton's administration in the 1990s. The EPA later extended the concept to the management of storm water runoff through waterways, wetlands, or drainage courses to mimic nature in controlling storm water, containing it at its source. The concept of green infrastructure as indicated also covers rain gardens, rain barrels and cisterns, blue and green roofs, permeable pavements, vegetated swales, pocket wetlands, and planted median strips. Some examples of green infrastructure are (Norman, 2008)

- Cincinnati, Ohio. According to the EPA, Cincinnati has one of the vastest plans for developing and implementing green infrastructure. The Metropolitan Sewer District of Greater Cincinnati (MSD) serves more than 800,000 people in 33 separate jurisdictions over 400 square miles (1036 km²) within Hamilton County. Approximately two-thirds of the MSD sewers are sanitary sewers and the rest is combined sewer. During heavy rainfall, the untreated overflow from the combined sewer enters creeks and rivers through nearly 300 discharge points throughout the county. In lieu of repair and building to capacity, which would cost $2 billion, the MSD is using green infrastructure options. Through a review of a series of maps showing areas of pervious, paved, soils, slopes, land use, and CSO locations, MSD selected favorable locations to implement green infrastructure.
- Milwaukee, Wisconsin. After spending over $3 billion in the 1980s and 1990s, this city began to consider green options. In 2003, Milwaukee adopted a downspout disconnection program to redirect roof downspouts into rain barrels, rain gardens, and pervious areas. Also, green roofs are being installed throughout the city.

---

* An American libertarian think tank, headquartered in Washington, DC, founded by Ed Crane in 1974.

- Portland, Oregon. This city is a pioneer in green infrastructure. City building codes require onsite storm water management for all new construction projects. New municipal buildings are required to have green roofs. Privately owned buildings with green roofs receive incentives. The city pays homeowners $53 for each downspout disconnected from storm drains.

Strategies to encourage incorporating green infrastructure solutions include

- Lowering storm water fees for homeowners who implement green storm water management practices, where applicable
- Private–public partnership
- Offset mitigation with credit trading

A number of cities, including Philadelphia, Washington, DC, and Portland, Oregon, now require redevelopment projects to minimize the onsite runoff volume using green infrastructure practices.

8.5 LEED AND GREEN BUILDINGS

Leadership in energy and environmental design, referred to as LEED, is a third-party ecology-oriented building certification program run under the auspices of the US Green Building Council (USGBC). This program was initiated in 1993 after then president Clinton announced a plan to "green" the White House. The USGBC website defines LEED as "a nationally accepted benchmark for the design, construction, and operation of high-performance green buildings" that provides building owners and operators with the tool they need to have an immediate and measurable impact on their buildings' performance. The built environment has a significant impact on our natural environment, economy, and health. In the United State alone, more than 2 million acres (8100 km²) of open space, wildlife habitat, and wetlands are developed each year. In 2002, buildings consumed approximately 78% of total electricity and 12.2% of the total amount of water (http://en.wikipedia.org/wiki/Green_Roof). So-called green buildings have environmental, economical, health, and community benefits. Their environmental benefits include

- Conserve natural resources
- Enhance and protect ecosystem and biodiversity
- Improve air and water quality
- Reduce solid waste

LEED programs cover schools; on-campus buildings; multiple buildings, including residential developments; commercial construction; department stores; commercial interior projects; operation and maintenance of existing buildings; and major renovation projects. Thus far, thousands of projects have been LEED certified in the United States, and state and local governments around the country are adopting LEED for public buildings of all kinds. LEED initiatives at the US Departments of Agriculture, Defense, Energy, and State drive actively at the federal land. In addition, various types of LEED projects are currently underway in over 40 other countries, including Canada, Brazil, India, and Mexico.

LEED performance is measured in five key areas: water efficiency, sustainable site development, energy efficiency, material selection, and indoor environmental quality. Storm water management credits for green buildings are covered under the LEED's sustainable site development. One credit point is given for quality control in storm water design. The intent is to limit disruption of natural hydrologic patterns by reducing impervious cover, increasing on-site infiltration, reducing or eliminating pollution from storm water runoff, and eliminating contaminants.

Two cases are included in the requirement. In case 1, existing imperviousness is less than 50%. In this case, a storm water management plan must avoid any increase in postdevelopment peak runoff rate and volume from those of predeveloped conditions for the 1- and 2-year, 24-hour storms. In case 2, which relates to existing imperviousness greater than 50%, the implemented storm water management plan must result in 25% reduction in the volume of storm water runoff from the 2-year, 24-hour design storm.

To acquire credit, the project site must be designed to maintain natural storm water flows by promoting infiltration using pervious pavement and other measures to minimize impervious surfaces and reusing storm water volumes for nonpotable uses, such as landscape irrigation and toilet and urinal flushing.

One point is awarded for storm water quality control design. The intent of this credit is to limit disruption and pollution of natural water through management of storm water runoff. The requirements for this credit include reducing impervious cover, promoting infiltration, and treating storm water runoff from the 90% of the average annual rainfall using acceptable BMPs. The said rainfall is 1.25 in. in New Jersey, 0.9 to 1.0 in. in Maryland, and 0.9 to 1.1 in. in New York State.

The provisioned BMPs to treat runoff must be capable of removing 80% of the average annual postdevelopment total suspended solids. To meet these criteria, the BMPs must be designed in accordance with standards and specifications of state or local agencies. Acceptable strategies include the use of nonimpervious surfaces, such as green roofs, pervious pavements, grid pavers, and nonstructural techniques such as rain gardens, vegetated swales, disconnection of impervious surfaces, and reuse of rainwater. Also included in acceptable strategies are sustainable design structures and techniques and environmentally sensitive design, such as low-impact development and constructed wetlands, vegetated filters, and grass channels to treat storm water runoff. Examples of LEED and green building include the following:

- The Sanctuary, a development constructed by Crescent Community of Raleigh, North Carolina, is a good example of a green project. This project, which comprises 187 homes on 1300 acres, bordering Lake Wylie in Charlotte, incorporates energy and water conservation measures. It also includes green storm water management measures, such as bioretention gardens and rain barrels. The Sanctuary's lodge was the first recreational facility in Charlotte to be LEED certified. Also, the Sanctuary was the first residential community in the world to receive Audubon International's Three Diamond designation, the highest level of certification in the Audubon's Gold Signature Program (Brzozowski, 2007).
- The California Academy of Sciences building located in the Golden Gate Park in San Francisco is the largest public structure in the world ever to achieve platinum certification in the United States Green Building Council's LEED. This building is also the only facility to combine an aquarium, a planetarium, a natural history museum, and science research. Because of its undulating 2.5 acre (1 ha) vegetated roof, its innovative environmental sustainability and its energy saving, this building has been called the greatest museum in the world. The vegetated building roof is covered with 50,000 coconut husk trays containing 6 in. of soil to nurture 1.7 million native Californian plants. This green roof helped the academy's new home attain platinum certification (Reid, 2009). A photo of this roof is included as Figure 8.24 in a later section in this chapter.
- West Michigan Environmental Action Council directed a green development project in which a vacant contaminated brownfield was transformed into the grounds for a 7200 ft^2 (670 m^2) multiple complex. The project included provision for rain gardens and green roofs to create Grand Rapids' first zero-storm water-discharge commercial site. The complex received the world's first US Green Building Council LEED double gold certification (Cunningham, 2009).
- Rickland College Science Building in Texas, 141,167 ft^2 (13,115 m^2) building and parking lots, is designed with a goal of receiving LEED Platinum status under the US Green Building Council's LEED rating system. The storm water management features include green roofs to partly retain the roof rain and slow the water, disconnecting downspouts from impervious surfaces, and bioinfiltration swales to allow time for infiltration, evaporation, and evapotranspiration. The green roofs improve energy efficiency and bioinfiltration swales improve water quality.

The paved parking lot is drained through curb openings onto landscape islands, which terminate at bioinfiltration swales. The water in the bioinfiltration swales flows down an engineered sand–soil matrix for rapid infiltration, which is underlain by a perforated underdrain system. The plants in the swale take up hydrocarbons and suspended solids before the water enters the underdrain. From the underdrain the cleansed water enters an underground Atlantic Raintank where it is stored and reused for outdoor landscape irrigation and nonpotable indoor water uses (Wilkins, 2008).

In 2008, the National Association of Home Builders (NAHB) initiated a plan for incorporating environmentally conscious products and practices for builders and homeowners. There is a basic difference between LEED and NAHB. LEED standards are mandatory to achieve points, but the NAHB plan is voluntary and has four levels. These levels are bronze, silver, gold, and emerald, the latter being the highest level of "green" achievement. The NAHB plan, designated as Green Building Standard, has been approved by the American National Standards Institute (ANSI). The plan includes seven standards, one of which relates to water efficiency and conservation. Others include lot preparation and design, resource efficiency, energy efficiency, occupancy comfort and indoor environmental quality, operation and maintenance, and homeowner education. In developing these standards, NAHB has coordinated with the International Code Commission of ANSI and also incorporated input from the public during the commentary period.

The standards on resources include indoor water saving through the use of low-flow faucets and showerheads and waterless toilets, and outdoor water conservation using landscape design that centers on native plant selection. Another potential outdoor conservation is utilizing an irrigation system zoned separately for lawn and for plants. The system could include drip irrigation, bubblers, and subsurface irrigation with buried lines. An example of NAHB is as follows:

- The Pearl River Tower in Guangzhou, China, opened March 2011, is the most energy efficient of all the world's super tall structures. Figure 8.2 shows images of this building. This 310 m (1016 ft) tall, 71-story building, which is the China National Tobacco Company's new headquarters, was designed by Chicago-based Skidmore, Owings and Merrill LLP

FIGURE 8.2 Images of Pearl River Tower. (a) Side view and (b) front view.

Engineering firm and Adrian Smith and Gordon Gill, architects. The design originally sought to construct a "net-zero energy" structure, but considering costs, the design was revised to achieve 60% reduction in energy consumption. As a technologically advanced superstructure, it will serve as a model for sustainable design and, in particular, CO_2 reduction; CO_2 emissions are widely believed to be one of the main culprits in climate change (Frechette and Gilchrist, 2009; Powell, 2009).

8.6 POROUS PAVEMENTS

A porous pavement represents a material through which water can flow. Porous pavements can be divided into a number of categories in terms of material and construction. These include open decks, open-celled paving grids, open-graded aggregate (gravel pavement), porous asphalt, pervious concrete, open-joint paving blocks (known as pavers), and synthetic turf (Ferguson, 2005). The words "pervious" and "permeable" are also interchangeably applied to porous pavements. "Pervious" means ability of a material to let water penetrate at its surface; "permeable" represents a material allowing water to infiltrate through it.

Decks are level or elevated wooden structures to bear foot traffic and are suitable for where they can be built around the existing environment, such as wetlands and boardwalks in coastal areas. Open-graded aggregate is one of the most permeable materials and the least costly. To avoid creating dust, aggregate is made of single sized angular material. This type of pavement has 30% to 40% void and tens of inches per hour permeability.

Because of reducing runoff due to infiltration, porous pavements, and particularly porous asphalt, pervious concrete and pavers are gaining great interest by site planners and public work departments. Regardless of terminology, some porous pavements, including porous asphalt, pervious concrete, and block pavers, have a common feature: a permeable surface and an underlying stone reservoir to temporarily retain surface runoff and infiltrate it into subsoil. Also, these pavements have the same design considerations as follows:

- Soils should have a permeability greater than 0.5 in./h (13 mm/h).
- The bottom of porous pavement should be at least 3 ft (1 m) above the seasonable high-water table.
- The bottom of the stone reservoir should be completely flat so that water can infiltrate through the entire surface.

In an interview, Ferguson (2009) sheds insight on answers to common questions regarding porous pavements.

8.6.1 OPEN CELL PAVING GRIDS

Open cell paving grids are cells with open spaces for gravel fill or grass to grow. The cells may be made of concrete or plastic. Open cell pavers are difficult to walk on, but they are suitable for low traffic areas, such as loading areas and fire and other emergency access lanes. Plastic paving grids consist of cells held together with ribs and filled with either aggregate or grass. Figure 8.3 shows a plan view and installation detail of a plastic grid cell made of recycled high-density polyethylene (HDPE), UV stabilized. This grid, trade-named "Grassy Paver™," is 15-1/4 in. × 13-3/16 in. × 1-3/4 in. thick (39 × 34 × 4.5 cm), covers 1.43 ft^2, and is manufactured by R.K. Manufacturing Inc. Figure 8.4 depicts an installed cross section of grassy pavers. Geo-Synthetics Inc. is another company that makes a Geoweb® cellular confinement system, named Presto® Geoweb. This system is available in three cell sizes—GW20V, GW30V, and GW40V—and four depths: 3, 4, 6, and 8 in. These come in folded panels that can be expanded to sizes ranging from 7.7 to 9.2 ft (2.3–2.8 m) wide and 32.2 to 68.5 ft (9.8–20.9 m) long. The cell openings are 44.8, 71.3, and 187 in.^{2} (289,

FIGURE 8.3 Plan view of one grid. Cells can be filled with gravel or soil and grass.

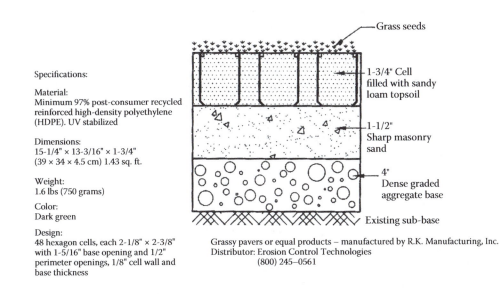

Specifications:

Material:
Minimum 97% post-consumer recycled
reinforced high-density polyethylene
(HDPE). UV stabilized

Dimensions:
15-1/4" × 13-3/16" × 1-3/4"
(39 × 34 × 4.5 cm) 1.43 sq. ft.

Weight:
1.6 lbs (750 grams)

Color:
Dark green

Design:
48 hexagon cells, each 2-1/8" × 2-3/8"
with 1-5/16" base opening and 1/2"
perimeter openings, 1/8" cell wall and
base thickness

Grass seeds

1-3/4" Cell
filled with sandy
loam topsoil

1-1/2"
Sharp masonry
sand

4"
Dense graded
aggregate base

Existing sub-base

Grassy pavers or equal products – manufactured by R.K. Manufacturing, Inc.
Distributor: Erosion Control Technologies
       (800) 245–0561

FIGURE 8.4 Grassy Paver™ details; installed cross section.

460, and 0.12 m²) for GW20, GW30, and GW40, respectively. Plastic geocells are used for emergency access lanes, auxiliary parking areas, pedestrian walks, golf cart path shoulders, and aprons. Geocells are also used for soil stabilization on steep slopes.

Flexi-Pave is the trade name of another permeable paver made of recycled tires. This product, manufactured by K.B. Industries, has high permeability and holds very well in freeze/thaw conditions (Brzozowski, 2009). Open cell pavers and plastic grids eliminate the need for retention/detention ponds and any drainage system.

Selection of grass pavers (plastic geogrids or concrete grids) depends on durability, ease of installation, and load-carrying capacity.

8.6.2 Porous Asphalt

Porous asphalt was first developed in the 1970s at the Franklin Institute in Philadelphia, Pennsylvania (Adams, 2003). This pavement, also referred to as "open-graded mix," "gap-graded mix," and "popcorn mix," contains the same elements as a conventional pavement; however, it has a different recipe. Porous asphalt is basically bituminous asphalt except that fine aggregate, smaller than

600 μm (no. 30 sieve) have been screened and reduced. A standardized mix of a porous pavement found to perform well is listed in Table 8.1. Porous asphalt is underlain by a choke course on an open-graded sub-base reservoir. A choke course is a permeable layer, typically 1.5–2.0 in. (40–50 mm) thick, of small size open-graded aggregate, which provides a level stabilized bed surface for the porous asphalt.

Open-graded sub-base reservoir is a thick layer of uniformly grade 1.0 to 2-1/2 in. (25–65 mm) clean stone to retain the design storm. However, based on availability, larger or smaller stone sizes may be used. Depending on the design frequency and the soil permeability, the sub-base reservoir depth can vary from over 12 in. (30 cm) to 36 in. (1 m). The soil beneath the stone sub-base should be minimally compacted to retain porosity and permeability. A 3–4 in. (75–100 mm) thick layer consisting of 3/8 to 3/4 in. (10–19 mm) stone may be placed between the chock layer and the sub-base. This high infiltration rate layer, which is intended to avoid migration of stones from the choke layer into the sub-base reservoir, also stores water.

In instances where the subgrade soil has a very low permeability, an underdrain may be installed in the sublayer to facilitate water removal. An underdrain is a small diameter perforated, typically PVC or HDPE, pipe. The use of a filter fabric between the sub-base reservoir and subgrade uncompacted soil is optional. Figure 8.5 shows a typical porous asphalt pavement section. Storm water drains through the asphalt and is held in the stone base, which has approximately 40% void space and percolates slowly into the underlying soil. A layer of geotextile filter fabric is placed below the stone to prevent the movement of soil into the stone bed.

The porous asphalt has been experimented on at the University of New Hampshire Stormwater Center (UNHSC) for a number of years. The asphalt studied in the UNHSC consisted of porous asphalt at the surface and a stone choker course (base course) on an underlying layer of finer material consisting of sand and gravel (Roseen and Ballestero, 2008).

A general misperception exists in the industry about functionality of porous asphalt (and pervious concrete, as well) in cold weather environments. Based on field observations, as well as the UNHSC experiments, freeze–thaw action and freezing of filter material is not an issue in the porous asphalt. By allowing the runoff to infiltrate, porous asphalt has fewer icing conditions and there is less need for snow plowing. While the filter may freeze, it does not freeze solid and still allows water to percolate. Research findings at the UNHSC show that, with the porous asphalt, salt application could be reduced by up to 75% based on snow and ice cover. According to Dr. Robert Roseen, the director of UNHSC (from 2004 to 2012), "With only 25% of the salt, the snow and ice cover on the porous asphalt was the same as on the dense-mix asphalt parking lot." He indicates that "even

TABLE 8.1
Standard Mix of Porous Asphalt[a]

US Standard Sieve Size, in. (mm)	Percent Passing
1/2 (12.5)	100
3/8 (10)	95
#4	35
#8	15
#16	10
#30	2

Note: Percent bituminous 5.75–6% by weight.

[a] See Appendix B at the end of the book for Unified Soil Classification System, US standard sieve sizes, and coarse aggregates.

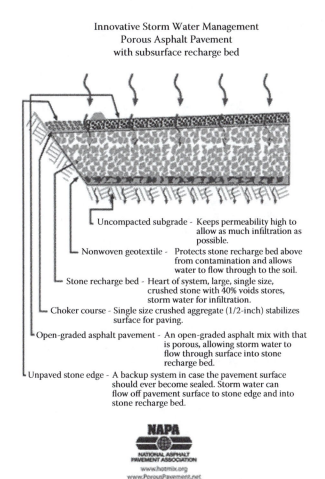

Innovative Storm Water Management
Porous Asphalt Pavement
with subsurface recharge bed

Uncompacted subgrade - Keeps permeability high to allow as much infiltration as possible.

Nonwoven geotextile - Protects stone recharge bed above from contamination and allows water to flow through to the soil.

Stone recharge bed - Heart of system, large, single size, crushed stone with 40% voids stores, storm water for infiltration.

Choker course - Single size crushed aggregate (1/2-inch) stabilizes surface for paving.

Open-graded asphalt pavement - An open-graded asphalt mix with that is porous, allowing storm water to flow through surface into stone recharge bed.

Unpaved stone edge - A backup system in case the pavement surface should ever become sealed. Storm water can flow off pavement surface to stone edge and into stone recharge bed.

NAPA
NATIONAL ASPHALT
PAVEMENT ASSOCIATION
www.hotmix.org
www.PorousPavement.net
www.PaveGreen.com

FIGURE 8.5   Porous asphalt section.

with no salt, porous asphalt has higher frictional resistance than dense-mix asphalt with the 100% of normal salt application."

Porous asphalt is well suited for parking lots, sidewalks, playgrounds, and low- to moderate-traffic areas. It has been used in hundreds of sites in the United States. One of the first porous asphalts, which was constructed in the parking lot of a corporate office park in East Whiteland Township, a suburb of Philadelphia, has been functioning well and has not been repaved for over 20 years. A parking lot in Walden Pond State Reservation in Massachusetts, constructed in 1977, is one of the best examples of porous asphalt. This parking lot has not been repaved, and even with the freeze–thaw cycles of New England winters, the porous pavement was reportedly holding in 2009. Among different mixes that were used in this lot, the one that performed best consisted of 2.5 in. (63 mm) porous asphalt on 1.5 in. (38 mm) type A stone on 10.5 in. (27 cm) type B stone on a gravel fill. Type A consisted of 3/8 and 1/2 in. (10–13 mm) stone and type B was a mix of stone 3/4 to 2.5 in. (19–64 mm) at certain percentage by weight. The porous asphalt included a stone mix ranging from 1/2 in. to sieve no. 200 and AC-20 viscosity grade asphalt (an acrylic latex sealant) 4.5% to 5.5% by weight.

Because of having less fine material, porous asphalt has smaller shear stress capacity than conventional pavements; therefore, it is not recommended for airport runways or slopes greater than 6%. Likewise, porous pavements are not suited where sand is applied for snow removal. Porous asphalt also is not recommended for truck stops or heavy industrial areas where there is a threat of groundwater contamination due to spills or leakage.

Different states have tried a variety of mix recipes. In a 1980 report titled "Porous Pavement Phase I Design and Operational Criteria," the EPA presented a standard pavement design. This report notes that the initial cost for a porous pavement can be up to 35–50% higher than the cost of conventional paving. However, this extra cost is more than offset by elimination or downsizing of conventional storm water management systems. The cost saving is even greater in urban settings where the land to install a detention basin or other storm water management facility is expensive. Also, based on life-cycle cost analysis, pervious pavement systems appear to be more economical in the longer run. "Normal parking lots made from impervious pavement typically last 12 to 15 years in northern climates where freeze–thaw is prevalent, while pervious pavement lots can last more than 30 years," Roseen says. Considering performance and storm water management benefits, the use of porous asphalt and other pervious pavements has increased rapidly in recent years and this trend is expected to grow even faster in the future.

In Portland, Oregon, one of the largest porous asphalt parking lots to store Hyundai cars was installed along the Port of Columbia River in 2006. This lot includes 35 acres (14.2 ha) of porous asphalt and 11 acres (4.45 ha) of standard dense graded asphalt, which suffer heavy traffic from delivery trucks. The pervious pavement consists of a 3 in. (7.5 cm) course of open-graded porous asphalt made of native river sand, and a 10 in. (25 cm) open-graded rock base. A reason for selecting porous asphalt was the availability of native river sand along the Columbia River. Although the porous pavement was somewhat more expensive than standard pavement, it eliminated the need for any drainage systems and storm water treatment devices, resulting in large cost savings. The 46 acre (18.6 ha) pavement cost $6.4 million (Brown, 2007). This amounts to $29/square yard ($34/m^2) of pavement on the average. The pavement is swept two to three times a year to avoid leaves and cottonwood seeds from clogging the voids and has been functioning well in infiltrating the runoff.

Figure 8.6 shows an experimental porous asphalt at the EPA laboratory in Edison, New Jersey. The asphalt fully infiltrates the sprinkler water. In Clark, New Jersey, porous asphalt was used at parking stalls on both sides of the parking lot of the municipal building to absorb the runoff from the lot. Figure 8.7 shows the porous asphalt at parking stalls abutting conventional asphalt driveways at this parking lot.

FIGURE 8.6   Experimental porous asphalt, EPA laboratory in Edison, New Jersey.

FIGURE 8.7 Porous asphalt at parking stalls, Municipal Building, Clark, New Jersey.

8.6.3 PERVIOUS CONCRETE

Pervious concrete is a mixture of cement, coarse-graded aggregate, and water. The main difference between pervious concrete, which was developed around 1970, and conventional concrete is that little or no sand is used in pervious concrete. Fibers and admixtures may be added to improve strength or other properties, but the common composition is the amount of water and the cementitious content and the absence of fine material. The result is a low slump mixture paste that forms a coating that bonds the aggregate together without flowing during mixing and placing. Using just enough paste produces a medium with interconnected voids.

Due to high void percentage, typically 15–25%, pervious concrete is relatively lightweight (Hun-Dorris, 2005), 100–120 lb/cf (16–19 kN/m^3). Upon placement, pervious concrete looks like sponge and provides an ultimate compression strength of 500–4000 psi (3400–27,600 kPa). Cured pervious concrete can pass 3–8 gal of water through each square foot (125–330 L/m^2) per minute. This reflects an infiltration rate of 4.8–13 in./min (12–33 cm/min). At this rate, pervious concrete can capture the most intense rainfalls. Thus, pervious concrete, similar to, yet better than, porous asphalt, reduces or eliminates the need for detention ponds. Likewise, it can downsize drainage systems and their costs. Figure 8.8 depicts infiltration capability of a small-scale pervious concrete model.

FIGURE 8.8 Running water draining into a pervious concrete model.

Pervious concrete has been used since the late 1970s in various applications, including driveways, sidewalks, and parking lots. In recent years, the use of pervious concrete has grown rapidly to address EPA and local storm water management regulations. In fact, the use of pervious concrete is recommended by the EPA as one of the best storm water management practices.

A benefit of pervious concrete is attributed to its light gray color, implying that it reflects more solar radiation and does not become as hot as the darker asphalt. In an experiment involving thermographic imaging, an asphalt roadway in Rio Verde, Arizona, was found to be 30°F warmer than an adjacent regular concrete parking lot at the same time of day. The lower temperature helps healthy growth of trees and plants, which in turn can reduce temperature in an urban setting.

Pervious concrete also offers a safety benefit in wet weather. A study by the University of Illinois showed a stopping distance of 162 ft (49 m) on dry concrete and 190 ft (58 m) on dry asphalt, for a given speed. The effect was even more significant in wet conditions, 316 ft (96 m) of stopping distance for concrete versus 440 ft (134 m) for asphalt. However, according to Dr. Roseen, porous asphalt functions better than pervious concrete in winter months. By absorbing more heat, porous asphalt will be warmer in winter, promoting more de-icing. A similar behavior was also observed by the author at his former residence, where the asphalt driveway abutted a concrete paver sidewalk.

Pervious concrete has been experimented with at the UNHSC in conjunction with the Northern New England Concrete Promotion Association. At the UNHSC, the makeup of the sub-base in an installed pervious concrete pavement is considerably different from most pervious pavement systems. The UNHSC pervious concrete consists of a thick stone base made of 1–3 in. (25–75 mm) stone, which provides considerable storage capacity and a sand layer that functions as a filter to improve water quality (Gunderson, 2008).

The sub-base should be thick enough to store an extreme storm event and the subsoil should be able to drain the water within 3 days to avoid standing water too long in a freeze–thaw environment. In soils of poor permeability, a subdrain pipe may be installed in the sub-base. Figure 8.9 shows a cross-sectional view of a typical pervious concrete pavement.

In terms of initial cost, the concrete industry claims that pervious concrete is comparable with porous asphalt, but it requires fewer repairs and has lower ownership costs in the long run. However, in a parking lot project at the University of North Carolina in Chapel Hill, where sections were constructed from pervious concrete and others from porous asphalt, pervious concrete pavement cost was four times as much as that of porous asphalt. A major problem with pervious concrete is difficulty of installation. Unlike porous asphalt, which can be handled by any qualified installer, placing pervious concrete generally requires a certified installer. This difficulty of preparation of a proper paste and installation was experienced in a 2003 pilot project on the Villanova University campus by the concrete manufacturer for the selection of a contractor to place pervious concrete (Traver et al., 2004, 2005). New Jersey Concrete and Aggregate Association is one of the companies that makes pervious concrete and can be visited at http://www.njconcrete.com for information on pervious concrete installers.

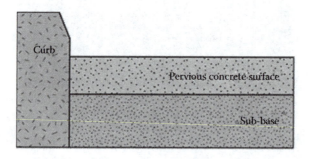

FIGURE 8.9   Cross-sectional view of typical pervious concrete pavement.

FIGURE 8.10　A pervious concrete sidewalk in Hogan Park, New Jersey.

Pervious concrete has long been used for sidewalks, driveways, and parking areas. However, there is a growing interest in using it in low- to moderate-traffic roadways, and this trend is expected to increase with more stringent storm water management regulations. The borough of Northvale, in northern New Jersey, is an example where pervious concrete was selected for paving walkways around four athletic fields in Hogan Park. Due to poor drainage, the sidewalks provided limited access for players and spectators following rainfalls. Pervious concrete was prepared in the Eastern Concrete Material Plant in Bogota, New Jersey, and placed by Let It Grow, Inc. of River Edge, New Jersey. In cross section, the pavement consists of 4.5 in. (11.5 cm) of pervious concrete over a 12 in.h (30 cm) layer of clean aggregate to store water that percolates through the concrete. The new pathway was 6–12 ft (2–4 m) wide and nearly a half mile (0.8 km) long and was constructed in 2008 (Justice, 2009). Figure 8.10 shows the pervious concrete sidewalk at Hogan Park. In a project reviewed by this author, pervious concrete was used to pave a parking lot of a commercial site in Englewood Cliffs, New Jersey. The pavement, which consisted of 4 in. (10 cm) thick pervious concrete over 24 in. (60 cm) stone reservoir, was installed at a cost of approximately $10/ft^2 ($110/ m^2) in 2012.

8.6.4　Glass Pave

Glass Pave is a permeable pavement and serves as an alternative to conventional pavement which consists of a coarse, granular sub-base, a finer base layer, and a top course, namely asphalt and concrete. As an alternative, pavements can be constructed using glass foam gravel and FilterPave™, both of which are made of recycled glass and are manufactured by Presto Geosystems (a construction product firm based in Appleton, Wisconsin). Glass foam gravel can be used as a substitute for sub-base and FilterPave™ can serve as an alternative for porous asphalt and pervious concrete and exhibits the best characteristics of both of these pavements. This material has 39% porosity, has stronger compressive strength than porous asphalt, and is more flexible and flexural than pervious concrete. The average comprehensive and flexural strengths of FilterPave™ are 8 MN/m^2 (1160 lb/ in.^{2}) and 3.5 MN/m^2 (508 lb/in.^{2}), respectively. As such, FilterPave™ is appropriate for low- and medium-duty traffic applications (Emersleben and Meyer, 2012).

8.6.5　Concrete Pavers

Concrete pavers were first developed in Europe over 50 years ago. In the United States the use of pavers for sidewalks, patios, driveways, and parking lots began nearly three decades ago. Pavers are a modern version of stone block pavements that were used by Romans and other old civilizations to construct roads for thousands of years. Figure 8.11 shows a stone paved parking lot with an adjoining green island in Budapest, Hungary. Figure 8.12 shows a color-striped cobblestone sidewalk with tree pods in Lisbon, Portugal, and Figure 8.13 shows a stone-paved street and sidewalk in another city in the same country. Figure 8.14 depicts pavers installed at a sidewalk in Van Saun Park in the borough of Paramus in New Jersey. Because of aesthetics and permeability, pavers are becoming increasingly popular in urban settings.

In Chicago alone, over several million square feet of permeable pavers with open-graded aggregate have been installed over the past 25 years. This city is at the forefront among large American cities in a green program to replace paved alleyways by permeable pavement to rectify vast flooding problems (Buranen, 2008b). A program, adopted in 2007, covers 20 alleys and incorporates three types of pervious materials: pervious concrete, porous asphalt, and pavers. Depending on the circumstance, an alley may be partly or fully covered with permeable pavement. For example, permeable concrete may be placed at the alley center with impervious concrete on either side.

Permeable pavers are installed by placing a layer of an open-graded aggregate and compacting it to provide a stable surface, placing a setting bed on top of a base course, and setting pavers either by hand or machine. The joints between the blocks are filled with finer material. No filter fabric is generally used in separating aggregate layers. Experience shows that separating aggregate layers

FIGURE 8.11   Cobblestone parking area in Budapest, Hungary. (Photo by the author, 2006.)

FIGURE 8.12   Striped cobblestone sidewalk with tree pods in Lisbon, Portugal. (Photo by the author, 2013.)

FIGURE 8.13 Stone-paved street and sidewalk in Portugal. (Photo by the author, 2013.)

FIGURE 8.14 Pavers at a sidewalk in Van Saun Park in Paramus, New Jersey. (Photo by the author.)

introduces clogging and causes more rapid degradation of the system. However, depending on the subsurface condition, a geogrid may be used below the base course aggregate to increase stability.

Depending on the application and the intended use, the aggregate layers may be from 6 to 12 in. (15–30 cm) thick. A 6 in. base is commonly used for patios and sidewalks, 8 in. (20 cm) in driveways, and 12 in. (30 cm) for significant exfiltration, light-traffic roads such as highway ramps, and parking lots. The thicker base also provides a larger retention capacity. The layers from bottom to top are as follows*:

Sub-base: Six to twelve inches (30 cm) of 1.5 in. (38 mm) crushed stone aggregate. This stone, which is called #4 aggregate, is compacted in 4–5 in. (10–13 cm) lifts. Railroad ballast, which is called #2 aggregate, may be used (DeLaria, 2008). Recycled concrete is also used—however, as a mix with dense graded aggregate (dga) or quarry-processed stone (known as QP).

* See Appendix B at the end of the book for Unified Soil Classification and AASHTO coarse aggregate sizing.

Base course: Four inches (10 cm) of 0.75 in. (18 mm) all-fracture-face aggregate, called #67. If #2 aggregate is used as the sub-base, #57 aggregate, which is 1 in. (25 mm) or smaller stone should be used as the base course in order not to fill pore space in the sub-base. The base course is also compacted.

Setting bed: One to one and one-half inch (2.5–4 cm) of coarse sand, also called concrete sand or stone dust. It is important that the sand has no more than 1% fines smaller than 0.075 mm (no. 200 sieve) to avoid slowing down drainage of the setting bed, also called bedding sand. This bed is not compacted.

In patios and sidewalks, which are not subjected to any traffic, 3/4 in. stone may be used for sub-base and the base course is eliminated. A paver patio at the author's former residence had been installed using a 6 in. (15 cm) QP, 1.5 in. (4 cm) thick stone dust (setting bed), and 6 × 9 in. (15.25 × 23.9 cm) concrete blocks. The pavers served superbly after 5 years and infiltrated the entire rainfall during light and moderate storms.

Like pervious concrete and porous asphalt pavements, pavers need maintenance, which basically involves sweeping or vacuuming annually to remove surface grit. Over a long period, the subsurface may accumulate enough silt to lose infiltration and detention capacity in any porous pavement. When this happens, the asphalt and concrete have to be removed completely and discarded; however, pavers can be deconstructed, the aggregate layers cleaned, and all materials reinstalled with minimal waste. Thus, pavers, while more expensive to construct than both pervious concrete and porous asphalt pavements, would cost far less in the long run. Another advantage of pavers compared with concrete and asphalt pavement is local repair, such as installation of a utility line. If one or more pavers are damaged or have to be removed, they can be replaced without being noticed. Also, pavers come in different sizes, colors, and shapes and can be arranged in different patterns.

The EPA accepts the use of pervious pavement as an alternative to traditional storm water pollution prevention (SWPP) BMPs such as grassy swales and trench infiltration systems. In the Chesapeake Bay visitor center in Annapolis, the worn-out asphalt pavement was retrofitted with Pine Hill brick pavers. The pavers infiltrated most of the runoff during Tropical Storm Nicole, which dropped slightly over 9 in. (23 cm) of rainfall in a 24-hour period in 2010. The pavers also functioned well during Hurricane Irene and Tropical Storm Lee in 2011.

CST, a paver manufacturing company, has introduced a paver called Aqua Paver, which is a 9-5/8 in. by 5 in. and 3-1/8 in. thick block (244.5 × 127 × 79.4 mm) providing 10% void area. This hardscape product provides sharp styling and aesthetic appeal and allows for effective infiltration of storm water runoff into the ground. Like other pavers, its applications range from parking areas to residential patios and driveways.

PaveDrain of Franklin, Washington, is a permeable articulating concrete block/mat system that, like other PCIPs, can be used in parking lots, driveways, and intersections. However, each block is thicker than other permeable pavers and shaped like an arch at the bottom for added retention storage. Each block is 12 in. × 12 in. and 5.65 in. (+1/8 in.) thick and weighs 45–49 lb. Unlike other pavers, PaveDrain is installed directly on stone base, eliminating the use of setting sand. The blocks provide 7% open space and 20% storage volume. Figure 8.15 shows a PaveDrain block. For more information, one may visit www.pavedrain.com.

Pavers can be installed at open joints to enhance infiltration. Figure 8.16 shows an unsanded paver at Lincoln Mall sidewalk in South Beach, Florida. Some pavers provide corner openings or large open cells, which can be filled with aggregate or soil and grass. Figure 8.17 shows a plan view and installation detail of a heavy-duty 3-1/8 in. thick paver, called Uni Eco-Stone, manufactured by Mutual Materials Company. Dimension of a block is also shown in this figure. The installation shown in the figure provides significant exfiltration and retention storage for rain. Uni Eco-Stone pavers are suitable for low-traffic areas such as parking lots and driveways.

XeriBrix is a recent generation of pavers manufactured by Xeripave Super Pervious Pavers of Vancouver, Washington. These pavers are made of pervious concrete and come in two different

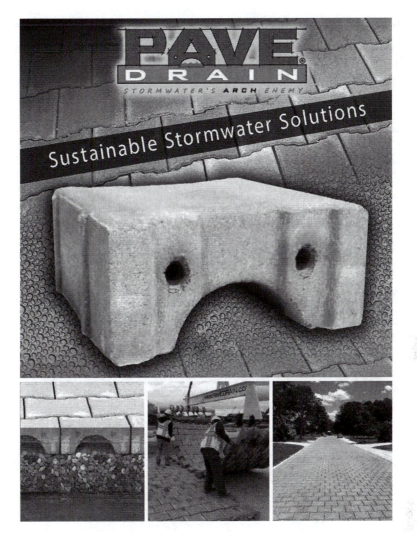

FIGURE 8.15   PaveDrain details.

FIGURE 8.16   Unsanded paver walk at Lincoln Mall in South Beach, Florida. (Photo by the author, 2012.)

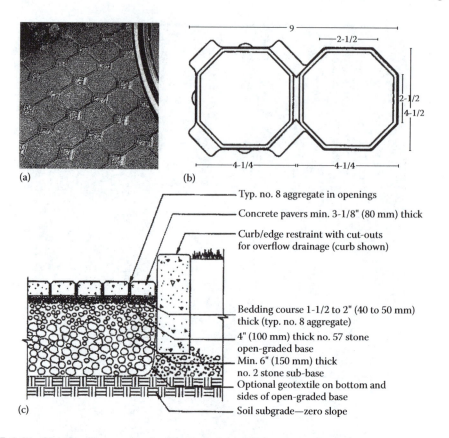

FIGURE 8.17 Uni Eco-Stone paver details. Typical installation for exfiltration. (a) Plan view, (b) paver detail, and (c) cross section.

models, Dupont and Montana. Each of these is 4.5 in. wide by 9.0 in. long by 2-3/8 in. thick (114 × 229 × 60 mm); Dupont weighs 6 lb (2.71 kg) and Montana 5 lb (2.2 6 kg). Figure 8.18 shows a picture of XeriBrix, its views, and information for its manufacturer. Since these pavers are made of pervious concrete, they infiltrate rainfall more effectively than traditional pavers. Figure 8.19 shows a typical cross section of installed XeriBrix pavers.

8.6.6 OPEN CELL PAVERS

Pavers are also manufactured in the form of large blocks or mats with openings to provide vegetated surface. An example of an open cell paver block is Turfstone, manufactured by CST Paving Stone and Versa-Lok Retaining Wall Company. Each block measures 23-1/8 in. long by 15-3/4 in. wide by 4 in. thick (59 × 40 × 10 cm) and block covers 2.53 ft² (0.235 m²). Figure 8.20 shows an open cell paver similar to Turfstone, which, however, is 27% larger than Turfstone and was used for a fire truck access driveway of a residential apartment building in Rochelle Park, New Jersey.

Grasscrete is a cast-in-place concrete paving system that is manufactured by GrassConcrete Limited, a UK company. This open cell paver is made by pouring concrete from a concrete mixer over "Formers," a mold that leaves open cells in the concrete. Once the cells are opened, they can be filled with a variety of porous materials such as vegetation or coarse stone. This paving system, like other pavers, has superior structural integrity; a long, useful life; and low maintenance costs. More information on this paving system can be viewed at www.sustainable pavingsystem.com.

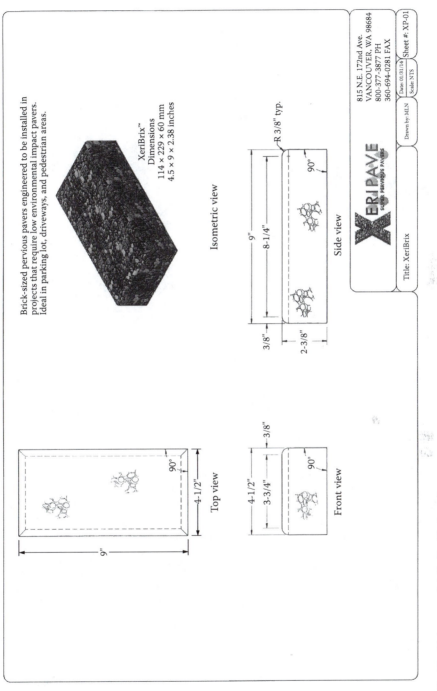

FIGURE 8.18 XeriBrix paver details (by Xeripave).

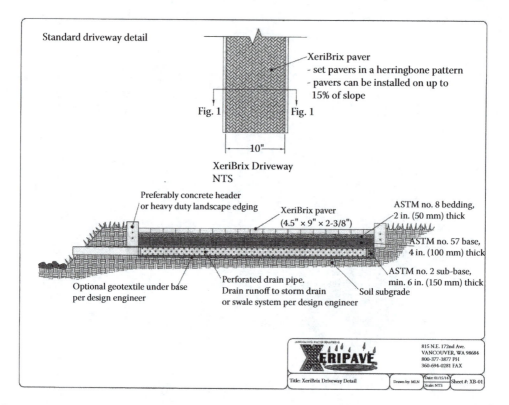

FIGURE 8.19   XeriBrix typical cross section.

A variation of an open cell concrete paving system is a 2 ft × 2 ft (60 cm × 60 cm) by 1.5 in. (3.8 cm) thick flexible mat commercially known as Drivable Grass® manufactured by Soil Retention Products, Inc. This mat consists of 36 (6 × 6) interlocked concrete cells and weighs 45 lb. Premanufactured holes and cracks allow for root penetration through the mat into the subgrade soil. This mat provides 61% planting area, but 90% of concrete base area (see Figure 8.21). Drivable Grass may be infilled with soil and grass or aggregate.

Drivable Grass pavers are estimated to provide 0.4 in. (10 mm) of retention storage at their surface and between 1 and 4 in./h (25–100 mm/h) infiltration for grass infill and up to 20 in./h or more (500 mm/h) for aggregate infill. This type of paver mat was installed at the parking area of the Oceanside fire station in California. The station is located adjacent to San Luis Bay River, where hundreds of gallons of contaminated water due to truck washing used to enter the river. By the use of pavers, the flow of contaminated water was eliminated. Figure 8.22 shows details of Drivable Grass pavement for light- and heavy-traffic application (http://www.soilretention.com). Appendix 8A includes technical specifications on Drivable Grass. Also included in this appendix are design guidelines for installation of Drivable Grass in various types of soil.

8.6.7   Nonconcrete Pavers

Recently, nonconcrete modular pavers have been introduced to the market. Terrecon, Inc., of Fountain Valley, California, makes modular blocks, 100% made of recycled material. This company offers three types of blocks. These are Terrewalks® made of 100% recycled plastic, Rubbersidewalks™ made of 100% recycled rubber tire, and Verlayo®, which is also made of 100% rubber tire. The Rubbersidewalks is the company's original nonconcrete sidewalks with smooth surface on the top and open spaces at the bottom to allow tree roots to grow in. The

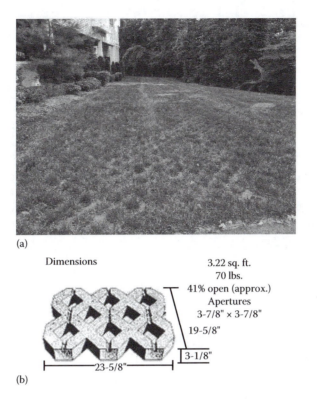

(a)

Dimensions   3.22 sq. ft.
                                                  70 lbs.
                                                  41% open (approx.)
                                                  Apertures
                                                  3-7/8" × 3-7/8"
                                                  19-5/8"

                                                  3-1/8"
                          23-5/8"

(b)

FIGURE 8.20   Open cell paver installed on the fire truck access driveway of a residential apartment building.

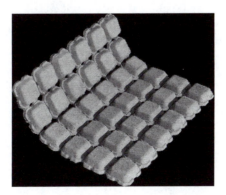

FIGURE 8.21   Drivable Grass paver (2 ft × 2 ft × 1.5 in.).

Verlayo is a thinner version of Rubbersidewalks, but has the same durability. The Terrewalks is the company's advanced paving material with a good appearance and designed for commercial and municipal applications. These pavers are eight times lighter than concrete pavers; Terrewalk comes in a gray concrete color, but Rubbersidewalk and Verlayo are available in a few colors. A 2 ft × 2 ft (0.6 × 0.6 m) Rubbersidewalks weighs about 25 lb (12 kg). These blocks click together, leaving small gaps, and are safe for walking on them in high-heel shoes. Figure 8.23 shows a picture of Terrewalks.

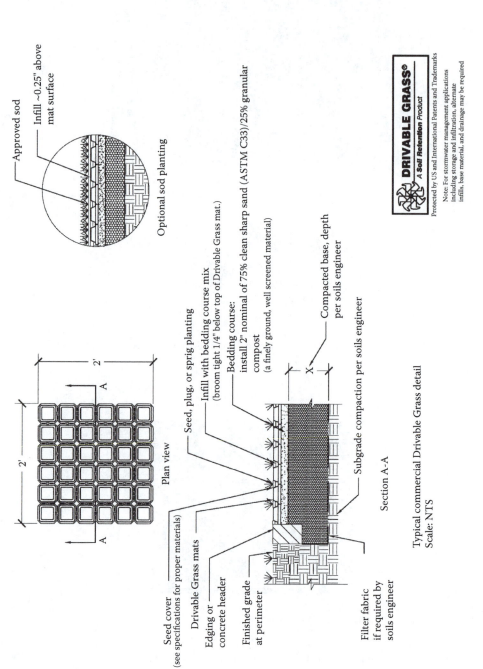

Approved sod

Infill ~0.25" above mat surface

Optional sod planting

DRIVABLE GRASS®
A Soil Retention Product

Protected by US and International Patents and Trademarks

Note: For stormwater management applications including storage and infiltration, alternate infills, base material, and drainage may be required

Plan view

2'

2'

A

A

Seed cover (see specifications for proper materials)

Drivable Grass mats

Edging or concrete header

Finished grade at perimeter

Filter fabric if required by soils engineer

Seed, plug, or sprig planting

Infill with bedding course mix (broom tight 1/4" below top of Drivable Grass mat.)

Bedding course: install 2' nominal of 75% clean sharp sand (ASTM C33)/25% granular compost (a finely ground, well screened material)

Compacted base, depth per soils engineer

Subgrade compaction per soils engineer

X

Section A-A

Typical commercial Drivable Grass detail
Scale: NTS

FIGURE 8.22 Drivable Grass® installation details.

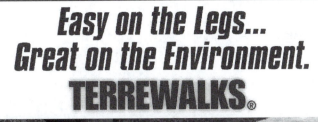

FIGURE 8.23 Terrewalks pedestrian pavement.

8.7 GREEN ROOFS

8.7.1 GREEN ROOF CONSTRUCTION

Green roofs were originated in Europe to reduce runoff from cities where nearly one-quarter of the total land is covered by roofs. The concept moved to the United States in the 1990s and the use of green roofs has since become increasingly popular. Now, all across the country, college and university administrators are raising green roofs on dormitories and department buildings. Examples of campus buildings with green roofs include the McIntyre School of Commerce at the University of Virginia; Cornell University, where an outdated dormitory was replaced with a green roof building; the Stephen M. Ross School of Business at the University of Michigan, which now has a 20,000 ft^2 (1860 m^2) green roof covered with 12 varieties of plants; and Pennsylvania State University, which has at least five green roofs.

The main components of a green roof typically include a lightweight growing medium with a high percentage of porous inorganic material, a root-repellent system, filter cloths, a drainage system, and a variety of native plants. The objective of the filter fabric is to avoid growing medium from clogging the drainage system that is intended to drain excess water from the roof and save plants from drowning. Green roof infrastructure can be loose lid, with each of the layers placed on each other, or modular, with a number of layers placed in a prefabricated tray. Because of ease of installation and lower cost, modular green roofs are more often used in practice.

In terms of construction, green roofs are divided into three types: extensive, semi-intensive, and intensive roofs. Extensive and intensive green roofs are also referred to as above grade and below grade, respectively.

Extensive green roofs require soil or lightweight medium less than 6 in. (15 cm) thick and are planted with some of the many varieties of the sedum family. The plant selection is critical in performance of green roofs and is dependent on local climate. While attractive, low-growing sedum is both drought tolerant and a good water absorbent.

Intensive green roofs, as their name implies, require more effort to establish and maintain. They are also thicker—at least 1 ft (30 cm) thick. This type of green roof can include trees and shrubs, which can be referred to as rooftop gardens, gardens of perennials, and/or annuals. The services of gardeners for regular weeding, mowing, pruning, and fertilizing or of maintenance crews and even a landscape architect are needed just as for a regular garden. Irrigation and drainage systems for intensive green roofs are more complex and the amount of fertilizer must be monitored so that pollution in runoff is not increased. One such green roof was constructed on portions of a new parking garage of Stevens Institute of Technology in Hoboken, New Jersey; the author served as a drainage consultant on this project. Figure 8.24 shows the undulating green roof of the California Academy of Sciences in San Francisco, which serves as a showpiece and has an educational objective.

Semi-intensive green roofs are composed of varying levels of growing plants—more plant variety than extensive system. This type of roof is used when the roof structure does not have the capacity to bear the weight of a full intensive green roof. All types of green roofs require water proofing membranes to avoid leakage into buildings. Also, an additional barrier layer to protect the membrane from root penetration and accidental damage by workers is required. Apart from extra materials and labor, a significant difference between the extensive and intensive green roofs is structural reinforcement to bear the extra weight of the latter roof.

The extensive green roof adds on average 7–8 lb/ft^2/in. (130–150 N/m^2/cm) of soil. Thus, an extensive green roof adds between 30 and 50 lb/ft^2 of dead load to a standard roof, which weighs

FIGURE 8.24 Undulating green roof of California Academy of Sciences in San Francisco. (Photo by the author, 2010.)

25–35 pounds per square foot (1200–1700 N/m^2). According to the EPA, an intensive green roof adds 80 to 150 lb/ft^2 (3800–7200 N/m^2) of load to the roof structure (http://www.epa.gov/neatisland /strategies/greenroofs.html).

Design of intensive and semi-intensive green roofs commonly includes a storage reservoir, which may be filled with stone. This reservoir also serves as a root barrier and improves drainage and aeration. Figure 8.25 shows a section of such a green roof.

Because of city living, green roofs are far more popular and practical in Europe than in the United States. In Germany 7% of buildings are covered with green roofs compared with 0.01 of 1% in the United States. One reason is that the majority of people in Germany live in multistory apartment buildings with flat roofs, while the majority of dwellings in the United States are single-family homes with steep roofs.

Green roofs retain a large portion of storm water during light and moderate storms but create no attenuation in peak runoff during sustained rainfalls. Their effectiveness in retention and delay is affected by the air temperature, antecedent moisture conditions, duration of rain, and plant uptake. Experiments show that green roofs can retain 50–60% of the annual precipitation in the Northeast. A study was conducted at the Lawrence Technological University in southeast Michigan where 13,000 ft^2 of green roof was installed in October 2010 and monitoring began in August 2011. After 9 months of testing, it was found that the green roof retained approximately 53% of the rainfall during that period (Brzozowski, 2012). An educational brochure prepared by Penn State's Center for Green Roof Research indicates that green roofs can retain 80–90% of the rain in May through September and 20–40% of the rain and snow from October through March. A typical 4 in. (10 cm) green roof can fully retain rains 0.6–0.7 in. (15–18 mm) that fall after a few dry days in summer. The retainage capacity drops if the rain occurs after moist weather. In a test in Belgium when, in a 24-hour period 15 mm (0.6 in.) of rain fell, the runoff from the green roof was measured to be 5 mm (0.2 in.). This reflects a 10 mm (0.4 in.) retention of rain for that rainfall. Assuming a minimum of 10 cm (4 in.) thick green roof, the retention capacity would be 1–1.5 mm/cm (0.1–0.15 in./1 in.) of soil media.

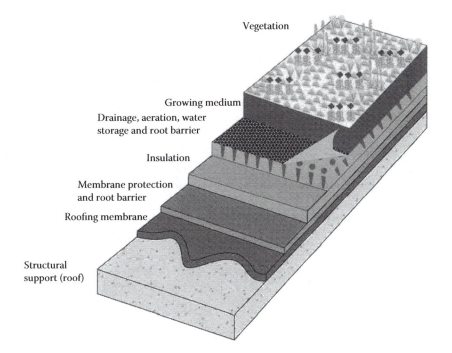

FIGURE 8.25 Green roof section (intensive and semi-intensive). (From Maryland Stormwater Design Manual, Figure 5.2, 2009.)

A 2005 modeling study included in the previously referenced EPA website indicates that installing green roofs on 20% of the buildings over 10,000 ft^2 in Washington, DC, would provide 23 million gal (87,000 m^3) of storage and reduce the runoff to the storm sewers by 300,000 million gal (1.1 million m^3) per year.

By dissipating heat, green roofs reduce the cost of heating and cooling and are also effective in mitigating the heat island effect of buildings in cities. This effect is highly dependent on the climate. A study in Canada modeled the heating and cooling energy saving of 32,000 ft^2 (2970 m^2) green roof of a one-story commercial building in Toronto. The analyses estimated that the green roof could save approximately 6% of total cooling energy and 10% in total heating, amounting to 21,000 kWh annually (http:www.epa.gov/heatislands/resources/pdf/GreenRoofsCompendium.pdf). Estimating the annual maintenance cost of a green roof at just $0.25/ft^2 and the energy cost at $0.20/kWh, the annual maintenance cost and energy savings would be calculated at $8000 and $4200, respectively. Thus, the energy savings do not even cover the annual maintenance cost of the green roof.

Green roofs also extend the useful life of roofs. In a 2007 paper, Stephen Peck, the founder and president of Green Roofs for Healthy Cities, reports that the standard expectation in Germany is that the membrane with a green roof above it will last 40 years; this may be an optimistic expectation.

Green roofs may have the ability to remove pollutants from runoff. The pollutant removal effect depends on the amount of fertilizer and compost application and the total amount of nitrogen in the rain. In a study at Penn State University, where just enough fertilizer was used, the nitrogen concentration in the runoff was nearly equal to that in the rain. This indicates a net reduction in total nitrogen since the amount of runoff is reduced. An analytical study at the University of Michigan and test results from a green roof at York University in Toronto indicated a 50% reduction in nitrate concentration compared to a conventional roof. However, the water quality results from the aforementioned Lawrence Technological University study were not conclusive.

An EPA website, updated February 9, 2009, reports estimated costs of installing a green roof starting from $10/ft^2 for simple extensive roofing to over $25/ft^2 for intensive roofs. Annual cost of maintenance of either type ranges from $0.75 to $1.50/ft^2 (http://www.epa.gov/heatisland/mitigation/greenroofs.htm). In comparison, traditional roofs cost $1.25/ft^2 and have little annual maintenance cost. Green roofs are increasingly used in the United States and a great deal of knowledge has been gained in their application during the past 10 years. In hot climates, such as the Carolinas, it is more challenging to get vegetation to fully establish quickly. In these states, as well as many other southwestern states such as Arizona, Nevada, and New Mexico, green roofs are impractical.

The previously indicated significant difference between the cost of traditional and green roofs discourages people from using green roofs. In lieu of green roofs, herb and vegetable planters and pots may be placed on roofs or balconies. These can be installed on a conventional flat roof or balcony and therefore cost far less than green roofs. Also, they provide fresh produce for consumption. Figure 8.26 shows basil planters on an apartment balcony in Laval, Canada, and Figure 8.27 depicts herb pods on a partial roof of Hilton Hotel in Cincinnati, Ohio.

Also in hot climates, apart from installation and maintenance expenses, green roofs, according to this author's estimates, cost from $5000 to $10,000 for every cubic meter of stored water. Therefore, if managing the roof runoff is the primary concern, green roofs are not the solution; the author believes that other, far more cost-effective alternatives are available. These include rain gardens, dry wells, and rain tanks; the latter will be discussed in Chapter 10. In addition to being far less expensive than green roofs, these other options are adaptable to single-family dwellings, which commonly have steep roofs that preclude raising of green roofs. The use of rain tanks can also conserve water for reuse.

8.7.2 STORM WATER MANAGEMENT ANALYSIS OF GREEN ROOFS

All of the previously referenced case studies are site specific and do not provide a universal method of estimating the volume and the peak runoff from green roofs. A number of jurisdictional agencies

FIGURE 8.26   Basil planters on an apartment balcony in Laval, Canada. (Photo by the author, 2012.)

FIGURE 8.27   Herb pots on a partial roof of the Hilton Hotel in Cincinnati, Ohio. (Photo by the author, 2013.)

recommend the application of SCS TR-55 method with a certain soil curve number (CN), for estimating the peak and/or volume of runoff discharged from green roofs. The Center for Neighborhood Technology, a sustainability organization located in the Midwest, for example, specifies CN = 75 for a green roof with no justification.

Fairfield County, in Virginia, uses CN values of 75 and 65 for extensive and intensive green roofs, respectively. An experimental study conducted at the University of Georgia between 2003 and 2004 showed CN = 86 for a given green roof (Moody, 2012). The Maryland Department of Environment, as indicated in Chapter 5, specifies the CN value range of 77 to 94 for 8 in. (20 cm) and 2 in. (5 cm) thick green roofs, respectively. Maryland's CN value represents an initial abstraction, I_a, of 0.6 in. (15 mm) for CN = 77 and 0.12 in. (3 mm) for CN = 94. These figures, unlike the previously referenced specified CN values, properly reflect the retention storage capacity of a green roof. However, using a CN value less than 100 (98 used in practice) implies deep percolation, which is nonexistent in green roofs. Therefore, using the SCS method underestimates the peak runoff from green roofs, which become saturated during the early hours of a 24-hour storm.

In arriving at a general method of analyzing green roofs in terms of storm water management capability, it should be realized that because of the impermeable liner, green roofs omit any percolation. Rather, a green roof functions as a small retention basin of limited capacity. Specifically, it retains water while the voids in the soil media are being filled, but, once saturated, discharges the entire rainfall that falls on it with no attenuation. The maximum retention capacity of a soil is the difference between its porosity and its wilting point (the moisture content below which vegetation wilts). The porosity and wilting point of a soil vary somewhat with the type of soil, though they can be taken as 35% and 15% by volume on the average. Therefore, the very maximum amount of rain that a green roof can retain is 0.20 cm/cm (0.2 in./in.) of soil thickness. Considering that for healthy vegetation, the soil moisture should be well above the wilting point, the retention capacity of green roofs would be closer to 15% than 20%. This implies that a 10 cm (4 in.) thick green roof can retain 1.5 cm (0.6 in.) of rain. Thus, on an annual basis, a green roof can retain a large percentage of rainfall, but it is rather ineffective for runoff control during heavy, sustained rainstorms.

The runoff depth from a green roof may be estimated from the following equation:

$$R = P - 0.15 \, MT \qquad (8.1)$$

where
P = rainfall depth, mm (in.)
R = runoff depth, mm (in.)
MT = soil media thickness, mm (in.)

This equation indicates that, for small rainfalls, green roofs may have no discharge. But, during a sustained rainfall, once the retention storage in the soil is exhausted, the green roof functions the same as a conventional roof. Therefore, the peak runoff from green roofs during heavy storms may be calculated using the following equations, which are based on the universal runoff model and resemble the rational method with a runoff coefficient, $C = 1.0$:

$$Q = I \cdot A/3600 \text{ (SI)} \qquad (8.2)$$

$$Q = I \cdot A/43200 \text{ (CU)} \qquad (8.3)$$

where
I = rainfall intensity, mm/h (in./h)
A = green roof area, m^2 (ft^2)
Q = peak discharge, L/s (cfs)

Example 8.1

An extensive green roof consists of 10 cm soil and covers 2000 m^2 of a 2200 m^2 roof. Calculate the retention storage volume of the green roof and the cost of retaining 1 m^3 of rainwater. The additional cost of roof structure is estimated at $150/m^2.

Solution

Estimate the effective porosity at 15%.

$$\text{Retention volume} = 2000 \times (10/100) \times 0.15 = 30 \text{ m}^3$$

$$\text{Added roof cost} = 2200 \times 150 = \$330,000$$

$$\text{Cost of water storage} = \$330,000/30 = \$11,000/\text{m}^3$$

8.8 BLUE ROOFS

A blue roof is a nonvegetated source control storm water management practice. It includes an impermeable structure that stores rain temporarily. The stored water is dissipated partly due to evaporation. By releasing the stored water after the storm has ended, blue roofs also mitigate the peak flows to storm drains or combined sewer systems. A variety of flow controls are available including modified inlets, check dams, or trays. Depending on the types of control devices used to control discharge of water that is stored temporarily, blue roofs can be classified as "active" or "passive." All types of flow control provide some detention effect. However, the tray and check dam systems are more effective than others. Blue roofs are more effective than green roofs in storing water. Also, because of absence of vegetation, they require far less maintenance than green roofs.

Blue roofs include open water surfaces, storage within or beneath a porous media or modular surface or below a raised decking cover. Apart from their storm water management benefits, blue roofs can provide storage for reuse such as irrigation or serve as recreation and water play areas. A blue roof can also be used to cool the roof of a building on hot days in order to reduce the HVAC load on the mechanical equipment and save energy costs.

Blue roofs add much less load on the roof than green roofs do and therefore do not require a significantly stronger structure than conventional roofs. Storing 40 mm (1.6 in.) of water provides the same retention storage as a 20–25 cm (8–10 in.) green roof but weighs nearly 10 times less. In addition, blue roofs, unlike green roofs, are practical in arid or semi-arid climates. In short, blue roofs are far more effective and less costly than green roofs.

A number of blue roof pilot projects have been conducted across the United States A significant blue roof pilot project was undertaken by the New York City Department of Environmental Protection between 2010 and 2012. This project was the first to utilize a novel passive blue roof tray design developed by Geosyntec Consultants. The design relies on the lateral transitivity of nonwoven filter fabric for drawdown control in a full-scale pilot. Monitoring of these systems has demonstrated their performance as an effective means for peak flow mitigation and lagging the peak flow in combined sewer systems. Coupled with light-color roofing material, the blue roofs provide energy cost savings through rooftop cooling.

8.9 STORM WATER WETLANDS

Storm water wetlands, also known as constructed wetlands, are structural storm water management practices that can both treat and store runoff. They are similar to wet ponds except that they are generally shallow and incorporate wetland plants in a shallow marsh area.

Natural and constructed wetlands have many beneficial functions, one of which is water filtration. As water flows through a wetland, it slows down and most of the suspended solids become trapped by vegetation and settle out. Other pollutants are transformed into less soluble forms and are taken up by plants or become ineffective. Wetland plants also create favorable conditions for microorganisms to live and reproduce. Through a series of complex processes, these microorganisms also transform and remove pollutants from water.

Natural wetland systems have often been described as the "earth's kidneys" because they filter pollutants from water that flows through on its way to receiving lakes, streams, and oceans. Because these systems can improve water quality, engineers and scientists construct systems that replicate the functions of natural wetlands. Constructed wetlands are treatment systems that use natural processes involving wetland vegetation, soils, and their associated microbial assemblages to improve water quality (EPA, 2004).

Storm water wetlands are designed specifically for storm water treatment and are one of the most effective storm water practices in that regard. They also provide aesthetic amenity and habitat value. Storm water wetlands are basically different from natural wetlands; they have less biodiversity than natural wetlands. Several design variations of storm water wetlands exist; these differ in the relative amount of shallow and deep water and dry storage.

Storm water wetlands are applicable in most regions in the United States, with the exception of arid climate, and also have limited applicability in highly urbanized settings. If storm water wetlands are used in high spots where land use generates highly contaminated runoff, they will need significant separation from groundwater. Storm water wetlands can be used in almost all types of soil and sites with slopes of up to 15%.

Constructed wetlands are generally built on uplands and outside flood plains in order to avoid drainage to natural wetlands and other aquatic systems. Wetlands are commonly constructed by excavation, backfilling, grading, berming, and, occasionally, installation of flow control structures. In highly permeable soils, an impermeable compacted clay liner is usually installed and the original soil placed over the liner. Wetland vegetation is then planted or allowed to establish naturally.

Storm water wetland design varies considerably depending on site constraints. However, some features, including pretreatment, conveyance, maintenance reduction, and landscaping, should be incorporated into most wetland designs. Design variations of storm water wetlands include shallow marsh, extended detention wetland, pond/wetland design, pocket wetland, and gravel-based

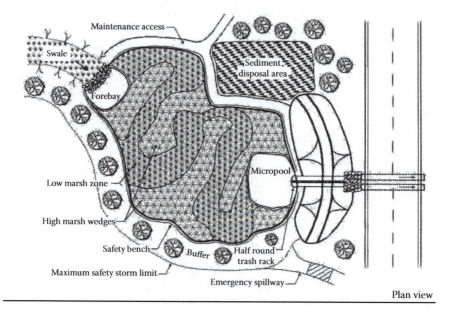

Plan view

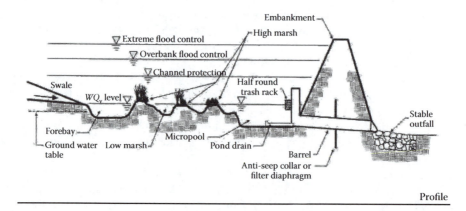

Profile

FIGURE 8.28 Pocket wetland. (From New York State Stormwater Management Design Manual, 2010.)

wetlands. Figure 8.28 shows a variation of pocket wetland. In the shallow marsh, most of wetland volume is in a shallow high or low marsh and the only deep portions are the forebay at the inlet to the wetland and the micropool at the end. Extended detention wetland is similar to the shallow marsh with additional storage above the surface of the marsh that provides extended detention basin. A pond/wetland system, as the name implies, combines a wet pond with shallow marsh. Storm water runoff enters the wet pond and then flows through the marsh. Since the pond can be 6–8 ft (1.8–2.4 m) deep, this type of wetland requires less surface area than shallow marsh. In the pocket pond design, the bottom of the wetland intersects the groundwater, which helps in maintaining the permanent pool. Because of groundwater flows, this type of wetland may not be as effective as others and should be used when the drainage area is not large enough to maintain a permanent pool. In gravel-based wetlands, water flows through a rock filter. At the surface, pollutants are removed through uptake by the plants and biological activity on the surface of the rocks. This type of wetland is basically different from other wetland designs and it is more similar to filtering systems.

Table 8.2, which is based on limited data points, presents a comparison of various types of storm water wetlands. Storm water wetlands are relatively inexpensive, but consume 2–4% of their drainage area, which is high compared to other storm water management practices.

Construction cost information for storm water wetlands is scarce, but it can be assumed that wetlands cost about 25% more than wet ponds of an equivalent storage volume. The cost of a wetland to store 1000 m³ (35,000 ft³) of a 10-year storm runoff may be estimated at $50,000.

Suggested design criteria for constructed wetlands are as follows:

- Depending on its depth, the surface area of storm water wetland should be between 2% and 4% of the contributing drainage area. For gravel-base wetlands, the area also varies with the thickness of the gravel.
- In a constructed wetland, contrary to a wet pond, only a small fraction (less than one-fifth, on average) of the surface area is deep open water.
- At a minimum, one-third of the surface area should be shallower than 15 cm (6 in.) and at least two-thirds of the surface area 50 cm (18 in.) or shallower.
- A forebay shall be located at the inlet, and a 1.2–1.5 m (4–6 ft) deep micropool that stores approximately 10% of WQ_v shall be located at the outlet to avoid the low-flow opening from clogging and sediment resuspension (Maryland Stormwater Management Manual, 2009).

TABLE 8.2
Typical Removal Rates (%) of Wetlands

	Storm Water Treatment Practice Design Variation			
Pollutant	Shallow Marsh	ED Wetland[a]	Pond/Wetland System	Submerged Gravel Wetland[a]
TSS	83 ± 51	69	71 ± 35	83
TP	43 ± 40	39	56 ± 35	64
TN	26 ± 49	56	19 ± 29	19
NOx	73 ± 49	35	40 ± 68	81
Metals	36–85	(80)–63	0–57	21–83
Bacteria	76[a]	NA	NA	78

Source: EPA, NPDES Stormwater Wetlands (http://cfpub.epa.gov/npdes/stormwater/menuofbmps/index.cfm?action =factsheet_).

[a] Data based on fewer than five data points.

- If extended detention is used in a storm water wetland, provide a minimum of 50% of the WQ_v in the permanent pool; the maximum water surface elevation of WQ_v-ED shall not extend more than 1 m (3 ft) above the permanent pool (New York State Stormwater Management Design Manual, 2010).
- Use of storm water wetlands on trout waters is strongly discouraged as these practices increase stream temperatures.

8.10   SUBSURFACE GRAVEL WETLANDS

A subsurface gravel wetland (SGW), also referred to as submerged gravel wetland, is a storm water management system that functions similarly to natural wetlands except that it is partly underground. This system, which was originally designed at the University of New Hampshire Stormwater Center (UNHSC), is based in part on the idea of multiple-pack reactors commonly used in wastewater treatment. The original SGW consists of two treatment cells, preceded by a forebay to retain coarse particles and gross debris. Both cells include a surface wetland and a crushed stone sublayer separated by a minimum 3 in. (8 cm) of graded aggregate filter to prevent wetland soil from entering the gravel sublayer. Figure 8.29 depicts the layout of the UNHSC subsurface gravel wetland. The water leaving the discharge pipe from the forebay is retained in the first cell and slowly moves down to the gravel sub-base through holes in a stand pipe. Any water in excess of the capacity of the first cell enters the second cell and is also filtered down through the stand pipe. The water in the gravel layer moves under the first cell to the second and is discharged through an outlet with its invert at the top of the gravel sub-base. The system may be preceded by a hydrodynamic separator device, swale, or forebay.

For permit compliance with the regulations, water quality monitoring was performed from July 2007 through October 2010 at the GSW of Greenland Meadows, a high-use commercial site. The discharge from the GSW enters Pickering Brook, which is an impaired waterway. The results from this project indicated that the treated runoff had a median total suspended solids (TSS) concentration of 3 mg/L (3 ppm by weight). This figure was lower than 5 ppm for preconstruction TSS and far less than the measured 53 ppm TSS concentration of Pickering Brook (Gunderson et al., 2012).

The long-term monitoring data of subsurface gravel wetland at the UNHSC has indicated an average annual nitrate removal of 75–85% during the growing season and 33% during winter. Annual average phosphate removal was measured to be 75% and total suspended solid removal was greater than 95%.

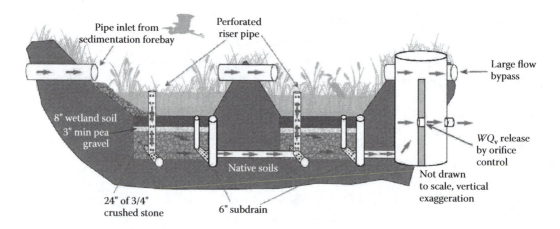

FIGURE 8.29   UNHSC subsurface gravel wetland profile.

The UNHSC's specifications for GSW are as follows:

- A subsurface water level is maintained through the design of the outlet invert elevation (invert just below the wetland soil surface).
- Retain and filter the entire water quality volume (WQ_v), 10% in the forebay and 45% above each of the respective treatment cells.
- Option to retain the channel protection volume (Cp_v) for 24–48 h.
- No geotextile or geofabric layers are used within this system, but may be used to line walls.
- If a native low-hydraulic conductivity soil is not present below the desired location of the SGW, a low-permeability liner or soil (hydraulic conductivity less than 10^{-5} cm/s = 0.03 ft/day) below the gravel layer should be used to minimize infiltration, preserve horizontal flow in the gravel, and maintain the wetland plants.
- Gravel length to width ratio of 0.5 (L:W) or greater is needed for each treatment cell with a minimum flow path (L) within the gravel substrate of 15 ft (4.6 m).
- There shall be 8 in. (20 cm) minimum thickness of a leveled wetland soil as the top layer.
- There shall be 3 in. (8 cm) minimum thickness of an intermediate layer of a graded aggregate filter to prevent the wetland soil from moving down into the gravel sublayer. Material compatibility between layers needs to be evaluated.
- There shall be 24 in. (0.6 m) minimum thickness of 3/4 in. (2 cm) crushed stone (gravel) sublayer. This is the active zone where treatment occurs.
- The primary outlet invert shall be located 4 in. (10 cm) below the elevation of the wetland soil surface to control groundwater elevation. Care should be taken to not design a siphon that would drain the wetland. The primary outlet location must be open or vented; the outlet can be a simple pipe.
- An optional high-capacity outlet at equal elevation or lower to the primary outlet may be installed for maintenance. This outlet would need to be plugged during regular operation. This optional outlet allows for flushing of the treatment cells at higher flow rates. If it is located lower, it can be used to drain the system for maintenance or repairs.
- The bypass outlet (emergency spillway or secondary spillway) is sized to pass designs flows (10-year, 25-year, etc.). This outlet is sized by using conventional routing calculations of the inflow hydrograph through the surface storage provided by the subsurface gravel wetland system. Local criteria for peak flow reductions are then employed to size this outlet to meet those criteria.
- The primary outlet structure and its hydraulic rating curve are based on a calculated release rate by orifice control to drain the WQ_v in 24–48 h.
- The minimum spacing between the subsurface perforated distribution line and the subsurface perforated collection drain at either end of the gravel in each treatment cell is 15 ft (4.6 m). There should be a minimum horizontal travel distance of 15 ft (4.6 m) within the gravel layer in each cell.
- Vertical perforated or slotted riser pipes deliver water from the surface down to the subsurface, perforated, or slotted distribution lines. These risers shall have a maximum spacing of 15 ft (4.6 m). Oversizing of the perforated or slotted vertical risers is useful to allow a margin of safety against clogging with a minimum recommended diameter of 12 in. (30 cm) for the central riser and 6 in. (15 cm) for end risers. The vertical risers shall not be capped, but rather covered with an inlet grate to allow for an overflow when the water level exceeds the WQ_v.
- Vertical cleanouts connected to the distribution and collection subdrains, at each end, shall be perforated or slotted only within the gravel layer and solid within the wetland soil and storage area above. This is important to prevent short-circuiting and soil piping.

- Berms and weirs separating the forebay and treatment cells should be constructed with clay or nonconductive soils and/or a fine geotextile, or some combination thereof, to avoid water seepage and soil piping through these earthen dividers.
- The system should be planted to achieve a rigorous root mat with grasses, forbs, and shrubs with obligate and facultative wetland species. In northern climates refer to the New York Stormwater Manual (http://www.dec.state.ny.us/website/dow/toolbox/swmanual/) or approved equivalent local guidance for details on local wetland plantings.

A subsurface gravel wetland does not necessarily have to be two cells like the UNHSC's SGW. Figure 8.30 depicts a submerged gravel wetland comprising only one cell. Shown in this figure are two options for inflow–outflow arrangement. As shown, the inflow can enter the body of water above the treatment media gravel or enter directly into the gravel media. The outlet pipe would be issued from the treatment media in the former case and the water body in the latter case.

The Maryland Department of the Environment specifies that subsurface gravel wetlands be used exclusively on C or D hydrologic soils. Design guidelines by the same agency indicate the following:

- SGW is best suited for soils of poorly drained or high water table.
- SGW implementation on a site should not exclude the use of other ESD options. Since SGW does not have drainage area limitations, using one SGW undermines the MEP mandate to mimic local hydrology and distribute runoff controls uniformly across the site.
- Only the aboveground storage volume in the SGW may be used for credit toward ESD treatment.

8.11   FILTER STRIPS

8.11.1   APPLICATION

A filter strip is a vegetated surface that is designed to treat sheet flow from adjacent pavements. Filter strips function by slowing runoff velocities and removing the sediment and other pollutants. The removal is achieved through filtration, absorption, and uptake by plants and infiltration. Although filter strips also reduce runoff volumes considerably, this effect is commonly ignored in practice. To be effective and stable, filter strips should receive runoff as sheet flow and have small or moderate slope, 15% at a maximum. To avoid concentrated flow to a filter strip, the NJDEP limits the length (along the flow) of filter strips to 100 ft (30 m); the same length is specified as 75 ft (25 m) in the New York Stormwater Management Design Manual (2010). Thus, filter strips are most suited for parking lots and alongside multilane roads. Filter strips can be natural or man-made and may be covered with a variety of vegetation such as woods, grass, herbaceous and bushy plants, or a combination thereof. Figure 8.31 shows a filter strip application.

Filter strips are applicable in most regions. However, in areas steeper than 15% or where the space is limited, filter strips are not feasible. The New Jersey Stormwater Best Management Practices Manual (2004) has adopted the maximum total suspended solids (TSSs) removal rate of vegetative filters as a function of vegetative cover as follows:

Turf grass	60%
Native grasses, meadow	70%
Indigenous woods	80%

With multiple vegetated cover, the composite TSS removal may be calculated based upon a weighted average of the previously indicated adopted rates. The required length of a filter strip to achieve the preceding TSS removal rates also depends on land slope and hydrologic soil group.

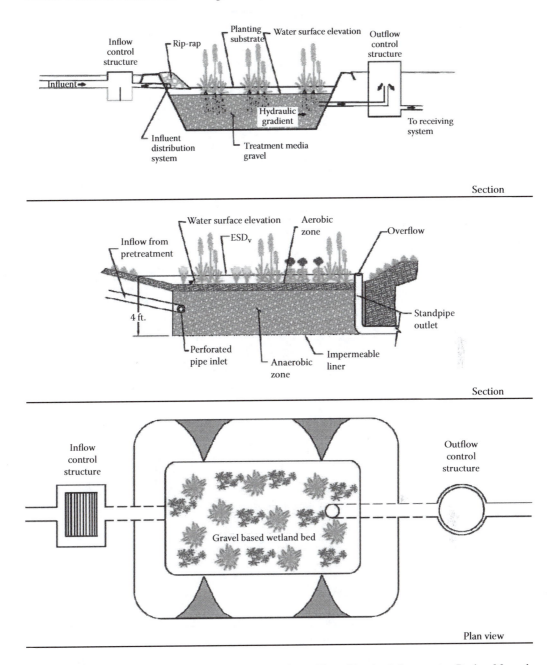

FIGURE 8.30 Submerged gravel wetland alternate plans. (From Maryland Stormwater Design Manual, Figure 5.10, 2009.)

Figure 8.32, which is an excerpt from the NJDEP, shows the required length of a filter strip as a function of slope and vegetative cover for hydrologic group B soil.

For a filter strip length (along the flow path) of at least one-half of that of the paved area to be treated, a TSS removal of 80% is reported by the NRCS of Illinois (Code 853, 1999). This is significantly larger than the aforementioned TSS removals adopted in New Jersey. Filter strips are also capable of removing other pollutants such as nitrate, total phosphorus, and heavy metals. However, removal of these other pollutants is not regulated by many jurisdictional agencies, as yet.

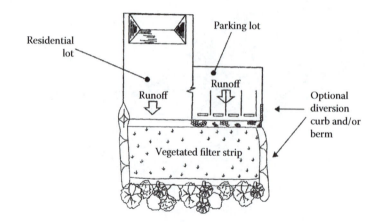

FIGURE 8.31   Filter strip.

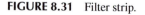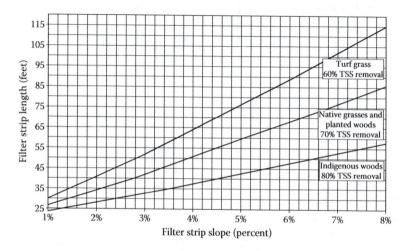

FIGURE 8.32   TSS removal for vegetated filter strips drainage area soil: loam, silt loam HGS:B. (Adapted from NJDEP, 2004, Chapter 9.) One of two figures for B soils.

Some typical locations for filter strip applications are

a. Adjacent to roadways, parking lots, and other impervious surfaces. Case Studies 7.4 and 7.5 in Chapter 7 indicated the application of filter strips for addressing water quality requirements of a New Jersey Garden State Parkway improvement project and a municipal parking lot expansion, respectively. The application of filter strip in the latter case study is elaborated through Case Study 8.1.

b. At roof downspouts, lawn and planted strips disperse and infiltrate roof runoff. Figure 8.33 shows a downspout terminating at a planted area at the author's residence. Discharges from this downspout, which carries runoff from a quarter of the dwelling roof, are fully retained and infiltrated by landscape, which is level with the lawn.

c. Buffer areas adjacent to streams, lakes, and ponds.

d. On construction sites and along disturbed land to filter sediment from bare soil.

FIGURE 8.33 Planted area (filter strip) at a roof downspout of the author's residence.

8.11.2 DESIGN CRITERIA

- The maximum drainage area to a filter strip should be 2 ha (5 acres).
- Slope of filter strip should be less than 15% (approximately 1 V to 7 H).
- Minimum filer strip length (dimension along the flow path should be longer than one-half of the unit area length, namely the ratio of drainage area to the width of filter strip. The width is the dimension perpendicular to the flow path. The width of a filter strip is commonly the same as that of the impervious area being treated. Therefore, the length of filter strip should be one-half of the flow length of the impervious surface.

CASE STUDY 8.1

This case study presents calculations for the TSS (total suspended solids) removal of the filter strip in Case Study 7.5 and the composite TSS removal of the filter strip and the detention basin in the parking lot of the Franklin Lakes municipal building.

SOLUTION

As was indicated in Case Study 7.5, a 30 ft (9.15 m) wide vegetated filter strip varying in slope from 2 to 2.5% was provisioned to supplement an extended detention basin in addressing the water quality requirements for the parking lot expansion. The strip is constructed alongside the parking area and is planted with native vegetation and small bushes. A review of the Bergen County soil survey maps shows that Dunellen soil, which is classified as a hydrologic soil group A, present at the site. For an average slope of 2.25%, Figure 8.34 indicates a minimum 70% TSS removal.

After passing through the filter strip, runoff from the parking area enters the detention basin, which, according to routing computations in the aforementioned case study, provides 12 hours of retention time. For this retention time, Figure 8.35 indicates a 40% TSS removal. Therefore, the composite TSS removal of the filter strip plus detention basin may be calculated as follows:

$$\text{TSS removal} = 70 + 40 - (70 \times 40/100) = 82\%$$

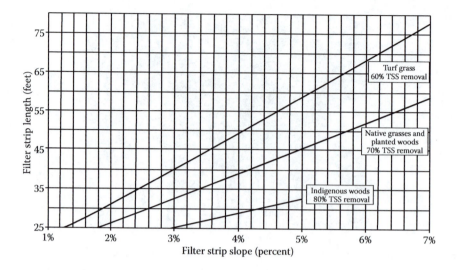

FIGURE 8.34 TSS removal of vegetated filter strip (Case Study 8.1), Sandy Soil, HSG A.

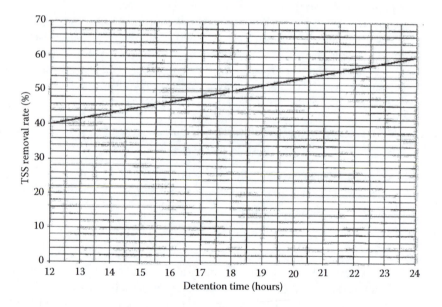

FIGURE 8.35 TSS removal rate (40%) of detention basin.

Therefore, the system provides a greater TSS removal than required. Since this site is situated in the PA1 (metropolitan planning area) and was previously disturbed, it is exempt from groundwater recharge.

8.12   BIORETENTION BASINS, SWALES, AND CELLS

These structures are commonly designed to treat water quality; however, they can also provide some runoff reduction.

8.12.1 Bioretention Basins

A bioretention basin, also known as bioinfiltration basin, consists of a planted soil bed over an underdrained layer of sand and gravel. Storm water runoff that enters a bioretention basin is first filtered by vegetation and then by the soil bed and sand layer before discharging through the underdrain. Bioretention basins are very effective in removing suspended solids, nutrients, metals, hydrocarbons, and bacteria from storm water runoff. Bioretention basins are similar to infiltration basins except for bed material and planting. Infiltration basins are covered with sand and have no plants, whereas bioretention basins are covered with a soil mix and are planted. Also, infiltration basins have no underdrain, but bioretention basins may be provided with an underdrain.

To avoid prolonged impoundment, bioretention basins are constructed 2–3 ft (0.67–1 m) deep. These basins are commonly employed to improve water quality. However, they also can attenuate the peak rates of runoff if they are designed as multipurpose, dual-stage systems. Figure 8.36 shows a multipurpose bioretention basin provided with an outlet structure. Labeled on this figure are the

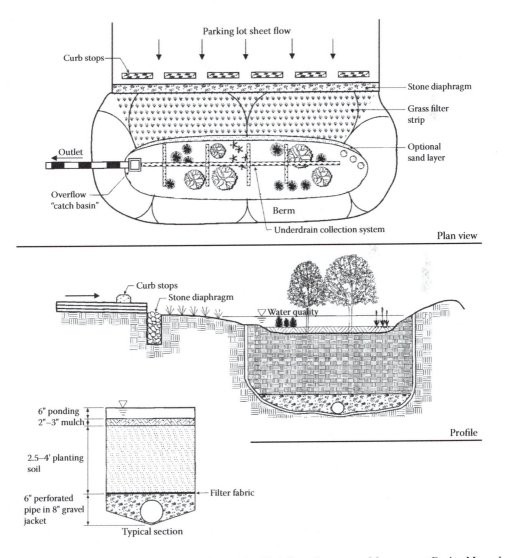

FIGURE 8.36 A bioretention basin variant. (From New York State Stormwater Management Design Manual, Figure 6.19, 2010.)

522   Urban Storm Water Management

suggested thickness of the soil bed and sand layer. A mulch layer may be placed on the surface of the planted soil bed in order to retain moisture and help plant growth.

Design parameters of a bioretention basin include its storage volume, the thickness and permeability of planting soil mix, and the discharge capacity of the underdrain pipe. The storage volume must be sufficient to contain, at a minimum, the design storm water runoff, generally the water quality storm. The soil and sand must be thick enough to effectively treat storm water and to drain the basin within 3, but preferably 2 days. Also, the underdrain should have enough capacity to convey the filtered water. The depth of soil mix can vary from 45 cm (18 in.) for shrubs to 1 m (±3 ft) for trees.

Bioretention basins may be constructed in parking lot islands, roadway medians, and lawns. To prevent premature clogging, a bioretention basin should not be placed in service until its tributary drainage area is fully stabilized.

To function properly, the bottom of a bioretention underdrain must be at least 30 cm (1 ft), but preferably 0.6–0.9 m (2–3 ft) above the seasonal high-water table. The function of bioretention basins is superior to that of infiltration basins in terms of water quality,* and they are also aesthetically more pleasing. However, a bioretention basin requires more maintenance than both infiltration basin and retention–infiltration basin. The fallen leaves should be raked and shrubs should be pruned at least once a year to avoid the return of nitrogen and phosphorus to soil (Minton, 2012). If privately owned, a bioinfiltration basin must be protected through easement or deed restrictions to ensure that they will not be adversely altered. More information on the design of bioretention basins can be found in the Maryland Stormwater Design Manual (2009) and NJDEP Stormwater Best Management Practices Manual (2004).

Specific designs of bioretention basins, like those of rain gardens and constructed wetlands, vary considerably depending on the preferences of designer and site conditions. There are some main features that, however, should be incorporated into bioretention design. These features, similar to those described for constructed wetlands, are pretreatment, treatment, conveyance, maintenance reduction, and landscaping. In arid climates, bioretention basins should be planted with drought-tolerant species. Bioretention basins have a few limitations: They cannot be used to treat runoff from large drainage areas and cannot be employed where the space is limited. However, they can be used as a storm water retrofit by modifying existing lawn or landscape areas. As indicated, bioretention basins are commonly sized for water quality storms. Therefore, to achieve channel protection and flood control, bioretention basins should be supplemented with other measures such as ponds or detention basins. Table 8.3 presents field test results of pollutant removal effectiveness of two bioretention basins in Maryland. Bioretention basins with subdrains can remove 80% or more of suspended sediment. The cost of bioretention basins varies considerably depending on design details. On an average basis, the cost of a bioretention basin increases almost linearly with its storage volume and may be twice as much as that for an infiltration basin.

In the design of any bioretention basin like that of any infiltration system, the soil permeability should be measured to determine whether or not a subdrain is needed. The calculation for drain time should be based on actual, rather than assumed percolation rate. A subdrain is provided in areas of low permeability, where the ponding period is longer than 48 or 72 hours. In some studies reported in literature (see, e.g., Jones, 2012), the percolation rate was assumed to be 0.3 in./h (7.6 mm/h) for hydrologic group A soil to 0.025 in./h (0.6 mm/h) for group D soil. The former rate grossly underestimated actual seepage losses through the ambient soil.

8.12.2 BIOSWALES

A bioswale, also referred to as bioretention swale, is similar to grass swales presented in Chapter 7. Therefore, only the differences are discussed herein. Unlike grass swales, bioretention swales are covered with shrubs and taller grass species and are not mowed. Design parameters for a bioretention

---

* The New Jersey DEP has accepted bioretention basins for a TSS removal of 90 percent.

TABLE 8.3
Pollutant Removal Effectiveness of Two Bioretention Basins in Maryland

Pollutant	Pollutant Removal
Copper	43–97%
Lead	70–95%
Zinc	64–95%
Phosphorus	65–87%
Total Kjeldahl nitrogen (TKN)	52–67%
Ammonium (NH_4)	92%
Nitrate (NO_3)	15–16%
Total nitrogen (TN)	40%
Calcium	27%

swale follow those of a grass swale; however, the Manning's n for the former is significantly greater than that for the latter. As a result, bioswales retard the flow and increase the time of concentration far better than grass swales. Also, dense, taller vegetation provides a higher pollutant removal than grass.

An early application of bioswales occurred in Portland, Oregon. In 1996 a total of 2330 linear feet (710 m) of bioswale were designed and installed in Willamette River Park in Portland to capture pollutants from runoff entering the Willamette River (http://en.wikipedia.org/wiki /Bioswale).

To improve infiltration, the bottom of bioswales may be covered with 0.35–0.9 m (1.2–3.0 ft) thick sandy loam over 15 cm (6 in.) of coarse sand. In areas of low permeability, a perforated subdrain may also be installed in the sand layer. To promote side stability, the side slopes should be no more than 1 by 4 (1V/4H).

Bioswale applications include

- Parking lot islands
- Highway medians or roadside swales
- Residential roadside channels

8.12.3 Bioretention Cells

Bioretention cells are compact bioretention basins for application at curbside or parking areas. These cells typically consist of a soil medium placed in a bottomless box and planted with a tree. A cell may have a grate or curb opening to capture runoff. Contech, as indicated in Chapter 6, used to manufacture UrbanGreen™ BioFilter. This biofilter consisted of an open-bottomed concrete box filled with soil mix at the site and planted with a tree. Since acquiring Filterra bioretention system (Filterra), Contech had stopped making UrbanGreen.

Filterra, formerly a division of AmeriCast (now a part of Contech), makes a biofilter called Filterra Tree Pod that is somewhat similar to Ultra Urban except that it receives runoff from curb openings. Filterra also makes boxless units named Filterra® Boxless™ (FTBXLS), which, as the name implies, has no box. In this type of filter, soil media are placed in an excavated area and planted with trees. This type of filter is particularly suited for parking areas where runoff from parking may be filtered through a short gravel berm before entering Filterra. This type of biofilter results in considerable cost savings and, for larger applications, may cost about the same as or less than bioretention basins.

Table 8.4 presents pollutant removal efficiency of Filterra and Table 8.5 lists the available Filterra Tree Box dimensions for contributory drainage areas of 1/6 to 1.21 acres (675 to 4900 m²).

A residential project in College Park, a city on the southern border of Atlanta, Georgia, installed 111 Filterra tree pods in 2009 to address water quality requirements for the project (Miller, 2011).

TABLE 8.4
Filterra Expected Pollutant Removal[a]

TSS removal	85%
Phosphorus removal	60–70%
Nitrogen removal	43%
Total copper removal	>58%
Dissolved copper removal	46%
Total zinc removal	>66%
Dissolved zinc removal	58%
Oil and grease	>93%

Note: Information on the pollutant removal efficiency of the filter soil/plant media is based on third-party lab and field studies.

[a] Ranges varying with particle size, pollutant loading, and site conditions.

TABLE 8.5
Filterra Quick Sizing Table for Type I and Type II Street Trees

Available Filterra Street/Shade Tree Box Sizes (ft)	Maximum Contributing Drainage Area (acres) Where C = 0.85
4 × 6 or 6 × 4	0.16
4 × 8 or 8 × 4	0.23
Standard 6 × 6	0.26
8 × 6 or 6 × 8	0.36
4 × 12 or 12 × 4	
10 × 6 or 6 × 10	0.46
12 × 6 or 6 × 12	0.56
13 × 7 or 7 × 13	0.71

Available Filterra Street/Shade Tree Box Sizes (ft)	Maximum Contributing Drainage Area (acres) Where C = 0.50
4 × 6 or 6 × 4	0.28
4 × 8 or 8 × 4	0.39
Standard 6 × 6	0.45
8 × 6 or 6 × 8	0.61
4 × 12 or 12 × 4	
10 × 6 or 6 × 10	0.78
12 × 6 or 6 × 12	0.95
13 × 7 or 7 × 13	1.21

Note: Typical street tree standards recommend a 1.5 in. to 2.5 in. caliper. To accommodate these size requirements, Filterra has appropriately sized each unit at a 5 ft, 2 in. depth (INV to TC) for type I and type II trees. (Three inches or greater caliper trees will require a 6 ft, 2 in. depth unit.) A standard schedule—40 pipe coupling is cast into the wall for easy connection to discharge drain.

Fabco Industries has recently introduced a bioinfiltration cell, named "Focal Point Biofiltration System." This system, unlike UrbanGreen and Filterra Tree Pod, has no soil media. Instead, it includes multilayers of filter elements: The top layer is made of clean, shredded hardwood mulch; the second, thicker layer is a performance biofiltration medium; and the third layer is clean stone on an open mesh separation fabric. These are placed on a hollow modular underdrain storage. Runoff entering the unit is filtered through the layered elements, which remove pollutants, and clean water enters the storage module. Fabco claims that FocalPoint removes up to 73% phosphorus, 43% total nitrogen, 85% total suspended solids, 90% of oil and grease, and 87% of bacteria.

8.13 RAIN GARDENS

A rain garden is a shallow planted bed on top soil, underlain by a sand base to retain the runoff from a paved area and infiltrate it into the ground. Though the concept of rain gardens is not new, the term *rain garden* has been introduced to storm water management practice in the recent past. Ancient civilizations had practiced directing runoff into vegetable gardens and, particularly, rice fields, which require ample watering. In this country it appears that the rain gardens originated in Maryland, developed in Minnesota, and quickly spread to other states.

Even in New Jersey, with an average monthly rainfall of approximately 4 in. (100 mm) during the growing season, rain gardens are becoming increasingly popular in retaining rainwater from dwelling roofs and driveways. Englewood Cliffs is one of the towns in New Jersey that formed a committee to develop and publicize rain gardens as well as other green storm water management measures in the borough. The author, who served as a member of this committee, reviewed and made site-specific recommendations for reducing runoff from developments.

A regular flower garden, a vegetable garden, or simply a lawn depression all function like a rain garden. However, a rain garden is commonly a shallow depression or series of shallow depressions covered with plants to retain the runoff it receives and infiltrate it over time. To discharge water in excess of its capacity, a rain garden may be provided with a riser and trash rack. Figure 8.37 shows a profile of a typical rain garden with an overflow system and Figure 8.38 depicts a deeper rain garden at the University of Connecticut (UCONN) campus in Storrs. This rain garden includes a riser to discharge storms beyond its capacity.

Rutgers University and the Native Plant Society of New Jersey have prepared a rain garden manual titled "Rain Garden Manual of New Jersey" (2010). This manual provides useful information on design and construction of rain gardens and selection of proper plants (http://www.npsnj.org).

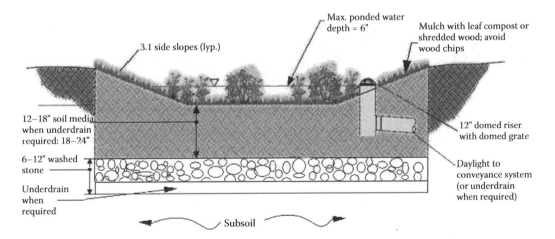

FIGURE 8.37 Typical rain garden section. (From New York State Stormwater Management Design Manual, Figure 5.42, 2010.)

FIGURE 8.38   A rain garden at UCONN. (Photo by the author, 2014.)

To improve infiltration as well as aesthetics, rain gardens are planted with a variety of native grasses, herbs, and woody plants that are adapted to the soil and site climate. The native plants have deeper root systems that help increase recharge of groundwater, are drought tolerant, and require no fertilizer. The root systems of plants in rain gardens act as a filter in removing pollutants from rain infiltrating through soil. Rain gardens are effective in reducing flooding and improving water quality and are becoming popular as a green solution to flooding and pollution problems.

The city of Aurora in Illinois performed a model study of pollutant removal effectiveness of rain gardens using the Source Loading and Management Model (SLAMM). Five hydraulically connected rain gardens, each 11 ft (3.35 m) wide, were placed between curb line and sidewalks along Spring Street. One of the rain gardens was 60 ft (18.29 m) long and the remaining four were 50 ft (15.24 m) long. All rain gardens included a substrate and plants that consisted of native shrubs, flowers, and plants (dogwood and nannyberry). Three of the gardens had 0.49 acre (1983 m²) drainage area. SLAMM modeling for these three gardens showed that they received 29,085 ft³ (824 m³) of runoff during the 2-year modeling period. Of this, 82% was infiltrated and the rest overflowed to a catch basin in the third rain garden. The modeling study showed that the three garden systems captured 85% of total solids, 89% of particulate phosphorus, 84% of total Kjeldah nitrogen, 85% of heavy metals (copper, lead, and zinc), and 83% of fecal coliform bacteria (Seth, 2011). No measurements, however, were taken to confirm these results.

In 2005, Johnson County officials launched a plan to construct 10,000 rain gardens in the Kansas City metropolitan area (Buranen, 2008a). This plan was intended to serve as a regional effort to educate the public about what each person can do to manage storm water runoff and improve water quality on personal and community property (http://www.rain.kc.com).

The Lab School in Cook County is one of the high schools in Chicago with a pilot program to educate students and the public about the role of rain gardens in managing storm water runoff and improving the water quality.

Figures 8.39 and 8.40 depict a lawn depression, 6–12 in. (15–30 cm) deep that the author designed to retain the runoff from the roof and parking lot of a fire station in a municipality in northern New Jersey. This lawn depression (which is basically the simplest, least expensive rain garden) measures approximately 450 m² (4840 ft²) and receives runoff from 2980 m² (32,080 ft²) of pavement consisting of 430 m² (4630 ft²) of roof and 2550 m² (27,450 ft²) of parking. This lawn depression has been functioning satisfactorily without any maintenance other than regular mowing since it was constructed in 2005.

Figure 8.41 shows a depressed garden (rain garden) at the backyard of the author's residence. This garden measures approximately 25 m² (270 ft²) on plan area and is up to 20 cm (8 in.) deep. It is

FIGURE 8.39 Rain basin retains runoff from roof and driveway of a North Jersey municipal fire station. The rocks at right protect the basin from erosion at the roof drain. (Photo by the author.)

FIGURE 8.40 Another view of lawn depression (a simple, inexpensive rain garden).

FIGURE 8.41 Depressed garden (rain basin) at the backyard receiving roof and yard runoff.

planted with hosta, pachysandra, a hydrangea plant, and five evergreen trees. The garden receives runoff from approximately 45 m² (480 ft²) of roof area (approximately one-fourth of the dwelling roof) plus a gazebo, cabana, and 35 m² (375 ft²) of concrete pavement. The planted depression has no stone/sand sub-base and in fact is underlain by a clay layer 25–30 cm below surface. It fills up 10–15 cm (4–6 in.) after heavy rains and drains out within 24 to 36 hours. The depression was filled approximately 15 cm (6 in.) deep after two consecutive thunderstorms, which dropped nearly 13 cm (5 in.) of rain on May 22 and 23, 2014, and drained completely a day and a half later. This garden has required no watering, whatsoever.

A rain garden does not have to be significantly depressed. It may be constructed as a nearly flat landscaped garden when it is downslope from its tributary area. Figure 8.42 shows a rain garden the author designed to capture uphill runoff entering the yard of a property in a north New Jersey town.

Despite the benefits of rain gardens, many people consider them to be unattractive and shy away from them. However, a rain garden, if properly constructed and decorated with ornamental plants, can be quite attractive. A case in point was a 200-acre mixed-use project in Lynnfield, Massachusetts. Another example worth visiting is a demonstration rain garden at the Nature Center of the Stony Brook-Millstone Watershed Association in New Jersey, constructed in the spring of 2007.

Rain garden plants may be selected based on the USDA Plant Hardiness Zone Map (US National Arboretum), which divides the country into 20 zones, designated as 1, 1a, 1b, 2a, 2b through 10a, 10b, and 11. These zones range in cold temperature from below –50°F for Zone 1 to above 40°F for Zone 11, in 5° intervals for intermediate zones. In New England states and the eastern United States down to North Florida, designated as zones 3a to 9b, the following plants may be selected for rain gardens: bur oak, switch grass, hosta, little bluestem, and butterfly milkwood for full sun and wild bergamot and hydrangeas for sun and shade. Figure 8.43 shows Aphrodite plantain lily, a hosta family plant. For selection of proper plants, one may consult with state plant societies or local nurseries or Monrovia online (http://www.monrovia.com).

The site of the rain garden should be tested to determine if good soil is present. Low permeability soils need to be removed and replaced with a good soil mix comprising 50% sand, 20% top soil, and 30% compost. If the soil contains less than 10% clay, it can be used in place of imported topsoil in the mix. Existing soil can also be improved by adding amendments such as lime, gypsum, and certain nutrients.

Rain gardens can be almost any size and shape. A typical residential rain garden ranges in size from 10 to 30 m² (100–330 ft²). Rain gardens smaller than 10 m² (100 ft²) cannot have desirable plant variety, and gardens larger than 30 m² (330 ft²) are not easy to make level. As a rule of thumb, a rain garden should be twice as long (perpendicular to slope) as it is wide. The depth of rain gardens is determined based on the ground slope. For a flat garden, the depth can vary from 7.5 to 15 cm (3–6 in.) for slopes less than 4% to 20 cm (8 in.) for slopes 8% to 12%.

The size of a rain garden is a function of its depth and its distance from downspout. Because of infiltration losses en route, the farther the rain garden is from a downspout, the smaller its size is. For soils of moderate permeability, the author suggests using an area factor ranging from 0.25 for a rain garden, 3 m (10 ft) from a downspout, to 0.05 for a rain garden more than 10 m (30 ft) away from it; the area factor is the ratio of rain garden area to the roof or pavement area being treated.

Rain gardens cost between $4 and $25 per square foot ($45–275/m²). The latter figure is associated with cases where there is need for other control structures such as storm drains, underdrains, and curbing. In the absence of any drainage structure, the cost of plants forms the majority of the overall cost of a rain garden. More detailed information on rain gardens can be found in "Rain Garden Manual for New Jersey" (Native Plant Society of New Jersey, 2010), the EPA website (http://www.epa.gov/nsp/toolbox/other/cwc_raingardenbrochure.pdf), and Wikipedia site (http://en.wikipedia.org/wiki/Rain_garden).

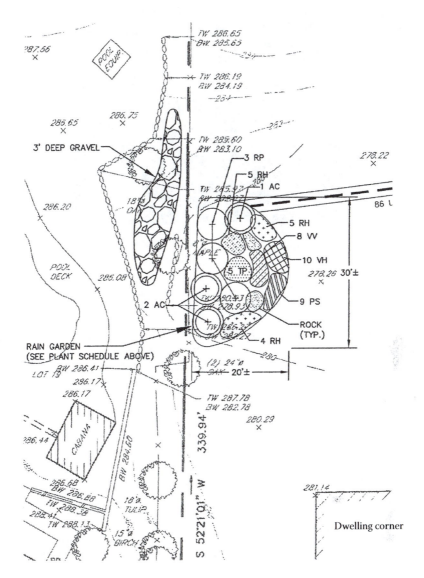

FIGURE 8.42   Rain garden for retainage of runoff entering a residential yard from an uphill property. Plant selection listed below.

Plant Schedule

Key	Quantity	Botanical Name	Common Name	Type
AC	3	*Viburnum teilobum*	American cranberry bush	Shrub
AV	6	*Andropogon virginicus*	Busy broom sedge	Grass
PS	9	*Phlox stolonifera*	Creeping phlox	Herb
RH	9	*Rudbeckia hirta*	Black-eyed Susan	Herb
RP	3	*Rosa palustris*	Swamp rose	Shrub
TP	5	*Thelypteris palustris*	March fern	Fern
WV	10	*Verbena hastata*	Blue varvain	Herb
VV	8	*Veronicastrum virginicum*	Culver's root	Herb

FIGURE 8.43 Aphrodite plantain lily, a hosta family plant, suitable for plant hardiness zone 7 (New Jersey included).

Example 8.2

A 15 m²/150 ft² rain garden receives runoff from a 100 m²/1000 ft² roof area. The rain garden consists of a 15 cm/6 in. stone base, 20 cm/8 in. soil mix and includes a 15 cm/6 in. depression. Calculate the rainfall depth that the rain garden can retain.

Solution

Calculations are performed separately in SI and English units. A 40% void in stone and 15% effective porosity (porosity less moisture content for soil mix) is assumed.

- Storage volume in stone and soil mix are

$$\forall_s = 15\times(15/100)\times0.4 = 0.9\,m^3;\ 150\times(6/12)\times0.4 = 30\,ft^3$$

$$\forall_{sm} = 15\times(20/100)\times0.15 = 0.45\,m^3;\ 150\times(8/12)\times0.15 = 15\,ft^3$$

- Dry storage above bed

$$\forall = 15\times15/100 = 2.25\,m^3;\ 150\times(6/12) = 75\,ft^3$$

Total storage = 0.9 + 0.45 + 2.25 = 3.6 m³; 30 + 15 + 75 = 120 ft³

Retainage depth = $V/(A_{roof} + A_{gar})$

Rainfall depth = 3.6/(100 + 15) = 0.031 m = 3.1 cm; (120/1150) × 12 = 1.25 in.

8.14 COST EFFECTIVENESS OF BMPs

To arrive at a meaningful comparison of various BMPs, their cost should be expressed per unit volume of runoff being retained. While the unit volume is commonly taken as gallons, the costs are expressed in cubic meters and cubic feet herein. The average cost of infiltration basins is $175/m³ ($5/ft³) of the stored water. Bioretention basins cost from $250 to $525/m³ ($7–$15/ft³). Rain gardens cost anywhere from \$275 to \$600/m³ (\$8–\$17/ft³) of retention storage depending on their depth and whether or not they are equipped with an overflow and discharge system. The cost of storm water wetlands ranges from \$75 to \$210/m³ (\$2–\$6/ft³) depending on their size; the larger the size is, the smaller is the unit cost.

The lawn depressions cost less than \$10/m² (\$0.90/ft²) of their surface area. This reflects an average unit cost of \$65/m³ (\$1.80/ft³) of stored runoff for a 15 cm (6 in.) depression. The cost of blue roofs is in the range of \$1000–\$2000/m³ (\$30–\$60/ft³) of the stored water. The costliest system is a green roof, which costs anywhere from \$125 to \$350/m² (\$12–\$30/ft²) of roof area in the United States. In terms of stored water, this reflects a cost of \$5000 to over \$10,000/m³ (\$140–\$285/ft³). In fact, these estimates are less than some reported costs. For example, Grey et al. (2013) report the cost of green roofs at \$10 to \$325/ft² and \$16 to \$522 per gallon of water managed. These figures represent \$4200 to over \$135,000 per cubic meter of stored water for green roofs.

The cost of porous pavements depends on the thickness of stone reservoir; the thicker the stone base is, the smaller is the unit cost of stored water. Pavers cost in the range of \$110–\$150/m² (\$10–14/ft²) and provide up to 6–8 cm (2.4–3.2 in.) of storage volume. As such, the cost of storing runoff is estimated at \$1750/m³ (\$50/ft³) on average. Both the construction cost and the retention cost of porous asphalt and pervious concrete are lower than those of pavers. The average construction costs of porous asphalt and pervious concretes are \$50/m² (\$5/ft²) and \$90/m² (\$8/ft²), respectively. These permeable pavements, when provided with 75 cm (30 in.) of stone reservoir, can hold up to 25 cm (10 in.) of water. Therefore, the cost of storing water is approximately \$200/m³ (\$5.70/ft³) for porous asphalt and \$350/m³ (\$10/ft³) for pervious concrete.

8.15   OTHER NONSTRUCTURAL MEASURES

As indicated previously, source reduction is more effective and a far better solution than structural BMPs in controlling both quantity and quality of runoff. A number of measures are available for source reduction, some of which are easy to implement. Measures can be grouped into common and specific, as follows.

8.15.1   COMMON (GENERAL) MEASURES

These measures, which can be applied to any type of development, include

a. Separating roof drains and downspouts from driveways and drainage systems, instead directing them to lawn/landscape areas and rain gardens (see Figure 8.46)
b. Pitching driveways slightly toward lawn and vegetated areas, where practicable
c. Minimizing land disturbance and grading
d. Collecting roof runoff in rain barrels/tanks and cisterns for reuse (see Chapter 10), or retaining roof runoff in dry wells, infiltration chambers or the like
e. Creating lawn/landscape depression and rain gardens to retain the runoff from roof/paved areas and infiltrating it over time
f. Constructing a raised planter or planted bed at the foot of a steep slope to retain/reduce runoff flowing onto downhill areas (see Figures 8.47 and 8.48)
g. Limiting the application of lawn chemicals to the minimum amount needed
h. Using water-based, rather than oil-based, paints and household cleaners
i. Conserving outdoor water use because overwatering does not make a healthier lawn but is only a waste
j. Minimizing the width of access roads, cul-de-sacs, and driveways; narrowing a roadway width from 22 to 18 ft reduces the peak rate and volume of road runoff by over 20%
k. Installing roadside swale and/or vegetation buffers in lieu of inlets and pipes; vegetation absorbs and infiltrates a large portion if not the entire runoff from paved roads

l. Substituting hardscape (porous pavements and, in particular, pavers) in lieu of impervious covers for parking lots, driveways, patios, and sidewalks
m. Covering pervious areas with native vegetation, shrubs, and flowering plants rather than high-maintenance, water-thirsty lawns
n. Sweeping driveways, sidewalks, and patios with a broom instead of using a hose
o. Starting a compost pile instead of disposing of yard waste
p. Picking up after pets
q. Disposing of chemicals properly to prevent polluted runoff
r. Checking one's car for leaks and recycling motor oil and antifreeze when changing these fluids
s. Using car wash facilities that recycle water to avoid adverse impacts to runoff due to detergents, grime, and brake dust

8.15.2 CLUSTERED DEVELOPMENTS

Clustered development is a specific measure for source reduction. This type of development tends to maximize conservation of trees and natural vegetation and minimize the impervious coverage for any given type of subdivision. For residential projects, for example, the objectives of clustered developments are

a. To maximize preservation of open space for recreational and scenic purposes
b. To provide a development pattern harmonious with natural land features
c. To offer a variety of choices and more economical housing
d. To use land efficiently with smaller networks of paved streets and utilities

This type of development creates a common green space to be maintained by the property owners association. Before the association is formed, the developer will be responsible for all maintenance. The common green space usually includes irreplaceable natural features located on the tract such as stream buffers, significant stands of trees, individual trees of significance, steep slopes, rock outcrops, open fields and meadows, any historic and archeological sites, and wetlands and their buffers. Submissions for a preliminary cluster residential development review by a municipality (similar to any type of development project) generally include

- Topographic maps
- Wetlands plan
- Preliminary plot plan showing the maximum number of lots allowed under regular subdivision ordinance
- Preliminary site plan showing the subdivision layout
- Calculations associated with the net residential acreage and density
- A written narrative explaining adherence of the proposed cluster subdivision with the ordinance

8.16 MINIMAL IMPACT DEVELOPMENTS

It is a common misperception that single-family homes with large lots have less adverse impact on the storm water runoff than multifamily residential developments, such as townhouses and condominiums. Basing the environmental impact on the amount of pavement and disturbance on a per-parcel of land or a development site is misleading. It should be realized that people need housing and therefore the impact criteria should be based not on how much of an impervious area is placed on 1 acre (hectare) of land, but rather on how much pavement is created to provide housing for one person. On this basis, as indicated in Chapter 1, large single-family homes, contrary to a general notion, disturb more land, create more storm water runoff, and have more adverse environmental impacts than condensed residential developments per capita.

Table 8.6 presents typical amounts of pavements for single-family homes on lots ranging from 675 m² (1/6 acre) to 8100 m² (2 acres) in the United States. This table indicates that larger homes, while having a smaller percentage of imperviousness, create far more impervious coverage than smaller homes. The access roads and streets are excluded from the listed impervious covers in this table. Table 8.7 shows impervious coverage for typical single-family homes, accounting for paved streets along each lot. This table also lists the per-capita impervious coverage for variously sized lots. Listed in the table are the calculated annual volume of runoff from impervious surfaces in each lot. The calculations are based on a rainfall amount of 760 mm, which represents the average annual precipitation in the United States and some European countries as well.

The impact of multifamily residential developments would be nil if the project is in a brownfield development, namely, developing on an abandoned industrial or commercial site. The following residential project exemplifies this case.

A residential development project, known as Cambridge Crossings, in Clifton, New Jersey, included construction of 47 three-story multifamily buildings, two club houses, swimming pools, and tennis courts on a 42.5 acre (10 ha) parcel of land. The site had been formerly occupied by American Cyanamid, a chemical industry, and was left abandoned for some time. Because of the development, the impervious surfaces inclusive of roads, buildings, driveways, and parking lot were reduced from 20.3 acres (8.22 ha) to 15.6 acres (6.31 ha). The development created 640 residential units, comprising 210 townhouses, 160 flats, and 270 condominiums. All units had two bedrooms,

TABLE 8.6
Lot Impervious Coverage: Single-Family Homes[a]

Lot Size, m² (ft²)	Roof, m² (ft²)	Driveway, m² (ft²)	Patio/Walks, m² (ft²)	Total Imp., m² (ft²)	% Cover[a]
675 (7270)	110 (1185)	65 (700)	45 (480)	220 (2370)	33.0%
1010 (10,870)	135 (1450)	95 (1020)	65 (700)	295 (3150)	29.0%
2025 (21,800)	185 (1990)	180 (1940)	110 (1180)	475 (5110)	23.5%
4050 (43,560)	280 (3010)	350 (3770)	160 (1720)	790 (8500)	19.5%
8100 (87,120)	370 (3980)	520 (5600)	205 (2210)	1095 (11,790)	13.5%

[a] Rounded to first decimal place.

TABLE 8.7
Overall Per-Capita Impervious Coverage: Single-Family Homes

Lot Size Acres, m² (ft²)	Total Imp., m² (ft²)	% Cover[a]	Per Capita[b], m² (ft²)	Avg. Annual Pavement Runoff[c], m³ (10³ cf)
675 (7270)	325 (3500)	48%	81 (875)	55.0 (1.94)
1010 (10,870)	430 (4630)	43%	107 (1160)	73.2 (2.84)
2025 (21,800)	675 (7270)	33%	169 (1820)	115.6 (4.08)
4050 (43,560)	1080 (11,630)	27%	216 (2330)	147.7 (5.21)
8100 (87,120)	1485 (15,980)	18%	297 (3200)	203.1 (7.17)

[a] Rounded to second decimal place.

[b] Based on family of four for smaller than 4500 m² (1 acre) and family of five for 4500 m² and 8100 m² (1 and 2 acre) lots.

[c] Based on 760 mm annual rainfall and 0.90 runoff/rainfall ratio from pavements.

occupied by two or three people, except for the condominiums, which were age restricted and had two residents. As such, the development potentially would provide housing for 1465 persons. Since two streams were traversing through the site, 14.1 acres (5.7 ha) of land remained intact to serve as stream buffer and conservation area. Of the remainder 28.4 acres (11.5 ha), 12.8 acres (5.2 ha) comprised green space including landscape islands and vegetated detention basins. Thus, the overall amount of land use and impervious coverage on a per-capita basis (assuming undisturbed land prior to the development) were approximately 844 square feet and 464 square feet (78 and 43 m^2), respectively. Table 8.8 provides a summary of pre- and postpavement areas and the per-capita impervious coverage. A sample photo of townhomes in this residential development is included as Figure 8.44.

The amount of impervious coverage per capita for midrise or high-rise apartment buildings would be even smaller than the previous development. Table 8.9 exemplifies a six-story apartment building in Laval, Quebec, Canada. Figure 8.45 depicts the apartment buildings (two in all) that were under construction in July 2012. As indicated by Table 8.9, the per-capita impervious coverage for this midrise apartment building is estimated at 175 ft^2 (16.3 m^2) per capita. For high-rise apartment buildings, the per-capita impervious coverage drops to 100 ft^2 (9.3 m^2) per capita or even less.

TABLE 8.8

Impervious Coverage Summary

Per-Capita Impervious Coverage
Multifamily Residential Developments

Example: Cambridge Crossings, Clifton, NJ

Site area = 42.5 acres

Development area = 28.4 acres

Impervious coverage = 15.6 acres[a] (including streets)

No. of units = 640 (one/two bedrooms)

Households (2.2 per unit) = 1465[b]

Per-capita impervious area = 483 ft^2/capita

[a] Pre-existing impervious coverage = 20.3 acres.
[b] 2 per condo; 2.3 persons, on the average, per townhomes and flats.

FIGURE 8.44   A typical depiction of townhomes in the Cambridge Crossings development. (Photo by the author, 2012.)

TABLE 8.9

Per-Capita Impervious Coverage for a Midrise Apartment Building

Per-Capita Impervious Coverage: Six-Story Apartment Buildings

Estimate: 1500 ft² apt. (50 ft × 30 ft flat)

Two household/apt.

Households: 12/1500 ft² footprint

Street: 12 × 50 = 600 ft²

Total imp. area = 2100 ft²

Per-capita imp. = 2100/12 = 175 ft²/capita

FIGURE 8.45 Two 6-story apartment buildings in Laval, Canada. (Photo by the author, 2013.)

In comparison, a typical midsized single-family home in a one-half acre (2000 m²) suburban subdivision creates over 8000 square feet (743 m²) of impervious surface, accounting for streets, sidewalks, driveway, patio, and roof. For a family of four, the suburban dwelling creates approximately 2000 square feet (186 m²) of impervious coverage per capita. Thus, the exemplified residential development and midrise apartment building have, respectively, over fourfold and 11-fold smaller impervious surfaces those of the previously indicated single-family home. Residential towers, like those in major cities in the United States and all over the world, result in even smaller land disturbance and impervious cover on the per-capita basis. Thus, city living has a far smaller adverse impact on the environment than suburbs do.

As indicated in a previous chapter, current storm water management regulations exclude small sites—in particular single-family homes. It is to be noted, however, that single-family homes, which form the majority of residential dwellings in the United States, are the culprit of storm water runoff and flooding problems. Though each single-family home has a small effect, collectively they create a significant increase in the peak and volume of runoff. To reduce these impacts, as indicated earlier, roof runoff should be directed to landscaped areas or rain gardens or drained into rain tanks/barrels or retention–infiltration systems such as dry wells (seepage pits) and chambers in stone trench. Figure 8.46 shows a downspout terminating at a front yard landscape area of the author's residence. The landscape area seldom overflows. The result is little runoff from the roof and the yard discharging to the municipal storm drain system. Figures 8.47 and 8.48 depict a landscape area in the form of a raised rain garden that the author had personally built at his former residence in West Milford, New Jersey, to capture the runoff from the hilly area at the back of the house. This raised landscape fully absorbed all of the uphill runoff and only overflowed during Tropical Storm Floyd in September 1999, when more than 10 in. of rain fell in less than a 24-hour period.

FIGURE 8.46   Roof leader terminating at landscaping at author's residence.

FIGURE 8.47   Raised planter for capturing runoff from uphill area; author's former residence.

FIGURE 8.48   A raised flowers bed the author built at his former residence to retain uphill runoff. The only time this planter (on the right) overflowed was during Tropical Storm Floyd in September 1999, when more than 10 in. of rain fell in less than 24 hours.

8.17   STORM WATER FEES

Addressing the current and ongoing storm water management regulations will require significant budgets for municipalities. To obtain the budget, a number of alternate solutions, such as real estate transfer fees, real estate tax adjustments, sales taxes, state revolving funds, voluntary offset programs, and storm water utility fees have been sought. Among these, the storm water utility fee is the most practical. While adjusting real estate taxes may compensate the cost of operation and maintenance of storm water programs, state and federal facilities do not pay local taxes. To be fair to private property owners, however, the option of tax increases should not be considered. To pay for the storm water management cost, many municipalities have established a storm water fee, just like the water, sewer, and energy fees. The use of storm water utility fees for financing urban storm water programs is growing. In the 1970s and 1980s, storm water utilities were first established in Washington state. To arrive at a more informed storm water management decision, the Center for Urban Water Resources at the University of Washington was formed.

In 2007 there were well over 600 storm water utilities in the United States. Over 2000 utilities were in place in the United States in 2010. This is an impressive change, considering that few such utilities existed just two decades ago. Communities see a utility fee as the steadiest and fairest way to pay for storm water systems. But getting started is still a hard battle for municipal officials. This is exemplified by a new utility proposed by the city of Hartsville in South Carolina to protect creeks and streams from storm water pollution. "We have never done anything larger in the past and we have never done anything that will be so controversial," says Mike Wetch, then the director of Public Works (Faile, 2008). As proposed, the utility charge will be a flat rate of $4 and $5 a month for residential and commercial properties, respectively.

The city of Ithaca in New York State is one of the regulated MS4s that established a storm water utility user fee to implement a storm water management program. The fee is set at $4.60 per equivalent residential unit, defined as 1976 ft^2 (184 m^2) of impervious area on the average (Zolezi, 2009). The future will tell "if these rates will be sufficient or inadequate"; the latter is more likely to be the case, the author believes.

Establishing a rate structure for storm water utilities is very complex. The complexity is partly due to insufficient data on long-term maintenance costs of storm water management facilities to address the increasingly stringent water quality standards. Many municipalities and governmental agencies have little experience with the maintenance of the storm water management systems required to address the regulations. The long-term cost of maintenance is more unknown even by many owners and operators of storm water management facilities. A storm water management system designed by the author for a roadway improvement program in a northern New Jersey municipality, constructed in 2010, was the first such system to be maintained by Bergen County in New Jersey.

A method of assessing storm water charges is to relate them with the size of the property and the improvements thereon (Hoag, 2004). However, even two nearly identical properties in a given municipality may generate different amounts of runoff due to the implemented storm water provisions on the property. To account for this difference, a credit may be given to onsite measures that reduce the impact on the downstream drainage system. In addition to quantitative aspects of storm water runoff, the impact on water quality is to be considered in the fee calculations and the appropriation of any credit as well. However, the latter impact is difficult to assess as it involves evaluating the sources of pollutants of concern and costs of their treatment. Apart from all these difficulties, this method does not account for the costs of maintenance of public drainage and storm water management facilities.

Unlike other utilities, such as gas, water, and electricity, which are metered and customers pay for their usage, the storm water fees are not related to any measured quantity. Because of lack of metering, it is hard for people to justify paying for something they can neither see nor use. With other utilities, such as water or energy, people can conserve to lower their fees; but with storm water, there is little choice and the public still has to pay monthly fees even if there is no rain. The majority of people and institutions understand the benefits of flood-free roads and clean water; however, they do not realize how much the management of storm water runoff costs.

While the concept of a storm water fee is gaining popularity in municipalities, legal opposition to utilities is getting more sophisticated as well (Kaspersen, 2004). The monthly storm water fee for a single-family home is generally a few dollars, not enough to encourage the average taxpayer to go to court. However, for schools, universities, and federal government installations, storm water utilities fees can run to hundreds or even thousands of dollars annually. Some tax-exempt institutions balk when they find they are not exempt from the utility fee.

There is also a general opposition to storm water fees by the public (Woolson, 2005). A question facing many municipalities is how they may overcome the public objections. One of the solutions is public awareness. As more people get to know what they are paying for, they may accept that utilities fees are worth paying. Savvy marketing comes into play as well. Also, as more utilities are created, the more acceptable they will become. As an incentive, bonuses or credits may be given to private property owners who implement measures to reduce the runoff, which in turn lowers pollution (Reese, 2007). This may produce more savings to the community by reducing the size of storm water management facilities and their maintenance.

After 15 years of planning, Philadelphia adopted a new approach for assessing nonresidential property owners' storm water fees on July 1, 2010. This fee structure is a parcel-based storm water billing system that charges owners based on the property area and impervious surface coverage. The Philadelphia Water Department acknowledges that a change from a meter-based rate would be a challenging concept to many customers. However, the change is intended to make the cost of service more equitable to customers (Cunningham, 2011).

A successful implementation of an equitable storm water fee program requires a clear understanding of key policy issues (Hoag, 2004; Kumar and White, 2008). It will take time and experience before equitable, fair, and practical storm water user-fee practices are developed. As more and more municipalities adopt a storm water utility program, more experience will be gained.

PROBLEMS

8.1 The construction of the Ambulance Corps in Hawthorne, Passaic County, New Jersey, affects 1.073 acres and increases the impervious coverage from 1400 square feet to 32,080 square feet. A rain basin in the form of lawn depression (see Figures 8.39 and 8.40) is designed to fully retain the entire runoff from the paved area during the 10-year, 60-minute storm, having 2 in./h intensity. Calculate the required storage volume of the rain basin.

8.2 A 20 m^2 rain garden receives runoff from a 150 m^2 roof area. The rain garden includes a 15 cm deep depression, 30 cm thick top soil and 15 cm stone sublayer. Calculate the maximum rainfall depth that the rain garden can retain.

8.3 Solve Problem 8.2 for a 200 ft^2 rain garden receiving runoff from a 1500 ft^2 roof area. The garden consists of 6 in. depression, 1 ft thick top soil, and 6 in. stone base.

8.4 Design a rain garden to retain the runoff from a 1200 ft^2 residential dwelling roof during a New Jersey water quality storm that is 1.25 in. of rainfall in 2 hours. The rain garden includes a 6 in. depression and 1 ft thick soil medium over a 6 in. layer of sand. Estimate porosity of the soil and sand at 0.2 and 0.35, respectively.

8.5 Solve Problem 8.4 for a 120 m^2 roof area, 15 cm depression, 30 cm thick soil, 15 cm thick sand, and 30 mm rainfall.

8.6 A 1500 ft^2 roof area drains to a 200 ft^2 rain garden that includes 12 in. thick soil, 6 in. thick stone, and allowable ponding of 3 in. Evaluate if this rain garden satisfies the WQ_v requirements of New York State (refer to Chapter 5). The 90% rainfall event in the area is 0.9 in. State any assumptions made.

8.7   Solve Problem 8.6 for a 150 m^2 roof area and 20 m^2 rain garden. The thicknesses of soil and drainage layers are 30 cm and 15 cm, respectively. The 90% rainfall event is 23 mm and the allowable ponding is 7.5 cm.

8.8   Size a bioretention basin for a 1300 ft^2 pavement and 2.0 in. of rainfall. The bioinfiltration basin includes 2.5 ft thick soil mix, 10 in. thick sand, 1.0 ft deep depression, and 3 in. thick mulch. Estimate the porosity of soil, sand, and mulch at 15%, 35%, and 40%, respectively. The basin abuts the pavement.

8.9   Solve Problem 8.8 if the bioinfiltration basin is 10 ft away from the pavement. Assume that 35% of the runoff from the pavement is captured by lawn en route to the basin.

8.10 Redo Problem 8.8 for a 120 m^2 pavement and 50 mm rainfall. The bioretention basin is 30 cm deep and the thicknesses of soil mix, sand base, and mulch are 75, 25, and 7.5 cm, respectively.

8.11 Solve Problem 8.10 if the bioretention basin is 3 m away from the pavement and 35% of pavement runoff is absorbed by the lawn between the bioretention basin and the depression.

8.12 A 25-acre single-family residential development comprises 1/3 acre lots and includes 35% impervious coverage. Calculate:
a. The impervious coverage per dwelling and per capita, in square feet
b. The annual runoff volume per dwelling and per capita
Base your calculations on a household of four persons per dwelling and 30 in. annual precipitation. State any assumptions made.

8.13 A 25-acre multiresidential development includes 600 condominium units and is 50% paved. Calculate:
a. The impervious coverage per condominium unit and per capita in square feet
b. The annual runoff volume per capita
Base your calculations on two persons per condo and 30 in. annual precipitation. State any assumptions you need to make.

8.14 A 10 ha residential development comprises 75 single-family homes and is 30% covered with impervious surfaces. Calculate:
a. The average impervious coverage per dwelling and per capita in square meters
b. The average annual runoff volume per dwelling and per capita
Base your calculations on a household of four persons per dwelling and 750 mm of annual precipitation. State your assumptions.

8.15 A multiresidential development includes 750 apartments on a 10 ha parcel of land that is 50% covered with pavements and roofs. Calculate:
a. The average impervious surface per apartment and per capita
b. The annual runoff volume per capita
Base your calculations on two people per apartment and 750 mm annual precipitation. State any assumptions made.

APPENDIX 8A: DRIVABLE GRASS® TECHNICAL SPECIFICATION GUIDE

Drivable Grass® Technical Specification Guide

Drivable Grass® is a permeable, flexible and plantable pavement system. Drivable Grass® is designed to be installed over a properly prepared subgrade and compacted aggregate base structural section. Drivable Grass® is intended to be used in areas that are exposed to traffic and / or areas that will have exposure to small drainage flows. Drivable Grass® is designed to facilitate planting which will produce a vegetated pavement section. Depending on the base and subgrade structural section, Drivable Grass® can be used for loadings resulting from both light and heavy-duty traffic areas. The vegetated Drivable Grass® and compacted aggregate base section can also be used for biofiltration and as an underground detention basin.

Recommended Uses. Recommended uses include, but are not limited to the following:

Light Duty Applications
- a. Golf Cart Paths
- b. Service Roads
- c. Dog Parks
- d. Irrigation Pathways
- e. Pump Stations
- f. Trail Reinforcement
- g. Roadway Shoulders
- h. Residential Driveways
- i. Parking Lots
- j. Concrete Swale Replacements
- k. Overflow Parking Areas
- l. RV and Boat Access Drives and Parking Areas
- m. Truck & Cart Wash-Down Areas
- n. Outdoor Shower & Drinking Fountain Runoff Areas

Heavy Duty Applications
- a. Fire Access Lanes
- b. Emergency Vehicle Access Drives
- c. Service Vehicle Utility Roads
- d. Truck Maintenance and Equipment Yards
- e. RV and Car Sales Centers

Non-Traffic Applications
- a. V-Ditch Lining
- b. Linings for Ditches
- c. Energy Dissipater Aprons
- d. Low-flow stream linings
- e. Lining for roadside drainage features
- f. Bioswales / Trickle Channels
- g. Erosion Control on Slopes

Non-Recommended Uses
- a. Surfacing for Athletic Fields (baseball diamonds, football field, soccer field, under playground equipment....)
- b. Support of tread driven equipment (tread driven military equipment, tread driven construction equipment.....)
- c. Use in high velocity streams, rivers or channels
- d. Very steep grades unless secured via pins/staples or regular spaced mow curbs or strips

Turf Maintenance Comments:
1. Avoid the use of aeration, roto-tilling, and de-thatching equipment in area where Drivable Grass® pavement is installed.
2. The need for de-thatching can be minimized by planting turf varieties that resist thatch build-up, collecting grass clippings and adopting deep watering techniques.

Soil Definitions
1. **Coarse Sandy Loam** – 25% or more very coarse and coarse sand and less than 50% any other one grade of sand.
2. **Sandy Loam** - 30% or more very coarse, coarse and medium sand, but less than 25% very coarse sand, and less than 30% very fine or fine sand.
3. **Fine Sandy Loam** – 30% or more fine sand and less than 30% very fine sand or between 15 and 30% very coarse, coarse, and medium sand.
4. **Loamy Coarse Sand** – 25% or more very coarse and coarse sand and less than 50% any other one grade of sand.

Drivable Grass® Technical Specification Guide   Page 2

5. **Loamy Sand** – 25% or more very coarse, coarse and medium sand and less than 50% fine or very fine sand.
6. **Loamy Fine Sand** – 50% or more fine sand or less than 25% very coarse, coarse, and medium sand and less than 50% very fine sand.
7. **Sand** – 25% or more very coarse, coarse and medium sand and less than 50% fine or very fine sand.
8. **Fine Sand** – 50% or more fine sand or less than 25% very coarse, coarse, and medium sand and less than 50% very fine sand.
9. **Very Fine Sand** – 50% or more very fine sand.

Drivable Grass® Installation Guidelines

1. **Delivery, Storage and Handling**
   a. Deliver materials to site in manufacturer's original palletized configuration with labels clearly identifying product style number, color, name and manufacturer.
   b. Check all materials upon delivery to assure that the proper type, grade, color, and certification have been received.
   c. Store materials in clean, dry area in accordance with manufacturer's instructions.
   d. Protect all materials from damage due to jobsite conditions and in accordance with manufacturer's recommendations. Damaged materials shall not be incorporated into the work.

2. **Sub-grade Preparation**
   a. Define boundary of proposed area to receive Drivable Grass® by using a string line, header board, existing hardscape, or other means of delineating the boundary shown on the construction drawings.
   b. Excavate to the lines and grades shown on the construction drawings.
   c. Proof roll foundation area as directed to determine if remedial work is required.
   d. Owner's representative shall inspect the excavation and approve prior to placement of base material or fill soils.
   e. Over-excavation and replacement of unsuitable sub-grade soils with approved compacted fill shall be compensated as agreed upon with the Owner.

3. **Installation of Filter Fabric**
   a. Install filter fabric on prepared sub-base. A filter-weave fabric by Mirafi Inc. or equal shall be used if required by contract documents.

4. **Installation of Aggregate Base and Sand Setting Bed**
   a. Install and compact base as required by the contract documents.
   b. Base aggregate shall consist of "Class II Permeable", "Crushed Miscellaneous Base" (CMB), crushed rock, or similar structural material normally used as a base course for pavement systems and meeting the gradation requirements shown on the construction drawings and specifications. Base layer shall be designed to carry the imposed loading as well as any stormwater storage considerations for the site. Base layer thickness to be determined by engineer of record for the project.
   c. Install subdrain as required by contract documents.
   d. Install, level and compact approximately 1" thick well graded sand bedding layer for non-planting applications. Install, level and compact approximately 1.5" thick well graded sand bedding layer for planting applications. Well graded sand to be comprised of a moderate percentage (20%) of organic or other plant nutrients for heavy duty applications and 30% organic material for light duty applications. A small amount of fertilizer may be added to facilitate grass growth.

5. **Install Permeable, Flexible and Plantable Pavement System**
   a. Install permeable, flexible, and plantable pavement system in accordance with the manufacturer's guidelines.
   b. Install system to the line, grades and locations required by the contract documents.
   c. Butt mats against each other leaving no significant gaps.
   d. Mats may be "fit" to the geometry of the site and obstructions by cutting with a concrete saw, or severing the polymeric reinforcement strands with a utility knife or other sharp cutting device.

Bending the mat over onto itself to expose the back side of the mats will facilitate exposure of reinforcing strands to be severed.

- e. Installations of Drivable Grass® (used for traffic loading) on grades steeper than 12% must be evaluated by a qualified engineer or architect.
- f. Use of geotextile pins or nails may be required for added stability of the mat on sloping terrain or as directed by contract documents. Industry practice suggests that 6 inch minimum length nails with 1.0 inch max diameter washers may work well in hard or rocky soils, while 12 to 18 inch long geotextile pins with 1.0 inch max diameter washers may work well in sandy soils. A qualified architect or engineer may be required to assess the need for securing devices.
- g. Anchoring frequency and pattern of securing devices (where required) should be as shown on the construction drawings or as specified by the engineer or architect. Anchoring is not required if mow strips / curbs are provided to confine the Drivable Grass® product.
- h. Installation of concrete mow strips / curbs: Mow curbs to be 4" x 4" (min.) thickness w/ (1) #4 rebar continuous or as directed by the specifications and drawings.

6. **Fill Grooves of Drivable Grass with Infill**

- a. Backfilling of Drivable Grass® must be conducted as soon after installation as practically possible. In no case shall Drivable Grass® be left un-filled for more than 30 days after installation unless specifically approved by the project architect or engineer.
- b. Backfill permeable, flexible and plantable pavement system with soil infill in accordance with the manufacturer's installation instructions. Soil in which grasses will be planted will have a moderate percentage of organic or other plant nutrients added to clean sand. Sand mixture to be 80% well graded sand and 20% organic material for heavy duty applications, and 70% well graded sand and 30% organic material for light duty applications. Infill not intended to support vegetation is likely to consist of decorative stone of varying color and quality, depending on application and aesthetic needs. Product may also be left bare. A layer of landscape fabric installed below sand setting bed is recommended to deter weed growth in non-planted systems.
- c. Prepare for planting by sweeping or otherwise spreading soil infill uniformly across the mats.

7. **Vegetate Mat System (Option 1 - Seeding)**

- a. Install lawn with the planting materials and manner as specified in the construction drawings.
- b. Broadcasting seed may be done by hand or mechanical spreading device. It is also recommended to mix seed into fill material. A topper may be used on top of the seed to facilitate germination.
- c. Set irrigation system (where required) such that complete and adequate irrigation coverage is provided for the installation area. Proper irrigation will promote healthy vegetation growth. The irrigation system may need to be installed before the base for large areas.
- d. Root barrier systems should be provided around the perimeter of trees that exist or may be planted near Drivable Grass installation to minimize the potential for future tree root damage.

8. **Vegetate Mat System (Option 2 - Top Dressing With Sod)**

- **a.** Lay sod on backfilled Drivable Grass® system being sure to cut out sod where sprinkler heads exist. An additional 1" of soil infill should be used between the top of the Drivable Grass and the sod.
- **b.** Sod should be laid in a staggered pattern to ensure a stable sod matrix.
- **c.** Refrain from traversing sodded areas for about 30 days or until sod has been established.
- **d.** Irrigate in-place sod and set watering schedule.

9. **Erosion Control**

- a. Provide dust and erosion control protection plan in accordance with the contract documents.

10. **Field Quality Control**

- a. The Owner shall engage inspection and testing services, including independent laboratories, to provide quality assurance and testing services during construction. This does not relieve the Contractor from securing the necessary construction control testing during construction when required by the contract documents.
- b. Qualified and experienced technicians and engineers shall perform testing and inspections services.
- c. As a minimum, quality assurance testing should include sub-grade soil inspection, aggregate base quality, thickness, and compaction, and observation of construction for general compliance with design drawings and specifications.

REFERENCES

Adams, M.C., 2003, Porous asphalt pavement with recharge beds 20 years and still working, *Stormwater*, May/June.

Americast, 1135 Virginia Precast Road, Ashland, VA 23005 (http://www.filtera.com).

Baker, L.A., 2007, Stormwater pollution, getting at the source, *Stormwater*, November/December, pp. 16–42.

Brown, D., 2007, Porous in Portland, parking lot incorporate porous asphalt for optimal storm water management, *Stormwater Solutions*, January/February, pp. 16–19.

Brzozowski, C., 2007, Green stormwater, *Stormwater*, October, pp. 52–59.

——— 2008, Private and public green, *Stormwater*, January/February, pp. 58–63.

——— 2009, Permeable pavers, *Stormwater*, September, pp. 82–90.

——— 2012, Testing the water part I, *Stormwater*, October, pp. 24–30.

Buranen, M., 2008a, Rain gardens reign, *Stormwater*, May, pp. 78–87.

——— 2008b, Chicago's green alleys, *Stormwater*, October, pp. 50–57.

Contech Stormwater Solutions, West Chester, Ohio (http://www.contechstorm water.com), Ph. 800-925-5240.

Cunningham, C., 2009, Wet-weather management blossoms, *Water and Wastes Digest*, editor's focus, July, pp. 12–13.

——— 2011, GIS for green, Philadelphia introduces parcel based storm water billing and corresponding online application, *Stormwater Solutions*, May/June, pp. 12–13.

DeLaria, M., 2008, What I learned in paver school, the role of permeable paver systems as a stormwater management technique. *Stormwater*, pp. 54–58.

Drivable Grass, Soil Retention, Ph. 800-346-7995 (http://www.soilrention.com).

Ecoloc/UniEco-Stone, Mutual Materials (http://www.mutualmaterials.com).

Emersleben, A. and Meyer, N., 2012, *Civil Engineering Magazine*, June, pp. 66–69.

EPA, 1980, Porous pavement, phase i design and operational criteria," report no. EPA-600/2-80-135, NTIS #PB8110-4796.

——— 2004, Constructed Treatment Wetlands, EPA 842-F-03-013, Office of Water, August (http://www.epa.gov/owow/wetlands/consructed/index.cfm).

——— 2008, Low impact development (LID): A literature review, EPA-841-B-00-005, October (http://www.epa.gov/owow/nps/lid/lid.pdf).

Fabco-Industries, 66 Central Avenue, Farmington, NY 11735, Ph. 631-393-6024 (http://www.fabco-industries.com).

Faile, J., 2008, New storm water utility to be created, *Morning News*, October 16.

——— 2009, Porous pavements Q&A, answers from the man who wrote the book on the subject, *Stormwater*, September, pp. 92–99.

Ferguson, B.K., 2005, *Porous pavements*, CRC Press, Boca Raton, FL.

Frechette, L.A., III and Gilchrist, R., 2009, Seeking zero energy, *ASCE Magazine*, January, pp. 38–47.

Geoweb Cellular Confinement System, Presto Products Company, 670 Perkings Street, PO Box 2399, Appleton, WI 54912-2399, Ph. 800-548-3424 (info@prestogeo.com).

Grassy Paver, R.K. Manufacturing, Inc.

Grey, M., Sorem, D., Alexander, C., and Boon, R., 2013, Low-impact development BMP installation, operation and maintenance cost in Orange County, CA, *Stormwater*, March/April, pp. 26–35.

Gunderson, J., 2008, Pervious pavements, *Stormwater*, September, pp. 62–71.

Gunderson, J., Rosen, R.M., Ballesters, T.P., Watts, A., Houle, J., and Farah, K., 2012, Subsurface gravel wetlands for stormwater management, *Stormwater*, November/December, pp. 8–17.

Hoag, G., 2004, Developing equitable storm water fees, *Stormwater*, January/February, pp. 32–39.

Hun-Dorris, T., 2005, Advances in porous pavement, *Stormwater*, March/April, pp. 82–88.

Jones, M., 2012, Design review and simulation for permeable pavement and bioretention, *Stormwater*, May, pp. 54–58.

Justice, K., 2009, Borough of Northvale chooses pervious concrete, *The Conveyor* (the official publication of the New Jersey Concrete and Aggregate Association), *Winter*, pp. 9–11.

Kaspersen, J., 2004, Selling storm water utilities, editor's comments, *Stormwater*, September/October.

Kumar, P. and White, A., 2008, Know your way, policy development in storm water-user-fee implementation, *Stormwater*, May, pp. 20–34.

Maryland Stormwater Design Manual, May 2009.

McLaughlin, J., Stein, J., Mehrotra, S., Leo, W., and Jones, M., 2012, Stormwater source control in New York City, implementing green infrastructure within the nation's largest urban landscape, *Stormwater*, September, pp. 40–47.

Miller, S.H., 2011, Growing business with BioFilters, tree-box technology aids neighborhood development, *Stormwater*, May/June, pp. 16–17.

Minton, G.R., 2012, Bioretention filters, part 2, what we know with respect to hydrologic and treatment performance, *Stormwater*, January/February, pp. 22–31.

Moody, J., 2012, Up in the air, LID for flood control? It depends who you ask, *Stormwater Solutions*, May/June, pp. 20–22.

Native Plant Society of New Jersey, Office of Continuing Professional Education, Cook College, 2010, Rain garden manual for New Jersey, 2010, 102 Ryders Lane, New Brunswick, NJ.

New Jersey Concrete and Aggregate Association, Pervious concrete, Ph. 609-393-3352 (http://www.njconcrete.com).

New Jersey DEP, 2004, Stormwater Best Management Practices Manual, February.

New York State Stormwater Management Design Manual, 2010.

Nisenson, L., 2004, Smart growth, guest editorial, *Stormwater*, November/December, pp. 8–12.

Norman, M., 2008, Cincinnati storm water, a new approach to an old problem, *Stormwater*, November/December, pp. 50–54.

NRCS, Illinois, 2003, Filterstrips, acre 393, June, 10 p. (http://www.efotg.sc.egov.usda.gov/references/public/IL/IL393.pdf).

PaveDrain, LLC, PMB292, 7245 S. 76th St., Franklin, WI, 53132-9041, Ph. 888-575-5339.

Peck, S., 2007, In the green, green-roof infrastructure gains popularity, *Stormwater*, January/February, pp. 24–27.

Powell, A.E., 2009, Editor's note, *ASCE Magazine*, January, p. 37.

Reese, A.J., 2007, Storm water utility user fee credits, *Stormwater*, November/December, pp. 56–66.

Reid, R.L., 2009, Under one green roof, *ASCE Magazine*, March, pp. 46–57.

Rogers, W.M. and Faha, M., 2007, Pulling together the pieces of the storm water puzzle, porous pavement and other techniques in Portland, *Stormwater*, September, pp. 92–100.

Roseen, R.M. and Ballestero, T.P., 2008, Porous asphalt pavements for storm water management in cold climates, *Hot Mix Asphalt Technology*, May/June.

Roy, S.P. and Braga, A.M., 2009, Saving Silver Lake, *ASCE, Civil Engineering*, February, pp. 72–79.

Seth, I., 2011, Runoff modeling for rain gardens, determining volume reduction and pollutant removal rates, *Stormwater*, July/August, pp. 36–39.

Terrecon, Inc., 10061 Talbot Avenue, #200, Fountain Valley, CA 92708, Ph. 714-964-1400 (http://terrecon.com).

Traver, R. et al., 2004, Porous concrete, *Stormwater*, July/August, pp. 30–45.

——— 2005, Lessons in porous concrete, an update on the Villanova urban storm water partnership's demonstration site, *Stormwater*, July/August, pp. 130–136.

Turfstone Open Cell Pavers, CST Paving Stone and Versa Lok Retaining Walls, Concrete Stone and Tile Corp., corp. office, 23 Ridge Road, PO Box 2191, Branchville, NJ 07826, Ph. 973-948-7193 (http://www.cstpavers.com).

UNHSC, 2009, Subsurface gravel wetland design specifications, University of New Hampshire Stormwater Center, June (http://www.unh.edu/eng/estev).

US National Arboretum, USDA plant hardiness zone map (http://www.usna.usda.gov/Hardzone/ushzmap.html).

Wilkins, R., 2008, Going platinum, *Stormwater Solutions*, May, pp. 11–14.

Woolson, E., 2005, The price of a utility, *Stormwater*, July/August, pp. 10–16.

Zolezi, C., 2009, Study examines feasibility of city's storm water utility user fee, *Urban Water Management*, April, pp. 17–19.

9 Installation, Inspection, and Maintenance of Storm Water Management Systems

This chapter presents an overview of installation, inspection, and maintenance of storm water conveyance systems, and structural and nonstructural storm water management facilities. Also presented herein is a brief description of soil erosion and sediment control measures, which are highly important in the reduction of storm water pollution during construction.

9.1 SOIL EROSION AND SEDIMENT CONTROL MEASURES

During construction, however short, a large quantity of sediment is created due to erosion of loose, unvegetated soil. Figure 9.1 depicts a woodland in natural condition. Figure 9.2 shows loose, disturbed soil, highly vulnerable to erosion and water pollution, at a construction site. To reduce the soil migration from a site to paved roads and drainage structures offsite, a stone blanket is installed at the entrance to the site. Also, a silt fence is installed downhill from the site and hay bales placed around catch basins and/or geofabric materials installed beneath inlet grates. More recently, filtersocks made of leaf compost have been employed as a replacement for plastic silt fencing. Also, filter pads are available to cover inlet grates. Figure 9.3 shows a commercially available filter mat, known as Inlet Filter, manufactured by Blocksom & Co. of Michigan City, Indiana.

To remove sediment from runoff, siltation basins are commonly built within the construction site. Often, the sites of permanent storm water management ponds/detention basins or infiltration basins are used for the construction of temporary siltation basins. It is imperative that these basins be fully cleaned of any silt and restored following the completion of site work and prior to the installation of inlet–outlet structures.

In a large construction site of a water supply reservoir in west central New Jersey, natural depressions were utilized as small sedimentation basins pursuant to the author's recommendation in 1986. This solution proved an effective sediment control measure. In such sites, the stored water in depressions can be used for sprinkling disturbed areas and watering temporary vegetation after the silt is allowed to settle between storms.

9.2 INSTALLATION OF PIPES

Presented in this section are installation of reinforced concrete pipes (RCPs) and high-density polyethylene (HDPE) pipes, which are employed far more than other makes of pipes for conveyance of storm water runoff. Corrugated metal pipes (CMPs) are also used nationwide. However, these pipes are far less durable and significantly more vulnerable to deformation, deterioration, and collapse than either RCP or HDPE pipes. The author has evidenced numerous cases of deformation, damage, and even total collapse of corrugated metal pipes that were constructed to serve as conveyance systems or underground detention basins in the states of New Jersey and New York (particularly in Albany County). Considering longevity, durability, and strength, both RCP and HDPE pipes are far superior to CMPs and are recommended for conveyance and detention system applications.

Contech manufactures high-quality PVC pipes for drainage and detention system application. These pipes, known as A-2000, are available at up to 36 in. diameter. Contech makes a

FIGURE 9.1   Woodland in natural condition.

FIGURE 9.2   Loose, highly erodible soil at a construction site; a potential source of water pollution.

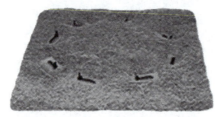

FIGURE 9.3   Commercially available filter mat Inlet Filter, manufactured by Blocksom & Co.

pipe, trade named DuroMaxx. This pipe is extruded of high-density polyethylene resin body embedded in ribs of high-yield steel for added strength. The pipe has a smooth internal surface and rib profile outer wall, resulting in exceptional strength and superior hydraulics suitable for drainage and detention basin applications. DuroMaxx pipe maintains stiffness with temperature, and in this regard, it has an advantage over HDPE pipe, which loses its stiffness at high temperatures. DuroMaxx pipes are currently manufactured in 14 and 20 ft sections of 24 to 120 in. sizes, in 12 in. intervals. Because of lower cost per unit of storage volume, the larger sizes are more suitable for underground detention basins. Considering structural integrity, light weight, and hydraulic efficiency, DuroMaxx pipes may be the next generation of drainage and detention pipes.

9.2.1 ROUND REINFORCED CONCRETE PIPES

RCPs have been used for over a century in the United States and are manufactured in various plants in many states. These pipes come in round, elliptical, and arch geometries. Arch concrete pipes have a much smaller market than elliptical pipes and are manufactured by a few companies in the United States. Hancock Concrete Products in Hancock, Minnesota, is the nearest manufacturer of arch concrete pipes to the northeastern United States. Because of transportation cost and delay in delivery, this type of pipe has little application in many parts of the country.

Table 9.1 shows the sizes and weight of round reinforced concrete pipes. RCPs are available in three ASTM classes: III, IV, and V. Class V pipes, being the strongest of all classes, are used only in unusual cases where the pipe does not have more than 6 in. cover. Tables 9.2 and 9.3 list the minimum allowable earth cover over round and elliptical pipes, respectively. These tables indicate that class IV pipes require no more than 1 foot of cover, and for class III pipes 36 in. and larger, a 6 in. cover is satisfactory.

Figure 9.4 shows a typical pipe trench for concrete pipes as specified by the NJDOT. The trench width is commonly specified equal to the pipe outside diameter (O.D.) plus 18 in. (0.5 m) for pipes 18 in. (450 mm) and smaller, O.D. plus 24 in. (0.6 m) for pipes 24–48 in. (60–1200 mm), and O.D. plus 36 in. (1 m) for larger pipes. The American Concrete Pipe Association in a recent publication presents a table of trench widths based on 1.25 times outside diameter of the pipe plus 1 ft (American Concrete Pipe Association, 2007–2014). The latter specified trench width is not significantly different from the former.

TABLE 9.1
Dimensions and Approximate Weights of Reinforced Concrete Pipes

Internal Diameter, in.	Wall A		Wall B[a]		Wall C	
	Minimum Wall Thickness, in.	Approximate Weight, lb/ft	Minimum Wall Thickness, in.	Approximate Weight, lb/ft	Minimum Wall Thickness, in.	Approximate Weight, lb/ft
12	1 3/4	79	2	93	–	–
15	1 7/8	103	2 1/4	127	–	–
18	2	131	2 1/2	168	–	–
21	2 1/4	171	2 3/4	214	–	–
24	2 1/2	217	3	264	3 3/4	366
27	2 5/8	255	3 1/4	322	4	420
30	2 3/4	295	3 1/2	384	4 1/4	476
33	2 7/8	336	3 3/4	451	4 1/2	552
36	3	383	4	524	4 3/4	654
42	3 1/2	520	4 1/2	686	5 1/4	811
48	4	683	5	867	5 3/4	1011
54	4 1/2	864	5 1/2	1068	6 1/4	1208
60	5	1064	6	1295	6 3/4	1473
66	5 1/2	1287	6 1/2	1542	7 1/4	1735
72	6	1532	7	1811	7 3/4	2015
78	6 1/2	1797	7 1/2	2100	8 1/4	2410
84	7	2085	8	2409	8 3/4	2660
90	7 1/2	2398	8 1/2	2740	9 1/4	3020
96	8	2710	9	3090	9 3/4	3355
102	8 1/2	3078	9 1/2	3480	10 1/4	3760
108	9	3446	10	3865	10 3/4	4160

[a] ASTM C76, AASHTO M170.

TABLE 9.2
Minimum Allowable Cover[a] over Round Reinforced Concrete Pipes

Pipe Diameter, in. (mm)	ASTM Class Pipe	Minimum Cover (Surface to Top of Pipe, in.)
12 (300)	III	15
	IV	10
	V	6
15 (375)	III	15
	IV	9
	V	6
18 (450)	III	12
	IV	6
21 (525)	III	11
	IV	6
24 (600)	III	9
	IV	6
36 (900) and larger	III	6
	IV	6

[a] Minimum cover as designated by the American Concrete Pipe Association.

TABLE 9.3
Minimum Allowable Cover over Elliptical RCPs

Pipe Size, in. (cm)	ASTM Pipe Class	Minimum Cover, in. (cm)
14 in. × 23 in.	III	12 (30)
(36 × 58)	IV	6 (15)
19 in. × 30 in.	III, IV	6 (15)
(48 × 76) and larger		

The American Concrete Pipe Association (2007) used to classify the trench bedding for reinforced concrete pipes into four classes: class A through class D. Class A is concrete cradle bedding and used only for circular pipes in stone trench. This type of bedding should be as wide as the outside pipe diameter and extend up to one-fourth of the outside diameter at the sides. Class B is used for a shaped subgrade with granular foundation. The bottom of excavation is shaped to conform with the pipe geometry and sufficiently wide to allow 6/10 of the outside diameter of circular pipe and 7/10 of the outside span of elliptical pipe to be bedded in finished granular fill placed in the shaped excavation. Densely compacted backfill should be placed at the sides of the pipe at least 1 ft (30 cm) above the top of the pipe.

Class C bedding is also used for shaped subgrade. The pipe is bedded with normal care in a soil foundation shaped to fit the lower one-half of the pipe diameter for circular pipes and 1/10 of the outside pipe rise for arch and elliptical pipes and box culverts. In trench installation, backfill should be placed at the sides and at least 6 in. (15 cm) above the pipe and lightly compacted. If pipe is installed in an embankment, no more than 9/10 of the height of the pipe should project above the bedding.

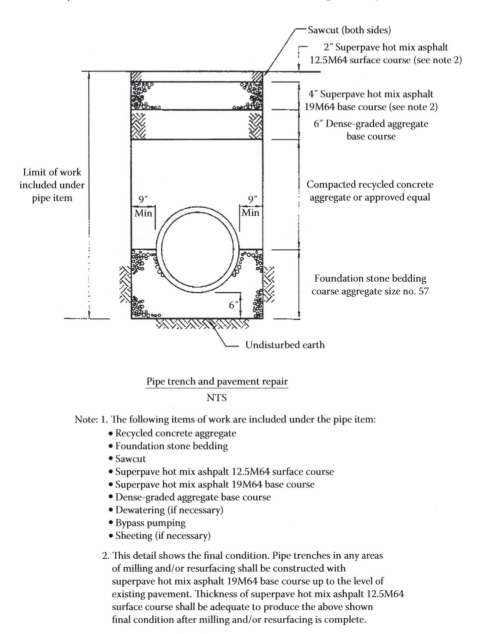

Pipe trench and pavement repair
NTS

Note: 1. The following items of work are included under the pipe item:
   - Recycled concrete aggregate
   - Foundation stone bedding
   - Sawcut
   - Superpave hot mix ashpalt 12.5M64 surface course
   - Superpave hot mix asphalt 19M64 base course
   - Dense-graded aggregate base course
   - Dewatering (if necessary)
   - Bypass pumping
   - Sheeting (if necessary)

2. This detail shows the final condition. Pipe trenches in any areas of milling and/or resurfacing shall be constructed with superpave hot mix asphalt 19M64 base course up to the level of existing pavement. Thickness of superpave hot mix ashpalt 12.5M64 surface course shall be adequate to produce the above shown final condition after milling and/or resurfacing is complete.

FIGURE 9.4 Typical concrete pipe trench under pavement.

Class D bedding only applies to circular pipes where granular material is used for the bedding. In this case, little or no care is taken for the bedding to fit the lower part of the pipe or to fill spaces around it. This class of bedding is also used for the case of pipe on rock foundation where either no earth is placed on the rock or the earth cushion is so thin that the pipe may come in contact with the rock under the load. The depth of bedding material below pipe in classes A, B, and C should be at least 3 in. (75 mm) for pipes 27 in. (686 mm) or smaller, 4 in. (100 mm) for pipes 30 to 60 in. (750 to 1500 mm), and 6 in. (150 mm) for pipes 66 in. (1650 mm) or larger.

Class C bedding is more widely used than other classes for concrete pipes (see Figure 9.4). Note that the NJDOT specifies 6 in. (15 cm) minimum bedding. Backfilling and compaction for reinforced concrete pipes are almost identical to that for other makes of pipes. More detailed information on

TABLE 9.4

Minimum Level of Compaction for Standard Installation Soils

Installation Type	Haunch and Outer Bedding	Lower Side
Type 1	95% Category I	90% Category I
		95% Category III
		100% Category III
Type 2	90% Category I	85% Category I
	95% Category II	90% Category II
		95% Category III
Type 3	85% Category I	85% Category I
	90% Category II	90% Category II
	95% Category III	95% Category III
Type 4	No compaction	No compaction
	Categories I and II	Categories I and II
	85% Category III	85% Category III

TABLE 9.5

Maximum Permissible Depth of Cover over Reinforced Concrete Pipes, ft

Pipe Diameter	C-76 Class III	C-76 Class IV	C-76 Class V
12 in.	9 ft	18 ft	50 ft
15 in.	10 ft	21 ft	50 ft
18 in.	11 ft	24 ft	50 ft
21 in.	12 ft	25 ft	50 ft
24 in.	10 ft	18 ft	50 ft
30 in.	11 ft	19 ft	50 ft
33 in.	11 ft	19 ft	49 ft
36 in.	12 ft	20 ft	49 ft
42 in.	12 ft	21 ft	50 ft
48 in.	12 ft	21 ft	47 ft
54 in.	11 ft	17 ft	31 ft
60 in.	11 ft	18 ft	32 ft
66 in.	11 ft	18 ft	32 ft
72 in.	12 ft	19 ft	33 ft
78 in.	12 ft	19 ft	33 ft
84 in.	12 ft	19 ft	33 ft

the installation of concrete pipes can be found in the American Concrete Pipe Association publications, "Concrete Pipe Handbook" (2005), "Concrete Pipe Design Manual" (2007), and "Concrete Pipe and Box Culvert Installation" (2007).

More recently, the American Concrete Pipe Association (2014) has specified four standard installations. These installations identify four principal zones within the lower half of the pipe. These four zones include the middle bedding, the outer bedding, the haunch, and the lower side.

The type of material and level of compaction of these zones vary with the installation type (1, 2, 3, and 4). In all these types (except type 4 which requires no bedding) the minimum bedding depth is specified as the outer pipe diameter divided by 24 ($D_o/24$) or 3 in. (7.5 cm), whichever is greater. In rock foundation the minimum bedding for all types is $D_o/12$ or 6 in. (15 cm). The level of compaction in the haunch and the lower side vary for soil categories I, II, and III. Soil category I is gravelly sand and includes SW, SP, GW, and GP.* Category II is sandy silt including GM, SM, and ML as well as GC and SC with less than 20% passing #200 sieve; category III is silty clay, which includes CL, MH, GC, and SC. The level of compaction for installation types 1 through 4 is summarized in Table 9.4.

Table 9.5 presents maximum permissible depth of cover over RCPs. The depths listed in this table are calculated assuming backfill material having a unit weight of 120 lb/ft³ (1920 kg/m³, 19 kN/m³). It is to be noted that reinforced concrete pipes have been installed at depths greater than 100 ft (30 m) with engineered backfill and compaction.

9.2.2 ELLIPTICAL CONCRETE PIPES

For elliptical concrete pipes, the middle bedding width should be at least equal to one-third of the outer diameter pipe, ($D_o/3$), where D_o is the diameter along the bedding (longer diameter for horizontal elliptical or shorter for vertical elliptical). The level of compaction for various soil types is the same as that of round pipes.

Both round and elliptical reinforced concrete pipes come in solid or perforated walls. Solid pipes are used for both conveyance and detention applications, and perforated pipes are employed only for underground retention–infiltration basins. The minimum spacing between round and elliptical pipes is generally specified as 12 in. (0.3 m) or one-half of the pipe diameter (longer axis for horizontal elliptical pipes), whichever is greater.

Arch reinforced concrete pipes are employed primarily as culverts under roads and railways. To ensure that pipes are installed in accordance with plans, they should be inspected during the entire construction period. Appendix 9C presents an overview of inspection during installation of any make of pipe.

9.2.3 PRESTRESSED CONCRETE PIPES

Concrete pipes are also made in prestressed forms for pressure flow applications. Price Brothers of Dayton, Ohio, used to manufacture prestressed concrete pipes (PCCPs) in sizes ranging from 400 mm (16 in.) to 3600 mm (144 in.). These pipes complied with the American Water Works Association (AWWA) standards C301 and were made in 20 or 16 ft nominal length. Pipes up to 108 in. were 20 ft long and pipes 114 to 144 in. were 16 ft. Hanson Pipe and Precast of Irvine, Texas, which is one of the largest manufacturers of concrete pipes and culverts, acquired Price Brothers in March 2007. Hanson prestressed concrete cylinder pipes, also complying with AWWA C301 standards, come in L-301 and E-301 series. Prestressing is achieved by helically wrapping steel wire under measured tension and uniform spacing around the concrete lined steel cylinder. The wire wrap compresses the steel cylinder and concrete core, allowing the pipe to withstand specified hydrostatic pressures and external loads comparable to other concrete pipes. Hanson prestress pipes also come in sizes 400 mm (16 in.) to 3600 mm (1400 in.), inside diameter. Tables 9.6 and 9.7 list dimensions and weights of Hanson prestressed concrete lined cylinder pipes for sizes 16 in. to 48 in. and for 54 in. to 144 in., respectively. Prestressed concrete pipes are commonly used in water and sewer applications. Because of having longer lengths and fewer joints than reinforced concrete pipes, prestressed pipes require fewer joints in underground detention basin applications. Figures 9.5 and 9.6 show the joint closure of Hanson L-301 and E-301 pipes, respectively.

---

* For soil designation, see Appendix B at the back of this book.

TABLE 9.6
AWWA C301 Pipe Data Sheet (For Lined Cylinder Pipe Made in U.S.)

Inside Pipe Diameter	Core Thickness Including Cylinder	Max. Outside Diameter at Bell	Weight per Lineal Foot	Standard Laying Length
16"	1"	21"	120#	20'–32'
18"	1 1/8"	23"	150#	20'–32'
20"	1 1/4"	25 1/2"	175#	20'–32'
24"	1 1/2"	30"	230#	20'–32'
27"	1 11/16"	33 1/2"	285#	20'–32'
30"	1 7/8"	37"	330#	20'–32'
33"	2 1/16"	40 1/2"	390#	20'–32'
36"	2 1/4"	43 1/2"	445#	20'–24'
39"	2 7/16"	47"	515#	20'–24'
42"	2 5/8"	50 1/2"	575#	20'–24'
48"	3"	57 1/2"	725#	16'–20'

Note: Availability of diameters and laying lengths varies by location. Contact your sales representative for more information.

TABLE 9.7
AWWA C301 Pipe Data Sheet (For Embedded Cylinder Pipe Made in U.S.)

Inside Pipe Diameter	Max. Outside Diameter at Bell	Weight per Lineal Foot	Standard Laying Length
54"	64"	1010#	20
60"	70 1/2"	1240#	20
66"	78"	1500#	16'/20'
72"	84 1/2"	1780#	20'/24'
78"	90 1/2"	2060#	20'
84"	96 1/2"	2390#	20'
90"	103 1/2"	2540#	20'
96"	111"	2700#	16'/20'
102"	118"	2990#	16'/20'
108"	124"	3150#	16'/20'
114"	131"	3530#	16'/20'
120"	138"	3930#	16'/20'
126"	144"	4450#	12'
132"	151"	4550#	12'
138"	158"	4990#	12'
144"	164"	5350#	12'/16'

Note: Availability of diameters and laying lengths varies by location. Contact your sales representative for more information.

Joint Closure: Hanson's circular O-ring gasket provides a highly dependable positive joint seal. Made of high-quality synthetic rubber, extruded to exacting tolerances and measured volumetrically, the gasket fits within an accurately shaped spigot groove.

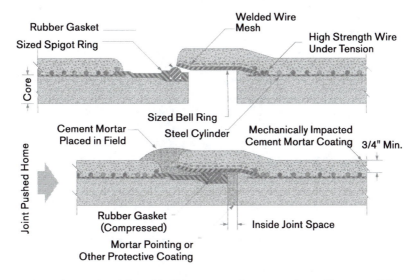

FIGURE 9.5   Dimensions and weights of L-301 prestressed concrete pipes. (Courtesy of Hanson Pressure Pipe.)

Joint Closure: Hanson's circular O-ring gasket provides a highly dependable positive joint seal. Made of high-quality synthetic rubber, extruded to exacting tolerances and measured volumetrically, the gasket fits within an accurately shaped spigot groove.

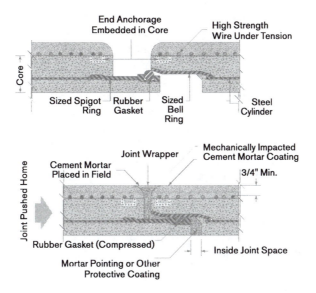

FIGURE 9.6   Dimensions and widths of Hanson E-301 prestressed concreted embedded cylinder pipe. (Courtesy of Hanson Pipe and Precast, Irvine, Texas.)

9.2.4 Concrete Box Culverts

Box culverts have a broad, flat surface and as such neither normally settle nor can be forced down to be level. Therefore, it is important to have a good level grade established before the box is placed. Base material should be fine to medium granular stone. It is best to place and compact a 6 in. (15 cm) thick medium granular material first and cover it with 2 in. (5 cm) thick fine granular material as a leveling course. The boxes should be aligned accurately prior to pulling in place. This is especially important for the first few sections as it will impact the line and grade of the sections that follow. To prevent the granular material from moving into the joint as the boxes are pulled together, the material should be removed 6 in. (15 cm) wide and 2–3 in. (5 to 7.5 cm) deep in front of the groove of the last installed section.

Any unstable material present at the bottom of the excavation should be removed to the depth and width specified by the engineer and replaced with granular material. Also, rock and boulders encountered at the bed must be removed at least 6 in. (15 cm) below the bottom of the box section and replaced with medium to fine granular material.

When putting two sections together their joint should be sealed. Place a joint sealant such as butyl sealant on the bottom half of the groove of the last section first and place the balance of joint material on the top half of the tongue (spigot) of the box to be set. The material should be placed about 1 in. (2.5 cm) from the leading edge of the groove and tongue. In cold weather the joint sealant may have to be heated prior to application. For waterproofing, an expanding waterstop secondary sealant should be applied after butyl sealant has been placed. Upon placing sections and waterproofing, lift inserts (holes) should be filled with grout flush with the top of the box culvert and backfill should be placed uniformly on each side and compacted, taking care that the box's alignment is maintained. The backfill material, the compaction level, and the depth of layers should follow contract specifications. More information on installation of concrete box culverts can be found at Geneva Pipe and Precast (http://info@genevapipe.com) and the concrete pipe handbook (1980).

9.2.5 HDPE Pipes

High-density polyethylene (HDPE) pipes are manufactured by a number of companies, including Advanced Drainage Systems (ADS) and J. M. Eagle, in the United States. After acquiring Hancor in September 2005, ADS is now the largest HDPE pipe company in the United States. HDPE pipes come in single wall and double walls. Single-wall pipes are available in 3 to 24 in. (75–600 mm) for highway drainage application. Double-wall pipes have smooth interiors for hydraulic efficiency and corrugated exteriors for added strength. HDPE double-wall pipes are available in solid and perforated walls, varying in size from 4 to 60 in. (100–1500 mm). ADS pipes come in 6 m (19.7 ft) long sections but are also available in 13 ft (4.25 m) lengths for smaller trench boxes. ADS also makes three-wall pipes made of polypropylene for added strength. These pipes have a smooth exterior, a smooth interior, and a corrugated structural core and are labeled high performance (HP). They are available in HP Storm, which comes in 12 to 30 in., and SaniTite HP in 30 to 60 in. for sanitary sewer applications.

HDPE pipes are lightweight and can be easily handled by two-man crews for sizes up to and including 18 in. Pipes of 24 and 30 in. are lifted by one sling and larger pipes by a two-point sling. HDPE pipes are nearly 20 times lighter than equivalent size concrete pipes. A 6 m (19.7 ft) section of 18 in. ADS double-walled pipe, for example, weighs approximately 126 lb (57 kg). In comparison, an 8 ft (2.4 m) section of 18 in. RCP weighs over 1400 lb. Table 9.8 presents nominal sizes and weights of double-wall HDPE pipes from 4 to 60 in. (100 to 1500 mm) in diameter. This table also lists approximate outside diameter of ADS pipes, rounded to the nearest whole inch.

For parallel pipe installations, a minimum spacing should be allowed between pipes. This minimum is 12 in. (0.3 m) or one-half of nominal pipe size (inside diameter), whichever is greater. It is to be noted that the said minimum spacing applies to solid pipes only. For perforated pipes in stone trench, the minimum spacing is generally wider than for solid pipes. The spacing for underground perforated HDPE pipes was listed in Table 7.1 in Chapter 7.

TABLE 9.8

Sizes and Approximate Weights of Dual Wall (N-12) HDPE Pipes

Inside Diameter		Outside Diameter		Weight	
in.	mm	in.	mm	lb/ft	kg/m
4	100	4.6	117	0.44	0.65
6	150	7.0	178	0.85	1.30
8	200	9.5	241	1.50	2.20
10	250	12.0	305	2.10	3.10
12	300	14.5	368	3.20	4.70
15	375	18.0	457	4.60	6.80
18	450	22.0	559	6.40	9.50
24	600	28.0	711	11.00	16.40
30	750	36.0	914	15.40	22.90
36	900	42.0	1067	19.80	29.40
42	1050	48.0	1219	26.40	39.30
48	1200	54.0	1372	31.30	46.60
54	1350	61.0	1549	34.60	51.50
60	1500	67.0	1702	45.20	67.30

Source: Advanced Drainage Systems.

HDPE pipes are designed for use under H-25 and E-80 live loads. They are also highly resistant to corrosion. High-density polyethylene drainage pipes have been in use since the late 1950s. Contractors have had over 60 years of satisfactory experience with HDPE pipes. During the last two decades some governmental agencies, such as the Federal Highway Administration (FHWA), have approved the use of HDPE pipes for federal highway drainage projects. Consequently, the use of HDPE pipes has grown rapidly and this trend is expected to continue in the future. Initially, the useful life of HDPE pipes had been estimated at 50 years; however, based on the condition of their HDPE pipes after 14 years in service, the Philadelphia Power and Light predicts that the pipes will last 100 years.

Storage and handling of HDPE pipes should follow the manufacturer's specifications. ADS recommends the following precautionary measures in handling and storage of HDPE pipes:

- Stack pipes less than 6 ft (1.8 m) high.
- Alternate bells for each row of pipe. Figure 9.7 shows stacked 24 in. solid HDPE pipes for an underground detention basin application.
- Do not drop the pipe from delivery trucks into an open trench or onto uneven surfaces.
- Avoid dragging the pipe across ground or striking the pipe against another pipe or object.
- Inspect the pipe and joining system before installation.

Ambient temperature extremes have little effect on the strength of polyethylene pipes. Depending on the product, either carbon black or titanium dioxide is added to the polyethylene to protect against ultraviolet light. However, it is advisable to avoid long-term storage under direct sunlight. At an 80°F (27°C) temperature, the pipe wall temperature can reach 110°F (43°C) if pipe is left in the sun.

Installation of HDPE pipes in trench is similar to, though it requires more care than, installing reinforced concrete pipes. Alignment of pipe is established by field survey, known as construction stakeout. The width of the trench depends on the pipe diameter, backfill material, and the method

FIGURE 9.7   Stacked 24 in. solid HDPE pipes.

of compaction. Generally, 6–8 in. (15–20 cm) on either side of the pipe is the minimum allowable trench width when compaction equipment is not needed. In poor native soils such as peat, muck, or highly expansive soils, a wider trench width than these minimums will be required.

Trench widths for small diameter pipes of all makes are dictated by the bucket size available for the excavation and, in many cases, exceed twice the pipe diameter. Table 9.9 lists the minimum trench width as recommended by ADS. Pipe trench should be excavated with sidewalls nearly vertical. In deep excavation or poor soil conditions, it may be necessary to excavate with sidewalls sloped adequately. If this results in an excessively wide trench, trench boxes will be used. The length of the box should be suitable for the pipe length. As indicated, standard length of reinforced concrete pipes is 8 ft (2.44 m) and ADS N-12 IB pipe is 6 m (19.7 ft); however, 14 ft (4.25 m) length can be ordered for shorter trench boxes.

To prevent disruptions of backfill envelope, the bottom of the trench box should be placed no higher than 24 in. (0.6 m) from the bottom of the trench. This may require raising the trench box during pipe installation to conform with the previously indicated spacing, which is an OSHA regulation. To properly backfill and compact the soil around the pipe, the trench box may be dragged along the trench—but only if it does not damage the pipe or disrupt the backfill. In some cases it may be necessary to move the trench box two or three times to achieve the required compaction of the soil envelope. Figure 9.8 shows a typical subtrench installation. For more information on the use of trench boxes, the reader is referred to technical note TN 5.01 (March, 2009) in the ADS Drainage Handbook (2014).

TABLE 9.9
Minimum Trench Width for HDPE Pipes

Nominal Pipe Diameter, in. (mm)	Minimum Trench, in. (m)	Nominal Pipe Diameter, in. (mm)	Minimum Trench, in. (m)
4 (100)	21 (0.5)	24 (600)	48 (1.2)
6 (150)	23 (0.6)	30 (750)	56 (1.4)
8 (200)	26 (0.7)	36 (900)	64 (1.6)
10 (250)	28 (0.7)	42 (1050)	72 (1.8)
12 (300)	30 (0.8)	48 (1200)	80 (2.0)
15 (375)	34 (0.9)	54 (1350)	88 (2.2)
18 (450)	39 (1.0)	60 (1500)	96 (2.4)

Source: ADS, Drainage Handbook, Installation 5-1, May 2012.

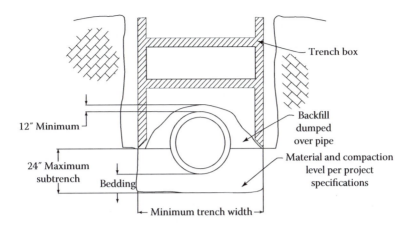

FIGURE 9.8 Subtrench installation.

When installing pipe in rocky trench, a minimum of 12 in. (0.3 m) of acceptable backfill material should be placed below the bottom of the pipe to provide a cushion between the pipe and the rock. In soft, unstable soils, the bedding should be at least 2 ft (0.6 m) thick. The minimum width of trench for installation in loose soil is pipe diameter plus 4 ft (1.2 m).

In suitable soil conditions, a minimum of 4–6 in. (10–15 cm) of bedding should be placed and compacted on the trench to equalize load distribution along the bottom of the pipe. Upon the placement of the pipe, backfill is placed in the trench. Acceptable backfill materials and compaction are very similar and, in many cases, identical to those for other types of pipes. Figure 9.9 shows a typical backfill structure. Appendix 9A provides a description of backfill material, placement, and compaction of HDPE pipes.

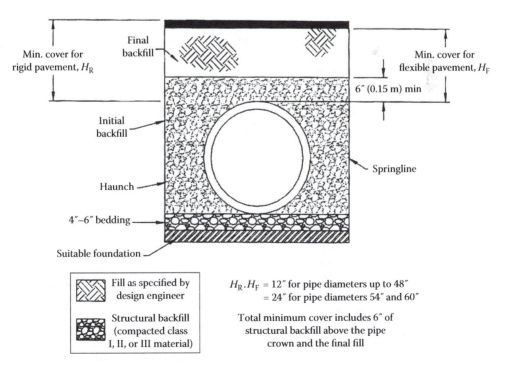

FIGURE 9.9 Typical backfill detail.

TABLE 9.10

Maximum Cover for ADS N-12, N-12 ST, and N-12 WT Pipes, ft (m)

Diameter in. (mm)	Class 1		Class 2			Class 3	
	Compacted	Dumped	95%	90%	85%	95%	90%
4 (100)	37 (11.3)	18 (5.5)	25 (7.6)	18 (5.5)	12 (3.7)	18 (5.5)	13 (4.0)
6 (150)	44 (13.4)	20 (6.1)	29 (8.8)	20 (6.1)	14 (4.3)	21 (6.4)	15 (4.6)
8 (200)	32 (9.8)	15 (4.6)	22 (6.7)	15 (4.6)	10 (3.0)	16 (4.9)	11 (3.4)
10 (250)	38 (11.6)	18 (5.5)	26 (7.9)	18 (5.5)	12 (3.7)	18 (5.5)	13 (4.0)
12 (300)	38 (11.6)	18 (5.5)	26 (7.9)	18 (5.5)	13 (4.0)	19 (5.8)	14 (4.3)
15 (375)	42 (12.8)	20 (6.1)	28 (8.5)	20 (6.1)	14 (4.3)	20 (6.1)	15 (4.6)
18 (450)	35 (10.7)	17 (5.2)	24 (7.3)	17 (5.2)	12 (3.7)	17 (5.2)	12 (3.7)
24 (600)	30 (9.1)	15 (4.6)	21 (6.4)	15 (4.6)	10 (3.0)	15 (4.6)	11 (3.4)
30 (750)	25 (7.6)	12 (3.7)	18 (5.5)	12 (3.7)	8 (2.4)	13 (4.0)	9 (2.7)
36 (900)	29 (8.8)	13 (4.0)	20 (6.1)	13 (4.0)	9 (2.7)	14 (4.3)	9 (2.7)
42 (1050)	27 (8.2)	13 (4.0)	19 (5.8)	13 (4.0)	8 (2.4)	13 (4.0)	9 (2.7)
48 (1200)	25 (7.6)	12 (3.7)	17 (5.2)	12 (3.7)	7 (2.1)	12 (3.7)	8 (2.4)
54 (1350)	26 (7.9)	12 (3.7)	18 (5.5)	12 (3.7)	7 (2.1)	12 (3.7)	8 (2.4)
60 (1500)	29 (8.8)	13 (4.0)	20 (6.1)	13 (4.0)	8 (2.4)	14 (4.3)	9 (2.7)

Source: Advanced Drainage Systems (TN 2.01, September 2014).

Notes:

1. Results based on calculations shown in the "Structures" section of the ADS Drainage Handbook (v20.2). Calculations assume no hydrostatic pressure and a density of 120 pcf (1926 kg/m^3) for overburden material.
2. Installation assumed to be in accordance with ASTM D2321 and the installation section of the Drainage Handbook.
3. For installations using lower quality backfill materials or lower compaction efforts, pipe deflection may exceed the 5% design limit; however, controlled deflection may not be a structurally limiting factor for the pipe. For installations where deflection is critical, pipe placement techniques or periodic deflection measurements may be required to ensure satisfactory pipe installation.
4. Backfill materials and compaction levels not shown in the table may also be acceptable. Contact ADS for further details.
5. Material must be adequately "knifed" into haunch and in between corrugations. Compaction and backfill material is assumed uniform throughout the entire backfill zone.
6. Compaction levels shown are for standard Proctor density.
7. For projects where cover exceeds the maximum values listed here, contact ADS for specific design considerations.
8. Calculations assume no hydrostatic pressure. Hydrostatic pressure will result in a reduction in allowable fill height. Reduction in allowable fill height must be assessed by the design engineer for the specific field conditions.
9. Fill height for dumped class I material incorporates an additional degree of conservatism that is difficult to assess due to the large degree of variation in the consolidation of this material as it is dumped. There are limited analytical data on its performance. For this reason, values as shown are estimated to be conservatively equivalent to class 2, 90% SPD.

ADS recommends a minimum of 1 ft (30 cm) cover on 4–48 in. (100–1200 mm) pipes and 2 ft (60 cm) for 54 and 60 in. pipes. Considering any deviation from the manufacturer's installation specification, a 1.5 ft (45 cm) minimum cover is recommended for pipes smaller than 48 in. Tables 9.10 and 9.11 specify the maximum cover on ADS N-12 pipe and single-wall, heavy-duty pipe, respectively. These tables include data for classes 1, 2, and 3 bedding.

When groundwater is present in the trench, it is necessary to dewater in order to maintain stability of the in situ and impacted bedding and backfill materials. The water level in the trench should be kept below the bedding during the pipe installation. A concern with HDPE (and other lightweight material) pipes in high water table areas is floatation due to uplift forces. The condition of stability should be considered when the pipe is empty. Assuming a soil friction angle of 36.8° and specific weight of 120 pcf (1920 kg/m^3, 19 kN/m^3) for backfill material, Hancor (acquired by ADS in 2005) has performed calculations for the amount of minimum cover; the results are presented in Table 9.12.

TABLE 9.11

Maximum Cover for ADS Single Wall Heavy Duty and Highway Pipes, ft (m)

Diameter, in. (mm)	Class 1		Class 2			Class 3		
	Compacted	Dumped	95%	90%	85%	95%	90%	85%
4 (100)	41 (12.5)	13 (4.0)	27 (8.2)	18 (5.5)	13 (4.0)	19 (5.8)	13 (4.0)	11 (3.9)
6 (150)								
8 (200)	38 (11.6)	12 (3.7)	25 (7.6)	17 (5.2)	12 (3.7)	18 (5.5)	12 (3.7)	10 (3.0)
10 (250)								
12 (300)								
15 (375)								
18 (450)								
24 (600)	32 (9.8)	11 (3.4)	21 (6.4)	15 (4.6)	11 (3.4)	16 (4.9)	11 (3.4)	9 (2.7)

TABLE 9.12

Minimum Recommended Cover to Prevent Flotation

Nominal Diameter, in. (mm)	Minimum Cover, in. (cm)[a]	Nominal Diameter, in. (mm)	Minimum Cover, in. (cm)
6 (150)	4 (10)	24	17 (43)
8 (200)	5 (13)	30	22 (56)
10 (250)	7 (18)	36	25 (64)
12 (300)	9 (23)	42	29 (74)
15 (375)	11 (28)	48	33 (84)
18 (450)	13 (33)	60	40 (101)

Note: For structural purposes, a minimum cover of 12 in. (30 cm) is required for 4–48 in. pipe, and 24 in. (60 cm) for 54–60 in. pipe.

[a] Rounded to nearest full centimeter.

More information on technical specification of HDPE pipes can be found in http://www.ads-pipe.com.

9.2.6 Dewatering

In areas of high water table, dewatering may be required to maintain a dry trench during pipe installation. Groundwater conditions should be investigated prior to excavation. Test borings may be performed to determine the depth, the rate, and the direction of groundwater flow. Groundwater is usually controlled by one of or a combination of the following methods:

- Drains
- Tight sheeting
- Pumping
- Well points

Drains and tight sheeting are practical when the water table is not significantly above the bottom of the excavation. Pumping or well point should be used when the bottom of the excavation is deeply submerged. Figure 9.10 shows a direct pumping from sump. In this dewatering method, the pump should be submersible or self-priming so that intermittent flows can be discharged. Centrifugal pumps are best suited for larger quantities of water. In mud slurry conditions, diaphragm pumps should be used. Regardless of the specific type of pump, a standby pump should be available in case the operating pump clogs.

Sump pumps lower the water table in a localized area (see Figure 9.11). For lowering the water table along the pipe alignment, a well point system is used (see Figure 9.12). This dewatering method provides an effective means of controlling groundwater in permeable soils and may eliminate the need for sheeting and shoring. A well point system consists of 1 or 2 in. pipes (well points), which are driven vertically into the wet soil. The well points are connected to a leader pipe by swing joints, which facilitate placement of well points. The spacing and number of well points are calculated based on the soil permeability and the amount of water to be pumped.

FIGURE 9.10   Sump pump dewatering pipe installation in a high water table site. (Photo by the author.)

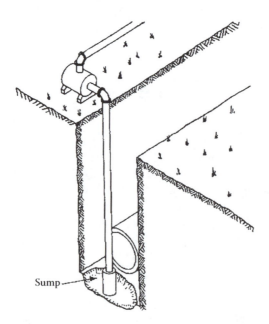

FIGURE 9.11   Dewatering by sump pump.

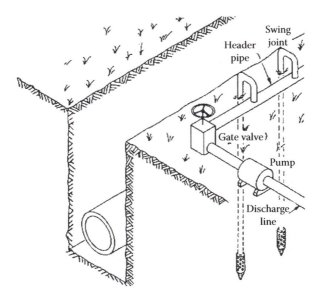

FIGURE 9.12   Well point system.

Spacing of 3 ft (1 m) of well points is common for a 6 in. (15 cm) diameter header pipe and 2 ft for 8 in. (20 cm) header pipe. In long header pipes, gate valves are installed at 100–200 ft (30–60 m) intervals. Significant lowering of the water table may result in subsidence of the ground in the surrounding area and structures in close proximity of the dewatered area. In such applications, a geotechnical investigation should be performed and adequate precautions taken.

9.3   WATERTIGHT JOINTS

To prevent leakage, pipe joints should be watertight. Also, the connection of pipes to manholes/inlets should be made leak free.

9.3.1   PIPE JOINTS

A variety of HDPE pipes come standard with watertight joints. HDPE double-walled smooth interior (N-12) pipes come in N-12 WT 1B and N-12 ST 1B, representing watertight and soil-tight joints, respectively. These joints are shown in Figure 9.13. A soil-tight joint is adequate for storm drains and watertight joints are used for both storm drains and sanitary sewers.

O-ring and bell and spigot are used to make a watertight joint in concrete pipes. Figure 9.14a and b show two variations of O-ring joints. In practice, mortar is also used at joints. Joints employing mortar sealants are rigid, and any movement or deflection after installation may cause cracks and leakage.

An external jointing system for concrete pipes utilizes external rubber and mastic sealing band conforming to ASTM 877. Figure 9.15 shows an external joint. Generally limited to a noncircular pipe with tongue and groove configurations, external sealing bands provide resistance to external loads normally encountered in drainage pipes.

Historically, watertight concrete pipes and structures have been associated only with sanitary sewer systems. Over the past two decades, environmental regulations have resulted in the need for watertight storm drains. However, mortar is still employed often at concrete pipe joints.

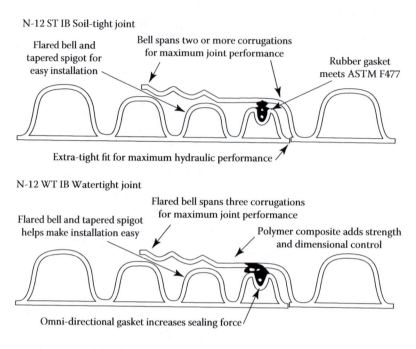

FIGURE 9.13   ADS watertight and soil-tight joints. (From Advanced Drainage Systems.)

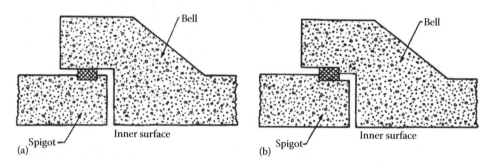

FIGURE 9.14   O-ring joints in concrete pipe: (a) opposing shoulder with O-ring and (b) spigot groove with confined O-ring. (From American Concrete Pipe Associations, Concrete Pipe Handbook.)

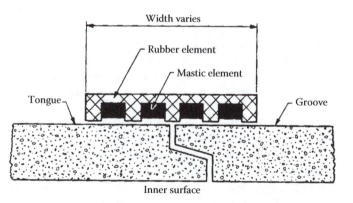

FIGURE 9.15   External sealing band for noncircular concrete pipes. (From American Concrete Pipe Association.)

9.3.2 Pipe Connection to Manhole/Inlet

Generally, there is a much larger gap at the connection of pipes to manholes or inlets than the pipe joints themselves. Therefore, it is important to provide a watertight connection between pipes and manholes or inlets.

In practice, the joints are filled or covered by mortar. Figure 9.16 shows a typical pipe-to-inlet connection in a shallow installation. Due to settlement, the mortar becomes cracked or broken and water begins to leak from the joint. Over time, the leakage washes the soil away and creates a cavity at the connection. The result is a collapse of pavement and sinking of the ground around the manhole inlet. This problem is especially the case in steep areas where the water flows at a high velocity in the pipe. Figure 9.17 depicts this problem in a fairly steep street in a suburban town in northern New Jersey. To avoid this common problem, a flexible gasket providing a watertight seal should be employed between concrete manhole/inlet and pipes of any make.

Press-Seal Gasket Corporation, headquartered in Fort Wayne, Indiana, is one of the larger manufacturers of flexible seal for both reinforced concrete and HDPE pipes. One such seal, known as "Cast-A-Seal 802," is a cast-in connector for straight wall application for any pipe make of 18 in. (375 mm) and larger. See Figure 9.18, A Lok Products, Inc., in Tullytown, Pennsylvania, is another company that manufactures all sizes and styles of connections for flat and curved walls. Figure 9.19 shows a curved wall connector. Trelleborg Pipe Seals of Milford, New Hampshire, is another company that manufactures flexible pipe to manhole connectors. This company offers

FIGURE 9.16   Pipe connection to an inlet using mortar. (Photo by the author.)

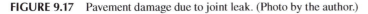

FIGURE 9.17   Pavement damage due to joint leak. (Photo by the author.)

FIGURE 9.18   Cast-A-Seal, boot-type pipe to structure connector. (Made by Press-Seal Gasket Corp.)

Curved wall field sleeve
with A·Lok xcel connector

FIGURE 9.19   A·Lok's curved wall connector. (From A-Lok Products, Inc.)

two types of rubber connections formerly made by NPC, one of which is called NPC Kor-N-Seal I, which is more often employed in sewer pipes than storm drains.

A seal, called ADS Pipe Adapter, is specifically made for ADS N-12 drainage (and sanitary) HDPE pipes up to 30 in. size. This adapter, also manufactured by Press-Seal Gasket Corporation, consists of a rubber ring with corrugated interior, which fits into the HDPE pipe corrugation, and a smooth exterior, which connects to a manhole or other structures using PSX-Direct Drive or other flexible connectors. Figure 9.20 shows this adapter.

Connectors cost well under 5% of total job cost, but save significant repair costs later. Stopping a leak alone can be expensive, but repairing a pavement failure is costly. A single pavement damage due to subsidence can cost several thousand dollars to repair, plus the danger and disruption of traffic. A flexible joint, not only at structures but also in pipes, is an important advantage. Sometimes the soil material around a joint settles, and with a grouted connection there is a risk of disconnection or crack of the pipe. To ensure a watertight storm drainage system, an infiltration/exfiltration test may be conducted as described next.

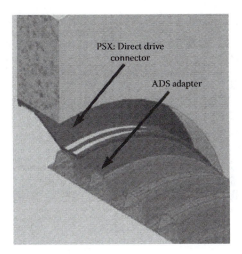

FIGURE 9.20   Rubber adapter for ADS N-12 drainage (and sanitary) corrugated pipes. (Made by Press-Seal Gasket Corp.)

9.3.3 INFILTRATION/EXFILTRATION TESTING

For pipes in watertight applications without specifying any ASTM specification for testing, an infiltration/exfiltration test is a simple and easy method of ensuring proper joint performance. In an infiltration/exfiltration test, a reach of piping is tested by filling the system with water from structure to structure (manhole or inlet), allowing the system to stabilize for 24 hours, measuring the water level, and then measuring the water level again after a period of time. The drop of water level can then be converted to gallons of leakage/inch pipe diameter/mile length of pipe/day and compared to the permissible level established for the project. In the absence of a specified level, 200 gal/in. dia./mi/day (18 L/mm dia./km/day) is commonly considered watertight for storm sewer applications.

9.4 CONSTRUCTION OF DETENTION BASINS/PONDS

In this section, construction of detention basins, ponds, infiltration basins, and grass swales, which are the most common structural storm water best management practices, is discussed.

9.4.1 DETENTION BASINS

Dry detention basins are commonly constructed through excavation below grade. Due to topography, an embankment may have to be erected on the downhill side of the detention basin. To avoid unnecessary jurisdictional compliance, the height of the embankment should be kept below the limit above which the embankment is classified as a dam.

The side slopes of the detention basin should be selected considering stability and maintenance. To facilitate mowing, grass-lined banks should be at slopes less than 3:1 (3 H, 1 V). In stone-lined or riprap-lined detention basins, the banks should be no steeper than the angle of response of granular material, but no more than 2:1. The bottom of detention basins should preferably be covered with vegetation to improve water quality and to enhance infiltration. (See Figure 7.1 in Chapter 7.)

Detention basins should be provided with a low-flow channel to convey low flows from the location of inflow pipe to outlet structure. Riprap stone is preferable to concrete for lining low-flow channels. By draining the bottom of the basin after a storm, riprap lining prevents soggy conditions

FIGURE 9.21   A detention basin covered with overgrown vegetation. (Photo by the author.)

in the basin. This will facilitate mowing the bottom and banks of the basin, which in turn avoids overgrowth of vegetation. Figure 9.21 depicts overgrowth of vegetation in a detention basin without a low-flow channel and neglected maintenance just 5 years after its construction. In soils of poor permeability, a subdrain may be installed below the bottom of the basin to help drain the basin.

All detention basins and ponds should be provided with an emergency outlet or spillway to safely release storms in excess of the design frequency. Also, an access driveway should be provided atop all basins for maintenance of outlet structure and forebays, if any. In addition, a maintenance right-of-way (ROW) or easement should be provided from a public or private road. Maintenance access should be at least 12 ft (4 m) wide and have less than 15% slope.

9.4.2 Infiltration Basins

Infiltration basins, similarly to detention basins, are constructed by excavation and erecting a low embankment, if necessary. Natural depressions can serve as infiltration basins. These basins may be placed online or off-line relative to the drainage route. In an online arrangement, the basin is designed to fully capture a selected storm, typically the water quality storm and overflow when a larger storm occurs. Off-line infiltration basins are designed to receive only the water quality storm with the larger storms bypassed either downstream or into a detention basin. In either case, the basin is relatively shallow, commonly 2–3 ft (0.6–0.9 m).

Infiltration basins may be designed to serve as a dual-purpose infiltration–detention basin. In this type of basin, the water quality storm is retained below the lowest opening in the outlet structure so that it is infiltrated slowly into the ground. The stored volume above the water quality storm level is discharged in a regulated manner through an outlet structure. Case Study 7.6 (Chapter 7) presented design calculations for this type of infiltration–detention basin.

Most failures in infiltration basins stem from lack of prior soil testing, poor construction practices resulting in over compaction, and premature silting. To avoid any loss of permeability due to compaction, excavation should be performed from the perimeter of the basin, avoiding heavy machinery entering the excavation site. Thus, the bottom width of aboveground infiltration basins should be limited to 30 ft (10 m) when a backhoe is used for excavation. For wider basins, draglines or other longer reaching arm machinery or, alternatively, light machinery should be employed. To prevent premature siltation, the use of infiltration basins as temporary siltation basins should be avoided to the maximum extent feasible. If due to a lack of room, an infiltration basin has to be used during construction, it must be thoroughly cleaned of silt and sediment before it is lined with a sand layer.

To be functional, infiltration basins should only be considered and constructed in areas where soil permeability is at least 1 in./h (25 mm/h). Also, the bottom of infiltration basins should be at least 2 ft, but preferably 3 ft (1 m) above the seasonal high water table. The bottom of infiltration basins may be covered with vegetation or sand. In the latter case, the sand layer should be at the minimum 6 in. (15 cm) but preferably 1 ft (30 cm) thick to enhance infiltration. A soil log and permeability test should be conducted at the location of any infiltration basin to determine that soil is sufficiently permeable and the water table will not be encountered.

Typical problems associated with infiltration basins are standing water, soggy surfaces, sedimentation, and inadequate access. The first two problems usually arise from either insufficient soil permeability or the loss of infiltration capacity due to deposition of silt on the bottom of the basin. In the latter case, the soil infiltration capacity may be restored by scraping the soil from the bed or removing and replacing the sand layer by light machinery. To access the basin for maintenance, ramps sloping 15% or less should be provided to the bottom of the basin. To prolong the longevity of an infiltration basin, it is recommended that runoff receive a pretreatment before entering the basin. The use of inlet filters, particularly in roadways and parking lots, offers a cost-effective means of pretreatment.

9.4.3   WET PONDS

Wet ponds are suitable in soils of very low permeability and/or high water table. In fact, in high water table areas and costal zones, a wet pond can be the most feasible structural BMP for attenuating the peak rate of runoff. In addition to peak flow reductions, wet ponds also improve water quality by slowing down the inflow runoff through the stagnant body of water in the pond. Thus, a wet pond can be far more effective than an extended detention basin in terms of removing suspended sediment matter and improving water quality. Where the water table is not shallow and soil is permeable, the bottom and sides of a wet pond should be lined with an impermeable geofabric material, or a clay liner. Wet ponds are commonly constructed through excavation with little or no embankment. Since the water fluctuates due to filling and drawdown following a storm, the banks of a wet pond should be constructed at a slope significantly less than the angle of repose of native soil. To stabilize soil, banks may be covered with stone lining. Figure 9.22 shows a pond with rock-lined banks designed by the author.

9.4.4   GRASS SWALES

A grass swale can be an effective means of removing sediment and pollutants provided that it is properly designed and maintained. To facilitate mowing, grass swales should be constructed having a parabolic section with sides at 3:1 (3 H, 1 V) slope or flatter. To prevent overgrowth of

FIGURE 9.22   A rock-lined pond (designed by the author).

vegetation, grass swales should be planted with grass suitable for mowing. However, grass should be left long enough to filter the water quality storm. Grass swales are more practical where there is sufficient rain during the growing season. In dry weather conditions, periodic irrigation is necessary for healthy grass growth. In the southwest and other arid or semiarid parts of the country, stone is far more practical than grass as a cover for roadside channels.

9.4.5 Dry Wells and Infiltration Chambers

A dry well, also called a seepage pit, generally consists of a hollow concrete cylinder installed in stone trenches (see Figure 9.23). Figure 9.24 shows installation of seepage pits in stone trench. Dry wells filled with stone are also used in practice. Figure 9.25 depicts this arrangement. However, filling a dry well results in a loss of its internal storage by nearly 60%.

Perforated pipes or chambers in stone trenches also serve the same purpose. Some of the most widely employed chambers include StormTech, StormChamber, Cultec, and Triton, which were discussed in Chapter 7. Dry wells are typically designed to accept the storm water runoff from roofs and other impervious areas. To prevent their premature clogging due to leaves, grass clippings, and debris, seepage pits should preferably be used for roof runoff and roof leaders equipped with a screen. In addition, seepage pits should be provided with a solid cover to prevent runoff from lawn and paved areas entering them. When seepage pits and infiltration chambers are used in a parking area, inlets should be equipped with filters such as FloGard+Plus® from Oldcastle® Stormwater Solutions or the like to remove silt, debris, oil, and grease from runoff. Design criteria for seepage pits and retention–infiltration basins were discussed in Chapter 7.

A soil log and percolation test should be performed at the location of a dry well or underground retention chamber to confirm the occurrence of suitable soil and absence of groundwater.

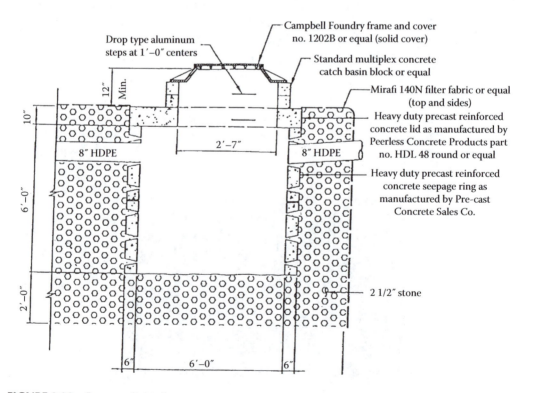

FIGURE 9.23 Seepage pit detail.

FIGURE 9.24   Seepage pits being installed in stone trench. (Photo by the author.)

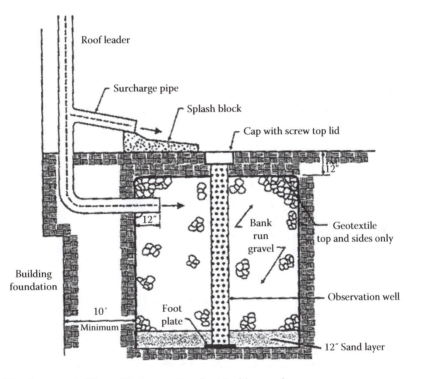

FIGURE 9.25   Seepage pit filled with stone and equipped with a surcharge.

Following the completion of construction, dry wells should be vacuum cleaned; if deemed necessary, they should be tested to ascertain their actual performance. If tests show that seepage pits fail to completely drain within 48 hours, other measures should be designed for the control of storm water runoff.

9.4.6   Outlet Structures

Outlet structures should be readily accessible for maintenance. They should have a minimum on-plain dimension of 4 × 4 ft (1.25 × 1.25 m) to enable a maintenance crew to enter. Upstream from

FIGURE 9.26   Trash rack with parallel bars on an outlet structure. (Designed by the author.)

an outlet, the basin should be lined with riprap, gabion mattress, or concrete to provide firm footing for maintenance operations.

Outlets for all types of storm water management facilities should be equipped with trash racks of properly sized openings. The openings should be small enough to collect debris that may block the outlet pipe, but must be sufficiently spaced to permit passage of leaves and small debris, which are not likely to clog the outlet pipe. This will minimize the frequency of cleaning and maintenance. If trash rack openings are too small, they become clogged by leaves and debris and cause frequent flooding. A review of outlet structures of detention basins is presented in an ASCE (1985) publication.

In a development project comprising detention basins and ponds, the contractor installed aluminum grids on the outlet structures pursuant to a request by the developer. After a heavy midfall rainstorm, the tree limbs and fallen leaves completely blocked trash racks, which had 2.5 in. square openings, and as a result, the basins were flooded out. With the buildup of pressure against the outlet, some of the racks were snapped off from the structures and carried through the outlet pipes downstream. Following the incident, typical trash racks of proper sizing were designed and installed on all the outlet structures. Figure 9.26 shows a typical trash rack designed by the author.

Prefabricated plastic trash racks are available for a variety of applications. Plastic Solutions, Inc., makes various HDPE trash racks, commercially known as StormRax™ (http://www.plastic -solutions.com). StormRax is also available at Contech.

To lessen maintenance, outlet structures should be designed with openings large enough to avoid frequent clogging. Regulations in many states neglect this important maintenance consideration, allowing use of small orifices in detention basins. In New Jersey, for example, orifices as small as 2.5 in. are allowed not only in aboveground but also in underground detention basins. Such small orifices are highly vulnerable to clogging in underground detention systems where the clogging can be left undetected. The detention basin may be partly, if not almost completely, filled with water when a storm occurs. Considering proper functioning and ease of maintenance, the author recommends using a 6 in. (150 mm) orifice as the smallest opening in any underground detention system. Following this recommendation, 6 in. orifices were substituted for 2.5 to 4 in. orifices in hundreds of underground detention basins in New Jersey.

9.5   SLOPE STABILIZATION

In steep slopes and sandy soils the land is vulnerable to erosion. As indicated in a previous chapter, erosion control blankets (ECBs) and turf reinforcement mats (TRMs) are employed to establish vegetation and control erosion in such cases. ArmorMax, manufactured by Propex, is a high-performance (HP) TRM with a life expectancy of 50 years or longer. This TRM serves as an effective erosion control measure and can also be applied in channels. Appendix 9B includes installation guidelines for ArmorMax.

9.6 INSPECTION AND MAINTENANCE

Addressing current storm water management regulations requires the provision of storm water management facilities in urban development projects in all parts of the country. Even single-family homes, when exceeding a threshold limit, are subject to these regulations in some states. Regardless of their size, type, and physical features, all storm water management facilities require regular cleaning and proper maintenance to function effectively. Also, installation of storm water management systems should be inspected to ensure that they are properly constructed. Appendix 9C includes a checklist of construction inspection for drainage systems.

9.6.1 Objectives of Inspection and Maintenance

The primary objective of regular inspection is to ensure that storm water management facilities operate satisfactorily and safely. Storm water management facilities should be inspected regularly by qualified personnel. The inspection should include all storm water management elements, including conveyance system, detention basins of any type, outlet structures, and water treatment devices. Nonstructural systems, such as pervious pavers, rain gardens, and green roofs, need to be inspected as well. Inspections provide information on the effectiveness of scheduled maintenance procedures and changes in scope and scheduling that are warranted. In addition to storm water management facilities, vegetative areas tributary to drainage systems and detention basins should be inspected and maintained regularly.

Conveyance systems generally include inlets, manholes, pipes, and grass swales/stone channels. All of these elements should be inspected regularly and cleaned of silt and floating debris such as leaves, paper, bottles, and cans, as necessary. The frequency of inspection depends on the location where the system is installed and the amount and type of impurities likely to be carried by runoff. Inspection should be conducted at least twice annually; however, in places such as shopping centers and supermarket parking lots, drainage systems should be inspected four times or more a year.

The purpose of maintenance is to keep the storm water management facilities effective, operational, and safe. Regular maintenance of storm water management facilities will also minimize the need for major, costly repairs. Maintenance may be divided into preventative (or regular) maintenance and repairs. Regular maintenance is intended to prevent a breakdown or failure and includes vegetative maintenance and facilities maintenance.

There is a general neglect in maintenance of storm water management systems; see Pazwash (1991) and Bryant (2004). Because of the lack of general understanding of the impact of urban developments on water quality, the public does not consider maintenance of storm water management facilities to be important. Also, because many of the storm water management facilities are hidden underground and out of site, they remain unattended. As a result, many private homeowners, homeowner associations, and even municipal officials neglect to maintain drainage and storm water management systems. Figure 9.27 shows an inlet grate in a park almost completely clogged by leaves and debris and Figure 9.28 depicts a municipal drainage pipe at a headwall, over two-thirds filled with silt. Figures 9.29 and 9.30 exemplify an unattended, totally ineffective sand filter in a residential development in New Jersey. This sand filter was installed to address the state's water quality requirements. However, due to a general lack of maintenance, the sand had become fully clogged, bypassing the runoff through an internal overflow grate without getting any treatment.

Delaware DOT's Nonpoint Pollutant Discharge Elimination System (DelDOT's NPDES) Group quantifies the benefits of good housekeeping practices and nonstructural BMPs. These practices

FIGURE 9.27   Inlet grate, clogged by leaves. (Photo by the author.)

FIGURE 9.28   Drainage pipe, over two-thirds filled with silt. (Photo by the author.)

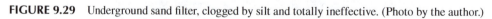

FIGURE 9.29   Underground sand filter, clogged by silt and totally ineffective. (Photo by the author.)

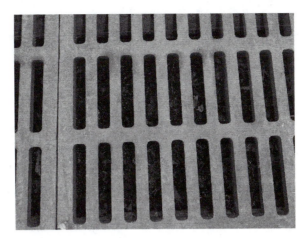

FIGURE 9.30 Grates on a sand filter. (Photo by the author.)

include cleaning inlets, pipes, and detention/retention basins; street sweeping; and inspection of infrastructures. In addition to maintenance, the NPDES group presents public education programs to convince citizens not to overwater their lawns or dump their used oil in storm drain systems. These measures help protect such important watersheds as the Delaware River and Chesapeake Bay (Keating, 2005).

The New Jersey storm water management regulations, adopted on February 2, 2004, included a municipal storm water program that required all MS4s municipalities (559 in all) to clean and maintain their drainage systems within 5 to 6 years depending on the size of their system. By June 2009, over 675,000 inlets had been cleaned and nearly 680,000 tons of trash, debris, and silt been removed from storm sewers. These figures demonstrate the need for and importance of good housekeeping and maintenance of drainage and storm water management facilities. A brief review of maintenance measures for storm water management facilities is presented in the following sections. For more information, the reader is referred to ASCE manuals (1992), ASCE standards (2006), Brzozowski (2004), and Barron and Lankford (2005).

9.6.2 Maintenance of Vegetative and Paved Areas

9.6.2.1 Lawns/Landscapes

Lawns, landscapes, trees, and shrubs that are located within the drainage area of a detention, retention, or infiltration basin or pond should be maintained to minimize soil erosion. This will decrease deposition of silt in inlets, pipes, and detention basins. Lawns should be mowed regularly during the growing season to maintain the grass at 2 to 3 in. high. Fertilizers should be applied at a minimum rate required and in no case more than 10 lb per 1000 square feet (5 kg/100 m^2). Damaged and dead lawn should be repaired and bare areas vegetated. If the season prevents the reestablishment of grass, exposed areas should be covered with salt, hay, mulch, or straw. Where a reseeding is not effective in establishing a nonerosive vegetative cover, soil should be protected by other materials including sod, erosion control blankets, riprap, or gravel.

The plant beds should be mulched with hardwood every 2 years in order to retain moisture around the root zones and to provide a growth medium for shrubbery. Shrubs and trees should be pruned to maintain shape and appearance: trees once a year (preferably in early spring) and shrubs regularly, as needed. For more information on maintenance frequency, the reader is referred to Reese and Presler (2005) and Kang et al. (2008).

FIGURE 9.31   Street dirt from 100 ft of curbside. (Photo by the author.)

9.6.2.2 Pavements

Management measures for parking lots and access driveways include the removal of sediment debris and other pollutants. Paved areas should be swept regularly (at least once every 3 months) and the swept material must be disposed of properly. Figure 9.31 shows dirt the author swept from approximately 30 m (100 ft) of curb along his property. This dirt, weighing nearly 20 kg (45 lb), was accumulated within 6 months.

Porous asphalt and concrete and permeable interlocking concrete pavers can become clogged with sediment over time and this reduces their infiltration rate and decreases their storage capacity. To restore their infiltration, the sediment should be removed from the surface of permeable pavers.

Pervious pavement of any type must be inspected for clogging and excessive sediment accumulation at least twice a year as well as after every intense rainstorm—1 in. of rainfall in less than 1 hour in New Jersey. It is best to inspect porous asphalt, pervious concrete, or permeable interlocking concrete pavements (PICPs) immediately after a heavy rainfall. Any standing water indicates the need for desilting the pavement. The permeability can also be measured by conducting an infiltration test using ASTM C1701 standard test method for infiltration rate of in-place pervious concrete. The Interlocking Concrete Pavement Institute (ICPI) recommends cleaning if the tested surface infiltration rate falls below 10 in./h (250 mm/h).

Sediments are most effectively removed by vacuum type street cleaning equipment without brooms and water sprayer. This equipment can remove the top 1 in. (25 mm) of sediment. Regenerative air sweepers (i.e., those that blow air across the pavement surface) are not recommended. Vacuuming should be performed at least twice a year, but preferably three times a year (once during dry, warm weather). Disposal of sediment should follow all applicable local and state waste regulations. The author uses water blasting (low-velocity pressure washing) to clean out the gaps in his backyard pavers once or twice a year.

9.6.3 Maintenance of Storm Water Drainage Systems

9.6.3.1 Restoration of Grass- and Riprap-Lined Swales

Grass and riprap swales should be inspected for erosion at least once a year and maintained as necessary. Maintenance of grass includes mowing on a regular basis and seeding/sodding eroded areas as needed. Regular mowing ensures that large storms do not create a soggy condition due to vegetation overgrowth. The stone-lined swale maintenance includes regrading banks and replenishing eroded areas with new stone.

FIGURE 9.32   Floatables trapped by curb piece. (Photo by the author.)

9.6.3.2   Snow and Ice Removal

Accumulation of snow and ice hinders the functioning of inlets, conveyance systems, and outlet structures. Snow should be removed from impervious areas to assure the facilities will be functional during the winter season.

9.6.3.3   Removal of Sediment and Floatables from Drainage Systems

Storm water management elements are expected to receive and trap sediment and trash. Inlets, pipes, drainage swales, outlet structures, and water quality filter inserts should be inspected at least twice a year as well as after every major or intense storm (1 in. in less than 1 hour in the East, South, Northwest, and Midwest and flash floods in the arid Southwest). The sediment and trash should be removed and disposed of in accordance with local regulations.

Removal of silt and debris from inlets and manholes may be performed either manually using a shovel or by a vacuum truck. Pipes and culverts can be cleaned by flushing, vacuuming, or a combination thereof. New designs of curb pieces in combination inlets trap floatables such as bottles, cans, paper cups, and the like that would otherwise enter the inlets (see Figure 9.32).

9.6.3.4   Control of Potential Mosquito Breeding Habitats

Stagnant bodies of water, such as ponds and wetlands, are potential habitats for mosquito breeding. Standing water in inlets and outlet structures is also a source of mosquito culture. Removal of all obstructions in inlets and drainage conveyance systems helps avoid creation of mosquito breeding areas. Eliminating potential breeding areas is preferable to application of chemicals to control mosquitoes.

9.6.4   Maintenance of Ponds/Detention Basins

9.6.4.1   Algae and Weed Control

Shallow ponds and detention basins with prolonged standing water are vulnerable to weed and algae growth. Excessive algae growth results in oxygen depletion, causing the development of anaerobic conditions. Low oxygen results in the emission of foul odors and an unpleasant scene. Figure 9.33 shows a stagnant irrigation pond fully covered with algae. Weeds associated with ponds and detention basins may be submergent, floating, or emergent. Submergent vegetation is the most difficult to detect and to control.

Ponds and detention basins should be inspected at least twice annually for algae and plant growth. Algae growth can often result from the misuse of fertilizers on lawns. Proper application of

FIGURE 9.33 Stagnant irrigation pond fully covered with algae. (Photo by the author.)

fertilizers lessens the algae problem. Aeration offers a practical solution to algae growth in ponds. Weeds that become a problem may be cleared by pond maintenance professionals.

9.6.4.2   Underground Detention Basins

Underground detention basins should be inspected for excessive deposition of sediment, grit, and debris at manholes, inspection ports thereon, and outlet structures, and within detention chambers/ pipes. Clogging of perforations in pipes and chambers prevents the utilization of the enveloping stone. The ground above and adjacent to underground detention basins should also be inspected for excessive settlement.

Sediment, debris, and trash should be removed from the underground detention basins and disposed of properly. It is recommended that the basin be dewatered before removing the sediment and debris from it. Entering into underground detention basins, if necessary, should be performed by a person who is trained for a confined space entry and in compliance with the latest OSHA confined space regulations. In addition, specific manufacturers' maintenance procedures should be followed when using underground pipes and chambers.

9.6.4.3   Wet Ponds

A wet pond functions like a continuous sedimentation basin. As more sediment accumulates in the pond, its effectiveness in removal of pollutants is diminished. Pollutants such as phosphorus, which attaches to sediment, can be chemically released under anaerobic conditions. Therefore, to maintain the effectiveness of this type of BMP, the sediment should be removed on a regular basis. Although guidelines generally call for removing sediment when 25% of the permanent pool volume has been lost, the author suggests removing the silt when it is accumulated to 1 ft thick, or 20% depth, whichever is smaller.

9.6.4.4   Outlet Structures

Outlet structures are the most important flow control element in any detention basin, whether aboveground or underground. These structures and their trash racks should be inspected at least four times annually and after every intense storm (1 in. in less than 1 hour in the northeastern United States and many parts of the country, and flash floods in the arid Southwest). Any floating debris, such as brush, leaves, tree limbs, paper, tumbleweeds, and plastic cans, should be removed from trash racks. Also, sediment and obstructions should be removed from outlet structures and, in particular, low-flow openings.

9.6.5   Maintenance of Water Treatment Devices

9.6.5.1   Catch Basin Inserts

Catch basin inserts should be inspected for clogging due to sediment and floatable trash accumulation and cleaned as necessary. It is advisable to perform the inspection initially after installation and two to four times annually during the first year as well as after any major storm event. In Encinitas, California, where 32 Kristar catch-basin inserts were used, it was found that semiannual maintenance is sufficient and that each insert costs approximately $40 to maintain annually (Brzozowski, 2004). Cleaning of catch basin inserts is relatively simple and can be performed by one person. Maintenance basically involves removing the filter insert, emptying its content, replacing oil absorbent, and placing everything back in the inlet. Alternatively, the insert may be cleaned using a vacuum truck. Specific manufacturer's specifications should also be followed. Appendix 9D contains inspection and maintenance specifications for FloGard+Plus® filter, formerly manufactured by Kristar Enterprises of Santa Rosa, California (now a part of Oldcastle Stormwater Solutions). The frequency of cleaning is highly dependent on the location. While in residential developments, it would be sufficient to clean catch basin inserts once or at the most twice a year, the inserts may have to be cleaned at least three or four times annually in shopping center and supermarket parking lots.

9.6.5.2   Manufactured Water Treatment Devices

Many structural storm water management facilities, such as detention basins and ponds, do not provide adequate water quality provisions to address the applicable storm water management requirements for a project. This is particularly the case for underground detention basins, which commonly are neither intended nor suited or accepted for water quality. As indicated in a previous chapter, numerous water treatment devices are available and more are being introduced to the market every year.

Water treatment units include filter media and nonmedia devices. Most nonmedia filter devices, such as CDS, Vortechnics, BaySaver, and many others, are not approved for 80% total suspended sediment removal in New Jersey and many other states. Many of these devices have to be placed off-line, bypassing flows beyond the water quality storm. Filter media devices, which are accepted for 80% TSS removal by many jurisdictional agencies, are commonly employed in connection with and placed past a detention basin. StormFilter, CDS Media Filtration System (MFS), Jellyfish, and AquaFilter are among filter media devices.

The latter devices are generally employed for treatment of small rates of flow, are far more expensive than nonfilter media devices, and require more maintenance. Manufacturers' recommended inspection and maintenance procedures should be followed for each of the water quality devices. Table 9.13 provides a comparison of maintenance costs of some of the filter media MTDs approved for 80% TSS removal by the NJDEP.

9.6.6   Repair of Storm Water Management Facilities

Inlets, pipes, detention basins, water quality units, and outlet structures that are damaged due to floods, settlement, vandalism, or other causes must be repaired promptly. The urgency of the repair depends on the effect of damage on functioning and safety of the element. Major structural repairs should be designed by a professional engineer. For deteriorated concrete and corrugated metal pipes, CentriPipe provides a cost-effective rehabilitation solution. CentriPipe is a trenchless method of slip lining pipes up to 144 in. (3600 mm) in diameter. The method involves centrifugally casting a specialty mortar and admixtures inside the pipe walls. The method can also be applied to elliptical pipe and old brick culverts. CentriPipe is offered by AP/M PermaForm, which also offers other trenchless solutions such as Permacast for ConShield, a biotech armor for sanitary sewer concrete pipes.

TABLE 9.13
Maintenance Cost Estimates of Some Filter Media MTDs

Product Model	StormFilter 96 in. Manhole or 6 × 12 Vault	Jellyfish 72 in. Manhole JF6-3-1	Bayfilter[a] 8 × 10 Vault	UpFlo Flter[a]	AquaFilter[a] AS4 and AF 4.4
					Note: 2 Structures
Flow (cfs) (assumed approx. 1/4 acre site)	0.6	0.6	0.6	0.6	0.6
Flow (gpm)	270	270	270	270	270
Required number of cartridges/bags[b]	12	4	9	14	56
Cost per replacement cartridge/bag	$150	$700	$750	$100	$50
Year 1					
Inspection[c]	$250	$250	$250	$500	$750
Vactor[d]	$–	$1000	$1000	$1000	$1000
Disposal of media/ materials[e]	$–		$1000	$1000	$1000
Replacement media/ materials	$–	$–	$6750	$1400	$2800
Rinse cartridges	$–	$500	$–	$–	$–
Year 2					
Inspection	$250	$250	$250	$500	$750
Vactor	$1000	$1000	$1000	$1000	$1500
Disposal of media/ materials	$1000		$1000	$1000	$1000
Replacement media/ materials	$1800	$–	$6750	$1400	$2800
Rinse cartridges	$–	$500	$–	$–	$–
Year 3					
Inspection	$250	$250	$250	$500	$750
Vactor	$–	$1000	$1000	$1000	$1500
Disposal of media/ materials	$–	$500	$1000	$1000	$1000
Replacement media/ materials	$–	$2800	$6750	$1400	$2800
Rinse cartridges	$–	$–	$–	$–	$–
3-Year cost total	$4550	$8050	$27,000	$11,700	$18,150
20-Year cost total	$43,000	$51,800	$180,000	$78,000	$121,000

Source: Imbrium Systems Corp.

[a] Costs and frequencies are assumed based on available information from these manufacturers. No claims are made to the accuracy of this information.

[b] Assumes largest/highest flow cartridge/bag for each manufacturer.

[c] Inspections assumed to be $250 per event.

[d] Vactor cost factors in the number of structures that need to be cleaned ($1000 for 1, $1500 for 2).

[e] Disposal of materials based on expected material volume and components.

9.6.7 NEGLECT IN MAINTENANCE

While the public can visualize the importance of controlling the runoff to prevent flooding, there is a general lack of understanding of the need for improving water quality. Consequently, many private homeowners do not consider inspection and maintenance of water quality units necessary. An inspection of 29 storm water quality practices revealed that approximately 70% of them were improperly installed and nearly the same percentage of these practices needed maintenance (Bryant, 2004). A statewide survey of various storm water management practices, conducted by the University of Minnesota, indicated that over 84% of the cities conduct routine maintenance once a year or less (Kang et al., 2008). These and other cases indicate a vast negligence in selection and upkeep of storm water management facilities. The following recommendation should help correct the neglect in maintenance (Pazwash, 1991):

- Designers and planners should give a more serious consideration to ease of maintenance of storm water management facilities in general and outlet structures in particular.
- Public and private owners of storm water management systems should pay more attention to their maintenance. We generally tend to neglect that these facilities need to be maintained. While maintaining our cars and homes, we pay no or little attention to drainage facilities we own.
- Homeowners should be educated about the need for assuming the maintenance responsibilities for ditches, swales, and sediment cleanup. A majority of homeowners even fail to consider that such responsibilities rest with them.
- The responsible party (private or public) should be clearly defined and the required maintenance measures should be specified in the design phase and/or upon the construction of a project.
- Practical means of enforcing the maintenance of storm water management infrastructure should be identified.

As indicated in Chapter 8, a large number of municipalities have established storm water management fees to implement necessary maintenance measures. Some others have prepared storm water management agreements to make sure that privately owned storm water management systems would be properly maintained. An example of the latter approach is Douglas County, Georgia, where the county water and sewer authority has created a formal development form to be signed by the owner and developer of every project before a land disturbance permit is issued (Barron and Lankford, 2005). Other means of developing and implementing maintenance programs are presented in a number of publications (see, for example, Reese and Presler, 2005; Kang et al., 2008, among others).

In some states, Pennsylvania included, a memorandum of understanding (MOU) is issued between the municipality and the county conservation district to serve as an additional layer of review of storm water management maintenance plans. An opinion held by some practitioners is that construction and maintenance of all detention and retention basins in general and infiltration basins in particular should be undertaken by the municipality to ensure proper functioning. In the majority of cases, storm water management facilities within the ROW of roadway are maintained by the department of public works of the municipality or county or by the state and federal highway agencies.

9.7 INSPECTION, OPERATION, AND MAINTENANCE MANUAL

Depending on the nature of the project, a specific storm water management inspection, operation, and maintenance manual should be prepared. The manual should include the name and full address of the project, the party responsible for maintenance and his or her name, phone and fax numbers, post office address, and, of course, e-mail address. Also included in the manual should be a brief description of provisioned storm water management facilities. In addition, each manual should include a separate inspection and maintenance checklist for record keeping. An estimate of annual cost of maintenance is also helpful.

CASE STUDY 9.1   Storm Water Management Operation and Maintenance, New Jersey Garden State Parkway Interchange 98

INTRODUCTION

The storm water management system for the Garden State Parkway roadway improvement project at Interchange 98 includes three infiltration basins. These basins have been designed to address the NJDEP's Stormwater Management Regulations relating to the peak rates of runoff. The infiltration basins are also designed to provide water quality in accordance with the regulations. The location of the infiltration basins is selected considering existing depressions, topographic features, and accessibility. This minimizes the number of structures needed and eases maintenance and inspection of the storm water management systems, as well. The site location is shown in Figure 9.34.

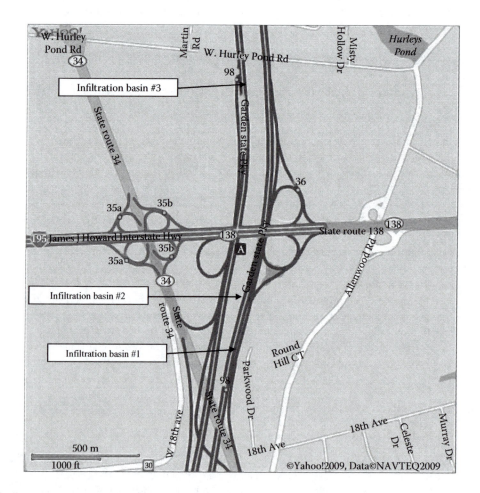

FIGURE 9.34   Location map of infiltration basins.

Soils in the area are predominately Downer, which is a sandy soil of high permeability. This type of soil is ideal for infiltration basins. The infiltration basins are located in highly visible areas that facilitate inspection. All three are also easily accessible from the southbound lanes to perform any needed maintenance.

This Operation and Maintenance Manual (O&M manual) is intended to provide a guidance for the inspection and maintenance of the infiltration basins by the New Jersey Turnpike Authority. The manual can also serve as a support document in connection with the permit application to the NJDEP.

DESCRIPTION

The infiltration basins are depicted on the grading and drainage plans. For brevity, only a location plan of the basins is enclosed herein. All three infiltration basins have sand bed and grass-lined side slopes. The sand beds are specified to be K-5 permeability rated. The sand layer is 6 in. (15 cm) thick in basins 1 and 2, and 12 in. (30 cm) thick in basin 3. In accordance with the New Jersey Stormwater Best Management Practices Manual, dated February 2004, all infiltration basins are designed to be no more than 2 ft (0.6 m) deep.

Located between the north- and southbound lanes, infiltration basin 1 is the most southern basin. The runoff from the roadway enters this basin as sheet flow, passing through grass and woods. Only an emergency spillway is incorporated in this basin to allow for discharge of runoff in excess of the design storm. Overflow from the spillway flows downhill overland toward existing wetlands.

Infiltration basin 2 is located about 600 ft (180 m) north of infiltration basin 1. Unlike infiltration basin 1, a conveyance system is included within this infiltration basin. Much of the runoff to the basin does occur as sheet flow; also a 15 in. RCP carries runoff from two existing inlets along the northbound lanes of the Garden State Parkway to the basin. A riprap apron is provided at the outlet of the 15 in. RCP to prevent erosion and scouring in the basin bottom. A concrete box with a flat grate is proposed to serve as an overflow structure and emergency spillway for this basin. This box will be constructed over an existing 24 in. RCP drainage pipe that traverses the southbound lanes in a westerly direction.

Infiltration basin 3 is located on the southbound side of the Garden State Parkway between the exit ramp for Interchange 98 and the southbound lanes. Runoff tributary to this basin travels overland, similar to infiltration basin 1. The grate atop an existing inlet, located within the basin, will be raised to serve as an emergency spillway. An existing 15 in. pipe emanates from this inlet, extends westerly below the exit lanes, and terminates at a swale. Since this basin is very shallow and abuts the roadway, a 3 in. orifice is provided at the bottom of the structure to drain the basin between rainstorms.

OPERATION AND MAINTENANCE RESPONSIBILITY

The owner and the responsible party for inspection, maintenance and repair of the storm water management systems is New Jersey Turnpike Authority. The designated person for inspection and maintenance is as follows:

Maintenance Office
New Jersey Turnpike Authority
740 US-46
Clifton, NJ 07013
Phone No.: (973) 478-8337

INSPECTION FREQUENCY

Following construction, the infiltration basins should be inspected regularly following rainfalls exceeding 1 in. in less than 1 hour and on a quarterly and annual basis. For record keeping, a quarterly inspection report form has been prepared. Also, an annual inspection report form is provided for those items that require less frequent inspection.

The following drainage elements should be inspected quarterly and after every storm exceeding 1 in. of rainfall in 1 hour:

- The infiltration basins shall be visually inspected for ponding and debris and sediment accumulation.
- The riprap apron shall be inspected for erosion, and significant debris and sediment accumulation.
- Established vegetation shall be inspected for vegetation health, density, and diversity. If greater than 25% of vegetation is damaged, the area shall be reestablished in accordance with the original specifications.
- Infiltration basins that do not infiltrate within 72 hours following a storm require immediate corrective measures.
- The conveyance system, inclusive of two inlets and drainage pipes, shall be inspected for sediment and debris accumulation.

The annual inspections include the following:

- Permeability rate of the soil below the basins shall be tested.
- Outlet structures and inlets shall be inspected for cracking, subsidence, spalling, erosion, and deterioration.
- Riprap aprons and emergency spillways shall be inspected for erosion.
- The basin's vegetation shall be inspected for unwanted growth and erosion/scouring.

The inspection forms are completed by checking "yes" or "no." Any item that is checked as "yes" requires maintenance and must be corrected as outlined in the next section.

INFILTRATION BASIN MAINTENANCE INSTRUCTIONS

The maintenance shall include the following:

- Removal of sediment and debris. This should take place when the infiltration basins are thoroughly dry. The sediment should be disposed of in compliance with all applicable local, state, and federal regulations.
- Grass should be mowed at least once a month during the growing season.
- Unwanted vegetation should be removed with minimum disruption to surrounding vegetation and basin subsoil.
- Reestablished vegetation should be inspected biweekly during the first growing season until the vegetation is established.
- Routine tilling should be performed with light equipment to maintain the infiltration capacity and break up clogged surfaces when necessary.
- Vegetation should be reestablished in eroded areas and riprap stone placed when erosion or scouring occurs at the emergency spillway.
- The outlet structure should be repaired or replaced, if necessary.
- Sediment and debris should be removed from inlets and conveyance systems and disposed of, in compliance with all applicable local, state, and federal regulations.

Case Study 9.1
INFILTRATION BASINS
Garden State Parkway Interchange 98
Township of Wall
Monmouth County, New Jersey

QUARTERLY INSPECTION REPORT

Date:_____   Time:_____   Weather Conditions:_____

	No	Yes
1. *Infiltration basin #1*		
I. Basin bottom:		
a. Sediment/debris accumulation	_____	_____
b. Clogging of sand/standing water	_____	_____
II. Vegetarian/grass:		
a. Areas of vegetation damage (>50%)	_____	_____
2. *Infiltration basin #2*		
I. Basin bottom:		
a. Sediment/debris accumulation	_____	_____
b. Clogging of sand/standing water	_____	_____
II. Vegetarian/grass:		
a. Areas of vegetation damage (>50%)	_____	_____
III. Riprap apron:		
a. Sediment/debris accumulation	_____	_____
b. Erosion/scouring	_____	_____
IV. Conveyance system:		
a. Sediment/debris accumulation	_____	_____
3. *Infiltration basin #3*		
I. Basin bottom:		
a. Sediment/debris accumulation	_____	_____
b. Clogging of sand/standing water	_____	_____
II. Vegetarian/grass:		
a. Areas of vegetation damage (>50%)	_____	_____
III. Outlet structure:		
a. Clogging of 3″ orifice	_____	_____

Additional Notes:

_____

_____

This report represents the conditions observed by: _____

_____

Signature

Case Study 9.1
INFILTRATION BASINS
Garden State Parkway Interchange 98
Township of Wall
Monmouth County, New Jersey

ANNUAL INSPECTION REPORT

Date:_____   Time:_____   Weather conditions:_____

		No	Yes
1. *Infiltration basin #1*			
I.	Basin bottom:		
	a. Permeability rate of sub soils (<4″/h)	_____	_____
II.	Vegetation/grass:		
	a. Unwanted growth (i.e., trees)	_____	_____
	b. Erosion/scouring	_____	_____
III.	Emergency spillway:		
	a. Erosion/scouring	_____	_____
2. *Infiltration basin #2*			
I.	Basin bottom:		
	a. Permeability rate of sub soils (<4″/h)	_____	_____
II.	Vegetation grass:		
	a. Unwanted growth (i.e., trees)	_____	_____
	b. Erosion/scouring	_____	_____
III.	Outlet structure:		
	a. Cracking/subsidence/spalling/erosion/deterioration	_____	_____
3. *Infiltration basin #3*			
I.	Basin bottom:		
	a. Permeability rate of sub soils (<4″/h)	_____	_____
II.	Vegetation/grass:		
	a. Unwanted growth (i.e., trees)	_____	_____
	b. Erosion/scouring	_____	_____
III.	Outlet structure:		
	a. Cracking/subsidence/spalling/erosion/deterioration	_____	_____

Additional Notes:

_____

_____

This report represents the conditions observed by: _____

_____

                                              Signature

Case Study 9.1
INFILTRATION BASINS
Garden State Parkway Interchange 98
Township of Wall
Monmouth County, New Jersey

MAINTENANCE LOG SHEET

Date	Location	Maintenance Performed	Performed By

PROBLEMS

9.1   Calculate the volume of a standard trench excavation per lineal feet for a 24 in. RCP.
       The average cover over the pipe is 18 in.

9.2   Redo Problem 9.1 for a 600 mm RCP and 45 cm average cover.

9.3   Calculate the excavation volume of the trench for a 30 in. HDPE pipe and 24 in. average
       cover.

9.4   Redo Problem 9.3 for a 750 mm HDPE pipe and 600 mm cover.

9.5   An underground detention system consists of four rows of 80 ft long, 42 in. solid RCP
       pipes under 18 in. cover. The pipes are capped at one end and terminate to a 4 ft wide
       by 6 ft deep chamber at the other end. Calculate:
       a. The storage volume of detention pipes
       b. The volume of excavation, assuming that the trench sides are at a slope of 1:5 (1 H, 5 V).

9.6   Redo Problem 9.5 for 1000 mm concrete pipes, each 24 m long and 45 cm cover; the
       chamber width and height are 1.2 and 1.8 m, respectively.

9.7   An underground detention system includes five rows of 750 mm solid HDPE pipes,
       60 m long and 750 mm headers pipes. Calculate:
       a. The detention storage volume
       b. The amount of excavation; the pipes are under 60 cm of cover. The trench is at 1:5
           side slope.

9.8   Redo Problem 9.7 for 180 ft long, 30 in. pipes under 2 ft of cover.

9.9   Estimate the annual cost of maintenance of treating storm water runoff at a flow rate of
       0.6 cfs for 80% TSS removal by an MTD of your choice.

9.10   Estimate the first year's and 3 years' maintenance of an MTD of your choice that treats
       30 L/s for 80% TSS removal. State any assumption you make.

9.11   In Problem 9.10, perform an estimate for 20 L/s.

APPENDIX 9A: INSTALLATION OF HDPE PIPE BY ADVANCED DRAINAGE SYSTEM (ADS)

9A.1 BACKFILL ENVELOPE CONSTRUCTION

ASTM D2321 serves as the basis for installation recommendations of HDPE pipes in traffic areas. Acceptable backfill materials and construction methods are very similar to those required for other types of pipe material. However, with a flexible pipe such as HDPE, the selection, placement, and compaction of backfill material are more important in its load carry capacity. Acceptable soil types and compaction for HDPE pipes are shown in Table 9A.1.

It is to be noted that the combination of the type of material and compaction level will determine the soil strength. When a variety of options will work in a particular installation, the final decision can depend on what is most available locally in order to keep the cost of the installation to a minimum. Native soil may be specified when it meets the backfill requirements of Table 9A.1. If the native material is not acceptable, then appropriate material will need to be brought in.

Controlled low-strength material (CLSM) or flowable fill is another, more specialized type of backfill material that is increasingly used throughout the country. This material is essentially very low strength concrete that is poured around the pipe. With CLSM, or flowable fill, the trench width can be reduced to a minimum of the outside diameter of the pipe plus 12 in.; however, it will misalign or float the pipe unless precautions, such as weighting the pipe or pouring the flowable fill in layers, are taken. Conventional compacted granular material creates structurally sound backfill that is easier to use and often less expensive to install.

TABLE 9A.1
Acceptable Backfill Material and Compaction Requirements

Description	Soil Classification			Minimum Standard Proctor Density, %	Maximum Compaction Layer Height, in. (m)
	ASTM D2321	ASTM[a] D2487	AASHTO M43		
Angular crushed stone or rock, crushed gravel	Class I (crushed stone)	–	5, 56 57, 6 67	Dumped	12 (0.3)
Well-graded sand, gravels and gravel/sand mixtures; poorly graded sand, gravels, and gravel/sand mixtures; little or no fines	Class II (gravelly sand)	GW GP SW SP GW-GC SP-SM	57 6 56 67 5	85%	12 (0.3)
Silty or clayey gravels, gravel/sand/silt or gravel and clay mixtures; silty or clayey sands, sand/clay or sand/silt mixtures	Class III (sandy silt)	GM GC SM SC	Gravel and sand (<10% fines)	90%	6 (0.15)
Inorganic silts and gravelly, sandy, or silty clays; some fine sands; low- to medium-plasticity clays	Class IVA	ML CL	–	Not recommended	–

Note: Layer heights should not exceed one-half the pipe diameter. Layer heights may also need to be reduced to accommodate compaction method.

[a] See Unified Soil Classification System in Appendix B at the back of the book.

9A.2   BACKFILL PLACEMENT

In soft or rocky soil, HDPE pipes should be placed on a firm foundation. Muck and other soft soil, which allow the pipe to settle, and rock protrusions, which apply point loads, can affect the hydraulics or structural integrity of the system. It is recommended that unsuitable foundation material be removed. Where a rock or unyielding or soft foundation is present, the design engineer or a geotechnical engineer shall be consulted to determine the extent to which the undesirable material is to be excavated and replaced with a layer of approved structural material. After unsuitable material is removed, the trench bed is leveled and backfilled as follows:

- Adequate bedding material 4 to 6 in. (10 to 15 cm) thick is placed, leveled, and compacted as shown in Figure 9.6.
- Backfill material is placed in layers to meet requirements of Table 9A.1. The first layer, called haunching, shall be placed in lifts of 4–6 in. (10–15 cm) and compacted by handheld equipment. Avoid impacting pipe with heavy equipment. Middle of bedding (one-third of pipe O.D.) should be loosely placed.
- Backfilling is continued in accordance with Table 9A.1. Pipes 4–60 in. (100–1500 mm) in H-25 traffic area will be backfilled a minimum of 6 in. above the crown of the pipe.
- Minimum cover may be reduced for areas with no traffic or infrequent, light traffic. These situations must first be reviewed by the pipe manufacturer.

Final backfill, which extends from the previously indicated backfill to the ground surface, shall be at least 6 in. (0.15 m) thick for pipes that are 48 in. (1200 mm) or smaller and 18 in. (0.5 m) for 54 and 60 in. (1350 and 1500 mm) diameter pipes. Therefore, the minimum cover on HDPE pipes is 12 in. for pipes smaller than or equal to 48 in. in diameter and 24 in. for 54 in. and 60 in. diameters. This does not include any pavement. Where flexible pavement is installed over the pipe, height of cover is measured from the crown of the pipe to the bottom of the flexible pavement. Where rigid pavement is installed over the pipe zone, height of cover is measured from the top of the pipe to the top of the rigid pavement. When no pavement will be installed, but vehicle traffic is expected (e.g., gravel driveway), a minimum cover of 18 in. (0.5 m) for 4 to 48 in. (100–1200 mm) diameters and 30 in. (0.8 m) for 54 and 60 in. (1350 and 1500 mm) diameters is recommended to minimize rutting. If roads or driveways will be crossing the pipe, a relatively high degree of compaction is needed to prevent pavement settlement. Excavated materials may be of adequate quality for final backfill, depending on the intended use at the surface. Selection, placement, and compaction of final backfill shall be as directed by the design engineer. Manufacturers' specifications should be followed using different soil classes.

9A.3   MECHANICAL COMPACTION EQUIPMENT

Compacting the haunching layer may be performed by handheld tampers. Tampers for horizontal layers shall not weigh more than 20 lb (9 kg) and the tamping face shall be limited to an area no larger than 6 by 6 in. (0.15 by 0.15 m).

Rammers or rammer plates (Figure 9A.1) use an impact action to force out air and water held between soil particles to consolidate the fill. This equipment works well on cohesive or high-clay-content soils. Care should be taken not to use rammer-type compactors directly on the pipe.

Static compactors are most suitable when used on noncohesive backfill away from the pipe. Vibrating compactors may be used near the pipe only if care is taken not to impact the pipe directly with a great deal of force.

Selecting the right equipment for the fill material is the key to achieving the most efficient compaction. For soil mixtures, the component having the highest percentage will dictate what type of compaction equipment is needed.

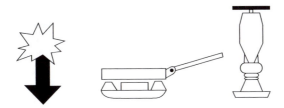

FIGURE 9A.1   Rammer compactors.

9A.4   Joints

The Advanced Drainage Systems, Inc. (ADS) N-12® ST IB integral bell and spigot joining systems meet or exceed soil-tight performance and ADS N-12 WT IB offers a 10.8 psi laboratory rated watertight joining system. Both ADS N-12 ST IB and ADS N-12 WT IB joins work well in all soil conditions and can be used in storm water management practice.

9A.5   Construction and Paving Equipment

Some of the construction vehicles and paving equipment used on site are not as heavy as the design traffic load of AASHTO HS-25. Table 9A.2 presents the minimum cover that can be permitted during construction for these vehicles.

It is recommended that 3 ft (0.9 m) of cover be used over the pipe in installations involving construction vehicles between 30 and 60T (267–534 kN). This cover can simply be mounded and compacted over the pipe during the construction phase and then graded following construction to provide the minimum required cover.

9A.6   Joining Different Pipe Types or Sizes

Drainage systems may involve connecting pipe of different materials or sizes. Options to make these transitions are often limited by the joint quality required. A common method of connecting different types of pipe of the same size, and in some cases different sizes, is through the use of a concrete collar. Another option may be using fittings or adapters specifically designed for this application. ADS offers a selection of fittings designed to make the transition from one material directly to another. Fittings may be more watertight than a concrete collar.

TABLE 9A.2

Temporary Cover Requirements for Light Construction Traffic

Vehicular Load at Surface, psi (kPa)	Temporary Minimum Cover, in. (mm), for 4–48 in. (100–1200 mm) Diameters	Temporary Minimum Cover, in. (mm), for 54 and 60 in. (1350 and 1500 mm) Diameters
75 (517)	9 (230)	12 (300)
50 (345)	6 (150)	9 (230)
25 (172)	3 (80)	6 (150)

9A.7 CURVILINEAR INSTALLATIONS

HDPE pipe can be laid on a curved alignment as a series of straight sections deflected horizontally at each joint. Typically, ADS N-12 ST IB and N-12 WT IB bell and spigot pipe joints can only accommodate small deflection angles (<1°). Split couplers and bell–bell couplers used to couple plain end N-12 pipes will also permit small deflection angles (approximately 1° to 3°). For larger angles, custom bends, manholes, or inlets are used.

9A.8 VERTICAL INSTALLATIONS

N-12 pipe is sometimes installed vertically for use as catch basins or manholes, meter pits, and similar applications. Vertical risers do not behave the same as pipe that is installed horizontally because the pipe/soil interaction is different. Installation requirements are especially important for vertical installations. Backfill shall extend a minimum of 1 ft (0.3 m) completely around the vertical structure. Backfill material recommendations are identical to those for a horizontal installation; compaction levels and maximum lift requirements must be strictly adhered to (refer to Table 9A.1).

Additional general applications limits include the following:

- Height of the vertical riser must not exceed 8 ft (2.4 m), unless the design is reviewed by the ADS Application Engineering Department.
- In traffic areas, a concrete collar or similar structure designed to transmit the load into the ground (away from the pipe) must be used at the surface.
- Cast iron frames, holding grates, or lids must be seated on a concrete collar or similar structure so that the weight of the frame and grate or lid is transferred into the ground, not to the vertical pipe.

Vertical installations of any fitting by ADS or its manufacturer should first be reviewed for suitability with ADS Application Engineering or its manufacturer. This includes, but is not limited to, tees, elbows, and reducers of any combination.

9A.9 STEEP SLOPE INSTALLATIONS

Where pipe slope is equal to or greater than 12%, precaution must be taken to ensure that application conditions will not adversely affect the pipe structure or flow characteristics. One design consideration is proper venting, to ensure negative pressure does not form inside the pipe. Venting can be provided along the pipe slope, at the head of the slope, or by designing the flow in the slope not to exceed 75% full in peak design flow conditions. Next, thrust blocks must be designed and constructed at all fittings and grade changes and, above all, a change in flow direction, which can cause excessive force against the pipe wall. Finally, consideration must be given to pipe slippage along the slope, which can result in slope failure of the surrounding soil, structural damage to the pipe wall, or compromising of joint quality for the overall system. Pipe should be restrained through the use of concrete blocks or pipe anchors.

Note: A grade of 12% is listed for reference purposes only; additional design consideration may be necessary for slopes less than 12%, where slope stabilization, negative pressure, or water hammer may be of concern.

9A.10 CAMBERED INSTALLATIONS

Pipe installation under high embankments may need to design for uneven settlement regardless of the backfill envelope quality and construction. In order to eliminate low pockets under the embankment, the pipe should be cambered. Cambering is the process of installing the pipe so that the expected settlement will create the design slope. It can be achieved by installing the upstream half of the pipe on a flat grade and the downstream half on a grade that is larger than design. A qualified soils engineer should be consulted for this specialized situation.

9A.11 SLIPLINING

Due to abrasive or corrosive environments, premature deterioration of some types of pipe may occur. In lieu of a total replacement, sliplining the existing pipe with HDPE pipe or PVC pipe is often an economical and efficient way to prolong a culvert's service life. CentriPipe (centrifugally cast-in-place lining), which was discussed earlier, provides a very effective solution. Typically, HDPE pipe can only be used for open-ended applications where the pipe does not need to be bent for installation. Other considerations during design and preconstruction should include the inside and outside diameter of the carrier pipe and HDPE pipe, length of installation, and grout installation.

APPENDIX 9B: INSTALLATION GUIDELINES FOR ARMORMAX

Channels	# INSTALLATION GUIDELINES FOR ARMORMAX™
	Anchored Reinforced Vegetation System (ARVS)

This document provides general installation guidelines for the ArmorMax™ anchored reinforced vegetation system used in non-structural applications including: Channels. These non-structural applications are typically designed using Type 2 anchors.

PRE-CONSTRUCTION

A pre-construction meeting shall be held with the construction team and a representative from Propex ®. This meeting shall be scheduled by the contractor with at least two weeks notice. Also, Propex suggests that installation monitoring of the ArmorMax System be performed by a qualified independent third party.

SITE PREPARATION

- Grade and compact area of ArmorMax installation as directed and approved by an Engineer. Subgrade shall be uniform and smooth. Remove all rocks, clods, vegetation or other objects so the installed mat will have direct contact with soil surface.
- Prepare seedbed by loosening the top 2-3 inches (50-75 mm) minimum of soil. This may be accomplished with a rotary tiller on slopes 3:1 or flatter.
- Perform a site specific soil test to determine what amendments such as lime and fertilizer need to be incorporated.
- Do not mulch areas where mat is to be placed.

SEEDING

- Keep seeded areas as moist as necessary to establish vegetation. When watering seeded areas, use fine spray to prevent erosion of seeds or soil. If as a result of a rain, prepared seedbed becomes crusted or eroded, or if eroded places, ruts or depressions exist for any reason, rework soil until smooth and reseed such areas.
- Apply 50% of the seed required for installation onto the soil surface before installing the High Performance Turf Reinforcement Mat (HPTRM).
- Disturbed areas should be reseeded.
- Consult project plans and/or specifications for seed types and application rates.

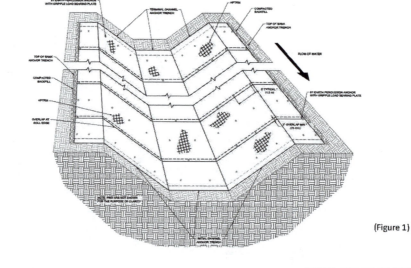

(Figure 1)

Propex™ Geotextile Systems

TESTED. PROVEN. TRUSTED.
www.geotextile.com

Propex Operating Company, LLC · 6025 Lee Highway, Suite 425 · PO Box 22788 · Chattanooga, TN 37422
ph 423 899 0444 · ph 800 621 1273 · fax 423 899 7619

GENERAL INSTALLATION GUIDELINES FOR A CHANNEL

- Refer to Figure 1 for a general layout of the installation in storm water channels. (Note: the details are for 8.5 ft wide HPTRM roll widths)

- Excavate an Initial Channel (IC) anchor trench a minimum of 12 in wide by 12 in deep (300 x 300 mm) across the channel at the downstream side of the project (see Figure 2/Figure 1). Deeper initial trench and/or hard armoring may be required for channels that have the potential for scour.

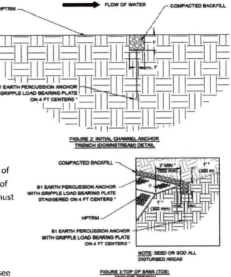

FIGURE 2: INITIAL CHANNEL ANCHOR TRENCH (DOWNSTREAM) DETAIL

- Excavate the Top of Bank (TOB) anchor trench a minimum of 12 in wide by 12 in deep (300 x 300 mm) along both sides of the installation (see Figure 3/Figure 1). Each TOB anchor must be a minimum of 3 ft (900 mm) over the crest of the bank (see Figure 3).

- Beginning at the downstream end of the channel, place HPTRM roll end into a TOB anchor trench and secure with Gripple Earth Percussion Anchors on 4 ft (1.2 m) centers (see Figure 3).

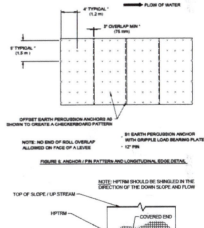

FIGURE 3: TOP OF BANK (TOB) ANCHOR TRENCH

- Unroll the HPTRM down the first channel bank, across and up the opposing channel bank. Terminate the roll end in the opposite TOB anchor trench securing it in a similar manner as the first one with Gripple Earth Percussion Anchors on 4 ft (1.2 m) centers (see Figure 3).

- Place the edge of the HPTRM into the IC anchor trench and secure it with Gripple Earth Percussion Anchors on 4 ft (1.2 m) centers (see Figure 2).

- Continue installation as follows:

  o Position adjacent rolls with a minimum of 3 in (75 mm) overlap (upstream mat on top) and secure in TOB anchor trench in same manner as the first roll (see Figure 2).

  o Unroll adjacent roll keeping the 3 in (75 mm) overlap constant. Secure overlap with one row of pins on 12 in (300 mm) centers and with one row of Gripple Earth Percussion Anchors on the designed anchor pin pattern detailed in Figure 6. A typical spacing on the overlapping seams for the Gripple Earth Percussion Anchors is 5 ft (1.5 m). When required, the Engineer is to create project details for transition to structures along the longitudinal edge or to address water flowing perpendicular to the seams.

  o In the case of a Roll End overlap: create a minimum of 6 in (150 mm) overlap with upslope mat on top. Secure with two rows of pins staggered 6 in (150 mm) apart on 12 in (300 mm) centers and with one row of Gripple Earth Percussion Anchors on 4 ft (1.2 m) centers (see Figure 9).

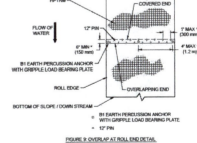

FIGURE 5: ANCHOR / PIN PATTERN AND LONGITUDINAL EDGE DETAIL

FIGURE 9: OVERLAP AT ROLL END DETAIL

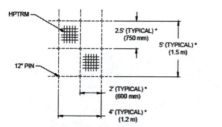

FIGURE 7: PIN PATTERN DETAIL SPECIFIC TO B1 ANCHOR RATIO OF 0.5 ANCHORS/YD² (FOR EARTH PERCUSSION ANCHOR RATIOS OTHER THAN THE ABOVE, PLEASE CONSULT WITH PROPEX ENGINEERING SERVICES AT 423-553-2450)

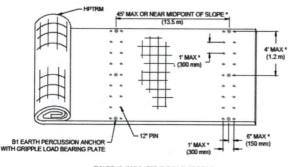

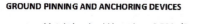

FIGURE 10: SIMULATED CHECK SLOT DETAIL

- Secure the rest of the mat using pins and Gripple Earth Percussion Anchors according to the pin and anchor patterns shown in Figure 6 and Figure 7.

- For channel bank heights or channel bottom widths greater than 45 ft (13.7 m), install simulated check slots per Figure 10. This method includes placing two rows of pins 12 in (300 mm) apart on 12 in (300 mm) centers and one row of Gripple Earth Percussion Anchors between the rows of pins on 4 ft (1.2 m) centers (see Figure 10). This pin/anchor pattern should be repeated every 45 ft (13.7 m) minimum or across the midpoint of the slope for slope lengths less than 60 ft (18.2 m) (see Figure 10).

- Excavate the Terminal Channel (TC) anchor trench a minimum of 12 in wide by 12 in deep (300 x 300 mm) across the channel at the upstream side of the project (see Figure 11/Figure 1). Deeper initial trench and/or hard armoring may be required for channels that have the potential for scour.

- Place the edge of the HPTRM into the TC anchor trench and secure it with Gripple Earth Percussion Anchors on 4 ft (1.2 m) centers (see Figure 11).

- Backfill and compact soil into each trench as directed and approved by and Engineer.

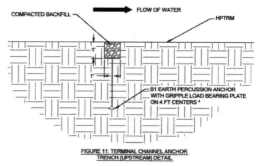

FIGURE 11: TERMINAL CHANNEL ANCHOR TRENCH (UPSTREAM) DETAIL

GROUND PINNING AND ANCHORING DEVICES

- Metal pins should be at least 0.20 in (5 mm) in diameter, made of steel, have a 1.5 in (38mm) diameter washer at the head, and be between 12 and 24 in (300-600 mm) long with sufficient ground penetration to resist pullout (see Figure 4). Longer pins may be required for looser soils. Heaver metal stakes may be required in rocky soils. Depending on soil pH and design life of the pin, galvanized or stainless steel pins may be required. Consult project plans and/or specifications for tie down device details.

- Gripple Earth Percussion Anchor assembly consists of an anchor head, stranded cable, gripping device and two crimping ferrules. Materials of each component have been selected to achieve an expected life of more than 50 years. The anchor head is made from die cast aluminum and is bullet nosed in shape to penetrate a turf mat without breaking strands of the mat. The cable is zinc-aluminum coated carbon steel and is of 1 x 19 construction. The ferrules are made from aluminum. The grip is die cast from zinc and uses a ceramic roller to clamp the cable in place. The one piece zinc top plate will have openings on the top to facilitate vegetative growth and the grip plate is approximately 0.2 in (5 mm) thick so it will only

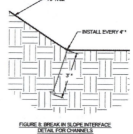

FIGURE 8: BREAK IN SLOPE INTERFACE DETAIL FOR CHANNELS

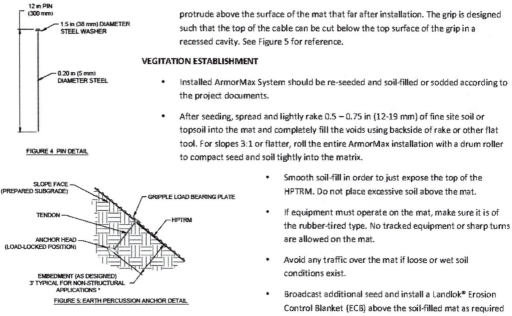

FIGURE 4 PIN DETAIL

FIGURE 5: EARTH PERCUSSION ANCHOR DETAIL

protrude above the surface of the mat that far after installation. The grip is designed such that the top of the cable can be cut below the top surface of the grip in a recessed cavity. See Figure 5 for reference.

VEGITATION ESTABLISHMENT

- Installed ArmorMax System should be re-seeded and soil-filled or sodded according to the project documents.

- After seeding, spread and lightly rake 0.5 – 0.75 in (12-19 mm) of fine site soil or topsoil into the mat and completely fill the voids using backside of rake or other flat tool. For slopes 3:1 or flatter, roll the entire ArmorMax installation with a drum roller to compact seed and soil tightly into the matrix.

- Smooth soil-fill in order to just expose the top of the HPTRM. Do not place excessive soil above the mat.

- If equipment must operate on the mat, make sure it is of the rubber-tired type. No tracked equipment or sharp turns are allowed on the mat.

- Avoid any traffic over the mat if loose or wet soil conditions exist.

- Broadcast additional seed and install a Landlok® Erosion Control Blanket (ECB) above the soil-filled mat as required by the Engineer. For levees or slopes steeper than 3:1, the addition of the ECB may be required or alternate methods of retaining the soil fill may be considered. Please contact the project engineer or Propex Engineering Services at (423) 553-2450.

- Irrigate as necessary to establish and maintain vegetation. Frequent, light irrigation will need to be applied to seeded areas if no natural rain events have occurred within two weeks of seeding and should continue until 75% of vegetation has established and has reached a height of 2 inches. Do not over irrigate.

SPECIAL TRANSITIONS

For applications that require special transitions (i.e. connections to riprap, concrete, T-walls, etc.), refer to the project specific drawings or consult with Propex Engineering Services at (423) 553-2450.

CONTRACTORS MAINTENANCE AND GUARENTEE PERIOD

It shall be the responsibility of the Owner to maintain all seed and ArmorMax areas after Engineer's acceptance. Maintenance shall consist of watering and weeding, repair of all erosion and any re-seeding as necessary to establish a uniform stand of the specified grasses. A minimum of 70 % of the seed area shall be covered with no bare or dead spots greater than 10 ft^2 (1 m^2). Seeded areas shall not be mowed prior to establishment of 70% vegetative density and a minimum grass growth of 4 in (100 mm). Throughout the duration of the project, the contractor shall be responsible for mowing to facilitate growth and shall not let the vegetation in the seeded areas exceed 18 in (450 mm). In addition, the Contractor shall water all grassed areas as often as necessary to establish satisfactory growth and to maintain its growth throughout the duration of the project.

Replanting is to be performed within 14 calendar days of notification by the Engineer.

APPENDIX 9C: CONSTRUCTION INSPECTION CHECKLIST: AN OVERVIEW

9C.1 SOIL EROSION AND SEDIMENT CONTROL MEASURES

- Check stone blanket at the entrance to construction site.
- Inspect silt fence/filter socks downhill of disturbed area.
- Regularly check the performance of siltation basins.
- Inspect hay bales/geofabric liner in inlets.
- Inform design engineer and contractor of any erosion control problems/deficiencies.
- Make sure corrective measures are implemented.

9C.2 EXCAVATION

- Frequently monitor the line and grade of the excavation for pipes.
- Check if excavated materials are placed at a minimum distance from the excavation edge.
- Make sure that unsuitable subgrade is overexcavated and stabilized.
- Classify rock excavation with contractor's representative.
- Keep a record of the excavated rock quantity.

9C.3 PIPE INSTALLATION

9C.3.1 Trenching

- Check if shoring is required and installed for trenches over 5 ft (1.5 m) deep.
- Check if necessary fences and barricades are installed to prevent accidental entry of persons or equipment into the trench.
- Check if escape ladders are placed per construction safety standards manual.
- Record soil types that deviate from the plans/specifications.

9C.3.2 Pipe Laying

- Check for proper size and class of pipe.
- Verify that pipe grade is checked frequently.
- Check if the joints (seals, couplings, gasket, bell, and groove) are installed per manufacturer's specifications.
- For concrete pipes, check that pulled holes are grouted and placed on the bottom.
- Verify that bedding material, backfill, and compaction follow specification.

9C.3.3 Manholes/Inlets

- Check for correct pipe grade and alignment at manholes/inlets.
- Check for proper location and size of inlet/manhole.
- Inspect the proper placement of filter gravel/bedding under inlet and outlet.
- Inspect for proper connection of pipes into inlet/manhole (intrusion length, outside joint seal, or mortar).
- Inspect backfilling and placement of the manhole cover/inlet grate.
- Ensure silt/debris are removed from inlets/manholes.
- Check if the outlet end of pipe is at proper location and grade.
- Check stone size and dimensions of riprap apron/scour hole at detention basin(s) or outfall to streams.

9C.3.4 Backfilling

- Ensure that proper measures are taken to prevent displacement of the pipe.
- Check that no organic matter, large rocks, or ice is placed within the specified distance from the pipe.

- Make sure that backfill is compacted as specified.
- Be sure that site restoration is done properly and on time.
- Record any site damage and its amount.
- Inspect any repairs to verify satisfactory condition.

9C.3.5   Repairs

- Make sure safety procedures are followed.
- Inspect resetting of grade and alignment of pipes.
- Check replacement of gravel envelope and backfill compaction.

9C.4   SITE RESTORATION

- Check debris and rock removal.
- Inspect inlets/manholes for cleanliness; make sure debris and silt are removed.
- Check site restoration.

APPENDIX 9D: GENERAL SPECIFICATIONS FOR MAINTENANCE OF FLOGARD+PLUS CATCH BASIN INSERT FILTERS

GENERAL SPECIFICATIONS FOR MAINTENANCE OF
FLOGARD +PLUS® CATCH BASIN INSERT FILTERS

SCOPE:

Federal, State and Local Clean Water Act regulations and those of insurance carriers require that stormwater filtration systems be maintained and serviced on a recurring basis. The intent of the regulations is to ensure that the systems, on a continuing basis, efficiently remove pollutants from stormwater runoff thereby preventing pollution of the nation's water resources. These specifications apply to the FloGard +Plus Catch Basin Insert Filter.

RECOMMENDED FREQUENCY OF SERVICE:

Drainage Protection Systems (DPS) recommends that installed FloGard +Plus Catch Basin Insert Filters be serviced on a recurring basis. Ultimately, the frequency depends on the amount of runoff, pollutant loading and interference from debris (leaves, vegetation, cans, paper, etc.); however, it is recommended that each installation be serviced a minimum of three times a year, with a change of filter medium once per year. DPS technicians are available to do an on-site evaluation, upon request.

RECOMMENDED TIMING OF SERVICE:

DPS guidelines for the timing of service are as follows:
1. For areas with a definite rainy season: Prior to, during and following the rainy season.
2. For areas subject to year-round rainfall: On a recurring basis (at least three times per year).
3. For areas with winter snow and summer rain: Prior to and just after the snow season and during the summer rain season.
4. For installed devices not subject to the elements (washracks, parking garages, etc.): On a recurring basis (no less than three times per year).

SERVICE PROCEDURES:

1. The catch basin grate shall be removed and set to one side. The catch basin shall be visually inspected for defects and possible illegal dumping. If illegal dumping has occurred, the proper authorities and property owner representative shall be notified as soon as practicable.
2. Using an industrial vacuum, the collected materials shall be removed from the liner. (Note: DPS uses a truck-mounted vacuum for servicing FloGard +Plus Catch Basin Insert Filters.)
3. When all of the collected materials have been removed, the filter medium pouches shall be removed by unsnapping the tether from the D-ring and set to one side. The filter liner, gaskets, stainless steel frame and mounting brackets, etc. shall be inspected for continued serviceability. Minor damage or defects found shall be corrected on the spot and a notation made on the Maintenance Record. More extensive deficiencies that affect the efficiency of the filter (torn liner, etc.), if approved by the customer representative, will be corrected and an invoice submitted to the representative along with the Maintenance Record.
4. The filter medium pouches shall be inspected for defects and continued serviceability and replaced as necessary and the pouch tethers re-attached to the liner's D-ring. See below.
5. The grate shall be replaced.

EXCHANGE AND DISPOSAL OF EXPOSED FILTER MEDIUM AND COLLECTED DEBRIS:

The frequency of filter medium pouch exchange will be in accordance with the existing DPS-Customer Maintenance Contract. DPS recommends that the medium be changed at least once per year. During the appropriate service, or if so determined by the service technician during a non-scheduled service, the filter medium pouches will be replaced with new pouches and the exposed pouches placed in the DOT approved container, along with the exposed debris. Once the exposed pouches and debris have been placed in the container, DPS has possession and must dispose of it in accordance with local, state and federal agency requirements.

DPS also has the capability of servicing all manner of catch basin inserts and catch basins without inserts, underground oil/water separators, storm water interceptors and other such devices. All DPS personnel are highly qualified technicians and are confined space trained and certified. Call us at (888) 950-8826 for further information and assistance.

REFERENCES

A·Lok Products Inc., P.O. Box 1647, 697 Main Street, Tullytown, PA 19007, Ph. 800-822-2565/215-547-3366 (http://www.a-lok.com).

ADS, 2014, Water management drainage handbook, Specification, Section 1, and Installation.

American Concrete Pipe Association, 1980 (8th printing, 2005), Concrete pipe handbook.

——— 2007, Concrete pipe design manual, 222 W. Las Colinas Blvd., Irving, Texas (http://www.concrete-pipe.org), Chapter 5, supplemental data.

——— 2007–2014, Concrete pipe and box culvert installation, resource no. 01-103, Ph. (972) 506-7216 (http://info@concrete-pipe.org).

ASCE (American Society of Civil Engineers), 1985, Stormwater detention—Outlet control structures, ISBN0-87262-480-3.

——— 1992, Manuals and reports of engineering practice no. 77, WEF manual of practice FD-20, 1992, Design and construction of urban stormwater management systems, WFE and ASCE, Alexandria, VA.

——— 2006, ASCE standard, Standards guidelines for the design of urban stormwater systems, ASCE/EWRI 45-05, Standard guidelines for installation of urban stormwater systems, ASCE/EWRI 46-05, Standard guidelines for the operation and maintenance of urban stormwater systems, ASCE/EWRI 47-05, ASCE, Reston, VA.

Barron J., and Lankford, M., 2005, Maintenance of privately owned stormwater infrastructure: One approach to enforcement, *Stormwater*, September/October, pp. 100–102.

Blocksom & Co., 450 St. John Rd., Suite 710, P.O. Box 2007, Michigan City, Indiana 46361-8007 (http://www.blocksom.com).

Bryant, G., 2004, Stormwater inspection and maintenance, The Sleeping Giant, *Stormwater*, May/June, pp. 8–10.

Brzozowski, C., 2004, Maintaining stormwater BMPs, *Stormwater*, May/June, pp. 36–51.

CentriPipe by AP/M PermaForm, Ph. 800-662-6465 (http://www.centripipe.com).

Contech Stormwater Solutions, 9025 Center Pointe Drive, West Chester, OH 45069, Ph. 800-338-1122 (http://www.contechstormwater.com).

FloGard+Plus Catch Basin Insert Filters by Kristar Enterprises, Inc. 360, Sutton Place, Santa Rosa, Ca 95407, Ph. 800-579-8819 (http://contactstormwater@oldcastle.com).

Geneva Pipe, Suggested procedure for installation of precast concrete box culvert, Geneva Pipe and Precast, 1465 W. 400 N., Orem, UT 84104 (http://info@genevapipe.com).

Hanson Pressure Pipe, Hanson Heidelberg Cement Groupe, 8505 Freeport Parkway, Irving, TX 75063, Ph. 972-262-3600 (http://www.hansonpressurepipe.com).

Imbrium Systems Corp., 605 Global Way, Suite 113, Lithium, MD 21090, Ph. 301-279-8827; Canada, International, 407 Fairview Drive, Whitby, ON LIN 3A9, Canada, Ph. (416) 960-9900.

J.M. Eagle, HDPE corrugated dual wall pipe manufacturer, Los Angeles, CA, Ph. 800-621-4404/973-535-1633 (http://www.JMEagle.com/EagleCorPE).

Kang, J.-H., Weiss, P.T., Wilson, C.B., and Gulliver, J.S. 2008, Maintenance of stormwater BMPs frequency, effort and cost, *Stormwater*, Nov./Dec., pp. 18–29.

Keating, J., 2005, Stormwater good housekeeping: Prevention worth a pound of cure, *Stormwater*, July/August, pp. 116–120.

Kristar Enterprises Inc., P.O. Box 6419, Santa Rosa, CA 954006-0419, Ph. 800-579-8819 (http://www.kristar.com).

NPC Inc., 250 Elm Street, P.O. Box 301, Milford, NH 03055, Ph. 603-673-8680 (800 626-2180; http://www.npc.com).

Pazwash, H., 1991, Maintenance of stormwater management facilities, neglects in practice, Proceedings of the ASCE National Conference on Hydraulic Engineering and International Symposium on Groundwater, Nashville, TN, July 29–August 2, 1991, pp. 1072–1077.

Press-Seal Gasket Corporation, 2424 W. State Blvd., Fort Wayne, IN 46808, Ph. 460-436-0521, 1-800-348-7235 (http://www.press-seal.com).

Price Brothers, Prestressed concrete pipe (PCCP), 333 W. First Street, Dayton, OH 45402. Note: Price Brothers has been bought by Hanson Pipe and Precast (http://www.hansonpipeandprecast.com).

Reese, A.J., and Presler, H.H., 2005, Municipal stormwater system maintenance, an assessment of current practices and methodology for upgrading programs, *Stormwater*, September/October, pp. 36–61.

StormRax by Plastic Solutions, Inc., Ph. 800-877-5727 (http://www.plastic-solutions.com).

Trelleborg Pipe Seals Milford, Inc., 250 Elm Street, P.O. Box 301, Milford, NH 03055, Ph. 800-626-2180 (http://www.trelleborg.com, milfordsales@trelleborg.com).

10 Water Conservation and Reuse

The world population is continually growing while our freshwater supplies are shrinking. This trend is placing a constraint on the use of water, not only in agriculture, but also in domestic needs. The southwest United States and many countries around the globe already experience a water shortage. As this trend continues, the need for conservation and reuse of water becomes a challenging reality.

10.1 TRENDS IN SUPPLY AND DEMAND

The US population has risen from approximately 190 million in 1920 to 249 million in 1990 to 320 million at the end of 2014. It rose 0.7% in 2013. At this rate, the US population is expected to increase by 30% to 410 million by the year 2050. Meanwhile, the municipal water demand is predicted to increase by 20–25%.* According to a study at the Environmental Protection Agency's (EPA's) National Risk Management Research Laboratory in Edison, New Jersey, the total gross water use in the United States currently exceeds the total available freshwater supply, particularly in Florida and the southwest.

Population growth will continue in the western and southern states and in urban areas. It is predicted that by 2025 two-thirds of the US population will live in the South and the West. In 2005, the Census Bureau projected that over 88% of population growth between 2000 and 2030 will occur in Sun Belt cities with far less precipitation than the rest of the country. In Las Vegas, Nevada, for example, where precipitation averages a mere 4 in. (100 mm) annually, the population grew from approximately 165,000 in 1980 to 478,000 in 2000 (Von Minden, 2013). Since 2000, this city grew even more rapidly; and by 2013, her population surpassed 2 million. The West is going to face a severe water shortage, if not a crisis, which will require challenging actions to balance supply and demand.

On a global basis, the water shortage is even gloomier (McCarthy, 2008). The world's population has grown over 215% since 1970 to 7.2 billion in mid-2014. It grew 1.1% in 2013, compared with 0.7% in the United States. At this rate, the world's population will be over 10 billion in 2050. In 2000 and 2001, the United Nations committed to meeting multiple objectives aimed at alleviating poverty and improving conditions of the world's poor by 2015. In 2012, the World's Health Organization (WHO) and United Nations International Children's Emergency Fund (UNICEF, later shortened to United Nations Children's Fund) announced that one of the goals, which was access to safe drinking water, had been met in 2010—namely, more than 2 billion people had gained access to improved drinking water sources since 1990. However, this good news was tempered by the 2011 data from the WHO and UNICEF. These data, which as of 2014 are the most recent available information, indicate that over three-quarters of a billion people around the world were unable to obtain safe drinking water and over 2.5 billion lacked access to adequate sanitation in 2011 (Landers, 2014). With the population growth, the situation may get even worse in the future.

Approximately 3% of the earth's water is fresh and about 70% of that is confined in glaciers and polar ice caps; thus, less than 1% of the water on earth is suitable for drinking and 0.08% of this water is accessible to humans. Because of overdraft, the freshwater supplies are depleting while, due to population growth, the demand is rising. Therefore, if the current supply and demand process is continued, soon we will be running out of water. To prolong the useful life of available supplies, we should reduce our water usage through conservation measures to be discussed later in this chapter.

Although conservation can help extend existing water supplies, it will not be enough to cope with population growth. Addressing the water shortage worldwide will be a looming challenge (McCarthy, 2008). Erecting dams to store surface water, though not favored by environmentalists,

---

* The rise in demand is estimated at 67% by others (see, e.g., Means et al., 2005). However, because of a general trend in conservation, this estimate appears unrealistic.

can partly address the water shortage. However, construction of water supply reservoirs, apart from environmental issues, can be too expensive. The cost of a reservoir to supply 185 million gal (700,000 m³) of water daily to restock drinking water well fields in Palm Beach, Florida, was estimated at $360,000. This reflects a unit cost of approximately $2 for every gallon ($0.55 per liter) of water captured. Therefore, measures are to be taken to further reduce shortages. A cost-effective and environmentally friendly solution is the reuse of storm water runoff, in general, and rainwater from roofs, in particular. As the water supply becomes smaller, there will be a greater need for water reuse. In fact, to ensure sufficient supplies, water reuse will be absolutely essential.

10.2 WATER CONSERVATION

Water conservation, meaning reduced use, prolongs the useful life of our water supplies. It also lowers the cost of treatment and distribution of domestic water. Some people are not familiar with measures to save water and many others are not concerned at all about water conservation.

One reason for the lack of concern is that water bills are many times smaller than electric and gas bills in the United States and many other countries. A 2004 survey by the American Water Works Association (AWWA) indicated the average cost of water to US customers at $19.11 per 1000 ft³ or $2.6 per 1000 gal ($0.69/1000 L). Since then, the prices have gone up. The current price of water varies from $3 to over $8 per 1000 gal (3785 L) in the United States.* Contrary to intuition, the rates tend to be higher in the northeastern parts of the country than in the south and western states, where there is a water shortage. The average price of water is nearly the same in the United States and Canada and three times more in Denmark and Germany.

In the past, the water purveyors did not promote, but rather opposed the idea of conservation so as not to reduce their revenues. Now, the water utility authorities embrace conservation because it delays, if not eliminates, a potential need to upgrade their treatment plants, which is costly (Brzozowski, 2012). Also, water purveyors across the United States are beginning to significantly raise user rates to help fund needed water supply repairs and replacements. The New York Water Board adopted a 13% raise in water rates in 2011, marking the fourth consecutive year the rates had increased by more than 10%. New Orleans water rates will more than double by 2020, going up 10% since 2012. In Paramus, New Jersey, where the author resides, the water rates were increased from $3.74 (2009) to $5.51 per 1000 gal ($1–$1.46/1000 L) in 2014, a jump of 47% in 5 years.

A raise in water rates results in a reduction in per-capita consumption. To educate customers about their water usage and the importance of conservation, the utilities should proactively leverage advanced metering infrastructure (AMI). Customer portals that integrate with the AMI system can access their water usage online and learn about rates, leak detection, and measures to conserve water.

The state of California in 2009 set a goal to reduce water consumption by 20% by 2020. By 2012, the average per-capita water consumption in Southern California had already dropped 18% from 177 gallons per day (gpd) to 150 gpd. Implementing a conservation program in Utah, the average daily water demand in selected municipalities that spanned the entire state dropped from 227 gal (860 L) per capita (gpc) to 193 gpc (730 Lpc) from 2000 to 2010.

The conservation of water as a means of water resources management is on the rise in the United States—especially in California and Florida, where the growth in population exceeds the national average. Many water districts are also considering water recycling as a water management technique. Water conservation, which was initiated in this country about three decades ago, is still in a stage of infancy in urban areas.

To conserve water, urban water agencies and environmental groups are forming water conservation councils around the country. In California, for example, over 100 municipal water agencies and environmental groups formed the California Urban Conservation Council in 1991. The council signed a memorandum of understanding pledging to develop and implement 14 comprehensive

---

* This is still less than a penny for a gallon of water, which is over 10 times cheaper than dirt.

practices for conservation. The council has since grown fourfold. In San Francisco, which is one of the three districts in California, the goal is to reduce water use by 4.5 million gpd by 2030.

A US EPA publication (July 2002), titled "Cases in water conservation," contains case studies of water conservation for 17 cities around the country. Water usage can be conserved both indoors and outdoors. The public is more aware of the indoor than outdoor conservation measures. Also, means of reducing indoor uses were developed many years before outdoor conservation was even considered. Water-efficient appliances and fixtures were introduced to the market over 30 years ago, but conservation of outdoor water uses, though more important than indoor water conservation, is still in development. An AWWA (2006) manual presents details of benefits of water conservation to the local community and the environment.

Water reuse is growing everywhere nationwide, including in the eastern states, which receive from 40 to 46 in. (1000–1170 mm) of rainfall annually. In Florida, Georgia, and the Carolinas, water reuse has been established for some time. With stringent environmental regulation it is difficult, if not unfeasible, to build new dams. Thus, with increases in population, it becomes necessary to conserve and reuse water. Water efficiency and reuse have become popular in recent years and are expected to be widespread for a long time into the future. Since 1999, two magazines on the subject of storm water management and recycling and reuse have originated. *Stormwater* magazine began its publication in June 1999 and the first issue of *Water Efficiency* magazine was published in September/October 2006. These magazines are available online (http://www.stormH2O.com and http://www.waterefficiency.net), respectively.

10.3 INDOOR CONSERVATION

10.3.1 RESIDENTIAL BUILDINGS

Indoor conservation is achieved through the use of water-efficient faucets and appliances including showerheads, flush toilets, dishwashers, and clothes washers. Among these, the low-flush toilets and low-flow showerheads result in a larger conservation than others. The average indoor use in a nonconserving single-family home in the United States is estimated at 262 L per capita per day (Lpcd) (69.2 gal per capita per day). In comparison, the average daily per-capita demand in a water conserving home is 160 Lpcd (42.4 gpcd). Thus, using water conserving fixtures reduces the indoor uses by nearly 39%. Table 10.1 presents a comparison of the water uses by various fixtures in a water conserving home and a nonconserving home.

TABLE 10.1
Average Indoor Water Uses in Liters (Gallons) per Capita per Day

Fixture	Nonconserving Home		Conserving Fixtures	
	Lpcd	gpcd	Lpcd	gpcd
Toilet	70	18.5	31	8.2
Shower	44	11.6	33	8.8
Faucet	41	10.8	31	8.2
Washing machine	57	15.0	38	10.0
Leak	36	9.5	15	4.0
Miscellaneous	14	3.8	12	3.2
Total	262	69.2	160	42.4

Note: These figures are based on 3.5 gal per flush (gpf) for two-piece gravity tank, 2.5 gpm (gallons per minute) for showerheads, and 2.2 gpm for kitchen faucets for nonconserving fixtures and 1.6 gpf for flushometer toilet tanks, 2.0 gpm for water-saver showerheads. Recently, toilet tanks have become available that use only 0.8 gpf. Niagara Conservation is a manufacturer of such toilet tanks. The same company manufactures 1.5 gpm showerheads and 0.5 gpm faucet aerators (http://www.NiagaraConservation.com).

A wealth of information on indoor water use and conservation, as well as outdoor water use, can be found in a book by Amy Vickers (2001). An easily read paperback book by Mark Obmascik (1993) presents many suggestions for water savings for homeowners.

According to an estimate by the EPA, Americans on the average flush 4.8 billion gal (18×10^6 m³) of water per day. Table 10.1 implies that the use of water-saver toilets alone can save over 57,000 L (15,000 gal) of water a year in a household of four. Using Niagara Conservation's 0.8 gal per flush (gpf) toilets increases this saving twofold. By retrofitting all of the fixtures, nearly 140,000 L (3650 gal) of water can be saved in the same household annually. Just replacing low-flow showerheads and faucet aerators in his former residence, the author has found the indoor water use to drop below 170 Lpcd (45 gpcd).

According to the AWWA, nearly 34 billion gal (129 million m³) of water is processed daily by more than 55,000 community water systems. The AWWA (2006) estimates that the production can be reduced by 5.4 billion gal (20.5×10^6 m³) per day by using such water conservation measures as updating plumbing systems or installing low-volume toilets.

While the indoor conservation measures began over 30 years ago in the United States, inefficient showerheads and toilet fixtures are still in use in many, mostly older, homes. As indicated, there are people in this country who are neither aware of nor concerned about water efficiency in their homes. This is even more the case in some other countries. A survey by City West Limited, a government-operated water and sewer authority in Melbourne, Australia, found that over two-thirds of people were unaware of water efficiency in their households (Johnstone 2008).

To conserve indoor use, many municipalities and water efficiency alliances are changing plumbing codes. An example is the Alliance for Water Efficiency in Chicago, Illinois, which has been actively working on a national level to improve the model plumbing codes. Some municipal utility authorities, either individually or collectively with industry, have formed alliances for water conservation partnerships. In California, for example, one such alliance was formed by the East Bay Municipal Utility District and Shapell Industries, Inc., a private developer and builder in California. The long-term water supply conservation goal of this partnership is to reduce average water demand to 48 million gal per day (mgd) (181,000 m³/day) by 2020, of which 14 mgd (53,000 m³/day) would be derived from recycling (Maddaus et al., 2008). A list of some other alliances and their websites includes the following:

- Metropolitan Water District of Southern California; http://www.bewaterwise.com
- California Urban Water Conservation Council; http://www.CUWCC.org
- Colorado WaterWise Council; http://www.coloradowaterwiser.org
- EPA Alliance for Water Efficiency; http://www.allianceforwaterefficiency.com
- San Antonio (Texas) Water System; http://www.SAWS.org/conservation
- Southern Nevada Water Authority; http://www.snwa.com
- Albuquerque (New Mexico); http://www.cabg.gov

To avoid the cost of upsizing their water treatment plants and water supply systems, water purveyors in some communities have given free aerators and low-flow showerheads to their customers. In the metropolitan Boston area, for example, the water demand started to surpass the safe yield of 300 million mgd in the early 1980s. The Massachusetts Water Resource Authority (MWRA), which is the water and sewer wholesaler serving 50 communities in the metropolitan Boston area, had predicted that if no measures were taken, the consumption could rise 450 mgd (1.7 million m³/day) in 20 years. To reduce consumption, MWRA began to identify system leaks, educate people about conserving water, and provide and install water-saving fixtures free of charge going door to door. The program of giving out efficient fixtures was implemented in a number of other cities around the country, including the Clearwater, Florida, and the Santa Clara Valley in California. In addition to sending out faucet aerators, toilet fixtures, and leak detection dyes, the Santa Clara Water District offered rebates to smart irrigation users. The water district also mailed a great amount of literature to customers with suggestions and ideas on landscaping (Hildebrandt, 2008).

By retrofitting indoor fixtures and toilet flushers in the city of Seattle, Washington, the total per-capita water consumption (indoor and outdoor) dropped from 150 gpcd in 1990 to less than 100 gpcd in 2008 (Brzozowski, 2009). As a result, while the population grew by 16%, the water consumption in the city declined by nearly 26% from 1990 to 2008. The San Francisco Public Utilities Commission has a long-term plan to reduce water use by 4.5 mgd (17,000 m^3/day) by 2030 through conservation (Brzozowski, 2008). This reflects an approximately 6% reduction from 2005 water demands, which is impressive considering population growth during a quarter century.

Indoor conservation is not limited to residential use. Water can also be conserved in public, commercial, and industrial buildings. In Massachusetts, where 28% of domestic water is used in schools, a savings of 14% of total water demand was achieved by repairing leaky toilets and replacing them with low-flow toilet flush (Brzozowski, 2008). According to Michigan statistics, a lodging facility can conserve 13.5 gal (51 L) of water per guest room if bath towels and linens are not replaced daily. In hotels with typical 218 gal (825 L) use of water per occupied room, the usage can be reduced by 30% using water-efficient fixtures.

10.3.2 Urinals in Nonresidential Buildings

One of the most wasteful fixtures is automatic flush urinals, which waste 1 gal (3.8 L) of water per flush. Using waterless and low-volume urinals in offices and commercial buildings and malls will result in significant water savings. Waterless urinals can save, on average, 40,000 gal (150,000 L) per urinal per year. Falcon Waterfree Technologies and Sloan Valve are two manufacturers of water-free urinals; US Sloan Valve also makes low-flow automatic and hand-operated urinals that use one-eighth of a gallon (approximately 0.5 L) per flush. Figure 10.1 shows a water-free urinal. In an office building, substituting a waterless urinal for a 1 gal per flush fixture saves more than 3 gal of water per capita per day. The savings are much larger per urinal in a commercial building such as retail stores, where each urinal is flushed hundreds of times a day. Thus, to save water, either

FIGURE 10.1   A Sloan water-free urinal. (Photo by the author.)*

---

* Falcon is another company which makes waterless urinals. The Falcon water-free urinals cost from approximately $300 for WES-4000 and WES-5000 to $450 for WES-1000 models. Cartridges for these urinals cost approximately $40 and, depending on application, need to be replaced two to three times a year.

water-free urinals, hand-operated low-flush, or timer jets should be used in lieu of the automatic 1 gpf valves. The timer can be set to operate at desired intervals and only when a building or facility is in use. Using such urinals in hotels, hospitals, malls, retail stores, and large office buildings also saves thousands of dollars in water bills, year after year.

10.3.3 OTHER INDOOR SAVING TIPS

In addition to water-conserving fixtures, the water can be saved in a number of ways as follows:

- Changing our habits. An example is turning off the faucet while brushing our teeth or shaving. This change alone can save a few gallons a day.
- Making simple repairs. A faucet with a slow drip can waste over 250 gal of water a week. Dripping faucets and leaky flapper valves in toilet tanks take only a few minutes to fix, which can save hundreds of gal a week.
- Using low-flow toilet tanks. Niagara Conservation now makes an ultra high efficiency toilet (UHET) flusher, named Stealth System. This flusher is the world's first 0.8 gpf (3 L) UHET and can save up to 40,000 gal (150,000 L) of water per toilet tank, a year.
- Retrofitting plumbing fixtures at offices, malls, and public buildings. Replacing 1 gpf urinals by either water-free urinals or timing devices save thousands of gallons of water per urinal annually.
- Placing a displacement bag in an old toilet tank. This can save up to 1 gpf. An empty detergent jar (or 1 gal milk jar) filled with water may be used in lieu of commercially available displacement bags.
- Replacing showerheads with high-efficiency heads. Niagara Conservation now offers a Tri-Max showerhead rated at 0.5/1.0/1.5 gpm (1.9, 3.8, and 5.7 Lpm).
- Insulating hot water pipes. This reduces the time the tap is left open for the water to get hot. The most important pipe sections to insulate are the first few feet of the line entering and exiting the hot water heater. The insulation will also reduce the heat loss when the water is running between the water heater and the tap.
- Installing heat traps (one-way valves in the cold- and hot-water lines) when the hot water heater is up for replacement. These valves prevent hot water rising out of the heater and cold water from falling into it. At a cost of about $30–$40, the valves pay for themselves in less than a year.
- Using energy-efficient (same as water-efficient) clothes washers (i.e., washing machines). The front-loading washing machines use approximately one-third less water than top loaders. Also, the setting of the water level should be proportional to the load. Washing a large load saves more water than washing small or medium loads.
- Incorporating more water-efficient industrial and commercial processes.
- Incentives to encourage conservation such as rebates for purchasing high-efficiency washing machines and dishwashers, water-free urinals, and low-flow toilets.

10.3.4 ECONOMY OF WATER-SAVER FIXTURES

Some water-saver fixtures are very inexpensive and easy to install. A faucet aerator costs under a dollar, but can cut the indoor water consumption by as much as 6%. It takes only minutes to replace an old showerhead with a new water-efficient nozzle. The new nozzle costs a few dollars but results in substantial savings in both water and energy bills. Thus, water-saver showerheads and aerators are the first conservation measures to be taken at every home.

Ultra low-flow toilets cost from $80 to over $500; however, they reduce the water use from 5 gpf (19 Lpf) for old fixtures or 3.5 gpf (13.2 Lpf) for two-piece tanks to 1.6 gpf (6 Lpf). The Niagara

UHET, as indicated, uses only 0.8 gpf (3 Lpf). Thus low-flow toilets save a family of four from 20,000 to 40,000 gal of water annually. As a result, new water-efficient toilets can pay for themselves in less than 2 years.

Using water-efficient (normally referred to as energy-efficient) washing machines can save over 7000 gal (26,500 L) of water annually in a household with a family of four. While it may not be economical to replace a good working washer with an energy-efficient one, it certainly makes sense to buy a water-efficient machine when the old one is up for replacement. A side loader washing machine should be considered for more savings.

The use of low-volume toilets in nonresidential buildings reduces the water use significantly more than in residential dwellings. The average daily water savings per 1.6 gpd toilet range from 57 gpd for wholesale to 16 gpd for hotels and motels. The intermediate savings are 47 gpd in restaurants, 37 gpd in retail, and 30 gpd in office buildings. In fact, replacing a 3.5 gpf toilet with a 1.6 gpf fixture will save an estimated 1.9 gpcd for males and 5.7 gpcd for females. Thus, in large office buildings, the water savings per each low-flow toilet are significantly greater than the average 30 gpd cited before. The use of 0.8 gpf toilets will save on the average 2.7 gpcd in offices and over 15 gpcd in homes.

10.4 OUTDOOR CONSERVATION

10.4.1 An Overview

Outdoor demands for water are far greater than indoor demands. As such, more water can be saved through outdoor conservation measures. Until recently, however, little consideration was given to conserving outdoor uses. Thus, agriculture, which is the largest water user, needs to implement a lot more conservation measures than other sectors. In the United States, outdoor conservation measures are just coming to life. The AWWA estimates that between 50% and 70% of our tap water still goes to outdoor uses such as lawns and gardens (Hildebrandt, 2006). Agriculture is the largest outdoor water user. In California, approximately 25.8 million ac-ft (32 billion m^3) of water is used for agriculture alone. This is slightly over 41% of the total annual water use in the state.

Outdoor water uses in a single-family home include lawn and landscape irrigation, car washing, decks and driveway cleaning, and filling swimming pools. Of these, over 85% goes to watering lawns and landscaped yards.

According to the US Geological Survey (USGS) estimates, the average daily residential water demand in the United States was 26 plus billion gallons (100 million m^3) in 1995 (Solley et al., 1998). This amounts to 101 gpcd (380 Lpcd). An AWWA study (Mayer et al., 1999) estimates indoor uses in a single-family home at 69.3 gpcd (270 Lpcd). The difference, which is 31.7 gpcd (110 Lpcd), is attributed to outdoor domestic uses. This average, however, is not representative of outdoor uses in a single-family home considering that nearly one-half of the people in the United States live in apartments, town houses, condominiums, and high-rise buildings, especially in large cities such as New York, Chicago, and Los Angeles, just to name a few.

It is evident that outdoor uses are much higher than the previously indicated average per-capita figure in large suburban family homes and especially affluent homes. Local climatic conditions and landscape design also result in outdoor residential water demand that differs significantly from the national average. In fact, actual outdoor water use can range from 1 to 2 gpcd in apartment buildings to more than 200 gpcd in a large single-family home. The average daily outdoor water use in the United States varies from 20 gpcd in Seattle, Washington, to 180 gpcd in Scottsdale, Arizona (Mayer et al., 1999). Figure 10.2 shows a survey of the average indoor and outdoor single-family residential water use for 14 cities in the United States and Canada.

It is to be noted that agriculture is by far the largest outdoor water user. Based on a US Geological Survey study (USGS, 2008), almost 60% of all the world's freshwater withdrawals are used for irrigation. In 2000, withdrawals in the United States were estimated at 137,000 mgd (518×10^6 m^3/day)

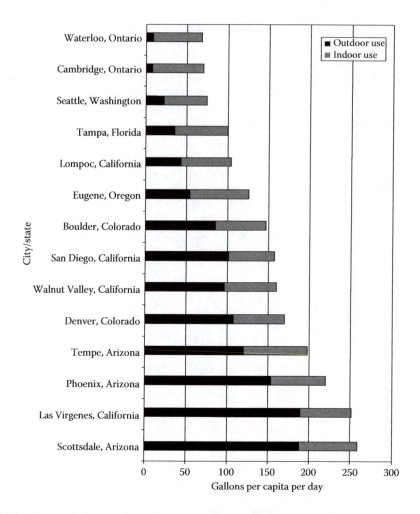

FIGURE 10.2 Average indoor–outdoor water use in single-family homes for certain North American cities. (From Mayer P.W. et al., AWWA Research Foundation and AWWA, Denver, 1999.)

or 153 million ac-ft (189 billion m³) per year. This represented 40% of total freshwater withdrawals and, when excluding water uses by the thermoelectric power industry, 65% of total withdrawals. Over 85% of the water uses for irrigation were in the 17 contiguous western states where the average annual precipitation is less than 4 to 20 in.—insufficient to support crops. California was the largest water user, consuming 22% of the total withdrawals. Figure 10.3 shows the trend in irrigation water uses and population between 1950 and 2000. The figure indicates that the withdrawals for irrigation had increased to approximately 150 billion gal per day (bgd) (567 million m³/day) by 1980, but since have been stabilized to 137 bgd (518 million m³/day). This represents an average annual irrigation use of approximately 30 in. (75 cm).

Also, about 90% of the water used for domestic or industrial needs is eventually returned to the environment, replenishing surface and groundwater supplies, which can be used again. However, only about one-half of the water used for irrigation is reusable. The rest is lost by evaporation or evapotranspiration into the air and vegetation growth. This indicates that measures to conserve water for irrigation, whether agricultural or domestic, will result in a significant water savings. Thus, it makes sense to give outdoor conservation priority.

According to the EPA's Water Sense Program, American homes, on average, use approximately 260 gal (980 L) of water daily. However, during summer months, the daily use can be as high as

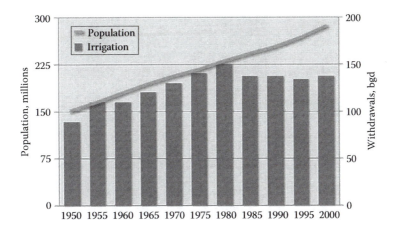

FIGURE 10.3 Population and irrigation withdrawal trends, 1950–2000. (From USGS [US Geological Survey], Irrigation water use, water science for schools [http://ga.water.usgs.gov/edu/wuïr.html], November 2008.)

1000–3000 gal (3800–11,400 L). It is estimated that at least 50% of the water used outdoors goes to waste due to evaporation, deep percolation, or runoff. Most water agencies are aware of this situation and many are working with municipal officials to offer rebates for more efficient irrigation systems such as rain sensors, weather-smart controllers, and drip irrigation. Some communities also offer rebates to property owners for replacing their conventional spray nozzles with low-flow rate nozzles (called low-precip nozzles in the irrigation industry).

Selection of plants is important in conserving irrigation water. Turf and lawn, which have shallow roots, require more watering than many other types of plants. Native grasses, bushy plants, and also deep rooted plants require far less watering than lawn. Also, landscape alternatives, including reduced lawn area, applying mulch around plants, and weed control, reduce irrigation needs. On a steep front yard in New Jersey, where the author planted junipers and pachysandra and placed mulch, the plants stabilized the slope and grew naturally. The plants formed a thick cover receiving only rainfall without any supplemental irrigation.

Soil composition also plays a key role in turf water consumption. Healthy soils produce healthy plants. A healthy soil has sufficient amounts of the organic matter that holds both water and nutrients, is not heavily compacted, and contains plenty of soil microbiology. Plant roots go deeper in the soil and require up to 20% less water. In Nevada and California, lawns and grass are being replaced by water stingy plants, such as cacti and succulents. Also, rocks are used in lieu of grass along public roads and highways.

Peter Landschoot, professor of turf grass science at Penn State University, foresees that the future advances in conservation will include grasses that require less water. However, grass species consideration in the turf grass industry has yet to be regulated and, also, people select species based on their own preferences and past experience. People in the United States may be overly obsessed with lush green grass and overwater their lawns, assuming that the more you water the lawn, the greener it gets. Watering twice a week is more than enough in many parts of the United States. Using sprinklers without rain sensors to irrigate lawns is another source of water waste; they turn on even when it rains. By watering the lawn in his former yard once every 3 dry days, the author had a healthier lawn than his neighbor's lawn, which was watered by sprinklers 1 hour every day, with excess water overflowing to streets and draining into a street inlet. In addition, automatic timer sprinklers, which are often not adjusted by homeowners with changes in season, water at the same rate in September and October, when the water need for irrigation is less than half that of July and August in many parts of the United States.

Irrigation technology has changed significantly during the past 10 or so years. Following an Irrigation Association conference in New Orleans in 2002, Smart water application technology (SWAT)

FIGURE 10.4   RainBird rotary nozzle.

protocols were created. The first self-adjusting smart box controllers were introduced by ET Water in 2005. Since then, over 20 major manufacturers like Toro, Hunter, Irrometer, and Rain Bird have improved their products. ET Water, headquartered in Novato, California, received a SWAT rating of nearly 100% effectiveness for its self-adjusting smart box controller in 2005 and it continues to advance its controllers (Corum, 2011). Smart box controllers can be installed in a new system or replace an existing one.

The newest technologies include the Toro precision spray nozzle and the Hunter MP rotary nozzle, both designed to improve water efficiency. The MP rotor emits multiple distinct streams of water at one-third of the rate of sprinkler heads. Smart box controllers have kept abreast with wireless technology. Controllers can now be linked to PCs and smart phones. Smart controllers come in weather-based and soil-based versions. The former creates an irrigation schedule based on evapotranspiration, rainfall, and radiation data taken from local weather stations; the latter controllers operate directly according to soil moisture readings of the sensor buried in the ground. Therefore, the latter can be more accurate than the former. In Frisco, Texas, which is one of the faster growing cities in the United States,* all new homes built since 2007 are required to install a smart controller from SWAT's approved list. The new smart controller systems are more efficient than sprinklers with timers and rain sensors. They provide significant conservation relative to the old sprinklers, especially those lacking any moisture sensors that would turn on even when it rained.

Nozzles now can generate larger droplets that are less affected by wind, minimizing airborne evaporation. Figure 10.4 depicts a rotary nozzle head manufactured by RainBird (http://www .rainbird.com) and Figure 10.5 shows the sprinkler in operation. Although low-precipitation nozzles use less water than before, they may be still using water inefficiently (Von Minden, 2013). Low-volume drip irrigation systems are designed to provide the needed amount of water to trees, bedding plants, street medians, and container plantings.

Easy-to-use remote control devices can regulate the amount of irrigation water based on seasonal and weather conditions and types of plants. An early study conducted by the University of Nevada, Las Vegas (UNLV), found that 20% water savings can be achieved by the use of "smart controllers" (Hildebrandt, 2006). The savings are expected to be much greater on the East Coast, in the Midwest, and in the Northwest with cooler fall weather than in Las Vegas. More recent studies indicate that combining these innovative technologies alone can reduce irrigation water usage by as

---

* Frisco's population grew from 1845 in 1970 to 110,000 in 2010.

FIGURE 10.5 Rotary sprinkler in operation.

much as 60%. Some states, including California and Florida, now have certification programs for landscape irrigation professionals (Poremba, 2009).

Efficient irrigation is not limited to residential landscaping, so are irrigating agricultural lands and public and private properties such as parks, golf courses, and commercial and industrial properties. As was noted, irrigation consumes over one-half of all water used in this country and many other places around the world. Therefore, recycled water and rainwater from roofs and pavements, rather than municipal potable water, should be used to the maximum extent practicable for irrigation.

Overwatering is perhaps the largest source of water waste in this country and abroad. In Colorado, where the drought during the past decade has raised public concern about the need for water conservation, overwatering has been found to be an important source of water waste in landscape irrigation. This is exemplified in the following:

Data on three residential developments in Colorado Front Range by a Water Engineering Company of Denver, Colorado, showed an average annual water use of 62.9 to 79.9 in. (160 to 203 cm) during a 2- to 3-year period between 1999 and 2001 at these developments (Clary et al., 2006). Compared with the estimated annual irrigation requirement of 29 in. (73.7 cm), the water consumption for irrigation was 2.2 to nearly 2.8 times larger than needed. Considering that over one-half of treated water along the Front Range is used for landscape irrigation, this overwatering represents a large waste. An estimated savings of more than 33% in potable water would be achieved by eliminating this waste.

Repair, replacement, and retrofitting of outdoor watering systems is another source of conservation. The Denver Zoo, for example, implemented a plan in 1999 to reduce the water usage of the flamingo pond. The plan, which consisted of repair, retrofitting, and replacement of the system, dropped the water usage from 300 million gal to 75 million gal in 2005 (Ramos, 2006).

During the 1990s a water-wise landscaping, called xeriscaping, was introduced in this country (Weinstein, 1999). This type of landscaping can reduce the irrigation water needs of a traditional landscaping by one-half. Xeriscaping is a water-wise landscaping. Not only does it use significantly smaller quantities of water than grass, but it also requires less maintenance and is more aesthetically pleasing. Americans are traditionally obsessed by lush green lawns. To gain acceptance, it is important for the people to see good examples of xeriscaping. Unless you need to play or walk on it, there is no reason to have lawn. Xeriscaping does not have to use a plant palette; even a traditional landscape incorporating lawn can be xeriscaped if it is properly arranged.

In Utah, which is the second driest state in the nation and more than 65% of the state's treated water is used for traditional turf, xeriscaping is growing in popularity in Salt Lake City (Ramos, 2007).

As more homeowners are becoming educated about xeriscaping, its application is expected to spread nationwide.

After an initial study of xeriscape water saving, the city of Austin in Texas initiated the "Xeriscape It!" rebate program in 1993 for residential water customers. The program offers $0.08 per square foot with up to a $240 rebate to participants who install drought-tolerant buffalo and Bermuda grass, water-stingy shrubs, and water-efficient ground covers in areas receiving more than 6 hours of sun daily (Fuller et al., 1995).

A highly water conserving method of irrigation used to be practiced in Persia centuries ago. In this practice, which may be referred to as jar irrigation, a hollow clay jar was placed near the roots of vegetable and fruit plants, such as melons, and filled with water when needed. As the soil moisture was taken up by plant roots, the soil became dry and extracted water from the clay jar through capillary suction. Thus, little water was lost through evaporation and none was wasted by irrigation. Thus, the Persians could cultivate in arid parts of the country where the rain was scarce and insufficient to grow vegetables and fresh produce.

It is to be noted that more important than innovative technologies to conserve water outdoors is to change public perception of water availability. Many of us are still under the illusion that existing water supplies are limitless. Also, at current water rates in the United States and many other countries, it is cheaper to waste water than to pay for an efficient, smart controller sprinkler. The best long-term solution to conservation is to educate people, especially at a young age.

To educate county residents about a potential water crisis and to change cultural attitudes about landscaping, San Diego County has constructed a variety of microgardens on a 4.2 acre (2 ha) of land owned by Cuyamaco Community College, east of San Diego (Corum, 2008). The garden includes, among others, a vegetative garden and a native plant garden. The exhibits in the garden illustrate the water savings of landscape versus lawn. One exhibit, for example, indicates that a stretch of lawn uses 25,000 gal (95 m³) of water per year. But, another exhibit on a water-wise landscape covered with a small patch of lawn surrounded by water-stingy shrubs, perennials, and a small tree needs just 6000 gal (23 m³) per year. Throughout the garden are 60 signs covered with educational information about low water demand landscaping and xeriscaping, which is a trade name for a landscaping method employing drought-resistant plants to conserve water.

10.4.2 Conservation of Outdoor Water Use: A Summary

Outdoor water uses can be reduced in a number of ways (Pazwash, 2002):

- Water when needed. Often water is applied when the grass does not need it. Automatic sprinklers without moisture sensors are a good example. They turn on even when it rains or the ground is still wet from a prior rain.
- Water in the early morning or late afternoon to reduce losses. Watering under the hot sun, as many do, causes the loss of over one-third of water to evaporation. Water before 9:00 a.m. in the east and after 6 p.m. in the west, when fungus is not a problem.
- Water at a low rate. Many sprinklers water at a rate beyond that which the soil can absorb and, as a result, the water flows onto sidewalks and streets and goes into storm drains. This is especially the case for steep slopes. To avoid this loss, water should be applied at a slow rate and preferably at several short cycles, rather than a single, long burst.
- Let the grass grow longer. Taller grass requires less water than shorter grass, as it gives the soil better cover, reducing sunrays and evaporation.
- Use mulch around plants. Mulch traps moisture beneath it, giving plants a steadier, longer water supply.
- Select landscape alternatives, using smaller lawn area. Perennials, wildflowers, bushes, and native gasses use a lot less water than water-thirsty lawn such as Kentucky bluegrass—better yet, xeriscape.

- Use drought-tolerant, low-maintenance plants in lieu of water-thirsty lawn.
- Avoid automatic timer sprinklers. These are one step backward for conservation. Many homeowners set them and forget to turn them off when watering is not needed. Turn them on manually, as needed.
- Install rain shut-off sensors on automatic sprinklers.
- Install water-efficient sprinklers equipped with rain sensors and "smart controllers." Note that sprinklers can be a step back from conservation if they lack rain sensors as they can turn on even when it rains.
- Adjust irrigation systems for seasonal changes.
- Use a broom, rather than a hose, to clean driveways, patios, sidewalks, and loading docks.
- Use treated grey water or wastewater for irrigating.
- Harvest rainwater. The roof rain collected in rain tanks and/or barrels provides an ideal source of water for sprinkling lawns and irrigating plants and bushes.

10.4.3 OTHER WATER CONSERVATION MEASURES

A water conservation measure known as the Water Conservation Hotel and Motel Program (Water CHAMP) began as a pilot study in 2002 in southwest Florida. By 2006, this program had expanded throughout 16 county districts. The program encourages hotel and motel guests to reuse their towels and linen during their stay to conserve water and reduce the amount of detergent wastewater. A survey of Water CHAMP indicated that the participants saved an average of 17 gal (65 L) per occupied room per day. In 2012, this savings was estimated at 149 million gal (563×10^6 L) of water. Likewise, a program named the Facilitating Agricultural Resource Management System (FARMS) has been set up to reduce the agricultural water use, which is a major water user throughout the district. Through the FARMS program, the district expedites the implementation of agricultural BMPs to reduce groundwater withdrawal from the upper Florida aquifer, improve water quality, and restore the area's ecology. The district projects that the agricultural industry could reduce groundwater use by 40 million gal per day through this project by the year 2025. Florida-Friendly Landscaping (FFL) and Florida Water Star (FWS) are other programs that encourage conserving water for landscaping and water efficiency in appliances, plumbing fixtures, and water recycling systems, respectively.

10.5 WATER REUSE

Water reuse involves reclamation, treatment, and recycling of wastewater, grey water, or storm water. The standards for reclamation and reuse in the United States are the responsibility of the state and local agencies. While there is no federal regulation for reuse, the EPA has developed guidelines for water reuse. The EPA guideline was first published as a research report in 1980 and was updated in 1992 and 2004. Recognizing the need for national guidance on water reuse regulations and planning, the EPA developed comprehensive up-to-date water guidelines in 2012. This document, titled *Guidelines for Water Reuse*, is over 640 pages and can be downloaded free of charge in PDF format at http://www.waterreuseguidelines.org. Appendix 10B includes a copy of Table 4.4 in this document, which covers urban, agricultural, industrial, and environmental water reuse and groundwater recharge. Table 10.2 here presents the state of New Jersey standards for the use of reclaimed water in irrigation and construction. This table is adapted from Appendix A: Effluent Reuse Treatment Guideline Table in a NJDEP Technical Manual titled "Reclaimed Water For Beneficial Reuse" (2005). In arid and semiarid parts of the United States that experience water shortages, the reuse of wastewater has been practiced for some time. Four states—Texas, California, Arizona, and Florida—account for over 80% of total water reuse in the United States. In the past few years, water reuse has gained popularity nationwide. A reason behind this trend is that some

TABLE 10.2
State of New Jersey Water Quality Standards for Water Reuse

Types of Reuse	Treatment and RWBR[a] Quality	RWBR Monitoring	Comments
RWBR public access systems: Examples include golf course spray irrigation, playground or park spray irrigation, vehicle washing, hydroseeding	Fecal coliform 2.2/100 mL, 7-day median, 14/100 mL maximum any one sample Minimum chlorine residual 1.0 mg/L after 15 min contact at peak hourly flow or design UV dose of 100 mJ/cm[2] under maximum daily flow 5 mg/L TSS maximum, 2 NTU maximum turbidity in UV applications Total nitrogen (NO_2+NH_3) 10 mg/L[b] Hydraulic loading rate 2 in. per week[c] Secondary[d] Filtration[e] Permit levels must be met	Continuous online monitoring of turbidity and CPO or UV criteria[f] Operating protocol required User/supplier agreement Annual usage report	• A chlorine residual of 0.5 mg/L or greater in the distribution system is recommended to reduce odors, slime, and bacterial regrowth • Chemical (coagulant and/or polymer) addition prior to filtration may be necessary • Loading rates can be increased based on a site-specific evaluation and department approval • Total nitrogen limitation can be less stringent if site evaluation submitted is approved by the NJDEP • Additional requirements dependent on application
RWBR for agricultural edible crops systems: Examples include irrigation of any edible crop that will be peeled, skinned, cooked, or thermally processed before consumption/commercially processed foods[i]	Fecal coliform 2.2/100 mL, 7 day-median, 14/100 mL maximum any one sample Minimum chlorine residual 1.0 mg/L after 15 min contact at peak hourly flow or design UV dose of 100 mJ/cm[2] under maximum daily flow 5 mg/L TSS maximum, 2 NTU maximum turbidity in UV applications Total nitrogen (NO_2+NH_3) 10 mg/L[b] Hydraulic loading rate 2 in. per week[c] Secondary[d] Filtration[e] Permit levels must be met	Continuous online monitoring of turbidity and CPO or UV criteria[f] Operating protocol required User/supplier agreement Annual usage report Annual inventory submittal on commercial operations using RWBR to irrigate edible crop	• A chlorine residual of 0.5 mg/L or greater in the distribution system is recommended to reduce odors, slime, and bacterial regrowth • Chemical (coagulant and/or polymer) addition prior to filtration may be necessary • Loading rates can be increased based on a site-specific evaluation and department approval • Total nitrogen limitation can be less stringent if site evaluation submitted is approved by the NJDEP • Additional requirements dependent on application

(Continued)

TABLE 10.2 (CONTINUED)
State of New Jersey Water Quality Standards for Water Reuse

Types of Reuse	Treatment and RWBR[a] Quality	RWBR Monitoring	Comments
RWBR restricted access systems and nonedible crops: Examples include irrigation of fodder crops or sod farms or other areas where public access is limited, such as landscaped areas within a secured perimeter	Fecal coliform 200/100 mL, monthly average, geometric mean 400/100 mL maximum any one sample[j]; Minimum chlorine residual 1.0 mg/L after 15 min contact at peak hourly flow or design UV dose of 75 mJ/cm² under maximum daily flow; TSS[g]; Total nitrogen 10 mg/L[b,j]; Hydraulic loading rate 2 in. per week[c,j]; Secondary[d]; Permit levels must be met	Submission of standard operations procedure that ensure proper disinfection[h]; User/supplier agreement; Annual usage report	• A chlorine residual of 0.5 mg/L or greater in the distribution system is recommended to reduce odors, slime, and bacterial regrowth • Loading rates can be increased based on a site-specific evaluation and department approval • Total nitrogen limitation can be less stringent if site evaluation submitted is approved by the NJDEP • Additional requirements dependent on application
RWBR for construction, and maintenance operations systems: Examples may include street sweeping, sewer jetting, parts washing, dust control, fire protection, and road milling	Fecal coliform 200/100 mL monthly geometric mean, 400/100 mL weekly geometric mean sample; TSS[g]; Secondary[d]; Permit levels must be met	Submission of standard operations procedure that ensure proper disinfection[h]; User/supplier agreement; Annual usage report	• Worker contact with RWBR shall be minimized • No windblown spray • Additional requirements dependent on application
RWBR industrial systems: Example includes closed loop systems (e.g., noncontact cooling water, boiler makeup water)	Permit levels must be met	Submission of standard operations procedures that ensure proper material handling; User/supplier agreement; Annual usage report	• Worker contact with RWBR shall be limited to individuals who have received specialized training to deal with the RWBR systems • Additional requirements dependent on application

a RWBR = reclaimed water for beneficial reuse.

b The Total Nitrogen (NO_3+NH_3) limit may be less stringent than 10 mg/L. See report/study requirements in Guidance Manual under Engineering Report Section.

c The Loading Rate may be greater than 2 inches per week. See report/study requirements in Guidance Manual under Engineering Report Section.

d Secondary treatment for the purpose of this manual, refers to the existing treatment requirements in the NJPDES permit, not including the additional RWBR treatment requirements.

e Filtration means the passing of wastewater through a filtration system in order to reduce TSS levels to below the 5 mg/L.

f The continuing monitoring for chlorine produced oxidant (CPO) or UV criteria and turbidity (in either case) is to ensure that all RWBR has been properly treated to the high-level disinfection requirements. The UV criteria include the continuous monitoring of lamp intensity, UV transmittance and flow rate.

g The TSS requirements in the application applies to the existing treatment requirements as specified in the NJPDES permit for the discharge.

h The Standard Operations Procedure is a written document on what methodology has been employed to ensure all the RWBR has been properly disinfected to the required RWBR treatment levels identified in this manual.

i Commercially processed food crops are those that, prior to final sale to the public or others, have undergone chemical or physical processing sufficient to destroy pathogens.

j Applicable limit for restricted spray irrigation applications.

states have adopted "needs analysis" or assessment prior to approval of discharge permits. Many more states are expected to establish the needs assessment as the national resources of freshwater supplies continue to shrink.

As indicated, between 50% and 70% of residential water demand goes to watering lawns and gardens. About 39,000 gal (150,000 L) of water is used, on average, to manufacture one automobile, and over 2000 gal (7500 L) of water is needed to produce one barrel (42 US gal, 159 L) of oil. These needs could be satisfied using nonpotable water derived from recycling of waste water, grey water, or, even better, storm water. For many applications, wastewater requires advanced treatment processes; however, the treatment level can be lowered using grey water and even more so using storm water by capturing the runoff.

10.5.1 Wastewater Reuse

Reuse of water involves collecting wastewater, treating it as necessary, and redistributing it for nonpotable uses. The reuse and reclamation of wastewater has been a common practice in Europe for over 60 years. In this country the recycling of water began in the 1970s and has been limited to the use of treated wastewater. In arid and semiarid regions, including Southern California, Nevada, Arizona, and New Mexico, where water supply is short, there are severe restrictions on water use. Water shortage is even felt in places such as Atlanta, Georgia, Florida, and some other southeastern states that are near the end of water reserves. In these places the recycling of wastewater is on a rapid rise. There is also a growing trend for water recycling in the populated states on the East Coast, including Massachusetts, New York, and New Jersey, with an annual precipitation of over 40 in. (1000 mm). By 2020 it is estimated that 36 states will face serious water shortages.

The reuse of treated wastewater rose from 1.5 billion gal per day (bgd) (5.7×10^6 m³/d) in 1990 (Mays, 1996) to well over 2 bgd in 2008. This trend is estimated to grow at a rate of 10% to 15% annually. In California alone over 500,000 ac-ft (617×10^6 m³) of water is recycled a year. The long-term goal is triple that amount by 2020.

Major applications of treated wastewater are as follows:

- Nationwide—88% irrigation, 11% industrial, 1% recharge and wetlands
- Arizona—irrigation, followed by recharge and industrial use
- California—irrigation, followed by industrial and then recharge
- Florida—irrigation, followed by recharge, then industrial and then wetlands

Examples of wastewater reuse are the following:

- City of Santa Monica: This city, like other municipalities in Southern California, receives 12 to 14 in. (300 to 450 mm) of rain annually, but water use for irrigation creates off-season flows that pick up and transport contaminants. To address this water quality and to reduce runoff, Santa Monica has established a comprehensive watershed-wide plan to maximize permeability throughout the city and increase infiltration. Santa Monica's watershed approach is dual purpose and includes an ordinance for harvesting runoff from new developments. The plan also includes harvesting 300,000 gal (1140 m³) a day of dry-weather runoff that the Santa Monica Urban Runoff Facility is designed to divert from the ocean. The collected all-dry-weather flow and some of wet-weather urban runoff are treated for irrigation of the city's two parks and the cemetery.
- Tucson, Arizona: In Pima County, Arizona, raw wastewater is delivered to a treatment plant, then returned to Tucson Water for direct tertiary treatment via pressure filtration or indirectly through aquifer recharge and recovery. Tucson Water delivers 16,000 ac-ft (19.7×10^6 m³) of water annually to 1000 customers, primarily golf courses, parks, and

schools. Also, 700 single-family homes use the reclaimed water for irrigation (Lovely, 2012). The treatment plant filters wastewater up to 10 mgd (37,800 m³/day). Filters are composed of a mixture of sand and anthracite coal. Chlorine is added before delivering water to customers. The recharge–recovery system consists of eight recharge basins that can produce 7500 ac-ft (9.25×10^6 m³) of water annually. Thus, the recharge–recovery system produces nearly one-half of the overall recycled water.

- The town of Cary, located in the heart of North Carolina, near Raleigh, treats wastewater by removing suspended solids as well as biological and chemical pollutants that consume oxygen. Nitrogen and phosphorous are also removed. The treated water is used for irrigation and cooling. Some hotels also use it for their climate control systems. Through reuse, Cary's goal is to provide a 20% reduction in water consumption by 2015 (Hildebrandt, 2007).

- In Linden, New Jersey, treated wastewater is reused at the Linden Combined Sewer Plant instead of being discharged into local streams. The reclaimed water from the Linden–Roselle Sewerage Authority Wastewater Treatment Plant is filtered through Hydro-Clear pulsed bed sand filters supplied by US Filter's Zimpro products of Rothschild, Wisconsin. The filters, which consist of seven cells in concrete tanks, treat effluent from the treatment plant at an average rate of 4200 gpm (16 m³/min). After filtration, the water is pumped approximately 1 mile to the power station and is treated further to prevent scaling and then made available for two 10-cell mechanical draft-cooling towers (see http://www .water-technology.net/project-printable.asp?Project_ID=2488).

- In Oklahoma City, Oklahoma, three out of four wastewater treatment plants can deliver up to 15 million gal (57×10^6 L) of recycled water per day to industrial customers. This saves the city over 1 billion gal (3.8×10^6 m³) of drinking water annually (Chavez, 2012).

- In the state of Sao Paulo, Brazil, which has a population of 41 million and is the world's seventh most populated area, drinking water is becoming increasingly scarce. Recognizing the importance of safeguarding drinking water to San Paulo inhabitants, the state government issued new regulations in 2011 to restrict the industrial use of potable water. Industries now have to use recycled water for their operations.

10.5.2 Recycled Wastewater Market

Reclaimed-water applications, depending on the level of treatment, range from landscape irrigation to industrial cooling processes, toilet flushing, vehicle washing, agriculture, groundwater recharge, and drinking water supply augmentation. The first small urban reuse was born with the irrigation of Golden Gate Park in 1912. Now over 100 years later, numerous communities rely on highly treated reclaimed water. Tucson Water, for example, has been producing reclaimed water for irrigation and nonpotable uses since 1985. Due to population growth, it will not take long before reclaimed water will be viewed as a practical and acceptable, rather than nontraditional and unfavorable, resource. In 1992, only the Southeast and Southwest practiced water reuse. Now there is quite a bit of reuse in other parts of the United States. In 2009 over 660,800 ac-feet (815×10^6 m³) of recycled water was produced, of which 37% was used in agricultural irrigation.

While wastewater recycling is rapidly on the rise in this country, ironically the reuse of storm water runoff, which is more plentiful and less costly to treat than wastewater, is generally neglected. One plausible reason is that since wastewater is continuous, it forms a reliable source of available water supply. Another, perhaps more important, reason is that recycling wastewater is a multimillion dollar industry.

A technical report by BCC Research (2006) estimated that the total value of the US water recycling and reuse industry was $2.2 billion in 2005 and that this value was expected to grow at an average annual rate of 8.8%, reaching nearly $3.3 billion in 2010. This estimate is broken down to nearly

70% filtration products and the balance in disinfection and demineralizing products. According to this report, landscape and agriculture irrigation are the largest consumers of the recycled water. The use of recycled water was growing at an annual rate of 12.4% and 10.3%, respectively, through 2010 for these users. Although the industrial market for water reuse is growing the most rapidly at 14.2%, this market forms a small share of the overall market.

The amount of water reuse for all applications in the United States was growing at 11.1% annually through 2010. The future growth will depend on droughts, future EPA regulations on wastewater and potable water, increased public awareness, the expected replacement of existing wastewater treatment plants (BCC Research), and climate change (Means et al., 2005).

The filtration alone is a multimillion dollar industry. Wastewater is commonly filtered by activated carbon systems, multimedia systems, membrane filtration, and zero liquid discharge (ZLD) systems. The combined market of these systems reached $620 million in 2004, and this market exceeded the billion dollar mark in 2010. Among these, the membrane filtration systems share was 61.2% followed by multimedia systems at 16.9%. Companies such as US Filter, AquaTech International, Parkson, F.P. Leopold, Pall, and Severn Trent Services have already begun to meet the new demand.

The main users of water recycling systems are municipalities and industries. The municipal market is growing more rapidly than industry, mainly because of a generation gap. Many of the water treatment facilities have been in service for 20 years or more and are in need of modernization. Also, municipalities view water reuse and recycling as a new avenue for revenues.

Wastewater recycling is a rapidly growing industry in this country. California, alone, has more than 300 water recycling plants operating since 2004. Forty-eight percent of recycled water is used for agriculture, 21% for landscape irrigation, 14% for groundwater recharge, and 19% for other uses. The California Recycled Water Task Force estimates that the state can recycle enough water to meet 30% to 50% of the domestic water needs of its projected growth (Grumbles, 2012). An example of a water recycling project in California is as follows:

Orange County in California appears to be on the forefront of wastewater recycling technology. This county, with a population of over 3 million, is the fifth most populated county in the United States (Duffy, 2008). The county has been undertaking a wastewater recycling project through groundwater recharge. This project, which is the world's largest water purification plant and known as the Groundwater Replenishment System (GWRS), has been a successful project. The GWRS is a jointly funded project operated by the Orange County Water District and the Orange County Sanitation District. Through this project the wastewater goes through a three-stage advanced treatment process of microfiltration, reverse osmosis, and ultraviolet light with hydrogen peroxide, producing water that exceeds all state and federal drinking water standards. The project, which has been operating since 2008, produces up to 70 million gal (265×10^6 L) of high-quality water daily to meet the needs of nearly 600,000 residents in north and central Orange County and Disneyland visitors. This project also uses less than half of the energy needed to pump water from northern California to Orange County and other parts of Southern California and uses less than one-third the energy required to desalinate ocean water. The wastewater recycling in Orange County is indirect in that a significant portion of GWRS is injected back into the ground to bar seawater intrusion and to recharge the aquifer, which becomes part of the region's drinking water supply. (Brzozowski, 2013)

To save energy, decentralized reclamation is a growing trend. Water purveyors use and distribute water at the lowest cost to their customers. With aging infrastructure, the biggest challenge in the future is to make sure people accept higher costs for their own safe, reliable water supply. Utilities will have to investigate where they are getting the water from and what treatment technology is required to achieve the desired water quality they want. Since less than 1% of treated water is digested, there is talk about smaller drinking water systems. When aging water lines must be replaced, then a dual water system warrants consideration.

10.5.3 Reuse of Grey Water

Grey water (also appears as gray water* in the literature) refers to residential wastewater other than that from toilets and kitchen sinks. These include showers, bathtubs, bathroom sinks and washing machines. Grey water gets its name from its cloudy appearance and from its status as being neither fresh nor heavily polluted. According to this definition, kitchen sinks, which may contain significant food residues or high concentrations of toxic chemicals from household cleaners, may be classified as dark grey or black water. Wikipedia, however, includes all residential wastewater other than toilets as grey water (http://en.wikipedia.org/wiki/Greywater).

Domestic wastewater is usually combined at the sewer so that grey and black waters are discharged together to rivers and streams after receiving treatment. Since the natural purification capacity of soil is millions of times more than that of water, dumping grey water directly on the soil is less ecologically damaging than sending highly treated grey water into natural waters.

Studies have established the presence of the same micro-organisms in grey water as those found in sewage, though in far lower concentrations. Due to lower levels of contamination, grey waters are much easier to treat than black waters. If collected using a separate plumbing system from that used for black water, grey water can be recycled directly within the home or garden or agriculture plot and used either immediately or treated first and stored. The simplest and least expensive means of recycling grey water is to direct it to garden or landscaping within a residential dwelling or a commercial facility. The diversion may be as simple as running a hose from a clothes washer out a window to a garden or it can be incorporated as a permanent part of house plumbing. When laundry grey water is delivered to a garden, a low-phosphate and -salt detergent must be chosen.

The use of grey water for landscape irrigation is becoming increasingly popular. Grey water may contain hair, detergents, pharmaceuticals, personal care pollutants, and small amounts of grease. Most of these impurities are degradable, but some may be sodium based, which can harm landscaping in arid climates.

Using grey water for toilet flushing requires a separate indoor plumbing line and may cause some potential problems with bacterial growth. For this reason, the use of grey water indoors is banned by many local health departments. To reduce risks to human health, residential grey water should be used outdoors only. Considering potential threats to human health and long-term impacts on plants, many states limit the reuse of grey water to landscaping irrigation.

Grey water comprises 50–80% of the overall wastewater. A single family of four generates between 100 and 160 gpd (380–600 L/d) of grey water. Where allowed, this water is enough to irrigate lawn and landscape in a single-family home. A 1999 study by the Soap and Detergent Association indicated that nearly 7% of US households were using grey water. Grey water is most popular in Southern California. In 1989, Santa Barbara County passed an ordinance requiring homes with lots larger than 2 acres to be provided with a grey water plumbing line.

Many western and southwestern states, including California, Nevada, Arizona, New Mexico, Idaho, and Utah, among others, have developed regulations or guidelines for grey water reuse. Some of these states, California, New Mexico, and Utah included, allow grey water for underground drip irrigation. In general, in those US states that adopt the International Plumbing Code, grey water can be used for underground irrigation and toilet flushing. And in those states that adopt the Uniform Plumbing Code, grey water can be disposed of in underground disposal fields.

A typical grey water recovery system includes an underground tank to settle sediments before the water gets into drainage lines. Fully engineered systems that incorporate a sump pump and sewage tank and deliver the water through subsurface drainage are available commercially. Among these are "Waterwise Greywater Gardener 230" (http://www.waterwisesystems.com/products/grey water-garden-230) and "Garden ResQ" (http://www.gardenresq.com/). Industrial facilities reuse their

---

* Gray water may be differentiated from grey water in that it also includes wastewater from kitchen sinks. Specifically, any wash water that has been used in the home, except water from toilets, is called gray water.

grey water to reduce the amount of wastewater discharged to municipal sewers to reduce sewer fees. Lee Valley Ice Center in Leyton, London, is the first example of a grey water system in an ice arena context, a concept that reduces the otherwise considerable water use.

10.5.4 TREATMENT OF WASTEWATER AND GREY WATER

Reuse of wastewater requires a fair amount of treatment. The level of treatment depends on the application and varies from secondary to advanced filtration and disinfection. The requirements for treatment and the permissible use of treated water vary to some extent from state to state.

Treated water is used directly for agricultural and landscape irrigation, and in industrial cooling towers, fire fighting, and toilet flushing. Indirect uses involve groundwater recharge through infiltration and injections to improve water quality. The water can be infiltrated through basins or injected to aquifers during the rainy season and pumped out in the dry season. In Canberra, Australia, where this type of indirect reuse has been practiced for some time, researchers have found that bacterial and viral pathogens have low rates of survival in aquifer environments (ASCE, 2002).

While grey water may be used with primary treatment, namely sedimentation, in certain irrigation applications wastewater requires, at a minimum, a secondary treatment. Even with secondary treatments, which involve biological oxidation and disinfection, it is only allowed for certain applications. Receiving advanced treatment, including chemical coagulations, filtration, or advanced disinfection, the treated wastewater can be used in more applications. The application of treated grey water and wastewater is as follows:

- Secondary treatment
  - Irrigation of orchards and vineyards
  - Nonfood crop irrigation
  - Underground landscape watering
  - Groundwater recharge of nonpotable aquifers
  - Industrial cooling towers
  - Wetlands and wildlife habitat augmentation
- Advanced treatment
  - Home gardening
  - Lawn and golf course irrigation
  - Food crop irrigation
  - Toilet flushing
  - Car washing
  - Paper mills
  - Construction activities such as concrete mixing
  - Indirect potable reuse through recharge of potable water aquifers

With advanced treatment, grey water can be used in applications other than those listed previously, including filling artificial lakes. The reuse of water for toilet flushing requires a separate plumbing and it is far more practical in public restrooms in malls, large office buildings, and the like than in single-family homes. Cape Coral, Florida, appears to be the first municipality that approved the reuse of recycled wastewater for household toilet flushing. This town has a dual water system comprising a separate line for nonpotable uses (mostly irrigation) that is supplemented by treated wastewater (Godman and Kuyk, 1997). In June 2008, the Oregon State Plumbing Board passed new standards that allow homeowners to install systems that reuse wastewater for flushing toilets. For a general in-depth study of wastewater treatment, the reader is referred to Tchobanoglous et al. (2003). Table 10.2 (referenced earlier) presents the state of New Jersey's water quality standards for the use of wastewater and grey water, as well as for applications varying from crops, very low health hazard potential, to irrigation of parks, playgrounds, and golf courses. A summary table from the EPA 2012 guidelines for water reuse is included in Appendix 10B.

10.6   REUSE OF RAINWATER AND STORM WATER RUNOFF

Storm water runoff is far more plentiful than wastewater here in this country and, for that matter, in most countries around the world. It is a vast water resource that can be and should be captured and used. Traditionally, storm water management practitioners viewed storm water runoff as a waste to be disposed of in a regulated manner. While this view is gradually changing, many practitioners have yet to incorporate storm water conservation and reuse measures in their storm water management planning.

Wastewater has been reused for nearly 40 years in this country and even longer in Europe while the reuse of storm water runoff has been ignored. Whatever the reason may be, it is unrelated to the quantity and/or availability of storm water runoff.

Rainwater harvesting has been practiced for thousands of years by ancient civilizations. As an organized industry, rainwater harvesting is still in its infancy. There exist no national standards regulating the use of rainwater, although various states and municipalities have begun adopting laws for its use. The rainwater harvesting industry has a national organization: the American Rainwater Catchment Systems Association.

Two states, Georgia and Texas, are frontiers in rainwater harvesting possibilities. In the last decade Georgia has published a set of *rainwater harvesting guidelines* and Texas published the *Texas Manual on Rainwater Harvesting* (2005), which provides guidance for many rainwater catchment agencies across the United States. A measurable impact of this manual was a 2-year educational effort in Brownwood, Texas, that resulted in development of a rainwater collection system with a capacity of over 100,000 gal (380,000 L). However, as of 2011, the city of Dallas still had no ordinance for rainwater harvesting; therefore, the installer had to go through a lengthy zoning application process. Now the installers have to meet minimal standards for small systems but a much longer review process for larger systems.

Depending on its intended use, treatment requirements of rainwater vary from none at all to sophisticated systems producing drinking water. Figure 10.6 shows a layout of an underground

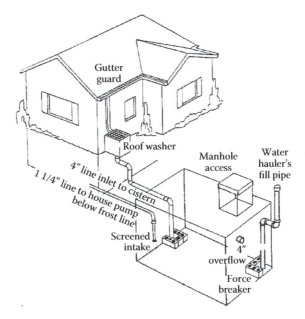

FIGURE 10.6   A roof catchment cistern system for outdoor use. (From Water Filtration Co., customer information brochure. Water Filtration Co, 108B Industry Rd., Marietta, OH 45750.)

cistern with roof washer, overflow pipe, and underground line to a pump for outdoor irrigation. A more sophisticated system may include a vertical filter that captures particles larger than 280 μm and an ultraviolet light for disinfection.

Apart from providing a water resource, rainwater harveting saves on water bills of consumers and reduces energy costs by water purveyors.

A Pepsico–Frito-Lay plant in Casa Grande, Arizona, has a system that demonstrates the potential of commercial water treatment operation. In this plant, water used in water production goes through extensive filtering and treatment processes, producing a purified water lower in metal and chemicals than the municipal water. The result is that 75% of the water supply in this plant comes from its water recovery operation and 25% comes from the city water (Goldberg 2013a,b).

Rainwater harvesting is most suited for roof rain, which is generally much cleaner than runoff from ground surfaces. In urban areas, and in particular cities, roof areas form a large portion of the overall impervious surfaces. If the runoff is harvested from pavements, it should be prefiltered. Without filtration, leaves and other organic debris decompose in the rain tank (or cistern) and create ammonic conditions, which support extensive bacteria growth. The silt also builds up in the bottom of the tank and the required regular maintenance to clean the tank may become cost prohibitive.

Rainwater harvesting is now a booming business; more than 10 new products come out every year. One such product is described herein. Rainwater Collection Solutions, a company in Alpharetta, Georgia, manufactures the original "Rainwater Pillow." This product is a horizontal flexible pillow that moves up when filled with rainwater and down when it empties. These pillows can be made with a capacity of 1000–20,000 gal (3800–76,000 L).

Collection of rainfall or runoff provides a vast supply of water for outdoor and indoor uses. It also mitigates adverse impacts of urbanization such as flooding, erosion, and pollution problems. The use of storm water runoff in general and roof rain in particular were introduced by the author in a 1994 paper (Pazwash and Tuvel, 1994). Subsequent papers by the author in 1997 (Pazwash and Boswell, 1997), 1999 (Pazwash and Boswell, 1999), and 2002 (Pazwash and Boswell, 2002) discussed the quantity of the roof rain (rainwater) and storm water runoff and presented suggestions for their collection and reuse. Examples of storm water and roof rain reuse are as follows:

Lakeland University has incorporated a recycling plan of runoff to create a decorative pond above an underground detention chamber next to a residential and food services building in Orillia, Ontario, Canada. The underground chamber receives runoff from an adjacent road after passing through a filtering system and also harvested grey water. The stored water is pumped up to the pond and is also used for toilet and urinal flushing. The construction began in the summer of 2009 and the building was opened over a year later. (Glist, 2010)

The city of Charlottesville, Virginia, uses captured rainwater for cleaning busses in the city's transit administration, maintenance and operations facilities, which opened in 2010. In this facility, the rain is harvested from approximately 26,000 ft² (2400 m²) of building roof. A similar recycling program has been employed in Clark, New Jersey. There, rainwater from the public works building is collected in a 5000 gal (18,900 L) tank. With the help of a booster pump, harvested water is used to wash cars on a concrete pad. The runoff from the concrete wash pad enters a rain garden located in the adjacent high school property to filter pollutants in runoff. Figures 10.7 and 10.8 depict the rain tank and the rain garden, respectively.

The Energy Coordination Agency (ECA), a nonprofit organization that was established in the 1980s and focuses on energy efficiency and weatherization, now has begun to get involved with storm water management. In the past few years ECA has installed a 3000 gal (11,300 L) cistern to collect the rain from its training center and use it for toilets and urinals and also for washing fleet vehicles. ECA has also been coordinating with Philadelphia Water Department on two different programs. One is the rain barrel program, through which they give free rain barrels to Philadelphia residents. EAC also offers workshops on installation and maintenance of rain barrels for residents. Over 3000 rain barrels had been given out as of the summer of 2013 (Goldberg, 2013b). The other program is the Rain Check Program, which trains two groups of contractors; one group identifies measures to reduce storm water runoff from residential homes and the other performs the installation. Five "green tools" have been identified for runoff reduction; these are downspout planter boxes, rain gardens, depaving of impervious surfaces, porous pavers, and yard trees.

FIGURE 10.7 Rain tank at public works building in Clark, New Jersey. (Photo by the author, 2014.)

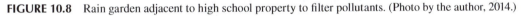

FIGURE 10.8 Rain garden adjacent to high school property to filter pollutants. (Photo by the author, 2014.)

10.6.1 QUANTITY OF URBAN RUNOFF

In the United States, with an average annual precipitation of 30 in. (760 mm), 1600 ac-ft of water falls on every square mile (760,000 m^3 per square km). Conservatively estimating the average annual runoff at 30% of the rainfall, more than 156 million gal of storm water runoff is discharged from every square mile of a typical suburban development. This runoff amounts to annual water needs of approximately 4285 persons, based on a daily per-capita demand of 100 gal. For a suburban community of a typical 0.5 acre lot size, this runoff is nearly 2.5 times greater than the outdoor water needs of the community. This estimate, as shown in the following, is based on a 50 gpd per-capita outdoor water need; an average of 0.25 acre of public lands, namely streets, sidewalks, and open space per lot; and an average of four households per dwelling.

Number of lots = 640/(0.5 + 0.25) = 853
Annual outdoor water need per home = 4 × 50 × 365 = 73,000 gal
Annual outdoor water demand per mi^2 = 73,000 × 853 = 62.3 × 10^6 gal

In metric units, the runoff volume and the outdoor water needs in every 1 km^2 of single-family home residential district with an average lot size equal to 2000 m^2 are calculated as follows:

Rainfall volume = 1,000,000 m^2 × 0.76 m = 760,000 m^3
Runoff volume = 760,000 × 0.3 = 228,000 m^3
Number of lots = 1,000,000 m^2/(2000 + 2000 × 0.5) = 333
Average daily outdoor water usage ≈ 190 L/d per capita
Annual outdoor water usage per home = 4 × 190 × 365 = 277,400 L = 277.4 m^3
Annual outdoor water usage per square kilometer = 277.4 × 333 ≈ 92,374 m^3

These calculations represent the average national condition on a conservative basis. In many parts of the country the amount of rainfall is greater than 30 in. and the outdoor demand is far less than 50 gpd per capita considering that the growing season lasts less than 6 months a year. Some other parts, such as southwestern states, receive significantly less rainfall than the national average, have a hot, dry climate, and have a growing season of nearly 12 months a year. In these places a combination of outdoor water conservation programs, drought-tolerant landscape, and aggressive collection of runoff will help supply water for outdoor demands. Calculations tailored to the local condition will identify water availability and demands. In New Jersey, where the author resides, the calculations are exemplified as follows.

In New Jersey, where the average annual precipitation is 46.6 in., approximately 243 million gal of runoff occurs per square mile each year. For the suburban development exemplified before, this runoff amounts to over 284,800 gal per lot. Estimating average annual outdoor water demand of a single-family home of four at 50 gpcd, each dwelling needs 73,000 gal annually. Thus, the runoff from the lot is sufficient to supply the outdoor demand nearly four times over. In metric units, the average annual runoff and the annual water demands per square kilometer of the development are calculated as shown:

Average annual runoff = 1,000,000 m^2 (46.6 × 25.4/1000) × 0.3 = 355,090 m^3
Average annual runoff per lot = 295,910/333 = 1066 m^3
Average per-capita outdoor demand = 190 L/day
Average annual outdoor usage per dwelling = 365(190 × 4) = 277,400 L = 277.4 m^3
Annual outdoor demand per square kilometer = 277.4 × 333 = 92,370 m^3

According to his records, the author estimated the average annual outdoor water use in his former home in West Milford, New Jersey, at 13,300 gal during the September 1989–August 2001 period. This figure, which translates to 9 gal per day per capita, is nearly sevenfold smaller than the previously calculated average outdoor water demand. However, as indicated, the author did not overwater lawn and landscapes and never watered the wooded area in the hilly backyard of his 3/4 acre lot. In his current residence in Paramus, New Jersey, where nearly 40% of the lot is covered with lawn and landscape, the outdoor use averages 190 gal (718 L) per day. This is less than 200 gal for a single-family home with four occupants on which the previous calculations are based. According to the author's records, the average outdoor water use during the growing season was approximately 350 gpd.

Evidently, the rainfall and outdoor water needs vary depending on climatic conditions, the type of development, and the nature of outdoor water demand. The calculations, however, indicate that outdoor demands in many parts of the country can be met by using urban runoff alone.

A study for the city of Austin indicates that collecting only 30% of runoff from paved areas provides sufficient water for 330 days of demand in the city (Hall, 2005). In Los Angeles, California, six projects capture 1.25 million gal of water for every inch of rain (1.86 × 10^6 L/cm). Tucson, Arizona, with 12 in. of annual rainfall, enacted the nation's first municipal rainfall harvesting ordinance for commercial projects, effective June 1, 2010. Tucson officials hope that diverting runoff from parking lots and roofs will supplement current municipal supplies (Cutright, 2009).

FIGURE 10.9   Underground cistern with aeration towers in Yezd, Iran.

In urban areas, the runoff can be directed to natural depressions or reservoirs after attenuation through storm water detention basins and ponds. It can also be injected into the ground to replenish groundwater supply. The use of large underground cisterns to store surface runoff was a common practice in Persia and many other ancient civilizations. There still are many of these cisterns in villages and rural areas. Some of these are constructed underground as architecturally appealing dome-shaped buildings with open windows on the sides and ceilings for natural aeration. Figure 10.9 shows an old underground cistern with four aeration towers in Yezd, an ancient city in the arid central plateau of Iran. The reuse of urban runoff has yet to be implemented in the United States.

10.7   RAINWATER HARVESTING

With an average annual precipitation of approximately 30 in. (760 mm), 18,700 gal of water falls on every 1000 ft^2 (76,000 L/100 m^2) of roof area in the United States At an average daily outdoor water demand of 50 gal per capita, this water is sufficient to satisfy the outdoor water needs of a single-family home of four for a period of over 3 months. It is ironic to see that such a large quantity of fairly pure water is wasted to storm drains, while wastewater is extensively treated and reused. Of course, due to the varied nature of precipitation, it is not feasible to collect all rainwater. However, collecting even a portion of this water produces a considerable water supply.

In New Jersey, the average annual precipitation varies from 51.8 in. (1316 mm) in Greenwood Lake, West Milford, to 40.3 in. (1024 mm) in Atlantic City. The state average is 46.6 in. (1184 mm). These rainfall figures are based on the National Oceanic and Atmospheric Administration's (NOAA's) published data of the monthly normal precipitation at 52 stations in New Jersey during three consecutive decades (the 1961–1990 period) (Owenby and Enzell, 1992). Based on the same data, the average rainfall during the growing season, namely the April 1–September 30 period, varies from approximately 28 in. (710 mm) at Long Valley to 20 in. (510 mm) at Cape May and Atlantic City. The state average is approximately 25 in. (635 mm).

A review of daily rainfall data at the New Milford, Newark Airport, and Atlantic City precipitation stations (which respectively represent north, central, and south Jersey) shows that the daily precipitation varies from less than 0.1 in. (3 mm) to over 6 in. (150 mm). The daily records during a normal, a dry (1995), and a wet (1996) year show that rainfalls of less than or equal to 1 in. (25 mm) account for 75% of the total depth of rainfall during the growing season, which is April 1 through September 30. Thus, collecting the rainfall of 1 in. (25 mm) or smaller from the roofs of residential buildings during the growing season alone provides over 11,700 gal from every 1000 ft^2 of roof area (47,600 L from every 100 m^2 of roof). On a statewide basis, capturing the roof runoff from residential buildings during the growing season provides nearly 54.1 billion gal (204 million m^3). This figure is based on 3,471,647 housing units (http://www.factfinder.census /servlet/ACSSAFFFacts) and an estimated 1000 ft^2 roof area per unit. Considering that a large percentage of the 8 million people in New Jersey live in apartment buildings and multifamily homes, the average outdoor need may be estimated at 20 gpcd (75 Lpcd). The rainwater can fully meet the outdoor water demand in this state.

10.7.1 HARVESTING ROOF RAIN

Roof rain can be stored in water tanks, which may be placed above ground or underground. Tanks are available in high-density polyethylene (HDPE), fiberglass, and stainless steel. Among these, fiberglass and HDPE tanks are the most economical for single-family homes. In addition to economy, these tanks have the following advantages over stainless steel tanks:

- Seamless construction, allowing easy cleaning and leak-free service
- Lightweight, less than one-half the weight of steel tanks
- Dent free and virtually maintenance free

HDPE tanks come in vertical, horizontal, and cone-bottom types (see Figure 10.10). These tanks are available through several manufacturers. Among these are Plasteel, Zerxes, Highland Tank and Manufacturing Company, and Snyder Industries. The price for HDPE tanks of up to 500 gal (1900 L) capacity is more or less the same regardless of the type. However, for larger capacities, vertical tanks are more economical than others. The price of Snyder vertical tanks, for example, ranges from $400 for a 500 gal (1900 L) tank to $1000 for a 2000 gal (7600 L) tank. The cost of shipping and installation must be added to these prices to arrive at the overall cost of a tank. Zerexe's fiberglass tanks are cylindrical and come in sizes varying from 600 to 50,000 gal (2.27–190 m^3). The 600 and 1000 gal (2.27–3.78 m^3) tanks are 4 ft (1.2 m) in diameter and the tanks up to 4000 gal (15.1 m^3) are 6 ft (1.8 m) in diameter. The length of these tanks ranges from 6 ft, 11-7/8 in. (2.29 m) to 21 ft, 11-1/2 in. (7.20 m) and they weigh from 500 to 1600 lb (226–723 kg).

The size of tank should be selected based on local variation of rainfall. As indicated, a tank sized for 1 in. (25 mm) of roof rain can collect a large portion of rain during the growing season in New Jersey. The same size tank is also suitable for all northeastern and eastern states from Maine to Virginia, where the rainfall distribution follows a similar pattern as that in New Jersey. Tentatively, a tank of nearly the same size appears appropriate for Midwestern states including Ohio, Indiana, Illinois, Missouri, Kansas, and Iowa. Alternately, the tanks may be sized for 90th percentile annual rainfall, which in New Jersey is 1.25 in. and, in the states of New York and Maryland, is 0.9–1.0 in.

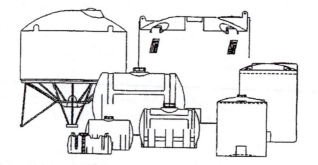

FIGURE 10.10 HDPE water tanks.

For a single-family dwelling with 1000 ft^2 or 100 m^2 of roof area, the size of water tank to retain 1 in. (25 mm) of rain is calculated as follows:

$$1000 \times (1 \text{ in.}/12) \times 7.48 \text{ gal/ft}^3 = 623 \text{ gal}$$
$$100 \times 25 \text{ mm}/1000 = 2.5 \text{ m}^3 = 2500 \text{ L}$$

Of course, sizing tanks for a larger rainfall results in a greater rainwater harvesting. Rain tanks may be sized for 0.4 in. (10 mm) rainfall in arid southwestern states, where rainfall is mostly light, and 1.25 in. (30 mm) for southeastern sates and the Gulf region, where rains are normally heavy. Table 10.3 lists suggested rain tank size and the estimated quantity of harvested roof rain per 100 m^2/1000 ft^2 dwelling in various parts of the United States. The following are examples of rain tank applications in the United States:

Oscar Smith Middle School in Chesapeake, Virginia, collects rain from its 220,000 ft^2 (20,440 m^2) building in four 65,000 gal ($\pm$250 m^3) cisterns. These cisterns are sized for 2 in. of rain and, as such, can hold more rainfall than that suggested by the author. Two of the cisterns supply water for indoor use (toilet and urinal flushing) and the other two cisterns provide outdoor use. Roof rain is first directed through nine large-capacity vortex filters that serve as first flush and mechanical filters. The water for indoor use is treated further with a 5 μm sediment filter and an ozone system to ensure that the water is clear and free of bacteria (Lawson, 2010).

The School of Global Sustainability for Global Solutions of the University of South Florida (USF) has installed a 30,000 gal (113 m^3) fiberglass tank to collect the rainwater from the building roof. The water is reused for urinals and toilets in this four-story 74,788 ft^2 ($\pm$6950 m^2) building that was constructed in September 2010. The university expects to harvest 506,000 gal (1.9 $\times$ 10^6 L) of rainwater, which is far more than the required 207,000 gal (780,000 L) per year to operate toilets and urinals. The rainwater, together with the condensation water collected from the building air conditioning system, is passed through a 200 μm vortex filter before it enters the cistern. After filtration, the water goes through ultraviolet (UV) treatment to eliminate bacteria. This cistern, which is sized to contain rainwater from a potential downpour, has eliminated the need for municipal water since it began operation (Cline, 2011).

Since roof rain is a fairly pure water, it can be used for all outdoor needs without any treatment. It can also be used for toilet flushing indoors. The plumbing line for toilets can be simply fed by a pump placed in an underground or aboveground tank. By placing the main rain tank on high ground or installing it on a platform, the water can be supplied for all outdoor needs without a need for any pumping. The tank may be drained at the end of the growing season to remove any sediment and the roof drains are bypassed throughout the cold season.

The use of roof rain for toilet flushing reduces one-fourth of residential indoor needs. The roof rain can save more water in commercial and office buildings, where the water can be used not only for toilet flushing but also in urinals and for washing cars, driveways, and any garage floor.

TABLE 10.3

Suggested Size of Rain Tanks and Estimated Amount of Water Saving per Dwelling during the Growing Season per 100 m^2/1000 ft^2 of Building Roofs

Region	Tank Size		Annual Water Saving	
	L	gal	L	gal
New Jersey	2500	625	48,000	12,000
Northeast	2500	625	48,000	12,000
Central Plains	2500	625	48,000	12,000
Southeast/Gulf states	3200	780	75,000	20,000
Arid Southwest	1000	250	20,000	5000

In addition to its conservation benefits, harvesting the rainwater from roofs will reduce the flow in municipal drainage systems and may also eliminate or reduce the size of storm water management facilities, such as detention basins and ponds. In a northern New Jersey town, the author suggested that the municipality consider offering rain tanks to homeowners, free of charge, to mitigate street flooding. Because of a general inadequacy of combined sewer systems in this town, overflow occurs after every heavy storm. The suggested solution would not only be less costly than replacing the municipal sewer system, but would also reduce the load on the treatment plant, saving overall operation and treatment costs.

Municipalities can take advantage of roof rain, which is a readily available and fairly pure source of water, either directly by collecting the roof rain from schools, and municipal and public buildings or indirectly by offering rebates or tax reduction incentives to homeowners. They can also require commercial and large residential developments to install a separate drainage line, fed by roof rain (or surface runoff) for irrigating lawn and landscape areas and supplying other outdoor uses. In multifamily residential, commercial, and industrial developments where the amount of roof rain may far exceed outdoor demands, the collected water can also be used for nonpotable indoor demands. In such developments storm water management measures may also include groundwater recharge through underground retention–infiltration basins.

The reuse of roof rain has been practiced in some cities in recent years and is gradually gaining popularity. While rain barrels were unheard of just 15 years ago, they are supplied in different shapes by several manufacturers today. Their prices range from $100 to $150 for a 50 gal barrel. Figure 10.11 shows a 50 gal (190 L) rain barrel made of polyethylene, wood grain look, that is commercially known as Achla RB03 rain catcher and measures $32 \times 23 \times 16$ in. ($81.3 \times 58.4 \times 40.6$ cm). Figure 10.12 shows another rain barrel, called Raintainer®, designed and marketed by Four Water LLC (Raintainer@insightbb.com).

As indicated, some municipalities already have a plan in place for the collection of roof rain. Examples are

- City of Austin in Texas: This city has a plan that offers up to $500 to residential and commercial water customers for installing a rainwater collection system. The rebate is given at a rate of $0.15 for every gallon of storage capacity for purchasing and installing qualified rain barrels/cisterns.
- City of San Antonio in Texas: The city water supply system offers rebates of $200 for every acre-foot of water saved over a 10-year period.

FIGURE 10.11   Achla, 50 gal (190 L) rain barrel.

FIGURE 10.12   Raintainer.

- Bloom Township in Illinois: The township district, together with two other school districts, supplies rain barrels to homeowners. The barrels are made available to Cook County residents for $40. The district has sold over 1500 rain barrels within 1 year after it began offering them in 2007. The barrels, connected to gutters at one end and hoses on the other, capture storm water runoff and supply a fresh clean water source for later use in yards and gardens.
- City of Adelaide, Australia: The south Australian city of Adelaide (population 1.1 million) has one of the largest storm water harvesting projects in the country. South Australia, like Southern California, faces a water shortage problem due to limited water supply and an expanding population.
- Coalition of five Arizona cities—Glendale, Mesa, Phoenix, Scottsdale, and Tempe—with a common wastewater treatment plant store leftover wastewater in aquifers to balance supply and demand (ASCE, 2004).
- Lady Bird Johnson Wildflower Center in San Antonio, Texas, appears to be the first public building where the reuse of roof rain is practiced. In this center, the rainwater is collected off roofs and terraces and is routed through a roof washer, consisting of a two-chamber, nonmechanical filtering system that removes leaves and pollen, and then into a series of surface cisterns that drain to an underground cistern. The water then goes through a filter before it is pumped up to a tank where it is stored for irrigation (O'Mally, 2007). According to the center's data, the collected rain from the 17,000 ft^2 (1580 m^2) of roof amounts to 10,200 gal (38,600 L) per inch of rainfall. Given an average annual rainfall of 30 in. (750 mm), the system can collect over 300,000 gal (1134 m^3) of rainwater annually.
- Destiny Mall in Syracuse, New York, captures roof runoff in a 90,000 gallon (340 m^3) underground tank. This water is filtered and reused to flush toilets, reducing the mall water demands by nearly 50%.
- Seaholm Power Plant in Austin, Texas, has installed a 10,000 gallon (37.8 m^3) water storage tank to collect the runoff from its approximately 35,000 ft^2 (3250 m^2) building roof

(Buranen, 2008). In this city, the average annual rainfall is 24 in. (610 mm) with monthly variation of 3.8 in. (96.5 mm) in March and 1.8 in. (45.7 mm) in May.

- A new Bank of America Tower in midtown, Manhattan (opened 2009), serves as an example where skyscraper water efficiency has been given a new dimension. In this 1200 ft (366 m) tall building, which may be the world's most environmentally responsible high-rise office building, rainwater, cooling tower makeup water, grey water, and even groundwater under the slab are tapped and recirculated. This ultragreen skyscraper and its developer, the Durst Organization, were striving for a US Green Building Council Leadership in Energy and Environmental Design (LEED) Platinum Certificate (Engle, 2007). In June 2010, the tower received the 2010 Best Tall Building Americas award by the Council on Tall Building and Urban Habitat. In this tower, rainfall is collected from the flat roof of the building and another 25,000 ft^2 (2320 m^2) podium roof and drained into four stacked water tanks positioned 10 ft (3 m) apart and arranged so that the upper tower roof flows into the top tank and then cascades down the lower tanks. From these tanks water flows by gravity to the building's 250–300 public toilets, 5 or 6 on each floor. In addition to roof runoff, the grey water from lavatories and other services, cooling water, and steam water condensation flow down into a large basement storage tank and are filtered, disinfected, and reused in this building.

Small-scale rainwater harvesting systems are commercially available to store from 100 to over 500 gal of water in aboveground tanks with pumps to transmit to sprinklers. The price of tanks varies from approximately $180 for 100 gal to $400 for 500 gal. More information can be viewed at http://www.rainharvest.com.

10.7.2 Problems with Rain Barrels

Rain barrels receive pollutants from the roof during the first flush. Grits from shingle roofs are carried during heavy rainfall. Roofs containing asbestos are not suitable for rainwater harvesting; likewise, gutters containing lead solder or lead-based paint should be avoided. A study report in January 2010 by the Texas Water Department Board indicated that the rainwater from various types of roofs did contain some contaminants above the USEPA drinking water standards. This indicates that harvested rainwater requires treatment for potable use. However, it does not need any treatment for sprinkling and other outdoor uses and some indoor uses, such as toilet flushing.

Standing water in rain barrels creates algae and is also a food source for mosquitoes. Excessive algae growth can occur rapidly in light-color or translucent rain barrels. Although many species of algae are not harmful themselves, algae blooms can deplete oxygen in water and release toxins that may be harmful to animals.

The EPA recommendations for algae control in rain barrels are as follows:

- Keep leaves out by regularly cleaning gutters and properly filtering the water from downspouts.
- Avoid barrels with open or screen tops that allow direct light inside; likewise, avoid light-colored barrels.
- Place barrels in shade; avoid full sun exposure.

Special consideration should be given to mosquito control in both aboveground and underground rainwater collection systems. If rainwater collection systems are improperly maintained, they can serve as breeding habitats for mosquitoes. Mosquito bites, apart from intense itching, can spread disease organisms. The most important mosquito-borne diseases in the United States are caused by viral pathogens, which include West Nile fever, St. Louis encephalitis, eastern or western equine encephalomyelitis, and Lacrosse encephalitis.

10.8 SUGGESTED ACTIONS FOR WIDESPREAD CONSERVATION AND REUSE

As individuals we should be concerned about our limited water supplies and take measures to collect storm water for our use. To achieve long-term water sustainability, local and state agencies and schools need to adopt challenging actions in leading the public to promote conservation and reuse of runoff in general, and roof rain in particular. Suggested actions include the following.

10.8.1 PUBLIC EDUCATION

A general lack of understanding exists among the public about the importance of water reuse. Education is the key to public awareness and is most effective when given earlier in life to students in preschools and elementary schools. Preparing concise pamphlets for indoor/outdoor uses of rainwater harvesting is another means of educating the public.

10.8.2 TASK FORCE

Task forces may be formed in schools and on campuses to promote measures for conservation and reuse. The University of Georgia, with nearly 33,000 students and 10,000 faculty, created a task force in late 2007 to conserve water. In just 4 months, from November 2007 to February 2008, water usage fell 21% compared to the same period a year before (Dendy and Freeland, 2008).

10.8.3 REACHING OUT

Reaching out to municipal agencies and officials provides a practical means of implementing and enforcing water-saving and reuse measures. If the public officials like a plan, they can, and most likely will, incorporate it into an ordinance that the community has to follow.

10.8.4 REWARD

Providing incentives for rain harvesting and rewarding innovative ideas about water reuse help conserve our valuable water resources and reduce runoff as well. Offering rain barrels and tanks, either free of charge or at reduced cost to homeowners, provides a practical solution to reuse of water. As indicated in a previous section, some cities, including Austin and Toronto, are doing just that. However, to effectively capture roof rain, barrels should be made larger following the suggested sizes of Table 10.3.

10.8.5 BLOCK PROGRAMS

As individuals we can solicit volunteers to form concerned citizen groups and block leaders to inform our neighbors of environmental and economical benefits of water saving, rain harvesting, and various measures to achieve this goal. Such a program has been adopted by the town of Cary in North Carolina to inform residents in their blocks about water conservation programs.

10.8.6 ENFORCEMENT

Malls, commercial buildings, schools, and university campuses should be required to implement measures for indoor/outdoor conservation and rainwater harvesting.

10.8.7 PILOT PROJECTS

Constructing a model or pilot project to demonstrate measures for and application of water harvesting and reuse is an effective means of educating the public. Involving the media to get the project/concept more publicity adds to public awareness.

10.8.8   Organizations/Alliances for Water Reuse

A sample of water conservation alliances was listed in Section 10.2. During the past few years, over 100 programs, water conservation and reuse nonprofit organizations, and alliances have been formed and the number is growing. Appendix 10A presents a partial list of programs and organizations (http://www.harvestH2O.com/resources.html).

10.8.9   Benefits of Water Conservation and Reuse

EPA's Office of Water had funded a study by the National Academy of Sciences (NAS) titled "Assessment of Water Reuse as an Approach for Meeting Future Water Supply Needs." The study was intended mainly to

- Revise EPA's 2004 water reuse guidelines
- Develop a comparison of performance, costs, energy requirements, and greenhouse gas releases for water reclamation/reuse versus desalination versus long-distance transport of water supplies versus pumping from deep aquifers
- Develop case examples of the use of reclaimed municipal effluent as an alternative water supply by various industries

Following the study, the EPA prepared a "Guidelines for Water Reuse" in 2012. (See Appendix 10B for a summary table.)

As far as water reuse is concerned, this publication mostly follows the general trend of water reuse, which is reclamation of wastewater. It is to be noted that rainfall and runoff are far more plentiful than wastewater and require less treatment. Therefore, water reuse programs in the future should logically pay more attention to collecting rainwater and storm water runoff.

Rooftops and driveways commonly account for 50% to 75% of total impervious area in an urban setting. Thus, retaining the rainwater from roofs alone through the use of rain barrels, tanks, or cisterns will provide a vast source of water supply. In addition, directing runoff from driveways to lawn and landscape areas serves as an effective measure both to conserve fresh water supplies and to reduce the urban storm water runoff. These measures also serve as decentralized BMPs offering viable source control solutions to storm water management.

Water reuse reduces both the average daily demand and the peak daily demand. Reductions in average daily demand affect the quantity of the water that must be developed and the size of facilities to import and store it. Reductions in peak daily demand impact the sizing of water treatment plants and their expansions and the size of treated water storage tanks. The size of pump stations and water distribution systems is also affected by water reuse.

Other advantages of water conservation are

- Lower freshwater withdrawal from lakes, rivers, and aquifers
- Reduced load to septic tanks
- Reduced energy use
- Reduced use of chemicals for treatment
- Topsoil nitrification and plant growth using recycled grey water

A comprehensive discussion of the benefits of water conservation to the local community and the environment is included in the American Water Works Association (AWWA) Manual M52, titled "Water Conservation Program, A Planning Manual" (2006).

PROBLEMS

10.1 In a single-family home of four households, the old fixtures are replaced with new water-saver fixtures. The flow ratings of the old and new fixtures are

Faucets:	2.5 gpm, old;	1.6 gpm, new
Showerheads:	2.5 gpm, old;	1.5 gpm, new
Toilet tanks:	3.6 gpf, old;	1.6 gpf, new

Calculate the annual water saving in this home. The per-capita daily uses are estimated as follows:

Faucets:	5 minutes
Shower:	5 minutes
Toilet flushes:	6 times

10.2 In Problem 10.1, calculate the annual savings in water costs assuming a $6.00 charge per CCF (100 ft^3) of municipal water.

10.3 Solve Problem 10.1 for the following data:

Old faucets:	10 Lpm;	showerheads: 10 Lpm;	flusher: 13.5 Lpf
New faucets:	6 Lpm;	showerheads: 6.0 Lpm;	flusher: 6.0 Lpf

10.4 In Problem 10.3 calculate the annual savings in water costs where the municipal water rate is $3.0/m^3.

10.5 The sprinkler system for a single-family home consists of 12 heads, each rated at 2.5 gpm. The system runs three times a week for 30 minutes. Calculate the total amount of water use during a 6-month growing season.

If this system is replaced by a self-adjusting smart box controller that reduces the irrigation water use by 50%, what will be the amount of water saving?

10.6 Solve Problem 10.5 for sprinkler heads rated at 9 L/min. Calculate the annual cost savings for a water rate of $3/m^3.

10.7 Calculate the annual amount of precipitation in ft^3 and gal that falls on a 40,000 ft^2 commercial building in New Jersey where the average annual precipitation is 46.6 in.

10.8 Calculate the annual amount of precipitation in m^3 and liters that falls on a 4000 m^2 roof of a commercial building in your area.

10.9 A rain tank is to be used for collecting the runoff from a 1750 ft^2 roof of a residential dwelling. Size the tank to fully retain a 1.25 in. rainfall.

10.10 An underground cistern is to be used to retain the rain from a 5000 m^2 roof of a commercial building for 30 mm rainfall. Size the cistern using:
   a. A cylindrical tank
   b. A prismatic tank

10.11 The cistern in Problem 10.10 is installed at a location where the average annual rainfall is 1200 mm, of which 80% is equal to or less than 30 mm. Calculate the amount of rainwater harvesting by the cistern.

APPENDIX 10A: LIST OF PROGRAMS AND NONPROFIT ORGANIZATIONS FOR WATER CONSERVATION AND REUSE

10A.1 WATER CONSERVATION

Before Your Harvest—Conserve—This article lists nearly 100 ways to save water, both inside and outside the house.

Clean Water Act—The primary federal law in the United States governing water pollution. Commonly abbreviated as the CWA, the Act established the symbolic goals of eliminating releases to water of toxic amounts of toxic substances.

Earth Works Institute—This nonprofit organization is dedicated to protecting the integrity of the natural environment by developing and promoting models of natural systems to create sustainable, self-sufficient communities.

Evaluation and Cost Benefit Analysis of Municipal Water Conservation Programs—This report evaluates the cost/benefits of various water conservation devices.

Global Water Futures—This report was published in 2005 by the Center for Strategic and International Studies and Sandia National Labs on the current state of water around the globe, with recommendations on what the United States can do to help solve the problem. The report is the output of the CSIS-SNL Global Water Futures Conference 2005.

Lifewater International—Lifewater International equips partner organizations and works with them to empower communities in developing countries to gain safe water, adequate sanitation, and effective hygiene.

Residential Water Conservation—This site is dedicated to water conservation. A section of the site has helpful information on conserving water with topics like "Why Should We Conserve?" "Conserving Water at Home," "Water Tips and Tools." There is a very small overview of rainwater harvesting in the tips and tools section.

United Nations Report on Water—If you wonder why you are thinking about a rainwater catchment system and need a little inspiration, read this report on the world's water condition and it should get you moving. This site has a very wide variety of water-related articles.

Water—Use it Wisely—This site includes a water audit tool to help you determine how much water you really use.

10A.2 WATER CONSERVATION AND REUSE—NONPROFIT ORGANIZATIONS

EPA WaterSense Program—The Environmental Protection Agency's voluntary program promotes water-efficient appliances.

Global Water Partnership—The Global Water Partnership is a working partnership among all those involved in water management: government agencies, public institutions, private companies, professional organizations, multilateral development agencies, and others committed to the principles initiated at the Dublin Conference on Water and the Environment in 1992.

Natural Resources Defense Council—This is one of the most effective environmental organizations in the United States working to protect the planet's wildlife and wild places, including clean water.

The Ocean Conservancy—This organization advocates for wild and healthy oceans. The Ocean Conservancy's International Coastal Cleanup is the largest and most successful volunteer event of its kind.

Save the Rain—This 501 (c) (3) nonprofit organization teaches people in water-starved areas to catch, store, clean, and use the rain as a sustainable water supply.

WaterAid—This international charity is dedicated to helping people escape the stranglehold of poverty and disease caused by living without safe water and sanitation.

WaterKeeper Alliance—This nonprofit organization's mission is that of a citizen watchdog for water resources. A WaterKeeper is a local, full-time, paid advocate responsible for keeping local waterways clean. The national organization provides a national and international voice for education, litigation and research, analysis, and review of water-related issues.

WaterReuse—This nonprofit organization's mission is to advance the beneficial and efficient use of water resources through education, sound science, and technology using reclamation, recycling, reuse, and desalination.

World Water Center—This nonprofit organization's function is to act as a clearinghouse for information related to water projects and activities on a worldwide basis and to provide a best practices rating system.

APPENDIX 10B: EPA 2012 GUIDELINES FOR WATER REUSE

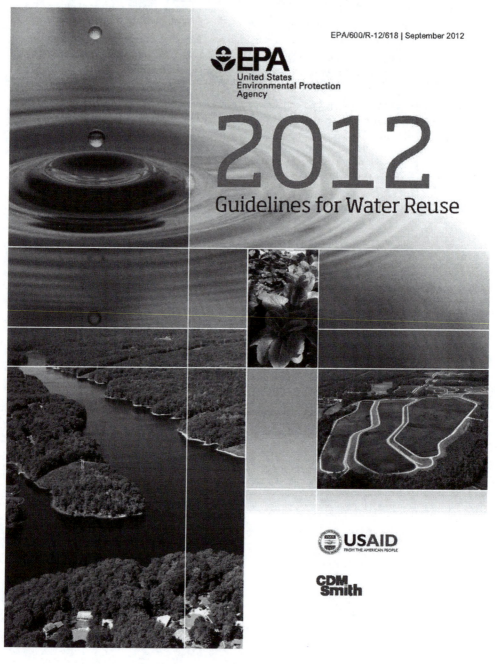

EPA/600/R-12/618 | September 2012

EPA
United States
Environmental Protection
Agency

2012
Guidelines for Water Reuse

USAID
FROM THE AMERICAN PEOPLE

CDM
Smith

Chapter 4 | State Regulatory Programs for Water Reuse

Table 4-4 Suggested guidelines for water reuse

Reuse Category and Description	Treatment	Reclaimed Water Quality [2]	Reclaimed Water Monitoring	Setback Distances [3]	Comments
Impoundments					
Unrestricted The use of reclaimed water in an impoundment in which no limitations are imposed on body-contact.	• Secondary [4] • Filtration [5] • Disinfection [6]	• pH = 6.0-9.0 • ≤ 10 mg/l BOD [7] • ≤ 2 NTU [8] • No detectable fecal coliform/100 ml [9,10] • 1 mg/l Cl2 residual (min.) [11]	• pH – weekly • BOD – weekly • Turbidity – continuous • Fecal coliform – daily • Cl2 residual – continuous	• 500 ft (150 m) to potable water supply wells (min.) if bottom not sealed	• Dechlorination may be necessary to protect aquatic species of flora and fauna. • Reclaimed water should be non-irritating to skin and eyes. • Reclaimed water should be clear and odorless. • Nutrient removal may be necessary to avoid algae growth in impoundments. • Chemical (coagulant and/or polymer) addition prior to filtration may be necessary to meet water quality recommendations. • Reclaimed water should not contain measurable levels of pathogens [12] • Higher chlorine residual and/or a longer contact time may be necessary to assure that viruses and parasites are inactivated or destroyed. • Fish caught in impoundments can be consumed. • See Section 3.4.3 in the 2004 guidelines for recommended treatment reliability requirements.
Restricted The use of reclaimed water in an impoundment where body-contact is restricted.	• Secondary [4] • Disinfection [6]	• ≤ 30 mg/l BOD [7] • ≤ 30 mg/l TSS • ≤ 200 fecal coliform/100 ml [8,13,14] • 1 mg/l Cl2 residual (min.) [11]	• pH – weekly • TSS – daily • Fecal coliform – daily • Cl2 residual – continuous	• 500 ft (150 m) to potable water supply wells (min.) if bottom not sealed	• Nutrient removal may be necessary to avoid algae growth in impoundments. • Dechlorination may be necessary to protect aquatic species of flora and fauna. • See Section 3.4.3 in the 2004 guidelines for recommended treatment reliability requirements.
Environmental Reuse					
Environmental Reuse The use of reclaimed water to create wetlands, enhance natural wetlands, or sustain stream flows.	• Variable • Secondary [4] and disinfection [6] (min.)	Variable, but not to exceed: • ≤ 30 mg/l BOD [7] • ≤ 30 mg/l TSS • ≤ 200 fecal coliform/100 ml [8,13,14] • 1 mg/l Cl2 residual (min.) [11]	• BOD – weekly • SS – daily • Fecal coliform – daily • Cl2 residual – continuous		• Dechlorination may be necessary to protect aquatic species of flora and fauna • Possible effects on groundwater should be evaluated. • Receiving water quality requirements may necessitate additional treatment. • Temperature of the reclaimed water should not adversely affect ecosystem. • See Section 3.4.3 in the 2004 guidelines for recommended treatment reliability requirements.
Industrial Reuse					
Once-through Cooling	• Secondary [4]	• pH = 6.0-9.0 • ≤ 30 mg/l BOD [7] • ≤ 30 mg/l TSS • ≤ 200 fecal coliform/100 ml [8,13,14] • 1 mg/l Cl2 residual (min.) [11]	• pH – weekly • BOD – weekly • TSS – weekly • Fecal coliform – daily • Cl2 residual – continuous	• 300 ft (90 m) to areas accessible to the public	• Windblown spray should not reach areas accessible to workers or the public
Recirculating Cooling Towers	• Secondary [4] • Disinfection [6] (chemical coagulation and filtration [5] may be needed)	Variable, depends on recirculation ratio • pH = 6.0-9.0 • ≤ 30 mg/l BOD [7] • ≤ 30 mg/l TSS • ≤ 200 fecal coliform/100 ml [8,13,14] • 1 mg/l Cl2 residual (min.) [11]	• Fecal coliform – daily • Cl2 residual – continuous	• 300 ft (90 m) to area accessible to the public. May be reduced if high level of disinfection is provided.	• Windblown spray should not reach areas accessible to workers or the public. • Additional treatment by user is usually provided to prevent scaling, corrosion, biological growths, fouling and foaming. • See Section 3.4.3 in the 2004 guidelines for recommended treatment reliability requirements
Other industrial uses – e.g. boiler feed, equipment washdown, processing power generation, and in the oil and natural gas production market (including hydraulic fracturing) have requirements that depends on site specific end use (See Chapter 3)					
Groundwater Recharge – Nonpotable Reuse					
The use of reclaimed water to recharge aquifers which are not used as a potable drinking water source.	• Site specific and use dependent • Primary (min.) for spreading • Secondary [4] (min.) for injection	• Site specific and use dependent	• Depends on treatment and use	• Site specific	• Facility should be designed to ensure that no reclaimed water reaches potable water supply aquifers. • See Chapter 3 of this document and Section 2.5 of the 2004 guidelines for more information. • For injection projects, filtration and disinfection may be needed to prevent clogging. • For spreading projects, secondary treatment may be needed to prevent clogging. • See Section 3.4.3 in the 2004 guidelines for recommended treatment reliability requirements

Table 4.4 Suggested guidelines for water reuse

Reuse Category and Description	Treatment	Reclaimed Water Quality²	Reclaimed Water Monitoring	Setback Distances³	Comments
Indirect Potable Reuse					
Groundwater Recharge by Spreading into Potable Aquifers	• Secondary[4] • Filtration[5] • Disinfection[6] • Soil aquifer treatment	Includes, but not limited to, the following: • No detectable total coliform/100 ml[8,10] • 1 mg/l Cl₂ residual (min.)[11] • pH = 6.5 – 8.5 • ≤2 NTU[8] • ≤2 mg/l TOC of wastewater origin • Meet drinking water standards after percolation through vadose zone	Includes, but not limited to, the following: • pH – daily • Total coliform – daily • Cl₂ residual – continuous • Drinking water standards – quarterly • Other[17] – depends on constituent • TOC – weekly • Turbidity – continuous • Monitoring is not required for viruses and parasites; their removal rates are prescribed by treatment requirements	• Distance to nearest potable water extraction well that provides a minimum of 2 months retention time in the underground.	• Depth to groundwater (i.e., thickness to the vadose zone) should be at least 6 feet (2m) at the maximum groundwater mounding point. • The reclaimed water should be retained underground for at least 2 months prior to withdrawal. • Recommended treatment is site-specific and depends on factors such as type of soil, percolation rate, thickness of vadose zone, native groundwater quality, and dilution. • Monitoring wells is necessary to detect the influence of the recharge operation on the groundwater. • Reclaimed water should not contain measurable levels of pathogens after percolation through the vadose zone.[12] • See Section 3.4.3 in the 2004 Guidelines for recommended treatment reliability requirements. • Recommended log-reductions of viruses, Giardia, and Cryptosporidium can be based on challenge tests or the sum of log-removal credits allowed for individual treatment processes. Monitoring for these pathogens is not required. • Dilution of reclaimed water with waters of non-wastewater origin can be used to help meet the suggested TOC limit.
Groundwater Recharge by Injection into Potable Aquifers	• Secondary[4] • Filtration[5] • Disinfection[6] • Advanced wastewater treatment[16]	Includes, but not limited to, the following: • No detectable total coliform/100 ml[8,10] • 1 mg/l Cl₂ residual (min.)[11] • pH = 6.5 – 8.5 • ≤2 NTU[8] • ≤2 mg/l TOC of wastewater origin • Meet drinking water standards	Includes, but not limited to, the following: • pH – daily • Turbidity – continuous • Total coliform – daily • Cl₂ residual – continuous • TOC – weekly • Drinking water standards – quarterly • Other[17] – depends on constituent • Monitoring is not required for viruses and parasites; their removal rates are prescribed by treatment requirements	• Distance to nearest potable water extraction well that provides a minimum of 2 months retention time in the underground.	• The reclaimed water should be retained underground for at least 2 months prior to withdrawal. • Monitoring wells is necessary to detect the influence of the recharge operation on the groundwater. • Recommended quality limits should be met at the point of injection. • The reclaimed water should not contain measurable levels of pathogens at the point of injection. • Higher chlorine residual and/or a longer contact time may be necessary to assure virus inactivation. • See Section 3.4.3 in the 2004 Guidelines for recommended treatment reliability requirements. • Recommended log-reductions of viruses, Giardia, and Cryptosporidium can be based on challenge tests or the sum of log-removal credits allowed for individual treatment processes. Monitoring for these pathogens is not required. • Dilution of reclaimed water with waters of non-wastewater origin can be used to help meet the suggested TOC limit.
Augmentation of Surface Water Supply Reservoirs	• Secondary[4] • Filtration[5] • Disinfection[6] • Advanced wastewater treatment[16]	Includes, but not limited to, the following: • No detectable total coliform/100 ml[8,10] • 1 mg/l Cl₂ residual (min.)[11] • pH = 6.5 – 8.5 • ≤2 NTU[8] • ≤2 mg/l TOC of wastewater origin • Meet drinking water standards	Includes, but not limited to, the following: • Drinking water standards – quarterly • Other[17] – depends on constituent • Monitoring is not required for viruses and parasites; their removal rates are prescribed by treatment requirements	• Site specific – based on providing 2 months retention time between introduction of reclaimed water into a raw water supply reservoir and the intake to a potable water treatment plant	• The reclaimed water should not contain measurable levels of pathogens.[12] • Recommended level of treatment is site-specific and depends on factor such as receiving water quality, time and distance to point of withdrawal, dilution and subsequent treatment prior to distribution for potable uses. • Higher chlorine residual and/or a longer contact time may be necessary to assure virus and protozoa inactivation. • See Section 3.4.3 in the 2004 Guidelines for recommended treatment reliability requirements. • Recommended log-reductions of viruses, Giardia, and Cryptosporidium can be based on challenge tests or the sum of log-removal credits allowed for individual treatment processes. Monitoring for these pathogens is not required. • Dilution of reclaimed water with water of non-wastewater origin can be used to help meet the suggested TOC limit.

Footnotes

[1] These guidelines are based on water reclamation and reuse practices in the U.S., and are specifically directed at states that have not developed their own regulations or guidelines. While the guidelines should be useful in may areas outside the U.S., local conditions may limit the applicability of the guidelines in some countries (see Chapter 9). It is explicitly stated that the direct application of these suggested guidelines will not be used by USAID as strict criteria for funding.

[2] Unless otherwise noted, recommended quality limits apply to the reclaimed water at the point of discharge from the treatment facility.

[3] Setback distances are recommended to protect potable water supply sources from contamination and to protect humans from unreasonable health risks due to exposure to reclaimed water

[4] Secondary treatment processes include activated sludge processes, trickling filters, rotating biological contractors, and may stabilization pond systems. Secondary treatment should produce effluent in which both the BOD and SS do not exceed 30 mg/l.

[5] Filtration means; the passing of wastewater through natural undisturbed soils or filter media such as sand and/or anthracite; or the passing of wastewater through microfilters or other membrane processes.

[6] Disinfection means the destruction, inactivation, or removal of pathogenic microorganisms by chemical, physical, or biological means. Disinfection may be accomplished by chlorination, ozonation, other chemical disinfectants, UV, membrane processes, or other processes.

[7] As determined from the 5-day BOD test.

[8] The recommended turbidity should be met prior to disinfection. The average turbidity should be based on a 24-hour time period. The turbidity should not exceed 5 NTU at any time. If SS is used in lieu of turbidity, the average SS should not exceed 5 mg/l. If membranes are used as the filtration process, the turbidity should not exceed 0.2 NTU and the average SS should not exceed 0.5 mg/l.

[9] Unless otherwise noted, recommended coliform limits are median values determined from the bacteriological results of the last 7 days for which analyses have been completed. Either the membrane filter or fermentation tube technique may be used.

[10] The number of total or fecal coliform organisms (whichever one is recommended for monitoring in the table) should not exceed 14/100 ml in any sample.

[11] This recommendation applies only when chlorine is used as the primary disinfectant. The total chlorine residual should be met after a minimum actual modal contact time of at least 90 minutes unless a lesser contact time has been demonstrated to provide indicator organism and pathogen reduction equivalent to those suggested in these guidelines. In no case should the actual contact time be less than 30 minutes.

[12] It is advisable to fully characterize the microbiological quality of the reclaimed water prior to implementation of a reuse program.

[13] The number of fecal coliform organisms should not exceed 800/100 ml in any sample.

[14] Some stabilization pond systems may be able to meet this coliform limit without disinfection.

[15] Commercially processed food crops are those that, prior to sale to the public or others, have undergone chemical or physical processing sufficient to destroy pathogens.

[16] Advanced wastewater treatment processes include chemical clarification, carbon adsorption, reverse osmosis and other membrane processes, advanced oxidation, air stripping, ultrafiltration, and ion exchange.

[17] Monitoring should include inorganic and organic compounds, or classes of compounds, that are known or suspected to be toxic, carcinogenic, teratogenic, or mutagenic and are not included in the drinking water standards.

[18] See Section 3.4.3.7 for additional precautions that can be taken when a setback distance of 100 ft (30 m) to potable water supply wells in porous media is not feasible.

REFERENCES

ASCE, 2002, Civil engineering news, *Civil Engineering Magazine*, May, p. 28.

———— 2004, Arizona Cities Move Forward with Groundwater Recharge Plan, civil engineering news, water resources, *Civil Engineering Magazine*, January, pp. 20–21.

———— 2006, Rainfall challenges to communities to treat, store, reuse and recharge, Special advertising section, civil engineering, *Civil Engineering Magazine*, August, pp. 73–79.

AWWA, 2006, Water conservation programs, a planning manual, manual 52.

BCC Research, 2006, Water recycling and reuse technologies and materials, report code, report no. RGB-331.

Brzozowski, C., 2008, Lessons in efficiency, *Water Efficiency*, March.

———— 2009, EPA promotional partner of the year—Seattle, *Water Efficiency*, March, pp. 28–31.

———— 2012, Double impact, utilities and water-intensive companies have begun to take another look at water reclamation and reuse, *Water Efficiency*, January/February, pp. 22–31.

———— 2013, Alternative Water Sources, *Water Efficiency*, November/December, pp. 29–37.

Buranen, M., 2008, Rain catcher's delight, *Water Efficiency*, September/October, pp. 40–44.

Chavez, A., 2012, Recycled water saves big, *Water and Wastes Digest*, September, pp. 30–32.

Clary, J., O'Brien, B., and Calomino, K., 2006, Working together to promote landscape water conservation, *Water Efficiency*, September/October, pp. 30–37.

Cline, K., 2011, On campus reuse, *Stormwater Solutions*, July/August, pp. 10–11.

Corum, L., 2008, A waterwise future, *Water Efficiency*, March, pp. 18–22.

———— 2011, Irrigation technology, smart water application technology comes of age, *Water Efficiency*, May/June, pp. 34–41.

Cutright, E., 2009, A first for rainwater harvesting, editorial, *Water Efficiency*, July.

Dendy, L. and Freeland, S., 2008, Water task force, *Water Efficiency*, November/December, pp. 46–49.

Duffy, D.P., 2008, The ultimate recycling program, *Water Efficiency*, May/June, pp. 24–25.

Engle, D., 2007, Green from top to bottom, *Water Efficiency*, March/April, pp. 10–15.

EPA (Environmental Protection Agency), 2002, Cases in water conservation, Office of Water (4204M), publication EPA832-B-02-003, July.

———— 2012, Guidelines for Water Reuse, EPA/600/R-12/616, Sept., http://www.waterreuseguidelines.org.

Falcon Waterfree Technologies, 4729 Division Ave., Suite C, Wayland, MI 49348, Ph. 866-975-0174 (http://www.falconwaterfree.com); International Headquarters: 2255 Barry Ave., Los Angeles, CA 90064, Ph. 310-209-7250 (info@falconwaterfree.com).

Fuller, F., Gregg, T., and Curry, L., 1995, Austin's Xeriscape It! replaces thirsty landscapes, *Opflow*, December, p. 3.

Glist, D., 2010, Pond on a pond water harvesting system meets aesthetic functional objectives, *Urban Water Management*, February, pp. 8–9.

Godman, R.R., and Kuyk, D.D., 1997, A dual water system for Cape Coral, *AWWA Journal* 89 (7): 45–53.

Goldberg, S., March/April 2013a, Rainwater harvesting, part 1. A growing industry with some challenges ahead, *Stormwater*, June, pp. 10–17.

———— 2013b, Rainwater harvesting, part 2, meeting specific challenges, *Stormwater*, May, pp. 40–45.

Grumbles, B.H., 2012, Rethink and reuse, *Water and Wastes Digest*, March, p. 12.

Hall, P.C., 2005, Municipal use of stormwater runoff, *Stormwater*, May/June, pp. 74–89.

Hildebrandt, P., 2006, Getting to the roots of water efficiency, *Water Efficiency*, September/October, pp. 44–48.

———— 2007, Reclaiming water in Cary, NC, *Water Efficiency*, January/February, pp. 26–30.

———— 2008, Communities in water conservation and efficiency share their secrets, *Water Efficiency*, January/February, pp. 14–19.

Johnstone, C., 2008, Waterwise for life kiosk, *AWWA Journal*, May, pp. 53–58.

Landers, J., 2014, What will it take? *Civil Engineering Magazine*, May, pp. 70–75, 79.

Lawson, S., 2010, Making the most of rain, a Virginia school harvests and treats storm water for indoor/outdoor reuse, *Stormwater Solutions*, January/February, pp. 28–30.

Lovely, L., 2012, H_2O, redux, *Water Efficiency*, May, pp. 34–41.

Maddaus, M.C., Maddaus, W.O., Torre, M., and Harris, R., 2008, Innovative water conservation supports sustainable housing development, *AWWA Journal*, May, pp. 104–111.

Mayer, P. W. et al., 1999, Residential end uses of water, AWWA Research Foundation and AWWA, Denver.

Mays, L.W., ed., 1996, Water reclamation and reuse, *Water resources handbook*, Chapter 21, McGraw-Hill, New York.

McCarthy, D., 2008, Water sustainability, a looming global challenge, *AWWA Journal*, September, pp. 46–47.

Means, E.G., III, West, N., and Patrick, R., 2005, Population growth and climatic change will pose tough challenges for water utilities, *AWWA Journal*, August, pp. 40–46.

Niagara Conservation, Ph. 800-831-8383, ext. 141, http://www.NiagaraConservation.com.

NJ Department of Environmental Protection, Division of Water Quality, 2005, Reclaimed water for beneficial reuse, appendix A.

Obmascik, M., 1993, *A consumer guide to water conservation*, 1st ed., AWWA, Denver, CO.

O'Mally, P.G., 2007, Stormwater harvesting: A project summary, *Water Efficiency*, March/April, pp. 51–56.

Owenby, J.R. and Enzell, D.S., 1992, Monthly station normals of temperature, precipitation and heating and cooling degree days, 1961–1990, New Jersey, Climatography of the United States, no. 81, US Department of Commerce, National Oceanic and Atmospheric Administration, January.

Pazwash, H. and Boswell, S.T., 1997, Management of roof runoff, conservation and reuse, *Proceedings of the ASCE 24th Annual Water Resources Planning and Management Conference*, Houston, TX, April 6–9, pp. 784–789.

———— 1999, Conservation of water, reuse of roof runoff, *Proceedings of 1999 International Water Resources Engineering Conference*, Seattle, WA, August 8–12.

———— 2002, Water conservation and reuse; an overview, *Proceedings of AWWA Water Sources Conference*, Las Vegas, NV, January 27–30.

Pazwash, H. and Tuvel, H.N., 1994, Conservation measures in urban stormwater management, *Proceedings of the 21st Annual Conference of the ASCE Water Resources Planning and Management*, Denver, CO, May 23–26, pp. 408–411.

Poremba, S.M., 2009, Target irrigation water waste, *Water Efficiency*, July/August, p. 23.

Ramos, A.R., 2006, Wild about water conservation, *Water Efficiency*, November/December, pp. 46–49.

———— 2007, Grass isn't always greener, *Water Efficiency*, May/June, pp. 46–49.

Sloan Valve Company, headquarters: 10500 Seymore Ave., Franklin Park, IL 60131, Ph. 847-671-4300/800-9VALVE9 (customer.service@sloanvalve.com); (international.411@sloanvalve.com).

Solley, W.B., Pierce, R.R., and Perlman, H. A., 1998, Estimated use of water in the United States in 1995, US Geological Survey Circular 1200, US Dept. of the Interior, USGS, Reston, VA.

Tchobanoglous, G. et al., 2003, *Wastewater engineering: Treatment and reuse*, McGraw-Hill, New York.

Texas Water Development Board, 2005, The Texas manual on rainwater harvesting, 3rd ed.

US Census Bureau, 2014, http://www.census.gov/servlet/SAFFPopulation.

USGS (US Geological Survey), 2008, Irrigation water use, water science for schools, November, http://ga.water.usgs.gov/edu/wuir.html.

Vickers, A., 2001, *Handbook of water use and conservation*, 1st ed., Water Plow Press, Amherst, MA.

Von Minden, L., 2013, Real-world watering, *Water Efficiency*, March/April, pp. 41–47.

Weinstein, G., 1999, *Xeriscape handbook*, Fulcrum Publishing, Golden, CO.

Glossary

Abstractions: Portions of the total rainfall that do not produce direct runoff. These include interception by vegetation, storage in depressions, and infiltration.

Adverse slope: Upward slope in the downstream direction in channel or pipe.

Algae: Aquatic organisms containing chlorophyll that grow in colonies and produce mats.

Angle of repose: Angle of slope formed by granular material under the critical equilibrium condition of incipient motion.

Antecedent moisture: Water stored in the soil before the start of rainfall.

Area-elevation curve: Curve relating surface area to elevation.

Arid: A climate associated with less than 10 in. (250 mm) of annual precipitation.

Artificial recharge: The intentional addition of water to an aquifer through injection or infiltration.

Atmospheric pressure: Force of air on unit area of a surface.

Auger hole test: Field test for measuring hydraulic conductivity.

Backwater curve: Plot of water depth along channel.

Bacteria: Single-cell microorganisms that reproduce by fission or by spores.

Bankfull flow: Flow conditions that fill a stream to the top of its banks.

Base flow: Flow in a channel or stream due to soil moisture or groundwater.

Best management practice (BMP): A structural or nonstructural measure designed to temporarily store and treat storm water runoff in order to avoid flooding, reduce pollution, and provide other amenities.

Biological oxygen demand (BOD): The oxygen taken by bacteria to oxidize soluble organic matter.

Bluegrass: Same as Kentucky bluegrass, a variety of cool-season turf grass.

Buffer: Zone of vegetated ground along both sides of natural streams or wetlands.

Bulk density: Ratio of the oven-dried mass of a sample to its original volume.

Bypass flow: Flow that bypasses an inlet on grade and is carried in the gutter to the next inlet.

Capillary rise: Height to which water rises under capillary action.

Capillary suction: Negative water pressure in soil above the water table.

Capillary water: Water in soil above the water table due to capillary forces.

Catch basin insert: Filters inserted in inlets to trap floatables, coarse and suspended solids. Some inserts also capture oil and grease.

Check dam: A small dam (berm) constructed in a gully or swale to decrease flow velocity.

Cistern: A tank or chamber to store rain or storm water.

Clustering: A development design technique that concentrates buildings on a part of the site to allow the remaining land to be used for agriculture, recreation, common open space, and preservation of environmentally sensitive features.

Combination inlet: An inlet with both a curb opening and a grate.

Combined sewer: A pipe that carries both sanitary sewage and storm water runoff.

Compaction: The closing of pore spaces among the particles of soil and rock, generally through heavy equipment during construction.

Concentration: The amount of a substance in a unit volume of solution.

Conduit: Any channel/pipe intended for conveyance of water.

Confluence: The location where two streams merge.

Constant head permeameter: Laboratory device for measuring hydraulic conductivity under a constant head.

Constructed wetland: A man-made freshwater wetland designed and constructed to serve a specific purpose.

Continuity equation: Equation based on conservation of mass; the product of cross-sectional area by flow velocity.

Contour (elevation): Line connecting points on a surface having the same elevation.

Control section: Cross section of a channel that has a controlling structure (bridge, free outfall, or weir).

Conveyance: Measure of amount of flow carried in a channel, defined from Manning's formula, related to n, A, and P.

Correlation coefficient: An index that represents the combined effects of soil characteristics, the land cover, the hydrologic condition, and antecedent soil moisture conditions.

Crest: Peak of a hydrograph; top of dam/spillway.

Critical depth: Depth of water at which specific energy is minimum.

Critical flow: Flow in an open channel at a minimum specific energy; Froude number = 1.0. Discharge through a channel for the minimum specific energy.

Critical slope: Channel slope at which uniform flow is critical.

Critical velocity: The velocity where stream flow passes from supercritical to subcritical conditions or vice versa.

Cross slope: The rate of change of roadway elevation with distance transverse to the direction of travel.

Crown: The inside top elevation of a conduit.

Culvert: A conduit used primarily to convey flow under highways and railroad embankments.

Cumulative mass curve: Graph of accumulated rainfall versus time.

Curb-opening inlet: An inlet with opening in the curb.

Curve number: An index related to soil characteristics, the land cover, and antecedent soil moisture conditions in the SCS (Soil Conservation Service—now NCRS) method.

Customary units (CU): Foot-pound system of units, also referred to as English units.

Dam: A barrier that impounds water for specific purpose(s).

Datum: Topographic reference elevation based on a benchmark (NGVD, 1929; NAVD, 1988; or arbitrary).

Dead storage: Storage in a reservoir or detention basin below the elevation of the principal outlet.

Density: Mass per unit volume of matter.

Depression storage: A depressed area that stores precipitation or runoff.

Depth of flow: Vertical distance from the bottom of a channel to the water surface.

Design discharge: Discharge associated with a selected return period.

Design storm: A hypothetical storm used in design.

Detention basin: A dry basin or other structure that detains runoff during a storm and releases it over time through an outlet structure.

Detention pond: A wet basin that temporarily stores storm water and releases the water at a controlled rate of flow.

Detention time: The time that it takes for design flow to be discharged from a detention facility.

Deterministic: Hydrologic model based on physical relations.

Development: Any use or change in the use of land due to construction of a structure or mining, excavation, landfill, or deposition, not including redevelopment.

Development density: The number of families, individuals, dwelling units, or households per unit area of land.

Dew point temperature: Temperature at which the air just becomes saturated with water.

Dimensionless hydrograph: A hydrograph presented in terms of ratios of flow and time—commonly, ordinates being the ratio of the discharge to the peak discharge and abscissae the ratio of time to the time to peak.

Direct runoff: The total runoff less losses.

Direct runoff hydrograph: Graph of direct runoff (rainfall—losses) versus time.

Directly connected impervious area: Impervious area that drains directly into a drainage system.

Disinfection: The process of removing or inactivating pathogens.

Diversion: Redirecting flow in a pipe or channel. Structure that redirects flow to another area. A channel or conduit to bypass flows above a certain elevation.

Domestic water use: Total of indoor and outdoor water uses in residential or industrial properties.

Drainage area: Area formed by topography that drains to a given point.

Drainage density: An index of the concentration of streams in a watershed, as measured by the ratio of the total length of streams to the drainage area.

Drainage divides: Topographic boundary that directs rainfall into different basins.

Drip irrigation: An irrigation system that delivers water to plants at low pressure and small drips through perforated plastic tubes.

Drought: An extended period of no or little rain.

Effective flow area: Portion of a cross-sectional area where water flows.

Effective precipitation: Total amount of rainfall minus evaporation and infiltration during rainfall period.

Effluent: Wastewater leaving the treatment plant.

Embankment: A man-made deposit of soil, rock, or other materials.

Emergency spillway: A structure that controls release of storm flows in excess of the design discharge from a detention basin or a reservoir.

Energy grade line: A line representing the sum of the pressure, velocity, and elevation heads.

Engineering fabric/filter fabric: Permeable textile placed below riprap to prevent piping and to permit natural seepage.

Erosion: The detachment, wearing away, or movement of soil or rock fragments by the action of water, wind, ice, or gravity. Erosion can be sheet, rill, or gully.

Erosion control blanket (ECB): A degradable material manufactured or fabricated into rolls. ECB is used to reduce soil erosion and helps growth and establishment of vegetation.

Erosive velocity: Velocity of water that is high enough to erode land surface.

Evaporation: Transformation of water from liquid to vapor. Loss of water from a surface due to vaporization.

Evapotranspiration: Water loss due to evaporation from soil and transpiration from plants.

Event: A single storm simulation.

Excavation: Any act by which soil or rock is dug, quarried, removed, or relocated.

Extended detention: Detention basins that hold and slowly release the storm water runoff following a storm event.

Falling head permeameter: Laboratory device with a tube of dropping water for measuring hydraulic conductivity.

Faucet aerator: A screen-like device with a tube of dropping water that is enclosed in a faucet to reduce flow volume.

Field capacity: Amount of water in soil after gravitational water is drained.

Fill: A man-made deposit of soil, rock, or other materials.

Filter blanket: One or more layers of graded noncohesive material placed below riprap to prevent soil piping and allow natural seepage.

Filter media: The sand, soil, or other material in a filtration device.

Filter strip: A strip of permanent vegetation to retard the flow of runoff.

Filtration: The process of screening suspended solids from storm water (also wastewater) through a porous medium.

Flood hazard zone: Area that will flood with a given probability. Such zones are commonly shown on FEMA or state flood maps.

Flood plain: Lands along a stream that would be inundated by a flood event.

Flow path: Common path of a fluid particle.

Force main: A pressurized conduit.

Freeboard: Vertical distance from the design water surface elevation to the top of the channel.

Frequency: Occurrence interval of a random variable.

Freshwater wetlands: An area that is inundated or saturated by surface water or groundwater.

Friction slope: Equal to the total energy slope in open channels.

Froude number: The ratio of inertia force to gravity force. Dimensionless parameter to characterize open channel flow regime.

Gabion: Rectangular wire mesh containers filled with stone.

Gabion mattress: A thin gabion, usually 6 to 9 in. thick, used to line channels.

Geographic information system (GIS): Computer application that displays numerous types of spatial data (such as land use, soil type, or topography) and links those data with a map.

Grading: Any stripping, cutting, filling, or any combination thereof.

Gradually varied flow: Open channel flow that changes gradually so that one-dimensional analysis can be applied in each reach.

Grate inlets: Inlets with transverse bars arranged to form an inlet structure.

Gravel: Aggregate of 1/4 in. to 3 in. stone mix.

Gravitational water: Water that will drain through the soil under the force of gravity.

Gray water: Water from showers and sinks (including kitchen sink; British and Australian practice).

Green development: A development that integrates ecology and real estate.

Green industry: Trades or stakeholders associated with landscape or irrigation or nonconventional management of runoff.

Grey water: Water from shower and sinks (excluding kitchen sink).

Groundwater: Water beneath the earth's surface.

Groundwater recharge: Replenishment of groundwater source, either naturally by percolation or artificially by injection.

Groundwater table/water table: The upper surface of an unconfined aquifer.

Gully: A channel cut by concentrated runoff.

Head: The height of water above any reference plane. In hydraulics, head may be potential (pressure and elevation) and kinetic (velocity).

Hotspot: Area where land use or activities produce highly contaminated runoff.

Hydraulic conductivity: Ratio of velocity to hydraulic gradient, indicating permeability of porous media.

Hydraulic grade line: A line showing the sum of the pressure and elevation heads.

Hydraulic jump: Sudden transition from supercritical flow to subcritical flow.

Hydraulic radius: The cross-sectional area of flow divided by wetted perimeter.

Hydraulically connected impervious area: Impervious area that drains directly into the drainage system.

Hydrograph: A graph showing variation of discharge with time.

Hydrologic cycle: A representation of the physical processes that control the distribution and movement of water.

Hydrologic routing: Computation relating inflow and outflow for a detention basin/reservoir that uses the continuity equation and a storage equation.

Hydrologic soil group (HSG): An SCS soil classification.

Hydrostatic pressure: Static pressure exerted at a depth below the water surface.

Hyetograph: Graph of rainfall intensity versus time.

Hygroscopic water: Moisture that is absorbed to the surface of soil grains in the unsaturated zone.

Infiltration: Movement of water from the surface into the soil.

Infiltration basin: A detention basin that impounds storm water and gradually exfiltrates it through the basin bed.

Infiltration capacity: Rate at which water can enter soil under surface inundation.

Infiltration rate: The rate of penetration of water through the soil surface.

Initial abstractions: The portion of rainfall that does not contribute to runoff.

Intensity: Depth of rainfall per unit of time.

Intensity–duration–frequency curve (IDF): A design chart that relates the rainfall intensity with its duration and return interval.

Invert: Bottom elevation of a channel or pipe.

Irrigation: The application of water to soil to meet water needs of plants, crops, turf, gardens, or wildlife to supplement rainfalls.

Karst: Carbonic rock, typified by the presence of limestone sinkholes and caverns.

Lag time: Time from the center of mass of rainfall to the peak of the hydrograph. In the Universal Method, developed by Pazwash, the lag time is defined as the beginning of rainfall to the onset of runoff.

Land cover/land use: The type of cover on the surface of the earth such as rooftop, pavement, grass, or tress.

Land disturbance: Any activity involving the clearing, cutting, excavation, grading, or filling of land that causes land to be exposed to erosion.

Landscape: A combination of turf and plants and mulches; may also include natural undisturbed areas.

Lining, composite: Combination of lining materials in a given cross section (e.g., riprap in low-flow channel and vegetated side slopes).

Lining, flexible: Lining material that can adjust to settlement; typically constructed of a porous material that allows infiltration and exfiltration.

Lining, rigid: Lining material that does not adjust to settlement, such as concrete or masonry.

Lining, temporary: Lining used for an interim condition (e.g., construction period).

Longitudinal slope: The rate of change of elevation with distance in the direction of travel or flow.

Low-flow faucet: A faucet that delivers no more than 2.5 gpm under 80 lb/in.^{2} pressure.

Low-flow shower head: A showerhead with no more than 2.5 gpm flow.

Low-flow urinal: Urinals with no more than 1 gallon per flush.

Low-impact development (LID): A development with minimal impacts on hydrology and water quality.

Low-volume toilet: Toilets with no more than 1.6 gallons per flush.

Main channels: Large channels that carry flow from collector channels to some outlet such as a lake or stream.

Mass rainfall curve: A plot of cumulative precipitation with time.

Median: The middle value in a set of values arranged in ascending or descending order.

Micro-irrigation: An irrigation system with small sprinkler heads or emitters that delivers small amounts of water either above or below ground.

Micropool: A smaller permanent pool incorporated into the design of larger stormwater pond.

Mild slope: Channel slope for which uniform flow is subcritical.

Mixed-use development: A tract of land with several different uses such as residential, office, manufacturing, retail, public, or entertainment.

Mulch: Bark, leaves, or straw placed around plants to reduce evaporation and weed growth.

Multifamily dwelling unit: Any building containing two or more dwelling units.

Native plants: Plants that are indigenous to an area and require little or no additional watering.

Natural drainage: Channels formed in the existing surface topography of the earth prior to changes made by man.

Natural ground surface: The existing surface of land prior to any land disturbance.

Nonpoint source pollutants: Pollutants generated from a spread-out area, rather than a localized point.

Nonstructural BMPs: Storm water treatment techniques that use natural measures to reduce run-off volume and pollution level.

Nonuniform flow: Flow of water through a channel that changes with distance.

Normal depth: Depth of a uniform channel flow.

One hundred-year storm: A storm event that occurs on the average once every 100 years.

Open channel: A natural or man-made conduit with free water surface.

Open channel flow: Flow of water through an open channel.

Open weave textile (OWT): A temporary degradable ECB made of natural or polymer yarns woven into a matrix. OWT provides erosion control and helps vegetation establishment.

Optimum channel cross section: Cross section that requires a minimum flow area.

Orifice: An opening through which water passes under pressure.

Orifice equation: An equation that relates the discharge through an orifice to the area of the orifice and the depth of water above the center of the orifice.

Orifice flow: Flow of water into a submerged opening due to water pressure.

Outfall: The point of water discharge from a conduit, drain, or detention basin.

Overland flow: Flow of water on the land surface in a down-slope direction

Oxidation: A chemical process in which a molecule or ion loses electrons to an oxidant.

Pathogens: Disease-producing microorganisms (bacteria, fungi, viruses).

Per-capita use: The amount of water used by one person during a specific time period, commonly 24 hours.

Percolation: The downward movement of water through soil.

Permeability: The capability of a soil or porous medium to transmit water; the rate of downward flow of water in soil.

Permissible shear stress: The force required to initiate movement of the channel bed or lining material.

Permissible velocity: The highest average velocity at which the water may be carried in a channel or discharged at an outfall and not cause scour.

Pervious/porous pavement: Any pavement that allows infiltration, such as porous asphalt, pervious concrete, and pavers.

Pocket pond: A storm water pond used for small drainage area (<5 acres) that relies on groundwater to maintain a permanent pool.

Point source: The location where wastewater or any polluted water is discharged.

Pollutant: Any contaminant at concentration high enough to pose a danger to public health or endanger aquatic environments.

Pond: A vernal or perennial body of standing water, smaller than a lake.

Population density: The total number of residents per total area of land, excluding water bodies.

Porosity: Ratio of volume of voids to total volume of soil sample.

Porous medium: Geologic material that will allow water to flow through it.

Potable water: Water suitable for drinking.

Potential evapotranspiration: Amount of water that can be lost to evapotranspiration if water is sufficiently available.

Precipitation: Water that falls to the earth in the form of rain, snow, hail, or sleet.

Pressure flow: Flow in a closed conduit with no free water surface. The flow occurs due to pressure forces.

Pretreatment: Techniques employed to remove coarse sediment before storm water enters a system.

Probability: Relative number of occurrences of an event after a large number of trials.

Rain sensor: A sensor that shuts off an irrigation system during rain.

Rainfall excess: Total rainfall minus the initial abstraction and losses.

Rapidly varied flow: Flow of water through a channel with rapidly changing characteristics.

Rating curve: A graph or equation that relates the stage and discharge.

Rational method: A simple, linear, rainfall–runoff relationship for estimating peak flow.

Recession curve: Portion of the hydrograph where runoff is from base flow.

Recharge: The addition of water to groundwater by precipitation or artificial infiltration and injection.

Record: A string of characters or groups of characters (fields) that are treated as a single unit in a file.

Recurrence interval: Time interval in which an event will occur once on the average.

Recycled water: Reuse of water repeatedly through a closed system; also used to refer to reclaimed water.

Redevelopment: The replacement or adaptive reuse of an existing structure, or of land from which previous improvements have been removed.

Relative humidity: Ratio of water vapor pressure to the saturated vapor pressure at the same temperature.

Reservoir: Man-made storage area for flood control or water supply.

Residential water use: Water use in homes, both indoor and outdoor.

Retardance classification: Qualitative measure of the resistance to flow by various types of vegetation.

Retention: Storage reservoir or pond that retains runoff or flood water without allowing it to discharge downstream.

Retention/detention facilities: Facilities that control the quantity and rate of runoff.

Retention pond or basin: A facility designed to retain storm water runoff on a development site.

Retrofit: To alter, adjust, or change plumbing fixtures or equipment or appliances to reduce water use.

Return period: Same as recurrence interval: time interval for which an event will occur once on the average.

Reuse: The additional use of previously used water.

Riprap: Broken rock, cobbles, or boulders placed on side slopes or the bottoms of channels to prevent erosion.

Riser: A vertical pipe or structure installed on a pond/detention basin as a flow control device.

Rising limb: Portion of the hydrograph where runoff is increasing.

Roadside channel: Stabilized drainage way to collect runoff from roadways and streets.

Rolled erosion control product (RECP): A degradable or nondegradable material fabricated into rolls. RECP is used to reduce soil erosion and to protect vegetation during establishment.

Root zone: Depth to which the vegetation draws water through its root system in soil.

Routing: The process of discharge of an inflow hydrograph through a detention system.

Runoff: Surface water from precipitation (rain or snow) and irrigation that is not absorbed by soil or retained in surface depressions.

Runoff coefficient: Ratio of runoff to precipitation.

Runoff curve number: Parameter used in the SCS method that accounts for soil type and land cover.

Saline water: Water containing salt at concentrations less than 35 g/L. (Slightly saline water contains 1–3 g/L, moderately saline water from 3–10 g/L, and highly saline water from 10–35 g/L.)

Salinity: The concentration of salts in water.

Sand filters: A structural water quality device to percolate runoff through a sand bed before discharge to drainage system.

Saturated zone: Zone of an aquifer in which the soil is under pressure greater than atmospheric pressure.

SCS county soil map: A book prepared by the Natural Conservation Resources Service (formerly Soil Conservation Service) of the USDA that presents maps and soil characteristics of a county.

Secondary treatment: With regard to wastewater and grey water, the biological process of removing suspended, colloidal, and dissolved organic matter in effluent from the primary treatment system.

Sediment: Soil and other granular materials carried by water and deposited into streams, lakes, reservoirs, and detention basins/ponds.

Sediment basin: A pond, basin, or other structure or measure that detains water to settle sediment.

Sewer: Any pipe or conduit used to collect and convey sewage or storm water runoff from its source to a treatment plant or receiving water body.

Shallow concentrated flow: Flow that has concentrated in rills or small gullies.

Shear stress: Force exerted by flow per unit of the wetted area of the channel; stress on the channel bottom due to the hydrodynamic forces of the flowing water.

Sheet flow: A shallow movement of runoff overland.

Side slope: Slope of the sides of a channel defined as the run per unit rise.

Site: A lot, tract, or parcel of land or a combination of contiguous lots, tracts, or parcels of land.

Slotted inlets: A section of pipe with openings along the longitudinal axis and transverse bars spaced to form slots.

Slug test: Field test for measuring permeability by adding or removing a volume of water in a single well and observing the drop or rise of water level.

Smart growth: Urban planning that achieves environmental, community, and economic improvements.

Soil amendment: Organic and inorganic materials added to soil to improve its texture, water-holding capacity, infiltration, and nutrients.

Soil moisture storage: Volume of water held in the soil.

Soil texture: Classification of soil based on percentage of sand, silt, and clay.

Specific energy: The sum of elevation head and velocity head at a cross section from the channel bed.

Sprawl: A pattern of development characterized by inefficient access between land uses or to public facilities or services and a lack of functional open space. Sprawl is typically an automobile-dependent, resource-consuming, discontinuous, low-density development pattern.

Spread: The lateral distance from the curb face to the edge of the water flowing in a gutter or on a roadway.

Sprinkler: A device for overhead delivery of water through small nozzles fastened to a hose. The sprinklers are commonly rotary or oscillating.

Stage–storage–discharge relationship: Variation of storage and discharge with stage for a detention basin/reservoir.

Steady flow: Flow that remains constant with respect to time.

Steep slope: Channel slope for which uniform flow is supercritical.

Storage–discharge relation: Relation between storage and outflow for a detention basin.

Storage routing: Flood routing in which discharge is uniquely related to the amount of storage.

Storm drain (storm sewer): A pipe that receives runoff from inlet(s) and conveys it downstream.

Storm drainage systems: Systems of pipes, inlets, and manholes that collect, convey, and discharge storm water.

Storm water management: The control and management of storm water runoff to minimize its detrimental effects on water quantity and quality.

Storm water wetlands: Shallow man-made ponds that treat storm water and allow for the growth of wetland-characterizing vegetation.

Streamflow: The fate of water flow at a section in streams or channels.

Streamlines: Flow lines that represent the direction of water movement in a flow.

Stripping: Any activity, such as clearing and grubbing operations, that removes or significantly disturbs vegetated or otherwise stabilized soil surface.

Sub-basins: Segments of a watershed with relatively homogeneous character.

Subcritical flow (tranquil flow): Flow of water at a velocity less than critical; Froude number less than 1.0.

Subdivision: The division of a lot, tract, or parcel of land into two or more lots or parcels.

Subsurface irrigation: The application of water below the soil surface, generally using a drip irrigation system.

Supercritical flow (rapid flow): Flow of water at a velocity greater than critical; Froude number greater than 1.0.

Superelevation: Local increase in water surface at the outer side of a channel bend.

Supplemental irrigation: The application of water to lawn or landscape area to supplement rainfall.

Surcharge: Condition in which the water level in a storm drain system rises above the grate of an inlet or manhole rim.

Sustainable development: Development that meets the needs of the present without jeopardizing the needs of future generations.

Synthetic design storm: Rainfall hyetograph obtained through statistical means.

Synthetic unit hydrograph: A unit hydrograph based on theoretical or empirical methods.

System International (SI): Also referred to as metric units; consist of meter, kilogram, or newton and second as basic dimensions.

Tailwater: Water surface level downstream of a conduit.

Technical release no. 20 (TR-20): A Soil Conservation Service (now NRCS) watershed hydrology computer model that performs runoff calculations and routing of storm through streams and ponds.

Technical release no. 55 (TR-55): A Soil Conservation Service hydrologic model that performs runoff calculations.

Time base: Total duration of direct runoff under the hydrograph.

Time of concentration: The time that it takes for runoff from all of a basin area to reach the outlet.

Time to peak: Time from the center of mass of rainfall to the peak of hydrograph.

Total dissolved solids (TDS): The quantity of dissolved material in water; usually expressed in mg/L or g/L.

Total dynamic head: The sum of static head, velocity head, and head losses.

Total energy: The sum of pressure, elevation, and velocity heads.

Total suspended solids (TSS): The quantity of suspended matter in water; expressed in mg/L or g/L.

Tractive force: Force developed due to the shear stress on the perimeter of a channel section.

Transient flow: Flow that varies with time.

Transpiration: The natural process of transferring water vapor to air by plants. Through this process the leaves and flowers keep cool and survive under the heat of the sun.

Trash rack: Grate, grill, or other device installed on an outlet structure or intake of pipe to remove large debris from entering the structure.

Turbidity: A suspension of fine particles in water that obscure light rays and take many days to settle.

Turf/turf grass: Hybridized grass that forms a dense growth of blades and roots.

Turf reinforcement mat (TRM): A nondegradable rolled erosion control product (RECP) made of synthetic fibers, filaments, or wire mesh processed into a three-dimensional matrix.

Uniform flow: Flow condition with a constant depth and velocity along the length of the channel.

Unit hydrograph: Graph of runoff versus time produced by a unit direct rainfall (1 in. or 1 cm) from a given duration storm.

Unit peak discharge: The peak discharge per unit area, with units of $m^3/s/km^2$, or cfs/mi^2.

Unsteady flow: Flow that changes with time.

Vadose zone: Zone of aeration that extends from the surface to the water table including the capillary fringe.

Vapor pressure: Partial pressure exerted by water vapor.

Varied flow: Flow where the flow rate and depth change along the channel.

Velocity head: Energy due to the velocity of water.

Velocity, mean: Discharge divided by the area of flow.

Velocity, permissible: Velocity that will not cause channel erosion.

Wastewater: The used water from residential, commercial, and industrial buildings/sites.

Wastewater conservation: Reductions in water use, loss, or waste.

Water conservation measures: Actions, behavioral changes, devices, improved design, technology, or process implemented to reduce water use, loss, and waste.

Water-efficient landscape: A landscape that minimizes the water need. See also xeriscape.

Water harvesting: The capture and use of rainfall or runoff.

Water quality inlets: Inlets with built-in filters that remove sediment, oil and grease, and floatables from runoff.

Water reclamation: The treatment of wastewater for reuse, commonly for nonpotable purposes.

Water reuse: The use of water that had previously been used for a specific purpose.

Watershed: Area of land that drains to a single outlet water course or stream and is separated from other watersheds by a topographic divide.

Watershed divide: Line that defines a watershed boundary.

Water surface profile: Plot of the depth of water along the length of a channel.

Water table: The top of saturated soil zone in an unconfined aquifer; the surface where the soil is saturated and water pressure is exactly atmospheric.

Water use: Water that is actually used for a specific purpose or customer such as residential or agricultural users.

Water year: A continuous 12-month period that begins on October 1 and ends on September 30th of the following year and is designated by the year in which it ends.

Weed: Any unwanted or troublesome plant.

Weir flow: Gravity flow over a horizontal barrier, such as a weir, roadway, or bridge.

Well: Vertical hole dug into the soil that penetrates an aquifer.

Wet ponds: A pond with a permanent pool.

Wetted perimeter: The length of contact between the flowing water and pipe or channel.

Wilting point: Moisture content below which plants cannot extract any water and will not survive.

Wing wall: Side wall extensions of a pipe/culvert used to prevent sloughing of channel or stream banks.

Withdrawal: Water delivered or extracted from a surface or groundwater source.

Xeriscape: A trademarked term representing a landscaping that involves the selection, placement, and maintenance of low-water-use plants, turf, shrubs, and trees.

Appendix A: System International (SI)

Dimensions describe physical quantities and units express their amounts. Units, however, vary from one system of units to another. The common system of units are metric, English, and SI. The metric and English systems of units have been in use for a long time (for example, the metric system since 1872). In 1960 an international system of units was proposed in Europe. In this system, abbreviated as SI (System International), mass, length, time, temperature, electrical current, luminous intensity, and the amount of matter are selected as dimensions. Any physical quantity can be described in terms of these dimensions. The units of these dimensions and their symbols appear in the following table. For temperature, the degree Celsius (°C) is more commonly used than kelvin in engineering practice. There is a general lack of familiarity with metric units in the United States. Figure A.1 serves as an example.

SI Dimensions and Units		
Dimensions	Units	Symbol
Length	Meter	m
Mass	Kilogram	kg
Time	Second	s
Temperature	Kelvin	K
Electrical current	Ampere	A
Luminous intensity	Candela	Cd
Amount of material	Mole	mol

In SI units there are also two supplementary units as follows:

Radian (rad), representing planar angles
Steradian (sr), representing solid angles (used in three dimensional space)

In hydraulic engineering, in general, and storm water management practice, in particular, three basic dimensions are used: length, mass, and time. Every hydraulic quantity can then be expressed in terms of these three basic dimensions, abbreviated as L, M, and T, respectively. For example, the relation for acceleration

$$a = \Delta V / \Delta t$$

shows that acceleration, a, has the dimensions of

$$a = (L/T)/T = LT^{-2}$$

Newton's law, which relates force and mass as

$$F = ma$$

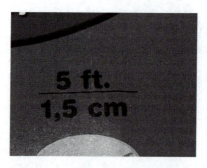

FIGURE A.1   An emblem on a cardboard box. The incorrect conversion indicates a greater need for familiarity with metric units. The emblem should read "1.5 m" rather than "1.5 cm."

may be used to establish the dimensions of F in terms of M, L, and T, with the following result:

$$F = MLT^{-2}$$

In hydraulics, we deal with fluid properties such as density ρ, specific weight γ, specific gravity $S_g = \gamma/\gamma_w$, viscosity μ (also called dynamic viscosity), kinematic viscosity $\nu = \mu/\rho$, and surface tension σ. Also encountered is pressure $p = F/A$ both in water at rest and in motion. In flowing water, moreover, such quantities as discharge Q, work ($W = FL$), power ($P = W/t$), angular velocity, momentum, and specific energy are involved. Dimensions of these quantities can be obtained from the laws of physics or mechanics.

As indicated, units of the aforementioned quantities depend on the system of units used. In the SI units, dimensions of quantities that are commonly used in hydraulic engineering are listed in the following table.

Quantity (Symbol)	Dimension	Units
Length (l)	L	Meter (m)
Mass	M	Kilogram (kg)
Time (s)	T	Second (s)
Velocity (V)	LT^{-1}	(m/s)
Acceleration (a)	LT^{-2}	(m/s^2)
Density (ρ)	ML^{-3}	(kg/m^3)
Force (weight) (F)	MLT^{-2}	Newton (N) = kg weight/9.81
Pressure (p)	$ML^{-1}T^{-2} = FL^{-2}$	Pascal (Pa) (N/m^2)
Discharge (Q)	L^3T^{-1}	m^3/s
Specific weight (γ)	$ML^{-2}T^{-2}$	N/m^3
Viscosity (μ)	$ML^{-1}T^{-2}$ ($FL^{-2}T$)	kg/m · s = N · s/m^2
Kinematic viscosity (ν)	L^2T^{-1}	m^2/s
Work, energy, heat (W)	$ML^2T^{-2} = (FL)$	N · m
Surface tension (σ)	$MT^{-2} = FL^{-1}$	N/m
Power (P)	$ML^2T^{-3} = (FLT^{-1})$	Watts (W), N · m/s
Momentum (mv) = M	MLT^{-1}	kg · m/s = N · s
Frequency (f)	T^{-1}	Hertz (Hz)

In this system, small or large quantities are described with the aid of a prefix, indicating the factor of 10 employed. The most common prefixes are as follows:

Prefix	Symbol	(Factor)
Tera	T	10^{12}
Giga	G	10^{9}
Mega	M	10^{6}
Kilo	k	10^{3}
Hecto	h	10^{2}
Deka	da	10^{1}
Deci	d	10^{-1}
Centi	c	10^{-2}
Milli	m	10^{-3}
Micro	μ	10^{-6}
Nano	n	10^{-9}
Pica	p	10^{-12}

On the basis of these symbols, for example, 10^5 pascal may be expressed as 10^2 kPa.

Conversion of some of the most commonly used units from English (also referred to as conventional units) to SI appear in the following table.

Conversion Factors from English to SI Units

Quantity	English Units	Metric Units	Factor
Length	inch (in.)	mm	$25.4\ e^a$
	foot (ft)	m	0.3048 e
	yard (yd)	m	0.9144 e
	mile	km	1.609
Area	in.^{2}	cm^2	6.452
	ft^2	m^2	0.0929
	yd^2	m^2	0.8361
	acre	m^2	4047
	acre	ha	0.4047
	mi^2	km^2	2.590
Volume	in.^{3}	cm^3	16.387
	ft^3	m^3	0.0283
	yd^3	m^3	0.7646
	ounce	cm^3	29.574
	quart	liters	0.9463
	gallon	liters	3.7853
	acre-ft	m^3	1233
Mass	pound (lb) mass	kg	0.4536
	kip (1000 lb)	ton	0.4536
	ton (2000 lb)	ton	0.9072
Mass density	lb/ft^3	kg/m^3	16.026
Force	lb	N	4.448
Pressure	lb/ft^2	Pa(N/m^2)	47.8803
	lb/in.^{2}	Pa	0.3325

(Continued)

Conversion Factors from English to SI Units (Continued)

Quantity	English Units	Metric Units	Factor
Viscosity	$lb \cdot s/ft^2$	$N \cdot s/m^2$	47.87
Kinematic viscocity	ft^2/s	m^2/s	0.0929
Power	ft-lb/s	watts	1.3558
	hp (horsepower)	watts	745.70
	Btu/h	watts	0.2931
Energy	1000 Btu	kWh	0.2931
Velocity	ft/s	m/s	0.3048 e
	mi/h	km/h	1.609
	knot (speed of sea vessels)	km/h	1.852

[a] e reflects exact conversion factor.

COMMON CONSTANTS

1. Water properties
   a. At standard conditions (4°C and 760 mm Hg)
      Specific weight: 9806 N/m³ (~1000 kg/m³) ≈ 62.4 lb/ft³
      Density: 1000 kg/m³ ≈ 1.94 slug/ft³
      Viscosity: 1.57×10^{-3} N · s/m² ≈ 3.28×10^{-5} lb · s/ft²
      Kinematic viscosity: 1.57×10^{-6} m²/s ≈ 1.69×10^{-5} ft²/s
   b. At normal temperature (20°C and 760 mm Hg)
      Specific weight: 9789 N/m³ ≈ 62.3 lb/ft³
      Density: 998.2 kg/m³ ≈ 1.94 slug/ft³
      Viscosity: ≈ 1.0×10^{-5} N · s/m² ≈ 2.09×10^{-5} lb · s/ft²
      Kinematic viscosity: ≈ 1.0×10^{-6} m²/s ≈ 1.08×10^{-5} ft²/s
      Standard atmospheric pressure = 101.4 kN/m² ≈ 14.71 lb/in²
      Gravitational constant = 9.81 m/s² = 32.2 ft/s²
   c. Freezing/boiling temperatures at sea level
      Freezing: 0°C = 32°F
      Boiling: 100°C = 212°F

REFERENCE

Pazwash, H., 2007, *Fluid mechanics and hydraulic engineering*, Tehran University Press, Tehran, Iran.

Appendix B: Unified Soil Classification System and Nominal Sizes of Coarse and Fine Aggregates

Soils seldom naturally exist as a single component such as sand and gravel. Rather, they occur as a mixture; each component contributes its characteristics to the soil. In the Unified Soil Classification System (USCS), soils are placed into three major classifications: coarse grained, fine grained, and highly organic. The USCS further divides soils into five major soil categories symbolized as

G: gravel
S: sand
M: silt
C: clay
O: organic

Most unconsolidated soils are represented by a two-letter symbol in this system. The first letter is one of the preceding five symbols. The second letter can be any one of the same symbols (except for peat) or one of the following letters describing the soil condition:

Letter	Definition
P	Poorly graded (mostly one size)
W	Well-graded (fine to coarse sizes)
H	Highly plastic
L	Low plasticity

The following is a list of various soil designations:

Symbol	Soil Group
GW	Well-graded gravel
GP	Poorly graded gravel
GM	Silty gravel
GC	Clayey gravel
SW	Well-graded sand
SP	Poorly graded sand
SM	Silty sand
SC	Clayey sand
ML	Silt of low plasticity
MH	Silt of high plasticity
CL	Clay of low plasticity
CH	Clay of high plasticity
OL	Organic soil of low plasticity
OH	Organic soil of high plasticity (organic silt/clay)
PT	Peat

For a further description of USCS, one is referred to the US Army (1997) Soils Engineering Field Manual 5-410 (http://www.adtdl.army.mil/cgi-bin/atdl.dll/fm/5-410/toc.htm) or Wikipedia (http://en.wikipedia.org/wiki/Unified_Soil_Classification_System).

A coarse-grained soil is defined as that in which at least half of the material is retained on a number 200 sieve (0.075 mm). In the United States, size designation of coarse aggregates is defined by AASHTO M43. Tables B.1 and B.2 present sizes of coarse aggregates in metric and English units, respectively.

Fine aggregates are designated based on the size of sieve opening passing the grains. Table B.3 shows the US standard sieve (mesh) number specified by the ASTM E11 and AASHTO M92 standards. According to this table, the 200 sieve, which differentiates coarse and fine aggregates, has a 0.003 in. opening. More information on AASHTO M43 can be found by searching for "703 Aggregate" and "AASHTO M43" and at many web addresses (e.g., http://www.odotnet.net/spec/703.htm).

TABLE B.1

Sizes of Coarse Aggregate (AASHTO M43) (Millimeters)

Size Number	Nominal Size Square Openings[a]	Amounts Finer than Each Laboratory Sieve (Square Openings), Percentage by Weight														
		100	90	75	63	50	37.5	25	19	12.5	9.5	4.75	2.36	1.18	300 μm	150 μm
1	90 to 37.5	100	90 to 100		25 to 60		0 to 15		0 to 5							
2	63 to 37.5			100	90 to 100	35 to 70	0 to 15		0 to 5							
24	63 to 19.0			100	90 to 100		25 to 60		0 to 10	0 to 5						
3	50 to 25.0				100	90 to 100	35 to 70	0 to 15		0 to 5						
357	50 to 4.75				100	95 to 100		35 to 70		10 to 30		0 to 5				
4	37.5 to 19.0					100	90 to 100	20 to 55	0 to 15		0 to 5					
467	37.5 to 4.75					100	95 to 100		35 to 70		10 to 30	0 to 5				
5	25.0 to 12.5						100	90 to 100	20 to 55	0 to 10	0 to 5					
56	25.0 to 9.5						100	90 to 100	40 to 75	15 to 35	0 to 15	0 to 5				
57	25.0 to 4.75						100	95 to 100		25 to 60		0 to 10	0 to 5			

(Continued)

TABLE B.1 (CONTINUED)
Sizes of Coarse Aggregate (AASHTO M43) (Millimeters)

Size Number	Nominal Size Square Openings[a]	Amounts Finer than Each Laboratory Sieve (Square Openings), Percentage by Weight														
		100	90	75	63	50	37.5	25	19	12.5	9.5	4.75	2.36	1.18	300 μm	150 μm
6	19.0 to 9.5							100	90 to 100	20 to 55	0 to 15	0 to 5				
67	19.0 to 4.75							100	90 to 100		20 to 55	0 to 10	0 to 5			
68	19.0 to 2.36							100	90 to 100		30 to 65	5 to 25	0 to 10	0 to 5		
7	12.5 to 2.36									90 to 100	40 to 70	0 to 15	0 to 5			
78	9.5 to 2.36								100	90 to 100	40 to 75	5 to 25	0 to 10	0 to 5		
8	9.5 to 1.18									100	85 to 100	10 to 30	0 to 10	0 to 5		
89	4.75 to 1.18									100	90 to 100	20 to 55	5 to 30	0 to 10	0 to 5	
9	4.75 to 1.18										100	85 to 100	10 to 40	0 to 10	0 to 5	
10	4.75 to 0[b]										100	85 to 100				10 to 30

Note: Where standard sizes of coarse aggregate designated by two- or three-digit numbers are specified, the specified gradation may be obtained by combining the appropriate single digit.

[a] In millimeters, except where otherwise indicated.

[b] Screenings.

TABLE B.2
Sizes of Coarse Aggregate (AASHTO M43) (Inches)

Size Number	Nominal Size Square Openings[a]	4	3-1/2	3	2-1/2	2	1-1/2	1	3/4	1/2	3/8	No. 4	No. 8	No. 16	No. 50	No. 100
					Amounts Finer than Each Laboratory Sieve (Square Openings), Percentage by Weight											
1	3-1/2 to 1-1/2	100	90 to 100		25 to 60		0 to 15		0 to 5							
2	2-1/2 to 1-1/2			100	90 to 100	35 to 70	0 to 15		0 to 5							
24	3-1/2 to 3/4			100	90 to 100		25 to 60		0 to 10	0 to 5						
3	2 to 1				100	90 to 100	35 to 70	0 to 15		0 to 5						
357	2 to no. 4				100	95 to 100		35 to 70		10 to 30		0 to 5				
4	1-1/2 to 3/4					100	90 to 100	20 to 55	0 to 15		0 to 5					
467	1-1/2 to no. 4					100	95 to 100	90 to 100	35 to 70		10 to 30	0 to 5				
5	1 to 1/2							90 to 100	20 to 55	0 to 10	0 to 5					
56	1 to 3/8							90 to 100	40 to 75	15 to 35	0 to 15	0 to 5				

(Continued)

TABLE B.2 (CONTINUED)
Sizes of Coarse Aggregate (AASHTO M43) (Inches)

Size Number	Nominal Size Square Openings[a]	Amounts Finer than Each Laboratory Sieve (Square Openings), Percentage by Weight														
		4	3-1/2	3	2-1/2	2	1-1/2	1	3/4	1/2	3/8	No. 4	No. 8	No. 16	No. 50	No. 100
57	1 to no. 4						100	95 to 100		25 to 60		0 to 10	0 to 5			
6	3/4 to 3/8							100	90 to 100	20 to 55	0 to 15	0 to 5				
67	3/4 to no. 4							100	90 to 100		20 to 55	0 to 10	0 to 5			
68	3/4 to no. 8							100	90 to 100		30 to 65	5 to 25	0 to 10	0 to 5		
7	1/2 to no. 4								100	90 to 100	40 to 70	0 to 15	0 to 5			
78	1/2 to no. 8								100	90 to 100	40 to 75	5 to 25	0 to 10	0 to 5		
8	3/8 to no. 8									100	85 to 100	10 to 30	0 to 10	0 to 5		
89	3/8 to no. 16									100	90 to 100	20 to 55	5 to 30	0 to 10	0 to 5	
9	No. 4 to no. 16										100	85 to 100	10 to 40	0 to 10	0 to 5	
10	No. 4 to 0[b]										100	85 to 100				10 to 30

Note: Where standard sizes of coarse aggregate designated by two- or three-digit numbers are specified, the specified gradation may be obtained by combining the appropriate single-digit standard size aggregates by a suitable proportioning device, which has a separate compartment for each coarse aggregate combined. The blending shall be done as directed by the laboratory.

[a] In inches, except where otherwise indicated. Numbered sieves are those of the US standard sieve series.
[b] Screenings.

TABLE B.3
US Standard Sieve Sizes

Alternative No.	Nominal Openings (in.)	Standard (mm/μm)
4 in.	4	100 mm
3-1/2 in.	3.5	90 mm
3 in.	3	75 mm
2-1/2 in.	2.5	63 mm
2.12 in.	2.12	53 mm
2 in.	2	50 mm
1-3/4 in.	1.75	45 mm
1-1/2 in.	1.5	37.5 mm
1-1/4 in.	1.25	31.5 mm
1.06 in.	1.06	26.5 mm
1 in.	1	25.0 mm
7/8 in.	0.875	22.4 mm
3/4 in.	0.75	19.0 mm
5/8 in.	0.625	16.0 mm
0.530 in.	0.53	13.2 mm
1/2 in.	0.5	12.5 mm
7/16 in.	0.434	11.2 mm
3/8 in.	0.375	9.50 mm
5/16 in.	0.312	8.00 mm
0.265 in.	0.265	6.70 mm
1/4 in.	0.25	6.30 mm
1/8 in.	0.125	3.17 mm
No. 3-1/2	0.233	5.66 mm
No. 4	0.187	4.75 mm
No. 5	0.157	4.00 mm
No. 6	0.132	3.35 mm
No. 7	0.111	2.80 mm
No. 8	0.0937	2.36 mm
No. 10	0.0787	2.00 mm
No. 12	0.0661	1.70 mm
No. 14	0.0555	1.40 mm
No. 16	0.0469	1.18 mm
No. 18	0.0394	1.00 mm
No. 20	0.0331	850 μm
No. 25	0.0278	710 μm
No. 30	0.0234	600 μm
No. 35	0.0197	500 μm
No. 40	0.0165	425 μm
No. 45	0.0139	355 μm
No. 50	0.0117	300 μm
No. 60	0.0098	250
No. 70	0.0083	212
No. 80	0.0070	180
No. 100	0.0059	150
No. 120	0.0049	125
No. 140	0.0041	106
No. 170	0.0035	90

(Continued)

TABLE B.3 (CONTINUED)
US Standard Sieve Sizes

Alternative No.	Nominal Openings (in.)	Standard (mm/μm)
No. 180	0.0033	80
No. 200	0.0029	75
No. 230	0.0025	63
No. 270	0.0021	53
No. 325	0.0017	45
No. 400	0.0015	38
No. 450	0.0012	32
No. 500	0.0010	25
No. 635	0.0008	20
No. 850	0.0004	10

Index

Page numbers followed by f, t and n indicate figures, tables and notes, respectively.

A

AbTech Industries, 14, 332, 333
Access holes, losses, 55–56, 56f
Achla RB03 rain catcher, 618
Acidification, 6
ACIS (applied climate information system), 74
ADS Drainage Handbook, 546
ADS N-12 drainage, 554, 555f
ADS Pipe Adapter, 554
Advanced drainage systems (ADS), 412, 415, 544–545
Advanced metering infrastructure (AMI) systems, 602
Airports, removing hydrocarbons and fuels, 14
Algae and weed control, 575–576; see also Ponds/
        detention basins, maintenance
A Lok Products, Inc., 563, 564f
AMC (antecedent moisture conditions), 117, 117t
American Concrete Pipe Association, 547–548, 550
American National Standards Institute (ANSI), 487
American Public Works Association, 327
American Society of Civil Engineers (ASCE), 326–327,
        570
American Water Works Association (AWWA), 541, 592,
        594, 597
    manual, 593, 622
Americast, 330, 370
Analytical testing, defined, 329
Antecedent moisture conditions (AMC), 117, 117t
A-2000 pipes, 535–536
Applied climate information system (ACIS), 74
Aprons, riprap, 206, 207–209, 208f, 210f
Aqua-Guardian™ catch basin insert, 333, 334f, 332t
Aqua-Swirl, 337, 337t
AquaTech International, 618
ArmorMax, 242, 571
    installation guidelines, 592–596
    physical properties, 256f–259f
Arrow Concrete Products, 413
ASTM C1701 standard test method, 574
Atlantic City, rainfall IDF data, 67, 71t–72t
Atlantis Water Management for Life, 420, 421f
Atmospheric dust
    defined, 15
    global dust-emission rates, 15
    particles and wind erosion, 15
Audubon International's Three Diamond designation, 486
Audubon's Gold Signature Program, 486
Average flow depth, defined, 225
AWWA (American Water Works Association), see
        American Water Works Association (AWWA)

B

Bacterra™ System, 350
Baffle box, Nutrient Separating Baffle Box, 347
Bare soil, 217–219, 218f, 218t, 219f

Barnegat Bay, 10
Basin development factor (BDF), 151
Baskets, gabion, 239–241, 240t
BayFilter, 344, 345
BaySaver Technologies, Inc., 340
BaySaver Treatment Technologies, 345
BaySeparator, 337, 340
Bazin formula, 35, 36t
Bend losses, 54
Bentley's Haestad Methods Pond Pack, 140
Benzene, toluene, ethyl benzene, and xylene (BTEX)
        compounds, 14
Berms, infiltration, 299
Best management practices (BMPs)
    cost effectiveness, 530–531
    database site, 326–326
    design, Maryland storm water management, 286–290
    International Stormwater BMP Database, 326–327
    publications on performance of storm water
        management, 326
BioClean Environmental Services of Oceanside, 333
Bio-Microbics of Shawnee, 350
Bioretention basin, 521–522
Bioretention cells, 350–352, 523–525
BioStorm® storm water treatment system, 350, 351f–352f
Bioswales, 302–303, 303f, 522–523; see also Swales
        Blocksom & Co., 333
Blocksom filters, 336, 336f
Blue roofs, 511
BMPs, see Best management practices (BMPs)
Box culverts, see Concrete box culverts
Box culverts, concrete, 554
BTEX (benzene, toluene, ethyl benzene, and xylene)
        compounds, 14
Buffalo Niagara International Airport, 14 Bureau of
        Nonpoint Pollution Control, 330

C

Calcium chloride (CaCl), 13
Calcium magnesium acetate (CMA), 13
Cambridge Crossings project, 533–535
Carpooling, 14
Cary Institute, 12
Cast-A-Seal 802, 563–564, 564f
Catch basin inserts, 330, 332–336
    Aqua-Guardian catch basin insert, 333, 336f
    FloGard+Plus® Filter
        description, 332, 333, 336f
        sponges, 333
    lawn inlet screen filter, 335
    REM Geo-Trap inserts, 333
    REM Triton Series, 333
    Smart Sponge, 333
    SNOUTs, 336, 336f

Ultra-Urban Filter
    description, 332, 333
    sponges, 333
Cato Institute, 484
CDS, *see* Continuous Deflective Separation (CDS)
Center for Watershed Protection, 306
Certification, of water quality devices, 327–333
    NJCAT, 328–333
        Bureau of Nonpoint Pollution Control, 330
        laboratory testing, 328–329
        by NJDEP, 333t
    NJDEP, letter for Filterra Bioretention Systems,
        370–371
    NJDEP Division of Science, Research and Technology
        (DSRT), 328
CGP, *see* Construction general permit (CGP)
Chamber Maxx, 417
Channel bends, 234–237
    bend factor, 235t
    example, 236–237
    superelevation
        defined, 234
        extra free board for, 235
Channel designs
    grass-lined channel design, 233–234
    gravel and stone-lined channels, 225
    HEC-15, NJ Standards for vegetal retardance D,
        226–227
    hydraulic parameters, NJ Standards, 230
    riprap, cobble, and gravel lining, 224
    shear stress and flow depth calculation, 228–229
    stable channel, hydraulic parameters, 229–230
    stone linings, 225, 226t
    stone size and, 225
    trapezoidal channel design, 231–234
Channel flow, flow path segment, 102, 103, 104, 105
Channel protection storage volume criteria, 284, 284t,
        285f
Channels, drainage, *see* Drainage channels
Chesapeake Bay, 10
Chezy equation
    average velocity, channel, 35
    Pavlosky, Prague Hydraulics Institute and Bazin
        formulas, 35, 36t
Chlordane, 10
Chlorpyritos, 10
Cholera, 12
Circular pipes
    flow depth calculation, 42–44
        hydraulic properties, 42, 43, 43t
        Manning formula, 42
        relative velocity, variation of, 43, 44f
    partly full, critical flow in, 31–34, 32t, 33f
Cisterns, 297
Clean Water Act (CWA)
    amendment, 265
    EPA and, 3, 5, 19–22, 265
CMA (calcium magnesium acetate), 13
CoCoRaHS (Community Collaborative Rain, Hale &
        Snow) Network, 74
Combination inlet, 160, 160f, 173, 174
Commercial development project, case study, 357–361
Community Collaborative Rain, Hale & Snow
        (CoCoRaHS) Network, 74

Composite lining, 237–238
Composite surfaces, equations for, 135–138
Concrete box culverts, 540
Concrete Pipe Design Manual, 194, 200
Concrete pipes
    elliptical, 551
    prestressed, 551
Conductivity, hydraulic, *see* Hydraulic conductivity
Conservation areas, sheetflow to, 296
Constant head permeameter, 93, 93f
Constructed wetlands, *see* Storm water wetlands
Construction and Development (C&D) rule, 267
Construction general permit (CGP), 266–268
Contech, 545–546, 570
Contech Construction Products Inc., 343
Contech Engineered Solutions, 337, 338f, 339, 340, 341f
Contech Stormwater Solutions, Inc., 345, 370
Continuous deflective separation (CDS)
    models, 339–340, 338f
        dimension and capacity of, 372t
        NJDEP approved rates for, 373t
Contractions, sudden, 53–54, 53f, 54t
Conveyance system design criteria, 314–315, 314f
Copper, pollutant constituent in urban runoff, 4
Cornell University Northeast Regional Climate Center
        website, 74
Corps of Engineers, 265
Critical Area Act, Maryland, 306
Critical flow, 26–34
    charts for round and elliptical pipes, 254f
    depth and discharge, calculation, 28–31, 34
    in partly full circular pipes, 31–34, 32t, 33f
    in rectangular channels, 28–29
    in trapezoidal channels, 29–31
CrystalClean Separator, 337
Crystal Springs Technologies, 337
CST, 498
Cultec chambers, 468
Cultec Inc., 468, 568
Cultec Recharger 280 HD chambers, 434
Cultec Recharger V8 chambers, 422
Culverts, hydraulic design of, 191–206
    calculations, 202
    capacity of, 191
    energy losses in, 50–59; *see also* Energy losses
    entrance loss coefficients, 200, 201t
    examples, 203–206
Curb opening inlet
    in combination inlets, 169
    design, 171–172
    overview, 160, 160f
    weir flow discharge, 171–172
CWA, *see* Clean Water Act (CWA)

D

Darcy's law, 82, 93, 94
Data, rainfall, 73–75, 73t, 74f
Dead zone, defined, 9
Debris clogging, grate inlets and, 165
De-icing, road salt for, 12–14
Delaware DOT's Nonpoint Pollutant Discharge Elimination
        System (DelDOT's NPDES) Group, 571–573
Delmarva unit hydrograph, 126n

Delta function model, *see* Green-Ampt model
Density-stiffness coefficient, 224, 224t
Depression, landscape, 297–298, 298f
Depression storage, 79–82
    defined, 131
    examples, 80–82
    rate of, 80
    ratio of runoff to rainfall, 81, 82
    runoff coefficient, 81
    typical values of, 80
    values, in universal runoff method, 132, 132t
    water volume, 80
Design, culverts, 191–206
    calculations, 202
    entrance loss coefficients, 200, 201t
    examples, 203–206
    inlet control
        capacity, 191, 194
        discharge, 196f, 198f, 200f
        headwater, 197f, 199f
        nomograph, 195f
        schemes, 193f
    Manning "n" values, 200, 201t
    outlet control
        capacity, 191, 193, 194
        discharge, 202, 202f
        schemes, 194f
    tabulation, 203f
Design, inlets, 165–176
    combination, 173, 174
    curb opening, 171–172
    grate, 165–171
        design, 165–171
        elements, 165
        frontal flow, defined, 165
        frontal flow to gutter flow ratio, 167, 167f
        gutter flow velocity, 168
        interception capacity, calculation, 170–171
        orifice flow, 175–176
        overall efficiency and intercepted flow, 169
        P-1-7/8 and P-1-7/8—four grates, 165, 166f
        reticuline, 165, 166f, 168
        on sag, 175–176
        side flow interception efficiency, 168
        splash-over velocity, 167, 168, 168t, 169f
        weir flow, 175–176
    New Jersey, 173–175, 174f, 175t
    slotted inlets, 172–173
Design, of storm drains, 184–191
    calculation processes, 184–186, 185f
    case studies
        HGL calculation, 191, 192f, 193f
    head loss calculations, 191, 192f
    runoff, peak rate of, 184
Design, swales, 288, 289f; *see also* Swales
Detention basins, 565–566
    Detention basins, preliminary sizing, 386–392
        adjusting detention storage volume estimation, 388–389
        rational and modified rational methods estimation, 386–387
        SCS TR-55 method estimation, 387–388
        universal method of storage volume estimation, 388
Detention basins/wet ponds, 375–385
    design, 375–377

flow routing, 377–378
outlet structure design, 379–385
    broad-crested weir, 382, 382f
    cipolleti weir, 381, 382f
    hydro-brake fluidic-cones, 383
    orifice, 379
    outflow grates, 382–383
    rectangular weir, 379–380, 380f
    stand pipes, 383, 383f
    thirsty duck, 383–384
    triangular weirs, 380–381, 381f
Developing Your Storm Water Pollution Prevention Plan, 266–267
Dewatering, 559–561
Dewatering time, 451
Diaphragm pumps, 560
Diazinon, 10
Dieldrin, 10
Division of Science, Research and Technology (DSRT), NJDEP, 326
Double ring infiltrometer, Rickly, 92, 93f
Downstream analysis, 313–314
    conditions, 314
    elements, 313–314
    10% rule, 313
    watershed, 313
Downstream Defender, 343
Drag coefficient, 8
Drainage channels, 213–238
    bare soil and stone lining, 217–219, 218f, 219f
        permissible shear stress, 217, 218t
        riprap-lined channels with side slopes, 219–220
    channel bends, 234–237
        bend factor, 235t
        example, 236–237
        shear stress, 235
        superelevation, 234, 235
    composite lining, 237–238
    flexible linings, 213
    grass lining, 221–224
        density-stiffness coefficient, 224, 224t
        grass cover factor, 223, 223t
        grass roughness coefficient, 224
        HEC-15, 222–223
        permissible shear stress of, 221
        retardance degree/classes, 221, 222, 222t
        soil roughness coefficient, 223
        vegetal retardance selection, 221t
    Manning's roughness coefficient variation, 224–234
        design charts, 226, 227f
        examples, 228–234
        flow depth, 224–225
        flow parameters and, 226, 227f
        flow velocity and hydraulic radius, 226, 226f
        grass-lined channel design, 233–234
        gravel and stone-lined channels, 225
        HEC-15, NJ Standards for vegetal retardance D, 226–227
        hydraulic parameters, NJ Standards, 230
        riprap, cobble, and gravel lining, 224
        shear stress and flow depth calculation, 228–229
        stable channel, hydraulic parameters, 229–230
        stone linings, 225, 226t
        stone size and, 225

trapezoidal channel design, 231–234
    permissible velocity concept, 213–215, 214f, 215t
        shear stress, 213, 215
        USSR data, cohesive and noncohesive soils, 214,
            214f
    rigid linings, 213
    side slope stability, 219–221
        angle of repose and angle of side slope, 220, 221f
        channel side shear stress to bottom shear stress
            ratio, 219–220, 220f
    tractive force method, 215–217, 216f
Drainage system, see Stormwater drainage systems
Drainage systems
    construction of, 1
    New Milford Avenue, case study, 186–190, 187f, 188f,
        189f, 190f
    roadway drainage analysis, 155–160
    storm, design of, see Storm drainage systems
Drivable grass, 502
    installation guidelines, 541–542
    technical specification guide, 540–541
Dry wells, 299–300
    and infiltration chambers, 568–569
DSRT (Division of Science, Research and Technology),
    NJDEP, 328
Dual-purpose detention basins, see Extended detention
    basins
Dual-Vortex Hydrodynamic Separator, 333
DuroMaxx pipes, 545–546
Dysentery, bacterial, 12

E

ECBs, see Erosion control blankets (ECBs)
Eco-curb piece, 160, 161, 162f
Effluent Limitations Guidelines, 267
EISA (Energy Independence and Security Act), section
    428, 271
Elliptical concrete pipes, 551
EMC (event mean concentration), 365–366, 366t
End-of-pipe practices, defined, 3
Energy equation, pipe and open channel flow, 25–26
    components, 25–26, 26f
    HGL and EGL, 26
    total energy head, 25–26
Energy grade line (EGL)
    defined, 26
    slope, 26, 34
Energy gradient, 35
Energy Independence and Security Act (EISA), section
    428, 271
Energy losses, in pipes and culverts, 50–59
    frictional losses, 50–51
    local losses, 51–59
        bend, 54
        defined, 50
        entrance and exit, 52–53, 52t
        example, 57–59, 58f
        head loss at transitions, 54, 55f, 55t
        junction, 55, 55f
        manholes, 55–56, 56f
        sudden expansions/contractions, 53–54, 53f, 54t
        velocity head times, 51–52
Enhanced filters, 296, 303–304, 304f

Enhanced phosphorus removal standards, 316
Enkamat Turf Reinforcement, 242
Entrance and exit losses, 52–53, 52t
Entrance loss coefficients, for culverts, 200, 201t
Environmental and Water Resources Institute (EWRI),
    327
Environmental Protection Agency (EPA) C&D rule, 267
    CGP, 266–268
    CWA of 1972, 5, 19–22, 265
    EISA Section 428, 271
    erosion and water pollution, 1
    fecal coliform concentrations, 12
    NPDES phase II program, 266–267
    NRC's report, 16, 19–22
    SWMM of, 83, 88, 140–141
    TCLP, 333, 366
    Terre Kleen verification, ETV program and, 363–367
        performance verification, 365–367, 365t
        technology description, 363–364
        testing description, 364–365
    water pollution, source of, 3
Environmental Retrofit Solutions (ERS), 161
Environmental site design (ESD), 290–304
    alternative surfaces, 294–295
        green roofs, 294, 295t
        permeable pavements, 294–295, 295t
        reinforced turf, 295
    implementing, 281
    Maryland's Stormwater Design Manual, 269
        microscale practices, 296–304
        dry wells, 299–300
        enhanced filters, 303–304, 304f
        infiltration berm, 299
        landscape infiltration (depression), 297–298, 298f
        microbioretention, 300–301, 301f
        performance standards, 296
        rain gardens, 301–302
        rainwater harvesting systems, 297
        submerged gravel wetlands, 297
        swales, 302–303
    nonstructural practices, 295–296
        conservation areas, sheetflow to, 296
        nonrooftop runoff disconnection, 296
        rooftop runoff, disconnection of, 296, 296t
    predevelopment runoff standards, 290
    runoff depth and storage volume calculation, 290, 291
    site development strategies, 290, 291t
    sizing requirement, Maryland's RCN for, 321t–323t
    storm water management requirements, 292–294, 293t
Environmental technology verification (ETV) program
    Terre Kleen, verification by EPA and NSF, 363–367
EPA, see Environmental Protection Agency (EPA)
Erosion control, outfalls, 206–213
    riprap aprons, 206, 207–209
        conduit outlet protection design, 206, 210f
        configuration, outlets, 208, 208f
    scour holes
        design tailwater depth calculation, New Jersey, 211
        EPA method, 212
        layout and section view, 209, 211f
        median stone diameter, 209
        SCS method, 212
Erosion control blankets (ECBs), 242–244, 570–571
    Landlok, 242, 243–244, 243t, 260f–263f

properties, 243–244, 243t
Propex, 242, 243, 260f–263f
ERS (Environmental Retrofit Solutions), 161
Escherichia coli, 12
ESD, *see* Environmental site design (ESD) ET Water, 610
Eutrophication, 10
Event mean concentration (EMC), 365–366, 366t
EWRI (Environmental and Water Resources Institute), 327
Executive Order 13504, 271
Expansions, sudden, 53–54, 53f, 54t
Extended detention basins
    defined, 375
Extreme flood control criteria
    Maryland storm water management regulations, 286
    New York storm water regulations, 312–313, 313f

F

Fabco Industries, 333, 337
Falcon Waterfree Technologies, 605
Falcon water-free urinals, 605n
Falling head permeameter, 93, 93f, 94
Fecal coliform, 11–12
Federal Clean Water Needs Survey, 480
Federal Highway Administration (FHWA), 555
Federal Register, 266
Federal regulations, storm water management, 265–268
    NPDES
        phase II program, 266–268
        phase I program, 265–266
Ferguson, 488
Fertilizers, from lawns, 10
FHWA, *see* Federal Highway Administration (FHWA) method
Field tests, 328
FilterPave™, 495
Filterra Bioretention Systems Bioretention device, 350, 353f
    certification for, 330
    cross section, 350, 353f
    NJDEP certification letter for, 370–371
    sizing table, 350, 354t
Filters
        BayFilter, 344, 345
        Blocksom, 336, 336f
        enhanced, 296, 303–304, 304f
    FloGard+Plus® Filter
        description, 332, 333, 334f
        sponges, 333
    Jellyfish, 343, 347, 349, 348f, 349t, 350t
    lawn inlet screen filter, 335
    media, 326, 344
    REM Geo-Trap™ catch basin filter, 333
    sand, 326
    sorbtive, 347
    StormFilter, 343, 347, 350, 357, 360, 361, 350t, 361f
    Ultra-Urban Filter
        description, 332, 333
        sponges, 333
    Up-Flow filter, 344
    Urban Green BioFilter, 350
Filter strips
    application, 516–518
        design criteria, 519
        Franklin Lakes municipal building, 519–520
Filter strips, vegetated
    NJDEP approved TSS removal rate for
        maximum slope, 318t
        required length, 318, 318f, 319f, 320f, 318t
Filtration water quality devices, media, 328, 340, 342–348
    BayFilter, 345
    BioStorm® storm water treatment system, 350, 351f–352f
    Infiltration StormFilter, 345, 346f
    Jellyfish filter, 343, 347, 349, 348f, 349t, 350t
    lists, 344t
    Sorbtive Filter, 347
    StormFilter, 345
    Up-Flow Filter, 344 First flush, concept, 269
Flexible linings, of channels, 213
Flexi-Pave, 489
Flexterra, 242
Floatables, 6, 7
FloGard+Plus®, 568
FloGard+Plus® Filter
    description, 332, 333, 334f
    installing inlet filters, 280
    sponges, 333
Florida-Friendly Landscaping (FFL) programs, 599
Florida Water Star (FWS) programs, 613
Flow, pipe and open channel
    classifications, 25
    depth, calculation, *see* Flow depth calculation
    design, 25
    energy equation, 25–26
    energy losses, 50–59
    frictional losses, 50–51; *see also* Local losses
    local losses, 51–59
    Manning equation, 35–41
    normal depth, 34–41; *see also* Normal depth
    round and elliptical, hydraulic properties, 62–65, 63f–65f, 62t, 63t
    specific energy, 26–34
Flow depth calculation, pipe and open channel, 42–50
    circular sections, 42–44
        hydraulic properties, 42, 43, 43t
        Manning formula, 42
        partly full, critical flow in, 31–34, 32t, 33f
        relative velocity, variation of, 43, 44f
    trapezoidal section, 44–50
        discharge calculation, 49–50
        horizontal elliptical RCP, 49
        hydraulic parameter and variation, 45–46, 45t, 46f
        Manning formula, 44
        roadside channel, 46–48
        slope calculation, 48–49
Flow path segments
    channel flow, 102, 103, 104, 105
    shallow concentrated flow, 102, 103f, 104–105, 104t
    sheet flow, 101–102, 102t, 103–104, 105
Flow quantity models, SWMM type, 140
Flow quantity-runoff quality models, SWMM type, 140
F.P. Leopold, 618
Franklin Lakes project, detention basin design, 444–445, 448
Free surface flow, defined, 25
Frictional losses, 50–51

Frontal flow
    defined, 165
    to gutter flow ratio, 167, 167f
Froude number, defined, 28
Futerra R45, 242

G

Gabion baskets and mattresses, 239–241
    design, 240
    drainage outfalls, tidal channel, 240
    eroded Glenwood Brook in Millburn, 241
    Keystone walls, 240, 241
    Reno, 208, 238, 239, 240f
    velocity, 240, 240t
    wall installation, 239, 240
    wire baskets, 239
Gap-graded mix, 489–492
Garden State Parkway nomograph, 99, 100f, 101
Gasoline, tetraethyl lead in, 11
Gastroenteritis, 12
Gauckler-Manning-Strickler formula, *see* Manning
    equation
Geological Survey Report, No. 32 (GSR-32) method, 275
Geo-Synthetics Inc., 488
Geoweb®, 488
Glass foam gravel, 495
Glycol, 13–14
Gradation analysis, soil, 94–96
    texture triangle, 94, 95f, 96t
    USDA soil classification, 94, 94t
GrassConcrete Limited, 500
Grass cover factor, 223, 223t
Grasscrete, 500
Grassed waterways, *see* Vegetative swales
Grass lining, of channels, 213, 221–224
    density-stiffness coefficient, 224, 224t
    grass cover factor, 223, 223t
    grass roughness coefficient, 224
    HEC-15, 222–223
    permissible shear stress of, 221
    retardance degree/classes, 221, 222, 222t
    soil roughness coefficient, 223
        vegetal retardance classification, 221, 222t
        permissible shear stresses, 222, 223t
        selection guide, 221, 221t
Grass pavers, 488–485
Grass roughness coefficient, 224
Grass swales, 302, 567–568
Grassy Paver™, 488
Grate inlets
    design, 165–171
    elements, 165 frontal flow
        defined, 165
        to gutter flow ratio, 167, 167f
    at grade, 165–171
    gutter flow velocity, 168
    inlet type, 160, 160f
    interception capacity, calculation
        equation for, 173
        gutter flow, 170
        side flow interception efficiency, 170–171
    orifice flow, 175–176
    overall efficiency and intercepted flow, 169

P-1-7/8 and P-1-7/8—four grates, 165, 166f
    reticuline, 165, 166f, 168
    on sag, 175–176
        orifice flow, 175–176
        weir flow, 175–176
    side flow interception efficiency, 168
    SI units, 168, 169f
    splash-over velocity, 167, 168, 168t, 169f
    weir flow, 175–176
Gravel wetland (SGW), subsurface, 297
    layout, 514
    Maryland Department of the Environment design
        guidelines, 516
    UNHSC's specifications, 515–516
Gray water, 619
Green-Ampt model, 83–87
    infiltration process, 83–84, 83f
    rainfall distribution, calculation process for, 86–87
    USDA soil texture classes, parameters, 84–85, 85t
    variation of infiltrated depth, 85
    water depth, 83–84, 85
Green Armor, 242
Green infrastructure, 480, 484–485
Green roofs, 294, 295t
    analysis, 508–510
    components, 505–510
    concept, 505
    construction types, 506–507
    installation cost, 508
    life and energy conservation, 507–508
    pollutant removal effect, 508
Grey water reuse, 619–620
Groundwater recharge standards, NJDEP, 275–278,
    276f–277f
GSR-32 (Geological Survey Report, No. 32) method, 275, 389
    annual groundwater recharge analysis, 403f–404f
Guidelines for Water Reuse, 613, 645–638
Gutter flow, roadway analysis, 155–160
    derivation of equation, 249–250
        partial, 249
        total, 249–250
        triangular, 249, 249f
        velocity, 249
    discharge vs. spread for composite gutter, 159f
    gutter flow for roadway of horizontal profile, 181–182,
        182f
    hydraulic capacity, 155–156
    Manning's n values, 157, 158t
    runoff and spread calculation, asphalt roadway, 159–160
    SI units, 156f–157f
    spread, 157, 158t, 159f
    triangular, 155–156, 156f–157f

H

Haestad Methods, 120 378
Hancock Concrete Products, 547
Hanson Pipe & Precast of Irvine, 551
HDPE (high-density polyethylene), 333
HDS (hydrodynamic sedimentation), 329, 337
Head losses, *see* Energy losses
Heavy metals, 11
HEC-1, 378
    model, SWMM, 140

HEC-15
    method, trapezoidal channel design, 228–229
    model, cohesive material in permissible shear stress, 255f, 255t
High-density polyethylene (HDPE), 333
High-density polyethylene (HDPE) pipes, 554–555
Highland Tank & Manufacturing Company, 626
High-performance TRM (HP-TRM), 242
Hillsborough Township, Somerset County Project; *see also* Extended detention basins
Horton equation, 87–90
Hotspots, storm water
    defined, 315
    land uses and activities, 315
    list of, 315t
HP Storm, 554
HP-TRM (high-performance TRM), 242
HSGs, *see* Hydrologic soil groups (HSGs)
Hunter MP rotary nozzle, 610
Hydraulic capacity, of gutter, 155–156
Hydraulic conductivity
    defined, 82, 83
    infiltration capacity and, 82–83, 84
    measuring, 93, 94
Hydraulic design of culverts, 193–206
    design calculations, 202
    entrance loss coefficients, 200, 201t
    examples, 203–206
    inlet control
        discharge, 196f, 198f, 200f
        headwater, 197f, 199f
        nomograph, 195f
        schemes, 193f
    Manning "n" values, 200, 201t
    outlet control
        discharge, 202, 202f
        schemes, 194f
    table, 203f
Hydraulic efficiency, grate inlets and, 165
Hydraulic grade line (HGL)
    calculation, 186, 191
    case study, 191, 192f, 193f
    defined, 26
    slope, 34
Hydraulic properties
    circular pipes
        flow depth calculation, 42, 43, 43t
        flowing full, CU, 62t
        flowing full, SI units, 62t
    elliptical concrete pipe, 63t
Hydraulic radius, defined, 35
HydroCAD, 140
Hydrocarbons, petroleum, 14–15
Hydrodynamic sedimentation (HDS), 329, 337
Hydrodynamic separation water quality devices, 330, 337–343
    Aqua-Swirl Model AS-5 CFD PCS, 340, 342f
    CDS units, 338–339, 341f
    NJDEP, 337, 340, 343
    oil absorption booms, 340
    partial list, 337t
    Stormceptor, 337, 343
    Terre Kleen, 340, 343f
    Vortechs®, 337, 339, 338f, 339t

Hydrograph(s), 96–109
    characteristics, 96
    defined, 1, 96
    Delmarva, 126n
    modified rational method, 113–115, 114f
    parameters, 96
    rational method, 110–112, 112f
    runoff, 75
    sheet flow length analysis, 108–109
    of single storm, 96, 96f
    snyder synthetic unit, 140, 148–149, 148f
    time of concentration, 97–108
        calculation, 101
        defined, 96, 97
        equations and nomographs, 97–101
        FHWA method, 103–108; *see also* Federal Highway Administration (FHWA) method
        Garden State Parkway nomograph, 99, 100f, 101
        Izzard equation, 97, 98, 99t
        Kirby equation, 98, 99, 99t
        Kirpich equation, 97, 98f
        SCS method, 101–103, 102t, 103f; *see also* Soil Conservation Service (SCS)
    urbanization on runoff, 2
    USGS Nationwide Urban Hydrograph, 150, 150t
    WinTR-55, 126
Hydro International, 345, 370
Hydrologic calculations, 67–152
    infiltration process, 82–92
        capacity and hydraulic conductivity, 82–83, 84
        Green-Ampt model, 83–87; *see also* Green-Ampt model
        Horton equation, 87–90; *see also* Horton equation
        indexes, 91–92, 92f
        natural and surface factors, 82
        percolation vs., 82
        Philip infiltration model, 90–91
        water movement through soil, 82
    initial abstractions, 76–82
        defined, 76
        depression storage, 79–82
        disposition of uniform rainfall, 78f
        interception, 78–79
        retention effect, vegetation, 76, 78
    permeability and infiltration, measurement, 92–96
        constant head permeameter, 93, 93f
        falling head permeameter, 93, 93f, 94
        infiltrometers, 92, 93f
        soil gradation analysis, 94–96, 94t, 95f, 96t
    rainfall process, 67–76
        daily precipitation data, 73–75, 73t, 74f
        hyetograph, 75–76, 75f, 76f
        IDF curves, 67–72, 68f, 69t–70t, 71t–72t
        SCS rainfall distributions, 75–76, 75f, 76f
        24-hour rainfall depth, New Jersey, 67, 71t–72t, 76, 77f, 78t
    runoff calculation methods, 109–131
        peak discharge calculations, 119–121; *see also* Peak discharge calculations
        rational method, 110–115; *see also* Rational method
        TR-55 method, SCS, 115–119; *see also* Technical Release No. 55 (TR-55) method

TR-55 method, limitations/drawbacks, 123–126;
    see also Technical Release No. 55 (TR-55)
    method
  unit hydrograph method, SCS, 121–123
  universal, *see* Universal runoff method
  WinTR-55 method, 126–131; *see also* WinTR-55
    method
snyder synthetic unit, 140, 148–149, 148f
SWMM, 140–141; SWMM 5, 141
  development, 141
  dynamic rainfall-runoff simulation model, 141
  of EPA, 83, 88, 140–141
  runoff component of, 141
universal runoff method, 131–140
  antecedent moisture condition, 133
  application to nonuniform rainfall, 139–140
  catchment area, 133–134
  composite surfaces, equations for, 135–138
  examples, 136–138
  excess rainfall, 131–132, 131f
  interception and depression storage, values of, 132,
    132t
  lag time, runoff equations for impervious surfaces
    and, 134–135
  lag time between rainfall and runoff, 131–133
  peak discharge, 133–135
  rainfall–runoff relation for impervious surfaces, 134f
  rational and SCS methods, drawbacks, 131
  runoff volume, 133–134, 134f
  universal rainfall–runoff relation, 131, 131f
USGS
  Nationwide Urban Hydrograph, 150, 150t
  regression equations for urban peak discharges,
    151, 151t
  StreamStats program, 140–141, 152, 152f
Hydrologic routing, 378
Hydrologic soil groups (HSGs)
  Lakehurst sand, 128
  runoff depth, calculation, 124–125
  SCS and, 117
  total site area within, 283
  vegetated filter strips for, 273
Hydrologic Solutions Inc., 419
Hyetograph, rainfall, 75–76, 75f, 76f

I

IDAs (intensely developed areas), 306
IDF, *see* Intensity-duration-frequency (IDF) curves
Imbrium Systems, 345, 347
Index(es)
  infiltration, 91–92, 92f
  tests, for RECPs (ECBs and TRMs), 243–244, 243t
Indoor water conservation
  economy of water-saver fixtures, 606–607
  other indoor saving tips, 606
  overview, 607–612
  residential buildings, 603–605
  urinals in nonresidential buildings, 605–606
Infiltration, 82–92
  capacity and hydraulic conductivity, 82–83, 84
  Green-Ampt model, 83–87
    infiltration process, 83–84, 83f
    rainfall distribution, calculation process for, 86–87

USDA soil texture classes, parameters, 84–85, 85t
    variation of infiltrated depth, 85
    water depth, 83–84, 85
  Horton equation, 87–90
    cumulative infiltration, 88–89, 90
    infiltration capacity, 88
    runoff amount, calculation, 90
    ultimate infiltration rates, values of, 88
  indexes, 91–92, 92f
  natural and surface factors, 82
  percolation vs., 82
  Philip infiltration model, 90–91
    cumulative depth, 90
    portion and runoff, rainfall ratio, 90–91
  urbanization on, 1, 2
  water movement through soil, 82
Infiltration and permeability, measurement, 92–96
  infiltrometers, 92, 93f
  permeameters
    constant head, 93, 93f
    falling head, 93, 93f, 94
  soil gradation analysis, 94–96
    texture triangle, 94, 95f, 96t
    USDA soil classification, 94, 94t
Infiltration basins, 448–450, 451–452, 566–567
  New Jersey residential project
    design of drainage system, 458
    drainage area maps, 452, 455
    infiltration basin design, 455–456, 458
    runoff calculations, 455
Infiltration berm, 299
Infiltration capacity
  defined, 82–83
  empirical equation for, 87–88
  hydraulic conductivity and, 82–83, 84
  rainfall intensity and, 84, 85, 86, 88
  for soil, 88
Infiltration StormFilter structure, 345, 346f
Infiltrometers, 92, 93f
Initial abstractions, 76–82
  defined, 76
  depression storage, 79–82
    examples, 80–82
    rate of, 80
    ratio, runoff to rainfall, 81, 82
    runoff coefficient, 81
    typical values of, 80
    water volume, 80
  disposition of uniform rainfall, 78f
  interception, 78–79
  retention effect, vegetation, 76, 78
  values, for soil curve numbers, 119, 121t
Inlets, in storm drainage systems
  combination, 160, 160f, 173, 174
  curb opening
    in combination inlets, 169
    overview, 160, 160f
    weir flow discharge, 171–172
  design, 165–176
    combination, 173, 174
    curb opening, 171–172
    grate, *see* Grate inlets
    New Jersey, 173–175, 174f, 175t
    slotted inlets, 172–173

grate, 165–171
  design, 165–171
  elements, 165
  frontal flow, defined, 165
  frontal flow to gutter flow ratio, 167, 167f
  gutter flow velocity, 168
  interception capacity, calculation, 170–171
  interception capacity of, 173
  orifice flow, 175–176
  overall efficiency and intercepted flow, 169
  P-1-7/8 and P-1-7/8—four grates, 165, 166f
  reticuline, 165, 166f, 168
  side flow interception efficiency, 168
  SI units, 168, 169f
  splash-over velocity, 167, 168, 168t, 169f
  weir flow, 175–176
hydraulic design charts for, 253, 253f
New Jersey
  design, 173–175
  efficiency, 175, 175t
  interception capacity, gutters, 174–175
  types of, 173, 174, 174f
roadways at 0% grade, 181–184
  effects of debris and clogging, 183
  flow equations, derivation of, 251–252, 251f
  gutter flow for roadway of horizontal profile,
      181–182, 182f
  maximum spacing for flat roadway, 183–184
  weir flow equation, 182
sags, grate
  orifice flow, 175–176
  weir flow, 175–176
slotted, 172–173
spacing, 176–181
  calculations, 177–178, 177f
  examples, 178–181
  maximum, for flat roadway, 183–184
  NJDOT criteria, 176
  NJDOT drainage design manual, 178
  P-1-1/8 FHA grates, 179–180
  procedure for, 177
types, 160–165
  ACO trench drain, 161, 163f–164f
  Campbell Foundry trench drains, 164, 165f
  combination, 160, 160f, 173, 174
  curb opening, *see* Curb opening inlet
  curb pieces, new and traditional, 161, 162f
  eco-curb piece, 160, 161, 162f
  foundries, 164
  grate, *see* Grate
  inlets median parking lot, 160, 161
  parking lot, 160
  slotted, *see* Slotted inlets
In situ permeability tests, 483
Intensely developed areas (IDAs), 306
Intensity duration-frequency (IDF) curves, 67–72
  data for Atlantic City, 71t–72t
  data for San Francisco, 69t–70t
  in New Jersey, 67, 68f
Interception, 78–79
Interlocking Concrete Pavement Institute (ICPI), 574
International Stormwater BMP Database, 326–327
Intestinal fever, 12
Irrigation Association conference, 609–610

Irrigation technology, 609–610
Irrometer, 610
Izzard equation, 97, 98, 99t

J

J. M. Eagle, 554
Jar irrigation, 612
Jellyfish filter, 343, 347, 349, 348, 349t, 350t
Jensen Precast, Inc., 330, 340, 345
Joint closure, 553f
Junction holes, losses, 55–56, 55f, 56f

K

Keystone walls, 240, 241
Kinematic wave model, 103
Kirby equation, 98, 99, 99t
Kirpich equation, 97, 98f
Kristar Enterprises, 332, 333
Kristar water quality devices, 332, 333
Kuichling, Emil, 110

L

Lacrosse encephalitis, 630
Lag time, in Pazwash's universal runoff method
  antecedent moisture condition, 133
  catchment area, 133–134
  excess rainfall, 131–132, 131f
  interception and depression storage, values of, 132,
      132t
  between rainfall and runoff, 131–133
  rainfall-runoff relation for impervious surfaces, 134f
  runoff equations for impervious surfaces and, 134
  universal rainfall-runoff relation, 131, 131f
Land disturbance, defined, 272
Landlok
  ECBs, 242, 243–244, 243t, 260f–263f
  TRMs, 242
Landscape infiltration (depression), 297–298, 298f
Land uses, pollutant and, 3, 4t
Lawn pollution, 481–482
Lawn(s)
  care habits, 11
  depression storage for, 80
  fertilizers from, 10
  inlet screen filter, 335
LDAs (limited development areas), 306
Lead, in aquatic systems, 11
LEED and green buildings
  environmental benefits, 485
  examples, 486–488
  performance measurement, 485–486
Lee Valley Ice Center in Leyton, 620
Level pool, 378
Limited development areas (LDAs), 306
Limit of disturbance (LOD), 304, 305
Linings, drainage channels
  composite, 237–238
  grass lining, *see* Grass lining
  Manning's roughness coefficient variation with, 224–234
    examples, 228–234
    flow parameters and, 226, 227f

product of flow velocity and, 226, 226f
stone linings, 225, 226t
rigid and flexible, 213
stone, 217–219, 218t, 219f
Local losses, 51–59
bend, 54
defined, 50
entrance and exit, 52–53, 52t
example, 57–59, 58f
head loss at transitions, 54, 55f, 55t
junction, 55, 55f
manholes, 55–56, 56f
sudden expansions/contractions, 53–54, 53f, 54t
velocity head times, 51–52
LOD (limit of disturbance), 304, 305
Losses, energy, 50–59
frictional losses, 50–51
local, 51–59
bend, 54
defined, 50
entrance and exit, 52–53, 52t
example, 57–59, 58f
head loss at transitions, 54, 55f, 55t
junction, 55, 55f
manholes, 55–56, 56f
sudden expansions/contractions, 53–54, 53f, 54t
velocity head times, 51–52
Low-impact development
Ipswich River, Massachusetts project, 482–483
objectives, 482–483

M

Maintenance cost, BMPs, 326
Major development, defined, 272
Manholes, losses at, 55–56, 56f
Manning formula
pipes and open channel, 35–41
Chezy and roughness coefficients, 35–36, 36t, 39t
circular sections, 42
discharge, 37
flow velocity, 36
geometric properties, 37, 38t–39t
nomographs, 37, 40f–41f
roughness coefficient, 35, 39t
trapezoidal section, 44
Manning's n values
for culverts, 200, 201t
gutter flow, roadway analysis, 157, 158t
of RECPs, 243
for sheet flow, 102t
Manning's roughness coefficient variation, 224–234
design charts, 226, 227f
examples, 228–234
flow depth, 224–225
flow parameters and, 226, 227f
grass-lined channel design, 233–234
gravel and stone-lined channels, 225
HEC-15, NJ Standards for vegetal retardance D, 226–227
hydraulic parameters, NJ Standards, 230
product of flow velocity and, 226, 226f
riprap, cobble, and gravel lining, 224
shear stress and flow depth calculation, 228–229
stable channel, hydraulic parameters, 229–230
stone linings, 225, 226t
stone size and, 225
trapezoidal channel design, 231–234
Manufactured water treatment devices (MTDs), 325–384
bioretention systems
cells, 350
Contech, 350, 353
Filterra, 350, 353f, 354t
Urban Green BioFilter, 350
BMPs
database site, 326–327
effectiveness of, 325
sand and media filters as, 325
case studies, 354–361
CDS
certification of water quality devices, 327
Contech and, 345
dimension and capacity of models, 372t
hydrodynamic separation water quality devices, 337–343, 341f
NJDEP approved rates for models, 373t
certification of, 325–329
CDS, 327
NJCAT, see New Jersey Corporation for Advanced Technology (NJCAT)
NJDEP Division of Science, Research and Technology (DSRT), 328
particle size, 327
pollutant removal rates, 327, 328
selection, 327
commercial development project, case study, 357–361
Filterra Bioretention Systems
Bioretention device, 350, 353f
certification for, 330
cross section, 350, 353f
NJDEP certification letter for, 370–371
sizing table, 350, 354t
overview, 325–327
parking lot expansion project, case study, 354–356
Terre Kleen
hydrodynamic separation water treatment devices, 338f, 339, 339t
verification, by EPA and NSF, ETV program, 363–367
catch basin inserts, 330, 332–336, 332t
hydrodynamic separation, 330, 337–343
media filtration, 330, 343–350
verification process, 368–369
Manufacturers Working Group (MWG), 368
Maryland, storm water management regulations, 280–306; see also State of Maryland storm water management regulations
Maryland's Stormwater Design Manual, 269
Microscale practices, addressing ESD, 296–304
dry wells, 299–300
enhanced filters, 303–304, 304f
infiltration berm, 299
landscape infiltration (depression), 297–298, 298f
microbioretention, 300–301, 301f
performance standards, 297
rain gardens, 301–302
rainwater harvesting systems, 297
submerged gravel wetlands, 297
swales, 302–303

Massachusetts Water Resource Authority (MWRA), 604

Mass first flush, defined, 269

Mats, TRMs, 241–242

Mattresses, gabion, 239–241, 240t; *see also* Gabion baskets and mattresses

Meadowlands Commission drainage standards, 464, 468

Mean flow depth, defined, 28

Media filtration water quality devices, 330, 343–350

  BayFilter, 345

  BioStorm® storm water treatment system, 350, 351f–352f

  as BMPs, 326

  Contech, 343, 345, 346f, 350

  filter surface calculation, 342

  Infiltration StormFilter, 345, 346f

  Jellyfish filter, 343, 347, 349, 348f, 349f, 350t

  lists, 344t

  Sorbtive Filter, 347

  StormFilter, 345

  Up-Flow Filter, 344

Median event mean concentration, for urban land uses, 4t

Microbioretention, 300–301, 301f

Minor losses, defined, 50

Modified rational method, 113–115

  hydrograph, 113–114, 114f

  longer storm durations, 113

  volume, calculation, 114–115

MS4s (municipal separate sewer storm water systems), 5, 265, 266

MTDs, *see* Manufactured water treatment devices (MTDs)

Municipal separate sewer storm water systems (MS4s), 5, 265, 266

Municipal storm water management plan (MSWMP), 279

Muskingum-Cunge method, 126

Mutual Materials Company, 498

MWG (Manufacturers Working Group), 368

N

N-12, 412

  storage capacities, 416t

National Academy of Science, 206

National Association of Home Builders (NAHB), 487

National Geodetic Vertical Datum (NGVD), 25

National Marine Fisheries Service, 265

National Motorist Association, 484

National oceanic and atmospheric administration (NOAA), 67, 73, 74

National pollutant discharge elimination system (NPDES), 5, 265, 479

  phase II program, 266–268

    CGP, 266–268

    EISA, 271

    MS4, 265, 266

    storm water discharges, classes of, 266

    SWPPP, 266–267

  phase I program, 265–266

    inventory report, 266

    NURP, 265

    water bodies, 265–266

National research council (NRC)

  EPA and, 16

  recommendation, 16

  report, 16–17, 19–22

National resources conservation service (NRCS), 126; *see also* Soil Conservation Service (SCS)

  charts for conduit outlet protection design, 209, 210f

  Conservation Practices Standard, 225

  hyetographs, development, 75

  Kirpich's equation, 97

  Manning's n for gravel and stone-lined channels, 225

  runoff calculation, NJ storm water management regulations, 278

  TR-55 method, 115

  website for hydrologic soil groups, 124

  WinTR-20 program, 126

National rivers and streams assessment report, 5, 266

National technical information service (NTIS), 4t

National urban runoff program (NURP), 4–5, 11, 265

National water quality inventory, 265

Neurological adverse effects, lead and, 11

Newark International Airport, 14

New Jersey, climate record for, 73t

New Jersey

  climate record for, 73t

  design tailwater depth calculation, 211

  IDF curves in, 67, 68f

  Monmouth County, 126–131, 127f, 128f, 129f, 129t, 130t

  New Milford Avenue, 186–190, 187f, 188f, 189f, 190f

  parking lot expansion project, 354–356, 355f, 356f

  rainfall IDF data, 67, 71t–72t

  24-hour rainfall depth, 67, 71t–72t, 76, 77f, 78t

New Jersey Corporation for Advanced Technology (NJCAT), certification, 328–331

  Bureau of Nonpoint Pollution Control, 330 laboratory testing, 328–329

  NJDEP Division of Science, Research and Technology (DSRT), 328

  procedure (2013), 368–369

  process for, 328

  Stormwater Management Unit, 330

  TSS removal efficiency by, 329–330

  verification from, 330

  water treatment devices certified by NJDEP, 331t

New Jersey Department of Environmental Protection (NJDEP)

  approved rates for CDS models, 373t

  approved TSS removal rate for vegetated filter strips

    maximum slope, 318t

    required length, 318, 318f, 319f, 320f, 318t

  certification letter for Filterra Bioretention Systems, 370–371

  curb pieces, replacing, 160–161

  DSRT, 328

  limiting application, 112

  municipal storm water regulation program, 268

  pollutant concentration in urban storm water, 6t

  process of verification for MTDs, 328

  storm water management regulations, 271–280; *see also* NJDEP storm water management regulations

  water treatment devices certified by, 331t

New Jersey Energy and Environmental Technology Verification Program, 328

New Jersey Geological Survey, 275

New Jersey Geological Survey (NJGS) spreadsheet GSR-32, 402

  annual groundwater recharge analysis, 403f–404f

New Jersey Highway Authority, 99
New Jersey inlets
  design, 173–175
  efficiency, 175, 175t
  interception capacity, gutters, 174–175
  types of, 173, 174, 174f
New Jersey pollutants discharge elimination system
      (NJPDES)
  defined, 272
  permits, 268
New Jersey Soil Erosion and Sediment Control Standards,
      112
New Jersey Storm Water Best Management Practices
      Manual, 273, 279
New Jersey Water Quality Standards for water reuse,
      614t–615t
New York State Codes Rules and Regulations (NYCRR),
      309
New York State Department of Transportation (NJDOT),
      12, 67, 103, 110, 547
  criteria, inlets spacing, 176
  drainage design manual, 178
  equation for inlet capacity, 178–180
  New Jersey inlet, design, 173
  tables for pipe sizing calculations, 184
New York State Storm Water Management Manual, 316
New York storm water regulations, 306–316; *see also* State
      of New York storm water regulations
NGVD (National Geodetic Vertical Datum), 25
Nitrate, 9
Nitrogen, nutrient for plants, 9–10
NJCAT, *see* New Jersey Corporation for Advanced
      Technology (NJCAT)
NJDEP, *see* New Jersey Department of Environmental
      Protection (NJDEP)
NJDEP storm water management regulations, 271–280,
      272t
  amended rules, 271
  goals, 271
  gravel and porous pavements, 272
  groundwater recharge standards, 275–278, 276f–277f
  improvement, 279–280
    grates on sand filter, 280
    runoff, 280
    TSS removal, 279–280
    water quality, 279–280
  MSWMP, 279
  municipalities, 279
  NJPDES, 272
  nonstructural strategies, 278–279
  planning agencies, 271, 272
  quality standards, 273–275
    1.25 in./2 h rainfall intensity curve, 273, 274f
    temporal distribution, 274t
    TSSs, 273, 275, 275t
  runoff
    calculation methods, 278
    quantity requirement, 272–273
  standards for structures, 278
NJ Geological Survey website, 275
NJPDES, *see* New Jersey Pollutants Discharge
      Elimination System (NJPDES)
NOAA (National Oceanic and Atmospheric
      Administration), 67, 73, 74

Nomographs
  inlet control, hydraulic design of culverts, 195f
  Manning equation, pipes and open channel and, 37,
      40f–41f
  for time of concentration, 97–101; *see also* Time of
      concentration
Nonpoint Source Management Program, 265
Nonpoint source (NPS) pollutant, 4–15
  atmospheric dust defined, 15
    global dust-emission rates, 15
    particles and wind erosion, 15
    concentration in urban storm water, 6t
  CWA of 1972, 3, 5, 19–22
  floatables, 6, 7
  heavy metals, 11
  NURP, 4–5
  nutrients and pesticides, 9–11
    effects, aquatic populations, 10
    gardening and lawn-care habits, 11
    phosphorus and nitrogen, 9–10
  pathogens and fecal coliform, 11–12
  petroleum hydrocarbons, 14–15
    concentrations, measurement, 14
    removing, airports and seaports, 14
    Smart Sponge, 14–15
  road salt, 12–14
    CaCl, CMA, and KA, 13
    de-icing, 12–14
    glycol, 13–14
    use and effects, 12–13
    Verglimit, 13
  sediment, 7–9
    falling velocity, of particle, 7–8
    load creation and discharge, 7
    measurements, 7
    settling velocity, spherical particles, 8–9, 9t
    trap efficiency, 7
  sources, 5t
  storm water, 5, 6t
  water quality, 3
Nonrooftop runoff disconnection, 296
Nonstructural practices, addressing ESD, 295–296
  disconnection of rooftop runoff, 296, 296t
  nonrooftop runoff disconnection, 296
  sheetflow to conservation areas, 296
Nonstructural storm water strategies, 278–279
Nonstructural strategies point system (NSPS),
      278–279
Nonuniform flow, 25
Normal depth, 34–41
  Chezy equation
    average velocity, channel, 35
    Pavlosky, Prague Hydraulics Institute and Bazin
        formulas, 35, 36t
  defined, 34
  Manning formula, 35–41
    circular sections, 42
    discharge, 37
    flow velocity, 36
    geometric properties, 37, 38t–39t
    nomographs, 37, 40f–41f
    roughness coefficient, 35, 39t
    trapezoidal section, 44
Northeast Regional Climate Center of University, 75

North Greenbush Town; *see also* Extended detention
    basins
Norton, Stephen, 13
NPC Kor-N-Seal I, 564
NPDES, *see* National Pollutant Discharge Elimination
    System (NPDES)
NPS, *see* Nonpoint source (NPS) pollutant NRC, *see*
    National research council (NRC)
NRCS, *see* National Resources Conservation Service
    (NRCS)
NSF International
    Terre Kleen verification, ETV program and,
        363–367
        performance verification, 365–367, 365t
        technology description, 363–364
        testing description, 364–365
NSPS (nonstructural strategies point system), 278–279
NTIS (National Technical Information Service), 4t
NURP (National Urban Runoff Program), 4–5, 11, 265
Nutrients, 9–11
Nutrient Separating Baffle Box, 347
NYCRR (New York State Codes Rules and Regulations),
    309

O

OldCastle Precast, 424
Oldcastle® Stormwater Solutions, 332, 333, 335f, 568
Open-graded mix, 489–492
Outdoor water conservation, measures, 612–613
Outfalls, erosion control, 206–213
    examples, 211–213
    riprap aprons, 206, 207–209
    scour holes, 209–213
Outlet structure design, 379–385; *see also* Detention
    basins/wet ponds
    broad-crested weir, 382, 382f
    cipolleti weir, 381, 382f
    hydro-brake fluidic-cones, 383
    orifice, 379
    outflow grates, 382–383
    rectangular weir, 379–380, 380f
    stand pipes, 383, 383f
    thirsty duck, 383–384
    triangular weirs, 380–381, 381f
Overbank flow control criteria
    design calculations for, 311–312
    storm, 311–312, 312f
Overbank protection volume criteria
    detention basins/ponds and underground chambers/
        vaults, 286
    rainfall depths, 284t, 286
    runoff, 286
Overwatering, 618

P

PAHs (polynuclear aromatic hydrocarbons), 14
Pall, 618
Parking lot expansion project, case study, 354–356
Parking lot inlet, 160
    median, 160, 161
Parkson, 618
Partly full circular pipes, critical flow in, 31–34, 32t, 33f

Pathogens, 11–12
PaveDrain of Franklin, 498
Pavements, permeable, 294–295, 295t
Pavlosky formula, 35
Pazwash universal method, 131–140
    application to nonuniform rainfall, 139–140
    composite surfaces, equations for, 135–138
    examples, 136–138
    lag time
        antecedent moisture condition, 133
        catchment area, 133–134
        excess rainfall, 131–132, 131f
        interception and depression storage, values of, 132,
            132t
        between rainfall and runoff, 131–133
        rainfall–runoff relation for impervious surfaces,
            134f
        runoff equations for impervious surfaces and,
            134
        universal rainfall–runoff relation, 131, 131f
    peak discharge, 133–135
    rational and SCS methods, drawbacks, 131
    runoff volume, 133–134, 134f, 136–137
Peak discharge calculations, SCS, 119–121
    tabular hydrograph method, 120
Peak discharges
    defined, 148
    in universal runoff method, 133–135
    urban, USGS regression equations for, 151, 151t
Percolation, infiltration process vs., 82
Permeability, measurement, 92–96; *see also* Infiltration
    and permeability
Permeable interlocking concrete pavements (PICPs), 574
Permeable pavements, 294–295, 295t
Permeameters
    constant head, 93, 93f
    falling head, 93, 93f, 94
Permissible shear stresses; *see also* Tractive force
    method
    for ambient soil, 206
    for bare soils and stone linings, 217, 218t
    of cohesive material in HEC-15, 255f, 255t
    of grass lining, 221
    of noncohesive and cohesive soils, 218f, 219f
    on RECP lining, 244–245
    on side of channel, 219
    for vegetal covers, 222, 223t
Permissible velocity concept, 213–215, 214f, 215t
Pervious pavement, 574
Pesticides, 9–11
Petroleum hydrocarbons, 14–15
Philip infiltration model, 90–91
Phosphorus, nutrient for plants, 9–10
Pipe and open channel flow, 25–65
    classifications, 25
    depth calculation, 42–50
        circular sections, 31–34, 42–44
        iterative process, 42
        trapezoidal section, 44–50
    design, 25
    energy equation, 25–26
        components, 25–26, 26f
        HGL and EGL, 26
        total energy head, 25–26

energy losses, 50–59
   frictional losses, 50–51
   local losses, 51–59
     bend, 54
     defined, 50
     entrance and exit, 52–53, 52t
     example, 57–59, 58f
     head loss at transitions, 54, 55f, 55t
     junction, 55, 55f
     manholes, 55–56, 56f
     sudden expansions/contractions, 53–54, 53f, 54t
     velocity head times, 51–52
Pipes installation
   concrete box culverts, 554
   dewatering, 559–561
   DuroMaxx, 545–561
   elliptical concrete pipes, 551
   high-density polyethylene (HDPE) pipes, 554–559
   prestressed concrete pipes, 551
   round reinforced concrete pipes (RCP), 547–551
   type of pipes, 545–546
Plasteel, 626
Plastic geocells, 489
Plastic Solutions, Inc., 570
Pollutant(s)
   absorption of, 325
   land uses and, 3, 4t
   NPS, *see* Nonpoint source (NPS) pollutant
   removal, 375–376
     device, verification of, 328
     particle size in, 325
Pollution
   storm water, 3
   urban, causes of, 4
Pollution Prevention Act, 265
Polynuclear aromatic hydrocarbons (PAHs), 14
PondPack, 378
Ponds/detention basins, maintenance
   algae and weed control, 575–576
   outlet structures, 576
   underground detention basins, 576
   wet ponds, 576
Popcorn mix, 489–492
Porosity, soil, percolation rate, 94
Porous pavements
   concrete pavers, 495–498, 500
   Glass Pave, 495
   nonconcrete modular pavers, 502–503
   open cell pavers, 500, 502
   open cell paving grids, 488–489
   pervious concrete, 493–495
   porous asphalt, 489–493
Potassium acetate (KA), 13
Prague Hydraulics Institute formula, 35
Prefabricated plastic trash racks, 570
Preformed scour holes, 209–213, 211f
Press-Seal Gasket Corporation, 563, 564
Pressure head, defined, 26
Presto®, 486, 495
Prestressed concrete pipes, 551
Price Brothers, 551
Propex, 571
   ECBs, 242, 243, 260f–263f
   TRMs, 242, 256f–259f

Public education on water conservation, 631
Pyramat, 242
Pyrethroid pesticides, 10–11

Q

Quick TR-55, 120

R

Rain barrels, 297
   problems with, 630
Rainfall duration, rainfall intensity on, 67
Rainfall intensity
   defined, 67
   on rainfall duration, 67
Rainfall process, 67–76
   daily precipitation data, 73–75, 73t, 74f
   hyetograph, 75–76, 75f, 76f
   IDF curves, 67–72
     data for Atlantic City, 71t–72t
     data for San Francisco, 69t–70t
     in New Jersey, 67, 68f
   SCS rainfall distributions, 75–76, 75f, 76f
   24-hour rainfall depth, New Jersey, 67, 71t–72t, 76, 77f, 78t
Rain gardens, 301–302, 525–526, 528
Rainstore, 422–423
Rain Water, 610
Rainwater and storm water runoff reuse
   overview, 621–622
   quantity of urban runoff, 623–625
Rainwater harvesting
   harvesting roof rain, 626–630
   rain barrels, problems with, 630
   systems, 297
Rainwater harvesting guidelines, 621
Rapid Repair Levee Break Laboratory, 242
Rational formula, defined, 110
Rational method, runoff calculation, 110–115
   drawbacks, 131
   formula, 110
   hydrograph, 111, 112f
   limitations, 112–113
   modified, 113–115
   volume, calculation, 114–115
   NJ storm water management regulations, 278
   peak rates, calculation, 112, 113
   runoff coefficient, values of, 110, 111t
   suggested runoff coefficients for Denver Area, 112–113, 113t
   time base, 111, 112t
RCAs (resource conservation areas), 306
Recession, defined, 96
Recharger 900 HD, 417
Recharge volume criteria
   calculation, 282, 282t, 283
   groundwater, 282, 283
   hydrologic soil group (HSG), 283
   percent volume and percent area method, 282t, 283
   water quality volume and, 283
Reclaimed-water applications, 617–618
Rectangular channels, critical flow in, 28–29
Recycled wastewater market, 617–618

Redevelopment, 304–305
    defined, 304
    New York storm water regulations
        alternative practices, 316
        water quality criteria, 316
        water quality volume, 316
    overview, 304
    policy, 304–305
Reduced curve numbers (RCNs), Maryland
    for green roofs, 294, 295t
    for ESD sizing requirement, 321t–323t
    for permeable pavements, 294–295, 295t
Regenerative air sweepers, 574
Regression equations, USGS, for urban peak discharges,
        151, 151t
Regulations, storm water management, *see* Storm water
        management regulations
Reinforced concrete pipe (RCP)
    critical discharge calculation, 34
    horizontal elliptical, 49
    minimum slope, calculation, 48–49
Reinforced turf, 295
REM Geo-Trap™ catch basin filter, 333
REM Triton Series, 333
Reno mattress, 208, 238, 239, 240f
Rensselaer County soil survey maps, 405
Reservoir routing, 378
Residential buildings and water conservation,
        603–605
Residential site improvement standards (RSIS), 178, 452,
        455, 458, 468
Resource conservation areas (RCAs), 306
Retardance degree/class
    defined, 221
    vegetal
        classification, 221, 222t
        permissible shear stresses, 222, 223t
        selection guide, 221, 221t
Retention-infiltration basin
    dry well (*see* page pit), 459–464
        changes in runoff, 461–463
        percolation losses, 461
        premature clogging, 459–461
    Secaucus, in Hudson County, New Jersey project,
        468–473
Reticuline grate, 165, 166f, 168
Revel Environmental Manufacturing, 333
Rickly double-ring infiltrometer, 92, 93f
Rigid linings, of channels, 213
Ring infiltrometer, 92, 93f
Riprap aprons, 206, 207–209, 208f, 210f
Riprap stone, 565
R.K. Manufacturing Inc., 488
Road salt, 12–14
    CaCl, CMA, and KA, 13
    de-icing, 12–14
    glycol, 13–14
    use and effects, 12–13
    Verglimit, 13
Roadway drainage analysis, 155–160
    gutter flow, 155–160
        derivation of equation, 249–250, 249f
        discharge vs. spread for composite gutter, 159f
        hydraulic capacity, 155–156

Manning's n values, 157, 158t
    runoff and spread calculation, asphalt roadway,
        159–160
    spread, 157, 158t, 159f
    triangular, 155–156, 156f–157f
Roadways at 0% grade, inlets on, 181–184
    derivation of flow equations for, 251–252, 251f
    effects of debris and clogging, 183
    gutter flow for roadway of horizontal profile, 181–182,
        182f
    maximum spacing for flat roadway, 183–184
    weir flow equation, 182
Rolled erosion control products (RECPs), 243–244, 243t;
        see also Erosion control
    example, 245–246
    lined channels, design, 244–246
    permissible shear stress on, 244–245
Roof rain, harvesting, 626–630
Rooftop runoff, disconnection of, 296, 296t
Roughness coefficients, Manning, 35–36, 36t, 39t
    variation with lining, 224–234
        design charts, 226, 227f
        examples, 228–234
Round reinforced concrete pipes (RCP), 547–551
Rubber sidewalks, 502–503
Runoff calculation methods, 109–131
    rational method, 110–115
        on assumptions of linearity and proportionality,
            110
        hydrograph, 111, 112f
        limitations, 112–113
        modified, 113–115
        peak rates, calculation, 112, 113
        runoff coefficient, values of, 110, 111t
        suggested runoff coefficients for Denver Area,
            112–113, 113t
        time base, 111, 112t
    TR-55 method, SCS, 115–119
        components, 115, 116f
        composite curve number, 120f
        curve numbers for urban areas, 117, 118t
        depths for AMC, 117, 117t
        examples, 124–126
        limitations/drawbacks of, 123–126
        potential and actual retention, 116
        soil curve number, 116–119
        soil groups, description of, 117, 117t
        Virgin Lands, curve numbers for, 117, 119t
        unit hydrograph method, SCS, 121–123
    universal, *see* Universal runoff method
    WinTR-55 method, 126–131
Runoff quantity regulations, NJDEP, 272–273

S

Salmonella group, 12
Sand filters, 442–444
    chambers, 442–443, 443f
    organic filter, 445f
    TSS removal, 444
SaniTite HP, 554
SBRs (statewide basic requirements), 268
Scour holes
    design tailwater depth calculation, New Jersey, 211

EPA method, 212
layout and section view, 209, 211f
median stone diameter, 209
riprap stone, 209
SCS method, 212
sizing, 209, 212–213
SCS, *see* Soil conservation service (SCS)
SCS TR-55 method estimation, 387–388; *see also*
Detention basins, preliminary sizing
Graphical method, 119–120
initial abstraction values, 119, 121t
peak runoff calculation, 119
pond and swamp adjustment factor, 119, 122t
unit peak discharge, 119, 122f
Seaports, removing hydrocarbons and fuels, 14
Sediment, NPS pollutant, 7–9
falling velocity, of particle, 7–8
trap efficiency, 7
Seepage pits, *see* Dry wells
Sefidrud Dam, 7
Sensitive waters, 305–306
Separation water quality devices, hydrodynamic, 330,
337–343
Aqua-Swirl Model AS-5 CFD PCS, 340, 339f
CDS® model 3945, 338–339, 341f
partial list, 337t
Vortechs®, 337, 338f, 339, 339t
Settling velocity, of spherical particles, 8–9, 9t
Severn Trent Services, 618
Shallow concentrated flow, flow path segment, 102, 103f,
104–105, 104t
Shear stresses
channel bends, 235
channel side, to bottom shear stress ratio, 219–220, 220f
permissible, *see* Permissible shear stresses
for straight channel, 217
variation in trapezoidal channel, 216, 216f
Sheet flow
to conservation areas, 296
flow path segment, 101–102, 102t, 103–104, 105
length analysis, 108–109
time of concentration in TR-55 method, 145–147
Shigella genus group, 12
Side flow interception efficiency
calculation, 168
defined, 168
Side slope stability, 219–221
angle of repose and angle of side slope, 220, 221f
channel side shear stress to bottom shear stress ratio,
219–220, 220f
Sieve analysis of soil samples, 472–473, 472f
Site, defined, 304
Sizing requirement, ESD Maryland's RCN for, 321t–323t
Skin Boss Filtration System, 347
Sloan Valve, 605
Slotted inlets
ACO trench drain, 161, 163f–164f
Campbell Foundry trench drain, 164, 165f
design, 172–173
overview, 160, 160f
Smart box controllers, 610
Smart growth, 483–484
Smart Sponge, 14–15, 333
Smart Sponge Plus, 333

Smart water application technology (SWAT), 609–610
SNOUT water quality inlet hood, 336, 336f
Snow and ice removal, 575; *see also* Storm water drainage
systems
Snyder Industries, 626
Snyder synthetic unit hydrograph, 140, 148–149, 148f
Soap and Detergent Association, 619
Sodium chloride, contaminant in urban runoff, 13
Soil(s)
bare, 217–219, 218f, 218t, 219f
noncohesive and cohesive, permissible shear stress,
218f, 219f
permissible velocities for, 206, 207t, 213–215, 214f,
215t
porosity, on percolation rate, 94
Soil and Water Assessment Tool (SWAT), 141
Soil composition and water consumption, 595
Soil conservation service (SCS), 101–103
TR-55 method, SCS, 115–119
components, 115, 116f
composite curve number, 120f
curve numbers for urban areas, 117, 118t
depths for AMC, 117, 117t
examples, 124–126
initial abstraction values for, 119, 121t
limitations/drawbacks of, 123–126
potential and actual retention, 116
soil curve number, 116–119
soil groups, description of, 117, 117t
Virgin Lands, curve numbers for, 117, 119t
unit hydrograph method, 121–123
peak runoff rate, 121
triangular distribution, 121, 123, 122f
Soil erosion and sediment control measures, 531
Soil gradation analysis, 94–96
texture triangle, 94, 95f, 96t
USDA soil classification, 94, 94t
Soil log and percolation test, 435
Soil Retention Products, Inc., 502
Solar reflectance index (SRI), 305
SOL (sum of loads) comparisons, 365–364, 366t
Sorbtive Filter, 347
Source control, *see* Source reduction
Source Loading and Management Model, 140
Source reduction, 479–480
benefits, 481–482
Source reduction measures
clustered development, 532
general, 531–532
Spacing, inlets, 176–181
calculations, 177–178, 177f
examples, 178–181
maximum, for flat roadway, 183–184
NJDOT criteria, 176
NJDOT drainage design manual, 178
P-1-1/8 FHA grates, 179–180
procedure for, 177
Specific energy
critical flow, 26–34
depth and discharge, calculation, 28–31, 34
in partly full circular pipes, 31–34, 32t, 33f
in rectangular channels, 28–29
in trapezoidal channels, 29–31
variation with depth, 26–27, 27f

SPP (storm water pollution plan), 281
Spread, gutter flow, 157, 158t, 159f
SRI (solar reflectance index), 305
St. Louis encephalitis, 630
State of Maryland storm water management regulations,
    280–306
  BMP design, 286–290
    filtration systems, 287–288
    infiltration trenches/basins, 287
    pretreatment storage, 288
    selection factors, 288, 290
    swales, 288, 289f
    treatment suitability, 288, 289t
    types of ponds, P-1 through P-5, 286
    types of storm water wetlands, W-1 through W-4,
        286–287
    watersheds, 286, 288
  channel protection storage volume criteria, 284, 284t,
        285f
  Department of Environment, 280–281
  ESD, 290–304
    addressing, 294–304
    alternative surfaces, 294–295
    microscale practices, 296–304
    nonstructural practices, 295–296
    predevelopment runoff standards, 290
    requirements, 292–294, 293t
    runoff depth and storage volume calculation, 290, 291
    site development strategies, 290, 291t
  extreme flood volume criteria, 286
  overbank protection volume criteria
    detention basins/ponds and underground chambers/
        vaults, 286
    rainfall depths, 284t, 286
    runoff, 286
  RCN for ESD sizing requirement, 321t–323t
  recharge volume criteria, 282–283
    calculation, 282, 282t, 283
    groundwater, 282, 283
    hydrologic soil group (HSG), 283
    percent volume and percent area method, 282t, 283
    water quality volume and, 283
  redevelopment, 304–305
    defined, 304
    overview, 304
    policy, 304–305
  sensitive waters, 305–306
  special criteria, 305–306
  SPP, 281
  unified storm water sizing criteria, 281t
    water quality volume
      calculation, 281–282
      overview, 281–282
    wetlands, waterways, and critical areas
      Critical Area Act, 306
      habitat, 306
      LDAs, 306
State of New York storm water regulations, 306–316
  conveyance system design criteria, 314–315, 314f
  downstream analysis, 313–314
    conditions, 314
    elements, 313–314
    10% rule, 313
    watershed, 313

enhanced phosphorus removal standards, 316
extreme flood control criteria, 312–313, 313f
hotspots, storm water
  defined, 315
  land uses and activities, 315
  list of, 315t
overbank flow control criteria
  design calculations for, 311–312
  storm, 311–312, 312f
overview, 306–307, 307t
redevelopment projects
  alternative practices, 316
  water quality criteria, 316
  water quality volume, 316
sizing criteria, 307t
stream channel protection volume requirement
      classification system, 309
  design storm, 309, 311f
  detention ponds/underground vaults, 309, 310t
  overview, 309, 311
  storage, 309, 311
water quality volume, 307–309
  calculation, 307
  impervious cover, 307, 308
  rainfall event, 307, 308f
  treatment practices, 308–309, 310t
Statewide basic requirements (SBRs), 268
Stealth System, 606
Steel mesh basket, Nutrient Separating Baffle Box, 347
Stokes's law, 8
Stone lining, 217–219
  Manning's n values, 225, 226t
  permissible shear stress for, 217, 218f, 218t, 219f
StormCAD, 120
Stormceptor, 327, 337, 343, 345, 347
StormChamber™, 415, 419, 568
Storm water drainage systems, 155–264
  critical flow charts for round and elliptical pipes, 254f
  culverts, hydraulic design, 191–206
    capacity of, 191
    charts for inlets, 253, 253f
    design calculations, 202
  design of, 184–191
    calculation processes, 184–186, 185f
    head loss calculations, 191, 192f
  gabion baskets and mattresses, 239–241
  gutter flow equation, derivation of, 249–250
  inlets; see also Inlets
    design, 165–176
  swales, see Swales
StormFilter, 343, 347, 350, 357, 360, 361, 350t, 361f
StormRax™, 570
StormSafe, 337
StormTank®, 422
StormTech, 415, 568
StormTrap, 420
StormVault, 340, 345, 331t
Storm water
  EPA and
    pollution study, 4–5
    program, 16
  management, 16–17
    BMPs, 16
    EPA's regulatory program, 16

NRC's report and compact developments, 16–17, 19–22
  regulations, 2–3
NPS pollutants, 5, 6t
pollution, 3
quantity, impacts on, 1–3
  construction of drainage systems, 1
  hydrograph, 1, 2
Stormwater 360, 345
Stormwater (magazine), 603
Stormwater antimicrobial treatment unit, 333
Storm water drainage systems, 574–575
  control of potential mosquito breeding habitats, 575
  removal of sediment and floatables from drainage systems, 575
  restoration of grass- and riprap-lined swales, 574
  snow and ice removal, 575
Storm water fee, 537–538
Storm water hotspots, 315, 315t
Storm Water Inspection & Maintenance Services, Inc., 333
Storm water management
  best management practices (BMPs)
    cost effectiveness, 530–531
    structural, 479
  bioretention basin, 521–522
  bioretention cells, 523–525
  bioswales, 522–523
  blue roofs, 511
  drivable grass, 540–542
  filter strips
    application, 516–518
    design criteria, 519
    Franklin Lakes municipal building, 519–520
  Green infrastructure, 484–485
  green roofs
    analysis, 508–510
    components, 505–506
    concept, 505
    construction types, 506–507
    installation cost, 508
    life and energy conservation, 507–508
    pollutant removal effect, 508
  LEED and green buildings environmental benefits, 485
    examples, 486–488
    performance measurement, 485–486
  low-impact development
    Ipswich River, Massachusetts project, 482–483
    objectives, 482–483
  minimal impact developments, 532–533
  porous pavements
    concrete pavers, 481–484, 486
    Glass Pave, 495
    nonconcrete modular pavers, 502–503
    open cell pavers, 500, 502
    open cell paving grids, 488–489
    pervious concrete, 493–495
    porous asphalt, 489–493
  rain gardens, 525–526, 528
  smart growth, 483–484
  source control, 480–481
    Green infrastructure techniques, 480
    New York City, 481
    Portland, 480–481

source reduction, 479–480
    benefits, 481–482
  source reduction measures
    clustered development, 532
    general, 531–532
  storm water wetlands
    benefits, 511
    construction cost, 513
    design criteria, 513–514
    design variations, 512–513
  subsurface gravel wetland (SGW)
    layout, 514
    Maryland Department of the Environment design guidelines, 516
    UNHSC's specifications, 515–516
  utility fee
    implementation, 538
    legal opposition, 538
    property size, 537
    rate structure, 537
Storm Water Management Act, 290
Storm Water Management Design Manual, 306
Storm water management model (SWMM), 140–141
  development, 141
  dynamic rainfall-runoff simulation model, 141
  EPA, 83, 88, 140–141
  runoff component of, 141
  SWMM 5, 141
  WinSLAMM with, 141
Storm water management program (SWMP), 266
Storm water management regulations, 265–323
  current, 268–271
    EISA section 428, 271
    input–output pollutant concentration, 270f
    pollutants removal, 269
    runoff quantity, 268–269
    TSS removal rate, 269–270
    volume of runoff, 269
    water quality, 268–269
  federal regulations, 265–268
  NJDEP, 271–280, 272t
    amended rules, 271
    approved TSS removal rate for vegetated filter strips, 318, 318f, 319f, 320f, 318t
    goals, 271
    gravel and porous pavements, 272
    groundwater recharge standards, 275–278, 276f–277f
    improvement, 279–280
    MSWMP, 279
    NJPDES, 272
    nonstructural strategies, 278–279
    planning agencies, 271–272
    quality standards, 273–275, 274f, 274t, 275t
    runoff calculation methods, 278
    runoff quantity requirement, 272–273
    standards for structures, 278
  NPDES
    phase II program, 266–268
    phase I program, 265–266
  overview, 265–268
State of Maryland, 280–306
State of New York, 306–316

Storm Water Inspection and Maintenance Services
     (SWIMS), 333
Storm water management systems
    construction
        detention basins, 565–566
        dry wells and infiltration chambers, 568–569
        grass swale, 567–568
        infiltration basins, 566–567
        outlet structures, 569–570
        wet ponds, 567
    inspection, operation and maintenance manual,
        579–580
    inspection and maintenance
        maintenance of water treatment devices, 577, 578t
        neglects, 579
        objectives, 571–573
        ponds/detention basins, 575–576
        repair of storm water management facilities, 577
    New Jersey Garden State Parkway Interchange 98, 99,
        100f, 101
    pipes installation
        concrete box culverts, 554
        dewatering, 559–561
        DuroMaxx, 545–546
        elliptical concrete pipes, 551
        high-density polyethylene (HDPE) pipes, 554–559
        prestressed concrete pipes, 551
        round reinforced concrete pipes (RCP), 547–551
    slope stabilization, 570–571
    soil erosion and sediment control measures, 545
    watertight joints
        infiltration/exfiltration testing, 565
        pipe connection to manhole/inlet, 563–564
        pipe joints, 561–562
Storm water pollution plan (SPP), 281
Storm water pollution prevention (SWPP) BMPs, 498
Storm water pollution prevention plan (SWPPP), 266–267
Storm water wetlands benefits, 511
    construction cost, 513
    design criteria, 513–514
    design variations, 512–513
Strahler-Horton methodology, 309
Stream channel protection volume requirement
    classification system, 309
    design storm, 309, 311f
    detention ponds/underground vaults, 309, 310t
    overview, 309, 311
    storage, 309, 311
StreamStats program, 140–141, 152, 152f
Structural BMPs, 325; *see also* Best management practices
     (BMPs)
Structural storm water management, standards for, 278
Subsurface gravel wetland (SGW), 297
    layout, 514
    Maryland Department of the Environment design
        guidelines, 516
    UNHSC's specifications, 515–516
Sudden expansions/contractions, 53–54, 53f, 54t
Sum of loads (SOL) comparisons, 365–364, 366t
Sump pumps, 560
Superelevation
    defined, 234
    extra free board for, 235
SuperGro, 242–243, 260f–263f

Surface depressions, *see* Depression storage
Swales, 302–303
    bioswales, 302–303, 303f, 508–509
    defined, 213
    design, 288, 289f
    drainage
        bare soil and stone lining, 217–219, 218f, 218t,
            219f
        channel bends, 234–237
        composite lining, 237–238
        flexible lining, 213
        grass lining, 221–224
        Manning's roughness coefficient variation,
            224–234
        permissible velocity concept, 213–215, 214f, 215t
        rigid linings, 213
        side slope stability, 219–221
        tractive force method, 215–217, 216f
    vegetative, 440–441
    wet, 303
SWAT (Soil and Water Assessment Tool), 141
SWIMS (Storm Water Inspection and Maintenance
     Services), 333
SWMM, *see* Storm water management model (SWMM)
SWMP (storm water management program), 266
SWPPP (storm water pollution prevention plan), 266–267
Synthetic triangular hydrograph, 121, 123, 122f

T

Tabular hydrograph method, 120
TAPE (technology assessment protocolecology) program,
     328, 350
TCLP (toxicity characteristic leach procedure), 333, 366
Technical Release No. 55 (TR-55) method, SCS, 115–123
    components, 115, 116f
    composite curve number, 120f
    soil curve numbers for urban areas, 117, 118t
    depths for AMC, 117, 117t
    description of soil groups, 117, 117t
    examples, 124–126
    graphical method, 119–120
        initial abstraction values, 119, 121t
        peak runoff calculation, 119
        pond and swamp adjustment factor, 119, 122t
        unit peak discharge, 119, 122f
    limitations/drawbacks of, 123–126
    potential and actual retention, 116
    for runoff calculation, NJ storm water management
        regulations, 278
    sheet flow time of concentration in, 145–147
    soil curve number, 116–119
    tabular hydrograph method, 120
    Virgin Lands, curve numbers for, 117, 119t
    WinTR-55 method, 126–131
        Monmouth County, New Jersey, 126–131, 127f,
            128f, 129f, 129t, 130t
Technology assessment protocolecology (TAPE) program,
     see TAPE
Terre Arch™/Terre Box™, 423, 424, 424f
Terrecon, Inc., 502–503
Terre Hill Concrete Products, 337
Terre Hill Stormwater Systems, 424
Terre Kleen

hydrodynamic separation water treatment devices, 338f, 339, 339t
verification, by EPA and NSF, ETV program, 328, 363–367
Terrewalks®, 502–503
Tetraethyl lead, 11
Texas Manual on Rainwater Harvesting, 611
Texture triangle, soil, 94, 95f, 96t
Time of concentration
  calculations
    TR-55 method, 101–108
    WinTR-55 method, 129, 130t
    Garden State Parkway nomograph, 99, 100f, 101
    Izzard equation, 97, 98, 99t
    Kirby equation, 98, 99, 99t
    Kirpich equation, 97, 98f
    Federal Highway Administration (FHWA) method, 103–108, 325
      calculation process, 103–108
      examples, 105–108
      flow path segments, 103–105, 104t
      intercept coefficient, 104, 104t
    sheet flow, in TR-55 method, 145–147
TMDL (total mean daily load) bacteria, 12
TN (total nitrogen), 270
Toro, 610
Toro precision spray nozzle, 596
Total energy, defined, 25
Total energy head, 25–26 defined, 25
Total mean daily load (TMDL) bacteria, 12
Total nitrogen (TN), 270
Total phosphorus (TP), 270
Total suspended solids (TSSs)
  removal
    of detention basin, 8, 9
    efficiency, 329–330
    by extended detention basins, 400–401, 400t, 402t, 401f, 431
  removal rate for vegetated filter strips, NJDEP
    approved maximum slope, 318t
    required length, 318, 318f, 319f, 320f, 318t
  removal rates for BMPs, 273, 275, 275t
  as surrogate pollutant parameter, 269–270
Toxicity characteristic leach procedure (TCLP), 333, 366
TP (total phosphorus), 270
TR-55, *see* Technical Release No. 55 (TR-55) method, SCS
Tractive force method
  defined, 213
  drainage channels, 215–217, 216f
Traffic safety, grate inlets and, 165
Transitions, head loss at, 54, 55f, 55t
Trap efficiency, defined, 7
Trapezoidal channel design
  critical flow in, 29–31
  flow depth calculation, 44–50
    discharge calculation, 49–50
    horizontal elliptical RCP, 49
    hydraulic parameter and variation, 45–46, 45t, 46f
    Manning formula, 44
    roadside channel, 46–48
    slope calculation, 48–49

Trash racks, 570
Trelleborg Pipe Seals, 563–564
Trench drains
  ACO, 161, 163f–164f
  Campbell Foundry, 164, 165f
Triton, 568
Triton StormWater Solution, 415, 420
TSSs, *see* Total suspended solids (TSSs)
Turf reinforcement mats (TRMs), 570–571
  manufacturing companies, 242
  overview, 241–242
  properties, 243–244, 243t
  Propex, 242, 256f–259f
Typhoid, 12

U
Ultra high efficiency toilet (UHET) flusher, 606
Ultra-Urban Filter, 332, 333
Underdrain, 490
Underground detention basins, 575–576; *see also* Ponds/detention basins, maintenance
  chambers, 415, 417–420
  plastic and concrete vaults, 420, 422–425
  solid and perforated pipes, 412, 415
  Woodcliff Lake Project, 426–431
Uni Eco-Stone, 498
Uniform flow, 25
United Nations International Children's Emergency Fund (UNICEF), 601
Unit hydrograph method, SCS, 121, 123, 122f
  peak runoff rate, 121
  triangular distribution, 121, 123, 122f
Unit peak discharge, for SCS, 119, 122f
Universal runoff method, 131–140
  application to nonuniform rainfall, 139–140
  composite surfaces, equations for, 135–138
  examples, 136–138
  lag time
    antecedent moisture condition, 133
    catchment area, 133–134
    excess rainfall, 131–132, 131f
    interception and depression storage, values of, 132, 132t
    between rainfall and runoff, 131–133
    rainfall–runoff relation for impervious surfaces, 134f
    runoff equations for impervious surfaces and, 134
    universal rainfall–runoff relation, 131, 131f
  peak discharge, 133–135
  rational and SCS methods, drawbacks, 131
  runoff volume, 133–134, 134f, 136–137
University of New Hampshire Stormwater Center (UNHSC), 490
Up-Flow Filter, 344
Urban Green BioFilter, 350
Urbanization impacts, runoff, 1–22
  infiltration, 1, 2
  NPS pollutants, 4–15
    atmospheric dust, 15
    concentration in urban storm water, 6t
    CWA of 1972, 3, 5, 19–22

floatables, 6, 7
heavy metals, 11
NURP, 4–5
nutrients and pesticides, 9–11
pathogens and fecal coliform, 11–12
petroleum hydrocarbons, 14–15
road salt, 12–14
sediment, 7–9, 9t sources, 5t
storm water, 5, 6t
storm water quantity and management, 1–3, 16–17
BMPs, 16
construction of drainage systems, 1
EPA's regulatory program, 16 hydrograph, 1, 2
NRC's report and compact developments, 16–17, 19–22
regulations, 2–3
water quality, 3–4
land uses and pollutant, 3, 4t
median event mean concentration, 4t
NPS pollutant, 3
storm water pollution, 3
Urban pollution, causes of, 4
Urinals in nonresidential buildings, 605–606; *see also* Indoor water conservation
US Army Corps of Engineers, 242
USDA, *see* US Department of Agriculture (USDA)
US Department of Agriculture (USDA)
Agricultural Research Services, 141
Snyder synthetic unit hydrograph and, 140
soil classification, 94, 94t
soil texture
Green-Ampt parameters for, 84, 85t
triangle, 94, 95f, 96t, 132
values, ultimate infiltration rates, 88, 89t
StreamStats, 140–141
US Department of Transportation, 103, 110, 184
US Filter, 618
US Fish and Wildlife Services, 265
US Geological Survey (USGS), 607–608
Nationwide Urban Hydrograph, 150, 150t
Quality of Our Nation's Water—Nutrients in Nation's Streams and Groundwater, 9
regression equations for urban peak discharges, 151, 151t
road salt, studies, 12
StreamStats program, 140–141, 152, 152f
urban storm water pollution, research study, 4
US Green Building Council (USGBC), 485
USGS, *see* US Geological Survey (USGS)
Utility fee
implementation, 538
legal opposition, 538
property size, 537
rate structure, 537

V

Vacuuming, 574
Vegetated buffers, 296
Vegetated filter strips
NJDEP approved TSS removal rate
maximum slope, 318t
required length, 318, 318f, 319f, 320f, 318t
Vegetation, retention effect, 76, 78

Vegetative and paved areas, 573–574
Vegetative swales, 428
Velocity head, defined, 26
Verglimit, anti-icing agent, 13
Verification
MTDs, process, 368–369
Vibrio cholerae, 12
Virginia Planning Board Commission, 326
Virgin Lands, curve numbers for, 117, 119t
V-notch weirs, 380–381
VortClarex, 345
Vortechemics, 327
Vortechnics Stormwater Management, 345
Vortechs®, 337, 339, 338f, 339t

W

Washington's Department of Ecology, 270
Wastewater and grey water, treatment of, 620–621
Wastewater market, recycled, 617–618; *see also* Water reuse
Wastewater reuse, 616–617; *see also* Water reuse
Water blasting, 574
Water conservation
benefits, 632
indoor conservation
economy of water-saver fixtures, 606–607
other indoor saving tips, 606
residential buildings, 603–605
urinals in nonresidential buildings, 605–606
non-profit organizations, 634–635
outdoor conservation
measures, 612–613
overview, 607–612
overview, 602–603
programs, 634–635
suggested actions
block programs, 631
enforcements, 631
organizations/alliances for water use, 632
pilot projects, 631
public education, 631
reaching out, 631
reward, 631
task force, 631
trends in supply and demand, 601–602
Water Conservation Hotel and Motel Program (Water CHAMP), 613
Water Efficiency (magazine), 603
Water Environment Research Foundation (WERF), 327
Water purveyors, 618, 622
Water quality, impacts on, 3–4
land uses and pollutant, 3, 4t
median event mean concentration for urban land uses, 4t
NPS pollutant, 3
storm water pollution, 3
Water Quality Act of 1987, 265
Water quality devices, *see* Manufactured water treatment devices (MTDs)
Water Quality Protection Center (WQPC), 363–367
Water quality storm
defined, 269
NJDEP, 273–275, 274f, 274t, 275t

Water quality volume
  State of Maryland storm water management regulations
    calculation, 281–282
    overview, 281–282
  State of New York storm water regulations calculation, 307
    impervious cover, 307, 308
    rainfall event, 307, 308f
    treatment practices, 308–309, 310t
Water reuse
  benefits, 632
  grey water reuse, 619–620
  New Jersey Water Quality Standards for, 614t–615t
  nonprofit organizations, 634–635
  rainwater and storm water runoff reuse
    overview, 621–622
    quantity of urban runoff, 623–625
  rainwater harvesting
    rain barrels, problems with, 630
    roof rain harvesting, 626–630
  recycled wastewater market, 617–618
  suggested actions
    block programs, 631
    enforcements, 617
    organizations/alliances for water reuse, 632
    pilot projects, 631
    public education, 631
    reaching out, 631
    reward, 631
    task force, 631
  treatment of wastewater and grey water, 620–621
  wastewater reuse, 616–617
Water-saver fixtures, 606–607
Water Sense Program, 608–609
Watertight joints
  infiltration/exfiltration testing, 565
  pipe connection to manhole/inlet, 563–565
  pipe joints, 561
Water treatment devices, see Manufactured water
    treatment devices (MTDs)
Water treatment devices, maintenance of
  catch basin inserts, 577
  manufactured water treatment devices, 577, 578t
  neglect in maintenance, 579
  repair of storm water management facilities, 577

Water treatment structures
  sand filters, 442–444
  chambers, 442–443, 443f
  organic filter, 445f
  TSS removal, 444
  vegetative swales, 440–441
Waterways, 306
WERF (Water Environment Research Foundation), 327
Westchester County Airport, 14
West Nile fever, 630
Wetlands
  for control of urban runoff quality, 5
  Maryland storm water management regulations, 306
Wet ponds, 567, 576; see also Ponds/detention basins,
    maintenance
Wet swales, 303; see also Swales
Wetting front, defined, 83
Wilson, Edward, Dr., 17
Wind erosion, 15
WinSLAMM, 140, 141
WinTR-55.exe, 126, 128, 129
WinTR-55 method, 126–131
  computed peak flows, 129, 130t
  curve number, calculations for, 128, 129t
  hydrographs, 126
  improvement elements, 126
  Monmouth County, New Jersey, 126–131, 127f, 128f,
    129f, 129t, 130t
WinTR-20 program, 126
WQPC (Water Quality Protection Center), 363t–367t

X

XeriBrix, 498–499
Xeripave Super Pervious Pavers, 498–499
Xeriscaping, 611

Z

Zero liquid discharge (ZLD) systems, 618
Zerxes, 626
Zinc, in storm water runoff, 11